T0344480

Boolean Functions

Written by prominent experts in the field, this monograph provides the first comprehensive and unified presentation of the structural, algorithmic, and applied aspects of the theory of Boolean functions.

The book focuses on algebraic representations of Boolean functions, especially disjunctive and conjunctive normal form representations. It presents within this framework the fundamental elements of the theory (Boolean equations and satisfiability problems, prime implicants and associated short representations, dualization), an in-depth study of special classes of Boolean functions (quadratic, Horn, shellable, regular, threshold, read-once functions and their characterization by functional equations), and two fruitful generalizations of the concept of Boolean functions (partially defined functions and pseudo-Boolean functions). Several topics are presented here in book form for the first time.

Because of the unique depth and breadth of the unified treatment that it provides and its emphasis on algorithms and applications, this monograph will have special appeal for researchers and graduate students in discrete mathematics, operations research, computer science, engineering, and economics.

Dr. Yves Crama is Professor of Operations Research and Production Management and the former Director General of the HEC Management School of the University of Liège, Belgium. He is widely recognized as a prominent expert in the field of Boolean functions, combinatorial optimization, and operations research, and he has coauthored more than seventy papers and three books on these subjects. Dr. Crama is a member of the editorial board of *Discrete Applied Mathematics, Discrete Optimization, Journal of Scheduling, and 4OR – The Quarterly Journal of the Belgian, French and Italian Operations Research Societies.*

The late Peter L. Hammer (1936–2006) was a Professor of Operations Research, Mathematics, Computer Science, Management Science, and Information Systems at Rutgers University and the Director of the Rutgers University Center for Operations Research (RUTCOR). He was the founder and editor-in-chief of the journals *Annals of Operations Research, Discrete Mathematics, Discrete Applied Mathematics, Discrete Optimization,* and *Electronic Notes in Discrete Mathematics.* Dr. Hammer was the initiator of numerous pioneering investigations of the use of Boolean functions in operations research and related areas, of the theory of pseudo-Boolean functions, and of the logical analysis of data. He published more than 240 papers and 19 books on these topics.

ENCYCLOPEDIA OF MATHEMATICS AND ITS APPLICATIONS

The titles below, and earlier volumes in the series, are available from booksellers or from Cambridge University Press at www.cambridge.org.

110 M.-J. Lai and L. L. Schumaker *Spline Functions on Triangulations*
111 R. T. Curtis *Symmetric Generation of Groups*
112 H. Salzmann et al. *The Classical Fields*
113 S. Peszat and J. Zabczyk *Stochastic Partial Differential Equations with Lévy Noise*
114 J. Beck *Combinatorial Games*
115 L. Barreira and Y. Pesin *Nonuniform Hyperbolicity*
116 D. Z. Arov and H. Dym *J-Contractive Matrix Valued Functions and Related Topics*
117 R. Glowinski, J.-L. Lions, and J. He *Exact and Approximate Controllability for Distributed Parameter Systems*
118 A. A. Borovkov and K. A. Borovkov *Asymptotic Analysis of Random Walks*
119 M. Deza and M. Dutour Sikirić *Geometry of Chemical Graphs*
120 T. Nishiura *Absolute Measurable Spaces*
121 M. Prest *Purity, Spectra and Localisation*
122 S. Khrushchev *Orthogonal Polynomials and Continued Fractions*
123 H. Nagamochi and T. Ibaraki *Algorithmic Aspects of Graph Connectivity*
124 F. W. King *Hilbert Transforms I*
125 F. W. King *Hilbert Transforms II*
126 O. Calin and D.-C. Chang *Sub-Riemannian Geometry*
127 M. Grabisch et al. *Aggregation Functions*
128 L. W. Beineke and R. J. Wilson (eds.) with J. L. Gross and T. W. Tucker *Topics in Topological Graph Theory*
129 J. Berstel, D. Perrin, and C. Reutenauer *Codes and Automata*
130 T. G. Faticoni *Modules over Endomorphism Rings*
131 H. Morimoto *Stochastic Control and Mathematical Modeling*
132 G. Schmidt *Relational Mathematics*
133 P. Kornerup and D. W. Matula *Finite Precision Numbers Systems and Arithmetic*
134 Y. Crama and P. L. Hammer *Boolean Models and Methods in Mathematics, Computer Science, and Engineering*
135 V. Berthé and M. Rigo *Combinatorics, Automata and Number Theory*
136 A. Kristály, V. D. Rădulescu, and C. Varga *Variational Principles in Mathematical Physics, Geometry, and Economics*
137 J. Berstel and C. Reutenauer *Noncommutative Rational Series with Applications*
138 B. Courcelle *Graph Structure and Monadic Second-Order Logic*
139 M. Fiedler *Matrices and Graphs in Geometry*
140 N. Vakil *Real Analysis through Modern Infinitesimals*
141 R. B. Paris *Hadamard Expansions and Hyperasymptotic Evaluation*

ENCYCLOPEDIA OF MATHEMATICS AND ITS APPLICATIONS

Boolean Functions

Theory, Algorithms, and Applications

YVES CRAMA

University of Liège, Belgium

PETER L. HAMMER

CAMBRIDGE
UNIVERSITY PRESS

CAMBRIDGE
UNIVERSITY PRESS

32 Avenue of the Americas, New York NY 10013-2473, USA

Cambridge University Press is part of the University of Cambridge.

It furthers the University's mission by disseminating knowledge in the pursuit of education, learning and research at the highest international levels of excellence.

www.cambridge.org
Information on this title: www.cambridge.org/9780521847513

© Yves Crama and Peter L. Hammer 2011

First published 2011

A catalogue record for this publication is available from the British Library

Library of Congress Cataloguing in Publication data
Crama, Yves, 1958–
Boolean functions / Yves Crama, Peter L. Hammer.
p. cm. – (Encyclopedia of mathematics and its applications)
Includes bibliographical references and index.
Contents: Theory, algorithms, and applications
ISBN 978-0-521-84751-3 (hardback)
1. Algebraic functions. 2. Algebra, Boolean. I. Hammer, P. L., 1936–2006. II. Title.
QA341.C73 2011
511.3'24–dc22   2011009690

ISBN 978-0-521-84751-3 Hardback

To Edith,
by way of apology for countless days
spent in front of the computer.
YC

Contents

Contributors   *page* xiii
Preface   xv
Acknowledgments   xix
Notations   xxi

Part I   Foundations

1   Fundamental concepts and applications   3
  1.1   Boolean functions: Definitions and examples   3
  1.2   Boolean expressions   8
  1.3   Duality   13
  1.4   Normal forms   14
  1.5   Transforming an arbitrary expression into a DNF   19
  1.6   Orthogonal DNFs and number of true points   22
  1.7   Implicants and prime implicants   24
  1.8   Restrictions of functions, essential variables   28
  1.9   Geometric interpretation   31
  1.10   Monotone Boolean functions   33
  1.11   Recognition of functional and DNF properties   40
  1.12   Other representations of Boolean functions   44
  1.13   Applications   49
  1.14   Exercises   65

2   Boolean equations   67
  2.1   Definitions and applications   67
  2.2   The complexity of Boolean equations: Cook's theorem   72
  2.3   On the role of DNF equations   74
  2.4   What does it mean to "solve a Boolean equation"?   78
  2.5   Branching procedures   80
  2.6   Variable elimination procedures   87

2.7	The consensus procedure	92
2.8	Mathematical programming approaches	95
2.9	Recent trends and algorithmic performance	103
2.10	More on the complexity of Boolean equations	104
2.11	Generalizations of consistency testing	111
2.12	Exercises	121

3   Prime implicants and minimal DNFs   123
Peter L. Hammer and Alexander Kogan

3.1	Prime implicants	123
3.2	Generation of all prime implicants	128
3.3	Logic minimization	141
3.4	Extremal and typical parameter values	159
3.5	Exercises	165

4   Duality theory   167
Yves Crama and Kazuhisa Makino

4.1	Basic properties and applications	167
4.2	Duality properties of positive functions	176
4.3	Algorithmic aspects: The general case	183
4.4	Algorithmic aspects: Positive functions	189
4.5	Exercises	198

Part II   Special Classes

5   Quadratic functions   203
Bruno Simeone

5.1	Basic definitions and properties	203
5.2	Why are quadratic Boolean functions important?	205
5.3	Special classes of quadratic functions	207
5.4	Quadratic Boolean functions and graphs	209
5.5	Reducibility of combinatorial problems to quadratic equations	218
5.6	Efficient graph-theoretic algorithms for quadratic equations	230
5.7	Quadratic equations: Special topics	243
5.8	Prime implicants and irredundant forms	250
5.9	Dualization of quadratic functions (Contributed by Oya Ekin Karaşan)	263
5.10	Exercises	266

6   Horn functions   269
Endre Boros

6.1	Basic definitions and properties	269
6.2	Applications of Horn functions	273
6.3	False points of Horn functions	277

6.4	Horn equations	281
6.5	Prime implicants of Horn functions	286
6.6	Properties of the set of prime implicants	292
6.7	Minimization of Horn DNFs	297
6.8	Dualization of Horn functions	306
6.9	Special classes	309
6.10	Generalizations	314
6.11	Exercises	321

7 Orthogonal forms and shellability — 326
7.1	Computation of orthogonal DNFs	326
7.2	Shellings and shellability	330
7.3	Dualization of shellable DNFs	336
7.4	The lexico-exchange property	338
7.5	Shellable quadratic DNFs and graphs	346
7.6	Applications	348
7.7	Exercises	349

8 Regular functions — 351
8.1	Relative strength of variables and regularity	351
8.2	Basic properties	355
8.3	Regularity and left-shifts	362
8.4	Recognition of regular functions	365
8.5	Dualization of regular functions	369
8.6	Regular set covering problems	377
8.7	Regular minorants and majorants	380
8.8	Higher-order monotonicity	391
8.9	Generalizations of regularity	397
8.10	Exercises	401

9 Threshold functions — 404
9.1	Definitions and applications	404
9.2	Basic properties of threshold functions	408
9.3	Characterizations of threshold functions	413
9.4	Recognition of threshold functions	417
9.5	Prime implicants of threshold functions	423
9.6	Chow parameters of threshold functions	428
9.7	Threshold graphs	438
9.8	Exercises	444

10 Read-once functions — 448
Martin C. Golumbic and Vladimir Gurvich
10.1	Introduction	448
10.2	Dual implicants	450

10.3   Characterizing read-once functions   456
10.4   The properties of P_4-free graphs and cographs   463
10.5   Recognizing read-once functions   466
10.6   Learning read-once functions   473
10.7   Related topics and applications of read-once functions   476
10.8   Historical notes   480
10.9   Exercises   481

11   Characterizations of special classes by functional equations   487
       Lisa Hellerstein
11.1   Characterizations of positive functions   487
11.2   Functional equations   488
11.3   Characterizations of particular classes   491
11.4   Conditions for characterization   495
11.5   Finite characterizations by functional equations   500
11.6   Exercises   506

Part III   Generalizations

12   Partially defined Boolean functions   511
       Toshihide Ibaraki
12.1   Introduction   511
12.2   Extensions of pdBfs and their representations   514
12.3   Extensions within given function classes   531
12.4   Best-fit extensions of pdBfs containing errors   547
12.5   Extensions of pdBfs with missing bits   551
12.6   Minimization with don't cares   558
12.7   Conclusion   561
12.8   Exercises   562

13   Pseudo-Boolean functions   564
13.1   Definitions and examples   564
13.2   Representations   570
13.3   Extensions of pseudo-Boolean functions   578
13.4   Pseudo-Boolean optimization   585
13.5   Approximations   593
13.6   Special classes of pseudo-Boolean functions   593
13.7   Exercises   607

A   Graphs and hypergraphs   609
A.1   Undirected graphs   609
A.2   Directed graphs   612
A.3   Hypergraphs   614

B Algorithmic complexity   615
   B.1 Decision problems   615
   B.2 Algorithms   617
   B.3 Running time, polynomial-time algorithms, and the class P   618
   B.4 The class NP   619
   B.5 Polynomial-time reductions and NP-completeness   620
   B.6 The class co-NP   621
   B.7 Cook's theorem   622
   B.8 Complexity of list-generation and counting algorithms   624

C JBool: A software tool   627
   Claude Benzaken and Nadia Brauner
   C.1 Introduction   627
   C.2 Work interface   628
   C.3 Creating a Boolean function   629
   C.4 Editing a function   632
   C.5 Operations on Boolean functions   633

Bibliography   635
Index   677

Contributors

Claude Benzaken
Laboratoire G-SCOP
Université Joseph Fourier
Grenoble, France

Endre Boros
RUTCOR – Rutgers Center for Operations Research
Rutgers University
Piscataway, NJ, USA

Nadia Brauner
Laboratoire G-SCOP
Université Joseph Fourier
Grenoble, France

Martin C. Golumbic
The Caesarea Rothschild Institute
University of Haifa
Haifa, Israel

Vladimir Gurvich
RUTCOR – Rutgers Center for Operations Research
Rutgers University
Piscataway, NJ, USA

Lisa Hellerstein
Department of Computer and Information Science
Polytechnic Institute of New York University
Brooklyn, NY, USA

Toshihide Ibaraki
Kyoto College of Graduate Studies for Informatics
Kyoto, Japan

Oya Ekin Karaşan
Department of Industrial Engineering
Bilkent University
Ankara, Turkey

Alexander Kogan
Rutgers Business School and RUTCOR
Rutgers University
Piscataway, NJ, USA

Kazuhisa Makino
Department of Mathematical Informatics
University of Tokyo
Tokyo, Japan

Bruno Simeone
Department of Statistics
La Sapienza University
Rome, Italy

Preface

Boolean functions, meaning $\{0,1\}$-valued functions of a finite number of $\{0,1\}$-valued variables, are among the most fundamental objects investigated in pure and applied mathematics. Their importance can be explained by several interacting factors.

- It is reasonable to argue that a multivariate function $f : A_1 \times A_2 \times \ldots \times A_n \to A$ is "interesting" only if each of the sets $A_1, A_2, \ldots, A_n$, and A contains at least two elements, since otherwise the function either depends trivially on some of its arguments, or is constant. Thus, in a sense, Boolean functions are the "simplest interesting" multivariate functions. It may even be surprising, actually, that such primitive constructs turn out to display a rich array of properties and have been investigated by various breeds of scientists for more than 150 years.
- When the arguments of a Boolean function are viewed as atomic logical propositions, the value of the function at a 0–1 point can be interpreted as the truth value of a sentence composed from these propositions. Carrying out calculations on Boolean functions is then tantamount to performing related logical operations (such as inference or theorem-proving) on propositional sentences. Therefore, Boolean functions are at the heart of propositional logic.
- Many concepts of combinatorial analysis have their natural Boolean counterpart. In particular, since every 0–1 point with n coordinates can be viewed as the characteristic vector of a subset of $N = \{1, 2, \ldots, n\}$, the set of points at which a Boolean function takes value 1 corresponds to a collection of subsets of N, or a "hypergraph" on N. (When all subsets have cardinality 2, then the function corresponds exactly to a graph.) Structural properties relating to the transversals, stable sets, or colorings of the hypergraph, for instance, often translate into interesting properties of the Boolean function.
- Boolean functions are ubiquitous in theoretical computer science, where they provide fundamental models for the most basic operations performed by

computers on binary digits (or bits). Turing machines and Boolean circuits are prime examples illustrating this claim. Similarly, electrical engineers rely on the Boolean formalism for the description, synthesis, or verification of digital circuits.

- In operations research or management science, binary variables and Boolean functions are frequently used to formulate problems where a number of "go – no go" decisions are to be made; these could be, for instance, investment decisions arising in a financial management framework, or location decisions in logistics, or assignment decisions for production planning. In most cases, the variables have to be fixed at values that satisfy constraints expressible as Boolean conditions and that optimize an appropriate real-valued objective function. This leads to – frequently difficult – Boolean equations ("satisfiability problems") or integer programming problems.
- Voting games and related systems of collective choice are frequently represented by Boolean functions, where the variables are associated with (binary) alternatives available to the decision makers, and the value of the function indicates the outcome of the process.
- Various branches of artificial intelligence rely on Boolean functions to express deductive reasoning processes (in the above-mentioned propositional framework), or to model primitive cognitive and memorizing activities of the brain by neural networks, or to investigate efficient learning strategies, or to devise storing and retrieving mechanisms in databases, and so on.

We could easily extend this list to speak of Boolean models arising in reliability theory, in cryptography, in coding theory, in multicriteria analysis, in mathematical biology, in image processing, in theoretical physics, in statistics, and so on.

The main objective of the present monograph is to introduce the reader to the fundamental elements of the theory of Boolean functions. It focuses on algebraic representations of Boolean functions, especially disjunctive or conjunctive normal form expressions, and it provides a very comprehensive presentation of the structural, algorithmic, and applied aspects of the theory in this framework.

The monograph is divided into three main parts.

Part I: *Foundations* proposes in Chapter 1: *Fundamental concepts and applications,* an introduction to the major concepts and applications of the theory. It then successively tackles three generic classes of problems that play a central role in the theory and in the applications of Boolean functions, namely, Boolean equations and their extensions in Chapter 2: *Boolean equations*, the generation of prime implicants and of optimal normal form representations in Chapter 3: *Prime implicants and minimal DNFs*, and various aspects of the relation between functions and their dual in Chapter 4: *Duality theory*.

Part II: *Special Classes* presents an in-depth study of several remarkable classes of Boolean functions. Each such class is investigated from both the structural and the algorithmic points of view. Chapter 5 is devoted to *Quadratic functions*, Chapter 6 to *Horn functions*, Chapter 7 to *Orthogonal forms and shellability*, Chapter 8 to

Regular functions, Chapter 9 to *Threshold functions*, and Chapter 10 to *Read-once functions*. Chapter 11: *Characterizations of special classes by functional equations* provides general conditions under which classes of functions can be "compactly" characterized.

Finally, Part III: *Generalizations* deals with two fruitful extensions of the concept of Boolean functions. Namely, Chapter 12: *Partially defined Boolean functions* deals with functions whose domain is restricted to a subset of all possible $\{0, 1\}$ points, and Chapter 13: *Pseudo-Boolean functions* proposes a brief overview of the theory of real-valued functions of binary variables.

In view of its emphasis on algorithms and applications, this monograph should appeal to researchers and graduate students in discrete mathematics, operations research, computer science, engineering, and economics. Although we believe that it is rather unique in its depth and breadth, our work has been influenced in various ways by many other books dealing with specialized aspects of the field, such as threshold logic, logical inference, operations research, game theory, or reliability theory. We like to mention, in particular, the classic monograph by P.L. Hammer and S. Rudeanu, *Boolean Methods in Operations Research and Related Areas* (Springer, Berlin, 1968). Although it focuses almost exclusively on Boolean models, rather than pseudo-Boolean ones, it can be seen as a distant follow-up to the 1968 monograph. We should also cite the influence of books by Anthony [25]; Brayton, Hachtel, McMullen, and Sangiovanni-Vincentelli [153]; Brown [156]; Chandru and Hooker [184]; Chang and Lee [186]; Hu [511, 512]; Jeroslow [533]; Kleine, Büning, and Lettmann [571]; Knuth [575]; Mendelson [680]; Muroga [698, 699]; Ramamurthy [777]; Rudeanu [795, 796]; Schneeweiss [811]; Störmer [849]; Truemper [871]; Wegener [902, 903]; and Winder [917], among others.

As a complement to the monograph, the reader is also advised to consult the collection of papers *Boolean Models and Methods in Mathematics, Computer Science and Engineering* (Y. Crama and P.L. Hammer, eds., Cambridge University Press, Cambridge, UK, 2010). Each chapter in that volume introduces the reader to specialized Boolean models and applications investigated in a particular field of science and provides a survey of important representative results.

Acknowledgments

The genesis of this book spread over many years, and over this long period, the authors have benefited from the support and advice provided by many individuals.

First and foremost, several colleagues have contributed important material to the monograph: Endre Boros, Marty Golumbic, Vladimir Gurvich, Lisa Hellerstein, Toshi Ibaraki, Oya Ekin Karaşan, Alex Kogan, Kaz Makino, and Bruno Simeone have coauthored several chapters and have provided input on various sections. Claude Benzaken and Nadia Brauner have developed a software package for manipulating Boolean functions that serves as a useful companion to the monograph. The contributions of these prominent experts of Boolean functions greatly enhance the appeal of the volume.

Comments, reviews, and corrections have been provided at different stages by colleagues and by RUTCOR students, including Nina Feferman, Noam Goldberg, Levent Kandiller, Shaoji Li, Tongyin Liu, Irina Lozina, Martin Milanic, Devon Morrese, David Neu, Sergiu Rudeanu, Gábor Rudolf, Jan-Georg Smaus, and Mine Subasi.

Special thanks are due to Endre Boros, who provided constant encouragement and tireless advice to the authors over the gestation period of the volume. Terry Hart provided the efficient administrative assistance that allowed the authors to keep track of countless versions of the manuscript and endless mail exchanges.

Finally, I am deeply indebted to my mentor, colleague, and friend, Peter L. Hammer, for getting us started on this ambitious project, many years ago. Peter spent much of his academic career stressing the importance and relevance of Boolean models in different fields of applied mathematics, and he was very keen on completing this monograph. It is extremely unfair that he did not live to see the outcome of our joint effort. I am sure that he would have loved it, and that he would have been very proud of this contribution to the dissemination of the theory, algorithms, and applications of Boolean functions.

Yves Crama
Liège, Belgium, September 2010

Notations

$\mathcal{B} = \{0,1\}, U = [0,1]$

$X = (x_1, x_2, \ldots, x_n), Y = (y_1, y_2, \ldots, y_n), \ldots$: components of points in $\mathcal{B}^n$

$x^\alpha = \begin{cases} x, & \text{if } \alpha = 1, \\ \overline{x}, & \text{if } \alpha = 0. \end{cases}$

$X \vee Y = (x_1 \vee y_1, x_2 \vee y_2, \ldots, x_n \vee y_n)$

$X \wedge Y = (x_1 \wedge y_1, x_2 \wedge y_2, \ldots, x_n \wedge y_n) = (x_1 y_1, x_2 y_2, \ldots, x_n y_n)$

$\overline{X} = (\overline{x}_1, \overline{x}_2, \ldots, \overline{x}_n)$

$X \leq Y$ (with $X, Y \in \mathcal{B}^n$) if and only if $x_i \leq y_i$ for $i = 1, 2, \ldots, n$

e_k : a unit vector $(0, \ldots, 0, 1, 0, \ldots, 0)$ of appropriate dimension, with 1 in kth position

e_A: the characteristic vector of $A \subseteq \{1, 2, \ldots, n\}$, that is, $e_A = \sum_{k \in A} e_k$; $e_\emptyset = 0$.

$supp(X)$: the support of $X \in \mathcal{B}^n$, that is, the set $\{i \in \{1, 2, \ldots, n\} \mid x_i = 1\}$

$T_{A,B} = \{X \in \mathcal{B}^n \mid x_i = 1 \text{ for all } i \in A \text{ and } x_j = 0 \text{ for all } j \in B\}$

$f, g, h, \ldots$: Boolean functions

$\phi, \psi, \theta, \ldots$: Boolean expressions

$\mathbf{1}_n$: the function that takes constant value 1 on $\mathcal{B}^n$

$\mathbf{0}_n$: the function that takes constant value 0 on $\mathcal{B}^n$

$T(f)$: the set of true points of function f

$F(f)$: the set of false points of function f

$minT(f)$: the set of minimal true points of a positive function f

$maxF(f)$: the set of maximal false points of a positive function f

f^d : the dual of function f

$|\phi|$: the (encoding) length, or size, of a Boolean expression ϕ; when ϕ is a DNF, $|\phi|$ is simply the number of literals appearing in ϕ

$|f|$: for a positive function f, $|f|$ denotes the size of the complete (prime irredundant) DNF ϕ of f, that is, $|f| \stackrel{\text{def}}{=} |\phi|$

$\|\phi\|$: the number of terms of a DNF ϕ

$(\omega_1, \omega_2, \ldots, \omega_n, \omega)$: the Chow parameters of a Boolean function on $\mathcal{B}^n$

$(\pi_1, \pi_2, \ldots, \pi_n, \pi)$: the modified Chow parameters of a Boolean function on $\mathcal{B}^n$

Part I

Foundations

1

Fundamental concepts and applications

The purpose of this introductory chapter is threefold. First, it contains the main definitions, terminology, and notations that are used throughout the book. After the introduction of our main feature characters – namely, Boolean functions – several sections are devoted to a discussion of alternative representations, or expressions, of Boolean functions. Disjunctive and conjunctive normal forms, in particular, are discussed at length in Sections 1.4–1.11. These special algebraic expressions play a very central role in our investigations, as we frequently focus on the relation between Boolean functions and their normal forms. Section 1.12, however, also provides a short description of different types of function representations, namely, representations over GF(2), pseudo-Boolean polynomial expressions, and binary decision diagrams.

A second objective of this chapter is to introduce several of the topics to be investigated in more depth in subsequent chapters, namely: fundamental algorithmic problems (Boolean equations, generation of prime implicants, dualization, orthogonalization, etc.) and special classes of Boolean functions (bounded-degree normal forms, monotone functions, Horn functions, threshold functions, etc.). Finally, the chapter briefly presents a variety of applications of Boolean functions in such diverse fields as logic, electrical engineering, reliability theory, game theory, combinatorics, and so on. These applications have often provided the primary motivation for the study of the problems to be encountered in the next chapters.

In a sense, this introductory chapter provides a (very) condensed digest of what's to come. It can be considered a degustation: Its main purpose is to whet the appetite, so that readers will decide to embark on the full course!

1.1 Boolean functions: Definitions and examples

This book is about Boolean functions, meaning: $\{0, 1\}$-valued functions of a finite number of $\{0, 1\}$-valued variables.

3

Definition 1.1. *A* Boolean function of *n* variables *is a function on B^n into B, where B is the set $\{0, 1\}$, n is a positive integer, and B^n denotes the n-fold carte-sian product of the set B with itself. A point $X^* = (x_1, x_2, \ldots, x_n) \in B^n$ is a* true point *(respectively,* false point*) of the Boolean function f if $f(X^*) = 1$ (respectively, $f(X^*) = 0$). We denote by $T(f)$ (respectively, $F(f)$) the set of* true points *(respec-tively,* false points*) of f. We denote by $\mathbf{1}_n$ the function that takes constant value 1 on B^n and by $\mathbf{0}_n$ the function that takes constant value 0 on B^n.*

It should be stressed that, in many applications, the role of the set B is played by another two-element set, like $\{\text{Yes}, \text{No}\}$, $\{\text{True}, \text{False}\}$, $\{\text{ON}, \text{OFF}\}$, $\{\text{Success}, \text{Failure}\}$, $\{-1, 1\}$ or, more generally, $\{a, b\}$, where a and b are abstract (uninterpreted) elements. In most cases, this distinction is completely irrelevant. However, it is often convenient to view the elements of B as *numerical quanti-ties* in order to perform arithmetic operations on these elements and to manipulate algebraic expressions like $1 - x, x + y - xy$, and so on, where x, y are elements of B.

As an historical aside, it is interesting to note that the ability to perform algebraic computations on logical symbols, in a way that is at least formally similar to what we are used to doing for numerical quantities, was one of the driving forces behind George Boole's seminal work in logic theory. Let us quote from Boole [103], Chapter V.6 (italics are Boole's):

> [...] any system of propositions may be expressed by equations involving symbols
> x, y, z, which, whenever interpretation is possible, are subject to laws identical in form
> with the laws of a system of quantitative symbols, susceptible only of the values 0 and
> 1. But as the formal processes of reasoning depend only upon the laws of the symbols,
> and not upon the nature of their interpretation, we are permitted to treat the above
> symbols, x, y, z, as if they were quantitative symbols of the kind above described. *We
> may in fact lay aside the logical interpretation of the symbols in the given equation;
> convert them into quantitative symbols, susceptible only of the values 0 and 1; perform
> upon them as such all the requisite processes of solution; and finally restore to them
> their logical interpretation.* And this is the mode of procedure which will actually be
> adopted [...]

In this book, we systematically follow Boole's prescription and adhere to the convention that $B = \{0, 1\}$, where 0 and 1 can be viewed as either abstract symbols or numerical quantities.

The most elementary way to define a Boolean function f is to provide its *truth table*.

Definition 1.2. *The* truth table *of a Boolean function on B^n is a complete list of all the points in B^n together with the value of the function at each point.*

Example 1.1. *The truth table of a Boolean function on B^3 is shown in Table 1.1.*   □

Of course, the use of truth tables becomes extremely cumbersome when the function to be defined depends on more than, say, 5 or 6 arguments. As a matter

Table 1.1. Truth Table for Example 1.1

(x_1, x_2, x_3)	$f(x_1, x_2, x_3)$
(0,0,0)	1
(0,0,1)	1
(0,1,0)	0
(0,1,1)	1
(1,0,0)	0
(1,0,1)	1
(1,1,0)	0
(1,1,1)	1

of fact, Boolean functions are often defined implicitly rather than explicitly, in the sense that they are described through a procedure that allows us, for any $0 - 1$ point in the domain of interest, to compute the value of the function at this point. In some theoretical developments, or when we analyze the computational complexity of certain problems, such a procedure can simply be viewed as a *black box oracle*, of which we can observe the output (that is, the function value) for any given input, but not the inner working (that is, the details of the algorithm that computes the output). In most applications, however, more information is available regarding the process that generates the function of interest, as illustrated by the examples below. (We come back to these applications in much greater detail in Section 1.13 and in many subsequent chapters of the book.)

Application 1.1. (Logic.) *In many applications (such as those arising in artificial intelligence), a Boolean function can be viewed as indicating the* truth value *of a sentence of propositional (or Boolean) logic. Consider, for instance, the sentence* S: *"If it rains in the morning, or if the sky is cloudy, then I carry my umbrella." Let us denote by* x_1, x_2, *and* x_3, *respectively, the subsentences "it rains in the morning," "the sky is cloudy," and "I carry my umbrella". Then,* S *can be identified with the sentence*

$$(x_1 \ OR \ x_2) \Rightarrow x_3.$$

It is easy to see that the function displayed in Table 1.1 computes the truth value of S for all possible values of x_1, x_2, x_3, *under the usual correspondence* True $\leftrightarrow$ 1, False $\leftrightarrow$ 0. □

Application 1.2. (Electrical engineering.) *In electrical or in computer engineering, a switching circuit is often abstracted into the following model, called a* combinational circuit. *The wiring of the circuit is described by an acyclic directed graph* $D = (V, A)$. *The vertices of D are the* gates *of the circuit. The indegree of each gate is at most 2. Each gate with indegree 2 is labeled either AND or OR, and each gate with indegree 1 is labeled NOT. The gates with indegree 0 are called* input gates *and are denoted* $v_1, v_2, \ldots, v_n$. *Also, all gates of D have outdegree 1, except for a single gate* f, *called* output gate, *which has outdegree 0.*

Every such circuit can be viewed as representing a Boolean function $f_D(x_1, x_2, \ldots, x_n)$. *First, for every* $(x_1, x_2, \ldots, x_n) \in \mathcal{B}^n$, *the* state $s(v)$ *of gate* $v \in V$ *is computed according to the following recursive rules:*

1. *For each input gate* v_i, $s(v_i) = x_i$ $(i = 1, 2, \ldots, n)$.
2. *For each AND-gate* $v \in V$, *if* $(u, v), (w, v) \in A$ *are the arcs entering* v, *then* $s(v) = \min(s(u), s(w))$.
3. *For each OR-gate* $v \in V$, *if* $(u, v), (w, v) \in A$ *are the arcs entering* v, *then* $s(v) = \max(s(u), s(w))$.
4. *For each NOT-gate* $v \in V$, *if* $(u, v) \in A$ *is the arc entering* v, *then* $s(v) = 1 - s(u)$. *Finally, we let* $f_D(x_1, x_2, \ldots, x_n) = s(f)$.

For instance, the circuit represented in Figure 1.1 computes the function given in Example 1.1. This can easily be verified by computing the state of the output gate (in this case, the OR-gate) for all possible 0–1 inputs. For example, if $(x_1, x_2, x_3) = (0, 0, 0)$, *then one successively finds that the state of each NOT-gate is* $1 (= 1 - 0)$; *the state of the AND-gate is* $1 (= \min(1, 1))$; *and the state of the output gate is* $1 (= \max(1, 0))$.

More generally, the gates of a combinational circuit may be "primitive" Boolean functions forming another class from the {AND,OR,NOT} collection used in our small example. In all cases, the gates may be viewed as atomic units of hardware, providing the building blocks for the construction of larger circuits. $\square$

Historically, propositional logic and electrical engineering have been the main nurturing fields for the development of research on Boolean functions. However, because they are such fundamental mathematical objects, Boolean functions have also been used to model a large number of applications in a variety of areas. To describe these applications, we introduce a few more notations.

Given a point $X \in \mathcal{B}^n$, we denote by $supp(X)$ the *support* of X, that is, $supp(X)$ is the set $\{i \in \{1, 2, \ldots, n\} \mid x_i = 1\}$. (Conversely, X is the *characteristic vector* of $supp(X)$.)

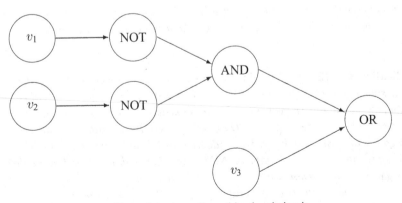

Figure 1.1. A small combinational circuit.

Application 1.3. (Game theory.) *Many group decision procedures (such as those used in legislative assemblies or in corporate stockholder meetings) can be viewed, in abstract terms, as decision rules that associate a single dichotomous "Yes–No" outcome (for instance, adoption or rejection of a resolution) with a collection of dichotomous "Yes–No" votes (for instance, assent or disagreement of individual lawmakers). Such procedures have been studied in the game-theoretic literature under the name of* simple games *or* voting games. *More formally, let* $N = \{1, 2, \ldots, n\}$ *be a finite set, the elements of which are to be called* players. *A simple game on N is a function* $v : \{A \mid A \subseteq N\} \to \mathcal{B}$. *Clearly, from our vantage point, a simple game can be equivalently modeled as a Boolean function* f_v *on* $\mathcal{B}^n$: *The variables of* f_v *are in 1-to-1 correspondence with the players of the game (variable i takes value 1 exactly when player i votes "Yes"), and the value of the function reflects the outcome of the vote for each point* $X^* \in \mathcal{B}^n$ *describing a vector of individual votes:*

$$f_v(X^*) = \begin{cases} 1 & \text{if } v(supp(X^*)) = 1, \\ 0 & \text{otherwise.} \end{cases}$$

□

Application 1.4. (Reliability theory.) *Reliability theory investigates the relationship between the operating state of a complex system S and the operating state of its individual components, say components* $1, 2, \ldots, n$. *It is commonly assumed that the system and its components can be in either of two states:* operative *or* failed. *Moreover, the state of the system is completely determined by the state of its components via a deterministic rule embodied in a Boolean function* f_S *on* $\mathcal{B}^n$, *called the* structure function *of the system: For each* $X^* \in \mathcal{B}^n$,

$$f_S(X^*) = \begin{cases} 1 & \text{if the system operates when all components in } supp(X^*) \text{ operate} \\ & \text{and all other components fail,} \\ 0 & \text{otherwise.} \end{cases}$$

A central issue is to compute the probability that the system operates (meaning that f_S *takes value 1) when each component is subject to probabilistic failure. Thus, reliability theory deals primarily with the stochastic theory of Boolean functions.*

□

Application 1.5. (Combinatorics.) *Consider a* hypergraph $\mathcal{H} = (N, \mathcal{E})$, *where* $N = \{1, 2, \ldots, n\}$ *is the set of* vertices *of* $\mathcal{H}$, *and* $\mathcal{E}$ *is a collection of subsets of N, called* edges *of the hypergraph. A subset of vertices is said to be* stable *if it does not contain any edge of* $\mathcal{H}$. *With* $\mathcal{H}$, *we associate the Boolean function* $f_{\mathcal{H}}$ *defined as follows: For each* $X^* \in \mathcal{B}^n$,

$$f_{\mathcal{H}}(X^*) = \begin{cases} 1 & \text{if } supp(X^*) \text{ is not stable,} \\ 0 & \text{otherwise.} \end{cases}$$

The function $f_{\mathcal{H}}$ *is the* stability function *of* $\mathcal{H}$.

□

Of course, the kinship among the models presented in Applications 1.3–1.5 is striking: It is immediately apparent that we are really dealing here with a single class of mathematical objects, in spite of the distinct motivations that originally justified their investigation.

Applications of Boolean functions will be discussed more thoroughly in Section 1.13, after we have introduced some of the fundamental theoretical concepts that underlie them.

Before we close this section, let us add that, in this book, our view of Boolean functions will be mostly combinatorial and algorithmic. For algebraic or logic-oriented treatments, we refer the reader to the excellent books by Rudeanu [795, 796] or Brown [156]. In these books, as in many related classical publications by other authors, Boolean functions are actually defined more broadly than in Definition 1.1, as (special) mappings of the form $f : \mathcal{A}^n \to \mathcal{A}$, where $\mathcal{A}$ is the carrier of an *arbitrary* Boolean algebra $(\mathcal{A}, \cup, \cap, \neg, 0, 1)$. By contrast, we shall essentially restrict ourselves in this book to the two-element Boolean algebra $(\mathcal{B}, \vee, \wedge, ^-, 0, 1)$, where $\mathcal{B} = \{0, 1\}$ (see Section 1.2). Brown [156], in particular, discusses in great detail the pros and cons of working with two-element, rather than more general, Boolean algebras. While acknowledging the relevance of his arguments, we feel that, at the risk of giving up some generality, our restricted framework is already sufficiently rich to model a variety of interesting applications and to allow us to handle a host of challenging algorithmic problems of a combinatorial nature. Also, the terminology introduced in Definition 1.1 has become sufficiently entrenched to justify its continued use, rather than the alternative terminology *switching functions* or *truth functions* which, though less liable to create confusion, has progressively become obsolete.

1.2 Boolean expressions

As the above examples illustrate, Boolean functions can be described in many alternative ways. In this section, we concentrate on a type of representation derived from propositional logic, namely, the representation of Boolean functions by *Boolean expressions* (see, for instance, [156, 680, 795, 848] for different presentations).

Boolean expressions will be used extensively throughout the book. In fact, the emphasis on Boolean expressions (rather than truth tables, circuits, oracles, etc.) can be seen as a main distinguishing feature of our approach and will motivate many of the issues we will tackle in subsequent chapters.

Our definition of Boolean expressions will be recursive, starting with three elementary operations as building blocks.

Definition 1.3. *The binary operation* $\vee$ *(disjunction, Boolean OR), the binary operation* $\wedge$ *(conjunction, Boolean AND), and the unary operation* $^-$ *(complementation, negation, Boolean NOT) are defined on* $\mathcal{B}$ *by the following rules:*

$$0 \vee 0 = 0, \ 0 \vee 1 = 1, \ 1 \vee 0 = 1, \ 1 \vee 1 = 1;$$

$$0 \wedge 0 = 0, \ 0 \wedge 1 = 0, \ 1 \wedge 0 = 0, \ 1 \wedge 1 = 1;$$

$$\overline{0} = 1, \ \overline{1} = 0.$$

For a Boolean variable x, we sometimes use the following convenient notation:

$$x^\alpha = \begin{cases} x, & \text{if } \alpha = 1, \\ \overline{x}, & \text{if } \alpha = 0. \end{cases}$$

Keeping in line with our focus on functions, we often regard the three elementary Boolean operations as defining Boolean functions on $\mathcal{B}^2$: $\text{disj}(x,y) = x \vee y$, $\text{conj}(x,y) = x \wedge y$, and on $\mathcal{B}$: $\text{neg}(x) = \overline{x}$. When the elements of $\mathcal{B} = \{0,1\}$ are interpreted as integers rather than abstract symbols, these operations can be defined by simple arithmetic expressions: For all $x, y \in \mathcal{B}$,

$$x \vee y = \max\{x,y\} = x + y - xy,$$

$$x \wedge y = \min\{x,y\} = xy,$$

$$\overline{x} = 1 - x.$$

Observe that the conjunction of two elements of $\mathcal{B}$ is equal to their arithmetic product. By analogy with the usual convention for products, we often omit the operator $\wedge$ and denote conjunction by mere juxtaposition.

We can extend the definitions of all three elementary operators to $\mathcal{B}^n$ by writing: For all $X, Y \in \mathcal{B}^n$,

$$X \vee Y = (x_1 \vee y_1, x_2 \vee y_2, \ldots, x_n \vee y_n),$$

$$X \wedge Y = (x_1 \wedge y_1, x_2 \wedge y_2, \ldots, x_n \wedge y_n) = (x_1 y_1, x_2 y_2, \ldots, x_n y_n),$$

$$\overline{X} = (\overline{x}_1, \overline{x}_2, \ldots, \overline{x}_n).$$

Let us enumerate some of the elementary properties of disjunction, conjunction, and complementation. (We note for completeness that the properties listed in Theorem 1.1 can be viewed as the defining properties of a general Boolean algebra.)

Theorem 1.1. *For all $x, y, z \in \mathcal{B}$, the following identities hold:*

(1) $x \vee 1 = 1$ *and* $x \wedge 0 = 0$;
(2) $x \vee 0 = x$ *and* $x \wedge 1 = x$;
(3) $x \vee y = y \vee x$ *and* $xy = yx$ (commutativity);
(4) $(x \vee y) \vee z = x \vee (y \vee z)$ *and* $x(yz) = (xy)z$ (associativity);
(5) $x \vee x = x$ *and* $xx = x$ (idempotency);
(6) $x \vee (xy) = x$ *and* $x(x \vee y) = x$ (absorption);
(7) $x \vee (yz) = (x \vee y)(x \vee z)$ *and* $x(y \vee z) = (xy) \vee (xz)$ (distributivity);
(8) $x \vee \overline{x} = 1$ *and* $x \overline{x} = 0$;
(9) $\overline{\overline{x}} = x$ (involution);
(10) $\overline{(x \vee y)} = \overline{x}\,\overline{y}$ *and* $\overline{(xy)} = \overline{x} \vee \overline{y}$ (De Morgan's laws);
(11) $x \vee (\overline{x} y) = x \vee y$ *and* $x(\overline{x} \vee y) = xy$ (Boolean absorption).

Proof. These identities are easily verified, for example, by exhausting all possible values for x, y, z. □

Building upon Definition 1.3, we are now in a position to introduce the important notion of *Boolean expression*.

Definition 1.4. *Given a finite collection of* Boolean *variables* $x_1, x_2, \ldots, x_n$, *a* Boolean expression (*or* Boolean formula) *in the variables* $x_1, x_2, \ldots, x_n$ *is defined as follows:*

(1) *The constants* $0, 1$, *and the variables* $x_1, x_2, \ldots, x_n$ *are Boolean expressions in* $x_1, x_2, \ldots, x_n$.

(2) *If* ϕ *and* ψ *are Boolean expressions in* $x_1, x_2, \ldots, x_n$, *then* $(\phi \vee \psi)$, $(\phi \psi)$ *and* $\overline{\phi}$ *are Boolean expressions in* $x_1, x_2, \ldots, x_n$.

(3) *Every Boolean expression is formed by finitely many applications of the rules* (1)–(2).

We also say that a Boolean expression in the variables $x_1, x_2, \ldots, x_n$ *is a* Boolean expression on $\mathcal{B}^n$.

We use notations like $\phi(x_1, x_2, \ldots, x_n)$ or $\psi(x_1, x_2, \ldots, x_n)$ to denote Boolean expressions in the variables $x_1, x_2, \ldots, x_n$.

Example 1.2. *Here are some examples of Boolean expressions:*

$$\phi_1(x) = x,$$
$$\phi_2(x) = \overline{x},$$
$$\psi_1(x, y, z) = (((\overline{x} \vee y)(y \vee \overline{z})) \vee ((xy)z)),$$
$$\psi_2(x_1, x_2, x_3, x_4) = ((x_1 x_2) \vee (\overline{x}_3 \overline{x}_4)).$$ □

Now, since disjunction, conjunction, and complementation can be interpreted as Boolean functions, every Boolean expression $\phi(x_1, x_2, \ldots, x_n)$ can also be viewed as generating a Boolean function defined by composition.

Definition 1.5. *The Boolean function* f_ϕ *represented (or expressed) by a Boolean expression* $\phi(x_1, x_2, \ldots, x_n)$ *is the unique Boolean function on* $\mathcal{B}^n$ *defined as follows: For every point* $(x_1^*, x_2^*, \ldots, x_n^*) \in \mathcal{B}^n$, *the value of* $f_\phi(x_1^*, x_2^*, \ldots, x_n^*)$ *is obtained by substituting* x_i^* *for* x_i ($i = 1, 2, \ldots, n$) *in the expression* ϕ *and by recursively applying Definition 1.3 to compute the value of the resulting expression.*

When $f = f_\phi$ *on* $\mathcal{B}^n$, *we also say that* f *admits the representation or the expression* ϕ, *and we simply write* $f = \phi$.

Example 1.3. *Consider again the expressions defined in Example 1.2. We can compute, for instance:*

$$f_{\phi_1}(0) = 0, \ f_{\phi_1}(1) = 1,$$
$$f_{\phi_2}(0) = \overline{0} = 1, \ f_{\phi_2}(1) = \overline{1} = 0,$$
$$f_{\psi_1}(0, 0, 0) = (((\overline{0} \vee 0)(0 \vee \overline{0})) \vee ((00)0)) = 1, \ldots$$

In fact, the expression ψ_1 in Example 1.2 represents the function f, where

$$f(0,0,1) = f(1,0,0) = f(1,0,1) = 0,$$
$$f(0,0,0) = f(0,1,0) = f(0,1,1) = f(1,1,0) = f(1,1,1) = 1.$$

Thus, we can write

$$f(x,y,z) = \psi_1(x,y,z) = (((\overline{x} \vee y)(y \vee \overline{z})) \vee ((xy)z)). \qquad \square$$

Remark. So that we can get rid of parentheses when writing Boolean expressions, we assume from now on a priority ranking of the elementary operations: Namely, we assume that disjunction has lower priority than conjunction, which has lower priority than complementation. When we compute the value of a parentheses-free expression, we always start with the operations of highest priority: First, all complementations; next, all conjunctions; and finally, all disjunctions. (This is similar to the convention that assigns a lower priority to addition than to multiplication, and to multiplication than to exponentiation when evaluating an arithmetic expression like $3x^2 + 5xy$.) Moreover, we also discard any parentheses that become redundant as a consequence of the associativity property of disjunction and conjunction (Theorem 1.1).

Example 1.4. *The expression ψ_1 in Example 1.2 (and hence, the function f in Example 1.3) can be rewritten with fewer parentheses as $f(x,y,z) = \psi_1(x,y,z) = (\overline{x} \vee y)(y \vee \overline{z}) \vee xyz$. Similarly, the expression ψ_2 in Example 1.2 can be rewritten as $\psi_2(x_1,x_2,x_3,x_4) = x_1 x_2 \vee \overline{x_3}\overline{x_4}$.* $\square$

The relation between Boolean expressions and Boolean functions, as spelled out in Definition 1.5, deserves to be carefully pondered.

On one hand, it is important to understand that every Boolean function can be represented by *numerous* Boolean expressions (see, for instance, Theorem 1.4 in the next section). In fact, it is easy to see that there are "only" 2^{2^n} Boolean functions of n variables, while there are infinitely many Boolean expressions in n variables. These remarks motivate the distinction we draw between functions and expressions.

On the other hand, since every Boolean expression ϕ represents a *unique* Boolean function f_ϕ, we are justified in interpreting ϕ itself as a function, and we frequently do so. The notation $f = \phi$ introduced in Definition 1.5, in particular, may initially seem confusing, since it equates a function with a formal expression, but this notational convention is actually innocuous: It is akin to the convention for real-valued functions of real variables, where it is usual to assimilate a function with its analytical expression and to write, for instance, equalities like

$$f(x,y) = x^2 + 2xy + y^2 = (x+y)^2. \qquad (1.1)$$

As a matter of fact, since Definition 1.5 implies that we write both $f = \psi$ and $f = \phi$ when ψ and ϕ represent the same function f (compare with Equation (1.1)), it also naturally leads to the next notion.

Definition 1.6. *We say that two Boolean expressions ψ and ϕ are* equivalent *if they represent the same Boolean function. When this is the case, we write $\psi = \phi$.*

Note that any two expressions that can be deduced from each other by repeated use of the properties listed in Theorem 1.1 are equivalent even though they are not *identical*.

Example 1.5. *The function $f(x,y,z)$ represented by $\psi_1(x,y,z) = (\overline{x} \vee y)(y \vee \overline{z}) \vee xyz$ (see previous examples) is also represented by the expression $\phi = \overline{x}\,\overline{z} \vee y$. Indeed,*

$$(\overline{x} \vee y)(y \vee \overline{z}) \vee xyz = (\overline{x}y \vee \overline{x}\,\overline{z} \vee yy \vee y\overline{z}) \vee xyz \quad (distributivity)$$
$$= \overline{x}y \vee \overline{x}\,\overline{z} \vee y \vee y\overline{z} \vee xyz \quad \begin{array}{l}(idempotency \\ and\ associativity)\end{array}$$
$$= \overline{x}\,\overline{z} \vee y \quad (absorption).$$

Thus, $\psi_1(x,y,z)$ and $\phi(x,y,z)$ are equivalent, that is, $\psi_1(x,y,z) = \phi(x,y,z)$. $\square$

A recurrent theme in Boolean theory concerns the transformation of Boolean expressions into equivalent expressions that display specific desirable properties. For instance, in the previous example, the expression ϕ is intuitively much "simpler" or "shorter" than ψ_1, even though these two expressions represent the same function. More generally, for algorithmic purposes, it is necessary to have a definition of the *length* of a Boolean expression.

Definition 1.7. *The* length *(or size) of a Boolean expression ϕ is the number of symbols used in an encoding of ϕ as a binary string. The length of ϕ is denoted by $|\phi|$.*

We refer to standard books on computational complexity for additional comments regarding this concept (see, for instance, [371, 725]). For most practical purposes, we can conveniently think of $|\phi|$ as the total number of symbols (constants, variables, operators, parentheses) occuring in the expression ϕ.

To conclude this important section on Boolean expressions, let us note that complex Boolean functions can be introduced by substituting functional symbols for the variables of a Boolean expression. That is, if $\psi(y_1, y_2, \ldots, y_m)$ is a Boolean expression on $\mathcal{B}^m$, and $f_1, f_2, \ldots, f_m$ are m Boolean functions on $\mathcal{B}^n$, then the Boolean function $\psi(f_1, f_2, \ldots, f_m)$ can be defined in the natural way: Namely, for all $X^* \in \mathcal{B}^n$,

$$\big(\psi(f_1, f_2, \ldots, f_m)\big)(X^*) = \psi(f_1(X^*), f_2(X^*), \ldots, f_m(X^*)), \qquad (1.2)$$

where we identify the expression ψ with the function f_ψ that it represents (thus, (1.2) simply boils down to function composition). In particular, if f and g are two

Boolean functions on $\mathcal{B}^n$, then the functions $f \vee g$, $f \wedge g$, and $\overline{f}$ are defined, for all $X^* \in \mathcal{B}^n$, by

$$(f \vee g)(X^*) = f(X^*) \vee g(X^*),$$
$$(f \wedge g)(X^*) = f(X^*) \wedge g(X^*),$$
$$\overline{f}(X^*) = \overline{f(X^*)}.$$

1.3 Duality

With every Boolean function f, the following definition associates another Boolean function f^d called the *dual* of f:

Definition 1.8. *The* dual *of a Boolean function f is the function f^d defined by*

$$f^d(X) = \overline{f(\overline{X})}$$

for all $X = (x_1, x_2, \ldots, x_n) \in \mathcal{B}^n$, where $\overline{X} = (\overline{x}_1, \overline{x}_2, \ldots, \overline{x}_n)$.

Example 1.6. *Let f be the 2-variable function defined by $f(0,0) = f(0,1) = f(1,1) = 1$ and $f(1,0) = 0$. Then the dual of f satisfies $f^d(0,0) = f^d(1,0) = f^d(1,1) = 0$ and $f^d(0,1) = 1$.* $\square$

Dual functions arise naturally in many Boolean models. We only describe here one simple occurence of this concept; more applications are discussed in Chapter 4.

Application 1.6. (Voting theory.) *Suppose that a voting procedure is modeled by a Boolean function f on $\mathcal{B}^n$, as explained in Application 1.3. Thus, when the players' votes are described by the Boolean point $X^* \in \mathcal{B}^n$, the outcome of the voting procedure is $f(X^*)$. What happens if all the players simultaneously change their minds and vote $\overline{X}^*$ rather than X^*? In many cases, we would expect the outcome of the procedure to be reversed as well, that is, we would expect $f(\overline{X}^*) = \overline{f(X^*)}$, or equivalently, $f(X^*) = \overline{f(\overline{X}^*)} = f^d(X^*)$. When the property $f(X) = f^d(X)$ holds for all $X \in \mathcal{B}^n$, we say that the function f (and the voting procedure it describes) is* self-dual. *Note, however, that some common voting procedures are not self-dual, as exemplified by the two-thirds majority rule.* $\square$

We first list some useful properties of dualization.

Theorem 1.2. *If f and g are Boolean functions, then*

(a) $(f^d)^d = f$ *(involution: the dual of the dual is the function itself);*
(b) $(\overline{f})^d = \overline{(f^d)}$;
(c) $(f \vee g)^d = f^d \, g^d$;
(d) $(fg)^d = f^d \vee g^d$.

Proof. Definition 1.8 immediately implies (a) and (b). For property (c), observe that

$$(f \vee g)^d(X) = \overline{(f \vee g)(\overline{X})}$$
$$= \overline{(f(\overline{X}) \vee g(\overline{X}))}$$
$$= \overline{f(\overline{X})}\,\overline{g(\overline{X})} \quad \text{(by de Morgan's laws)}$$
$$= f^d(X) g^d(X).$$

Property (d) follows from (a) and (c). $\square$

Observe that, in view of property (a), dualization defines a bijective correspondence on the space of Boolean functions.

It is natural to ask how the Boolean expressions of a function relate to the expressions of its dual. To settle this question, we introduce one more definition.

Definition 1.9. *The* dual *of a Boolean expression ϕ is the expression ϕ^d obtained by exchanging the operators $\vee$ and $\wedge$, as well as the constants 0 and 1, in ϕ.*

Example 1.7. *If $\phi(x,y,z) = (\overline{x} \vee y)(y \vee \overline{z}) \vee xyz$, then $\phi^d(x,y,z) = (\overline{x}y \vee y\overline{z})$ $(x \vee y \vee z)$.* $\square$

For our notations and terminology to be consistent, ϕ^d should represent the dual of the function represented by ϕ. This is indeed the case.

Theorem 1.3. *If the expression ϕ represents the Boolean function f, then the expression ϕ^d represents f^d.*

Proof. Let t denote the total number of conjunction, disjunction and negation operators in ϕ. We prove the theorem by induction on t. If $t = 0$, then ϕ is either a constant or a literal, and the statement is easily seen to hold.

Assume now that $t > 0$. Then, by Definition 1.4, ϕ takes either the form $\psi \vee \theta$, or the form $\psi\theta$, or the form $\overline{\psi}$. Assume, for instance, that $\phi = \psi \vee \theta$ (the other cases are similar). Then, by Definition 1.9, $\phi^d = \psi^d \theta^d$. Let g be the function represented by ψ and let h be the function represented by θ. By induction, ψ^d and θ^d represent g^d and h^d, respectively. So, ϕ^d represents $g^d h^d$, which is equal to f^d by Theorem 1.2. $\square$

Duality is an important concept in Boolean theory and we shall return to this topic many times in subsequent chapters of this book. Chapter 4, in particular, is fully devoted to duality.

1.4 Normal forms

In this section, we discuss some classes of Boolean expressions of special interest. Let us adopt the following notations: If $\{\phi_k | k \in \Omega\}$ is a family of Boolean expressions indexed over the set $\Omega = \{k_1, k_2, \ldots, k_m\}$, then we denote by $\bigvee_{k \in \Omega} \phi_k$ the expression $(\phi_{k_1} \vee \phi_{k_2} \vee \ldots \vee \phi_{k_m})$, and we denote by $\bigwedge_{k \in \Omega} \phi_k$ the expression $(\phi_{k_1} \wedge \phi_{k_2} \wedge \ldots \wedge \phi_{k_m})$. By convention, when Ω is empty, $\bigvee_{k \in \Omega} \phi_k$ is equivalent to the constant 0 and $\bigwedge_{k \in \Omega} \phi_k$ is equivalent to the constant 1.

Definition 1.10. *A* literal *is an expression of the form x or $\overline{x}$, where x is a Boolean variable. An* elementary conjunction (*sometimes called* term, *or* monomial, *or* cube) *is an expression of the form*

$$C = \bigwedge_{i \in A} x_i \bigwedge_{j \in B} \overline{x}_j, \quad \text{where } A \cap B = \emptyset,$$

and an elementary disjunction (*sometimes called* clause) *is an expression of the form*

$$D = \bigvee_{i \in A} x_i \vee \bigvee_{j \in B} \overline{x}_j, \quad \text{where } A \cap B = \emptyset,$$

where A, B are disjoint subsets of indices.

A disjunctive normal form (DNF) *is an expression of the form*

$$\bigvee_{k=1}^{m} C_k = \bigvee_{k=1}^{m} \left(\bigwedge_{i \in A_k} x_i \bigwedge_{j \in B_k} \overline{x}_j \right),$$

where each C_k ($k = 1, 2, \ldots, m$) is an elementary conjunction; we say that each conjunction C_k is a term *of the DNF.*

A conjunctive normal form (CNF) *is an expression of the form*

$$\bigwedge_{k=1}^{m} D_k = \bigwedge_{k=1}^{m} \left(\bigvee_{i \in A_k} x_i \vee \bigvee_{j \in B_k} \overline{x}_j \right),$$

where each D_k ($k = 1, 2, \ldots, m$) is an elementary disjunction; we say that each disjunction D_k is a clause *of the CNF.*

In particular, 0 is an elementary (empty) disjunction, 1 is an elementary (empty) conjunction, and any elementary disjunction or conjunction is both a DNF and a CNF. Additional illustrations of normal forms are provided in the next example.

Example 1.8. *The expression $\phi(x, y, z) = \overline{x}\,\overline{z} \vee y$ is a disjunctive normal form; its terms are the elementary conjunctions $\overline{x}\,\overline{z}$ and y. It is easy to check that ϕ is equivalent to the CNF $(\overline{x} \vee y)(y \vee \overline{z})$ with clauses $(\overline{x} \vee y)$ and $(y \vee \overline{z})$.*

The expression $\psi_2(x_1, x_2, x_3, x_4) = x_1 x_2 \vee \overline{x}_3 \overline{x}_4$ is a DNF; it is equivalent to the CNF $(x_1 \vee \overline{x}_3)(x_1 \vee \overline{x}_4)(x_2 \vee \overline{x}_3)(x_2 \vee \overline{x}_4)$. $\square$

Bringing together the observations in Examples 1.5 and 1.8, we see that we have obtained three different expressions for the same Boolean function f:

$$f(x, y, z) = (\overline{x} \vee y)(y \vee \overline{z}) \vee xyz \tag{1.3}$$

$$= \overline{x}\,\overline{z} \vee y \tag{1.4}$$

$$= (\overline{x} \vee y)(y \vee \overline{z}). \tag{1.5}$$

In particular, we have been able to derive both a DNF representation (1.4) and a CNF representation (1.5) of the original expression (1.3) (which is not a normal form). This is not an accident. Indeed, we can now establish a fundamental property of Boolean functions.

Theorem 1.4. *Every Boolean function can be represented by a disjunctive normal form and by a conjunctive normal form.*

Proof. Let f be a Boolean function on $\mathcal{B}^n$, let T be the set of true points of f, and consider the DNF

$$\phi_f(x_1, x_2, \ldots, x_n) = \bigvee_{Y \in T} \left(\bigwedge_{i \mid y_i = 1} x_i \bigwedge_{j \mid y_j = 0} \overline{x_j} \right). \tag{1.6}$$

If we interpret ϕ_f as a function on $\mathcal{B}^n$, then a point $X^* \in \mathcal{B}^n$ is a true point of ϕ_f if and only if there exists $Y = (y_1, y_2, \ldots, y_n) \in T$ such that

$$\bigwedge_{i \mid y_i = 1} x_i^* \bigwedge_{j \mid y_j = 0} \overline{x_j^*} = 1. \tag{1.7}$$

But condition (1.7) simply means that $x_i^* = 1$ whenever $y_i = 1$, and $x_i^* = 0$ whenever $y_i = 0$, that is, $X^* = Y$. Hence, X^* is a true point of ϕ_f if and only if $X^* \in T$, and we conclude that ϕ_f represents f.

A similar reasoning establishes that f is also represented by the CNF

$$\psi_f(x_1, x_2, \ldots, x_n) = \bigwedge_{Y \in F} \left(\bigvee_{j \mid y_j = 0} x_j \vee \bigvee_{i \mid y_i = 1} \overline{x_i} \right), \tag{1.8}$$

where F is the set of false points of f. $\qquad\square$

Note that, alternatively, the second part of Theorem 1.4 can also be derived from its first part by an easy duality argument. Indeed, in view of Theorem 1.3, the function f is represented by the CNF

$$\bigwedge_{(A,B) \in \Omega} \left(\bigvee_{i \in A} x_i \vee \bigvee_{j \in B} \overline{x_j} \right) \tag{1.9}$$

exactly when its dual f^d is represented by the DNF

$$\bigvee_{(A,B) \in \Omega} \left(\bigwedge_{i \in A} x_i \bigwedge_{j \in B} \overline{x_j} \right). \tag{1.10}$$

Let us now illustrate Theorem 1.4 by an example.

Example 1.9. *The set of true points of the function f represented by the expression (1.3) is $T = \{(0,0,0), (0,1,0), (0,1,1), (1,1,0), (1,1,1)\}$, and its set of false points is $F = \{(1,0,0), (0,0,1), (1,0,1)\}$ (see Example 1.3). Thus, it follows from the proof of Theorem 1.4 that f is also represented by the DNF*

$$\phi_f = \overline{x}\,\overline{y}\,\overline{z} \vee \overline{x}\,y\,\overline{z} \vee \overline{x}\,y\,z \vee x\,y\,\overline{z} \vee x\,y\,z$$

and by the CNF

$$\psi_f = (\overline{x} \vee y \vee z)(x \vee y \vee \overline{z})(\overline{x} \vee y \vee \overline{z}).$$

□

The expressions (1.6) and (1.8) have a very special structure that is captured by the following definitions:

Definition 1.11. *A* minterm *(respectively,* maxterm*) on $\mathcal{B}^n$ is an elementary conjunction (respectively, disjunction) involving exactly n literals.*

Let f be a Boolean function on $\mathcal{B}^n$, let $T(f)$ be the set of true points of f, and let $F(f)$ be its set of false points. The DNF

$$\phi_f(x_1, x_2, \ldots, x_n) = \bigvee_{Y \in T(f)} \left(\bigwedge_{i|y_i=1} x_i \bigwedge_{j|y_j=0} \overline{x_j} \right) \tag{1.11}$$

is the minterm *expression (or canonical DNF) of f, and the terms of ϕ_f are the* minterms *of f. The CNF*

$$\psi_f(x_1, x_2, \ldots, x_n) = \bigwedge_{Y \in F(f)} \left(\bigvee_{j|y_j=0} x_j \vee \bigvee_{i|y_i=1} \overline{x_i} \right) \tag{1.12}$$

is the maxterm *expression (or canonical CNF) of f, and the terms of ψ_f are the* maxterms *of f.*

Observe that Definition 1.11 actually involves a slight abuse of language, since the minterm (or the maxterm) expression of a function is unique only up to the order of its terms and literals. In the sequel, we shall not dwell on this subtle, but usually irrelevant, point and shall continue to speak of "the" minterm (or maxterm) expression of a function.

With this terminology, the proof of Theorem 1.4 establishes that every Boolean function is represented by its minterm expression. This observation can be traced all the way back to Boole [103]. In view of its unicity, the minterm expression provides a "canonical" representation of a function. In general, however, the number of minterms (or, equivalently, of true points) of a function can be very large, so that handling the minterm expression often turns out to be rather impractical.

Normal form expressions play a central role in the theory of Boolean functions. Their preeminence is partially justified by Theorem 1.4, but this justification is not sufficient in itself. Indeed, the property described in Theorem 1.4 similarly holds for many other special classes of Boolean expressions. For instance, it can be observed that, besides its DNF and CNF expressions, every Boolean function also admits expressions involving only disjunctions and complementations, but no conjunctions (as well as expressions involving only conjunctions and complementations, but no disjunctions). Indeed, as an immediate consequence of De Morgan's laws, every conjunction xy can be replaced by the equivalent expression $\overline{(\overline{x} \vee \overline{y})}$, and similarly, every disjunction $x \vee y$ can be replaced by the expression $\overline{(\overline{x}\,\overline{y})}$.

More to the point, and as we will repeatedly observe, normal forms arise quite naturally when one attempts to model various problems within a Boolean framework. For this reason, normal forms are ubiquitous in this book: Many of the problems to be investigated will be based on the assumption that the Boolean functions at hand are expressed in normal form or, conversely, will have as a goal constructing a normal form of a function described in some alternative way (truth table, arbitrary Boolean expression, etc.).

However, it should be noticed that DNF and CNF expressions provide closely related frameworks for representing or manipulating Boolean functions (remember the duality argument invoked at the end of the proof of Theorem 1.4). This dual relationship between DNFs and CNFs will constitute, in itself, an object of study in Chapter 4.

Most of the time, we display a slight preference for DNF representations of Boolean functions over their CNF counterparts, to the extent that we discuss many problems in terms of DNF rather than CNF representations. This choice is in agreement with much of the classical literature on propositional logic, electrical engineering, and reliability theory, but is opposite to the standard convention in the artificial intelligence and computational complexity communities. Our preference for DNFs is partially motivated by their analogy with real polynomials: Indeed, since the identities $x \wedge y = xy$ and $\bar{x} = 1 - x$ hold when x, y are interpreted as numbers in $\{0, 1\}$, every DNF of the form

$$\bigvee_{k=1}^{m} \left(\bigwedge_{i \in A_k} x_i \bigwedge_{j \in B_k} \bar{x}_j \right)$$

can also be rewritten as

$$\bigvee_{k=1}^{m} \left(\prod_{i \in A_k} x_i \prod_{j \in B_k} (1 - x_j) \right),$$

a form that is reminiscent of a multilinear real polynomial like

$$\sum_{k=1}^{m} c_k \prod_{i \in T_k} x_i.$$

Ultimately, however, because of the above-mentioned "duality" between DNFs and CNFs, our preference can also simply be viewed as a matter of taste and habit, and so we will make no further attempts to justify it.

Before we close this section, we introduce some additional terminology (inspired by the analogy of DNFs with polynomials over the reals) and notation that will be useful in our dealings with DNFs.

Definition 1.12. *The* degree *of an elementary conjunction* $C = \bigwedge_{i \in A} x_i \bigwedge_{j \in B} \bar{x}_j$ *is the number of literals involved in* C, *namely,* $|A| + |B|$. *If* $\phi = \bigvee_{k=1}^{m} C_k$ *is a DNF, then the* degree *of* ϕ *is the maximum degree of the terms* C_k *over all* $k \in \{1, 2, \ldots, m\}$. *A DNF is called* linear *(respectively,* quadratic, cubic, ...*) if its degree is at most 1 (respectively, at most 2, 3, ...).*

Note that the (encoding) length $|\phi|$ of a DNF ϕ, as introduced in Definition 1.7, comes within a constant factor of the number of literals appearing in ϕ. Therefore, we generally feel free to identify these two measures of the size of ϕ (especially when discussing asymptotic complexity results). We denote by $||\phi||$ the number of terms of a DNF ϕ.

1.5 Transforming an arbitrary expression into a DNF

How difficult is it to transform an arbitrary expression ϕ into an equivalent DNF? Clearly, the construction given in the proof of Theorem 1.4 is not algorithmically efficient, as it requires the enumeration of all the true points of ϕ. On the other hand, a very simple procedure may come to mind immediately: Given the expression ϕ, the properties listed in Theorem 1.1 (especially, De Morgan's laws and the distributivity laws) can be repeatedly applied until a DNF is obtained.

Example 1.10. *The expression* $\phi(x_1,x_2,x_3,x_4) = (x_1 \vee x_4)\overline{(x_1 \vee (x_2\overline{x}_3))}$ *can be successively transformed into:*

$$\phi = (x_1 \vee x_4)(\overline{x}_1(\overline{x}_2 \vee x_3)) \quad \textit{(De Morgan and involution)}$$
$$= x_1\overline{x}_1\overline{x}_2 \vee x_1\overline{x}_1x_3 \vee \overline{x}_1\overline{x}_2x_4 \vee \overline{x}_1x_3x_4 \quad \textit{(distributivity and commutativity)}$$
$$= \overline{x}_1\overline{x}_2x_4 \vee \overline{x}_1x_3x_4.$$

$\square$

The problem with this method (and, actually, with *any* method that transforms a Boolean expression into an equivalent DNF) is that it may very well require an exponential number of steps, as illustrated by the following example:

Example 1.11. *The function represented by the CNF*

$$\phi(x_1,x_2,\ldots,x_{2n}) = (x_1 \vee x_2)(x_3 \vee x_4)\ldots(x_{2n-1} \vee x_{2n})$$

has a unique shortest DNF expression; call it ψ *(this will result from Theorem 1.23 hereunder). The terms of* ψ *are exactly those elementary conjunctions of n variables that involve one variable out of each of the pairs* $\{x_1,x_2\}, \{x_3,x_4\}, \ldots, \{x_{2n-1},x_{2n}\}$. *Thus,* ψ *has* 2^n *terms. Writing down all these terms requires exponentially large time and space in terms of the length of the original formula* ϕ. $\square$

Example 1.11 essentially shows that there is no hope of transforming an arbitrary expression (or even a CNF) into an equivalent DNF in polynomial time.

In Chapter 4, we shall return to a finer discussion of the following, rather natural, but surprisingly difficult question: Given a DNF expression ψ of the function f, what is the complexity of generating a CNF expression of f? Or, equivalently, what is the complexity of generating a DNF expression of f^d? For now, we are going to present a procedure that achieves a less ambitious goal in polynomial (indeed, linear) time. This procedure is essentially due to Tseitin [872] (see also Blair, Jeroslow,

and Lowe [98]). With an arbitrary Boolean expression $\phi(X) = \phi(x_1, x_2, \ldots, x_n)$
on $\mathcal{B}^n$, the procedure associates a DNF $\psi(X,Y) = \psi(x_1, x_2, \ldots, x_n, y_1, y_2, \ldots, y_m)$
(where $(y_1, y_2, \ldots, y_m)$ are additional variables, and possibly $m = 0$) and a distinguished literal z among the literals on $\{x_1, x_2, \ldots, x_n, y_1, y_2, \ldots, y_m\}$. These
constructs have the properties that, for all $X^* \in \mathcal{B}^n$, there is a (unique) point
$Y^* \in \mathcal{B}^m$ such that $\psi(X^*, Y^*) = 0$. Moreover, in every solution (X^*, Y^*) of the
equation $\psi(X,Y) = 0$, the distinguished coordinate of the point Y^* takes the value
$z^* = \phi(X^*)$.

The DNF $\psi(X,Y)$ can be regarded as providing an *implicit DNF representation*
of the function $\phi(X)$, in the following sense: in order to compute the value of ϕ
at a point $X^* \in \mathcal{B}^n$, one can solve the equation $\psi(X^*, Y) = 0$ and read the value
of z in the unique solution of the equation. We will encounter some applications
of this procedure to the analysis of switching circuits in Section 1.13.2 and to the
solution of Boolean equations in Chapter 2.

The procedure recursively processes each of the subexpressions of ϕ and then
recombines the resulting DNFs into a single one using additional variables. Intuitively, each additional variable y_i ($i = 1, 2, \ldots, m$) represents the value of one of the
subexpressions occurring in ϕ. The formulation of the recombination step depends
on whether the outermost operator in ϕ is a complementation, a disjunction, or a
conjunction.

Before we give a formal statement of this procedure, let us illustrate it on a
small example.

Example 1.12. *Consider the expression $\phi = (x_1 \vee x_4)\overline{(x_1 \vee (x_2\overline{x}_3))}$ (see Example
1.10). When working out an example by hand, it is easiest to apply the recur-
sive procedure "from bottom to top." So, we start at the lowest level, with the
subexpression $\phi_1 = x_1 \vee x_4$. This subexpression gives rise to an associated DNF*

$$\psi_1(x_1, x_4, y_1) = x_1\overline{y}_1 \vee x_4\overline{y}_1 \vee \overline{x}_1\overline{x}_4\, y_1,$$

*where y_1 is the distinguished literal associated with ψ_1. We will explain later how
this DNF ψ_1 has been constructed, but the reader can already verify that, in every
solution of the equation $\psi_1(x_1, x_4, y_1) = 0$, there holds $y_1 = \phi_1 = x_1 \vee x_4$, so that
the literal y_1 can be viewed as implicitly representing the subexpression ϕ_1.*

*Let us now proceed with the remaining subexpressions of ϕ. The subexpression
$\phi_3 = x_1$ yields the trivial expansion $\psi_3(x_1) = 0$, with x_1 itself as distinguished
literal, while the subexpression $\phi_4 = x_2\overline{x}_3$ expands into*

$$\psi_4(x_2, x_3, y_4) = \overline{x}_2 y_4 \vee x_3 y_4 \vee x_2\overline{x}_3\, \overline{y}_4,$$

*with y_4 as distinguished literal. Note again that, in every solution of
$\psi_4(x_2, x_3, y_4) = 0$, we have $y_4 = \phi_4 = x_2\overline{x}_3$.*

Combining ψ_3 and ψ_4, we obtain the DNF expansion of $\phi_2 = (x_1 \vee (x_2\overline{x}_3))$ as

$$\psi_2(x_1, x_2, x_3, y_2, y_4) = x_1\overline{y}_2 \vee y_4\overline{y}_2 \vee \overline{x}_1\overline{y}_4 y_2 \vee \overline{x}_2 y_4 \vee x_3 y_4 \vee x_2\overline{x}_3\overline{y}_4,$$

*where y_2 is the distinguished literal associated with ψ_2. Here again, one can verify
that the equality $y_2 = \phi_2 = (x_1 \vee (x_2\overline{x}_3))$ holds in every solution of the equation
$\psi_2 = 0$.*

The same DNF ψ_2 is also the DNF expansion of $\overline{(x_1 \vee (x_2 \overline{x}_3))}$, this time with $\overline{y}_2$ as associated literal.

Finally, putting all the pieces together, we obtain the desired expression of ψ:

$$\psi(x_1, x_2, x_3, y_1, y_2, y_4, z) = \overline{y}_1 z \vee y_2 z \vee y_1 \overline{y}_2 \overline{z}$$
$$\vee x_1 \overline{y}_1 \vee x_4 \overline{y}_1 \vee \overline{x}_1 \overline{x}_4 y_1$$
$$\vee x_1 \overline{y}_2 \vee y_4 \overline{y}_2 \vee \overline{x}_1 \overline{y}_4 y_2$$
$$\vee \overline{x}_2 y_4 \vee x_3 y_4 \vee x_2 \overline{x}_3 \overline{y}_4$$

with distinguished literal z. We leave it as an easy exercise to check that, for every $X = (x_1, x_2, x_3, x_4) \in \mathcal{B}^4$, the unique solution of the equation $\psi = 0$ satisfies

$$y_1 = \phi_1(X) = x_1 \vee x_4$$
$$y_2 = \phi_2(X) = x_1 \vee (x_2 \overline{x}_3)$$
$$y_4 = \phi_4(X) = x_2 \overline{x}_3$$
$$z \ = \phi(X).$$

$\square$

Figure 1.2 presents a formal description of the procedure EXPAND. Let us now establish the correctness of this procedure.

Theorem 1.5. *With every Boolean expression $\phi(X)$ on $\mathcal{B}^n$, the procedure EXPAND associates a DNF $\psi(X, Y)$ on $\mathcal{B}^{n+m}$ ($m \geq 0$) and a distinguished literal z among the literals on $\{x_1, x_2, \ldots, x_n, y_1, y_2, \ldots, y_m\}$ with the property that, for each $X^* \in \mathcal{B}^n$, there is a unique point $Y(X^*) \in \mathcal{B}^m$ such that $\psi(X^*, Y(X^*)) = 0$; moreover, in this point, the distinguished literal z is equal to $\phi(X^*)$. EXPAND can be implemented to run in linear time.*

Proof. We proceed by induction on the number of symbols in the expression ϕ. The statement trivially holds if ϕ contains only one literal. If ϕ contains more than one literal, then it must be of one of the types identified in EXPAND. Let us concentrate on the case $\phi = (\phi_1 \vee \phi_2 \vee \ldots \vee \phi_k)$ (the other cases are similar). Let $\psi_j(X, Y_j) = $ EXPAND(ϕ_j), where Y_j is a Boolean vector of appropriate dimension, and let z_j denote the distinguished literal of ψ_j, for $j = 1, 2, \ldots, k$. Then, by construction,

$$\psi := z_1 \overline{y} \vee z_2 \overline{y} \vee \ldots \vee z_k \overline{y} \vee \overline{z}_1 \overline{z}_2 \ldots \overline{z}_k y \vee \psi_1(X, Y_1) \vee \psi_2(X, Y_2) \vee \ldots \vee \psi_k(X, Y_k).$$

Fix $X^* \in \mathcal{B}^n$. By induction, there exist k points $Y_1^*, Y_2^*, \ldots, Y_k^*$, each of them uniquely defined, such that $\psi_j(X^*, Y_j^*) = 0$ and $z_j^* = \phi_j(X^*)$ for $j = 1, 2, \ldots, k$. It is then straightforward to verify that the condition $\psi(X^*, Y_1, Y_2, \ldots, Y_k, y) = 0$ holds for a unique choice of $(Y_1, Y_2, \ldots, Y_k, y)$, namely, for $Y_j = Y_j^*$ ($j = 1, 2, \ldots, k$), and for

$$y = z_1^* \vee z_2^* \vee \ldots \vee z_k^* = \phi_1(X^*) \vee \phi_2(X^*) \vee \ldots \vee \phi_k(X^*) = \phi(X^*).$$

The time complexity of the procedure is easily established by induction. $\square$

Procedure EXPAND(ϕ)

Input: A Boolean expression $\phi(x_1, x_2, \ldots, x_n)$.
Output: A DNF $\psi(x_1, x_2, \ldots, x_n, y_1, y_2, \ldots, y_m)$, with a distinguished literal z among the literals on $\{x_1, x_2, \ldots, x_n, y_1, y_2, \ldots, y_m\}$.

begin
    if $\phi = x_i$ for some $i \in \{1, 2, \ldots, n\}$
        then return $\psi(x_1, x_2, \ldots, x_n) = \mathbf{0}_n$ and the distinguished literal x_i
    else if $\phi = \overline{\phi_1}$ for some expression ϕ_1 **then**
        begin
            let $\psi_1 := \text{EXPAND}(\phi_1)$ and let z be the distinguished literal of ψ_1;
            return $\psi := \psi_1$ and the distinguished literal $\overline{z}$;
        end
    else if $\phi = (\phi_1 \vee \phi_2 \vee \ldots \vee \phi_k)$ for some expressions $\phi_1, \phi_2, \ldots, \phi_k$ **then**
        begin
            for $j = 1$ **to** k **do** $\psi_j := \text{EXPAND}(\phi_j)$;
            let z_j be the distinguished literal of ψ_j, for $j = 1, 2, \ldots, k$;
            create a new variable y;
            return $\psi := z_1\overline{y} \vee z_2\overline{y} \vee \ldots \vee z_k\overline{y} \vee \overline{z}_1\overline{z}_2 \ldots \overline{z}_k y \vee \psi_1 \vee \psi_2 \vee \ldots \vee \psi_k$
            and the distinguished literal y;
        end
    else if $\phi = (\phi_1 \phi_2 \ldots \phi_k)$ for some expressions $\phi_1, \phi_2, \ldots, \phi_k$ **then**
        begin
            for $j = 1$ **to** k **do** $\psi_j := \text{EXPAND}(\phi_j)$;
            let z_j be the distinguished literal of ψ_j, for $j = 1, 2, \ldots, k$;
            create a new variable y;
            return $\psi := \overline{z}_1 y \vee \overline{z}_2 y \vee \ldots \vee \overline{z}_k y \vee z_1 z_2 \ldots z_k \overline{y} \vee \psi_1 \vee \psi_2 \vee \ldots \vee \psi_k$
            and the distinguished literal y;
        end
end

Figure 1.2. Procedure EXPAND.

1.6 Orthogonal DNFs and number of true points

A classical problem of Boolean theory is to derive an *orthogonal disjunctive normal form* of an arbitrary Boolean function. In order to define this concept, consider a DNF

$$\phi = \bigvee_{k=1}^{m} \left(\bigwedge_{i \in A_k} x_i \bigwedge_{j \in B_k} \overline{x}_j \right), \tag{1.13}$$

where $A_k \cap B_k = \emptyset$ for all $k = 1, 2, \ldots, m$.

Definition 1.13. *A DNF of the form* (1.13) *is said to be* orthogonal, *or to be a sum of disjoint products, if* $(A_k \cap B_\ell) \cup (A_\ell \cap B_k) \neq \emptyset$ *for all* $k, \ell \in \{1, 2, \ldots, m\}, k \neq \ell$.

Definition 1.13 simply states that every two terms of an orthogonal DNF must be "conflicting" in at least one variable; that is, there must be a variable that appears complemented in one of the terms and uncomplemented in the other term. This property is easy to test for any given DNF.

Note also that a DNF is orthogonal if and only if, for every pair of terms $k, l \in \{1, 2, \ldots, m\}, k \neq l$, and for every $X^* \in \mathcal{B}^n$,

$$\left(\bigwedge_{i \in A_k} x_i^* \bigwedge_{j \in B_k} \overline{x^*}_j \right) \left(\bigwedge_{i \in A_l} x_i^* \bigwedge_{j \in B_l} \overline{x^*}_j \right) = 0.$$

The terminology "orthogonal" is quite natural in view of this observation, the proof of which is left to the reader.

Example 1.13. *The DNF* $\phi = \overline{x}_1 \overline{x}_2 x_4 \vee \overline{x}_1 x_3 x_4$ *is not orthogonal since the point* $X^* = (0, 0, 1, 1)$ *makes both of its terms equal to 1. But ϕ is equivalent to the DNF* $\psi = \overline{x}_1 \overline{x}_2 x_4 \vee \overline{x}_1 x_2 x_3 x_4$, *which is orthogonal.* □

As this example illustrates, the following specialization of Theorem 1.4 holds.

Theorem 1.6. *Every Boolean function can be represented by an orthogonal DNF.*

Proof. It suffices to observe that the minterm expression (1.6) used in the proof of Theorem 1.4 is orthogonal. □

Let us now establish a remarkable property of orthogonal DNFs that reinforces several of our earlier comments about the usefulness of interpreting the elements of $\mathcal{B} = \{0, 1\}$ as numbers (see Section 1.1) and the similarity between DNFs and polynomials over the reals (see the end of Section 1.4).

Theorem 1.7. *If the Boolean function f on $\mathcal{B}^n$ is represented by an orthogonal DNF of the form* (1.13), *and if the elements of $\mathcal{B}$ are interpreted as numbers, then*

$$f(X) = \sum_{k=1}^{m} \left(\prod_{i \in A_k} x_i \prod_{j \in B_k} (1 - x_j) \right), \tag{1.14}$$

for all $X = (x_1, x_2, \ldots, x_n) \in \mathcal{B}^n$.

Proof. Since the terms of (1.14) are pairwise orthogonal, at most one of them takes value 1 at any point $X \in \mathcal{B}^n$. □

One of the main motivations for the interest in orthogonal DNFs is that, for functions expressed in this form, computing the number of true points turns out to be extremely easy.

Theorem 1.8. *If the Boolean function f on $\mathcal{B}^n$ is represented by an orthogonal DNF of the form* (1.13), *then the number of its true points is equal to*

$$\omega(f) = \sum_{k=1}^{m} 2^{n - |A_k| - |B_k|}.$$

Proof. The DNF (1.13) takes value 1 exactly when one of its terms takes value 1. Since the terms are pairwise orthogonal, $\omega(f) = \sum_{k=1}^{m} \alpha_k$, where α_k denotes the number of true points of the k-th term. The statement follows easily. □

At this point, the reader may be wondering (with some reason) why anyone would ever want to compute the number of true points of a Boolean function. We present several applications of this concept in Section 1.13. For now, it may be sufficient to note that determining the number of true points of a function f is a roundabout way to check the consistency of the Boolean equation $f = 0$.

Chow [194] introduced several parameters of a Boolean function that are closely related to the number $\omega(f)$ defined in Theorem 1.8.

Definition 1.14. *The* Chow parameters *of a Boolean function f on $\mathcal{B}^n$ are the $n+1$ integers $(\omega_1, \omega_2, \ldots, \omega_n, \omega)$, where $\omega = \omega(f)$ is the number of true points of f and ω_i is the number of true points X^* of f such that $x_i^* = 1$:*

$$\omega_i = | \{X^* \in \mathcal{B}^n \mid f(X^*) = 1 \text{ and } x_i^* = 1\} |, \quad i = 1, 2, \ldots, n.$$

The same reasoning as in Theorem 1.8 shows that the Chow parameters of a function represented in orthogonal form can be efficiently computed: For ω, this is just a consequence of Theorem 1.8; for ω_i ($1 \le i \le n$), this follows from the fact that the DNF obtained by fixing x_i to 1 in an orthogonal DNF remains orthogonal.

Example 1.14. *The function f represented by the orthogonal DNF $\psi = \overline{x}_1\overline{x}_2x_4 \vee \overline{x}_1x_2x_3x_4$ has Chow parameters $(\omega_1, \omega_2, \omega_3, \omega_4, \omega) = (0, 1, 2, 3, 3)$. Indeed, f has exactly three true points, $x_1 = 0$ and $x_4 = 1$ in all true points, $x_2 = 1$ in exactly one true point, and $x_3 = 1$ in exactly two true points.* $\square$

Chow parameters, and variants thereof, have been independently rediscovered by several researchers; in particular, up to scaling and shifting, they are identical to the so-called degree-0 and degree-1 *Fourier coefficients* of a Boolean function. Chow parameters have found applications in as diverse fields as electrical engineering (Chow [194], Winder [920]), game theory (Banzhaf [52], Dubey and Shapley [279]), reliability theory (Birnbaum [91], Barlow and Proschan [54]), cryptography (Carlet [170]), and theoretical computer science (see Ben-Or and Linial [60], Bruck [157], Kahn, Kalai and Linial [543]); see also O'Donnell [716] for an overview of applications. We return to Chow parameters in Section 1.13 and in subsequent chapters, especially in Chapter 9. Orthogonal forms are further discussed in Chapter 7.

1.7 Implicants and prime implicants

Definition 1.15. *Given two Boolean functions f and g on $\mathcal{B}^n$, we say that f* implies *g (or that f is a* minorant *of g, or that g is a* majorant *of f) if*

$$f(X) = 1 \implies g(X) = 1 \quad \text{for all } X \in \mathcal{B}^n.$$

When this is the case, we write $f \le g$.

This definition extends in a straightforward way to Boolean expressions, since every such expression can be regarded as a Boolean function.

The terminology "f *implies* g" is obviously borrowed from logic: If f and g model, respectively, the truth value of propositional sentences S_f and S_g, then $f \leq g$ holds exactly when $S_f \Rightarrow S_g$. On the other hand, the terms "minorant," and "majorant," as well as the notation "$f \leq g$" are easily motivated by looking at f and g as integer-valued functions. Also, as suggested by the notation, the equality $f = g$ holds if and only if $f \leq g$ and $g \leq f$ hold simultaneously.

The following alternative forms of Definition 1.15 are frequently useful.

Theorem 1.9. *For all Boolean functions f and g on $\mathcal{B}^n$, the following statements are equivalent:*

(1) $f \leq g$;
(2) $f \vee g = g$;
(3) $\overline{f} \vee g = \mathbb{1}_n$;
(4) $f g = f$;
(5) $f \overline{g} = \mathbb{0}_n$.

Proof. It suffices to note that each of the assertions (1)–(5) fails exactly when there exists $X \in \mathcal{B}^n$ such that $f(X) = 1$ and $g(X) = 0$. $\square$

Let us record a few additional properties of the implication relation.

Theorem 1.10. *For all Boolean functions f, g, and h on $\mathcal{B}^n$,*

(1) $\mathbb{0}_n \leq f \leq \mathbb{1}_n$;
(2) $f g \leq f \leq f \vee g$;
(3) $f = g$ if and only if $(f \leq g$ and $g \leq f)$;
(4) $(f \leq h$ and $g \leq h)$ if and only if $f \vee g \leq h$;
(5) $(f \leq g$ and $f \leq h)$ if and only if $f \leq g h$;
(6) if $f \leq g$ then $f h \leq g h$;
(7) if $f \leq g$ then $f \vee h \leq g \vee h$;

Proof. All these properties are easily verified. $\square$

When two Boolean functions f and g are represented by arbitrary Boolean expressions, it can be quite difficult to check whether or not f implies g. Definition 1.15 does not suggest any efficient way to perform this task, except by complete enumeration of all the points in $\mathcal{B}^n$, nor does Theorem 1.9 help in this respect. We will come back to this point in Chapter 2, when we discuss the complexity of solving Boolean equations.

For elementary conjunctions, however, implication takes an especially simple, easily verifiable form: Indeed, an elementary conjunction implies another one if and only if the latter results from the former by deletion of literals (the "longer" conjunction implies the "shorter" one). More formally:

Theorem 1.11. *The elementary conjunction $C_{AB} = \bigwedge_{i \in A} x_i \bigwedge_{j \in B} \overline{x}_j$ implies the elementary conjunction $C_{FG} = \bigwedge_{i \in F} x_i \bigwedge_{j \in G} \overline{x}_j$ if and only if $F \subseteq A$ and $G \subseteq B$.*

Proof. Assume that $F \subseteq A$ and $G \subseteq B$ and consider any point $X = (x_1, x_2, \ldots, x_n) \in \mathcal{B}^n$. If $C_{AB}(X) = 1$, then $x_i = 1$ for all $i \in A$ and $x_j = 0$ for all $j \in B$, so that $x_i = 1$ for all $i \in F$ and $x_j = 0$ for all $j \in G$. Hence, $C_{FG}(X) = 1$ and we conclude that C_{AB} implies C_{FG}.

To prove the converse statement, assume for instance that F is not contained in A. Set $x_i = 1$ for all $i \in A$, $x_j = 0$ for all $j \notin A$ and $X = (x_1, x_2, \ldots, x_n)$. Then, $C_{AB}(X) = 1$ but $C_{FG}(X) = 0$ (since $x_k = 0$ for some $k \in F \setminus A$), so that C_{AB} does not imply C_{FG}. □

Definition 1.16. *Let f be a Boolean function and C be an elementary conjunction. We say that C is an* implicant *of f if C implies f.*

Example 1.15. *Let $f = xy \vee x\overline{y}z$. Then xy, $x\overline{y}z$ and xz are implicants of f.* □

We can now formulate an easy observation.

Theorem 1.12. *If ϕ is a DNF representation of the Boolean function f, then every term of ϕ is an implicant of f. Moreover, if C is an implicant of f, then the DNF $\phi \vee C$ also represents f.*

Proof. For the first statement, notice that, if any term of ϕ takes value 1, then ϕ, and hence f, take value 1. For the second statement, just check successively that $\phi \vee C \leq f$ and $f \leq \phi \leq \phi \vee C$. □

Example 1.16. *By Theorem 1.12, the function $f = xy \vee x\overline{y}z$ (see Example 1.15) admits the DNF expression $xy \vee x\overline{y}z \vee xz = xy \vee xz$ (the last equality is easily verified to hold).* □

Example 1.16 illustrates an important point: With a view toward simplification of Boolean expressions, it makes sense to replace "long" implicants by "short" ones in DNF representations of a Boolean function. The meaning of "long" and "short" can be clarified by reference to Theorem 1.11. This line of reasoning leads to the following definitions (see Quine [766, 768]).

Definition 1.17. *Let f be a Boolean function and C_1, C_2 be implicants of f. We say that C_1* absorbs *C_2 if $C_1 \vee C_2 = C_1$ or, equivalently, if $C_2 \leq C_1$.*

Definition 1.18. *Let f be a Boolean function and C_1 be an implicant of f. We say that C_1 is a* prime implicant *of f if C_1 is not absorbed by any other implicant of f (namely, if C_2 is an implicant of f and $C_1 \leq C_2$, then $C_1 = C_2$).*

Example 1.17. *Consider again the function f defined in Example 1.15. It is easy to verify that xy and xz are prime implicants of f, whereas $x\overline{y}z$ is not prime (since $x\overline{y}z \leq xz$). As a matter of fact, f has no prime implicants other than xy and xz.* □

Prime implicants play a crucial role in constructing DNF expressions of Boolean functions. This role is best described by the next theorem (compare with Theorem 1.4).

Theorem 1.13. *Every Boolean function can be represented by the disjunction of all its prime implicants.*

Proof. Let f be a Boolean function on B^n, and let $P_1, P_2, \ldots, P_m$ be its prime implicants (notice that m is finite because the number of elementary conjunctions on n variables is finite). Consider any DNF representation of f, say $\phi = \bigvee_{k=1}^{r} C_k$. By Theorem 1.12, the DNF

$$\psi = \left(\bigvee_{k=1}^{r} C_k \right) \vee \left(\bigvee_{j=1}^{m} P_j \right)$$

also represents f. Consider any term C_k of ϕ ($1 \le k \le r$). Since C_k is an implicant of f, it is absorbed by at least one prime implicant of f, say, by P_j (where possibly $C_k = P_j$). Then, it follows that $C_k \vee P_j = P_j$, from which we deduce $\psi = \bigvee_{j=1}^{m} P_j$. $\qquad\square$

The DNF representation introduced in Theorem 1.13 will be used repeatedly throughout this book, and therefore deserves a special name.

Definition 1.19. *The disjunction of all prime implicants of a Boolean function is called the* complete DNF *(or the* Blake canonical form*) of this function.*

Note that the complete DNF is only unique up to the order of its terms and literals. However, just as we did in the case of minterm expressions, we shall disregard this subtlety and simply look at the complete DNF as being uniquely defined.

An interesting corollary of Theorem 1.13 is that each Boolean function is uniquely identified by the list of its prime implicants. Equivalently, two Boolean functions are equal if and only if they have the same complete DNF. Let us stress, however, that it is not always necessary to know *all* the prime implicants of a function to know the function, and that it is not always necessary to take the disjunction of *all* the prime implicants to obtain a correct DNF representation of the function.

Example 1.18. *The function $g = x\overline{y} \vee \overline{x}y \vee x\overline{z}$ has four prime implicants, namely, $x\overline{y}$, $\overline{x}y$, $x\overline{z}$, and $y\overline{z}$.* $\qquad\square$

More generally, let us introduce the following terminology.

Definition 1.20. *Let f be a Boolean function on B^n and let $\phi = \bigvee_{k \in \Omega} C_k$ be a DNF representation of f. We say that ϕ is a* prime DNF *of f if each term C_k ($k \in \Omega$) is a prime implicant of f. We say that ϕ is an* irredundant DNF *of f if there is no $j \in \Omega$ such that $\bigvee_{k \in \Omega \setminus \{j\}} C_k$ represents f; otherwise, we say that f is* redundant.

So, a redundant DNF expression can be turned into a shorter equivalent DNF by dropping some of its terms. For instance, Example 1.18 shows that the complete DNF of a Boolean function is not necessarily irredundant. Similarly, if a DNF is not prime, then at least one of its terms can be replaced by a prime implicant that absorbs it (remember Theorem 1.12 and the comments following it). Therefore, prime irredundant DNFs provide the shortest possible DNF representations of Boolean functions. In Chapter 3, we return to the study of prime irredundant DNFs in detail.

Of course, the concepts of implicants and prime implicants have their natural disjunctive counterparts.

Definition 1.21. *Let f be a Boolean function and D be an elementary disjunction. We say that D is an* implicate *of f if f implies D. We say that the implicate D is* prime *if it is not implied by any other implicate of f.*

Similarly to Theorem 1.13, we obtain:

Theorem 1.14. *Every Boolean function can be represented by the conjunction of all its prime implicates.*

Proof. The proof is a straightforward adaptation of the proof of Theorem 1.13. $\square$

Example 1.19. *The function g considered in Example 1.18 has four implicates, namely,* $(x \vee y)$, $(x \vee y \vee z)$, $(x \vee y \vee \overline{z})$, *and* $(\overline{x} \vee \overline{y} \vee \overline{z})$. *However, only the first and the last implicates in this list are prime, and we conclude that* $g = (x \vee y)$ $(\overline{x} \vee \overline{y} \vee \overline{z})$. $\square$

1.8 Restrictions of functions, essential variables

We now introduce the concept of *restriction* (sometimes called *projection*) of a Boolean function.

Definition 1.22. *Let f be a Boolean function on* $\mathcal{B}^n$, *and let* $k \in \{1, 2, \ldots, n\}$. *We denote by* $f_{|x_k=1}$ *and* $f_{|x_k=0}$, *respectively, the Boolean functions on* $\mathcal{B}^{n-1}$ *defined as follows: For every* $(x_1, \ldots, x_{k-1}, x_{k+1}, \ldots, x_n) \in \mathcal{B}^{n-1}$,

$$f_{|x_k=1}(x_1, \ldots, x_{k-1}, x_{k+1}, \ldots, x_n) = f(x_1, \ldots, x_{k-1}, 1, x_{k+1}, \ldots, x_n),$$

$$f_{|x_k=0}(x_1, \ldots, x_{k-1}, x_{k+1}, \ldots, x_n) = f(x_1, \ldots, x_{k-1}, 0, x_{k+1}, \ldots, x_n).$$

We say that $f_{|x_k=1}$ *is the* restriction *of f to* $x_k = 1$ *and that* $f_{|x_k=0}$ *is the* restriction *of f to* $x_k = 0$.

Even though $f_{|x_k=1}$ and $f_{|x_k=0}$ are, by definition, functions of $n-1$ variables, we can also look at them as functions on $\mathcal{B}^n$ rather than $\mathcal{B}^{n-1}$, via the following convention: For every $(x_1, x_2, \ldots, x_n) \in \mathcal{B}^n$, we simply let

$$f_{|x_k=1}(x_1, x_2, \ldots, x_n) = f(x_1, \ldots, x_{k-1}, 1, x_{k+1}, \ldots, x_n),$$

and similarly for $f_{|x_k=0}(x_1,x_2,\ldots,x_n)$. This slight abuse of definitions is innocuous and we use it whenever it proves convenient. Also, we use shorthand like $f_{|x_1=0,x_2=1,x_3=0}$ instead of the more cumbersome notation $\left(\left(f_{|x_1=0}\right)_{|x_2=1}\right)_{|x_3=0}$.

The link between representations of a function and representations of its restrictions is straightforward.

Theorem 1.15. *Let f be a Boolean function on B^n, let ψ be a representation of f, and let $k \in \{1,2,\ldots,n\}$. Then, the expression obtained by substituting the constant 1 (respectively, 0) for every occurrence of x_k in ψ represents $f_{|x_k=1}$ (respectively, $f_{|x_k=0}$).*

Proof. This is an immediate consequence of Definitions 1.22 and 1.5. □

Example 1.20. *Consider the function $f = (xz \vee y)(x \vee \overline{z}) \vee \overline{x}\,\overline{y}$. After some easy simplifications, we derive the following expressions for $f_{|y=1}$ and $f_{|y=0}$:*

$$f_{|y=1} = (xz \vee 1)(x \vee \overline{z}) \vee \overline{x}\,\overline{1} = x \vee \overline{z},$$

$$f_{|y=0} = (xz \vee 0)(x \vee \overline{z}) \vee \overline{x}\,\overline{0} = xz \vee \overline{x} = z \vee \overline{x}.$$

□

We now prove a trivial, but useful identity.

Theorem 1.16. *Let f be a Boolean function on B^n, and let $k \in \{1,2,\ldots,n\}$. Then,*

$$f(x_1,x_2,\ldots,x_n) = x_k f_{|x_k=1} \vee \overline{x}_k f_{|x_k=0} \tag{1.15}$$

for all $(x_1,x_2,\ldots,x_n) \in B^n$.

Proof. This is immediate by substitution of the values $x_k = 0$ or $x_k = 1$ in (1.15). □

The right-hand side of the identity (1.15) is often called the *Shannon expansion* of the function f with respect to x_k, by reference to its use by Shannon in [827], although this identity was already well-known to Boole [103]. It can be used, in particular, to construct the minterm DNF of a function (Theorem 1.4 and Definition 1.11). More interestingly, by applying the Shannon expansion to a function and to its successive restrictions until these restrictions become either 0, or 1, or a literal, we obtain an orthogonal DNF of the function (this is easily proved by induction on n). Not every orthogonal DNF, however, can be obtained in this way.

Example 1.21. *Consider again the function f in Example 1.20. The Shannon expansion of $f_{|y=1}$ with respect to x is*

$$x f_{|y=1,x=1} \vee \overline{x} f_{|y=1,x=0} = x \vee \overline{x}\,\overline{z}.$$

Observe that $f_{|y=1,x=1}$ is identically 1 and $f_{|y=1,x=0} = \overline{z}$ is a literal, so we terminate here the expansion $f_{|y=1}$.

Similarly, the Shannon expansion of $f_{|y=0}$ with respect to z (for a change) is

$$z f_{|y=0,z=1} \vee \overline{z} f_{|y=0,z=0} = z \vee \overline{z}\,\overline{x}.$$

Putting the pieces together, we obtain

$$f(x,y,z) = y\,(x \vee \overline{x}\,\overline{z}) \vee \overline{y}\,(z \vee \overline{z}\,\overline{x}) = x\,y \vee \overline{x}\,y\,\overline{z} \vee \overline{y}\,z \vee \overline{x}\,\overline{y}\,\overline{z},$$

which is an orthogonal DNF of f.

Another orthogonal DNF of f is $x\,y \vee \overline{x}\,\overline{z} \vee \overline{y}\,z$. But this DNF cannot be obtained from successive Shannon expansions, since, when applying this procedure, we necessarily produce a DNF in which one of the variables appears in all the terms.   □

Let us now turn to the concept of essential variables.

Definition 1.23. *Let f be a Boolean function on B^n, and let $k \in \{1,2,\ldots,n\}$. We say that the variable x_k is inessential for f, or that x_k is a dummy for f, or that f does not depend on x_k, if $f_{|x_k=1}(X) = f_{|x_k=0}(X)$ for all $X \in B^{n-1}$. Otherwise, we say that x_k is essential.*

If a function has a representation in which some specific variable x_k does not appear, then, as a consequence of Theorem 1.15, the function does not depend on x_k. The converse statement is slightly less obvious but nevertheless valid.

Theorem 1.17. *Let f be a Boolean function on B^n, and let $k \in \{1,2,\ldots,n\}$. The following statements are equivalent:*

(1) *The variable x_k is inessential for f.*
(2) *The variable x_k does not appear in any prime implicant of f.*
(3) *f has a DNF representation in which the variable x_k does not appear.*

Proof. The second statement implies the third one by Theorem 1.13, and the third statement implies the first one by Theorem 1.15. Let us now assume that x_k is inessential for f, and let us consider an arbitrary implicant of f, say, $C_{AB} = \bigwedge_{i \in A} x_i \bigwedge_{j \in B} \overline{x}_j$. Assume, for instance, that $k \in A$ (the argument would be similar for $k \in B$) and consider the conjunction C obtained by deleting x_k from C_{AB}:

$$C = \left(\bigwedge_{i \in A \setminus \{k\}} x_i \right) \left(\bigwedge_{j \in B} \overline{x}_j \right).$$

We claim that C is an implicant of f: This will in turn entail that the prime implicants of f do not involve x_k. To prove the claim, let $X = (x_1, x_2, \ldots, x_n)$ be any point in B^n such that $C(X) = 1$, and let us show that $f(X) = 1$. Since neither C nor f depend on x_k, we may as well suppose that $x_k = 1$. Then, $C(X) = C_{AB}(X) = 1$ and hence $f(X) = 1$, as required.   □

It should be obvious, however, that any particular representation of a function may involve a variable on which the function does not depend. So, for instance, the minterm expression introduced in Definition 1.11 involves n variables for every function on $\mathcal{B}^n$ (except the null function $\mathbf{0}_n$), even when the function depends on much fewer than n variables.

Example 1.22. *The DNF* $\phi(x_1, x_2, x_3, x_4) = x_1 x_2 \vee x_1 \overline{x}_2 \vee \overline{x}_1 x_2 \vee \overline{x}_1 \overline{x}_2$ *represents the constant function* $\mathbf{1}_4$ *on* $\mathcal{B}^4$. *In particular,* ϕ *does not depend on any of its variables.* $\square$

We will prove later that, for a function represented by an arbitrary DNF expression, it is generally difficult to determine whether any given variable is essential or not (see Theorem 1.32 in Section 1.11).

Finally, let us mention an interesting connection between the concept of essential variables and of Chow parameters.

Theorem 1.18. *Let* f *be a Boolean function on* $\mathcal{B}^n$, *let* $(\omega_1, \omega_2, \ldots, \omega_n, \omega)$ *be its vector of Chow parameters, and let* $k \in \{1, 2, \ldots, n\}$. *If the variable* x_k *is inessential for* f, *then* $\omega = 2\omega_k$.

Proof. The sets $A = \{X \in \mathcal{B}^n \mid f(X) = 1, x_k = 1\}$ and $B = \{X \in \mathcal{B}^n \mid f(X) = 1, x_k = 0\}$ partition the set of true points of f, and $|A| = \omega_k$, $|B| = \omega - \omega_k$. If x_k is inessential, then A and B are in one-to-one correspondence, so $\omega = 2\omega_k$. $\square$

The converse of Theorem 1.18 is not valid since the function $f(x_1, x_2) = x_1 \overline{x}_2 \vee \overline{x}_1 x_2$ has Chow parameters $(1,1,2)$, and both variables x_1, x_2 are essential.

1.9   Geometric interpretation

Most of the concepts introduced in the previous sections have simple, but frequently useful, geometric interpretations. First, the points of $\mathcal{B}^n$ can be identified with the vertices of the unit hypercube

$$U^n = \{X \in \mathbb{R}^n \mid 0 \le x_i \le 1 \text{ for } i = 1, 2, \ldots, n\}.$$

Every Boolean function defines a partition of the vertices of U^n into true points and false points. Conversely, this partition completely characterizes the function.

Consider an arbitrary elementary conjunction of the form $C_{AB} = \bigwedge_{i \in A} x_i$ $\bigwedge_{j \in B} \overline{x}_j$. The set of true points of C_{AB} is

$$T_{AB} = \{X \in \mathcal{B}^n \mid x_i = 1 \text{ for all } i \in A \text{ and } x_j = 0 \text{ for all } j \in B\}.$$

Geometrically, the points in T_{AB} are exactly the vertices contained in a face of U^n. Every such face is itself a hypercube of dimension $n - |A| - |B|$ containing $2^{n-|A|-|B|}$ vertices, and will therefore be referred to as a *subcube*. (Some authors, especially in the electrical engineering literature, actually use the term "cube" instead of "elementary conjunction.")

Consider now a Boolean function f. In view of the previous observation, each implicant of f corresponds to a subcube of U^n that contains no false points of f. The implicant is prime if the corresponding subcube is maximal with this property.

Let $\phi = \bigvee_{k=1}^m C_k$ be an arbitrary DNF expression of the function f. The set of true points of f coincides with the union of the sets of true points of the

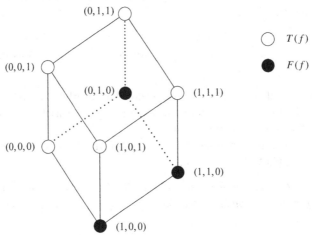

Figure 1.3. A 3-dimensional view of the Boolean function of Example 1.23.

Table 1.2. A Karnaugh map

		(x_2, x_3)			
		00	01	11	10
(x_1)	0	1	1	1	0
	1	0	1	1	0

terms C_k. In other words, a DNF expression of f can be viewed as a collection of subcubes of U^n that cover all the true points of f and none of its false points. In particular, an orthogonal DNF is one for which the subcubes in the collection are pairwise disjoint. This observation motivates the terminology "sum of disjoint products" mentioned in Definition 1.13 and may provide an alternative insight into Theorem 1.8.

The classical representation of Boolean functions by *Karnaugh maps* is directly inspired by the geometric point of view. Although Karnaugh maps may be useful for visual inspection of functions involving a small number of variables (up to 5 or 6, at most), they are inadequate for algorithmic purposes and thus have become obsolete. We illustrate them here with only a simple example and refer the interested reader to Maxfield [678] or to Mendelson [680] for more details.

Example 1.23. *Consider again the function given by Table 1.1 in Section 1.1.*

A Karnaugh map for this DNF is given by the matrix displayed in Table 1.2. The rows of the map are indexed by the values of the variable x_1; its columns are indexed by the values of the pair of variables (x_2, x_3); and each cell contains the value of the function in the corresponding Boolean point. For instance, the cell in the second row, fourth column, of the map contains a 0, since $f(1, 1, 0) = 0$.

Because of the special way in which the columns are ordered, two adjacent cells always correspond to neighboring vertices of the unit hypercube U; that is, the corresponding points differ in exactly one component. This remains true if we think of the Karnaugh map as being wrapped on a torus, with cell $(0,10)$ adjacent to cell $(0,00)$. Likewise, each row of the map corresponds to a 2-dimensional face of U, and so do squares formed by 4 adjacent cells, like $(0,01)$, $(0,11)$, $(1,01)$, and $(1,11)$.

Now, note that every cell of the map containing a 1 can alternatively be viewed as representing a minterm of the function f. For instance, the cell $(0,01)$ corresponds to the minterm $\bar{x}_1\bar{x}_2x_3$. Moreover, any two adjacent cells with value 1 can be combined to produce an implicant of degree 2. So, the cells $(0,01)$ and $(0,11)$ generate the implicant $\bar{x}_1x_3$, and so on. Finally, each row or square containing four 1's generates an implicant of degree 1; e.g., the cells $(0,01)$, $(0,11)$, $(1,01)$, and $(1,11)$ correspond to the implicant x_1.

So, in order to derive from the map a DNF expression of f, we just have to find a collection of subsets of adjacent cells corresponding to implicants of f and covering all the true points of f. Each such collection generates a different DNF of f. For instance, the pairs of cells $((0,00),(0,01))$, $((0,01),(0,11))$, and $((1,01),(1,11))$ simultaneously cover all the true points of f and generate the DNF

$$\phi = \bar{x}_1\bar{x}_2 \vee \bar{x}_1x_3 \vee x_1x_3.$$

Alternatively, the true points can be covered by the pair $((0,00),(0,01))$ and by the square $((0,01),(0,11),(1,01),(1,11))$, thus giving rise to the DNF

$$\psi = \bar{x}_1\bar{x}_2 \vee x_3.$$

□

Karnaugh maps have been mostly used by electrical engineers to identify short (irredundant, prime) DNFs of Boolean functions of a small number of variables. Extensions of this problem to arbitrary functions will be discussed in Section 3.3.

1.10 Monotone Boolean functions

In this section, we introduce one of the most important classes of Boolean functions, namely, the class of monotone functions, which subsumes several other special classes of functions studied further in this book. We establish some of the fundamental properties of monotone functions and of their normal forms. Many other properties of monotone functions will be uncovered in subsequent chapters (see also Korshunov [580] for a long survey devoted to monotone functions).

1.10.1 Definitions and examples

"Monotonically increasing" and "monotonically decreasing" real-valued functions are classical objects of study in elementary calculus. The following definition attempts to capture similar concepts in a Boolean framework.

Definition 1.24. *Let f be a Boolean function on $\mathcal{B}^n$, and let $k \in \{1,2,\ldots,n\}$. We say that f is* positive *(respectively,* negative) *in the variable x_k if $f_{|x_k=0} \leq f_{|x_k=1}$ (respectively, $f_{|x_k=0} \geq f_{|x_k=1}$). We say that f is* monotone *in x_k if f is either positive or negative in x_k.*

Thus, when f is positive in x_k, changing the value of x_k from 0 to 1 (while keeping the other variables fixed) cannot change the value of f from 1 to 0.

Definition 1.25. *A Boolean function is* positive *(respectively,* negative) *if it is positive (respectively, negative) in each of its variables. The function is* monotone *if it is monotone in each of its variables.*

Example 1.24. *The function $f(x_1,x_2,x_3) = \overline{x}_1\overline{x}_2 \vee x_3$, whose truth table was given in Example 1.1, is negative in x_1, negative in x_2, and positive in x_3. Hence, f is monotone, but it is neither positive nor negative.*
   The function $h(x,y) = x\overline{y} \vee \overline{x}y$ is neither monotone in x nor monotone in y. For instance, to see that h is not positive in x, observe that $h(0,1) = 1$, whereas $h(1,1) = 0$, and hence $h_{|x=0} = y \not\leq h_{|x=1} = \overline{y}$. $\square$

Application 1.7. (Voting theory.) *Remember the decision-making situation sketched in Application 1.3. Voting rules are usually designed in such a way that the outcome of a vote cannot switch from "Yes" to "No" when any single player's vote switches from "No" to "Yes". For this reason, simple games are most adequately modeled by positive Boolean functions.* $\square$

Application 1.8. (Reliability theory.) *In the context described in Application 1.4, it is rather natural to assume that a currently working system does not fail when we replace a defective component by an operative one. Therefore, a common hypothesis in reliability theory is that structure functions of complex systems are positive Boolean functions.* $\square$

Application 1.9. (Graphs and hypergraphs.) *The stability function of a hypergraph, as defined in Application 1.5, is a positive Boolean function since every subset of a stable set is stable.* $\square$

As is too often the case in the Boolean literature, the terminology established in Definitions 1.24 and 1.25 is not completely standardized, and authors working in different fields have proposed several variants. So, for example, monotone functions are also called *unate* or *1-monotone* in the electrical engineering and threshold logic literature. Computer scientists usually reserve the qualifier "monotone" for what we call "positive" functions, and so forth.

Notice that, in many applications, the distinction between positive and negative variables (and hence, between positive and monotone functions) turns out to be irrelevant. This holds by virtue of the following fact.

Theorem 1.19. *Let f be a Boolean function on $\mathcal{B}^n$, and let g be the function defined by*

$$g(x_1,x_2,\ldots,x_n) = f(\overline{x}_1,x_2,\ldots,x_n)$$

for all $(x_1, x_2, \ldots, x_n) \in \mathcal{B}^n$. *Then,* g *is positive in the variable* x_1 *if and only if* f *is negative in* x_1.

Proof. This is a trivial consequence of Definition 1.24. $\qquad\square$

So, when a monotone function is neither positive nor negative (as in the preceding Example 1.24), it can always be brought to one of these two forms by an elementary change of variables. This suggests that, in many cases, it is sufficient to study the properties of positive functions to understand the properties of monotone functions. This is our point of view in the next sections.

Let us give a characterization of positive functions that can be seen as a simple restatement of Definitions 1.24 and 1.25. For two points $X = (x_1, x_2, \ldots, x_n)$ and $Y = (y_1, y_2, \ldots, y_n)$ in $\mathcal{B}^n$, we write $X \le Y$ if $x_i \le y_i$ for all $i = 1, 2, \ldots, n$.

Theorem 1.20. *A Boolean function* f *on* $\mathcal{B}^n$ *is positive if and only if* $f(X) \le f(Y)$ *for all* $X, Y \in \mathcal{B}^n$ *such that* $X \le Y$.

Proof. The "if" part of the statement is trivial, and the "only if" part is easily established by induction on the number of components of X and Y such that $x_i < y_i$. $\qquad\square$

1.10.2 DNFs and prime implicants of positive functions

Let us now try to understand the main features of positive functions in terms of their DNF representations and their prime implicants. To this effect, we first introduce some remarkable classes of disjunctive normal forms.

Definition 1.26. *Let* $\psi(x_1, x_2, \ldots, x_n)$ *be a DNF, and let* $k \in \{1, 2, \ldots, n\}$. *We say that*

- ψ *is positive (respectively, negative) in the variable* x_k *if the complemented literal* $\overline{x}_k$ *(respectively, uncomplemented literal* x_k*) does not appear in* ψ;
- ψ *is monotone in the variable* x_k *if* ψ *is either positive or negative in* x_k;
- ψ *is positive (respectively, negative) if* ψ *is positive (respectively, negative) in each of its variables;*
- ψ *is monotone if* ψ *is either positive or negative in each of its variables.*

Example 1.25. *Every elementary conjunction is monotone (since each variable appears at most once in it). The DNF* $\phi(x, y, z) = xy \vee x\,\overline{y}\,\overline{z} \vee xz$ *is positive in* x *and neither positive nor negative in* y *and* z. *The DNF* $\theta(x, y, z) = xy \vee x\,\overline{z} \vee y\,\overline{z}$ *is monotone (as it is positive in* x, y *and negative in* z*), but it is neither positive nor negative. The DNF* $\psi(x, y, z, u) = xy \vee xzu \vee yz \vee yu$ *is positive.* $\qquad\square$

It is important to realize that a nonpositive (or even nonmonotone) DNF may very well represent a positive function. The DNF ϕ in Example 1.25 provides an example: Indeed, this DNF can be checked to represent the monotone function $f(x, y, z) = x$. The following result spells out the relation between positive functions and positive DNFs.

Theorem 1.21. *Let f be a Boolean function on $\mathcal{B}^n$, and let $k \in \{1, 2, \ldots, n\}$. The following statements are equivalent:*

(1) *f is positive in the variable x_k.*
(2) *The literal $\overline{x}_k$ does not appear in any prime implicant of f.*
(3) *f has a DNF representation in which the literal $\overline{x}_k$ does not appear.*

Proof. To see that the first assertion implies the second one, consider any prime implicant of f, say, $C_{AB} = \bigwedge_{i \in A} x_i \bigwedge_{j \in B} \overline{x}_j$, and assume that $k \in B$. Since C_{AB} is prime, the conjunction C obtained by deleting $\overline{x}_k$ from C_{AB}, namely,

$$C = \bigwedge_{i \in A} x_i \bigwedge_{j \in B \setminus \{k\}} \overline{x}_j,$$

is not an implicant of f. Therefore, there exists a point $X^* \in \mathcal{B}^n$ such that $C(X^*) = 1$ and $f(X^*) = 0$. Since C_{AB} is an implicant of f, this implies that $C_{AB}(X^*) = 0$, and hence, $x_k^* = 1$. Consider now the point $Y^* \in \mathcal{B}^n$ defined by $y_i^* = x_i^*$ for $i \neq k$ and $y_k^* = 0$. Then, $C_{AB}(Y^*) = 1$ implies $f(Y^*) = 1$. This establishes that f is not positive in the variable x_k, as required.

By Theorem 1.13, the second assertion implies the third one.

Assume now that the third assertion holds, and let $\phi = \bigvee_{j=1}^{m} C_j$ be any DNF of f in which the literal $\overline{x}_k$ does not appear. Recall from Theorem 1.15 that the expression obtained by substituting 1 (respectively 0) for every occurence of x_k in ϕ represents $f_{|x_k=1}$ (respectively $f_{|x_k=0}$). Now, if a term C_j does not involve x_k, then the substitution has no effect on this term. On the other hand, if C_j involves x_k (in uncomplemented form, by hypothesis), then this term vanishes when we substitute x_k by 0. This directly implies that $f_{|x_k=0} \leq f_{|x_k=1}$, and hence, f is positive in x_k. $\square$

As an immediate corollary of Theorem 1.21, the prime implicants of a positive Boolean function do not involve any complemented variables; therefore, every positive function has at least one positive DNF. This property can actually be stated more accurately. Before we do so, we first establish a result that facilitates the comparison of positive DNFs.

Theorem 1.22. *Let ϕ and ψ be two DNFs and assume that ψ is positive. Then, ϕ implies ψ if and only if each term of ϕ is absorbed by some term of ψ.*

Proof. We suppose, without loss of generality, that ϕ and ψ are expressions in the same n variables. The "if" part of the statement holds even when ψ is not positive, as an easy corollary of Theorem 1.11. For the converse statement, let us assume that ϕ implies ψ, and let us consider some term of ϕ, say, $C_k = \bigwedge_{i \in A} x_i \bigwedge_{j \in B} \overline{x}_j$. Consider the characteristic vector of A, denoted e_A. There holds $C_k(e_A) = \phi(e_A) = 1$. Thus, $\psi(e_A) = 1$ (since $\phi \leq \psi$), and therefore, some term of ψ must take value 1 at the point e_A: Denote this term by $C_j = \bigwedge_{i \in F} x_i$ (remember that ψ is positive).

Now, since $C_j(e_A) = 1$, we conclude that $F \subseteq A$, and hence, C_j absorbs C_k as required. $\square$

As a consequence of Theorem 1.22, one can easily check in polynomial time whether an arbitrary DNF ϕ implies a positive DNF ψ (the same question is much more difficult to answer when both ϕ and ψ are arbitrary DNFs; see the comments following Theorem 1.10).

Beside its algorithmic consequences, Theorem 1.22 also allows us to derive one of the fundamental properties of DNFs of positive functions (remember Definitions 1.19 and 1.20):

Theorem 1.23. *The complete DNF of a positive Boolean function f is positive and irredundant; it is the unique prime DNF of f.*

Proof. Let f be a positive function, let $P_1, P_2, \ldots, P_m$ be its prime implicants, and let $\phi = \bigvee_{k=1}^{m} P_k$ denote the complete DNF of f. By Theorem 1.21, ϕ is positive. Consider now an arbitrary prime expression of f, say, $\psi = \bigvee_{k=1}^{r} P_k$, where $1 \le r \le m$. Since $f = \phi = \psi$, we deduce from Theorem 1.22 that each term of ϕ is absorbed by some term of ψ. In particular, if $m > r$, then P_m must be absorbed by some other prime implicant P_k with $k \le r$. This, however, contradicts the primality of P_m. Hence, we conclude that $r = m$, which shows that ϕ is irredundant and is the unique prime DNF of f. $\square$

Theorem 1.23 is due to Quine [767]. It is important because it shows that the complete DNF provides a "canonical" shortest DNF representation of a positive Boolean function: Since the shortest DNF representation of a Boolean function is necessarily prime and irredundant (see the comments following Definition 1.20), no other DNF representation of a positive function can be as short as its complete DNF. Notice that this unicity result does not hold in general for nonpositive functions, as illustrated by the example below.

Example 1.26. *The DNFs $\psi_1 = x\,y \vee \overline{y}\,z \vee \overline{x}\,\overline{z}$ and $\psi_2 = x\,z \vee y\,\overline{z} \vee \overline{x}\,\overline{y}$ are two shortest (prime and irredundant) expressions of the same function.* $\square$

We conclude this section with a useful result that extends Theorem 1.23: This result states that the complete DNF of a positive function can be obtained by first dropping the complemented literals from any DNF representation of the function and then deleting the redundant implicants from the resulting expression.

Theorem 1.24. *Let $\phi = \bigvee_{k=1}^{m} \left(\bigwedge_{i \in A_k} x_i \right) \left(\bigwedge_{j \in B_k} \overline{x}_j \right)$ be a DNF representation of a positive Boolean function f. Then, $\psi = \bigvee_{k=1}^{m} \left(\bigwedge_{i \in A_k} x_i \right)$ is a positive DNF representation of f. The prime implicants of f are the terms of ψ which are not absorbed by other terms of ψ.*

Proof. Clearly, $f = \phi \le \psi$ (see Theorem 1.22). To prove the reverse inequality, consider any point $X^* = (x_1^*, x_2^*, \ldots, x_n^*) \in B^n$ such that $\psi(X^*) = 1$. There is a term of ψ that takes value 1 at the point X^*, or equivalently, there exists $k \in \{1,2,\ldots,m\}$ such that $x_i^* = 1$ for all $i \in A_k$. If e_{A_k} is the characteristic vector of A_k, then $\phi(e_{A_k}) = f(e_{A_k}) = 1$. Moreover, $e_{A_k} \leq X^*$ and therefore, by positivity of f, $f(X^*) = 1$. This establishes that $\psi \leq f$, and thus $f = \psi$ as required.

For the second part of the statement, consider the complete DNF of f, say, ψ^*. Since $\psi = \psi^*$ and ψ is positive, Theorem 1.22 implies that every term of ψ^* is absorbed by some term of ψ. However, the terms of ψ are implicants of f, and the terms of ψ^* are prime implicants of f. Hence, all prime implicants of f must appear among the terms of ψ. This completes the proof. $\qquad\square$

Example 1.27. *As already observed in the comments following Example 1.25, the DNF $\phi(x,y,z) = xy \vee x\,\overline{y}\,\overline{z} \vee xz$ represents a positive function; call it f. An alternative representation of f is derived by deleting all complemented literals from ϕ. In this way, we obtain the redundant DNF $\psi = xy \vee x \vee xz$, and we conclude that x is the only prime implicant of f.* $\qquad\square$

1.10.3 Minimal true points and maximal false points

The definition of the true points (respectively, false points) of an arbitrary Boolean function has been stated in Definition 1.1: These are simply the points in which the function takes value 1 (respectively, 0). Let us now consider a further refinement of these concepts.

Definition 1.27. *Let f be a Boolean function on B^n, and let $X \in B^n$. We say that X is a minimal true point of f if X is a true point of f and if there is no true point Y of f such that $Y \leq X$ and $X \neq Y$. Similarly, we say that X is a maximal false point of f if X is a false point of f and if there is no false point Y of f such that $X \leq Y$ and $X \neq Y$. We denote by $minT(f)$ (respectively, $maxF(f)$) the set of minimal true points (respectively, maximal false points) of f.*

Minimal true points and maximal false points have been defined for arbitrary Boolean functions. However, these concepts are mostly relevant for positive functions, as evidenced by the following observation:

Theorem 1.25. *Let f be a positive Boolean function on B^n and let $Y \in B^n$.*

(1) *Y is a true point of f if and only if there exists a minimal true point X of f such that $X \leq Y$.*
(2) *Y is a false point of f if and only if there exists a maximal false point X of f such that $Y \leq X$.*

Proof. The "only if" implication is trivial in both cases (and is independent of the positivity assumption). The converse implications are straightforward corollaries of Theorem 1.20. $\qquad\square$

As a consequence of Theorem 1.25, positive functions are completely characterized by their set of minimal true points (or maximal false points). More precisely, if
S is a subset of B^n such that every two points in S are pairwise incomparable with
respect to the partial order $\leq$, then there is a unique positive function which has S
as its set of minimal true points and there is a unique positive function which has
S as its set of maximal false points.

At this point, it is interesting to remember that, as we discussed in Section
1.7, Boolean functions are similarly characterized by their list of prime implicants
(see Theorem 1.13 and the comments following it). This analogy is not fortuitous:
Indeed, there exists a simple, but fundamental, one-to-one correspondence between
the minimal true points and the prime implicants of a positive function.

Theorem 1.26. *Let f be a positive Boolean function on B^n, let $C_A = \bigwedge_{i \in A} x_i$ be
an elementary conjunction and let e_A be the characteristic vector of A. Then,*

(1) *C_A is an implicant of f if and only if e_A is a true point of f.*
(2) *C_A is a prime implicant of f if and only if e_A is a minimal true point of f.*

Proof. Consider the first statement. If C_A is an implicant of f, then clearly e_A is
a true point of f (this holds even if f is not positive). Conversely, if e_A is a true
point of f, then $\bigwedge_{i \in A} x_i \bigwedge_{j \notin A} \overline{x}_j$ is an implicant of f. Then, by positivity of f, C_A
also is an implicant of f (by the same argument as in the proof of Theorem 1.24).

For a proof of the second statement, consider an elementary conjunction
$C_B = \bigwedge_{i \in B} x_i$ and the characteristic vector e_B of B. Observe that $C_A \leq C_B$ if
and only if $B \subseteq A$, that is, if and only if $e_B \leq e_A$. This observation, together with
the first statement, implies that C_A is a prime implicant of f if and only if e_A is a
minimal true point of f.   □

Example 1.28. *Consider the positive function $f(x, y, z, u) = xy \vee xzu \vee yz$. From
Theorem 1.26, we conclude that the minimal true points of f are $(1,1,0,0), (1,0,1,1)$
and $(0,1,1,0)$.*

*Theorem 1.26 crucially depends on the positivity assumption. To see this, con-
sider the function $g(x, y, z) = x y \vee \overline{x}\,\overline{z}$. The point $(1,1,0)$ is a true point of g derived
from the prime implicant $x y$, as explained in Theorem 1.26. However, $(0,0,0)$ is
the unique minimal true point of g.*   □

A similar one-to-one correspondence holds between the maximal false points
and the prime implicates of a positive function.

Theorem 1.27. *Let f be a positive Boolean function on B^n, let $D_A = \bigvee_{i \in A} x_i$
be an elementary disjunction, and let $e_{N \setminus A}$ be the characteristic vector of $N \setminus A$.
Then,*

(1) *D_A is an implicate of f if and only if $e_{N \setminus A}$ is a false point of f.*
(2) *D_A is a prime implicate of f if and only if $e_{N \setminus A}$ is a maximal false point
     of f.*

Proof. It suffices to mimic the proof of Theorem 1.26. Alternatively, Theorem 1.27 can be derived as a corollary of Theorem 1.26 via De Morgan's laws or simple duality arguments.   □

Example 1.29. *The function f given in Example 1.28 has four prime implicates, namely, $(x \vee y)$, $(x \vee z)$, $(y \vee z)$ and $(y \vee u)$. Accordingly, it has four maximal false points, namely, $(0,0,1,1)$, $(0,1,0,1)$, $(1,0,0,1)$ and $(1,0,1,0)$.*   □

1.11 Recognition of functional and DNF properties

In this section, we concentrate on the broad algorithmic issue of deciding whether a given function or expression belongs to a particular class. We refer the reader to Appendix B for a brief primer on computational complexity, and for a reminder of concepts like NP-completeness, NP-hardness, and so on.

If $\mathcal{C}$ is a set of Boolean expressions, we define the decision problem:

DNF MEMBERSHIP IN $\mathcal{C}$
Instance: A DNF expression ϕ.
Question: Is ϕ in $\mathcal{C}$?

Similarly, if $\mathcal{C}$ is a set of Boolean functions, then we define the decision problem:

FUNCTIONAL MEMBERSHIP IN $\mathcal{C}$
Instance: A DNF expression of a function f.
Question: Is f in $\mathcal{C}$?

Roughly speaking, a DNF membership problem bears on the DNF itself, whereas a functional membership problem bears on the *function represented by* the DNF. The distinction between both types of problems, however, is not as clear-cut as it may seem, since every functional membership problem can be viewed as a DNF membership problem of a special type: Indeed, if ϕ is a DNF expression of the function f, then f is in the class $\mathcal{C}$ if and only if ϕ is in the class $\mathcal{C}^* = \{\psi \mid \psi$ is a DNF representation of a function in $\mathcal{C}\}$.

Consider, for instance, the following classes of Boolean expressions and Boolean functions:

- The class $\mathcal{TDNF}$ of all DNF expressions of the constant function $\mathbf{1}$ (on an arbitrary number of arguments) and the class $\mathcal{T}$ of all constant functions $\{\mathbf{1}_n \mid n \in \mathbb{N}\}$.
- The class $\mathcal{ZDNF}$ of all DNF expressions of the constant function $\mathbf{0}$ (on an arbitrary number of arguments) and the class $\mathcal{Z}$ of all constant functions $\{\mathbf{0}_n \mid n \in \mathbb{N}\}$.

Since $T\mathcal{DNF} = T^*$ and $Z\mathcal{DNF} = Z^*$, it is obvious that the functional membership problems associated with T and Z are equivalent to the DNF membership problems associated with the classes $T\mathcal{DNF}$ and $Z\mathcal{DNF}$, respectively.

On the other hand, define now

- the class $\mathcal{D}_+$ of all positive DNFs;
- the class $\mathcal{F}_+$ of all positive functions.

The relationship between $\mathcal{D}_+$ and $\mathcal{F}_+$ is not trivial, since a positive function may very well be represented by a nonpositive DNF. In particular, $\mathcal{D}_+ \neq \mathcal{F}_+^*$ and the DNF membership problem associated with the class $\mathcal{D}_+$ does not reduce to a functional membership problem. ($\mathcal{D}_+$ is, in fact, defined by a purely syntactical property.)

Similarly, consider

- the class $\mathcal{D}_k$ of all DNFs of degree at most k ($k \in \mathbb{N}$),
- the class $\mathcal{F}_k$ of all functions representable by a DNF in $\mathcal{D}_k$ ($k \in \mathbb{N}$).

Here again, $\mathcal{D}_k \neq \mathcal{F}_k^*$, and the DNF membership problem associated with $\mathcal{D}_k$ is not equivalent to the functional membership problem associated with $\mathcal{F}_k$.

Now, as one may expect, the difficulty of membership problems depends to a large extent on the specification of the class $\mathcal{C}$. For instance, it is quite easy to test whether a DNF is identically 0, or has degree at most k, or is positive.

Theorem 1.28. DNF MEMBERSHIP IN $Z\mathcal{DNF}$ and FUNCTIONAL MEMBERSHIP IN Z can be tested in constant time. DNF MEMBERSHIP IN $\mathcal{D}_+$ and DNF MEMBERSHIP IN $\mathcal{D}_k$ ($k \in \mathbb{N}$) can be tested in linear time.

Proof. Every elementary conjunction takes value 1 in at least one point. Consequently, a DNF ϕ is identically 0 if and only if it has no term, a condition which can be tested in constant time.

Furthermore, computing the degree of a DNF, or checking whether a DNF is positive, only requires linear time in the size of the DNF.   □

As illustrated by Theorem 1.28, many DNF membership problems are easy to solve. By contrast, however, functional membership problems tend to be difficult: Intuitively, we might say that this is because most properties of Boolean functions are not reflected in a straightforward way in their normal form representations. As a first manifestation of this phenomenon, we can formulate a result that is a simple restatement of Cook's fundamental theorem on NP-completeness [208]. The restatement applies to the so-called *tautology problem*, that is, to the functional membership problem in T:

Theorem 1.29. *The tautology problem* FUNCTIONAL MEMBERSHIP IN T *is co-NP-complete.*

Proof. Cook's theorem was originally stated and proved in the following form (see also Theorem 2.1 in Chapter 2): Given a CNF $\psi = \bigwedge_{k=1}^m \left(\bigvee_{i \in A_k} x_i \vee \bigvee_{j \in B_k} \overline{x}_j \right)$ in

n variables, it is NP-complete to decide whether there exists a point $X^* \in \mathcal{B}^n$ such that $\psi(X^*) = 1$. Trivially, the answer to this decision problem is affirmative if and only if the DNF $\overline{\psi} = \bigvee_{k=1}^{m} \left(\bigwedge_{i \in A_k} \overline{x}_i \bigwedge_{j \in B_k} x_j \right)$ is *not* in $\mathcal{T}$. $\square$

We now extend Theorem 1.29 to a broad category of functional membership problems. This result can be found in Hegedűs and Megiddo [481]; it relies on a simple extension of an argument originally proposed by Peled and Simeone [735].

Theorem 1.30. *Let $\mathcal{C}$ be any class of Boolean functions with the following properties:*

(a) *There exists a function g such that $g \notin \mathcal{C}$.*
(b) *For all $n \in \mathbb{N}$, the constant function $\mathbf{1}_n$ is in $\mathcal{C}$.*
(c) *$\mathcal{C}$ is closed under restrictions; that is, if f is a function in $\mathcal{C}$, then all functions obtained by fixing some variables of f to either 0 or 1 are also in $\mathcal{C}$.*

Then, the problem FUNCTIONAL MEMBERSHIP IN $\mathcal{C}$ *is NP-hard.*

Proof. Let $g \notin \mathcal{C}$ be a function of m variables, and let γ be an arbitrary DNF representation of g. We are going to reduce the problem FUNCTIONAL MEMBERSHIP IN $\mathcal{T}$ to FUNCTIONAL MEMBERSHIP IN $\mathcal{C}$. Let ϕ be a DNF in n variables (defining an instance of FUNCTIONAL MEMBERSHIP IN $\mathcal{T}$) and let us construct a new DNF ψ on $\mathcal{B}^{n+m}$:

$$\psi(X, Y) = \phi(X) \vee \gamma(Y),$$

where X and Y are disjoint sets of n and m variables, respectively. Notice that ψ can be constructed in time polynomial in the length of ϕ, since γ is fixed independently of ϕ. We claim that ϕ represents a tautology (that is, $\phi = \mathbf{1}_n$) if and only if ψ represents a function in $\mathcal{C}$.

Indeed, if $\phi = \mathbf{1}_n$, then $\psi = \mathbf{1}_{n+m}$ and, by virtue of condition (b), ψ represents a function in $\mathcal{C}$. Conversely, if ϕ is not identically 1, then there exists a point $X^* \in \mathcal{B}^n$ such that $\phi(X^*) = 0$, so that $\psi(X^*, Y) = \gamma(Y)$ for all $Y \in \mathcal{B}^m$. Thus, γ is a restriction of ψ and condition (c) implies that ψ does not represent a function in $\mathcal{C}$ (remember that γ represents $g \notin \mathcal{C}$). This proves the claim and the theorem. $\square$

In spite of its apparent simplicity, Theorem 1.30 is a very general result that can be applied to numerous classes of interest due to the weakness of its premises. Indeed, condition (a) is perfectly trivial because the membership question would be vacuous without it. Condition (b) is quite weak as well: It is fulfilled by all the classes introduced earlier in this section, except by $\mathcal{Z}$ (remember Theorem 1.28). Condition (c) is stronger than the first two, but it arises naturally in many situations. In particular, the condition holds again for all the classes of functions discussed above.

Without further knowledge about the class $\mathcal{C}$, Theorem 1.30 does not allow us to draw conclusions about NP-completeness or co-NP-completeness of the membership problem. In any specific application of the theorem, however, we may know that the problem is in NP or in co-NP, and we may strengthen the conclusions accordingly. Several examples will be encountered further in the book. Some of the results presented in Chapter 11 (characterizations by finite sets of functional equations), in particular, imply that certain classes of functional membership problems are in co-NP. Also, Aizenstein et al. [13] investigate various relations between the FUNCTIONAL MEMBERSHIP problem and *query learnability* of Boolean formulas, which allow them to derive a general criterion for FUNCTIONAL MEMBERSHIP to be in co-NP.

Let us illustrate these comments on a few simple examples.

Theorem 1.31. *The problem* FUNCTIONAL MEMBERSHIP IN $\mathcal{F}_+$ *is co-NP-complete. The problem* FUNCTIONAL MEMBERSHIP IN $\mathcal{F}_k$ *is co-NP-complete for all* $k \in \mathbb{N}$.

Proof. As already observed, $\mathcal{F}_+$ and $\mathcal{F}_k$ ($k \in \mathbb{N}$) fulfill conditions (a)–(c) in Theorem 1.30.

The problem FUNCTIONAL MEMBERSHIP IN $\mathcal{F}_+$ is in co-NP: Indeed, to certify that a function f is not in $\mathcal{F}_+$, it suffices to exhibit two points X and Y such that $X \leq Y$, $f(X) = 1$ and $f(Y) = 0$.

The problem FUNCTIONAL MEMBERSHIP IN $\mathcal{F}_k$ is also in co-NP when $k \leq 2$, as follows from Theorem 11.4 and Theorem 11.5 in Chapter 11.

A similar argument does not apply when $k \geq 3$ (because $\mathcal{F}_k$ cannot be characterized by a finite set of functional equations; see the comments at the end of Chapter 11). However Aizenstein et al. [13] were able to establish that $\mathcal{F}_k$ is in co-NP for all $k \in \mathbb{N}$.   □

Let us finally observe that, even though they are not direct corollaries of Theorem 1.30, some related complexity results may sometimes be derived from it as well. For instance:

Theorem 1.32. *Given a DNF expression* $\psi(x_1, x_2, \ldots, x_n)$ *of a function* f *and an index* $i \in \{1, \ldots, n\}$, *it is NP-complete to decide whether the variable* x_i *is essential for* f, *and it is co-NP-complete to decide whether* f *is positive in* x_i.

Proof. If there exists a polynomial algorithm to check whether a variable is essential or not, then the same algorithm can be applied repeatedly (for every variable) to decide in polynomial time whether a given function is identically 1 or not. Thus, detecting essential variables is NP-hard. Moreover, to show that x_i is essential, it is enough to exhibit two points X and Y that differ only in their i-th component such that $f(X) \neq f(Y)$. This establishes that the problem is in NP. A similar reasoning shows that testing the positivity of individual variables is co-NP-complete.   □

1.12 Other representations of Boolean functions

As we mentioned earlier, we mostly concentrate in this book on Boolean functions represented by Boolean expressions, in particular by DNF and CNF expressions. The applications presented in Section 1.13 demonstrate that this class of representations is extremely rich and allows us to model and tackle a wide variety of interesting problems.

However, many other representations of Boolean functions also exist and have proved useful in various contexts. We briefly mention here some of the most important ones.

1.12.1 Representations over GF(2)

Definition 1.28. *The* exclusive-or function, *or* parity function, *is the Boolean function* $\oplus: \mathcal{B}^2 \to \mathcal{B}$ *defined by*

$$\oplus(x_1, x_2) = x_1 \overline{x}_2 \vee \overline{x}_1 x_2$$

for all $x_1, x_2 \in \mathcal{B}$. *We usually write* $x_1 \oplus x_2$ *instead of* $\oplus(x_1, x_2)$.

It is easy to check that, when viewed as a binary operator, $\oplus$ is commutative and associative, that is, $x_1 \oplus x_2 = x_2 \oplus x_1$ and $(x_1 \oplus x_2) \oplus x_3 = x_1 \oplus (x_2 \oplus x_3)$ for all $x_1, x_2, x_3 \in \mathcal{B}$. Also, for every $n \in \mathbb{N}_0$, the function $f(x_1, x_2, \ldots, x_n) = x_1 \oplus x_2 \oplus \ldots \oplus x_n$ takes value 1 exactly when the number of ones in the point $(x_1, x_2, \ldots, x_n) \in \mathcal{B}^n$ is odd. Actually, the operation $\oplus$ defines addition modulo 2 over the Galois field $GF(2) = (\{0, 1\}, \oplus, \wedge)$.

It is well-known that every Boolean function can be represented uniquely as a *sum-of-products modulo 2*. Namely, if we let $\mathcal{P}(N)$ denote the power set of $N = \{1, 2, \ldots, n\}$, then:

Theorem 1.33. *For every Boolean function f on $\mathcal{B}^n$, there exists a unique mapping* $c: \mathcal{P}(N) \to \{0, 1\}$ *such that*

$$f(x_1, x_2, \ldots, x_n) = \bigoplus_{A \in \mathcal{P}(N)} c(A) \prod_{i \in A} x_i. \qquad (1.16)$$

Proof. We provide a constructive proof from first principles. To establish the existence of the representation, we use induction on n. A representation of the form (1.16) clearly exists when $n = 0$, or when $n = 1$ (since $\overline{x} = x \oplus 1$). For $n > 1$, the existence of the representation directly follows from the trivial identity (note the analogy with the Shannon expansion (1.15)):

$$f = f_{|x_n=0} \oplus x_n f_{|x_n=0} \oplus x_n f_{|x_n=1}. \qquad (1.17)$$

Indeed, by induction, both $f_{|x_n=0}$ and $f_{|x_n=1}$ can be expressed in the form (1.16). Substituting these expressions in (1.17) yields a sum-of-products modulo 2 that may contain pairs of identical terms. In this case, these pairs of terms can be removed using the identity $x \oplus x = 0$.

To prove uniqueness, it suffices to observe that there are exactly 2^{2^n} expressions of the form (1.16) and that this is also the number of Boolean functions on $\mathcal{B}^n$. $\square$

Representations of Boolean functions over GF(2) are sometimes called Reed-Muller expansions, or Zhegalkin polynomials, or algebraic normal forms. They are a common tool in algebra (see, for instance, Pöschel and Rosenberg [752]), in cryptography and coding theory (see, for instance, McWilliams and Sloane [642] or the survey by Carlet [170]), and in electrical engineering (see, for instance, Astola and Stanković [35]; Davio, Deschamps and Thayse [259]). The concept of *Boolean derivative* (introduced by Reed [782]) also plays a useful role in these applications.

Definition 1.29. *Let f be a Boolean function on $\mathcal{B}^n$, and let $k \in \{1, 2, \ldots, n\}$. The (Boolean) derivative of f with respect to x_k is the function $\partial f / \partial x_k : \mathcal{B}^{n-1} \to \mathcal{B}$ defined by*

$$\frac{\partial f}{\partial x_k} = f_{|x_k=0} \oplus f_{|x_k=1}. \tag{1.18}$$

Comparing (1.18) with (1.17), we see that $\partial f / \partial x_k$ acts indeed like a formal derivative. Also, it is quite obvious that a function f depends on its k-th variable if and only if $\partial f / \partial x_k \neq 0$. A complete theory of Boolean differential calculus can be built on the basis of Definition 1.29; see [35, 259, 795, 862].

1.12.2 Representations over the reals

A *pseudo-Boolean function of n variables* is a function on $\mathcal{B}^n$ into $\mathbb{R}$, that is, a real-valued function of Boolean variables. Pseudo-Boolean functions provide a far-reaching extension of the class of Boolean functions and are discussed in more detail in Chapter 13. For now, we simply need the following fact (compare with Theorem 1.33):

Theorem 1.34. *For every pseudo-Boolean function f on $\mathcal{B}^n$, there exists a unique mapping $c : \mathcal{P}(N) \to \mathbb{R}$ such that*

$$f(x_1, x_2, \ldots, x_n) = \sum_{A \in \mathcal{P}(N)} c(A) \prod_{i \in A} x_i. \tag{1.19}$$

Proof. This result will be established in Chapter 13; see Theorem 13.1. $\square$

In particular, every Boolean function has a unique representation as a multilinear polynomial over the reals.

Example 1.30. *The function $f = x y \vee \overline{x}\,\overline{z} \vee \overline{y} z$ can be expressed as*

$$f = x y + (1 - x)(1 - z) + (1 - y) z$$

(by orthogonality) or, after some rewriting, as

$$f = 1 - x + x y + x z - y z.$$

$\square$

For Boolean functions, we could actually have observed the existence of the representation (1.19) while discussing orthogonal expressions in Section 1.6. Indeed, the existence of a polynomial representation is an immediate corollary of Theorem 1.7. The latter result also underlines that Boolean functions admit various representations over the reals.

However, the *uniqueness* of the multilinear polynomial (1.19) makes it especially attractive, as it provides a canonical representation of every Boolean function. Note that checking whether a given multilinear polynomial represents a Boolean function, rather than an arbitrary pseudo-Boolean function, is quite easy. Indeed, a pseudo-Boolean function f is Boolean if and only if $f^2(X) = f(X)$ for all $X \in \mathcal{B}^n$. This condition can be checked efficiently due to the unicity of expression (1.19).

Finally, we note that, when the Boolean function f is viewed as a function on the domain $\{-1,+1\}^n$ and taking its values in $\{-1,+1\}$, then f obviously admits an alternative polynomial representation of the form (1.19), sometimes called the *Fourier expansion* of f. Although the Fourier expansion is perfectly equivalent to the multilinear polynomial in 0-1 variables, one of these two expressions may occasionally prove more useful than the other, depending on the intended purpose. Applications of the Fourier expansion in the theoretical computer-science literature are numerous; some illustrations can be found, for instance, in [163, 543, 714, 716]; we refer to Bruck [157] for an introduction to this very fruiful topic. (See also Carlet [170] for uses of the pseudo-Boolean representation and of the Fourier expansion of Boolean functions in crytpography and in coding theory.)

1.12.3 Binary decision diagrams and decision trees

We have already discussed the analogy between the representation of Boolean functions by combinational circuits and Boolean circuits investigated in complexity theory. Another graphical representation of Boolean functions is provided by *binary decision diagrams*. A binary decision diagram (or BDD) on n variables consists of an acyclic directed graph $G = (V, A)$ in which exactly one vertex (the *root*) has indegree 0, and all vertices have outdegree either 0 (the *leaves*) or 2 (the *inner vertices*), together with a labeling of the vertices and of the arcs. The inner vertices are labeled by variables from $\{x_1, x_2, \ldots, x_n\}$, while the leaves get labels from $\{0, 1\}$. One of the arcs leaving each inner vertex is labeled by 0, the other by 1.

The BDD $G = (V, A)$ represents a Boolean function f_G on $\mathcal{B}^n$, in the following sense. For each point $X^* \in \mathcal{B}^n$, the value of $f_G(X^*)$ is computed recursively by traversing G, starting from its root. If vertex v is reached during the traversal, and v is labeled by variable x_i, then the traversal leaves v along the arc labeled by x_i^*. The value of $f_G(X^*)$ is the label of the leaf reached at the end of the computation. (Note that by switching the labels on the leaves, we can similarly view G as providing a representation of $\overline{f_G}$, the complement of f_G.)

Example 1.31. *A binary decision diagram is displayed in Figure 1.4. It is easy to verify that it represents the Boolean function* $f(x_1, x_2, x_3) = \overline{x}_2 \vee x_1 x_3 \vee \overline{x}_1 \overline{x}_3$. □

Some special classes of BDDs have been more thoroughly investigated in the literature. A BDD is a *decision tree* if its underlying graph is a tree. A BDD is *ordered*, or is an OBDD, if there exists a permutation $\pi = (\pi_1, \pi_2, \ldots, \pi_n)$ of $\{1, 2, \ldots, n\}$ with the following property: $\pi_i < \pi_j$ for each arc $(u, v) \in A$ such that u is labeled by x_i and v is labeled by x_j. Note that, in an OBDD, the variables that appear on a path from the root to a leaf form a subsequence of π, and each input variable is read at most once while evaluating the value of the function at any given point.

Example 1.32. *The BDD in Figure 1.4 is ordered by the permutation* $\pi = (2, 1, 3)$. □

BDDs have become popular in the engineering community, mostly since Bryant [160] established the efficiency of OBDDs for performing several operations on Boolean functions (evaluation, solution of Boolean equations, etc.). Decision trees are widely used in artificial intelligence, where they provide a tool for the solution of various machine learning and classification problems (e.g., Quinlan's ID3 method; see [770] and Section 12.2.5 in Chapter 12). BDDs have also been studied in the theoretical computer-science literature, under the name of *branching programs*, in connection with the derivation of lower bounds on the computational complexity of structured Boolean functions. A very thorough account of the literature on BDDs is found in Wegener's book [903] and in the survey paper by Bollig et al. [100].

Although we do not intend to discuss BDDs in great detail, we nevertheless establish a few connections between the concepts introduced in this section and

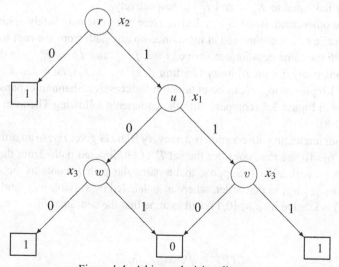

Figure 1.4. A binary decision diagram.

Procedure DECISION TREE(f)
Input: A Boolean function $f(x_1, x_2, \ldots, x_n)$.
Output: A decision tree $D(f) = (V, A)$ representing f.

begin
   if f is constant **then**
      $D(f)$ has a unique vertex $r(f)$ (which is both its root and its leaf);
      $r(f)$ is labeled with the constant value of f (either 0 or 1);
   else
      let $f_0 := f_{|x_i=0}$ and $f_1 := f_{|x_i=1}$;
      run DECISION TREE(f_0) to build $D(f_0)$ with root $r(f_0)$;
      run DECISION TREE(f_1) to build $D(f_1)$ with root $r(f_1)$;
      introduce a root $r(f)$ labeled by x_1;
      make $r(f_0)$ the right son and $r(f_1)$ the left son of $r(f)$;
      label the arc $(r(f), r(f_0))$ by 0 and the arc $(r(f), r(f_1))$ by 1;
      return $D(f)$;
end

Figure 1.5. Procedure DECISION TREE

in the remainder of the chapter. Each vertex $u \in V$ of a BDD-graph $G = (V, A)$ can in fact be viewed as defining a Boolean function f^u: Rather than starting the computation at the root, as we did when defining f_G, simply start it at vertex u. So, if r is the root of G, we have $f_G = f^r$. Alternatively, if u is labeled by variable x_i, and if v, w are the children of u, then it is easy to see that $f^u = x_i f^v \vee \overline{x}_i f^w$ (where we assume that the arc (u, v) is labeled by 1, and the arc (u, w) is labeled by 0). The similarity of this construction with the Shannon expansion (1.15) is rather obvious. Note, however, that f^v and f^w may depend on x_i and, thus, are generally not equal to $f^u_{|x_i=1}$ and $f^u_{|x_i=0}$, respectively.

On the other hand, when G is a decision tree, then one may safely assume that each variable x_i is encountered at most once on any path from the root to a leaf. Thus, with the same notations as above, $f^v = f^u_{|x_i=1}$ and $f^w = f^u_{|x_i=0}$ in decision trees. Conversely, for an arbitrary function $f(x_1, x_2, \ldots, x_n)$, an ordered decision tree $D(f)$ representing f can be obtained by successive Shannon expansions as described in Figure 1.5 (compare with the comments following Theorem 1.16 in Section 1.8).

Another interesting observation is that every BDD G gives rise to an orthogonal DNF of f_G. To see this, consider the set $\mathcal{P}$ of all directed paths from the root r to the leaves with label 1. Suppose that a particular path P contains the vertices $u_1(= r), u_2, \ldots, u_p$, in that order, where u_i is labeled by variable $x_{k(i)}$ and the arc (u_i, u_{i+1}) is labeled by $a_i \in \{0, 1\}$, and assume that the conjunction

$$C(P) = \bigwedge_{i \mid a_i = 1} x_{k(i)} \bigwedge_{j \mid a_j = 0} \overline{x}_{k(j)}$$

is not identically 0 (that is, a variable x_k and its complement $\overline{x}_k$ do not simultaneously appear in the conjunction). Then, $C(P)$ is an implicant of f_G. Moreover, $\bigvee_{P \in \mathcal{P}: C(P) \neq 0} C(P)$ is an orthogonal DNF of f_G.

Of course, by applying the same procedure to the paths from the root to the 0–leaves of G, one can similarly compute an orthogonal DNF of $\overline{f_G}$ and of the dual function f_G^d.

Example 1.33. *When we apply this procedure to the binary decision diagram in Figure 1.4, we obtain the orthogonal DNF $\psi = \overline{x}_2 \vee x_1 x_2 x_3 \vee \overline{x}_1 x_2 \overline{x}_3$ for the function f represented by the BDD, and the orthogonal DNF $\phi = \overline{x}_1 x_2 x_3 \vee x_1 x_2 \overline{x}_3$ for its complement $\overline{f}$.* □

For arbitrary BDDs, the above procedure may be inefficient because the number of paths in $\mathcal{P}$ may be exponentially large in the size of G. When G is a decision tree, however, we obtain a stronger result:

Theorem 1.35. *Let f be a Boolean function represented by a decision tree D, let L be the number of leaves of D and let δ be the depth of D, that is, the length of a longest path from root to leaf in D. Then, an ODNF of f and an ODNF of f^d with degree δ can be computed in time $O(\delta L)$.*

Proof. When D is a decision tree, there is exactly one path from the root to each leaf of D. Hence, the number of terms in the ODNF is at most L, and each term can be built in time $O(\delta)$. □

Finally, we note the following corollary:

Theorem 1.36. *Under the assumptions of Theorem 1.35, the prime implicants of f and the prime implicants of f^d can be generated in time $O(\delta L)$ when f is a positive function.*

Proof. This follows from Theorem 1.35 and Theorem 1.24. □

1.13 Applications

In this section, we return to some of the areas of application that we briefly mentioned earlier in this chapter: propositional logic, electrical engineering, game theory, reliability, combinatorics, and integer programming. We sketch how the basic Boolean concepts arise in these various frameworks and introduce some of the problems and concepts investigated in subsequent chapters. We stress again, however, that Boolean functions and expressions play a role in many other fields of science. We have already mentioned their importance in complexity theory (see, for instance, Krause and Wegener [583], Papadimitriou [725], Wegener [902]); in coding theory or in cryptography (see Carlet [170], McWilliams and Sloane [642]); and we could cite a variety of additional applications arising in social sciences (qualitative analysis of data; see Ragin [775]); in psychology (human concept learning; see Feldman [326, 327]); in medicine (diagnostic, risk assessment; see Bonates and Hammer [102]); in biology (genetic regulatory networks; see Kauffman [553], Shmulevich, Dougherty and Zhang [831], Shmulevich and Zhang [832]), and so on.

Beyond the specific issues arising in connection with each particular application, we want to stress that the unifying role played by Boolean functions and, more generally, by Boolean models, should probably provide the main motivation for studying this book (it certainly provided one of the main motivations for writing it). This theme will be recurrent throughout subsequent chapters, where we will see that the same basic Boolean concepts and results have repeatedly been reinvented in various areas of applied mathematics.

1.13.1 Propositional logic and artificial intelligence

As suggested in Application 1.1, propositional logic is essentially equivalent to the calculus of Boolean functions (see, e.g., Stoll [848], Urquhart [882]). Besides its fundamental role as a theoretical model of formal reasoning, propositional logic has found practical applications in several domains of artificial intelligence. For more information on this topic, we refer the reader to classic texts by Chang and Lee [186], Gallaire and Minker [359], Loveland [627], Kowalski [582], Jeroslow [533], Anthony and Biggs [29], and so on. We only briefly touch here upon the surface of this topic.

Consider three propositional variables, say x, y, z. The exact interpretation of these variables is not relevant here, but, for the sake of the discussion, one may think of them as representing elementary propositions such as:

x : The patient shows symptom X.

y : Test Y is negative.

z : Diagnosis Z applies.

The knowledge base of an expert system is a list of *rules* expressing logical relationships of the "if-then-else" type between the propositional variables of interest. For instance, a knowledge base may contain the following rules:

Rule 1 : "If x is false and y is true then z is true."
Rule 2 : "If x is false and y is false then z is false."
Rule 3 : "If z is true then x is false."
Rule 4 : "If y is true then z is false."

Let us associate a Boolean expression $\phi(x, y, z)$ with the above knowledge base:

$$\phi(x, y, z) = \overline{x}\, y\, \overline{z} \vee \overline{x}\, \overline{y}\, z \vee x\, z \vee y\, z, \tag{1.20}$$

where each term of ϕ corresponds in a straightforward way to one of rules 1–4. The interpretation of ϕ is easy: A 0–1 point (x, y, z) is a false point of ϕ if and only if the corresponding assignment of True–False values to the propositional variables does not contradict any of the rules in the knowledge base. Thus, in the terminology of logic theory, the set of *solutions* of the Boolean equation $\phi(x, y, z) = 0$ is exactly the set of *models* of the knowledge base. In particular, the set of rules is not "self-contradictory" if and only if ϕ is not identically 1, that is, if and only if the Boolean equation $\phi = 0$ admits at least one solution (which is easily seen to be the case for our small example).

The main purpose of an expert system is to draw inferences and to answer queries involving the propositional variables, such as: "Is the assignment $z = 1$ consistent with the given set of rules?" (that is, "Does diagnosis Z apply under at least one imaginable scenario?"). This question can be answered by plugging the value $z = 1$ into ϕ and checking whether the resulting Boolean equation $\phi_{|z=1} = 0$ remains consistent. For our example, this procedure yields the equation

$$\overline{x}\,\overline{y} \vee x \vee y = 0,$$

which is clearly inconsistent. Thus, $z = 1$ is not possible in the world described by the above knowledge base.

This short discussion illustrates how simple questions pertaining to the atomic propositions involved in a knowledge base can be reduced to the solution of Boolean equations. The solution of Boolean equations by algebraic techniques has been an ongoing topic of research ever since Boole's original work appeared in print 150 years ago. We return to Boolean equations in much greater detail in Chapter 2 of this book.

In actual expert systems, for pragmatic reasons of computational efficiency, it is usual to restrict the rules incorporated in the knowledge base to so-called *Horn clauses*, namely, to rules of the form

if x_{i_1} is true and x_{i_2} is true and ... and x_{i_k} is true, then $x_{i_{k+1}}$ is true,

or

either x_{i_1} is false or x_{i_2} is false or ... or x_{i_k} is false,

where $x_{i_1}, x_{i_2}, \ldots, x_{i_k}, x_{i_{k+1}}$ are arbitrary variables. When all the rules are Horn clauses, then the associated Boolean expression ϕ is a DNF with terms of the form $x_{i_1} x_{i_2} \ldots x_{i_k} \overline{x}_{i_{k+1}}$ or $x_{i_1} x_{i_2} \ldots x_{i_k}$. This leads to the following definition:

Definition 1.30. *A DNF is a Horn DNF if each of its terms contains at most one complemented variable.*

We will show in Chapter 6 that, when ϕ is a Horn DNF, the Boolean equation $\phi(X) = 0$ can be solved easily, more precisely, in linear time. This single fact suffices to explain the importance of Horn DNFs in the context of expert systems, where large Boolean equations must be solved repeatedly. Moreover, we also discover in Chapter 6 that Horn DNFs possess a host of additional remarkable properties making them a worthwhile object of study.

Before we close this section, we must warn the reader that our view that (propositional) knowledge bases define Boolean expressions and, concomitantly, Boolean functions, is quite unorthodox in the artificial intelligence literature, where rules are more traditionally regarded as forming a "loose" collection of Boolean clauses rather than a single function. We claim, however, that our point of view has definite advantages over the traditional one. Indeed, it allows us to take advantage of the huge body of knowledge regarding Boolean functions and to draw inspiration from concepts and properties pertaining to such functions.

As an example of this general claim, let us go back to the small knowledge base just given and to the corresponding DNF ϕ displayed in equation (1.20). It should be clear from our previous discussion that, as far as drawing inferences goes, all the information contained in the knowledge base is adequately translated in ϕ. More precisely, any Boolean expression representing the same Boolean function as ϕ provides the same information as the original knowledge base. Indeed, if ψ is any expression such that $\psi = \phi$, then the set of models of the knowledge base is in one-to-one correspondence with the set of false points of ψ, which coincides with the set of false points of ϕ. This observation implies that Boolean transformations can sometimes be applied in order to obtain a simpler, but equivalent, representation of the knowledge base. The simplification of Boolean expressions is one of the main topics of Chapter 3. For now, however, the discussion in Section 1.7 already suggests that the prime implicants of ϕ may play an interesting role in this respect. For our example, it turns out that ϕ only has two prime implicants, namely, $\overline{x} y$ and z. By way of consequence (recall Theorem 1.13), $\phi = \overline{x} y \vee z$, so that the original rules 1–4 are equivalent to the conjunction of the following two rules:

Rule 5 : "either x is true or y is false."
Rule 6 : "z is false."

(Note that Rule 6 provides a confirmation of our previous conclusion, according to which z can never be true.)

Recently, the Boolean formalism has found a very central role in another area of artificial intelligence, namely, in *computational learning theory*. In intuitive terms, many of the fundamental questions in this field take the following form: Given a class C of Boolean functions and an unknown function f in C, how many rows of the truth table of f is it necessary to query to be "reasonably confident" that f is known with "sufficient accuracy?" Another type of question would be: Given a class C of Boolean functions and two subsets (or "samples") $T, F \subseteq \mathcal{B}^n$, is there a function $f \in C$ such that f takes value 1 on T and value 0 on F? Related issues will be tackled in Chapter 12 of this book. For more information on computational learning theory, we refer the reader to the textbook by Anthony and Biggs [29] and to survey papers by Anthony [26] and Sloan, Szörényi, and Turán [838].

1.13.2 Electrical and computer engineering

We have already mentioned that every switching or combinational circuit can be viewed as a device computing the value of a Boolean function f (see Application 1.2). Given a description of an {AND,OR,NOT}–circuit, an expression of f can be constructed recursively. Indeed, let us assume that the output gate of the circuit is an OR-gate. If we delete this gate, then we obtain two subcircuits that compute two functions, say f_1 and f_2, for which we can (recursively) construct the representations ϕ_1 and ϕ_2. Then, the expression $\phi_1 \vee \phi_2$ represents f.

Example 1.34. *The circuit displayed in Figure 1.6 computes the function*

$$\phi = (x_1 \vee x_4)\overline{(x_1 \vee (x_2\overline{x}_3))}. \tag{1.21}$$

□

It can be much more difficult, however, to obtain a DNF of the function associated to a given circuit. Fortunately, for many applications, it is sufficient to have an implicit representation of f via a DNF $\psi(X, Y, z)$ similar to the DNF produced by the procedure EXPAND (see Section 1.4). In the DNF $\psi(X, Y, z)$, the vector X represents the inputs of the circuit, z represents its output, and Y can be viewed as a vector of variables associated with the outputs of the "hidden" gates of the circuit (in the physical realization of a switching circuit, the input and ouput signals can be directly observed, whereas the value of all other signals cannot, hence the qualifier "hidden" applied to these internal gates). On every input signal X^*, the circuit produces the output z^*, where (X^*, Y^*, z^*) is the unique solution of the equation $\psi(X^*, Y, z) = 0$. (See Abdulla, Bjesse, and Eén [1] for a more detailed contribution along similar lines.) Let us illustrate this construction on an example.

Example 1.35. *Consider again the circuit displayed in Figure 1.6 and the corresponding expression ϕ given by (1.21). We have already shown in Example 1.12*

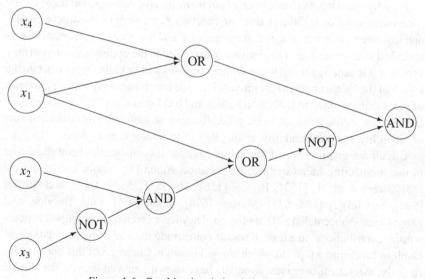

Figure 1.6. Combinational circuit for Example 1.34.

that, when applying EXPAND *to* ϕ, *we obtain the expression*

$$\psi = \overline{y}_1 z \vee y_2 z \vee y_1 \overline{y}_2 \overline{z} \vee x_1 \overline{y}_1 \vee x_4 \overline{y}_1 \vee \overline{x}_1 \overline{x}_4 y_1$$

$$\vee x_1 \overline{y}_2 \vee y_4 \overline{y}_2 \vee \overline{x}_1 \overline{y}_4 y_2 \vee \overline{x}_2 y_4 \vee x_3 y_4 \vee x_2 \overline{x}_3 \overline{y}_4. \qquad (1.22)$$

For every point $(x_1^*, x_2^*, x_3^*, x_4^*)$ *describing the input signals, the output of the circuit is given by the value of* z *in the unique solution of the equation* $\psi(x_1^*, x_2^*, x_3^*, x_4^*, y_1, y_2, y_4, z) = 0$. *Similarly, as discussed in Example 1.12, the value of* y_1 *in this solution indicates the state of the first (topmost) OR-gate produced by the inputs* $(x_1^*, x_2^*, x_3^*, x_4^*)$; *the value of* y_2 *indicates the state of the second OR-gate; and the value of* y_4 *indicates the state of the first AND-gate.*

Consider, for instance, the input $(x_1^*, x_2^*, x_3^*, x_4^*) = (0, 1, 1, 1)$. *In this point, the equation* $\psi = 0$ *boils down to*

$$\overline{y}_1 z \vee y_2 z \vee y_1 \overline{y}_2 \overline{z} \vee \overline{y}_1 \vee y_4 \overline{y}_2 \vee \overline{y}_4 y_2 \vee y_4 = 0,$$

which has the unique solution $(y_1^*, y_2^*, y_4^*, z^*) = (1, 0, 0, 1)$. *Thus, the output signal of the circuit is* $z^* = 1$, *which is indeed equal to* $\phi(0, 1, 1, 1)$. $\square$

We now turn to the opposite problem of constructing a circuit that computes a given Boolean function. Notice that, as an easy consequence of Theorem 1.4, such a circuit exists for every Boolean function. Actually, if we allow AND-gates and OR-gates to have indegree larger than two, then every DNF can even be computed by a switching circuit involving at most four layers and one OR-gate, with all input gates in the first layer, all NOT-gates in the second layer, all AND-gates in the third layer, and the OR-gate in the fourth layer (in the role of output gate). By the same reasoning, every positive function corresponds to a circuit involving at most three layers and one OR-gate, since NOT-gates are superfluous in this case.

Broadly speaking, the basic issue of *circuit design* (or *network synthesis*) can be formulated as follows: Given a Boolean function f, we want to construct a combinational circuit of minimal size that computes f and that satisfies a number of pre-specified side constraints. The measure of size used in the optimality criterion may vary, but it is usually related to the number of gates and/or to the depth (that is, the length of the longest path) of the circuit. The side constraints may restrict the types of gates that are allowed (only AND-gates and NOT-gates, no NOT-gates, etc.) or the indegree of the gates, or, more generally, may be motivated by considerations of reliability, manufacturability, availability of technology, and so on.

Circuit design problems of this nature have for several decades been addressed in the engineering literature; see, for instance, Adam [5]; Astola and Stanković [35]; Brayton et al. [153]; Brown [156]; Hu [511, 512]; Kunz and Stoffel [590]; McCluskey [634, 635]; Muroga [698]; Sasao [804]; Villa, Brayton, and Sangiovanni-Vincentelli [891], and so on. They have given rise, among other noteworthy contributions, to a host of results concerning the size of representations of Boolean functions, a topic to which we will return in Chapter 3 of this book. More recently, theoretical computer scientists have shown renewed interest for similar questions arising in the framework of computational complexity. Although their research stresses asymptotic measures of performance ("Is it possible to compute all functions in a given class by circuits of polynomial size?") rather than engineering or economic considerations, the issues they investigate remain very much akin to those studied in electrical engineering. We refer the reader to the monographs by Wegener [902] and Vollmer [892] or to the survey by Krause and Wegener [583] for a wealth of information on this line of research, which largely falls outside the scope of our book.

Starting in the late 1950s, electrical engineers have devoted a lot of attention, from both an applied and a theoretical perspective, to combinational circuits built from a remarkable type of switching gates called *threshold gates*. For our purpose (and brushing aside all technicalities involved in their implementation), threshold gates are electronic devices that compute a special class of Boolean functions called *threshold functions*.

Definition 1.31. *A Boolean function f on $\mathcal{B}^n$ is a* threshold *(or linearly separable) function if there exist n weights $w_1, w_2, \ldots, w_n \in \mathbb{R}$ and a threshold $t \in \mathbb{R}$ such that, for all $(x_1, x_2, \ldots, x_n) \in \mathcal{B}^n$,*

$$f(x_1, x_2, \ldots, x_n) = 0 \text{ if and only if } \sum_{i=1}^{n} w_i x_i \leq t.$$

In geometric terms, threshold functions are precisely those Boolean functions for which the set of true points can be separated from the set of false points by a hyperplane, namely, the *separator* $\{X \in \mathbb{R}^n \mid \sum_{i=1}^{n} w_i x_i = t\}$. It is easy to see that elementary conjunctions and disjunctions are threshold functions. As a consequence, every Boolean function can be realized by a circuit involving only threshold gates. The problem of designing optimal circuits of threshold gates has generated a huge body of literature. The concept of Chow parameters, for instance, has originally been introduced with the purpose of providing a numerical characterization of threshold functions (see Chow [194] and Winder [920] or books by Dertouzos [269], Hu [511], Muroga [698], etc.). We devote two chapters (Chapters 8 and 9) of this book to an investigation of the properties of threshold and related functions.

1.13.3 Game theory

As introduced in Applications 1.3 and 1.7, a simple game (or voting game) v on a set of players $N = \{1, 2, \ldots, n\}$ can be modeled as a positive Boolean function f_v on $\mathcal{B}^n$. This concept was introduced by von Neumann and Morgenstern [893] (albeit in set-theoretic, rather than Boolean terminology), in the seminal book that laid the foundations of game theory, and was further developed in Shapley [828]. More recent discussions can be found in several books, for instance [79, 777, 850].

Many of the notions introduced in previous sections have natural interpretations in a game-theoretic setting. Consider for instance an implicant $\bigwedge_{i \in A} x_i$ of the function f_v, where $A \subseteq N$, and consider any point $X^* \in \mathcal{B}^n$ such that $x_i^* = 1$

for all $i \in A$ (all players in A cast a "Yes" vote). By definition of an implicant, $f_v(X^*) = 1$ and, in view of the translation rules proposed in Application 1.3, $v(supp(X^*)) = 1$, where $supp(X^*)$ is the set of players who voted "Yes." Thus, the set A can be viewed as a group of players who, when simultaneously voting in favor of an issue, have the power to determine the outcome of the vote *irrespective* of the decision made by the remaining players. In game theory, such a decisive group of players is called a *winning coalition*. Clearly then, a prime implicant simply corresponds to an (inclusion-wise) minimal winning coalition, that is, to a subset A of players such that $v(A) = 1$, but $v(B) = 0$ for all subsets $B \subset A, B \neq A$.

It is well-known that every simple game is completely determined by the collection of its minimal winning coalitions, a fact we can regard as an immediate corollary of the results established in previous sections (see, for instance, Theorem 1.13 and Theorem 1.23).

A straightforward counterpart of minimal winning coalitions is provided by *maximal losing coalitions*, that is, by those subsets A of players such that $v(A) = 0$, but $v(B) = 1$ for all supersets $B \supset A, B \neq A$. It is easy to see that the maximal losing coalitions of v are in one-to-one correspondence with the prime implicates of f_v, in the sense that A is a maximal losing coalition of v if and only if $\bigvee_{i \in N \setminus A} x_i$ is a prime implicate of f_v (see also Theorem 1.27). The question of characterizing the collection $\mathcal{L}$ of maximal losing coalitions in terms of the collection $\mathcal{W}$ of minimal winning coalitions, or even of generating $\mathcal{L}$ from $\mathcal{W}$, arises quite naturally in this setting. We shall tackle this type of issues in Chapter 4, in the broader framework of *duality theory*.

The most common voting rules used in legislative or corporate assemblies are modeled by the class of so-called *weighted majority games*. In such a game, each player i carries a positive weight $w_i \in \mathbb{R}$: When he votes in favor of an issue, player i contributes his full weight w_i toward the issue ($i = 1, 2, \ldots, n$). The issue is adopted if the sum of the weights cast in its favor exceeds a predetermined threshold t. Comparing this definition with Definition 1.31, it is not too hard to see that weighted majority games correspond exactly to positive threshold functions. As a consequence, the theory of threshold functions has been thoroughly investigated in the game-theoretic literature. This happened at about the same time threshold functions were also attracting the attention of electrical engineers (starting mostly in the late fifties), so that many properties have been independently (re)discovered by researchers active in these two fields.

Another main theme of study in game theory is the computation of the "share of power" held by the players of a game, and several definitions of "power indices" coexist in the literature. Many of these indices are closely related to the Chow parameters of the associated Boolean functions (see Dubey and Shapley [279] and Felsenthal and Machover [329] for detailed presentations). In fact, most indices are naturally expressed in terms of the so-called *modified Chow parameters* of the function, which we now introduce.

Definition 1.32. *The* modified Chow parameters *of a Boolean function* $f(x_1, x_2, \ldots, x_n)$ *are the* $(n+1)$ *numbers* $(\pi_1, \pi_2, \ldots, \pi_n, \pi)$ *defined as* $\pi = \omega - 2^{n-1}$ *and* $\pi_k = 2\omega_k - \omega$ *for* $k = 1, 2, \ldots, n$, *where* $(\omega_1, \omega_2, \ldots, \omega_n, \omega)$ *are the Chow parameters of* f.

Note that there is a bijective correspondence between the vectors of Chow parameters and those of modified Chow parameters. Modified Chow parameters have been considered both in threshold logic (see [698, 920]) and in game theory (see [279, 329]). In the terminology of Dubey and Shapley [279], $\pi_1, \pi_2, \ldots, \pi_n$ are the *swing numbers*, or *raw Banzhaf indices*, of the function. The name "swing number" refers to the following concept.

Definition 1.33. *Let* f *be a positive Boolean function on* $\mathcal{B}^n$, *and let* $k \in \{1, 2, \ldots, n\}$. *A swing of* f *for variable* k *is a false point* $X^* = (x_1^*, x_2^*, \ldots, x_n^*)$ *of* f *such that* $X^* \vee e_k$ *is a true point of* f, *where* $e_k = (0, \ldots, 0, 1, 0, \ldots, 0)$ *denotes the* k-*th unit vector.*

The relation between swings and modified Chow parameters is simple.

Theorem 1.37. *If* f *is a positive Boolean function on* $\mathcal{B}^n$ *with modified Chow parameters* $(\pi_1, \pi_2, \ldots, \pi_n, \pi)$, *then* π_k *is the number of swings of* f *for* k, $k = 1, 2, \ldots, n$.

Proof. Let Y^* be any true point of f such that $y_k^* = 1$ (note that there are ω_k such points), and write $Y^* = X^* \vee e_k$, where $x_k^* = 0$. Then, either X^* is a swing for k, or X^* is a true point of f, but not both. Moreover, all swings of f for k and all true points of f whose k-th component is zero can be obtained in this way. Denoting by s_k the number of swings for k, we conclude that $\omega_k = s_k + (\omega - \omega_k)$ or, equivalently, $s_k = \pi_k$. $\qquad\square$

In voting terms, a swing for variable (that is, player) k corresponds to a losing coalition (namely, the coalition $\{i \in N \mid x_i^* = 1\}$) that turns into a winning coalition when player k joins it. Intuitively, then, player k retains a lot of power in the game v if f_v has many swings for k, since this means that k plays a "pivotal" role in many winning coalitions.

Accordingly, many authors define power indices as functions of the number of swings or, equivalently, of the modified Chow parameters. Banzhaf [52], for instance, made a proposal which translates as follows in our terminology (see also Penrose [739] for pioneering work on this topic).

Definition 1.34. *If* f *is a positive, nonconstant Boolean function on* $\mathcal{B}^n$ *with modified Chow parameters* $(\pi_1, \pi_2, \ldots, \pi_n, \pi)$, *then the* k-*th (normalized) Banzhaf index of* f *is the quantity*

$$\beta_k = \frac{\pi_k}{\sum_{i=1}^{n} \pi_i},$$

for $k = 1, 2, \ldots, n$.

The Banzhaf index ranks among the most extensively studied and widely accepted power indices for voting games. In spite of some fundamental drawbacks, it agrees on many accounts with what we would intuitively expect from a reasonable measure of power (see Dubey and Shapley [279]; Felsenthal and Machover [329]; and Straffin [850] for an axiomatic characterization and extensive discussions of the relation between Banzhaf and other power indices). Note, for instance, that, in view of Theorem 1.18, the Banzhaf index of an inessential player is equal to zero. The converse statement also holds for positive Boolean functions (the proof is left to the reader as an exercise). We return to this topic in Chapter 9. Many other connections between the theory of Boolean functions and the theory of simple games will also be established in the monograph.

Finally, it is interesting to observe that Boolean functions provide useful models for investigating certain types of nonsimple games, for example, 2-player positional games in normal form. We do not further discuss this topic now but refer the reader to Chapter 10 and to Gurvich [421, 423, 424, etc.] for more information.

1.13.4 Reliability theory

As explained in Applications 1.4 and 1.8, reliability theory models every complex system S by a positive Boolean function f_S called the *structure function* of S. To rule out trivial cases, it is often assumed that all variables of f_S are essential. When this is the case, the system and its structure function are said to be *coherent*. This framework was introduced by Birnbaum, Esary, and Saunders [92] and is further discussed in Barlow and Proschan [54], Colbourn [205, 206], Ramamurthy [777], or Provan [759]. Colbourn [206], in particular, examines in depth the interplay between combinatorial and Boolean reliability models.

Let $N = \{1, 2, \ldots, n\}$ be the set of components. If $\bigwedge_{i \in A} x_i$ is an implicant of the function f_S, then the whole system S operates whenever the components in A operate, irrespectively of the state of the remaining components. In reliability parlance, the set A is called a *pathset* of S. If no subset of A is itself a pathset, then A is called a *minimal pathset*. Thus, we see that the (minimal) pathsets of S correspond exactly to the (prime) implicants of f_S.

As mentioned in Application 1.4, the fundamental problem of reliability theory is to compute the probability that the system S operates when its components fail randomly. Assume for the sake of simplicity that the components work or fail independently of each other, and let p_i denote the probability that component i works, for $i = 1, 2, \ldots, n$. Thus, we have $p_i = \text{Prob}[x_i = 1]$ and we want to compute $\text{Rel}_S(p_1, p_2, \ldots, p_n) = \text{Prob}[f_S = 1]$, which is the probability that the system S operates.

If $\phi = \bigvee_{k=1}^{m} \bigwedge_{i \in A_k} x_i \bigwedge_{j \in B_k} \overline{x}_j$ is an orthogonal DNF of f_S, then Theorem 1.7 can be used to compute $\text{Rel}_S(p_1, p_2, \ldots, p_n)$ (see [49, 205, 206, 619, 759, etc.]). Indeed, denoting by $E[f_S]$ the expected value of the random variable

$f_S(x_1, x_2, \ldots, x_n)$, we successively derive:

$$\text{Rel}_S(p_1, p_2, \ldots, p_n) = \text{Prob}[f_S = 1]$$

$$= E[f_S] \qquad \text{(since } f_S \text{ is a Bernoulli random variable)}$$

$$= \sum_{k=1}^{m} E[\prod_{i \in A_k} x_i \prod_{j \in B_k} (1 - x_j)] \quad \text{(by Theorem 1.7)}$$

$$= \sum_{k=1}^{m} \prod_{i \in A_k} p_i \prod_{j \in B_k} (1 - p_j) \text{ (by the independence assumption).}$$

When viewed as a function from $[0,1]^n$ to $[0,1]$, Rel_S is called the *reliability function* or *reliability polynomial* of S (see, e.g., [54, 205, 206, 777]). Observe that the polynomial Rel_S extends the Boolean function $f_S : \{0,1\}^n \to \{0,1\}$ over the whole unit cube $U^n = [0,1]^n$. As a matter of fact, if f_S is viewed as a pseudo-Boolean function, and if it is represented as a multilinear polynomial over the reals (see Section 1.12.2)

$$f_S(x_1, x_2, \ldots, x_n) = \sum_{A \in \mathcal{P}(N)} c(A) \prod_{i \in A} x_i,$$

then we similarly conclude that

$$\text{Rel}_S(p_1, p_2, \ldots, p_n) = \sum_{A \in \mathcal{P}(N)} c(A) \prod_{i \in A} p_i.$$

Similar observations have also been made in the game theory literature (see [456, 720, 777]).

When $p_i = \frac{1}{2}$ for $i = 1, 2, \ldots, n$, all vertices of $\mathcal{B}^n$ are equiprobable with probability 2^{-n}. As a result,

$$\text{Rel}_S\left(\frac{1}{2}, \ldots, \frac{1}{2}\right) = \text{Prob}[f_S = 1] = \frac{\omega(f_S)}{2^n},$$

where $\omega(f_S)$ denotes as usual the number of true points of f_S. Similarly, if $p_k = 1$ for some component k and $p_i = \frac{1}{2}$ for $i = 1, 2, \ldots, n, i \neq k$, then

$$\text{Rel}_S\left(\frac{1}{2}, \ldots, \frac{1}{2}, 1, \frac{1}{2}, \ldots, \frac{1}{2}\right) = \frac{\omega_k}{2^{n-1}},$$

where ω_k is the k-th Chow parameter of f_S. These observations show that computing the Chow parameters of a positive Boolean function is just a special case of computing the reliability of a coherent system. Also, similarly to what happened for simple games, variants of the modified Chow parameters have been used in the literature to estimate the "importance" of individual components of a coherent system. Ramamurthy [777] explains nicely how Banzhaf and other power indices (like the Shapley-Shubik index) have been rediscovered in this framework.

1.13.5 Combinatorics

Relations between Boolean functions and other classical combinatorial constructs, such as graphs, hypergraphs, independence systems, clutters, block designs, matroids, colorings, and so forth, are amazingly rich and diverse. Over time, these relations have been exploited to gain insights into the constructs themselves (see, e.g., Benzaken [64]), to handle algorithmic issues related to the functions or the constructs (see, e.g., Hammer and Rudeanu [460]; Aspvall, Plass, and Tarjan [34]; Simeone [834]) and to introduce previously unknown classes of combinatorial objects (see, e.g., Chvátal and Hammer [201]). These are but a few examples, and we will encounter plenty more throughout this book. In this section, we only mention a few useful connections between the study of hypergraphs and the concepts introduced so far.

The stability function $f = f_{\mathcal{H}}$ of a hypergraph $\mathcal{H} = (N, \mathcal{E})$ was introduced in Application 1.5. We observed in Application 1.9 that $f_{\mathcal{H}}$ is a positive function. In fact, if $N = \{1, 2, \ldots, n\}$, then it is easy to see that

$$f_{\mathcal{H}}(x_1, x_2, \ldots, x_n) = \bigvee_{A \in \mathcal{E}} \bigwedge_{j \in A} x_j. \qquad (1.23)$$

It is important to realize that the function $f_{\mathcal{H}}$ does not completely define the hypergraph $\mathcal{H}$. Indeed, consider two hypergraphs $\mathcal{H} = (N, \mathcal{E})$ and $\mathcal{H}' = (N', \mathcal{E}')$. If $\mathcal{E} \subseteq \mathcal{E}'$, and if every edge in $\mathcal{E}' \setminus \mathcal{E}$ contains some edge in $\mathcal{E}$, then $\mathcal{H}$ and $\mathcal{H}'$ have exactly the same stable sets, so that $f_{\mathcal{H}} = f_{\mathcal{H}'}$. Thus, the expression (1.23) of $f_{\mathcal{H}}$ can be rewritten as

$$f_{\mathcal{H}}(x_1, x_2, \ldots, x_n) = \bigvee_{A \in \mathcal{P}} \bigwedge_{j \in A} x_j,$$

where $\mathcal{P}$ is the set of minimal edges of $\mathcal{H}$. Putting this observation parallel with Theorem 1.23, we see that the terms $\bigwedge_{j \in A} x_j$ ($A \in \mathcal{P}$) are nothing but the prime implicants of $f_{\mathcal{H}}$.

Obviously, the minimal edges of a hypergraph $\mathcal{H}$ form a *clutter* (or a *Sperner family*), namely, a subhypergraph $(N, \mathcal{P})$ of $\mathcal{H}$ with the property that

$$A \in \mathcal{P}, B \in \mathcal{P}, A \neq B \Rightarrow A \not\subseteq B.$$

Conversely, any clutter can also be viewed as defining the collection of minimal edges of a hypergraph or, equivalently, the collection of prime implicants of a positive Boolean function.

Many operations on hypergraphs or clutters are natural counterparts of operations on Boolean expressions. For instance, if $\mathcal{H} = (N, \mathcal{E})$ is a clutter and $j \in N$, the clutter $\mathcal{H} \setminus j$ is defined as follows: $\mathcal{H} \setminus j = (N \setminus \{j\}, \mathcal{F})$, where $\mathcal{F} = \mathcal{E} \setminus \{A \in \mathcal{E} \mid j \in A\}$ (*deletion* of j; see, e.g., Seymour [821] or the literature on matroid theory). Thus, $f_{\mathcal{H} \setminus j}$ is simply the restriction of $f_{\mathcal{H}}$ to $x_j = 0$.

Similarly, the clutter $\mathcal{H}/j$ is defined as $\mathcal{H}/j = (N \setminus \{j\}, \mathcal{G})$, where $\mathcal{G}$ is the collection of minimal sets in $\{A \setminus \{j\} \mid A \in \mathcal{E}\}$ (*contraction* of j). We see that

$f_{\mathcal{H}/j}$ is the restriction of $f_{\mathcal{H}}$ to $x_j = 1$. We shall come back to these operations in Chapter 4, when we discuss duality theory.

A (simple, undirected) graph $G = (V, E)$ is a special type of hypergraph such that $|e| = 2$ for all edges $e \in E$ (we adopt a well-entrenched convention and denote edges of a graph by lowercase letters). Thus, graphs are in one-to-one correspondence with *purely quadratic* positive Boolean functions, that is, positive functions with prime implicants of degree 2 only. This, and related connections between graphs and quadratic functions, will be exploited repeatedly in later chapters (see, in particular, Chapter 5). For now, let us just illustrate its use in deriving the following observation due to Ball and Provan [49] (we refer to [371, 725, 883] and Appendix B for a definition of #P-completeness):

Theorem 1.38. *Computing the number of true points of a Boolean function f expressed in DNF is #P-complete, even if f is purely quadratic and positive.*

Proof. Let $f(x_1, x_2, \ldots, x_n) = \bigvee_{\{i,j\} \in E} x_i x_j$, and let G be the corresponding graph, namely, $G = (N, E)$. We denote by $s(G)$ the number of stable sets of G, and by $\omega(f)$ the number of true points of f. Valiant [883] proved that computing $s(G)$ is #P-complete. Since $s(G) = 2^n - \omega(f)$, the result follows. $\square$

As a corollary of this theorem, we can also conclude that computing the Chow parameters of a quadratic positive function is #P-hard. Observe that these results actually hold independently of the representation of f. Indeed, *if we know* in advance that f is purely quadratic and positive, then the complete DNF of f can easily be obtained by querying $O(n^2)$ values of f: For all pairs of indices $i, j \in N$, compute $f(e_{\{i,j\}})$, where $e_{\{i,j\}}$ is the characteristic vector of $\{i, j\}$. Those pairs $\{i, j\}$ such that $f(e_{\{i,j\}}) = 1$ are exactly the edges of G.

We conclude this section by mentioning one last connection between combinatorial structures and positive Boolean functions. In 1897, Dedekind asked for the number $d(n)$ of elements of the free distributive lattice on n elements. This famous question is often referred to as *Dedekind's problem* [572]. As it turns out, $d(n)$ is equal to the number of positive Boolean functions of n variables. The number $d(n)$ grows quite fast and its exact value is only known for small values of n; see Table 1.3 based on Berman and Köhler [73]; Church [196]; Wiedemann [908]; and sequence A000372 in Sloane [840]. Kleitman [572] proved that $\log_2 d(n)$ is asymptotic to the middle binomial coefficient $\binom{n}{\lfloor n/2 \rfloor}$ (see also [542, 573, 578, 579] for extensions and refinements of this deep result).

We should warn the reader, however, that the relations between combinatorics and Boolean theory are by no means limited to the study of positive Boolean functions. Later in the book, we shall have several opportunities to encounter nonpositive Boolean functions linked, in various ways, to graphs or hypergraphs.

Table 1.3. The number of positive Boolean
functions of n variables for $n \leq 8$

n	$d(n)$
0	2
1	3
2	6
3	20
4	168
5	7581
6	7828354
7	2414682040998
8	56130437228687557907788

1.13.6 Integer programming

Consider a very general 0–1 integer programming problem $\mathcal{P}$ of the form

$$\text{maximize} \quad z(x_1, x_2, \ldots, x_n) = \sum_{i=1}^{n} c_i x_i \tag{1.24}$$

$$\text{subject to} \quad (x_1, x_2, \ldots, x_n) \in F, \tag{1.25}$$

where $c_1, c_2, \ldots, c_n$ are integer coefficients and $F \subseteq \mathcal{B}^n$ is a set of feasible 0–1 solutions. Following Granot and Hammer [410], we call *resolvent* of F the Boolean function $f_F(x_1, x_2, \ldots, x_n)$ that takes value 0 on F and value 1 elsewhere (see also Hammer and Rudeanu [460], where the function $\overline{f_F}$ is called the *characteristic function* of F, or Granot and Hammer [411], where f_F is implicitly described). So, problem $\mathcal{P}$ is equivalent to

$$\text{maximize} \quad z(x_1, x_2, \ldots, x_n) = \sum_{i=1}^{n} c_i x_i \tag{1.26}$$

$$\text{subject to} \quad f_F(x_1, x_2, \ldots, x_n) = 0 \tag{1.27}$$

$$(x_1, x_2, \ldots, x_n) \in \mathcal{B}^n. \tag{1.28}$$

Let us assume for a moment that we have somehow obtained a DNF expression of the resolvent f_F. Then, problem $\mathcal{P}$ can be rewritten as a linear 0–1 programming problem with a very special structure. Indeed, as observed by Balas and Jeroslow [43] and by Granot and Hammer [410, 411]:

Theorem 1.39. *If*

$$\psi = \bigvee_{k=1}^{m} \left(\bigwedge_{i \in A_k} x_i \bigwedge_{j \in B_k} \overline{x}_j \right) \tag{1.29}$$

is a DNF expression of the resolvent f_F, then problem $\mathcal{P}$ is equivalent to the generalized covering problem

$$\text{maximize} \quad z(x_1, x_2, \ldots, x_n) = \sum_{i=1}^{n} c_i x_i \tag{1.30}$$

$$\text{subject to} \quad \sum_{i \in A_k} x_i - \sum_{j \in B_k} x_j \leq |A_k| - 1, \quad k = 1, 2, \ldots, m \tag{1.31}$$

$$(x_1, x_2, \ldots, x_n) \in \mathcal{B}^n. \tag{1.32}$$

Proof. We must show that the set of false points of f_F coincides with the set of solutions of (1.31). Let X^* be a false point of f_F. For each $k = 1, 2, \ldots, m$, since $\psi(X^*) = 0$, either there is an index $i \in A_k$ such that $x_i^* = 0$, or there is an index $j \in B_k$ such that $x_j^* = 1$. In either case, we see that X^* satisfies the k-th inequality in (1.31). The converse statement is equally easy.   □

Theorem 1.39 takes an especially interesting form when f_F is positive. Indeed, remember that a *set covering problem* is a linear programming 0–1 problem of the form

$$\text{minimize} \quad \sum_{i=1}^{n} w_i y_i$$

$$\text{subject to} \quad \sum_{i \in S_k} y_i \geq 1, \quad k = 1, 2, \ldots, m$$

$$(y_1, y_2, \ldots, y_n) \in \mathcal{B}^n,$$

where $S_1, S_2, \ldots, S_m$ are subsets of $\{1, 2, \ldots, n\}$ (see, e.g., Nemhauser and Wolsey [707]).

Now, if we assume that (1.29) is a positive DNF of f_F, namely, if $B_k = \emptyset$ for $k = 1, 2, \ldots, n$, then we obtain (Granot and Hammer [411]):

Theorem 1.40. *If the resolvent f_F is positive, and if*

$$\psi = \bigvee_{k=1}^{m} \bigwedge_{i \in A_k} x_i \tag{1.33}$$

is a positive DNF of f_F, then problem $\mathcal{P}$ is equivalent to the following set covering problem $\mathcal{SCP}$:

$$\text{minimize} \quad z'(y_1, y_2, \ldots, y_n) = \sum_{i=1}^{n} c_i y_i \tag{1.34}$$

$$\text{subject to} \quad \sum_{i \in A_k} y_i \geq 1, \quad k = 1, 2, \ldots, m \tag{1.35}$$

$$(y_1, y_2, \ldots, y_n) \in \mathcal{B}^n. \tag{1.36}$$

Proof. By Theorem 1.39, $\mathcal{P}$ is equivalent to

$$\text{maximize} \quad z(x_1, x_2, \ldots, x_n) = \sum_{i=1}^{n} c_i x_i$$

$$\text{subject to} \quad \sum_{i \in A_k} x_i \leq |A_k| - 1, \quad k = 1, 2, \ldots, m$$

$$(x_1, x_2, \ldots, x_n) \in \mathcal{B}^n.$$

For $i = 1, 2, \ldots, n$, it is now sufficient to replace variable x_i by a new variable $y_i = 1 - x_i$ in this formulation. $\qquad\square$

Note that the latter result motivates the terminology "generalized covering" used in Theorem 1.39.

Another way to look at Theorem 1.40 is suggested by the connections established in Section 1.13.5. Indeed, when f_F is positive, the feasible solutions of $\mathcal{P}$ are exactly the stable sets of a hypergraph, and the feasible solutions of $\mathcal{SCP}$ are the transversals of this hypergraph. So, Theorem 1.40 simply builds on the well-known observation that stable sets are exactly the complements of transversals (see, e.g., Berge [72]).

Algorithms based on the transformations described in Theorems 1.39 and 1.40 have been proposed in [408, 409, 410]. Several recent approaches to the solution of Boolean equations also rely on this transformation (see, e.g., [184]).

We shall come back to integer programming problems of the form $\mathcal{P}$ in subsequent chapters of the book (see, in particular, Sections 4.2, 8.6, and 9.4). For now, we conclude this section with a discussion of the complexity of computing a DNF expression of the resolvent. For this question to make sense, we must first specify how the set F is described in (1.25). In the integer programming context, F would typically be defined as the solution set of a system of linear inequalities in the variables $x_1, x_2, \ldots, x_n$, say,

$$\sum_{i=1}^{n} a_{ki} x_i \leq b_k, \quad k = 1, 2, \ldots, s. \tag{1.37}$$

When this is the case, there is generally no low-complexity, practically efficient algorithm for computing a DNF of f_F. More precisely, in Section 9.5 of Chapter 9 we show that the size of every DNF of the resolvent may be exponentially large in the input size of (1.37), even when $s = 1$, that is, when $\mathcal{P}$ is a so-called *knapsack problem*. (Observe that when $s = 1$, the resolvent is a threshold function.) We also see in Chapter 9 that, in this context, the resolvent still turns out to be a useful concept.

On the other hand, there are examples of combinatorial optimization problems for which the resolvent of F is directly available in DNF. The most obvious example, in view of Theorem 1.40, is when $\mathcal{P}$ is a set-covering problem.

Finally, one should also notice that, as long as the description of F is in NP, Cook's theorem [208] guarantees the existence of a polynomial-time procedure which, given any instance of F, produces an integer $t \geq n$ and a DNF expression $\phi(y_1, y_2, \ldots, y_t)$ such that $X^* \in F$ if and only if $\phi(y_1^*, y_2^*, \ldots, y_t^*) = 0$ for some $(y_1^*, y_2^*, \ldots, y_t^*) \in \mathcal{B}^t$. However, although the DNF ϕ bears some resemblance with the resolvent of F, it usually involves a large number of additional variables beside the original variables $x_1, x_2, \ldots, x_n$ (compare with the DNF produced by the procedure EXPAND in Section 1.5).

1.14 Exercises

1. Compute the number of Boolean functions and DNF expressions in n variables, for $n = 1, 2, \ldots, 6$.
2. Show that $\overline{(ax \vee b\overline{x})} = \overline{a}x \vee \overline{b}\overline{x}$ for all $a, b, x \in \mathcal{B}$.
3. Prove that every Boolean function has an expression involving only disjunctions and negations, but no conjunctions, as well as an expression involving only conjunctions and negations, but no disjunctions.
4. The binary operator *NOR* is defined by $NOR(x, y) = \overline{x}\,\overline{y}$. Show that every Boolean expression is equivalent to an expression involving only the *NOR* operator (and parentheses). Show that the same property holds for the *NAND* operator defined by $NAND(x, y) = \overline{x} \vee \overline{y}$. (See, e.g., [752] for far-reaching extensions of these observations.)
5. A Boolean function f is called *symmetric* if $f(x_1, x_2, \ldots, x_n) = f(x_{\sigma_1}, x_{\sigma_2}, \ldots, x_{\sigma_n})$ for all permutations $(\sigma_1, \sigma_2, \ldots, \sigma_n)$ of $\{1, 2, \ldots, n\}$.
   (a) Prove that f is symmetric if and only if there exists a function $g : \{0, 1, \ldots, n\} \to \mathcal{B}$ such that, for all $X \in \mathcal{B}^n$, $f(x_1, x_2, \ldots, x_n) = g(\sum_{i=1}^n x_i)$.
   (b) For $k = 0, 1, \ldots, n$, define the Boolean function r_k by $r_k(X) = 1$ if and only if $\sum_{i=1}^n x_i = k$. Prove that f is symmetric if and only if there exists $A \subseteq \{0, 1, \ldots, n\}$ such that $f = \bigvee_{k \in A} r_k$.
   (c) Prove that the set of all symmetric functions is closed under disjunctions, conjunctions, and complementations.
   (d) What is the complexity of deciding whether a given DNF represents a symmetric function?
6. Design a data structure to store a DNF ϕ in which
   (a) ϕ can be stored in $O(|\phi|)$ space built in $O(|\phi|)$ time;
   (b) finding a term of η of a given degree requires $O(1)$ time;
   (c) finding a negative linear term of ϕ requires $O(1)$ time;
   (d) adding/deleting a term of degree k requires $O(k)$ time;
   (e) fixing/reactivating a literal occurring l times in ϕ requires $O(l)$ time.
7. Show that the degree of a DNF expression of a Boolean function may be strictly smaller than the degree of its complete DNF.
8. For an arbitrary Boolean function f on $\mathcal{B}^n$, define the *influence* of variable k $(k = 1, 2, \ldots, n)$ to be the probability that $f_{|x_k=1}(X) \neq f_{|x_k=0}(X)$, where

X is drawn uniformly at random over $\mathcal{B}^{n-1}$ (see Kahn, Kalai, and Linial [543]). Show that, when f is positive, the influence of variable k is equal to $\frac{\pi_k}{2^{n-1}}$, where π_k is the k-th modified Chow parameter of f.

9. Show that the binary operator $\oplus$ is commutative and associative, that is, $x_1 \oplus x_2 = x_2 \oplus x_1$ and $(x_1 \oplus x_2) \oplus x_3 = x_1 \oplus (x_2 \oplus x_3)$ for all $x_1, x_2, x_3 \in \mathcal{B}$.

10. The *parity function* on $\mathcal{B}^n$ is the function $p^n(x_1, x_2, \ldots, x_n) = x_1 \oplus x_2 \oplus \ldots \oplus x_n$.
    (a) Write a DNF expression of p^n.
    (b) Compute the Chow parameters of p^n.

11. Assume that f is represented either as a sum-of-products modulo 2 of the form (1.16) or as a multilinear polynomial over the reals of the form (1.19). In each case, show how to efficiently solve the equation $f(X) = 0$.

12. Show that, if f is a Boolean function on $\mathcal{B}^n$, and f has an odd number of true points, then
    (a) every orthogonal DNF of f has degree n;
    (b) every decision tree for f contains a path of length n from the root to some leaf.

13. Prove that every Boolean function f on $\mathcal{B}^n$ has a unique *largest positive minorant* f_- and a unique *smallest positive majorant* f^+, where
    (a) f_- and f^+ are positive functions on $\mathcal{B}^n$;
    (b) $f_- \leq f \leq f^+$;
    (c) if g and h are any two positive functions such that $g \leq f \leq h$, then $g \leq f_-$ and $f^+ \leq h$.

14. Prove that every Boolean function has the same maximal false points as its largest positive minorant and the same minimal true points as its smallest positive majorant (see previous exercise).

15. Consider the 0-1 integer programming problem (1.26)–(1.28) in Section 1.13.6. Prove that, when $c_j > 0$ for $j = 1, 2, \ldots, n$, (1.26)–(1.28) has the same optimal solutions as the set covering problem obtained upon replacing the resolvent f_F by its largest positive minorant (see previous exercises, and Hammer, Johnson, and Peled [443]).

Question for thought

16. (Open-ended). Characterize those multilinear polynomials over the reals that represent Boolean functions. (See Section 1.12.2 and Nisan and Szegedy [714].)

2

Boolean equations

The solution of Boolean equations is arguably the most fundamental problem arising in the theory of Boolean functions. Actually, the quote at the beginning of Chapter 1 shows that an important aspect of Boole's original research program was essentially to reduce logic to the solution of Boolean equations. Although his hopes eventually proved overly optimistic, it will become clear in subsequent chapters of this book that Boolean equations often arise as subproblems to be solved in the course of tackling more complex problems. Therefore, their solution is a cornerstone of many Boolean algorithms.

In this chapter, we present some representative models involving Boolean equations and describe various algorithmic procedures for their solution: branching, variable elimination, the consensus method, and mathematical programming approaches. In view of the importance of this topic, we spend quite a lot of time discussing the details of classical procedures, their interrelations, respective merits, and complexity. In the last section, we generalize the basic consistency-testing problem in several ways: We examine the problems of counting and of generating all solutions of a Boolean equation and briefly discuss the maximum satisfiability (MAX SAT) problem.

2.1 Definitions and applications

Definition 2.1. *A* Boolean equation *is an equation of the form* $\phi(X) = \psi(X)$, *where* $X = (x_1, x_2, \ldots, x_n)$ *is a vector of Boolean variables, and* ϕ, ψ *are Boolean expressions in these variables. A* solution *of the equation is a point* $X^* \in B^n$ *such that* $\phi(X^*) = \psi(X^*)$. *A Boolean equation is called* consistent *if it has a solution; otherwise, it is called* inconsistent.

For reasons to be discussed in Section 2.3, much of the literature on Boolean equations focuses on DNF equations.

Definition 2.2. *A* DNF equation *is a Boolean equation of the form* $\phi(X) = 0$, *where* ϕ *is a DNF. The* degree *of the DNF equation* $\phi(X) = 0$ *is the degree of* ϕ.

Boolean equations not only play a fundamental role in propositional logic and in theoretical computer science but also occur directly and naturally in many applications, such as artificial intelligence, electrical engineering, mathematical programming, and so on. Here are brief outlines of some typical applications.

Application 2.1. (Propositional logic, artificial intelligence.) *In propositional logic, a formula (or a Boolean expression) ϕ is called* satisfiable *if the equation $\phi(X) = 1$ is consistent, and it is called a* contradiction *otherwise. The formula is* valid, *or is a* tautology, *if ϕ is identically equal to 1, that is, if the equation $\phi(X) = 0$ is inconsistent. These classical concepts play a central role in (propositional) logic and in all applications of artificial intelligence in which propositional logic is used to model knowledge.*

To illustrate, consider a knowledge base of rules involving the propositional variables $x_1, x_2, \ldots, x_n$, and let $\phi(x_1, x_2, \ldots, x_n)$ be the Boolean expression associated with the knowledge base, as in Section 1.13.1 (Chapter 1). Then, as we have seen, the set of solutions of the equation $\phi = 0$ describes the set of models of the knowledge base, that is, the set of truth assignments that satisfy all the rules. In particular, the equation $\phi = 0$ is consistent if and only if the collection of rules is not self-contradictory. Also, questions relative to the atomic propositions – e.g., questions of the form, "Is $x_i = 1$ consistent with the given rules?" – are directly reducible to the solution of Boolean equations.

Similar principles are used in many other areas of artificial intelligence, notably in automated theorem proving. Assume, for instance, that a theorem proving system must prove or disprove a general implication of the form

$$\forall X \in \mathcal{B}^n; \; (\phi(X) = 0) \implies (\psi(X) = 0), \tag{2.1}$$

where ϕ and ψ are arbitrary Boolean expressions (the premise $\phi(X) = 0$ could express the axioms of the theory as well as a number of more specific hypotheses). The usual way to attack this question is to reason by contradiction and to solve the equation

$$\phi(X) \vee \overline{\psi}(X) = 0.$$

If this equation is consistent, then any of its solutions yields a counter-example to the conjecture (2.1). Conversely, if the equation is inconsistent, then the implication (2.1) is a theorem.

Our discussion focused on propositional logic. However, testing the validity of formulas in first-order predicate logic, even though an undecidable problem, can, in principle, be "reduced" to the solution of an infinite number of Boolean equations through an application of Herbrand's theorem. This type of reduction is used, either explicitly or implicitly, in many theorem-proving procedures for first-order logic; see, for example, Gilmore [380], Davis and Putnam [261], Robinson [787], Chang and Lee [186], Jeroslow [533], Thayse [863]. Boolean equations also find applications in solving decision problems from modal logic, as discussed in [384, 510]. □

Application 2.2. (Electrical engineering.) *Boolean equations play a central role in the design and analysis of logic circuits. We sketch here only some representative applications arising in this field and refer the reader to the specialized literature for more information; see, for instance, Abdulla, Bjesse, and Eén [1]; Brayton, Hachtel, McMullen, and Sangiovanni-Vincentelli [153]; Brown [156]; Herbstritt [490]; Kunz and Stoffel [590]; Schneeweiss [811]; Stephan, Brayton, and Sangiovanni-Vincentelli [846]; or, the surveys by Clarke, Biere, Raimi, and Zhu [204]; Gu, Purdom, Franco, and Wah [418]; Jiang and Villa [535]; or Villa, Brayton, and Sangiovanni-Vincentelli [891].*

When a Boolean function is to be physically realized by a VLSI circuit, it is usually desirable to first transform the original expression of the function into another equivalent expression. This is because the original expression, which arose from a functional specification of the circuit, may not be best suited for implementation purposes. Circuit designers would thus typically seek an expression requiring fewer gates, fewer contacts, and so on in order to reduce the size of the circuit and increase its speed and reliability. The transformed expression can be obtained by algebraic manipulations based on the elementary rules spelled out in Chapter 1 or, possibly, by other means. In particular, the Boolean minimization and dualization problems discussed in Chapters 3 and 4 of this book, mostly arise in this context; we shall see that the solution of Boolean equations is a basic subproblem in this framework. (Brayton et al. [153] developed the well-known computer program ESPRESSO-II *for logic design; according to the authors (page 64): "Answering the tautology question (deciding if $f \equiv 1$) is the most fundamental Boolean operation required by* ESPRESSO-II*.")*

Whatever means are used, the validity of the transformation of a given expression, say $\phi(X)$, to another expression, say $\psi(X)$, has to be carefully established in a so-called verification *phase before one can proceed with the actual implementation of the circuit. Verification can (in principle) be carried out by solving the Boolean equation $\phi(X) = \overline{\psi(X)}$. Indeed, $\phi(X)$ and $\psi(X)$ are equivalent expressions if and only if this equation is inconsistent.*

We saw in Chapter 1, Section 1.13.2, that the correct operation of a combinational circuit can be described by a Boolean equation $\psi(X, Y, z) = 0$, where $\psi(X)$ is a DNF, $X = (x_1, x_2, \ldots, x_n)$ is the vector of variables associated with the input signals of the circuit, z corresponds to the output signal of the circuit, and $Y = (y_1, y_2, \ldots, y_m)$ is a vector of variables associated with the outputs of the internal, "hidden" gates of the circuit.

In reality, a circuit may malfunction for any of a number of reasons, and the problem of detecting such malfunctions is crucial in VLSI engineering. Various techniques can be used for this purpose, depending on the type of faults that are expected. We briefly discuss the detection of stuck-at *faults. A stuck-at fault occurs when, due to some physical defect, one of the gates of the circuit produces a constant output, independent of the values of its inputs. The gate could be stuck at 1, meaning that it always produces a 1, or stuck at 0, meaning that it always outputs a 0. Since the hidden gates of the circuit are not directly observable, one can only* *infer stuck-at faults from the observed input and ouput signals of the circuit. In general terms, the test generation problem for stuck-at faults can be expressed as follows: Generate an input vector X^* (or possibly several) such that the output of the circuit is incorrect on that input when certain gates have stuck-at faults.*

To make this more explicit, let us focus on the test generation problem for diagnosing whether a specific OR-gate, say, gate k, is stuck at 1. (In practice, one may often safely assume that only a few gates are faulty in a circuit. It is even common to posit the "single fault hypothesis" according to which one gate at most could be faulty.) Let $\psi(X,Y,z)$ be the Boolean expression modeling the combinational circuit as explained in Section 1.13.2, let y_1, y_2 model the inputs of gate k, and let y_k model its output. So, in the expression ψ, we can isolate the terms associated with gate k by rewriting $\psi(X,Y,z)$ as

$$\psi(X,Y,z) = \phi(X,Y,z) \vee y_1 \overline{y}_k \vee y_2 \overline{y}_k \vee \overline{y}_1 \overline{y}_2 y_k. \tag{2.2}$$

Observe that the role of the last three terms of ψ in (2.2) is only to describe the correct operation of the OR-gate k (all three terms must be 0 when the gate is operating properly).

To model the behavior of the circuit when gate k is stuck at 1, we introduce a new variable w, representing the output of the faulty circuit, and a new vector of variables $V = (v_1, v_2, \ldots, v_m)$, where v_i represents the output signal of gate i $(i = 1, 2, \ldots, m)$ in the faulty circuit. Applying the same reasoning as in the absence of any fault, we can state: In every solution (X^*, V^*, w^*) of the equation $\phi_{|v_k=1}(X,V,w) = 0$, the variable associated with each gate represents the output of that gate on the input signal X^* on the assumption that gate k is stuck at 1 (note that the terms linking y_1, y_2, and y_k are absent from this equation).

It is then easy to conclude that every solution $(X^*, Y^*, z^*, V^*, w^*)$ of the equation

$$\psi(X,Y,z) \vee \phi_{|v_k=1}(X,V,w) \vee zw \vee \overline{z}\overline{w} = 0 \tag{2.3}$$

has the following property: On the input signal X^*, the correct circuit described by ψ produces the output z^*, while the faulty circuit in which gate k is stuck at 1 produces the output $w^* = \overline{z}^*$. In other words, a valid test vector for the stuck-at-1 fault at gate k can be generated by solving the Boolean equation (2.3).

Example 2.1. *Let us illustrate this procedure for the detection of a stuck-at-1 fault at the second (lower) OR-gate of the circuit displayed in Figure 1.6. The expression $\psi(X,Y,z)$ associated with this circuit is given by equation (1.22), where the output of the OR-gate under consideration is represented by variable y_2. As in equation (2.2), we can rewrite*

$$\psi(X,Y,z) = \phi(X,Y,z) \vee x_1\overline{y}_2 \vee y_4\overline{y}_2 \vee \overline{x}_1\overline{y}_4 y_2,$$

with

$$\phi(X,Y,z) = \overline{y}_1 z \vee y_2 z \vee y_1\overline{y}_2\overline{z} \vee x_1\overline{y}_1 \vee x_4\overline{y}_1 \vee \overline{x}_1\overline{x}_4 y_1 \vee \overline{x}_2 y_4 \vee x_3 y_4 \vee x_2\overline{x}_3\overline{y}_4.$$

Then, after some simplifications, equation (2.3) reduces to

$$x_1 \vee \overline{x}_4 \vee \overline{y}_1 \vee y_2 \vee y_4 \vee \overline{z} \vee \overline{v}_1 \vee w \vee x_2\overline{x}_3 \vee \overline{x}_2 v_4 \vee x_3 v_4 = 0.$$

The conclusion is that any input vector X^* satisfying $x_1^* = 0$, $x_4^* = 1$, and $x_2^* \overline{x}^*_3 = 0$ is a valid test vector for a stuck-at-1 fault at the lower OR-gate. Any such vector produces the output $w^* = 0$ in the faulty circuit, when it should produce the output $z^* = 1$ in the correct circuit (it is not very difficult to check that it is indeed so, by direct verification).   □

    Larrabee [599] has demonstrated that a Boolean approach to test pattern generation, based on the formulation just described, is extremely effective in practice and produces excellent results on well-known benchmark problems. In her experiments, the approach proved competitive with alternative structural approaches proposed in the specialized literature (see, e.g., [178, 590]).

    In more recent work, Clarke et al. [204] describe successful reformulations of other verification problems as Boolean DNF equations. They observe that this approach, known as bounded model checking, *appears to be remarkably efficient and robust on industrial systems that would be difficult for the more traditional model checking techniques based on binary decision diagrams; see also Jiang and Villa [535].*   □

Application 2.3. (Combinatorics.) *Many properties of graphs and hypergraphs can be easily expressed by means of Boolean equations. Theorem 2.1 in Section 2.2 provides a more precise statement of this claim, and Hammer and Rudeanu [460] give several explicit Boolean formulations of combinatorial problems. More examples will appear in subsequent chapters. So, we only present here a simple illustration.*

    Let $\mathcal{H} = (N, \mathcal{E})$ be a hypergraph, where $N = \{1, 2, \ldots, n\}$, and recall the terminology in Appendix A. We say that $\mathcal{H}$ is 2-colorable if N can be partitioned into two stable sets of $\mathcal{H}$. Equivalently, $\mathcal{H}$ is 2-colorable if each of its vertices can be assigned one of two colors, say blue or red, so that no edge of $\mathcal{H}$ is entirely blue or entirely red. Introduce now n Boolean variables $x_1, x_2, \ldots, x_n$ with the interpretation that vertex i is colored blue (respectively, red) if $x_i = 1$ (respectively, $x_i = 0$). Then, $\mathcal{H}$ is 2-colorable if and only if the following DNF equation is consistent:

$$\phi(x_1, x_2, \ldots, x_n) = \left(\bigvee_{A \in \mathcal{E}} \bigwedge_{j \in A} x_j \right) \vee \left(\bigvee_{A \in \mathcal{E}} \bigwedge_{j \in A} \overline{x}_j \right) = 0.$$

This straightforward observation seems to be part of the folklore of the field. Remark that, with the notations of Section 1.1, $\phi(X) = f_{\mathcal{H}}(X) \vee f_{\mathcal{H}}(\overline{X})$.

    Conversely, Linial and Tarsi [615] showed that testing the consistency of any DNF equation of the form

$$\phi(x_1, x_2, \ldots, x_n) = \bigvee_{k=1}^{m} \left(\bigwedge_{i \in A_k} x_i \bigwedge_{j \in B_k} \overline{x}_j \right) = 0, \tag{2.4}$$

can be very simply transformed to a hypergraph 2-colorability problem. To see this, let us define $V = \{x_1, x_2, \ldots, x_n, \overline{x}_1, \overline{x}_2, \ldots, \overline{x}_n, 1\}$. We build a hypergraph

$\mathcal{H} = (V, \mathcal{E})$ *on the vertex-set* V, *where*

- *for each* $i \in \{1, 2, \ldots, n\}$, $\{x_i, \overline{x}_i\}$ *is an edge in* $\mathcal{E}$;
- *for each* $k \in \{1, 2, \ldots, m\}$, $\{x_i \mid i \in A_k\} \cup \{\overline{x}_j \mid j \in B_k\} \cup \{1\}$ *is an edge in* $\mathcal{E}$.

It is an easy exercise to check that equation (2.4) is consistent if and only if $\mathcal{H}$ *is 2-colorable. As a consequence, any algorithm for testing the 2-colorability of hypergraphs can also be used to solve (2.4) (see [615] for the description of such an algorithm).*

Another property of this construction is that the equation (2.4) is consistent if and only if the hypergraph $\mathcal{H}$ *has the so-called Kőnig-Egerváry property, that is, if the maximum number of pairwise disjoint edges of* $\mathcal{H}$ *is equal to the minimum cardinality of a transversal of* $\mathcal{H}$. *A closely related result was previously established by Simeone [833, 834], who relied on this characterization to propose a linear time algorithm for the solution of quadratic DNF equations (see Chapter 5).*   □

Application 2.4. (Integer programming.) *In the course of solving linear or non-linear 0–1 optimization problems, logical relations can often be deduced between the values assumed by certain variables in every, or in some, optimal solutions. This happens typically, though not exclusively, in the preprocessing phase of the solution procedure. Suppose, for instance, that we are somehow able to derive that two variables* x *and* y *can never be simultaneously 0 in a feasible solution of the problem. Then, we know that the Boolean relation* $\overline{x}\,\overline{y} = 0$ *must hold. Similarly, if at most one of* x, y, *and* u *can take value 1 in a feasible solution, then* $xy \vee xu \vee yu = 0$ *must hold. Collecting several such relations and taking them simultaneously into account leads to a Boolean equation* $\phi(x, y, u, \ldots) = 0$, *which is consistent if and only if the the optimization problem is feasible. This observation can be used to set up feasibility tests in a branch-and-bound procedure or to accelerate heuristics (see, e.g., Granot and Hammer [410], Hammer and Nguyen [454], Hammer and Hansen [439], Jaumard [526], Boros, Hammer, and Sun [133]).*   □

Several researchers have recently reported encodings of a large variety of industrial problems in the form of Boolean equations, and solutions of these problems by general purpose algorithms. Besides the references already cited earlier, let us also mention the synthesis of small circuits for partially defined Boolean functions [546], the verification of the validity of an automated safety procedure implemented by Dutch railway stations [412], an application to product data management in the automotive industry [586], the analysis of data encryption standards [675], planning problems in logistics [556], and so on (see also the survey [418]).

2.2 The complexity of Boolean equations: Cook's theorem

In the previous section, we discovered several prominent applications of Boolean equations. We could have extended this list of applications to encompass several hundreds of questions. Indeed, it has been observed for a long time that numerous problems of a combinatorial nature can be reduced to the solution of Boolean equations (see, for instance, Fortet [342, 343]; Hammer and Rudeanu [460]). This statement was given a more precise and very dramatic formulation by Cook [208], who proved that *each and every* decision problem in a broad class of problems (namely, the so-called class NP) can be transformed *in polynomial time* into an equivalent Boolean equation. In order to express Cook's theorem in the usual format of complexity theory, we first pose the problem of solving Boolean equations as a decision problem (see Appendix B).

BOOLEAN EQUATION
Instance: Two Boolean expressions $\phi(X)$ and $\psi(X)$.
Question: Is the equation $\phi(X) = \psi(X)$ consistent?

A restricted version of this problem is:

DNF EQUATION
Instance: A DNF expression $\phi(X)$.
Question: Is the equation $\phi(X) = 0$ consistent?

Observe that the answer to DNF EQUATION is "No" exactly when ϕ is a tautology, as discussed in Application 2.1.

Theorem 2.1. (Cook [208]) *The problem* BOOLEAN EQUATION *is NP-complete, even when restricted to* DNF EQUATION *and to DNF expressions of degree* 3.

A proof of this deep and fundamental theorem requires the introduction of formal machinery from complexity theory, for example, a definition of models of computation, computing time, polynomial algorithms, reductions, and so on. We refer the interested reader to Appendix B for a succinct introduction to these concepts and for a proof of Theorem 2.1 (see also Theorems 2.3 and 2.4 following). For readers who are not familiar with complexity theory, Appendix B also provides valuable additional insights into the relevance of Cook's theorem.

It is important to observe that the DNF equation

$$\phi(X) = \bigvee_{k=1}^{m} \left(\bigwedge_{i \in A_k} x_i \bigwedge_{j \in B_k} \overline{x}_j \right) = 0 \tag{2.5}$$

has exactly the same set of solutions as the equation

$$\psi(X) = \bigwedge_{k=1}^{m} \left(\bigvee_{i \in A_k} \overline{x}_i \vee \bigvee_{j \in B_k} x_j \right) = 1, \tag{2.6}$$

where ψ is now a CNF. As a matter of fact, Cook's theorem is frequently stated (and was originally proved) in its dual form involving CNF rather than DNF equations.

More precisely, let us define the following decision problem:

SATISFIABILITY
Instance: A CNF $\psi(X)$.
Question: Is the equation $\psi(X) = 1$ consistent?

In view of Theorem 2.1 and of the equivalence of (2.5) and (2.6), we immediately conclude that SATISFIABILITY is NP-complete, even when each clause of the CNF ψ involves at most three literals (3-SATISFIABILITY or 3-SAT problem).

Boolean equations have been frequently stated as satisfiability problems in the artificial intelligence and computational complexity literatures. On the other hand, DNF formulations are more commonly used in electrical engineering and in propositional logic. In this book, we mostly deal with DNF equations rather than satisfiability problems, but it should be clear that this is purely a matter of convention.

The reader should also be aware that, in contrast to the foregoing comments, equations of the form $\phi = 1$, where ϕ is a DNF, are extremely easy to solve (and thus rather uninteresting). To see this, simply remember that a DNF takes value 1 if and only if at least one of its terms takes value 1.

Finally, it should be noted that the bound on the degree of the equation in Theorem 2.1 is tight, in the sense that the equation $\phi(X) = 0$ can be solved in polynomial time if ϕ is a quadratic DNF, as we shall see in Chapter 5. Numerous extensions of Theorem 2.1, of the form BOOLEAN EQUATION *is NP-complete, even when restricted to equations satisfying condition C*, have been established in the literature. We refer to [371, 571] for a discussion of such extensions, and we propose some of them as end-of-chapter exercises.

2.3 On the role of DNF equations

In subsequent sections, we frequently concentrate on solution techniques for DNF, rather than arbitrary Boolean equations. There are several reasons for this focus. Clearly, Theorem 2.1 claims an important role for DNF equations in the theory of computational complexity. But DNF equations also occur naturally in many practical settings, as illustrated by several of the applications presented in Section 2.1. Moreover, most solution techniques for Boolean equations actually start by reducing the given equation to a DNF equation.

The practical relevance of DNF equations is probably better understood when one realizes that (2.5) is in fact equivalent to the system of equations

$$\bigwedge_{i \in A_k} x_i \bigwedge_{j \in B_k} \overline{x}_j = 0, \quad k = 1, 2, \ldots, m.$$

Thus, a DNF equation is the natural expression of a system of conditions of the form "at least one of the variables in A_k must be 0, or at least one of the variables in B_k must be 1," all of which must be simultaneously satisfied. For instance, the production rules used in the knowledge base of an expert system frequently constitute a system of conditions of this type; see Applications 1.13.1 and 2.1. This is also the case for the Boolean equation associated with a logic circuit, as explained in Applications 1.13.2 and 2.2.

More generally, systems of (possibly complex) Boolean conditions also arise when instantiation techniques based on Herbrand's theorem are used to prove the validity of first-order logic formulas. Davis and Putnam [261] argued that, in this framework, it is quite natural and efficient to work with DNFs. Their argument goes as follows (for the sake of clarity, we replace the word "conjunctive" by "disjunctive" in the authors' original statement, without altering its meaning):

> That the disjunctive normal form can be employed follows from the remark that to put a whole system of formulas into disjunctive normal form we have only to put the individual formulas into disjunctive normal form. Thus, even if a system has hundreds or thousands of formulas, it can be put into disjunctive normal form "piece by piece", without any "multiplying out" (Davis and Putnam [261]).

In the remainder of this section, we show how an arbitrary system of Boolean conditions (equations and inequalities) can be efficiently transformed into an equivalent DNF equation. Let us first define what we mean by a "system of Boolean conditions."

Definition 2.3. *A* Boolean system *on $\mathcal{B}^n$ is a collection of Boolean equations and inequalities of the form*

$$\phi_k(X) = \psi_k(X) \quad k = 1, 2, \ldots, p, \tag{2.7}$$

$$\phi_k(X) \leq \psi_k(X) \quad k = p+1, p+2, \ldots, p+q, \tag{2.8}$$

where ϕ_k and ψ_k are Boolean expressions on $\mathcal{B}^n$, for $k = 1, 2, \ldots, p+q$. A solution *of the system is a point $X^* \in \mathcal{B}^n$ such that $\phi_k(X^*) = \psi_k(X^*)$ for $k = 1, 2, \ldots, p$ and $\phi_k(X^*) \leq \psi_k(X^*)$ for $k = p+1, p+2, \ldots, p+q$.*

An easy, but fundamental, result due to Boole [103] allows us to transform any Boolean system into a single Boolean equation.

Theorem 2.2. *The Boolean system (2.7)–(2.8) has the same set of solutions as the Boolean equation*

$$\bigvee_{k=1}^{p} \left(\phi_k(X) \overline{\psi_k}(X) \vee \overline{\phi_k}(X) \psi_k(X) \right) \vee \bigvee_{k=p+1}^{p+q} \left(\phi_k(X) \overline{\psi_k}(X) \right) = 0. \tag{2.9}$$

Proof. It suffices to observe that the system (2.7) is equivalent to the system

$$\phi_k(X) \leq \psi_k(X) \quad k = 1, 2, \ldots, p,$$

$$\phi_k(X) \geq \psi_k(X) \quad k = 1, 2, \ldots, p,$$

and that each inequality of the form $\phi_k(X) \leq \psi_k(X)$ is in turn equivalent to the equation $\phi_k(X) \overline{\psi_k}(X) = 0$. $\qquad\square$

In view of Theorem 2.2, it only remains to show that every Boolean equation of the form $\phi(X) = 0$ can be efficiently transformed into an equivalent DNF equation. A polynomial time transformation could of course be read from the proof of Cook's theorem (Theorem 2.1), but the resulting procedure would be too cumbersome to be of practical interest.

On the other hand, since every Boolean function has a DNF expression, the left-hand side of (2.9) could, in principle, be rewritten as an equivalent DNF. However, we have already observed that this may lead to an exponential explosion in the size of the problem (see Example 1.11). As a matter of fact, Example 1.11 essentially shows that there is no hope of achieving the desired polynomial time transformation of an arbitrary equation into an equivalent DNF equation, unless one is willing to introduce additional variables in the picture.

Definition 2.4. *Consider two Boolean systems, say $S_1(X)$ and $S_2(X,Y)$, where S_1 involves only the variables $(x_1, x_2, \ldots, x_n)$, whereas S_2 involves $(x_1, x_2, \ldots, x_n)$ and possibly additional variables $(y_1, y_2, \ldots, y_m)$. We say that S_1 and S_2 are equivalent if the following two conditions hold:*

    (a) *For every solution of S_1, say $X^* \in B^n$, there exists $Y^* \in B^m$ such that (X^*, Y^*) is a solution of S_2.*

    (b) *For every solution of S_2, say, $(X^*, Y^*) \in B^{n+m}$, X^* is a solution of S_1.*

So, when S_1 and S_2 are equivalent, the solution set of S_1 is the projection on B^n of the solution set of S_2. In particular, if S_2 only involves the X-variables, then S_1 and S_2 are equivalent if and only if they have the same solution set.

We are now ready for the main result of this section (Tseitin [872]; see also [78] for a broader discussion and for extensions of this result to first-order predicate logic).

Theorem 2.3. *Every Boolean system can be reduced in linear time to an equivalent DNF equation.*

Proof. First, Theorem 2.2 can be used to rewrite (in linear time) the system as a single equation of the form $\phi(X) = 0$. Then, apply the procedure EXPAND described in Section 1.5 to the expression $\phi(X)$. The output of EXPAND is a DNF $\psi(X,Y)$ and a distinguished literal z among the literals on (X,Y), with the property that the equation $\phi(X) = 0$ is equivalent to the DNF equation $\psi_{|z=0}(X,Y) = 0$. $\square$

Actually, we do not need the full power of the procedure EXPAND in order to establish Theorem 2.3. Indeed, we leave it to the reader to verify that the procedure EXPAND* in Figure 2.1, which introduces fewer additional variables and produces shorter DNFs than EXPAND, also achieves the required transformation (we refer, for instance, to Blair, Jeroslow and Lowe [98], Clarke, Biere, Raimi and Zhu [204], Eén and Sörensson [290], Jeroslow [533], Plaisted and Greenbaum [750], and Wilson [914], for descriptions and applications of related procedures).

Procedure EXPAND$^*(\phi)$

Input: A Boolean expression $\phi(X)$ on $\mathcal{B}^n$.
Output: A DNF $\psi(X,Y)$ on $\mathcal{B}^{n+m}$ such that the equations $\phi(X)=0$ and $\psi(X,Y)=0$ are equivalent.

begin

    if ϕ is a DNF **then** $\psi := \phi$;
    else if $\phi = \overline{\overline{\alpha}}$ for some expression α **then return** EXPAND$^*(\alpha)$;
    else if $\phi = \overline{(\overline{\phi_1 \vee \phi_2})}$ for some expressions ϕ_1, ϕ_2 **then return** EXPAND$^*(\overline{\phi_1}\,\overline{\phi_2})$;
    else if $\phi = \overline{(\overline{\phi_1 \phi_2})}$ for some expressions ϕ_1, ϕ_2 **then return** EXPAND$^*(\overline{\phi_1} \vee \overline{\phi_2})$;
    else if $\phi = (\phi_1 \vee \phi_2 \vee \ldots \vee \phi_k)$ for some expressions $\phi_1, \phi_2, \ldots, \phi_k$ **then**
        begin
            for $j = 1$ **to** k **do** $\psi_j := $ EXPAND$^*(\phi_j)$;
            return $\psi := \psi_1 \vee \psi_2 \vee \ldots \vee \psi_k$;
        end
    else if $\phi = (\phi_1 \phi_2 \ldots \phi_k)$ for some expressions $\phi_1, \phi_2, \ldots, \phi_k$ **then**
        begin
            for $j = 1$ **to** k **do** $\psi_j := $ EXPAND$^*(\phi_j)$;
            create k new variables, say $y_1, y_2, \ldots, y_k$;
            return $\psi := \overline{y}_1 \psi_1 \vee \overline{y}_2 \psi_2 \vee \ldots \vee \overline{y}_k \psi_k \vee y_1 y_2 \ldots y_k$;
        end

end

Figure 2.1. Procedure EXPAND*.

The next result underlines the special role played by DNF equations of degree 3. It can be seen as a strengthening of the second half of Theorem 2.1.

Theorem 2.4. *Every DNF equation can be reduced in linear time to an equivalent DNF equation of degree* 3.

Proof. Consider a DNF equation of the form (2.5) and assume that $|A_1| + |B_1| > 3$. Select two distinct indices in $A_1 \cup B_1$, say, $h, \ell \in A_1$ (similar arguments apply if one of the indices is in B_1). Let y be an additional Boolean variable, different from $x_1, x_2, \ldots, x_n$, and define

$$\psi(X,y) = \left(\bigwedge_{i \in A_1 \setminus \{h,\ell\}} x_i \bigwedge_{j \in B_1} \overline{x}_j \right) y \vee \bigvee_{k=2}^{m} \left(\bigwedge_{i \in A_k} x_i \bigwedge_{j \in B_k} \overline{x}_j \right) \vee x_h x_\ell \overline{y} \vee \overline{x_h}\, y \vee \overline{x_\ell}\, y.$$

We claim that the equations $\phi(X) = 0$ and $\psi(X,y) = 0$ are equivalent. To see this, consider any point $(X^*, y^*) \in \mathcal{B}^{n+1}$. It is easy to see that the expression $x_h^* x_\ell^* \overline{y^*} \vee \overline{x_h^*}\, y^* \vee \overline{x_\ell^*}\, y^*$ is equal to 0 if and only if $y^* = x_h^* x_\ell^*$. This implies that, for all solutions (X^*, y^*) of the equation $\psi(X,y) = 0$, there also holds $\phi(X^*) = 0$. And conversely, every solution X^* of $\phi(X) = 0$ gives rise to a solution (X^*, y^*) of the equation $\psi(X,y) = 0$, by simply setting $y^* = x_h^* x_\ell^*$. Thus, the equations are equivalent.

Note that the degree of the first term of ψ is equal to $|A_1| + |B_1| - 1$. Thus, repeatedly applying this reduction eventually yields a DNF equation of degree 3. It can be checked that the total number of additional variables and terms introduced by this transformation is $O\left(\sum_{k=1}^{m}(|A_k|+|B_k|)\right)$. We leave to the reader a more complete analysis of the complexity of this procedure.   □

Relying on Theorem 2.3 (and Theorem 2.4), the remainder of this chapter mostly concentrates on the solution of DNF equations. The reader should be aware, however, that the transformation of an arbitrary Boolean equation into a DNF equation typically introduces a large number of new variables into the picture, even when procedure EXPAND* is used, rather than EXPAND. Hence, in some cases, this transformation may artificially increase the difficulty of the problem at hand. Since some Boolean equations naturally arise in non-DNF form (e.g., equations of the form $\phi(X) = \psi(X)$ arising in logic circuit verification; see Application 2.2), it may sometimes be desirable to develop procedures capable of dealing directly with these alternative forms, rather than blindly relying on the general techniques discussed earlier.

To illustrate this comment, let us consider an equation of the form $\phi(X) = \psi(X)$, where ϕ and ψ are DNFs. According to our previous discussion, one way of handling this equation is to rewrite it as $\phi(X)\overline{\psi}(X) \vee \overline{\phi}(X)\psi(X) = 0$, and next to apply EXPAND* to the latter equation. However, a more efficient approach can be used here. First, check whether the system $\phi(X) = 1, \psi(X) = 1$ has a solution. Since ϕ and ψ are both DNFs, this system turns out to be very easy to solve (we leave this for the reader to check). If it is consistent, then we can stop right away. Otherwise, the original equation $\phi(X) = \psi(X)$ has been reduced to the system $\phi(X) = 0, \psi(X) = 0$, which is, in turn, equivalent to the DNF equation $\phi(X) \vee \psi(X) = 0$. Clearly, this approach usually involves much less work than the "standard" procedure.

It will also be easy to see that some of the equation-solving techniques presented in the following sections (e.g., the enumeration techniques) can be modified in a straightforward way to handle non-DNF equations. Other techniques have been generalized in a more sophisticated way with the same goal in mind, for example, the consensus technique (see Thayse [863], Van Gelder [885]) or local search heuristics (see Stachniak [844]).

2.4 What does it mean to "solve a Boolean equation"?

The phrase "solving a Boolean equation" can be interpreted in various ways. It is worthwhile to briefly clarify this issue before proceeding.

While discussing Cook's theorem in Section 2.2, we formalized Boolean equations as decision problems: Given a Boolean equation, the task was simply to *decide* whether the equation was consistent or not. Any algorithm for the solution of Boolean equations should be able, as a minimal requirement, to give an answer to this decision problem.

Now, any algorithm that tests the consistency of an equation can also be used, in principle, to *compute* a solution when there is one. To understand this, consider any such consistency-testing algorithm, say $\mathcal{A}$. Given an equation $\phi(x_1, x_2, \ldots, x_n) = 0$, we first use $\mathcal{A}$ to decide whether the equation is consistent. If the answer is No, then we can stop. Otherwise, we run $\mathcal{A}$ again in order to decide whether $\phi(x_1, x_2, \ldots, x_{n-1}, 0) = 0$ is consistent, where the expression $\phi(x_1, x_2, \ldots, x_{n-1}, 0)$ is obtained by substituting 0 for x_n in ϕ. If the answer is Yes, then we restrict our attention to solutions where $x_n = 0$, that is, we fix x_n to 0 in the equation. If the answer is No, then we know that x_n must be 1 in all solutions, and, accordingly, we fix x_n to 1 in the equation. Thus, in either case, we have reduced the original equation to an equation in $n - 1$ variables. Proceeding iteratively, we see that $n + 1$ calls on the algorithm $\mathcal{A}$ suffice to construct a solution of the equation, when there is one.

Fortunately, this roundabout way of computing solutions will usually not prove necessary. Indeed, it is difficult to imagine an algorithm that would simply test whether an equation is consistent that would not also, implicitly or explicitly, *find* a solution of the equation when there is one. As a result, all of the algorithms described in the coming sections will provide an answer to the latter "constructive" version of the problem.

But there are still other ways of interpreting, or of generalizing the task of "solving a Boolean equation." First, we may want to list *all* solutions of the given equation. This is, of course, a formidable requirement, since a Boolean equation may well have an exponential number of solutions. We discuss various ways of handling this problem, either explicitly, in Section 2.11.2, or implicitly (by giving a parametric representation of the set of solutions), in Section 2.11.3.

We may also be interested in *counting* the number of solutions of the equation. We have already briefly mentioned this question in Sections 1.6 and 1.13, for instance, in connection with the problem of computing the reliability of a complex system. We return to it in Section 2.11.1.

We may want to compute the *optimal* solution of the given equation according to a variety of numerical criteria. Such formulations bring us into the realm of integer programming. They have already been evoked in Section 1.13.6, where we have seen that they can be transformed into equivalent generalized covering problems, and we return to them several times in subsequent chapters.

Finally, even when the equation is inconsistent, we may want to compute a point *that comes as close as possible* to satisfying the equation. For instance, the famous *maximum satisfiability problem* (or MAX SAT problem) is of this nature. Indeed, it is equivalent to the following question: Given a DNF ϕ, find a point that cancels as many terms of ϕ as possible. This problem and some of its variants have been thoroughly investigated in the computational complexity literature. We discuss them in Section 2.11.4.

For now, let us turn to the fundamental task of testing the consistency of a Boolean equation. There is a huge field to cover, and we shall primarily concentrate on exact Boolean approaches, as opposed, for instance, to heuristics and to numerical methods. For additional information, we refer to the books [508, 571], to the collections of papers [278, 377, 537], to the surveys [209, 418, 881], and so on.

2.5 Branching procedures

Branching procedures (sometimes called splitting procedures) represent the most elementary and most natural approach to the solution of Boolean equations. Yet, in spite (or because) of their simplicity, they have established themselves as very efficient, reliable, and versatile methods. Therefore, they deserve special attention in this chapter. They also provide a general framework in which many useful algorithmic ideas can easily be explained and implemented.

The starting point of most branching procedures is the following obvious observation:

Theorem 2.5. *The Boolean equation* $\phi(x_1,\ldots,x_{n-1},x_n) = 0$ *is consistent if and only if either*

$$\phi(x_1,\ldots,x_{n-1},0) = 0$$

or

$$\phi(x_1,\ldots,x_{n-1},1) = 0$$

is consistent.

Theorem 2.5 suggests that we can solve the equation $\phi(x_1,\ldots,x_{n-1},x_n) = 0$ using a *branching*, or *enumerative*, procedure similar in spirit to the branch-and-bound methods developed for integer programming problems. We are now going to describe the basic scheme of such a procedure, first informally, and then more rigorously. We restrict ourselves to a depth-first search version of the procedure, partly for the sake of simplicity, and also because many efficient implementations fall under this category (the reader will easily figure out what a more general branching scheme may look like).

The procedure can be viewed as growing a binary enumeration tree (or "semantic tree"), where each node of the tree corresponds to a partial assignment of values to the variables. More precisely, each node is associated with a subproblem that we denote by (ϕ, T, F), where T and F are two disjoint subsets of $\{1,\ldots,n\}$. This subproblem is defined as follows: *Find a solution* $X^* = (x_1^*, x_2^*, \ldots, x_n^*)$ *of the equation* $\phi(X) = 0$ *such that* $x_i^* = 1$ *for all* $i \in T$, *and* $x_i^* = 0$ *for all* $i \in F$, *or decide that no such solution exists.* The root of the tree corresponds to the subproblem $(\phi, \emptyset, \emptyset)$, meaning that all variables are initially free.

The branching procedure uses a subroutine PREPROCESS(ϕ, T, F) which could perform a variety of preprocessing operations on the subproblem (ϕ, T, F). We simply assume that this subroutine always returns one of three possible outputs:

   (i) Either a solution X^* satisfying the conditions of the subproblem (ϕ, T, F).
  (ii) Or the answer No, if the procedure is able to establish conclusively that (ϕ, T, F) has no solution.
 (iii) Or a subproblem of the form (ψ, S, G) with the property that (ϕ, T, F) has a solution if and only if (ψ, S, G) has a solution.

To simplify our presentation, we further assume that ψ is defined on the same set of variables as ϕ, and that $T \subseteq S$ and $F \subseteq G$. (Typically, though not necessarily, (ψ, S, G) would be obtained by determining the value that certain variables must take in all solutions of (ϕ, T, F), and by simplifying ϕ and extending (T, F) accordingly.) Finally, let us agree that PREPROCESS always returns either a solution or the answer No in the trivial case where $T \cup F = \{1, \ldots, n\}$, that is, when values have been assigned to all variables.

Now consider an arbitrary node of the enumeration tree and the corresponding subproblem (ϕ, T, F). The branching procedure makes a first attempt at solving (ϕ, T, F) by calling the subroutine PREPROCESS(ϕ, T, F). If PREPROCESS succeeds in finding a solution X^* (case (i)), then the search stops, since X^* is by definition a solution of $\phi(X) = 0$. If PREPROCESS reports that (ϕ, T, F) is inconsistent (case (ii)), then the procedure backtracks by moving to another node of the search tree. Finally, if PREPROCESS returns the problem (ψ, S, G) (case (iii)), then the procedure resorts to Theorem 2.5: That is, a variable x_i is selected such that $i \notin S \cup G$, and the subproblems $(\psi, S, G \cup \{i\})$ and $(\psi, S \cup \{i\}, G)$ are recursively solved (this amounts to fixing x_i first to 0, then to 1). The subproblem (ϕ, T, F) is only reported to have no solution if both $(\psi, S \cup \{i\}, G)$ and $(\psi, S, G \cup \{i\})$ are eventually found to be inconsistent.

Figure 2.2 presents a more formal, recursive description of the branching procedure. The Boolean equation $\phi = 0$ can be solved by calling the procedure BRANCH$(\phi, \emptyset, \emptyset)$. The correctness of the procedure directly follows from Theorem 2.5 and from our previous discussion.

Procedure BRANCH(ϕ, T, F).

Input: A Boolean expression $\phi(x_1, x_2, \ldots, x_n)$ and two subsets T, F of $\{1, \ldots, n\}$ such that $T \cap F = \emptyset$.

Output: A solution $X^* = (x_1^*, x_2^*, \ldots, x_n^*)$ of the equation $\phi(X) = 0$ such that $x_i^* = 1$ for all $i \in T$, and $x_i^* = 0$ for all $i \in F$, if such a solution exists; No otherwise.

begin

    if PREPROCESS$(\phi, T, F) = X^*$ **then return** X^*;

    if PREPROCESS$(\phi, T, F) = $ No **then return** No;

    if PREPROCESS$(\phi, T, F) = (\psi, S, G)$ **then**

    {comment: branch }

    begin

        select an index $i \in \{1, \ldots, n\} \setminus (S \cup G)$;

        {comment: fix x_i to 0}

        if BRANCH$(\psi, S, G \cup \{i\}) = X^*$ **then return** X^*

        {comment: fix x_i to 1}

        else return BRANCH$(\psi, S \cup \{i\}, G)$;

    end

end

Figure 2.2. Procedure BRANCH.

Of course, BRANCH cannot really be called an algorithm until we specify the rules applied to selecting the branching variable, as well as the specific features of the subroutine PREPROCESS. In practice, as demonstrated, for instance, in [239, 281, 476], the efficiency of BRANCH hinges critically on these factors (as well as on the strategy used to explore the search tree; see Section 2.9). We now proceed with a discussion of these topics. We observe that, although no assumption has been formulated so far regarding the nature of the input expression ϕ, much of the literature has concentrated on the special (but important) case of DNF equations. In particular, branching rules and preprocessing operations have been mostly investigated for DNF equations, and from now on, we restrict our attention to such equations.

2.5.1 Branching rules

Let us concentrate on the situation arising at the root of the search tree, when the subproblem to be solved is the original (DNF) equation $\phi = 0$ (the situation at every other node is similar). When branching is necessary, the branching variable can be chosen according to various strategies. Most strategies tend to give higher priority to variables presenting a "large" number of occurrences in the DNF and/or to variables occurring in "many" terms of "low" degree (the idea being, in both cases, to reduce as much as possible the size of the DNF after branching). Some typical suggestions are listed hereunder. In order to describe them, let $h_i(u)$ denote the number of terms of degree i that contain the literal u in the DNF ϕ, for $i = 1, 2, \ldots, n$.

- Davis and Putnam [261] propose branching first on any literal appearing in a term of smallest degree (theoretical properties of this rule have been investigated by Chao and Franco [187] and Chvátal and Reed [202]).
- A popular variant of this rule consists in selecting a literal with the highest number of occurrences among the terms of smallest degree: Select u that maximizes $h_{min(\phi)}(u)$, where $min(\phi)$ is the minimum degree of any term in ϕ (see Cook and Mitchell [209]; Dubois, André, Boufkhad and Carlier [281]; Van Gelder and Tsuji [886], etc.).
- In the computer code ESPRESSO-II, Brayton et al. [153] branch first on a literal with the highest number of occurrences in the formula, namely, a literal u that maximizes $\sum_{i=1}^{n} h_i(u)$.
- Jeroslow and Wang [534] combine the above ideas by giving more weight to shorter terms: They suggest branching first on any literal u that maximizes

$$W(u) = \sum_{i=1}^{n} w_i \, h_i(u), \qquad (2.10)$$

where $w_i = 2^{-i}$. Jeroslow and Wang [534] and Harche, Hooker, and Thompson [476] obtained good computational results with this branching rule, but Dubois et al. [281] report that other choices of the weights w_i may be more effective. The performance of the branching rule has been investigated in depth by Hooker and Vinay [504], who also challenge its rationale and propose more efficient alternatives.

- Several researchers have successfully used branching rules of the following form: Select a variable x that maximizes

$$h_{min(\phi)}(x) + h_{min(\phi)}(\overline{x}) + \alpha \min\left(h_{min(\phi)}(x), h_{min(\phi)}(\overline{x})\right), \quad (2.11)$$

where $h_{min(\phi)}$ is defined as above and α is a numerical parameter; see, for instance, Buro and Kleine Büning [166], Dubois et al. [281], Pretolani [758]. Intuitively, this type of rule favors variables that not only appear frequently in short terms but also tends to pick variables for which the subtrees created after branching are roughly balanced (this provides the motivation for the last term in (2.11)).

Other practical branching rules are discussed in [58, 166, 239, 281, 418, 534, 613, 886], and so on. Dubois et al. [281], in particular, stress the fact that the branching strategies that prove most efficient on *consistent* instances may be different from those that perform well on *inconsistent* instances.

In order to improve the effectiveness of branching, several authors have suggested focusing on *control sets*, where a control set is a set S of indices such that, after branching on all the variables with indices in S, in any arbitrary order, the remaining equation is always "easy" to solve (that is, the subproblem (ϕ, T, F) is "easy" for every partition $T \cup F$ of S). This type of strategy appears, for instance, in publications by Brayton et al. [153], Chandru and Hooker [183], Boros et al. [116], Truemper [871], and so on. Crama, Ekin, and Hammer [229] proved that finding a smallest control set is NP-hard for a broad range of specifications of what constitutes an "easy" equation. Closely related concepts have recently been reexamined by Williams, Gomes, and Selman [913] under the name *backdoor sets*; see also [568, 581, 715, 854] and the discussion of relaxation schemes in Section 2.5.2 hereunder.

The branching rules described earlier may lead to ties that can be broken either deterministically (e.g., by choosing the variable with smallest index among the candidates) or by *random selection*. Implementations of sophisticated randomized branching rules are found, for instance, in Bayardo and Schrag [58] and Crawford and Auton [239]. Interestingly, Gomes et al. [403] provide evidence that randomized selection may noticeably influence the performance of branching procedures (namely, the variance of the running time is usually large when the randomized procedure is applied several times to a single instance).

Departing from the basic algorithm BRANCH, some authors have suggested branching on *terms* of the current DNF rather than on its variables. For instance, Monien and Speckenmeyer [690] proposed the following approach: If

PREPROCESS(ϕ, T, F) returns (ψ, S, G) at some node of the enumeration tree, then

(a) choose a term of ψ, say a term of the form $(\bigwedge_{i=1}^{r} x_i)(\bigwedge_{j=r+1}^{p} \overline{x}_j)$;
(b) create p subproblems, where, in the k-th subproblem:
   - If $1 \le k \le r$, then $x_1 = \ldots = x_{k-1} = 1$ and $x_k = 0$.
   - If $r + 1 \le k \le p$, then $x_1 = \ldots = x_r = 1$, $x_{r+1} = \ldots = x_{k-1} = 0$ and $x_k = 1$.

Thus, the subproblems created in the search tree correspond to mutually exclusive ways of setting the term $(\bigwedge_{i=1}^{r} x_i)(\bigwedge_{j=r+1}^{p} \overline{x}_j)$ to zero. In their computational experiments, Gallo and Urbani [365] and Bruni and Sassano [159] found this rule to perform well.

2.5.2 Preprocessing

Let us now discuss some of the possible ingredients that may go into the subroutine PREPROCESS. We assume again, for the sake of simplicity, that the current subproblem is the equation DNF $\phi = 0$. We successively handle rewriting rules, the Davis-Putnam rules, general heuristics, and relaxation schemes.

Rewriting rules

Any rewriting operation that replaces ϕ by an equivalent DNF can be applied. Examples of such operations are the removal of duplicate terms or, more generally, the removal of any term of ϕ that is absorbed by another term. Several authors have also experimented with rules which replace ϕ by an equivalent DNF of the form $\phi \vee C_1 \vee C_2 \vee \ldots \vee C_r$, where $C_1, C_2, \ldots, C_r$ are (prime) implicants of ϕ. The consensus procedure (see Section 2.7) can be interpreted in this framework; related ideas are found in [599, 886].

Davis-Putnam rules

In an oft-cited paper, Davis and Putnam [261] proposed a number of simple preprocessing rules that have attracted an enormous amount of attention in the literature on Boolean equations and that are implemented in most of the efficient equation solvers (strictly speaking, Davis and Putnam's suggestions were formulated in the framework of elimination algorithms – to be discussed in Section 2.6 – rather than branching algorithms; the application of these rules within branching procedures was popularized by Davis, Logemann, and Loveland [260] and Loveland [627]).

The Davis-Putnam rules identify various special circumstances under which a variable x_i can be fixed to a specific value without affecting the consistency of the equation. The rules fall into two categories: *unit literal rules* (sometimes called unit clause rules, unit deduction rules, forward chaining rules, etc.) and *monotone literal rules* (sometimes called pure literal rules, affirmative-negative rules, etc.).

To state them, it is convenient to assume that the terms of the DNF ϕ have been grouped as follows:

$$\phi = \overline{x}_i \, \phi_0 \vee x_i \, \phi_1 \vee \phi_2,\tag{2.12}$$

where ϕ_0, ϕ_1, and ϕ_2 are DNFs which do not involve x_i.

Unit literal rules: For $i = 1,2,\ldots,n$,

(a) if ϕ has the form $\overline{x}_i \vee x_i \vee \phi_2$, then return No: the equation $\phi = 0$ is inconsistent;
(b) if ϕ has the form $\overline{x}_i \vee \phi_2$, then fix x_i to 1;
(c) if ϕ has the form $x_i \vee \phi_2$, then fix x_i to 0.

The unit literal rules are obviously valid; that is, the equation obtained after applying the rules is consistent if and only if the original equation is consistent. Within branching algorithms, they are usually applied in an iterative fashion until their premises are no longer satisfied. At this point, either a complete solution of the equation $\phi = 0$ has been found, or an equivalent, but simpler equation has been derived. In the artificial intelligence literature, this procedure sometimes goes by the name of *unit resolution*, *clausal chaining*, or *Boolean constraint propagation* (BCP) (see, e.g., [186, 533, 627, 670, 693]).

The unit literal rules can be implemented to run in linear time and are computationally efficient. It is worth noting that they are somehow redundant with most of the branching rules described in the previous subsection, in the sense that these branching rules tend to select a variable appearing in a term of degree 1 when such a term exists (since the branching rules often give priority to variables appearing in short terms). Thus, many branching rules can be seen as automatically enforcing the unit literal rules when they are applicable, and as generalizing these rules to terms of higher degree otherwise. Separately handling the unit literal rules, however, usually allows for more efficient implementations.

Let us now turn to the monotone literal rules.

Monotone literal rules: For $i = 1,2,\ldots,n$,

(a) if x_i occurs only uncomplemented in ϕ, that is, if ϕ has the form $x_i \, \phi_1 \vee \phi_2$, then fix x_i to 0;
(b) if x_i occurs only complemented in ϕ, that is, if ϕ has the form $\overline{x}_i \, \phi_0 \vee \phi_2$, then fix x_i to 1.

The monotone literal rules are valid in the sense that $\phi = 0$ has a solution if and only if the equation obtained after applying the rules has a solution. From a practical viewpoint, they can be implemented to run in linear time but seem to have only a marginal effect on the performance of branching procedures. Generalizations of these rules have been investigated in [126].

Heuristics

Any heuristic approach to consistency testing can be used within the branching framework. For instance, Jeroslow and Wang [534] implement a "greedy" heuristic, which essentially consists in iteratively fixing to 0 any literal u that maximizes the expression $W(u)$ defined by (2.10). This process is repeated until either a solution X^* of $\phi(X) = 0$ has been produced or a contradiction has been detected. In the latter case, PREPROCESS simply returns the original equation. Jaumard, Stan, and Desrosiers [532] similarly rely on a tabu search heuristic at every node of the branching tree.

Relaxation schemes

An interesting approach to preprocessing has been initiated by Gallo and Urbani [365] (who also credit Minoux [unpublished] with a similar idea) and exploited by several other researchers in various frameworks. This approach makes use of a basic ingredient of enumerative algorithms: the notion of relaxation of a problem.

We define here a *relaxation scheme* as an operator that associates with every (DNF) equation $\phi(X) = 0$ another (DNF) equation $\psi(X,Y) = 0$ (its *relaxation*), with the property that $\phi(X) = 0$ is inconsistent whenever $\psi(X,Y) = 0$ is inconsistent.

Given a relaxation scheme, the subroutine PREPROCESS can proceed along the following lines:

For the current subproblem $\phi(X) = 0$,

(a) generate the relaxation $\psi(X,Y) = 0$, and solve it;
(b) if the relaxation is inconsistent, then return No; otherwise, let (X^*, Y^*) be a solution of the relaxation;
(c) if $\phi(X^*) = 0$, then return the solution X^*; otherwise, return the original equation.

Thus, solving the relaxation $\psi = 0$ either proves that the original equation $\phi = 0$ is inconsistent (in step (b)) or produces a candidate (heuristic) solution of $\phi = 0$ (in step (c)).

Generally speaking, the art consists in choosing the relaxation scheme in such a way that the relaxed equation $\psi(X,Y) = 0$ is "easy" to solve, while remaining sufficiently "close" to the original equation. One way of defining a relaxation scheme is to construct ψ so that $\psi(X,Y) \leq \phi(X)$ for all (X,Y), which can be achieved by removing a subset of terms from ϕ. In this framework, the goal is to remove as few terms as possible from ϕ (so that ψ remains "close" to ϕ) until the equation $\psi = 0$ becomes "easy" to solve. (This idea is related to the notion of control set introduced in Section 2.5.1.) Crama, Ekin, and Hammer [229] have investigated the computational complexity of several versions of this problem.

Gallo and Urbani [365] use Horn equations as relaxations of arbitrary DNF equations. Horn equations are precisely those DNF equations in which each term contains at most one complemented variable (recall Definition 1.30 in Section

1.13.1). As we will see in Chapter 6, Horn equations can be solved in linear time (essentially, by repeated application of the unit literal rules). A DNF equation $\phi = 0$ can be relaxed to a Horn equation by dropping from ϕ any term that contains more than one complemented variable. More elaborate schemes are discussed in Gallo and Urbani [365] or Pretolani [758].

Other authors have similarly proposed to relax the given DNF equation to a quadratic equation (quadratic equations, like Horn equations, are easily solved in linear time; see Chapter 5). Buro and Kleine Büning [166]; Dubois and Dequen [283]; Groote and Warners [412]; Jaumard, Stan, and Desrosiers [532]; Larrabee [599]; and Van Gelder and Tsuji [886] report on computational experiments relying on (variants of) such schemes. As Larrabee observed [599], one may expect these approaches to perform particularly well when the equation contains a relatively high number of quadratic terms, as is the case with the equations arising from stuck-at fault detection in combinational circuits (see Application 2.2).

Finally, we note that the decomposition techniques described by Truemper [871] share some similarities with relaxation schemes.

2.6 Variable elimination procedures

In this section, we discuss *variable elimination techniques* for the solution of Boolean equations. Variable elimination procedures apply to Boolean equations of the form

$$\phi(x_1, \ldots, x_{n-1}, x_n) = 0, \tag{2.13}$$

where ϕ is an arbitrary Boolean expression, not necessarily in disjunctive normal form. They rely on the following result.

Theorem 2.6. *The equation $\phi(x_1, \ldots, x_{n-1}, x_n) = 0$ is consistent if and only if the equation*

$$\phi(x_1, \ldots, x_{n-1}, 0)\, \phi(x_1, \ldots, x_{n-1}, 1) = 0 \tag{2.14}$$

is consistent.

Theorem 2.6 can be viewed as a trivial restatement of Theorem 2.5. It should be noted, however, that contrary to Theorem 2.5 which only holds in the two-element Boolean algebra, Theorem 2.6 holds (nontrivially) in general Boolean algebras as well, so that variable elimination techniques extend directly to such algebras. Theorem 2.6 and variable elimination procedures can actually be traced to the classical works of several 19th-century logicians (see, e.g., Boole [103], Chapter VII, Proposition 1; see also Kuzicheva [591] or Rudeanu [795] for historical accounts).

Equation (2.14) is an equation in $n - 1$ variables, which we view as resulting from (2.13) by *elimination* of variable x_n (the operation that associates equation (2.14) to equation (2.13) is sometimes called *variable splitting*; see e.g. [186, 261]).

By successive elimination of a subset of variables, a necessary and sufficient condition for the consistency of (2.13) can be obtained in terms of the remaining variables. This technique turns out to be useful in applications where some of the variables are not immediately relevant, but have rather been introduced in the equation in order to facilitate the formulation of a problem. For instance, in the Boolean equation $\phi(X, Y, z) = 0$ describing the correct functioning of a switching circuit (see Application 1.13.2), the variables Y associated with the output of the hidden gates are usually not of direct interest. In this application, eliminating the Y-variables from $\phi = 0$ leads to an equation whose solution set describes the relation between the input signals X and the output signal z (viz., the function computed by the circuit).

More specifically, successive elimination of *all* variables of the equation (2.13) eventually provides a straightforward consistency test for this equation. Before we make this more precise, however, we would like to address the following question: Suppose that the equation (2.14) is consistent, and that we know one of its solutions, say $(x_1^*, \ldots, x_{n-1}^*)$; how can we use this knowledge to produce a solution of the original equation (2.13)? The next result provides a constructive answer to this question.

Theorem 2.7. *If $(x_1^*, \ldots, x_{n-1}^*)$ is a solution of (2.14), if $x_n^* = \phi(x_1^*, \ldots, x_{n-1}^*, 0)$, and $x_n^{**} = \overline{\phi}(x_1^*, \ldots, x_{n-1}^*, 1)$, then both $(x_1^*, \ldots, x_{n-1}^*, x_n^*)$ and $(x_1^*, \ldots, x_{n-1}^*, x_n^{**})$ are solutions of (2.13).*

Proof. The validity of this statement can be verified by direct substitution. But the following proof provides more insight into the nature of the elimination technique. The Shannon expansion of the function ϕ is

$$\phi(x_1, \ldots, x_{n-1}, x_n) = x_n \, \phi(x_1, \ldots, x_{n-1}, 1) \vee \overline{x}_n \, \phi(x_1, \ldots, x_{n-1}, 0). \qquad (2.15)$$

Therefore, if $x_n^* = \phi(x_1^*, \ldots, x_{n-1}^*, 0)$, it follows from (2.15) that

$$\phi(x_1^*, \ldots, x_{n-1}^*, x_n^*) = \phi(x_1^*, \ldots, x_{n-1}^*, 0) \, \phi(x_1^*, \ldots, x_{n-1}^*, 1), \qquad (2.16)$$

which is zero by definition of $(x_1^*, \ldots, x_{n-1}^*)$. The same reasoning applies to x_n^{**}. $\square$

Let us now illustrate the use of Theorems 2.6 and 2.7 on a small example.

Example 2.2. *Consider the DNF equation $\phi_3(x_1, x_2, x_3) = 0$, where*

$$\phi_3 = \overline{x}_1 x_2 \overline{x}_3 \vee x_1 \overline{x}_2 x_3 \vee \overline{x}_1 \overline{x}_2 \vee \overline{x}_1 x_3 \vee x_2 x_3.$$

By Theorem 2.6, the equation $\phi_3(x_1, x_2, x_3) = 0$ is consistent if and only if the equation $\phi_2(x_1, x_2) = 0$ is consistent, where

$$\phi_2(x_1, x_2) = \phi_3(x_1, x_2, 0)\phi_3(x_1, x_2, 1) = (\overline{x}_1 x_2 \vee \overline{x}_1 \overline{x}_2)(x_1 \overline{x}_2 \vee \overline{x}_1 \overline{x}_2 \vee \overline{x}_1 \vee x_2).$$

Applying once again Theorem 2.6, $\phi_2(x_1, x_2) = 0$ is consistent if and only if $\phi_1(x_1) = 0$ is consistent, where

$$\phi_1(x_1) = \phi_2(x_1, 0)\phi_2(x_1, 1) = \overline{x}_1.$$

Finally, eliminating x_1 yields

$$\phi_0 = \mathbf{0}.$$

Procedure ELIMINATE(ϕ)

Input: A Boolean expression $\phi(x_1,\ldots,x_n)$.
Output: A solution $(x_1^*,\ldots,x_n^*)$ of the equation $\phi(X)=0$ if the equation is consistent; No otherwise.

begin

  $\phi_n := \phi(x_1,\ldots,x_n)$;
  {comment: begin successive variable elimination}
  for $j := n$ **down to** 1 **do** $\phi_{j-1}(x_1,\ldots,x_{j-1}) := \phi_j(x_1,\ldots,x_{j-1},0)\phi_j(x_1,\ldots,x_{j-1},1)$;
  {comment: consistency check}
  if $\phi_0 = \mathbf{1}$ **then return** No;
  if $\phi_0 = \mathbf{0}$ **then** {comment: the equation is consistent; begin backtracking}
    for $j := 1$ **to** n **do** $x_j^* := \phi_j(x_1^*,\ldots,x_{j-1}^*,0)$;
  return $(x_1^*,\ldots,x_n^*)$;
end

Figure 2.3. Procedure ELIMINATE.

The equation $\phi_0 = 0$ is clearly consistent, and therefore, we can conclude at this point that the original equation $\phi_3 = 0$ is consistent, too. Using iteratively Theorem 2.7, we now proceed to compute a solution (x_1^*,x_2^*,x_3^*) of $\phi_3 = 0$. First, we let $x_1^* = \phi_1(0) = 1$. Since $\phi_2(x_1^*,0) = \phi_2(1,0) = 0$, we next set $x_2^* = 0$. And finally, since $\phi_3(x_1^*,x_2^*,0) = \phi_3(1,0,0) = 0$, we let $x_3^* = 0$. Thus, we conclude that $(x_1^*,x_2^*,x_3^*) = (1,0,0)$ is a solution of $\phi_3 = 0$. $\qquad\square$

Figure 2.3 presents a formal statement of the procedure ELIMINATE(ϕ) for the solution of Boolean equations of the form (2.13). The correctness of the procedure is an immediate consequence of Theorems 2.6 and 2.7. It should be noted, however, that ELIMINATE can be implemented in a variety of ways. More precisely, the meaning of the assignment

$$\phi_{j-1} := \phi_j(x_1,\ldots,x_{j-1},0)\phi_j(x_1,\ldots,x_{j-1},1) \qquad (2.17)$$

in this procedure is not entirely determined. It leaves open an important question: *What expression of ϕ_{j-1} should we carry over to the next step of the algorithm?* Also, there is no reason to stick to the original ordering $(x_1,\ldots,x_n)$ of the variables in the elimination phase of the procedure. Rather, we may want to decide at each step, in a dynamic fashion, what variable to eliminate next. The answer to these questions may determine the efficiency of ELIMINATE to a large extent, and we now proceed to discuss them briefly.

Let us first consider the question of what expression to use for ϕ_{j-1} at each step of the elimination procedure. If we simply write ϕ_{j-1} as the conjunction of the expressions $\phi_j(x_1,\ldots,x_{j-1},0)$ and $\phi_j(x_1,\ldots,x_{j-1},1)$, without transforming the resulting expression any further, then we eventually obtain the following expression:

$$\phi_0 = \bigwedge_{X^* \in \mathcal{B}^n} \phi(X^*).$$

Successive elimination then amounts to the complete enumeration of all points of $\mathcal{B}^n$, and the necessary and sufficient condition for consistency, viz., $\phi_0 = \mathbf{0}$, becomes trivial (in the two-element Boolean algebra).

By contrast, transforming the expression (2.17) in each (or some) iteration(s) of ELIMINATE allows in general an increase of the efficiency of the algorithm. In particular, simplifying the expression ϕ_{j-1} sometimes allows us to immediately detect that ϕ_{j-1} is identically 0 or identically 1. The elimination procedure can then be curtailed: Indeed, if ϕ_{j-1} is constant, then clearly $\phi_{j-1} = \phi_0$, and ELIMINATE can immediately proceed with the consistency check.

To discuss this point more concretely, let us concentrate on the special case in which $\phi = \phi_n$ is a DNF (recall that no such assumption has been made so far). When this is the case, we can rewrite ϕ_n in the form

$$\phi_n = \overline{x}_n \psi_0 \vee x_n \psi_1 \vee \psi_2, \tag{2.18}$$

where ψ_0, ψ_1 and ψ_2 are DNFs involving the variables $x_1,\ldots,x_{n-1}$, but not x_n. Then,

$$\phi_n(x_1,\ldots,x_{n-1},0) = \psi_0 \vee \psi_2$$

and

$$\phi_n(x_1,\ldots,x_{n-1},1) = \psi_1 \vee \psi_2,$$

so that

$$\phi_{n-1} = \phi_n(x_1,\ldots,x_{n-1},0)\,\phi_n(x_1,\ldots,x_{n-1},1) = \psi_0 \psi_1 \vee \psi_2. \tag{2.19}$$

The expression (2.19) can be used to rewrite ϕ_{n-1} as a DNF. Indeed, by distributivity, the conjunction $\psi_0 \psi_1$ has a DNF expression ψ, each term of which is simply the conjunction of a term of ψ_0 with a term of ψ_1. This DNF can be further simplified by deleting any term that is identically 0 or is absorbed by another term. These straightforward rules yield a DNF equivalent to ϕ_{n-1}.

Since a DNF is identically zero if and only if it has no terms, this approach sometimes allows us to detect consistency early in the elimination procedure, thus reducing the number of iterations required by ELIMINATE and speeding up termination.

Example 2.3. *Consider the equation*

$$\phi_4 = x_1 x_2 \overline{x}_4 \vee \overline{x}_1 x_2 x_3 \overline{x}_4 \vee \overline{x}_2 x_4 \vee \overline{x}_1 \overline{x}_3 x_4.$$

By elimination of x_4, we get

$$\phi_3 = (x_1 x_2 \vee \overline{x}_1 x_2 x_3)(\overline{x}_2 \vee \overline{x}_1 \overline{x}_3).$$

Using distributivity, ϕ_3 is directly seen to be identically zero. Thus, we conclude that $\phi_3 = \phi_2 = \phi_1 = \phi_0 = \mathbf{0}$, and that the equation $\phi_4 = 0$ is consistent. The solution $(x_1^*,\ldots,x_4^*) = (0,\ldots,0)$ can be computed using Theorem 2.7. $\square$

We now turn to a brief discussion of the elimination ordering. As noted earlier, there is no compelling reason to eliminate the variables in the order $x_n, \ldots, x_1$ rather than in any other order. We may even want to determine *dynamically* (that is, on the run) which variable x_i to eliminate next. In some situations, an obvious choice can be made for this next variable. For instance, if the current DNF $\phi_j(x_1, \ldots, x_j)$ contains x_i as a term of degree 1, or if the variable x_i appears only uncomplemented in ϕ_j, then eliminating x_i is tantamount to fixing x_i to 0 in ϕ_j (we leave this for the reader to check). Similarly, if $\phi_j(x_1, \ldots, x_j)$ contains a term $\overline{x}_i$, or if the variable x_i appears only complemented in ϕ_j, then eliminating x_i is tantamount to fixing x_i to 1 in ϕ_j. It is easy to recognize in this description an alternative statement of the Davis-Putnam rules (see Section 2.5), cast here in terms of variable elimination.

Example 2.4. *Consider the DNF equation* $\phi_6(x_1, \ldots, x_6) = 0$, *where*

$$\phi_6 = \overline{x}_1 x_2 \overline{x}_3 \vee x_1 \overline{x}_2 x_3 \vee \overline{x}_1 \overline{x}_2 x_4 \vee \overline{x}_1 x_3 \vee x_2 x_3 x_4 \vee x_4 x_5 x_6 \vee \overline{x}_4 \overline{x}_5 \overline{x}_6 \vee \overline{x}_4 \vee x_3 x_5 \overline{x}_6.$$

Applying the unit literal rule, we see that x_4 *can be fixed to 1. This reduces* ϕ_6 *to*

$$\overline{x}_1 x_2 \overline{x}_3 \vee x_1 \overline{x}_2 x_3 \vee \overline{x}_1 \overline{x}_2 \vee \overline{x}_1 x_3 \vee x_2 x_3 \vee x_5 x_6 \vee x_3 x_5 \overline{x}_6.$$

Variable x_5 *only appears in uncomplemented form in this DNF. By the monotone literal rule, we can set* x_5 *to 0, thus reducing the original problem to the equation solved in Example 2.2.*   □

Davis and Putnam's original algorithm [261] is in fact a variant of the classical procedure ELIMINATE, especially tailored for the solution of DNF (or CNF) equations. The additional rules proposed by these authors consist in maintaining the DNF format throughout the procedure and in computing dynamically an effective variable elimination ordering. Since both the unit literal rules and the monotone literal rules lead to a simplification of the current DNF ϕ_j, it makes sense to apply them first in the elimination algorithm. When the rules are no longer applicable, Davis and Putnam [261] suggest proceeding with the elimination of any variable that appears in a shortest term of ϕ_j (recall our discussion of branching rules in Section 2.5).

Even with these refinements, however, the main computational hurdle of the elimination method remains: Namely, the number of terms in the equation tends to explode in the initial phases of the procedure, before it eventually decreases with the number of variables. As a result, computer implementations of elimination procedures rapidly face memory space problems, similar in nature to those encountered by other dynamic programming algorithms. In effect, these problems are often serious enough to prohibit the solution of equations involving many variables. This difficulty was first noticed by Davis, Logemann, and Loveland [260] and led them to replace the original form of the Davis-Putnam algorithm by a branching procedure of the type discussed in Section 2.5. We will see in Sections 2.11.2 and 2.11.3, however, that elimination procedures are well suited for generating all solutions or for computing parametric solutions of Boolean equations.

2.7 The consensus procedure

The *consensus procedure* has a long history in the Boolean literature. It was originally designed as a method generating (that is, listing) all prime implicants of a
Boolean function given in DNF and was repeatedly discovered in this form by several independent researchers; see Blake [99], Samson and Mills [801], Quine [768],
as well as Chapter 3. Brown [156] gives a interesting historical account of this line
of research.

As a solution method for *CNF* equations, the consensus method mostly owes
its fame to Robinson [787]. In his seminal paper, Robinson introduced an inference principle (which he calls the *resolution principle*) for first-order logic. The
resolution principle subsequently became the cornerstone of many algorithmic
techniques used by automated reasoning systems (see, e.g., Wos et al. [925]).
When specialized to propositional equations and translated from the CNF format favored by Robinson into the equivalent DNF framework adopted here, the
resolution method becomes essentially identical to the consensus method, and it
immediately follows from earlier works that resolution provides a correct solution
procedure for Boolean equations (although Robinson [787] used a direct, ad hoc
argument to establish this important result).

In this section, we explain how a consensus-based procedure can be used to
solve DNF Boolean equations. A general version of the consensus method, allowing the enumeration of all prime implicants of DNF expressions, is discussed
more extensively in Chapter 3. The essence of consensus procedures lies in the
following observation. (Note the similarity of this statement with the statement of
Theorem 2.6 when the latter is specialized to DNF equations; see (2.18), (2.19).)

Theorem 2.8. *Let $\phi(x_1, x_2, \ldots, x_n)$ be a Boolean expression of the form*

$$\phi = x_i \, C \vee \overline{x}_i \, D \vee \psi, \tag{2.20}$$

*where $i \in \{1, 2, \ldots, n\}$, and C, D are elementary conjunctions. Then, $\phi = \phi \vee CD$,
so that the equation $\phi = 0$ is consistent if and only if the equation $\phi \vee CD = 0$ is
consistent.*

Proof. The claims simply follow from the observation that, in every solution of
$\phi = 0$, either C or D must be 0. □

Theorem 2.8 motivates the following definition.

Definition 2.5. *If $x \, C$ and $\overline{x} \, D$ are two elementary conjunctions such that CD is
not identically 0, then we say that CD is the* consensus *of these two conjunctions,
and we say that CD is derived from $x \, C$ and $\overline{x} \, D$ by consensus on x.*

One interpretation of Theorem 2.8 is that, whenever $x_i \, C = 0$ and $\overline{x}_i \, D = 0$
express conditions to be satisfied by the variables $(x_1, x_2, \ldots, x_n)$, then $CD = 0$
expresses another such condition (note that this condition is uninteresting if CD
is identically 0). Therefore, we can see consensus derivation as the application

of an inference rule (namely, the classical *syllogism*) that allows us to draw the conclusion CD from the premises $x_i C$ and $\overline{x}_i D$.

This view of consensus derivation, as an operation producing new elementary conjunctions from existing ones, leads to a natural extension of the previous concepts.

Definition 2.6. *The elementary conjunction C can be derived by consensus from a set S of elementary conjunctions if there exists a finite sequence $C_1, C_2, \ldots, C_p$ of elementary conjunctions such that*

(1) $C_p = C$, *and*
(2) *for $i = 1, \ldots, p$, either $C_i \in S$ or there exist $j < i$ and $k < i$ such that C_i is the consensus of C_j and C_k.*

We are now ready to state the fundamental result that motivates the consideration of consensus derivation.

Theorem 2.9. *The DNF equation $\phi = 0$ is inconsistent if and only if the (empty) elementary conjunction 1 can be derived by consensus from the set of terms of ϕ.*

Proof. As mentioned earlier, this theorem can be viewed as an immediate corollary of the results in Chapter 3 (see Theorem 3.5 and its Corollary 3.4). For the sake of completeness, we prove it here from first principles.

The "if" part of the statement follows directly from Theorem 2.8. For the "only if" part, we assume that the DNF equation $\phi(x_1, x_2, \ldots, x_n) = 0$ is inconsistent, and we proceed by induction on the number n of variables. The result is trivial if $n = 1$. For $n > 1$, write ϕ as

$$\phi = \overline{x}_n \psi_0 \lor x_n \psi_1 \lor \psi_2, \qquad (2.21)$$

where ψ_0, ψ_1 and ψ_2 do not depend on x_n. Theorem 2.6 implies that the equation $\psi_0 \psi_1 \lor \psi_2 = 0$ is inconsistent. Now use distributivity to rewrite $\psi_0 \psi_1 \lor \psi_2$ as a DNF of the form $\psi = \bigvee_{k=1}^{m} C_k$, where each term C_k is either a term of ϕ or the conjunction of a term of ψ_0 with a term of ψ_1, namely, the consensus (on x_n) of two terms of ϕ. Since ψ depends on $n - 1$ variables, we know by induction that the constant 1 can be derived by consensus from $\{C_k \mid k = 1, 2, \ldots, m\}$. This, however, implies that 1 can be derived by consensus from the set of terms of ϕ. $\square$

A procedure for testing the consistency of DNF equations can now be stated as in Figure 2.4. The correctness of the procedure is an immediate corollary of Theorem 2.9 (note that the **while**-loop eventually terminates, since the number of elementary conjunctions on n variables is finite).

Example 2.5. *Consider the DNF equation $\phi(x_1, x_2, x_3, x_4) = 0$, where*

$$\phi = \overline{x}_1 x_2 \overline{x}_3 \lor x_1 \overline{x}_4 \lor \overline{x}_1 \overline{x}_2 \lor \overline{x}_1 x_3 \lor x_4.$$

From the terms $\overline{x}_1 x_2 \overline{x}_3$ and $\overline{x}_1 \overline{x}_2$, we can derive the consensus $\overline{x}_1 \overline{x}_3$. This new term together with $\overline{x}_1 x_3$ yields the consensus $\overline{x}_1$. On the other hand, the term

---

Procedure CONSENSUS(ϕ)

Input: A DNF expression $\phi(x_1,\ldots,x_n) = \bigvee_{k=1}^{m} C_k$.
Output: Yes if the equation $\phi = 0$ is consistent; No otherwise.

begin
    $S := \{C_k \mid k = 1,2,\ldots,m\}$;
    while there exist two terms $x_i\, C$ and $\overline{x}_i\, D$ in S such that
    $x_i\, C$ and $\overline{x}_i\, D$ have a consensus and CD is not in S **do**
        if $CD = 1$ **then return** No
        else $S := S \cup \{CD\}$;
    return Yes;
end

Figure 2.4. Procedure CONSENSUS.

x_1 *can be derived from* $x_1\overline{x}_4$ *and* x_4. *Combining now the derived terms* $\overline{x}_1$ *and* x_1, *we can produce the constant* 1, *and we conclude that the equation* $\phi = 0$ *is inconsistent.* $\square$

Two features of the consensus procedure deserve further attention. First, CONSENSUS does not produce a solution of the DNF equation when there is one. Second, CONSENSUS is not completely defined, since we did not specify how the terms $x_i\, C$ and $\overline{x}_i\, D$ are to be chosen in the **while**-loop. We now successively tackle these two points.

Consider first the fact that CONSENSUS only delivers a consistency verdict for DNF equations, but no solution. This is, from a theoretical viewpoint, no serious problem. Indeed, as explained in Section 2.4, CONSENSUS can easily be used as a subroutine to produce a solution of consistent equations.

But the situation is actually even better here. Indeed, we shall prove in Chapter 3 that, when the procedure CONSENSUS(ϕ) halts and returns the answer Yes, the set S contains all prime implicants of the function represented by the DNF ϕ. The knowledge of these prime implicants is, by itself, sufficient to produce a solution of the equation $\phi = 0$, as will also be explained in Chapter 3 (see Corollary 3.4).

Let us also notice, as a final remark on this topic, that the consensus procedure and its various extensions have been mostly used as equation-solving techniques within the field of automated theorem proving. As previously mentioned, many applications in this particular field do not require the explicit finding of solutions, since only inconsistent equations are "interesting" (because they correspond to theorems). On the other hand, what is valuable in this context is an explicit argument showing why a theorem is true (i.e., a proof of the theorem). A consensus derivation of inconsistency provides such an argument (although sometimes insufficiently clear; see [881, 925] for a more detailed discussion).

We now take up the second issue mentioned above: How are the terms $x_i\, C$ and $\overline{x}_i\, D$ to be selected in the **while**-loop of the consensus procedure? This question is closely related to the question of selecting the next variable to branch upon in branching procedures, and some of the available strategies should be by now very familiar.

A first strategy is to replace the condition "CD is not in S" in the **while**-statement by the stronger condition "CD is not absorbed by any term in S." The procedure remains correct under this modification, as easily follows from the proof of Theorem 2.9.

Another strategy, much in the spirit of the Davis-Putnam unit literal rule, is to give priority to so-called *unit consensus steps*, namely, to pairs of terms $\{x_i C, \overline{x}_i D\}$ such that either $x_i C$ or $\overline{x}_i D$ is of degree 1. Note, for instance, that the consensus of x_i and $\overline{x}_i D$ is simply D, which absorbs $\overline{x}_i D$. Thus, unit consensus steps can be implemented without increasing the cardinality of the set S. If we restrict the procedure CONSENSUS to the use of unit consensus steps, then the procedure becomes extremely fast. But, unfortunately, it can fail to detect inconsistent equations. Nevertheless, equation solving heuristics based on this approach are widely used in automated reasoning procedures.

Similarly, a substantially accelerated heuristic algorithm is obtained when we restrict consensus formation to pairs of terms of the form $\{x_i C, \overline{x}_i CD\}$; indeed, such a pair produces the term CD, which absorbs $\overline{x}_i CD$.

If CONSENSUS starts by selecting all pairs of terms having a consensus on x_n, as long as they are available, and proceeds next to pairs of terms having a consensus on $x_{n-1}, x_{n-2}, \ldots, x_1$, then CONSENSUS becomes essentially identical to the elimination procedure.

Other specialized forms of consensus used in automated reasoning are the so-called *set of support strategy*, *linear consensus*, *input refutation*, and so on. Some of these variants will be introduced in subsequent chapters (e.g., in Chapter 6). We also refer to [186, 571, 925] and to the exercises at the end of this chapter for more information.

2.8 Mathematical programming approaches

The approaches surveyed in this section are characterized by their treatment of Boolean variables as numerical quantities and by the transformation of Boolean equations into equivalent mathematical programming problems. This is in sharp contrast with the methods discussed in previous sections, which rely on a purely symbolic treatment of the variables. The idea of identifying the Boolean symbols 0 and 1 with numbers and reducing problems of logic to optimization problems goes back a long time (see, among others, Fortet [342, 343], Hammer and Rudeanu [460]); the interest in such approaches has been revived in recent years.

2.8.1 Integer linear programming

The basic observation underlying integer linear programming approaches can be phrased as follows:

Theorem 2.10. *The DNF equation*

$$\phi(x_1, x_2, \ldots, x_n) = \bigvee_{k=1}^{m} \left(\bigwedge_{i \in A_k} x_i \bigwedge_{j \in B_k} \overline{x}_j \right) = 0 \tag{2.22}$$

has the same set of solutions as $IS(\phi)$, where $IS(\phi)$ is the following system of linear inequalities in 0-1 variables:

$$\sum_{i \in A_k} (1 - x_i) + \sum_{j \in B_k} x_j \geq 1, \quad k = 1, 2, \ldots, m;$$

$$x_i \in \{0, 1\}, \qquad\qquad i = 1, 2, \ldots, n.$$

In particular, the following statements are equivalent:

(a) *The equation $\phi = 0$ is consistent.*
(b) *The system $IS(\phi)$ is feasible.*
(c) *The optimal value of $IP(\phi)$ is 0, where $IP(\phi)$ is the integer programming problem:*

$$\text{minimize} \quad z$$

$$\text{subject to} \quad z + \sum_{i \in A_k} (1 - x_i) + \sum_{j \in B_k} x_j \geq 1, \quad k = 1, 2, \ldots, m;$$

$$x_i \in \{0, 1\}, \qquad\qquad\qquad i = 1, 2, \ldots, n;$$

$$z \in \{0, 1\}.$$

Proof. The first claim is just a restatement of Theorem 1.39 (see Section 1.13.6) and the second one is an immediate corollary. □

In principle, any algorithm for handling 0-1 linear programming problems can be used to solve $IS(\phi)$ or $IP(\phi)$, thereby simultaneously solving the Boolean equation $\phi = 0$ (see [707] for an overview of integer progamming methods). Such approaches have been taken up and developed by several researchers, following, in particular, some early work by Jeroslow and his coworkers; see, for example, [98, 533], Williams [911, 912] and Hooker [498, 499, 500, 501], and so on. The book by Chandru and Hooker [184] covers these developments in great detail, so that we shall content ourselves with a brief survey of the basic ideas (see also Hooker [502] for a discussion of logic-based methods in optimization).

Blair, Jeroslow and Lowe [98] adopt a straightforward approach: They simply feed the formulation $IS(\phi)$ to standard integer linear programming codes that attempt to solve $IS(\phi)$ by branch-and-bound. Let us see what this approach amounts to.

First, consider the linear relaxation $LP(\phi)$ of problem $IP(\phi)$, namely, the linear programming problem:

minimize z

subject to $z + \sum_{i \in A_k} (1 - x_i) + \sum_{j \in B_k} x_j \geq 1, \quad k = 1, 2, \ldots, m,$

$$0 \leq x_i \leq 1, \qquad\qquad\qquad i = 1, 2, \ldots, n,$$

$$0 \leq z \leq 1.$$

Applying a basic branch-and-bound procedure to solve IP(ϕ) is tantamount to solving the equation $\phi = 0$ by the procedure BRANCH described in Section 2.5, with a subroutine PREPROCESS-LP which performs the following steps:

(a) Solve LP(ϕ) and let (X^*, z^*) be its optimal solution;
(b) If $z^* > 0$, then the optimal value of IP(ϕ) must be 1 and hence the equation $\phi = 0$ is inconsistent;
(c) If $z^* = 0$ and X^* is integral, then X^* is a solution of the equation $\phi = 0$.

When neither case (b) nor case (c) applies, then one of the variables assuming a fractional value in X^* can be selected for branching.

How effective is this particular version of BRANCH? Let us say that a variable x_i is fixed to the value 0 (respectively, 1) by unit consensus on ϕ if the term x_i (respectively, $\bar{x}_i$) can be derived from the terms of ϕ by a sequence of unit consensus steps (i.e., if the linear term x_i, respectively $\bar{x}_i$, arises after iterated applications of the unit literal rule on ϕ). Also, let us say that unit consensus detects that $\phi = 0$ is inconsistent if some variable x_i can be fixed both to 0 and to 1 by unit consensus. The next result is due to Blair, Jeroslow, and Lowe [98].

Theorem 2.11. (a) *If unit consensus does not detect that $\phi = 0$ is inconsistent, then there is a feasible solution (X^*, z^*) of LP(ϕ) in which $z^* = 0$ and $x_i^* = 1/2$ for each variable x_i that is not fixed by unit consensus ($i = 1, 2, \ldots, n$).*
(b) *For each $i = 1, 2, \ldots, n$, if x_i is fixed to the value $u \in \{0, 1\}$ by unit consensus on ϕ, then $x_i^* = u$ in all those feasible solutions (X^*, z^*) of LP(ϕ) for which $z^* = 0$.*
(c) *The optimal value of LP(ϕ) is strictly positive if and only if unit consensus detects that $\phi = 0$ is inconsistent.*

Proof. The theorem follows from the fact that, if all terms of ϕ have degree at least 2, then setting $z^* = 0$ and $x_j^* = 1/2$ for $j = 1, 2, \ldots, n$ defines a feasible solution of LP(ϕ). And conversely, if ϕ contains a term of degree 1, say, the term x_i, then LP(ϕ) contains the constraint

$$z + (1 - x_i) \geq 1,$$

so that $x_i^* = 0$ in every feasible solution (X^*, z^*) of LP(ϕ) for which $z^* = 0$. Statements (a) and (b) easily follow from these observations, by induction on the number of variables fixed by unit consensus. Statement (c) is a corollary of the previous ones.   □

It follows from Theorem 2.11 that, when applied to problem IP(ϕ) (or IS(ϕ)), a branch-and-bound algorithm does not detect inconsistency faster than the unit literal rules. One may still hope that, in the course of solving the linear relaxation LP(ϕ), integer solutions may be produced by "sheer luck," thus accelerating the basic branching procedure in the case of consistent equations. While this is true to some extent, computational experiments indicate that this approach is rather inefficient and that special-purpose heuristics tend to outperform this general-purpose LP-based approach (see [98, 534] and Section 2.5.2).

The integer programming framework, however, also offers insights of a more theoretical nature into the solution of Boolean equations. Let us first recall some definitions from [197, 211, 812]. (We denote by $\lceil x \rceil$ the smallest integer not smaller than x.)

Definition 2.7. *Let $A \in \mathbb{Z}^{m \times n}, b \in \mathbb{Z}^m$, and consider the system of linear inequalities $\mathcal{I}: (Ax \geq b, x \geq 0)$ for $x \in \mathbb{R}^n$. A* Chvátal cut *for $\mathcal{I}$ is any inequality of the form $cx \geq \delta$, where $c \in \mathbb{Z}^n$ and $\delta \in \mathbb{R}$, such that for some $d \in \mathbb{R}$, $\lceil d \rceil \geq \delta$, the inequality $cx \geq d$ can be obtained as a nonnegative linear combination of the inequalities in $\mathcal{I}$.*

It should be clear that every *integral* vector $x \in \mathbb{Z}^n$ that satisfies all the inequalities in $\mathcal{I}$ also satisfies every Chvátal cut for $\mathcal{I}$. Let us now consider the set of all the inequalities that can be obtained by iterated computations of Chvátal cuts.

Definition 2.8. *The inequality $cx \geq d$ is in the* Chvátal closure *of $\mathcal{I}: (Ax \geq b, x \geq 0)$ if there exists a finite sequence of inequalities $c_i x \geq d_i$ $(i = 1, 2, \ldots, p)$ such that*

(1) $c_p = c$, $d_p = d$, *and*
(2) *for $i = 1, \ldots, p$, either the inequality $c_i x \geq d_i$ is in $\mathcal{I}$, or it is a Chvátal cut for the system of inequalities $(c_j x \geq d_j : 1 \leq j < i)$.*

A deep theorem of Chvátal [197] asserts that, if the solution set of $\mathcal{I}$ is bounded, then *every* linear inequality $cx \geq \delta$ ($c \in \mathbb{Z}^n, \delta \in \mathbb{R}$) that is satisfied by *all* integral solutions of $\mathcal{I}$ is in the Chvátal closure of $\mathcal{I}$ (see also [812, 211]). In particular, if the system $\mathcal{I}$ has no integral solution, then the inequality $0 \geq 1$ must be in its Chvátal closure. We are now ready to apply these concepts to the solution of Boolean equations.

Theorem 2.12. *The DNF equation $\phi(X) = 0$ is inconsistent if and only if the inequality $0 \geq 1$ is in the Chvátal closure of the system*

$$\sum_{i \in A_k} (1 - x_i) + \sum_{j \in B_k} x_j \geq 1, \qquad k = 1, 2, \ldots, m, \qquad (2.23)$$

$$0 \leq x_i \leq 1, \qquad i = 1, 2, \ldots, n. \qquad (2.24)$$

Proof. By Theorem 2.10, we know that $\phi = 0$ is inconsistent if and only if the system IS(ϕ) is infeasible, that is, if and only if the system (2.23)–(2.24) has no integral solution. So, the statement follows from Chvátal's theorem. $\square$

As observed by Cook, Coullard, and Turán [210], Definition 2.8 and Theorem 2.12 suggest a purely algebraic *cutting-plane proof system* for establishing the inconsistency of DNF equations. The next result, proved in Cook, Coullard, and Turán [210]; Hooker [499, 500] and Williams [911], establishes a connection between this approach and the consensus method.

Theorem 2.13. *Let* $k \in \{1,...,n\}$; *let* A_1, A_2, B_1, B_2 *be subsets of* $\{1,...,n\} \setminus \{k\}$ *such that* $(A_1 \cup A_2) \cap (B_1 \cup B_2) = \emptyset$; *and consider the system of inequalities*

$$(1 - x_k) + \sum_{i \in A_1} (1 - x_i) + \sum_{j \in B_1} x_j \geq 1, \tag{2.25}$$

$$x_k + \sum_{i \in A_2} (1 - x_i) + \sum_{j \in B_2} x_j \geq 1, \tag{2.26}$$

$$0 \leq x_i \leq 1, \qquad i = 1, 2, ..., n. \tag{2.27}$$

Then, the inequality

$$\sum_{i \in A_1 \cup A_2} (1 - x_i) + \sum_{j \in B_1 \cup B_2} x_j \geq 1 \tag{2.28}$$

is a Chvátal cut for (2.25)–(2.27).

Proof. Take the sum of (2.25) and (2.26). Add $(1 - x_i) \geq 0$ to the resulting inequality for each i that appears in exactly one of A_1, A_2 and add $x_j \geq 0$ for each j that appears in exactly one of B_1, B_2. Divide both sides of the resulting inequality by 2. These operations yield the valid inequality

$$\sum_{i \in A_1 \cup A_2} (1 - x_i) + \sum_{j \in B_1 \cup B_2} x_j \geq \frac{1}{2}, \tag{2.29}$$

which shows that (2.28) is a Chvátal cut for (2.25)–(2.27). $\square$

Observe that (2.25) represents the elementary conjunction $C = x_k (\bigwedge_{i \in A_1} x_i \bigwedge_{j \in B_1} \overline{x}_j)$, (2.26) represents the elementary conjunction $D = \overline{x}_k (\bigwedge_{i \in A_2} x_i \bigwedge_{j \in B_2} \overline{x}_j)$, and (2.28) represents the consensus of C and D. Therefore, (2.28) can appropriately be called a *consensus cut* derived from (2.25)–(2.26).

Example 2.6. *Consider two terms of a DNF* ϕ, *say,* $C = x_1 x_2 \overline{x}_3 \overline{x}_4$ *and* $D = \overline{x}_1 x_2 \overline{x}_3 x_5$. *In the system IS($\phi$), these terms give rise to the inequalities*

$$(1 - x_1) + (1 - x_2) + x_3 + x_4 \geq 1,$$

$$x_1 + (1 - x_2) + x_3 + (1 - x_5) \geq 1.$$

The consensus cut (2.28) derived from these inequalities is

$$(1 - x_2) + x_3 + x_4 + (1 - x_5) \geq 1,$$

which is also the inequality associated to the consensus of C and D, viz.
$x_2 \overline{x}_3 \overline{x}_4 x_5.$ □

Comparing Definitions 2.6 and 2.8 in light of Theorem 2.13, we conclude that the consensus procedure can be interpreted as a special type of cutting-plane procedure (Cook, Coullard, and Turán [210]). Note in particular that, for an inconsistent equation $\phi = 0$, the sequence of consensus steps required to derive the empty conjunction must be at least as long as the number of cuts required to derive the inequality $0 \geq 1$. This observation raises hope that a cutting-plane approach to the solution of Boolean equations may be practically efficient (see also Section 2.10.1).

Hooker [500] attacked IP(ϕ) by a cutting-plane procedure based on Theorem 2.13. In the simplest approach, the procedure PREPROCESS-LP described earlier in this section is augmented by the following step:
(d) Try to derive one or more consensus cuts violated by X^* from the inequalities in IP(ϕ). If such cuts are found, then add them to IP(ϕ) and go back to step (a).

Finding violated consensus cuts can in principle be implemented by sequentially considering all pairs of inequalities in IP(ϕ) and checking the corresponding consensus cuts. This inefficient approach, however, can be accelerated in various ways. We refer to Chandru and Hooker [184] for details and for additional theoretical developments, and to Chai and Kuehlmann [177] or Manquinho and Marques-Silva [666] for recent computational work along similar lines.

Hooker [503] presents further results about the integer programming approach to logic.

2.8.2 Nonlinear programming

Several attempts have been made to model and to solve Boolean equations as nonlinear programming problems, either discrete or continuous.

One possible approach consists in minimizing the objective function $\sum_{i=1}^{n} x_i (1 - x_i)$ subject to the constraints (2.23)–(2.24), which define the continuous relaxation of IS(ϕ). Note that the optimal value of this problem is 0 if and only if the equation $\phi = 0$ is consistent. Kamath, Karmarkar, Ramakrishnan, and Resende [545, 546], for instance, propose an interior-point algorithm to solve a closely related model (they first perform the change of variables $y_i = 2x_i - 1$, which replaces the 0-1 variables $x_1, x_2, \ldots, x_n$ by new variables $y_1, y_2, \ldots, y_n$ taking values in $\{-1, +1\}$, and they maximize $\sum_{i=1}^{n} y_i^2$).

Another line of attack exploits the following observation:

Theorem 2.14. *Consider the DNF*

$$\phi(x_1, x_2, \ldots, x_n) = \bigvee_{k=1}^{m} \left(\bigwedge_{i \in A_k} x_i \bigwedge_{j \in B_k} \overline{x}_j \right) \tag{2.30}$$

and the real-valued function

$$f(x_1, x_2, \ldots, x_n) = \sum_{k=1}^{m} c_k \left(\prod_{i \in A_k} x_i \prod_{j \in B_k} (1 - x_j) \right), \tag{2.31}$$

where $c_1, c_2, \ldots, c_m$ are arbitrary positive coefficients. The following statements are equivalent:

(a) *The equation $\phi(x_1, x_2, \ldots, x_n) = 0$ is consistent.*
(b) *The minimum of $f(x_1, x_2, \ldots, x_n)$ over $\{0, 1\}^n$ is equal to zero.*
(c) *The minimum of $f(x_1, x_2, \ldots, x_n)$ over $[0, 1]^n$ is equal to zero.*

Proof. The equivalence of statements (a) and (b) is obvious. Their equivalence with statement (c) follow from the claim that $\min_{X \in [0,1]^n} f(X) = \min_{X \in \{0,1\}^n} f(X)$ (as observed by Rosenberg [789], this property actually holds for every multilinear function f; see also Theorem 13.12 in Section 13.4.3). To see this, consider an arbitrary point $X^* \in [0,1]^n$ and assume that one of its components, say, x_1^*, is not integral. The restriction of f to $x_i = x_i^*$ for $i \geq 2$, namely, the function $g(x_1) = f(x_1, x_2^*, \ldots, x_n^*)$, is affine. Hence, $g(x_1)$ attains its minimum at a 0-1 point $\hat{x}_1$. This implies in particular that $f(\hat{x}_1, x_2^*, \ldots, x_n^*) \leq f(x_1^*, x_2^*, \ldots, x_n^*)$. Continuing in this way with any remaining fractional components, we eventually produce a point $\hat{X} \in \{0, 1\}^n$ such that $f(\hat{X}) \leq f(X^*)$, which proves the claim and the theorem. $\square$

Any algorithm for *nonlinear* 0-1 *programming* can be used to optimize $f(X)$ over B^n (see Chapter 13 and the survey [469]). Hammer, Rosenberg, and Rudeanu [458, 460], for instance, have proposed a variable elimination algorithm (inspired from ELIMINATE) for minimizing functions of the form (2.31) over B^n. A streamlined version and an efficient implementation of this algorithm are described by Crama, Hansen, and Jaumard [235], who also observe that this algorithm is applicable to the solution of Boolean equations. The algorithm described in [235] relies on numerical bounding procedures to control (to a certain extent) the combinatorial explosion inherent to elimination procedures (see Section 2.6).

The coefficients c_k are arbitrary in (2.31), and the performance of any optimization algorithm based on Theorem 2.14 may be influenced by the choice of these coefficients. Wah and Shang [894] propose a discrete Lagrangian algorithm for minimizing (2.31), which can be viewed as starting with $c_k = 1$ for all k and dynamically adapting these values.

Recently, several authors have experimented with semidefinite programming reformulations of Boolean equations based on extensions of Theorem 2.14; see, for instance, Anjos [23, 24] and de Klerk, Warners, and van Maaren [266]. Gu [417] combines various continuous global optimization algorithms with backtracking techniques to compute the minimum of (2.31), or of closely related functions, over $[0, 1]^n$. Other nonlinear programming approaches to the solution of Boolean equations, including Lagrangian techniques and heuristics, are surveyed by Gu, Purdom, Franco, and Wah [418].

2.8.3 Local search heuristics

In recent years, several groups of researchers have experimented with *heuristics*, or *incomplete* methods, which are not *guaranteed* to solve the given equation $\phi(X) = 0$, but which do so (experimentally) with high probability.

As a matter of fact, heuristic methods for equation solving have been used for a long time in the artificial intelligence literature (see [186, 533, 627]). We have already mentioned, for instance, that unit consensus is sometimes viewed as providing such an incomplete method. Linear consensus (see Exercise 9 in Section 2.12) is another example. Both unit consensus and linear consensus may prove either consistency or inconsistency but may sometimes terminate without any conclusion.

By contrast, a more recent trend of research has turned to heuristics which are unable to prove inconsistency, and which simply concentrate on the quest for solutions of the equation. These approaches are typically based on Theorem 2.14 and attempt to minimize the pseudo-Boolean function

$$f(X) = \sum_{k=1}^{m} \left(\prod_{i \in A_k} x_i \prod_{j \in B_k} (1 - x_j) \right) \qquad (2.32)$$

by a descent algorithm enhanced with some local search ingredients. Pioneering work along these lines includes the work of Gu [415, 416]; Selman, Levesque, and Mitchell [820]; Selman, Kautz, and Cohen [818, 819] (a very similar scheme was implemented by Hansen and Jaumard [468] for solving the maximum satisfiability problem). The GSAT algorithm in [820], for instance, starts with a random point $X^* \in \mathcal{B}^n$ and repeats a number of times the following step: If $f(X^*) \neq 0$, then switch that component i (namely, replace x_i^* by $\overline{x^*}_i$) which results in the largest decrease of $f(X)$. The decrease may be negative or 0. If no solution of $\phi(X) = 0$ is found after a predetermined number of switches, then the process is restarted from scratch and, after a number of restarts, the equation is declared (perhaps wrongly) inconsistent. GSAT was found to perform surprisingly well on a variety of experimental benchmark problems. It can be improved even further, however, if the variable to be switched is picked more carefully. Algorithms in the WalkSAT family [818, 819], for instance, select a term randomly among all terms of (2.32) that are not canceled by the current assignment X^*, and then switch a variable within that term, either at random or greedily. Variations on this theme have been explored by several researchers.

Gu et al. [418] and Hoos and Stützle [507, 508] provide a wealth of details about heuristic approaches to SAT and about their practical performance. Finally, we note that local search algorithms have also been proposed as *exact* solution methods for DNF equations, as in Dantsin et al. [255].

2.9 Recent trends and algorithmic performance

Over the last 15 years, there has been an unprecedented flurry of algorithmic developments around the solution of Boolean equations. These developments came in fast, successive waves, with each wave bringing new computational breakthroughs and new insights into what the "ultimate" solution method might eventually look like.

In spite of its simplicity, the basic branching scheme enhanced by some additional features (such as a smart branching rule or a tight relaxation) has repeatedly proved to provide one of the most effective ways of solving DNF Boolean equations. State-of-the-art implementations are described in several of the references cited earlier (see also Hoos and Stützle [506] and the Web site http://www.satlive.org). It should be noted, however, that some of the most recent implementations depart in various ways from the basic scheme described in Section 2.5.

For instance, in their RELSAT algorithm, Bayardo and Schrag [58] have incorporated *look-back* strategies inspired from the constraint satisfaction literature: RELSAT is no longer restricted to performing a depth-first traversal of the search tree, but is allowed to backtrack in more intelligent ways. For this purpose, the authors, at every node of the search tree, rely on information which is derived from past branchings by implicitly applying consensus operations on carefully selected pairs of terms. Thus, their approach provides an interesting, and extremely effective link between branching-based and consensus-based techniques.

In another paper, Gomes et al. [403] argued convincingly that the performance of branching procedures can be further enhanced if randomized branching is used in conjunction with *rapid randomized restarts* (RRR). In RRR, if the branching procedure does not stop after a small number of backtracks, then the run is terminated and restarted from the root (since the branching rule is randomized, two successive runs of the procedure usually behave differently). In particular, the authors show that RRR further improves the performance of efficient algorithms like RELSAT ([58]) and SATZ ([613]).

Thus, in conclusion, today's most successful methods for the solution of Boolean equations are a mixture of a broad variety of ingredients. They are often elaborations of branching methods à la Davis-Putnam, augmented by smart branching rules spiced with a subtle touch of randomization, and they may rely on look-ahead or look-back techniques based on the solution of easy subproblems, on local search optimization or on partial consensus derivations. The highly efficient algorithms implemented by Bayardo and Schrag [58], Dubois and Dequen [283], Eén and Sörensson [289], Goldberg and Novikov [392], Gomes et al. [403], Hoos and Stützle [507, 508], Marques-Silva and Sakallah [670] or Moskewicz et al. [693] are good examples of these trends. The developments concerning the fast solution of Boolean equations have certainly not come to a halt yet and, in years to come, we should still witness much progress in the solution of this venerable problem.

2.10   More on the complexity of Boolean equations

We now briefly return to some theoretical complexity issues. We refer the reader
to the surveys [209, 344, 418, 881, 882] and to the book [571] for additional details
and references.

2.10.1   Complexity of equation-solving procedures

Let us first compare the relative complexity of the solution procedures described
in previous sections. Rather than viewing these procedures as precisely defined
algorithms, we look at them as broad algorithmic frameworks, or *proof systems*.
For instance, we do not want to specify how the next branching variable or how the
next consensus pair is selected. Moreover, we only consider the simplest versions
of the procedures, without any fancy preprocessing or additional heuristics.

We focus on the number of computational steps required to prove the inconsistency of a DNF equation $\phi(x_1, x_2, \ldots, x_n) = 0$ (this is in a sense the more difficult
half of the problem, since proving consistency only requires us to exhibit a solu-
tion). Loosely speaking, we say that algorithm $\mathcal{A}$ is *stronger* than algorithm $\mathcal{B}$ if,
for some implementation of $\mathcal{A}$, the number of steps required by $\mathcal{A}$ for proving the
inconsistenty of $\phi(x_1, x_2, \ldots, x_n) = 0$ is no larger than the number of steps required
by any implementation of $\mathcal{B}$ (see Urquhart [882] for a more rigorous statement of
this definition).

Theorem 2.15. *Cutting-plane procedures are stronger than consensus pro-*
cedures, which are stronger than both branching and variable elimination
procedures.

Proof. The relative strength of consensus and cutting-plane procedures was examined in Section 2.8.1 (see the comments following Theorems 2.12 and 2.13). We
noted at the end of Section 2.7 that consensus is more powerful than variable elimination, since eliminating variable x_j can be viewed as performing all possible
consensus steps on x_j, for $j = 1, 2, \ldots, n$. Thus, it only remains to establish that
the consensus procedure is stronger than branching.

More precisely, we want to prove that, if a branching tree contains β nodes
and eventually demonstrates the inconsistency of $\phi = 0$, then there is a consensus
derivation of the constant 1 in β steps. The proof is by induction on the number of
variables. Suppose that

$$\phi = \overline{x}_n \phi_0 \vee x_n \phi_1 \vee \phi_2, \tag{2.33}$$

where ϕ_0, ϕ_1 and ϕ_2 are DNFs which do not involve x_n, and suppose that the
first branching takes place on x_n. Two subtrees are created, corresponding to the
equations $\phi_0 \vee \phi_2 = 0$ and $\phi_1 \vee \phi_2 = 0$. Say these trees have sizes β_0 and β_1,
respectively, where $\beta_0 + \beta_1 = \beta - 1$. Since both equations are inconsistent, the
constant 1 can be derived by consensus from each of them in, at most, β_0 and β_1
steps, respectively. Now, apply the same consensus steps to the terms of ϕ (note

that each term of $\phi_0 \vee \phi_2$ or $\phi_1 \vee \phi_2$ corresponds to a term of ϕ). Either these consensus steps yield the constant 1, or they must respectively yield the terms $\overline{x}_n$ and x_n. Then, one more consensus step produces the constant 1, and the total length of this derivation is at most β. $\qquad\qquad\square$

Since solving Boolean equations is NP-hard, one may expect any solution procedure to take an exponential number of steps on some classes of instances. Identifying bad instances for any particular method, however, is not an easy task. The so-called pigeonhole formulae have played an interesting role in this respect. These formulae express that it is impossible to assign $n + 1$ pigeons to n holes without squeezing two pigeons into a same hole. In Boolean terms, this rather obvious fact of life translates into the inconsistency of the DNF equation

$$\bigvee_{i=1}^{n+1} \left(\bigwedge_{k=1}^{n} \overline{x}_{ik} \right) \vee \bigvee_{i=1}^{n} \bigvee_{j=i+1}^{n+1} \bigvee_{k=1}^{n} \left(x_{ik} x_{jk} \right) = 0, \qquad (2.34)$$

where variable x_{ik} takes value 1 if the i-th pigeon is assigned to the k-th hole. In a famous breakthrough result, Haken [433] showed that any consensus proof of inconsistency has exponential length for the pigeonhole formulae. Other hard examples for consensus (and hence, for branching and variable elimination) were later provided by Urquhart [880] (see also Section 2.10.2).

It can be shown, however, that cutting-plane derivations of length $O(n^3)$ are sufficient to prove the inconsistency of (2.34) (see [210]). Exponential lower bounds for cutting-plane proofs are provided by Pudlák [761]. Let us also mention that an extended version of consensus has been introduced by Tseitin [872], and is known to be at least as strong as cutting-plane proofs [210]. Interestingly, no exponential lower bound has been established for this extended consensus algorithm. We refer to Urquhart [882] for a discussion of the complexity of other proof systems.

A number of authors have examined *upper bounds* on the number of steps required to prove the inconsistency of a DNF equation $\phi(x_1, x_2, \ldots, x_n) = 0$. Branching procedures trivially require $O(2^n)$ steps. Monien and Speckenmeyer [690] have improved this bound by proving that a variant of the branching procedure solves DNF equations of degree k in at most $O(\alpha_k^n)$ steps, where α_k is the largest root of the equation

$$x^k = 2x^{k-1} - 1,$$

for $k = 1, 2, \ldots, n$. One computes: $\alpha_3 = 1.618$, $\alpha_4 = 1.839$, $\alpha_5 = 1.928$, and so on. Note that $\alpha_k < 2$ for all k, but that α_k quickly approaches 2 as k goes to infinity. It is an open question whether DNF equations in n variables can be solved in $O(\alpha^n)$ steps for some constant $\alpha < 2$.

The above bounds have been subsequently improved by several authors, see for instance Kullmann [588]; Schiermeyer [808]; Paturi, Pudlák, Saks, and Zane [731]; Dantsin et al. [255]. In particular, the algorithm in [255] requires $(2 - 2/(k+1))^n$ steps for equations of degree k and $O(1.481^n)$ steps for cubic DNF equations.

Van Gelder [885] described an algorithm requiring at most $O(1.093^{|\phi|})$ steps, where $|\phi|$ is the input length of the DNF (his analysis yields a bound of $O(1.189^{|\phi|})$ for arbitrary, non-DNF equations). Hirsch [492] strengthened the bound to $O(1.074^{|\phi|})$ and to $O(1.239^m)$, where m is the number of terms of the DNF. Yamamoto [930] slightly improved the latter bound to $O(1.234^m)$.

Crama, Hansen, and Jaumard [235] proved that a variable elimination algorithm for nonlinear 0-1 programming runs in time $O(n\,2^{tw(\phi)})$, where $tw(\phi)$ is the so-called *tree-width* of a graph associated to ϕ; their arguments are easily adapted to show that the same bound applies to the Boolean procedure ELIMINATE. We refer to their paper for details.

2.10.2 Random equations

A large body of literature has been devoted to the investigation of *random Boolean expressions* and *random Boolean equations*. This approach allows, for instance, a better understanding of the distinctive features of hard versus easy equations, the analysis of the behavior of algorithms over various distributions of instances, or nonconstructive proofs of the existence of certain types of expressions.

We limit our discussion to one particular distribution of random expressions (see, e.g., [209, 418, 763] for other probabilistic models).

Definition 2.9. *Let n,m and k be positive integers. A* random (n,m,k)-DNF *is a DNF $\phi(x_1,x_2,\ldots,x_n) = \bigvee_{j=1}^{m} C_j$ whose terms $C_1,C_2,\ldots,C_m$ are drawn independently and uniformly from among all elementary conjunctions of degree k on $x_1,x_2,\ldots,x_n$. A* random (n,m,k)-equation *is an equation $\phi = 0$, where ϕ is a random (n,m,k)-DNF.*

Note that all terms of a random (n,m,k)-DNF have degree *exactly* k, and that the definition allows for repeated terms but not for terms which are identically 0.

Since adding terms increases the probability of introducing inconsistencies, one can expect "long" equations, that is, equations where m is large relative to n, to be inconsistent with high probability, and "short" equations to be consistent with high probability. More precise versions of these statements can actually be established. We start with an easy observation due to Franco and Paull [345].

Theorem 2.16. *Let $\phi = 0$ be a random (n,m,k)-equation, where $m = cn$ for some constant c. If $c > -1/\log_2(1-2^{-k})$, then the equation is inconsistent with probability tending to 1 as n goes to infinity.*

Proof. Let the random equation be $\phi(X) = \bigvee_{j=1}^{m} C_j = 0$, and consider an arbitrary point X^* in $\mathcal{B}^n$. For $j = 1,2,\ldots,m$, the probability that $C_j(X^*) = 0$ is $1 - 2^{-k}$, and hence the probability that $\phi(X^*) = 0$ is $\left(1-2^{-k}\right)^m$. Therefore, the expected number of solutions of the equation is $2^n \left(1-2^{-k}\right)^m$. If $m = cn$ and $c > -1/\log_2(1-2^{-k})$, then this expected number goes to 0 as n goes to infinity, which proves the statement. $\qquad\square$

For instance, by setting $k = 2$ or $k = 3$ in the theorem, we conclude that almost all quadratic equations with more than $2.5n$ terms, and almost all cubic equations with more than $5.191n$ terms, are inconsistent.

A simple counting argument shows that very short equations are almost always consistent.

Theorem 2.17. *If $\sum_{j=1}^{m} 2^{-|C_j|} < 1$, then the DNF equation $\phi = \bigvee_{j=1}^{m} C_j = 0$ is consistent. In particular, every (n,m,k)-equation with $m < 2^k$ is consistent.*

Proof. Each term $C_j(X)$ takes value 1 in exactly $2^{n-|C_j|}$ points of $\mathcal{B}^n$, for $j = 1, 2, \ldots, m$. So, $\phi(X)$ takes value 1 in at most $\sum_{j=1}^{m} 2^{n-|C_j|}$ points of $\mathcal{B}^n$. If $\sum_{j=1}^{m} 2^{-|C_j|} < 1$, then $\phi(X)$ takes value 1 in less than 2^n points, which implies that $\phi(X) = 0$ is consistent. The second statement is an immediate corollary of the first one. $\square$

Of course, there is nothing really probabilistic about the previous result. In order to improve the bound on m, however, several researchers have analyzed algorithms which quickly find a solution of random equations with high probability. Following previous work by Chao and Franco [187], Chvátal and Reed [202] were able to show that, when $k \geq 2$ and $c < 2^k/4k$, random (n,m,k)-equations with $m = cn$ terms are consistent with probability approaching 1 as n goes to infinity (this is to be contrasted with the lower bound on c in Theorem 2.16, which grows roughly like $2^k \ln 2$).

These results motivate the following conjecture (see [209, 349, 418]).

Threshold conjecture. *For each $k \geq 2$, there exists a constant c^* such that random (n, cn, k)-equations are consistent with probability approaching 1 as n goes to infinity when $c < c^*$ and are inconsistent with probability approaching 1 as n goes to infinity when $c > c^*$.*

Despite its appeal, the considerable experimental evidence for its validity, and the existence of similar zero-one laws for other combinatorial structures, the threshold conjecture has only been established when $k = 2$. In this case, Chvátal and Reed [202] and Goerdt [390] were able to show that the conjecture holds for the threshold value $c^* = 1$. This result was subsequently sharpened by several researchers; see in particular the very tight results by Bollobás, Borgs, Chayes, Kim, and Wilson [101].

For $k = 3$, experiments indicate the existence of a threshold around the value $c^* = 4.2$, but at the time of this writing, the available bounds only imply that, if c^* exists, then $3.26 < c^* < 4.506$ (see Achlioptas and Sorkin [4]; Dubois, Boufkhad, and Mandler [282]; Janson, Stamatiou, and Vamvakari [525], etc.). In a remarkable breakthrough, however, Friedgut [348] proved that a weak form of the threshold conjecture holds for all k when c^* is replaced by a function depending on n only. Achlioptas and Peres [3] established that the conjecture holds asymptotically when $k \to +\infty$ with $c^* = 2^k \log 2 - O(k)$; see also Frieze and Wormald [350].

From an empirical point of view, it has been repeatedly observed that very long and very short equations are easy for most algorithms, whereas hard nuts occur in the so-called *phase transition region*, near the *crossover point* at which about half the instances are (in)consistent. These observations clearly have important consequences for the design of experiments aimed at assessing the quality of equation solvers. They have progressively led researchers to focus their computational experiments on the solution of special subclasses of random equations, or on structured equations derived from the encoding of hard combinatorial problems (see, e.g., [57, 239, 505, 687, etc.]).

The concept of random equations has also been used to analyze the efficiency of solution algorithms. In a far-reaching extension of the results of Haken [433] and Urquhart [880] (see Section 2.10.1), Chvátal and Szemerédi [203] proved that for all fixed integers c and $k \geq 3$, there exists $\varepsilon > 0$ such that, for large n, *almost no* random (n, cn, k)-equations have consensus proofs of inconsistency of length less than $(1 + \varepsilon)^n$. In view of Theorems 2.15 and 2.16, this result actually implies that almost all cubic equations with more than $5.191 n$ terms are hard for branching, variable elimination, and consensus algorithms.

For more information on the analysis of random equations, we refer to an extensive survey by Franco [344].

2.10.3 Constraint satisfaction problems and Schaefer's theorem

A *constraint satisfaction problem* (CSP) is a class of Boolean equations which can be formulated by imposing a finite number of constraints on Boolean variables. If the 0-1 points that satisfy the i-th constraint are modeled as the set of solutions of an "elementary" Boolean equation $f_i(X) = 0$, for $i = 1, 2, \ldots, q$, then the corresponding CSP is simply the equation

$$f_1(X) \vee f_2(X) \vee \ldots f_q(X) = 0. \tag{2.35}$$

Interestingly, and rather unexpectedly, the complexity of CSP can be characterized quite precisely: In an appropriate setting to be described later, it is possible to classify every CSP of the form (2.35) as either "easy" or "hard," depending only on the nature of the individual constraints $f_i(X) = 0$, which are used as building blocks of the problem.

This line of research has been initiated in a seminal paper by Schaefer [807] and pursued by several researchers after him; the book by Creignou, Khanna, and Sudan [243] contains a detailed account of their results. In this section, we give a precise statement of Schaefer's theorem without going into the intricacies of its proof.

Let us start with a formal definition of constraint satisfaction problems.

Definition 2.10. *A (Boolean) constraint set $\mathcal{F}$ is a finite set of Boolean functions, where f_i is defined on $\mathcal{B}^{n_i}$, $n_i \geq 1$, and f_i is not identically $\mathbf{1}_{n_i}$ $(i = 1, 2, \ldots, r)$. In this context, each of the functions f_i is called a* constraint.

If X^i is an n_i-dimensional vector of Boolean variables, then the pair (f_i, X^i) is called an application *of the constraint f_i to X^i.*

The constraint satisfaction problem *associated to the constraint set* $\mathcal{F}$, *or* $CSP(\mathcal{F})$, *is the (infinite) collection of Boolean equations of the form*

$$f_{i_1}(X^{i_1}) \vee f_{i_2}(X^{i_2}) \vee \ldots \vee f_{i_q}(X^{i_q}) = 0, \tag{2.36}$$

where $f_{i_1}, f_{i_2}, \ldots, f_{i_q}$ *are functions in the constraint set* $\mathcal{F}$, *and* $(X^{i_1}), (X^{i_2}), \ldots,$ (X^{i_q}) *are vectors of Boolean variables of appropriate lengths. So, an instance of* $CSP(\mathcal{F})$ *is defined by the list of applications* (f_{i_j}, X^{i_j}), $j = 1, 2, \ldots, q$.

Let us give a few examples of constraint satisfaction problems.

Example 2.7. *Consider the constraint set* $\mathcal{F}^{QUAD} = \{f_1, f_2, f_3, f_4, f_5\}$ *in which the constraints are represented by the following expressions:*

$$f_1(x) = x, \ f_2(x) = \overline{x}, \ f_3(x_1, x_2) = x_1 x_2, \ f_4(x_1, x_2) = x_1 \overline{x}_2, \ f_5(x_1, x_2) = \overline{x}_1 \overline{x}_2.$$

An instance of $CSP(\mathcal{F}^{QUAD})$ *is for example the equation:*

$$f_3(x_1, x_2) \vee f_3(x_2, x_3) \vee f_4(x_1, x_3) \vee f_4(x_1, x_4) \vee f_4(x_4, x_1) \vee f_4(x_4, x_3)$$
$$\vee f_5(x_2, x_3) = 0,$$

or, equivalently,

$$x_1 x_2 \vee x_2 x_3 \vee x_1 \overline{x}_3 \vee x_1 \overline{x}_4 \vee \overline{x}_1 x_4 \vee \overline{x}_3 x_4 \vee \overline{x}_2 \overline{x}_3 = 0.$$

Clearly, $CSP(\mathcal{F}^{QUAD})$ *is exactly the class of all (nontrivial) quadratic DNF equations. As we mentioned in Section 2.2, such equations can be solved in polynomial time (see also Chapter 5).*

Note that an immediate generalization of this example would show that, for every fixed integer k, the class of DNF equations of degree k can be represented as a constraint satisfaction problem. This problem is NP-complete for all k > 2, as stated by Cook's theorem.

Consider next the set $\mathcal{F}^{3NAE} = \{g\}$ *in which g is represented by the DNF*

$$g(x_1, x_2, x_3) = x_1 x_2 x_3 \vee \overline{x}_1 \overline{x}_2 \overline{x}_3.$$

Note that in any point $(x_1^*, x_2^*, x_3^*) \in \mathcal{B}^3$, $g(x_1^*, x_2^*, x_3^*) = 1$ *if and only if* $x_1^* = x_2^* = x_3^*$. *Therefore, we call g the "cubic not-all-equal" constraint, a name which is in turn reflected in the notation* $\mathcal{F}^{3NAE}$. *The constraint satisfaction problem* $CSP(\mathcal{F}^{3NAE})$ *is NP-complete (see the exercises at the end of the chapter).*   □

So, depending on the class $\mathcal{F}$, the problem $CSP(\mathcal{F})$ may be either easy or hard, as illustrated by the classes introduced in the previous example. Schaefer's theorem very accurately separates those classes for which $CSP(\mathcal{F})$ is polynomially solvable from those for which it is NP-complete. Before we can state this result, however, we need a few more definitions.

Extending Definitions 1.12 and 1.30 in Chapter 1, we say that a Boolean function is quadratic if it can be represented by a DNF in which each term contains at most two variables, and that the function is Horn if it can be represented by a DNF in which each term contains at most one complemented variable. Similarly, we say that a function is *co-Horn* if it can be represented by a DNF in which each term contains at most one noncomplemented variable. It will follow from the results in Chapter 6 that CSP($\mathcal{F}$) is polynomially solvable when all the constraints in $\mathcal{F}$ are Horn, or when they are all co-Horn.

Finally, we define a Boolean function f on $\mathcal{B}^n$ to be *affine* if the set of false points of f is exactly the set of solutions of a system of linear equations over GF(2), that is, if f can be represented by an expression of the form

$$f(x_1, x_2, \ldots, x_n) = \bigvee_{A \in \mathcal{E}_0} (\bigoplus_{i \in A} x_i) \vee \bigvee_{A \in \mathcal{E}_1} (1 \oplus \bigoplus_{i \in A} x_i), \qquad (2.37)$$

where $\mathcal{E}_0, \mathcal{E}_1$ are families of subsets of $\{1, 2, \ldots, n\}$ (compare with (1.16)). We now show that systems of linear equations over GF(2) can be solved by an elimination procedure closely resembling the classical Gaussian elimination process.

Theorem 2.18. *Systems of linear equations over GF(2) can be solved in polynomial time.*

Proof. Consider the system $f(X) = 0$, where f has the form (2.37), and assume that the first term of (2.37) defines the equation

$$a \oplus x_1 \oplus x_2 \oplus \ldots \oplus x_n = 0,$$

where $a \in \{0, 1\}$. This equation can be rewritten as

$$x_n = a \oplus x_1 \oplus x_2 \oplus \ldots \oplus x_{n-1},$$

which can be used to eliminate variable x_n from all subsequent equations.

We leave it as an exercise to work out the remaining details of this elimination procedure and to verify that it can be implemented to run in polynomial time. $\square$

We are finally ready to present Schaefer's result [807].

Theorem 2.19. *If $\mathcal{F}$ satisfies either one of the conditions (1)–(6) hereunder, then CSP($\mathcal{F}$) is polynomially solvable; otherwise, it is NP-complete.*

(1) *For every function $f \in \mathcal{F}$, $f(0, 0, \ldots, 0) = 0$.*
(2) *For every function $f \in \mathcal{F}$, $f(1, 1, \ldots, 1) = 0$.*
(3) *Every function $f \in \mathcal{F}$ is quadratic.*
(4) *Every function $f \in \mathcal{F}$ is Horn.*
(5) *Every function $f \in \mathcal{F}$ is co-Horn.*
(6) *Every function $f \in \mathcal{F}$ is affine.*

Proof. The first half of the theorem is easy: CSP($\mathcal{F}$) is trivial under conditions (1) and (2), and we have already discussed conditions (3)–(6). The NP-completeness statement, therefore, is the hard nut to crack: we refer to Schaefer [807] or to Creignou, Khanna, and Sudan [243] for a complete proof.   □

Theorem 2.19 underlines the special role played by quadratic, Horn, and affine functions in Boolean theory. Chapter 5 and Chapter 6 contain a thorough discussion of quadratic and Horn functions, respectively. Affine functions will not be further handled in the book. The monograph [243] contains additional facts about these functions, as well as several extensions and refinements of Theorem 2.19; see also Creignou and Daudé [241, 242] for probabilistic extensions.

Finally, we note that Boros, Crama, Hammer and Saks [116] established another theorem separating NP-hard from polynomially solvable instances of BOOLEAN EQUATION (see Section 6.10.2). Although the nature of their classification result is very different from Schaefer's classification, it also stresses the importance of quadratic and Horn equations.

2.11   Generalizations of consistency testing

In this section, we return to some of the extensions of consistency testing which we briefly introduced in Section 2.4.

2.11.1   Counting the number of solutions

Counting the number of solutions of a Boolean equation, or equivalently, the number of false points of a Boolean function, is an old problem with applications in reliability theory (see Section 1.13.4), in game theory (see Section 1.13.3), in artificial intelligence (see for instance [792]), etc. We have already observed in Theorem 1.38 that this problem is #P-complete even for quadratic positive functions. It generalizes in an obvious way the consistency question, and its solution has actually been used by some authors to attack Boolean equations indirectly (see [145, 522, etc.]).

Of course, counting the number of true points of a function on B^n is equivalent to counting the number of its false points, since the sum of these two numbers is exactly 2^n. Now, the set of true points of a DNF $\phi = \bigvee_{k=1}^{m} C_k$ is just $\bigcup_{k=1}^{m} T_k$, where T_k is the set of true points of the k-th term C_k, for $k = 1, 2, \ldots, m$. Hence, counting the number of true points of a function expressed in DNF can also be viewed as determining the size of a union of sets, a problem which is frequently attacked by inclusion-exclusion techniques. These links are explicitly stated and exploited by several authors, see [144, 145, 522, 551, 614, 630, 759, etc.].

As discussed in Section 1.6, another way of counting the number of true points of a function f consists in producing an orthogonal DNF of f and in applying Theorem 1.8. In view of the relationship between BDDs and orthogonal forms (cf. Section 1.12.3), related approaches can also be cast in a branching framework; see [48, 49, 90, 111, 280, 619, 759, etc.].

Recently, a number of specialized counting algorithms have been proposed in
[56, 257, 584, 802, etc.].

2.11.2 Generating all solutions

When the objective is to generate all solutions of a Boolean equation, we face
the additional difficulty that the number of solutions, and hence the length of
the output, can be exponential in the input size of the equation. Therefore, the
complexity of any algorithm solving this problem is most meaningfully analyzed
in terms of its input size *and* its output size (see Appendix B.8).

We first show that, in a sense, generating all solutions of an equation is not
much harder than testing its consistency.

Theorem 2.20. *There is an algorithm which, given a Boolean expression ϕ on n
variables, produces all solutions of the equation $\phi = 0$ by solving $q + 1$ Boolean
equations of size at most $|\phi| + nq$, where q is the number of solutions of $\phi = 0$.
If $t(L)$ is the complexity of solving a Boolean equation with input size at most L,
then the running time of this algorithm is polynomial in $|\phi|$, q and $t(|\phi| + nq)$.*

Proof. We describe an algorithm which performs $(q + 1)$ iterations and outputs a
new false point of ϕ at each of the first q iterations. To describe a generic iteration,
assume that we have already produced $k \leq q$ false points $X^1, X^2, \ldots, X^k$. For
$i = 1, 2, \ldots, k$, let C_i be the unique elementary conjunction such that $C_i(X) = 1$ if
and only if $X = X^i$, and solve the equation

$$\phi(X) \vee \bigvee_{i=1}^{k} C_i(X) = 0. \tag{2.38}$$

Clearly, $X^* \in \mathcal{B}^n$ is a solution of (2.38) if and only if X^* is a false point of
ϕ which differs from $X^1, X^2, \ldots, X^k$. Thus, if we find such a solution, we let
$X^{k+1} := X^*$ and proceed with the next iteration. Otherwise, we stop. Since each
iteration can be carried out in time $O(t(|\phi| + nq))$, the proof is complete. $\square$

Note that we did not assume anything about the expression ϕ in Theorem 2.20,
and hence, this result can be used to generate all solutions of a general equation
of the form $\phi(X) = \psi(X)$. Approaches of the type described in the proof of
Theorem 2.20 have also been used in the machine learning literature; see, for
instance, Angluin [21].

Other approaches rely on ad hoc modifications of the equation-solving procedures described in previous sections in order to generate all solutions of the
given equation. For instance, straightforward extensions of the branching procedure BRANCH can be used to handle the problem. We describe here another
approach, based on an extension of the variable elimination technique and the
following simple observations.

Theorem 2.21. *The one-variable equation $\phi(x) = a\,x \vee b\,\overline{x} = 0$, where a,b are Boolean constants, is consistent if and only if $a\,b = 0$, or, equivalently, if and only if $b \leq \overline{a}$. When this is the case, the solutions of the equation are the values of x that satisfy $b \leq x \leq \overline{a}$, namely,*

$$x^{(0)} = b = \phi(0), \tag{2.39}$$
$$x^{(1)} = \overline{a} = \overline{\phi(1)}. \tag{2.40}$$

Proof. This can be checked directly.   □

Theorem 2.22. *The point $(x_1^*, x_2^*, \ldots, x_n^*)$ is a solution of the equation $\phi(x_1, x_2, \ldots, x_n) = 0$ if and only if $(x_1^*, \ldots, x_{n-1}^*)$ is a solution of the equation*

$$\phi(x_1, \ldots, x_{n-1}, 0)\,\phi(x_1, \ldots, x_{n-1}, 1) = 0,$$

and x_n^* is a solution of the one-variable equation

$$\phi(x_1^*, \ldots, x_{n-1}^*, x_n) = 0.$$

Proof. This is trivial.   □

Taken together, Theorems 2.21 and 2.22 provide a slight generalization of Theorem 2.7, and they allow us to produce all solutions of the equation $\phi(x_1, x_2, \ldots, x_n) = 0$ by the following recursive procedure: First, we successively compute all expressions $\phi_n, \phi_{n-1}, \ldots, \phi_0$, as in the procedure ELIMINATE. If $\phi_0 = \mathbf{1}$, then the original equation is inconsistent. Otherwise, for $j = 1, 2, \ldots, n$ and for each solution $(x_1^*, \ldots, x_{j-1}^*)$ of the equation $\phi_{j-1}(x_1, \ldots, x_{j-1}) = 0$, we compute the solutions of the one-variable equation $\phi_j(x_1^*, \ldots, x_{j-1}^*, x_j) = 0$, and we use Theorem 2.22 to produce all solutions of the equation $\phi_j(x_1, \ldots, x_j) = 0$.

This procedure is reasonably efficient in the sense that, once the equation has been found to be consistent, all its solutions are produced in quick succession (exactly how efficient the procedure is depends on the size of the intermediate expressions $\phi_j, j = 1, 2, \ldots, n$). For special classes of Boolean equations, however, it may be possible to achieve a better performance; for instance, Feder [322] describes a polynomial-delay algorithm to generate all solutions of a quadratic DNF equation; see Section 5.7.1.

Finally, we observe that all solutions of the equation $\phi = 0$ are immediately available if a CNF expression of ϕ is at hand, because a CNF is equal to 0 exactly when (at least) one of its terms is 0. Of course, obtaining a CNF of ϕ (or equivalently, dualizing ϕ) is generally quite difficult. We return to this problem in Chapter 4.

2.11.3 Parametric solutions

Rather than explicitly generating all solutions of a Boolean equation, we may want to obtain an implicit representation of these solutions.

Definition 2.11. *A* parametric solution *of the Boolean equation* $\phi(x_1, x_2, \ldots, x_n) = 0$ *is a mapping* $\sigma \colon \mathcal{B}^m \to \mathcal{B}^n$ *with the property that, for all* $(x_1^*, x_2^*, \ldots, x_n^*) \in \mathcal{B}^n$, $(x_1^*, x_2^*, \ldots, x_n^*)$ *is a solution of the equation if and only if there exists* $(p_1, p_2, \ldots, p_m) \in \mathcal{B}^m$ *such that* $\sigma(p_1, p_2, \ldots, p_m) = (x_1^*, x_2^*, \ldots, x_n^*)$. *The parametric solution* σ *is called* reproductive *if* $n = m$ *and* $\sigma(p_1, p_2, \ldots, p_n) = (p_1, p_2, \ldots, p_n)$ *whenever* $(p_1, p_2, \ldots, p_n)$ *is a solution of the equation.*

In other words, a parametric solution is a surjective mapping $\sigma \colon \mathcal{B}^m \to F$, where F is the set of false points of ϕ, and a reproductive solution is the identity on F. Parametric solutions of Boolean equations have been investigated for a very long time (see Hammer and Rudeanu [460] or Rudeanu [795] for references). Their connection with the concept of "Boolean unification" has been recently examined, for instance, in [167, 672], where the authors point out their relevance for manipulating hardware descriptions (e.g., for verifying and testing digital circuits).

A classical result due to Löwenheim [628, 629] allows the construction of a parametric solution of the equation $\phi = 0$ once a *particular* solution of the equation is known (this is reminiscent of the solution of differential equations in calculus).

Theorem 2.23. *Let* $X^* = (x_1^*, x_2^*, \ldots, x_n^*)$ *be a particular solution of the Boolean equation* $\phi(X) = 0$, *and consider the functions* $\sigma_i \colon \mathcal{B}^n \to \mathcal{B}^n$ *defined by*

$$\sigma_i(p_1, p_2, \ldots, p_n) = x_i^* \, \phi(p_1, p_2, \ldots, p_n) \vee p_i \, \overline{\phi}(p_1, p_2, \ldots, p_n)$$

for $i = 1, 2, \ldots, n$. *Then,* $\sigma = (\sigma_1, \sigma_2, \ldots, \sigma_n)$ *is a reproductive parametric solution of* $\phi(X) = 0$.

Proof. Let $(p_1, p_2, \ldots, p_n) \in \mathcal{B}^n$, and let $X = (x_1, x_2, \ldots, x_n) = \sigma(p_1, p_2, \ldots, p_n)$. If $\phi(p_1, p_2, \ldots, p_n) = 1$, then $x_i = x_i^*$ for all i, so $X = X^*$ is a solution of the equation. If $\phi(p_1, p_2, \ldots, p_n) = 0$, then $x_i = p_i$ for all i, so $\phi(x_1, x_2, \ldots, x_n) = \phi(p_1, p_2, \ldots, p_n) = 0$. This implies that σ is a reproductive parametric solution which maps every true point to X^*. $\square$

Example 2.8. *Let us return to the equation* $\phi_3 = \overline{x}_1 x_2 \overline{x}_3 \vee x_1 \overline{x}_2 x_3 \vee \overline{x}_1 \overline{x}_2 \vee \overline{x}_1 x_3 \vee x_2 x_3 = 0$ *which was examined in Example 2.2. We found there that* $(x_1^*, x_2^*, x_3^*) = (1, 0, 0)$ *was a solution of this equation. Using Theorem 2.23, we obtain the parametric solution*

$$\sigma_1 = \phi_3(p_1, p_2, p_3) \vee p_1 \overline{\phi}_3(p_1, p_2, p_3),$$

$$\sigma_2 = p_2 \overline{\phi}_3(p_1, p_2, p_3),$$

$$\sigma_3 = p_3 \overline{\phi}_3(p_1, p_2, p_3).$$

Some additional manipulations show that this parametric solution can be alternatively represented as $(\sigma_1, \sigma_2, \sigma_3) = (1, p_1 \, p_2 \, \overline{p}_3, 0)$, *and that it correctly describes the two solutions of* $\phi_3 = 0$, *namely, the points* $(1, 0, 0)$ *and* $(1, 1, 0)$. $\square$

Another type of parametric solution can be derived from the variable elimination principle. Note first that a parametric solution of a consistent one-variable equation $\phi(x) = ax \vee b\overline{x} = 0$ is given by $\sigma(p) = p\overline{a} \vee \overline{p}b = p\overline{\phi(1)} \vee \overline{p}\phi(0)$ (compare with Theorem 2.21). This observation leads to the following reformulation of Theorem 2.22.

Theorem 2.24. *The point* $(x_1^*, x_2^*, \ldots, x_n^*)$ *is a solution of the equation* $\phi(x_1, x_2, \ldots, x_n) = 0$ *if and only if* $(x_1^*, \ldots, x_{n-1}^*)$ *is a solution of the equation* $\phi_{n-1}(x_1, \ldots, x_{n-1}) = 0$ *and*

$$x_n^* = p_n \overline{\phi(x_1^*, \ldots, x_{n-1}^*, 1)} \vee \overline{p}_n \phi(x_1^*, \ldots, x_{n-1}^*, 0)$$

for some parameter $p_n \in \mathcal{B}$.

Proof. This is an immediate consequence of Theorem 2.22 and the previous observation. □

Theorem 2.24 allows us to compute recursively a parametric solution of the equation $\phi = 0$: If $(\sigma_1, \sigma_2, \ldots, \sigma_{i-1})$ is a parametric solution of $\phi_{i-1}(x_1, x_2, \ldots, x_{i-1})$, then Theorem 2.24 indicates that σ_i can be obtained as

$$\sigma_i(p_1, p_2, \ldots, p_i) = p_i \overline{\phi_i(\sigma_1, \sigma_2, \ldots, \sigma_{i-1}, 1)} \vee \overline{p}_i \phi_i(\sigma_1, \sigma_2, \ldots, \sigma_{i-1}, 0).$$

This process yields a parametric solution in "triangular form," where σ_i depends on $(p_1, p_2, \ldots, p_i)$, but not on $(p_{i+1}, p_{i+2}, \ldots, p_n)$, for $i = 1, 2, \ldots, n$. Furthermore, this solution can be shown to be reproductive (we leave the proof of this claim as an exercise).

Example 2.9. *Let us again return to Example 2.2. Using Theorem 2.24 and the expression* $\phi_1 = \overline{x}_1$ *derived in Example 2.2, we find* $\sigma_1 = p_1 \overline{\phi_1(1)} \vee \overline{p}_1 \phi_1(0) = p_1 \vee \overline{p}_1 = 1$.

Next, $\sigma_2 = p_2 \overline{\phi_2(\sigma_1, 1)} \vee \overline{p}_2 \phi_2(\sigma_1, 0) = p_2 \overline{\phi_2(1, 1)} \vee \overline{p}_2 \phi_2(1, 0)$. *In view of Example 2.2,* $\phi_2(1, 0) = \phi_2(1, 1) = 0$, *so that* $\sigma_2 = p_2$.

Finally, $\sigma_3 = p_3 \overline{\phi_3(\sigma_1, \sigma_2, 1)} \vee \overline{p}_3 \phi_3(\sigma_1, \sigma_2, 0) = p_3 \overline{\phi_3(1, p_2, 1)} \vee \overline{p}_3 \phi_3(1, p_2, 0)$. *From the expression of* ϕ_3, *we find immediately that* $\sigma_3 = 0$.

Note that this solution $(\sigma_1, \sigma_2, \sigma_3) = (1, p_2, 0)$ *is in triangular form, as opposed to the solution derived in Example 2.8, and that it is reproductive.* □

More information on parametric solutions can be found in Hammer and Rudeanu [460], Martin and Nipkow [672], and Rudeanu [795, 796].

2.11.4 Maximum satisfiability

Definition 2.12. *If* $\phi(X) = \bigvee_{k=1}^{m} C_k$ *is a DNF on* $\mathcal{B}^n$, *and if positive real weights* $w_1, w_2, \ldots, w_m$ *are associated with the terms* $C_1, C_2, \ldots, C_m$, *then the (weighted) maximum satisfiability problem, or* MAX SAT, *asks for a point* $X^* \in \mathcal{B}^n$ *that*

maximizes the total weight of the terms canceled by X^*. In other words, MAX SAT
is the optimization problem

$$maximize \sum_{k=1}^{m} \{w_k \mid C_k(X) = 0\} \quad subject\ to\ X \in \mathcal{B}^n.$$

The name MAX SAT refers more properly to a dual version of the problem
in which the objective is to maximize the number of *satisfied* clauses of a CNF
$\psi(X) = \bigwedge_{k=1}^{m} D_k$, where a clause k is satisfied if it takes value 1. Clearly, both
versions of the problem are equivalent. To be consistent with the remainder of
the book, we carry on the discussion in terms of DNFs; but the terminology MAX
SAT is so deeply entrenched that we prefer to apply it to this DNF version as well,
rather than inventing some neologism like "maximum falsifiability problem."

MAX SAT is a natural generalization of DNF equations, viewed as collections
of logical conditions $C_1(X) = 0, C_2(X) = 0, \ldots, C_m(X) = 0$. When the equation
$\phi(X) = 0$ is inconsistent, we may be happy to find a model X^* that satisfies as
many of the conditions as possible. Applications are discussed, for instance, in
Hansen and Jaumard [468].

Let us call MAX d-SAT the restriction of MAX SAT to DNFs of degree d. In view
of Cook's theorem, MAX d-SAT is NP-hard for all $d \geq 3$. But a stronger statement
can actually be made (Garey, Johnson, and Stockmeyer [372]).

Theorem 2.25. MAX 2-SAT *is NP-hard, even when* $w_1 = w_2 = \ldots = w_m = 1$.

Proof. The problem of solving the DNF equation $\phi(x_1, x_2, \ldots, x_n) = \bigvee_{k=1}^{m} C_k = 0$
is NP-complete even when all terms of ϕ have degree exactly 3 (see [208, 371],
Theorem 2.4 and Exercise 4). With such an equation, we associate an instance of
MAX 2-SAT on $\mathcal{B}^{n+m}$, as follows. First, we introduce m new variables $y_1, y_2, \ldots, y_m$.
Next, for all $k = 1, 2, \ldots, m$, if $C_k = u_1 u_2 u_3$ is the kth term of ϕ, where u_1, u_2, u_3
are distinct literals, we create a subformula ψ_k consisting of 10 terms:

$$\psi_k = u_1 \vee u_2 \vee u_3 \vee \overline{u}_1 \overline{u}_2 \vee \overline{u}_1 \overline{u}_3 \vee \overline{u}_2 \overline{u}_3 \vee y_k \vee u_1 \overline{y}_k \vee u_2 \overline{y}_k \vee u_3 \overline{y}_k.$$

Finally, the instance of MAX 2-SAT is the DNF $\psi = \bigvee_{k=1}^{m} \psi_k$, with weight 1 on
each term. We claim that $\phi = 0$ is consistent if and only if the optimal value of this
MAX 2-SAT instance is at least $7m$.

Indeed, suppose that $\phi(X^*) = 0$ for some point $X^* \in \mathcal{B}^n$, and consider a term
$C_k = u_1 u_2 u_3$. Either 1, 2, or 3 of the literals u_1, u_2, u_3 take value 0 at X^*. If only
one of the literals is 0, then set $y_k^* = 1$; otherwise, set $y_k^* = 0$. The resulting point
(X^*, Y^*) cancels 7 terms of each DNF ψ_k, for $k = 1, 2, \ldots, m$, and hence it cancels
$7m$ terms of ψ.

Conversely, assume that the point (X^*, Y^*) cancels $7m$ terms of ψ. For
$k = 1, 2, \ldots, m$, it is easy to see that no assignment of values to u_1, u_2, u_3, y_k cancels
more than 7 terms of ψ_k. Moreover, if $u_1 = u_2 = u_3 = 1$, then at most 6 terms of
ψ_k can be cancelled. Therefore, (X^*, Y^*) must cancel exactly 7 terms of each DNF
ψ_k, and X^* must be a solution of the equation $\phi(X) = 0$. $\square$

The following extension of Theorem 2.10 and Theorem 2.14 will be useful in the sequel (see also Theorem 13.13 in Section 13.4.3).

Theorem 2.26. *If*

$$\phi(x_1, x_2, \ldots, x_n) = \bigvee_{k=1}^{m} \left(\bigwedge_{i \in A_k} x_i \bigwedge_{j \in B_k} \overline{x}_j \right), \tag{2.41}$$

then the optimal value of MAX SAT *is equal to the optimal value of the 0-1 linear programming problem*

$$\text{maximize } \sum_{k=1}^{m} w_k z_k \tag{2.42}$$

$$\text{subject to } \sum_{i \in A_k} (1 - x_i) + \sum_{j \in B_k} x_j \geq z_k, \qquad k = 1, 2, \ldots, m; \tag{2.43}$$

$$x_i \in \{0, 1\}, \qquad\qquad i = 1, 2, \ldots, n; \tag{2.44}$$

$$z_k \in \{0, 1\}, \qquad\qquad k = 1, 2, \ldots, m, \tag{2.45}$$

as well as to the maximum over $\{0, 1\}^n$ *and over* $[0, 1]^n$ *of the real-valued function*

$$f(X) = \sum_{k=1}^{m} w_k \left(1 - \prod_{i \in A_k} x_i \prod_{j \in B_k} (1 - x_j) \right). \tag{2.46}$$

Proof. In any optimal solution $(X^*, Z^*) \in \{0, 1\}^{n+m}$ of (2.42)–(2.45), variable z_k^* takes value 1 if and only if $C_k(X^*) = 0$, since $w_k > 0$ ($k = 1, 2, \ldots, m$). This proves the first statement.

Similarly, for every $X^* \in \{0, 1\}^n$, the expression $(1 - \prod_{i \in A_k} x_i^* \prod_{j \in B_k} (1 - x_j^*))$ takes value 1 if and only if $C_k(X^*) = 0$. This proves that the maximum of $f(X)$ over $\{0, 1\}^n$ coincides with the optimal value of MAX SAT.

Finally, we claim that, if we view $f(X)$ as a function on $[0, 1]^n$, then $\max_{X \in [0,1]^n} f(X) = \max_{X \in \{0,1\}^n} f(X)$. The proof of this claim is similar to the proof of Theorem 2.14 (see also Theorem 13.12 in Section 13.4.3). $\square$

So, MAX SAT can be seen as either a linear or a nonlinear optimization problem in 0-1 variables and can, in principle, be solved by any 0-1 programming algorithm (see, e.g., [707] and Chapter 13). Rather than diving into the details of specific implementations, we restrict ourselves here to a few elegant results concerning the performance of approximation algorithms for MAX SAT, as these results tie in nicely with previous sections of the chapter. We begin with a definition.

Definition 2.13. *Let* $0 < \alpha \leq 1$. *An* α-*approximation algorithm for* MAX SAT *is a polynomial-time algorithm which, for every instance of* MAX SAT, *produces a point* $\hat{X} \in \mathcal{B}^n$ *such that*

$$\sum_{k=1}^{m} \{ w_k \mid C_k(\hat{X}) = 0 \} \geq \alpha \max_{X \in \mathcal{B}^n} \sum_{k=1}^{m} \{ w_k \mid C_k(X) = 0 \}.$$

The parameter α is called the performance guarantee *of the algorithm.*

Of course, it is not a priori obvious that there should exist an α-approximation algorithm for MAX SAT, for some $\alpha > 0$. Johnson [536], however, was able to establish the existence of such an algorithm (in this and the following statements, we assume that the DNF ϕ has no empty terms).

Theorem 2.27. *For all $d \geq 1$, there is a $(1 - \frac{1}{2^d})$-approximation algorithm for the restriction of MAX SAT to DNFs in which every term has degree at least d. In particular, there is a $\frac{1}{2}$-approximation algorithm for MAX SAT.*

Proof. Let $d \geq 1$ be the minimum degree of a term of ϕ, let $X^H = (\frac{1}{2}, \frac{1}{2}, \ldots, \frac{1}{2})$ denote the center of the unit hypercube, and consider the value assumed by the function (2.46) at the point X^H. Then,

$$f(X^H) = \sum_{k=1}^{m} w_k \left(1 - (\frac{1}{2})^{|A_k| + |B_k|} \right) \geq (1 - \frac{1}{2^d}) \sum_{k=1}^{m} w_k \geq (1 - \frac{1}{2^d}) W^{MS}, \quad (2.47)$$

where W^{MS} is the optimal value of MAX SAT. Therefore, starting from the point X^H and proceeding as in the last part of the proof of Theorem 2.14, we can produce a point $\hat{X} \in \{0, 1\}^n$ such that $f(\hat{X}) \geq f(X^H) \geq (1 - \frac{1}{2^d}) W^{MS}$. This procedure clearly runs in polynomial time, and hence, the algorithm that returns $\hat{X}$ is a $(1 - \frac{1}{2^d})$-approximation algorithm. $\square$

Note that the proof actually establishes a little bit more than what we claimed: Namely, the algorithm always returns an assignment with value at least $\frac{1}{2} \sum_{k=1}^{m} w_k$. This shows in particular that, for any DNF equation $\phi = 0$, there exists a point that cancels at least half of the terms of ϕ.

Our proof of Theorem 2.27 is inspired from a probabilistic argument due to Yannakakis [934]. In this approach, each variable x_i is independently set to either 0 or 1 with probability $\frac{1}{2}$, and $f(\frac{1}{2}, \frac{1}{2}, \ldots, \frac{1}{2})$ is interpreted as the expected objective value of this random assignment. Then, the method of conditional probabilities is used to "derandomize" the procedure. The above proof translates this probabilistic method into a purely deterministic one (but not every probabilistic algorithm can be so easily derandomized; see, for instance, [689] for a brief introduction to probabilistic algorithms).

Theorem 2.27 has been subsequently improved by several authors, but the first real breakthrough came with a $\frac{3}{4}$-approximation algorithm proposed by Yannakakis [934] (note that Johnson's algorithm has a performance guarantee equal to $\frac{3}{4}$ for DNFs without linear terms). Goemans and Williamson [389] later proposed another, simpler $\frac{3}{4}$-approximation algorithm, which we now describe. We need some preliminary results.

Define the sequence

$$\beta_t = 1 - (1 - \frac{1}{t})^t, \quad t \in \mathbb{N}.$$

Lemma 2.1. *Let A, B be subsets of $\{1, 2, \ldots, n\}$ with $A \cap B = \emptyset$ and $|A| + |B| = n$. All solutions of the system*

$$\sum_{i \in A}(1 - x_i) + \sum_{j \in B} x_j \geq z, \tag{2.48}$$

$$(x_1, x_2, \ldots, x_n, z) \in [0, 1]^{n+1} \tag{2.49}$$

satisfy the inequality

$$1 - \prod_{i \in A} x_i \prod_{j \in B}(1 - x_j) \geq \beta_n z. \tag{2.50}$$

Proof. Assume without loss of generality that $|A| = n$ and $B = \emptyset$. The arithmetic-geometric mean inequality yields

$$\sqrt[n]{\prod_{i=1}^{n} x_i} \leq \frac{\sum_{i=1}^{n} x_i}{n},$$

or equivalently

$$\prod_{i=1}^{n} x_i \leq \left(\frac{\sum_{i=1}^{n} x_i}{n}\right)^n.$$

From (2.48), $\sum_{i=1}^{n} x_i \leq n - z$. Hence,

$$\prod_{i=1}^{n} x_i \leq \left(\frac{n-z}{n}\right)^n = \left(1 - \frac{z}{n}\right)^n.$$

The function $h(z) = 1 - (1 - \frac{z}{n})^n$ is concave on $[0, 1]$, $h(0) = 0$ and $h(1) = \beta_n$. Thus, $h(z) \geq \beta_n z$ on $[0, 1]$, and the lemma follows. $\qquad\square$

Goemans and Williamson [389] proved:

Theorem 2.28. *There is a β_d-approximation algorithm for* MAX d-SAT, *for all $d \geq 1$. In particular, there is a $(1 - \frac{1}{e})$-approximation algorithm for* MAX SAT.

Proof. Consider the linear relaxation of (2.42)–(2.45), that is, the linear programming problem obtained after replacing the integrality constraints (2.44)–(2.45) by the weaker constraints $x_i \in [0, 1]$ $(i = 1, 2, \ldots, n)$ and $z_k \in [0, 1]$ $(k = 1, 2, \ldots, m)$. Call this problem $LPMS$. Let $(X^{LP}, Z^{LP}) \in [0, 1]^{n+m}$ be an optimal solution of $LPMS$, with value $W^{LP} = \sum_{k=1}^{m} w_k z_k^{LP}$, and let W^{MS} denote the optimal value of MAX SAT. Note that (X^{LP}, Z^{LP}) can be computed in polynomial time and that $W^{LP} \geq W^{MS}$.

Consider now the value taken by the function f defined by (2.46) at the point X^{LP}. Since each term of ϕ has degree at most d, and the sequence β_t is decreasing with t, Lemma 2.1 implies

$$f(X^{LP}) \geq \sum_{k=1}^{m} w_k \beta_{|A_k| + |B_k|} z_k^{LP} \geq \beta_d W^{LP} \geq \beta_d W^{MS}. \tag{2.51}$$

As in the proof of Theorem 2.14, we can find in polynomial time a point $\hat{X} \in \{0,1\}^n$ such that $f(\hat{X}) \geq f(X^{LP}) \geq \beta_d \, W^{MS}$.

The second part of the statement follows from $\lim_{t \to \infty} \beta_t = 1 - \frac{1}{e}$. $\square$

Since $\beta_2 = 0.75$, Theorem 2.28 establishes the existence of a $\frac{3}{4}$-approximation algorithm for MAX 2-SAT. This result is in a sense complementary to Theorem 2.27, since Johnson's algorithm has performance guarantee equal to 0.75 when each term of ϕ has degree *at least* 2, while the new algorithm has performance guarantee equal to 0.75 when each term of ϕ has degree *at most* 2. This observation led Goemans and Williamson [389] to the following stronger result (note that $1 - \frac{1}{e} \approx 0.632$):

Theorem 2.29. *There is a $\frac{3}{4}$-approximation algorithm for* MAX SAT.

Proof. Let $f_1 = f(\frac{1}{2}, \frac{1}{2}, \ldots, \frac{1}{2})$, let $f_2 = f(X^{LP})$, where (X^{LP}, Z^{LP}) is an optimal solution of the linear relaxation $LPMS$, and let W^{LP} be the optimal value of $LPMS$. In order to prove the theorem, we only have to establish that

$$\max(f_1, f_2) \geq \frac{f_1 + f_2}{2} \geq \frac{3}{4} W^{LP} \geq \frac{3}{4} W^{MS}, \qquad (2.52)$$

and to conclude as usual.

The first and last inequalities in (2.52) are trivial. For the middle one, notice that (2.47) implies

$$f_1 = \sum_{k=1}^{m} w_k \left(1 - (\frac{1}{2})^{|A_k| + |B_k|} \right) \geq \sum_{k=1}^{m} w_k \left(1 - (\frac{1}{2})^{|A_k| + |B_k|} \right) z_k^{LP}.$$

Adding this to (2.51) yields

$$f_1 + f_2 \geq \sum_{k=1}^{m} w_k \left(1 - (\frac{1}{2})^{|A_k| + |B_k|} + \beta_{|A_k| + |B_k|} \right) z_k^{LP}.$$

Let $\gamma_t = 1 - (\frac{1}{2})^t + \beta_t$ for $t \in \mathbb{N}$. Then, $\gamma_1 = \gamma_2 = 1.5$, and, for $t \geq 3$, $\gamma_t \geq \frac{7}{8} + 1 - \frac{1}{e} \geq 1.5$. The middle inequality in (2.52) follows immediately. $\square$

Several further improvements on the performance guarantee of $\frac{3}{4}$ have been subsequently reported in the literature, and more will certainly follow in years to come. Several of these approaches rely on reformulations of MAX SAT as a semidefinite programming problem. We refer the reader to Asano and Williamson [31] for a 0.7846-approximation algorithm for MAX SAT; to Avidor, Berkovitch, and Zwick [38] for a 0.7968-approximation algorithm for MAX SAT; to Feige and Geomans [325] for a 0.931-approximation algorithm for MAX 2-SAT; to Lewin, Livnat, and Zwick [611] for a 0.9401-approximation algorithm for MAX 2-SAT; and to Karloff and Zwick [549] for a 0.875-approximation algorithm for MAX 3-SAT.

By contrast, Håstad [478] has proved that, unless P = NP, no approximation algorithm for MAX 2-SAT can achieve a better guarantee than $21/22 \cong 0.9545$, and no algorithm for MAX 3-SAT (and, a fortiori, for MAX SAT) can achieve a better guarantee than 0.875. Hence, the performance guarantee in [549] is best possible for MAX 3-SAT, but a small gap remains between the known upper and lower bounds for MAX 2-SAT. Khot et al. [567] have shown that, if the so-called "Unique Games Conjecture" holds, then it is NP-hard to approximate MAX 2-SAT to within any factor greater than 0.943, a bound that is extremely close to the approximation ratio of 0.9401 due to Lewin, Livnat, and Zwick [611].

Escoffier and Paschos [315] analyze the approximability of MAX SAT under a different type of metric, namely the *differential approximation ratio*. Creignou [240] and Khanna, Sudan, and Williamson [563] investigate and classify some generalizations of MAX SAT; see also Creignou, Khanna, and Sudan [243] for a complete overview.

We have concentrated in this section on the approximability of MAX SAT. On the computational side, numerous algorithms have been proposed for the solution of MAX SAT problems. Most of these algorithms rely on generalizations of techniques described in previous sections, especially in Section 2.8. We do not discuss these approaches in detail, and we refer instead to early work by Hansen and Jaumard [468]; to the papers [55, 540, 557, 785] in the volume edited by Du, Gu, and Pardalos [278]; to the book by Hoos and Stützle [508], and so on. Recent efficient algorithms are proposed by Ibaraki et al. [515] or Xing and Zhang [926]. De Klerk and Warners [265] examine the computational performance of semidefinite programming algorithms for MAX SAT. We also refer to Chapter 13 for a more general discussion of pseudo-Boolean optimization.

2.12 Exercises

1. Given an undirected graph G and an integer K, write a DNF ϕ such that the equation $\phi = 0$ is consistent if and only if G is K-colorable.
2. Prove that BOOLEAN EQUATION can be solved in polynomial time when restricted to DNF equations in which every variable appears at most twice.
3. Complete the proof of Theorem 2.18.
4. Prove that every Boolean equation can be transformed in linear time into an equivalent DNF equation in which all terms have degree exactly equal to 3.
5. Prove that BOOLEAN EQUATION is NP-complete, even when restricted to DNF equations in which every variable appears at most three times.
6. Prove that BOOLEAN EQUATION is NP-complete, even when restricted to cubic DNF equations in which every term is either positive or negative.
7. Let $\psi(X, Y)$ be the DNF produced by the procedure EXPAND* when running on the expression $\phi(X)$. Prove that, for all $X^* \in \mathcal{B}^n$, $\phi(X^*) = 0$ if and only if there exists $Y^* \in \mathcal{B}^m$ such that $\psi(X^*, Y^*) = 0$. Show that Y^* is not necessarily unique.
8. Prove that the following problem is NP-hard: Given a DNF $\phi = \bigvee_{k \in A} T_k$, find a largest subset of terms, say, $B \subseteq A$, such that the "relaxed" DNF $\psi = \bigvee_{k \in B} T_k$ is monotone (see Section 2.5 and [229]).

9. *Linear consensus* is the restricted form of CONSENSUS in which the pair $\{x_i C, \overline{x}_i D\}$ considered in each step of the **while**-loop (after the first one) involves the consensus generated in the previous step. Show that the empty conjunction 1 can be derived by linear consensus whenever it can be derived by consensus (i.e., whenever a DNF equation is inconsistent; see, e.g., [186, 571]).

10. *Input consensus* is the restricted form of CONSENSUS in which a pair of conjunctions $\{x_i C, \overline{x}_i D\}$ can be used to derive a consensus only if one of $x_i C, \overline{x}_i D$ is among the terms $\{C_k \mid k = 1, 2, \ldots, m\}$ of the original DNF ϕ.
    (a) Show that input consensus can fail to derive the empty conjunction 1 when $\phi = 0$ is inconsistent.
    (b) Prove that the empty conjunction 1 can be derived by input consensus if and only if it can be derived by unit consensus steps ([186, 571]).

11. Let X^* satisfy the inequalities (2.25)–(2.27) in the statement of Theorem 2.13. Show that, if X^* violates (2.28), then $0 < x_k^* < 1$ and the left-hand side of (2.25) and (2.26), when evaluated at X^*, is stricly less than 2 (see Hooker [500]).

12. Prove that the inconsistency of the n-th pigeonhole formula has a cutting-plane proof of length $O(n^3)$. *Hint:* Show how the inequality $\sum_{i=j}^{j+r} x_{ik} \leq 1$ can be generated by Chvátal cuts, for $r = 1, 2, \ldots, n$, $j = 1, \ldots, n-r$, $k = 1, 2, \ldots, n$.

13. Prove that, with probability tending to 1, random $(n, m, 1)$-equations of degree 1 are consistent whenever $m n^{-1/2} \to 0$ and inconsistent whenever $m n^{-1/2} \to \infty$ (see Chvátal and Reed [202]).

14. Prove that the parametric solution derived from the statement of Theorem 2.24 is reproductive.

3

Prime implicants and minimal DNFs

Peter L. Hammer and Alexander Kogan

This chapter is dedicated to two of the most important topics in the theory of Boolean functions. The first is concerned with the basic building blocks of a Boolean function, namely, its prime implicants. The set of prime implicants of a Boolean function not only defines the function, but also provides detailed information about many of its properties. In this chapter, we discuss various applications and basic properties of prime implicants and describe several methods for generating all the prime implicants of a Boolean function.

The second deals with problems related to the representation of a Boolean function by a DNF; that is, as a disjunction of elementary conjunctions. Since a Boolean function may have numerous DNF representations, the question of finding an "optimal" one plays a very important role. Among the most commonly considered optimality criteria, we discuss in detail the minimization of both the number of terms and the number of literals in a DNF representation of a given function. We explain the close relationship between these "logic minimization" problems and the well-known set covering problem of combinatorial optimization; we describe several efficient DNF simplification procedures; we establish the computational complexity of logic minimization problems; and we present a "greedy" procedure as an efficient and effective approximation algorithm for logic minimization.

3.1 Prime implicants

Let us first recall some of the notations and definitions introduced in Chapter 1 (see Section 1.7, in particular). For a Boolean variable x, we let

$$x^\alpha = \begin{cases} x, & \text{if } \alpha = 1, \\ \overline{x}, & \text{if } \alpha = 0. \end{cases}$$

An elementary conjunction $C_{PN} = \bigwedge_{i \in P} x_i \bigwedge_{j \in N} \overline{x}_j$ is an *implicant* of a Boolean function $f(x_1, x_2, \ldots, x_n)$ if $C_{PN} = 1$ implies $f = 1$, or equivalently, if $C_{PN} \leq f$. Clearly, every term of any DNF representing a Boolean function f is an implicant of f. We say that a term C *covers* a point X if $C(X) = 1$.

An implicant is called *prime* if it is not absorbed by any other implicant. In other words, an implicant is prime if each elementary conjunction obtained by eliminating an arbitrary literal from it is not an implicant. A DNF consisting only of prime implicants is called a *prime DNF*.

A remarkable property of prime implicants is the fact that the disjunction of all prime implicants represents the function. This expression is called the *complete DNF* of the function. We have already noted, however, that a disjunction of a (sometimes small) subset of prime implicants may already represent the function. For example, the prime DNF

$$x\bar{y} \vee yz$$

represents the function whose complete DNF is

$$x\bar{y} \vee yz \vee xz.$$

The prime implicants of a Boolean function can be viewed as its "building blocks." Indeed, a function is not only completely described by its prime implicants, but also, as will be seen later, the set of all prime implicants reveals many important properties of the function.

3.1.1 Applications to propositional logic and artificial intelligence

As already discussed in Chapter 1 (Section 1.13), Boolean functions find numerous applications in the knowledge bases of expert systems. Such knowledge bases are usually huge collections of propositional implication rules that formally represent the expert knowledge in a particular domain. As was shown in Section 1.13, a system of rules can be transformed in a straightforward way to a DNF expression of a Boolean function, and vice versa. Recall, for instance, that the rule system

Rule 1: If x is false and y is true then z is true
Rule 2: If x is false and y is false then z is false
Rule 3: If z is true then x is false
Rule 4: If y is true then z is false

corresponds to the DNF

$$\phi(x,y,z) = \bar{x}\,y\,\bar{z} \vee \bar{x}\,\bar{y}\,z \vee xz \vee yz. \tag{3.1}$$

This DNF is logically equivalent to the prime DNF $\phi = \bar{x}\,y \vee z$, so that the original rules 1–4 are equivalent to the conjunction of the following two rules:

Rule 5: If y is true then x is true
Rule 6: z is false

The two rule systems are logically equivalent in the sense that any logical deduction that follows from one system also follows from the other one.

The foregoing example shows how the application of the notion of prime implicants allows the simplification of an arbitrary system of rules. Moreover, any implicant of the associated DNF corresponds to a rule that can be deduced from the rule system, and vice versa. For example, the term $\bar{x}z$ is an implicant of the DNF $\phi(x,y,z)$, and therefore the rule

Rule 7: If z is true then x is true

can be deduced from the rule system.

Note that Rule 7 is not very interesting, since a more general rule (namely, Rule 6: "z is false") can also be deduced. Since z is a prime implicant of $\phi(x,y,z)$, it is impossible to deduce a more general rule than the latter one. It is therefore natural to consider the so-called irredundant rules, which correspond to the prime implicants of the associated Boolean function. The complete DNF of the associated Boolean function will provide all the irredundant rules that can be deduced from the given rule system. While some of these rules may be present in the original rule system or can be obtained by generalizing the rules of the original system (i.e., by removing some literals from them), some other irredundant rules may bear no evident similarity to any of the initial rules. Such rules can reveal some possibly interesting logical implications that are "hidden" in the original system.

3.1.2 Short prime implicants

In many cases, the most "important" prime implicants of a Boolean function are the shortest ones. The presence of short prime implicants may allow to simplify various problems concerning a Boolean function.

First of all, the constant 1, which is the only elementary conjunction of degree 0, is an implicant of a Boolean function $f(x_1,\ldots,x_n)$ if and only if $f(x_1,\ldots,x_n)$ is a tautology (namely, if $f(x_1,\ldots,x_n) = 1$ for all Boolean vectors $(x_1,\ldots,x_n)$), and in this case, it is its only prime implicant. In other words, the constant 1 is an implicant (which cannot be but prime) of f if and only if there is no solution to the equation $f(x_1,\ldots,x_n) = 0$.

Similarly, a prime implicant of degree 1 is a literal; such implicants will be called *linear*. If a function f has a linear prime implicant x (or a linear prime implicant $\bar{x}$), then no other prime implicant of f contains either x or $\bar{x}$. Indeed, if x is an implicant of f, then every other implicant of f containing the literal x is absorbed by the implicant x, and therefore is not prime. On the other hand, if $\bar{x}C$ is an implicant of f, where C is an elementary conjunction, then C is also an implicant of f, since

$$C = xC \vee \bar{x}C \leq x \vee \bar{x}C \leq f.$$

This reasoning, together with Theorem 1.13, easily leads to the following result:

Theorem 3.1. *If $x_i^{\alpha_i}$, $i = 1,2,\ldots,m$, are prime implicants of the Boolean function $f(x_1,\ldots,x_n)$, then there exists a Boolean function $g(x_{m+1},\ldots,x_n)$ such that*

$$f(x_1,\ldots,x_n) = \bigvee_{i=1}^{m} x_i^{\alpha_i} \vee g(x_{m+1},\ldots,x_n).$$

Moreover, an elementary conjunction different from $x_i^{\alpha_i}$, $i = 1, \ldots, m$, is a prime implicant of $f(x_1, \ldots, x_n)$ if and only if it is a prime implicant of $g(x_{m+1}, \ldots, x_n)$.

The decomposition provided by Theorem 3.1 allows the reduction of many problems involving Boolean functions to the case of Boolean functions without linear implicants.

Prime implicants of degree 2, also called *quadratic* prime implicants, define a partial order among certain literals. Indeed, if $x_1^{\alpha_1} x_2^{\alpha_2}$ is an implicant of a Boolean function $f(x_1, \ldots, x_n)$, then the inequality

$$x_1^{\alpha_1} \leq x_2^{\bar{\alpha}_2},$$

or equivalently, the inequality

$$x_2^{\alpha_2} \leq x_1^{\bar{\alpha}_1},$$

holds in every false point of f.

Example 3.1. *Consider the Boolean function*

$$f = xy \vee y\bar{z} \vee xwz.$$

Since xy and $y\bar{z}$ are prime implicants of f, it follows that $x \leq \bar{y}$ and $y \leq z$ in every false point of f. In other words, if a false point has $y = 1$, then it must have $x = 0$ and $z = 1$.   □

If $x_1 x_2$ is a prime implicant of f, then neither $x_1\bar{x}_2$ nor $\bar{x}_1 x_2$ can be an implicant of f, since otherwise, either x_1 or x_2 would also be an implicant, and hence $x_1 x_2$ would not be prime. If both $x_1\bar{x}_2$ and $\bar{x}_1 x_2$ are prime implicants of f, then the variables x_1 and x_2 are logically equivalent in the sense that, in every false point of f, the value of x_1 and the value of x_2 are the same. The next theorem shows that in this case x_1 and x_2 behave in a perfectly "symmetric" way in the prime implicants of f.

Theorem 3.2. *If both $x_1\bar{x}_2$ and $\bar{x}_1 x_2$ are prime implicants of the Boolean function $f(x_1, x_2, \ldots, x_n)$, then*

(1) *no other prime implicant of f depends on both x_1 and x_2;*
(2) *if an elementary conjunction C depends neither on x_1 nor on x_2, then $x_1^\alpha C$ is a prime implicant of f if and only if $x_2^\alpha C$ is a prime implicant of f;*
(3) *$f(x_1, x_2, x_3 \ldots, x_n) = x_1\bar{x}_2 \vee \bar{x}_1 x_2 \vee g(x_1, x_3, \ldots, x_n)$, where g is obtained from f by substituting x_1 for x_2 in f, that is, $g(x_1, x_3, \ldots, x_n) = f(x_1, x_1, x_3 \ldots, x_n)$;*
(4) *an elementary conjunction different from $x_1\bar{x}_2$ or $\bar{x}_1 x_2$ is a prime implicant of $f(x_1, x_2, \ldots, x_n)$ if and only if*
   - *it is a prime implicant of $g(x_1, x_3, \ldots, x_n)$, or*
   - *it is of the form $x_2^\alpha C$, where the elementary conjunction C is such that $x_1^\alpha C$ is a prime implicant of $g(x_1, x_3, \ldots, x_n)$.*

Proof. (1) Let $x_1^\beta x_2^\gamma C$ be a prime implicant of f. Clearly, $\gamma = \beta$ since otherwise $x_1^\beta x_2^\gamma C$ is absorbed by $x_1 \overline{x}_2$ or $\overline{x}_1 x_2$. However, in this case $x_1^\beta C$ is an implicant, since

$$x_1^\beta C = x_1^\beta x_2^\beta C \vee x_1^\beta x_2^{\overline{\beta}} C \leq x_1^\beta x_2^\gamma C \vee x_1^\beta x_2^{\overline{\beta}} \leq f.$$

Therefore, $x_1^\beta x_2^\gamma C$ is not prime since it is absorbed by $x_1^\beta C$. This proves statement 1.

(2) If $x_1^\alpha C$ is an implicant of f, then $x_2^\alpha C$ is an implicant of f, since

$$x_2^\alpha C = x_1^\alpha x_2^\alpha C \vee x_1^{\overline{\alpha}} x_2^\alpha C \leq x_1^\alpha C \vee x_1^{\overline{\alpha}} x_2^\alpha \leq f.$$

Therefore, by symmetry, $x_1^\alpha C$ is an implicant of f if and only if $x_2^\alpha C$ is an implicant of f.

If $x_1^\alpha C$ is a prime implicant of f, but the implicant $x_2^\alpha C$ is not prime, then, since C is not an implicant, there must exist a $C' < C$ such that $x_2^\alpha C'$ is an implicant of f. Then $x_1^\alpha C'$ is also an implicant of f, contradicting the assumption that $x_1^\alpha C$ is prime. Hence, by symmetry, $x_1^\alpha C$ is a prime implicant of f if and only if $x_2^\alpha C$ is a prime implicant of f. This proves statement 2.

(3) If we use Theorem 1.13 to represent f as its complete DNF, then statement 3 follows from the identity

$$x_1 \overline{x}_2 \vee \overline{x}_1 x_2 \vee x_2^\alpha C = x_1^\alpha x_2^{\overline{\alpha}} \vee x_1^\alpha x_2^\alpha C \vee x_1^{\overline{\alpha}} x_2^\alpha \vee x_1^{\overline{\alpha}} x_2^\alpha C = x_1^{\overline{\alpha}} x_2^\alpha \vee x_1^\alpha (x_2^{\overline{\alpha}} \vee x_2^\alpha C)$$

$$= x_1^{\overline{\alpha}} x_2^\alpha \vee x_1^\alpha (x_2^{\overline{\alpha}} \vee C) = x_1 \overline{x}_2 \vee \overline{x}_1 x_2 \vee x_1^\alpha C$$

applied to each prime implicant depending on x_2.

(4) To prove that statement 4 holds, let us first note that every implicant of g is an implicant of f.

Let us now show that any implicant C of f that does not depend on x_2 is an implicant of g. Indeed, if $C(X^*) = 1$ for some point X^*, then $f(X^*) = 1$, and since C does not depend on x_2, $f(x_1^*, x_1^*, \ldots, x_n^*)$ must also be 1, and therefore $g(X^*) = 1$. It follows now, just as in the proof of statement 2, that every prime implicant of g is a prime implicant of f, and conversely, every prime implicant of f that does not depend on x_2 is a prime implicant of g. This fact, together with statement 2, shows that a prime implicant of f that is not a prime implicant of g is of the form $x_2^\alpha C$, where $x_1^\alpha C$ is a prime implicant of g. This proves statement 4. $\square$

Remark 3.1. If $x_1 x_2$ and $\overline{x}_1 \overline{x}_2$ are prime implicants of a Boolean function f, then a valid statement analogous to Theorem 3.2 is obtained after replacing x_2 by $\overline{x}_2$, showing that in this case x_1 and $\overline{x}_2$ behave in a perfectly symmetric way in the prime implicants of f. $\square$

3.2 Generation of all prime implicants

In this section, we consider the problem of generating all prime implicants of a Boolean function, that is, the algorithmic problem:

PRIME IMPLICANTS
Instance: An arbitrary expression of a Boolean function f.
Output: The complete DNF of f or, equivalently, a list of all prime implicants of f.

This problem has been intensively investigated in the literature since the early 1930s. Its complexity depends very much on the expression of f. We shall successively handle the cases in which f is given by a list of its true points, or by an arbitrary DNF, or by a CNF.

3.2.1 Generation from the set of true points

Let us first assume that the input of PRIME IMPLICANTS takes the form of a list of all true points of the function f or, equivalently, of a truth table, or of a minterm expression of f. The results presented here have been known for a very long time and seem to belong to the "folklore" of the field.

It is useful to associate an elementary conjunction with a pair of points in a Boolean cube, as defined next.

Definition 3.1. *Given two Boolean points* $Y = (y_1, y_2, \ldots, y_n)$ *and* $Z = (z_1, z_2, \ldots, z_n)$, *the* hull *of* Y *and* Z *(denoted by* $[Y, Z]$*) is the elementary conjunction defined by*

$$[Y, Z] = \bigwedge_{i: y_i = z_i = 1} x_i \bigwedge_{j: y_j = z_j = 0} \overline{x}_j.$$

Example 3.2. *If* $Y = (1, 0, 1, 0, 1)$ *and* $Z = (0, 0, 1, 1, 1)$, *then* $[Y, Z] = \overline{x}_2 x_3 x_5$.   □

Clearly, for any two Boolean points Y and Z,

$$[Y, Z] = [Z, Y]$$

and

$$[Y, Z](Y) = [Y, Z](Z) = 1.$$

In fact, the set of true points of $[Y, Z]$ is the smallest subcube covering both Y and Z.

Theorem 3.3. *If* C *is an implicant of a Boolean function* f, *then for any true point* Y *of* f *such that* $C(Y) = 1$, *there exists a unique true point* Z *of* f *such that* $C(Z) = 1$ *and* $C = [Y, Z]$.

Proof. Let $C = \bigwedge_{i \in P} x_i \bigwedge_{j \in N} \overline{x}_j$. Since $C(Y) = 1$, the point $Y = (y_1, y_2, \ldots, y_n)$ is such that $y_i = 1$ for $i \in P$ and $y_j = 0$ for $j \in N$. Let us define the point $Z = (z_1, z_2, \ldots, z_n)$ in the following way:

$$z_i = \begin{cases} y_i, & \text{if } i \in P \cup N, \\ \overline{y}_i, & \text{if } i \notin P \cup N. \end{cases}$$

Clearly, $C(Z) = 1$, and therefore Z is a true point of f. Moreover, it follows easily from Definition 3.1 that $C = [Y, Z]$ and that $C \neq [Y, W]$ when $W \neq Z$. □

Corollary 3.1. *If a Boolean function f has m true points, then the number of (prime) implicants of f does not exceed $\binom{m}{2} + m$.*

Proof. By Theorem 3.3, every implicant of f is the hull of some pair of (possibly identical) true points of f, and every pair of true points generates in this way at most one implicant of f. □

We now describe an efficient way of generating all (prime) implicants of a Boolean function f when the set $T(f)$ of its true points is given. To generate all implicants of f, it is sufficient to examine all the pairs of true points of f, and check for every pair whether its hull is an implicant of f.

Given the set $T(f)$ of all true points of a Boolean function f on $\mathcal{B}^n$, and an elementary conjunction C of degree d, one can easily check whether C is an implicant of f. Indeed, one can simply count the number of points $X \in T(f)$ such that $C(X) = 1$: Obviously, C is an implicant of f if and only if this count equals 2^{n-d}. For every $X \in T(f)$, both evaluating $C(X)$ and incrementing the counter can be done in $O(n)$ time. Therefore, for every pair of true points, it can be checked in $O(n|T(f)|)$ time whether or not its hull is an implicant of f, and all the implicants of f (possibly with repetitions) can be generated in $O(n|T(f)|^3)$ time.

Now, given a list (possibly with repetitions) of all the implicants of a Boolean function f, one can generate all the prime implicants of f by eliminating from the list those implicants that are absorbed by some other ones. If the list contains M elementary conjunctions, a naive approach requires $\binom{M}{2}$ pairwise comparisons, each taking $O(n)$ time. In this way, the list of all prime implicants is generated in $O(nM^2)$ time. Since M is typically much larger than n, one may want to reduce the generation time by making use of the fact that all the implicants are present in the list. We now describe how to achieve this time reduction.

Let $C = \bigwedge_{i \in L} x_i^{\alpha_i}$, and let us denote $C^j = \bigwedge_{i \in L \setminus \{j\}} x_i^{\alpha_i}$ for $j \in L$. If C is an implicant of f, then C is prime if and only if no C^j (for $j \in L$) is an implicant of f. To be able to find out efficiently whether an elementary conjunction is present in the list of all implicants of f, we need to order the implicants in such a way that C^j always appears "before" C in the list.

Let us introduce such a linear order on the set of all elementary conjunctions. We place C' "before" C'' if the degree of C' is lower than the degree of C''. When

C' and C'' have the same degree, then their order is the lexicographic order induced by the linear order of literals, whereby x_i is before $\overline{x}_i$, $\overline{x}_j$ is before than $*$ (meaning "not present"), and x_i is before x_j if $i < j$.

A comparison of two elementary conjunctions according to this order can be performed in $O(n)$ time, and the set of implicants of f can be linearly ordered in $O(nM \log M)$ time. Ordering the list also allows us to eliminate possible repetitions. Then, using binary search, one can check whether a conjunction is present in the list by doing at most $\log M$ comparisons, that is, in $O(n \log M)$ time. For any implicant C, at most n conjunctions C^j need to be checked. Therefore, all the nonprime implicants can be eliminated from the list in $O(n^2 M \log M)$ time. When M is sufficiently large, this bound is better than the naive $O(nM^2)$ bound.

The arguments above prove the following statement:

Theorem 3.4. *If a Boolean function f of n variables is represented by the set $T(f)$ of its true points, then*

  (a) *all implicants of f can be generated in $O(n|T(f)|^3)$ time;*
  (b) *all prime implicants of f can be generated in $O(n|T(f)|^2(|T(f)| + n \log |T(f)|))$ time.* $\qquad\square$

Note that for those Boolean functions whose number of true points is sufficiently large, the additional expense of reducing the list of all implicants and keeping only the prime ones is asymptotically negligible compared with the time required to generate all the implicants.

3.2.2 Generation from a DNF representation: The consensus method

We now turn to the PRIME IMPLICANTS problem when its input is in disjunctive normal form. The best-known method of solving this problem is the *consensus method*. Recall that we introduced this fundamental procedure in Chapter 2, Section 2.7, for the solution of DNF equations. The consensus method, however, has been initially proposed, and repeatedly rediscovered, as a method of generating all prime implicants of a function represented in DNF. The most frequently cited references in this framework include Blake [99], Samson and Mills [801], and Quine [768]; see Brown [156] for a historical perspective on the development of the consensus method.

Given an arbitrary DNF $\phi(x_1,\ldots,x_n)$, the *consensus procedure* transforms ϕ by repeatedly applying the operations of *absorption* and *consensus*, as displayed in Figure 3.1 (recall Definition 2.5 and compare with Figure 2.4 in Section 2.7; the present section contains significant overlap with Section 2.7, but we find it advisable to repeat some of those definitions and concepts here for the sake of clarity). In the description of this procedure, we use the shorthand $\phi \setminus D$ to denote the DNF obtained by removing a term D from the DNF ϕ.

Procedure CONSENSUS*(ϕ)

Input: A DNF expression $\phi(x_1,\ldots,x_n) = \bigvee_{k=1}^{m} C_k$ of a Boolean function f.
Output: The complete DNF of f, that is, the disjunction of all prime implicants of f.

begin
      while one of the following conditions applies **do**
          if there exist two terms C and D of ϕ such that C absorbs D
             then remove D from ϕ: $\phi := \phi \setminus D$;
          if there exist two terms $x_i\, C$ and $\overline{x}_i\, D$ of ϕ such that $x_i\, C$ and $\overline{x}_i\, D$
          have a consensus and CD is not absorbed by another term of ϕ
             then add CD to ϕ: $\phi := \phi \vee CD$;
      end while
      return ϕ;
end

Figure 3.1. Procedure CONSENSUS*

The consensus procedure stops when

(1) the absorption operation cannot be applied, and
(2) either the consensus operation cannot be applied, or all the terms that can
be produced by consensus are absorbed by other terms in ϕ.

We shall say that a DNF is *closed under absorption* if it satisfies the first condition above, and that it is *closed under consensus* if it satisfies the second condition.

Note that the consensus procedure always terminates and produces a DNF closed under consensus and absorption in a finite number of steps: Indeed, the number of terms in the given variables is finite, and once a term is removed by absorption, it will never again be added by consensus.

Example 3.3. *Consider the following DNF:*

$$\phi(x_1,x_2,x_3,x_4) = x_1\overline{x}_2 x_3 \vee \overline{x}_1 \overline{x}_2 x_4 \vee x_2 x_3 x_4.$$

Note that absorption cannot be applied to ϕ. The application of consensus to the first two terms of ϕ transforms it into

$$\phi'(x_1,x_2,x_3,x_4) = x_1\overline{x}_2 x_3 \vee \overline{x}_1 \overline{x}_2 x_4 \vee x_2 x_3 x_4 \vee \overline{x}_2 x_3 x_4.$$

Again, absorption cannot be applied to ϕ'. The application of consensus to the last two terms of ϕ' transforms it into

$$\phi''(x_1,x_2,x_3,x_4) = x_1\overline{x}_2 x_3 \vee \overline{x}_1 \overline{x}_2 x_4 \vee x_2 x_3 x_4 \vee \overline{x}_2 x_3 x_4 \vee x_3 x_4.$$

Now the last term of ϕ'' absorbs the two previous terms, and ϕ'' is transformed into

$$\phi'''(x_1,x_2,x_3,x_4) = x_1\overline{x}_2 x_3 \vee \overline{x}_1 \overline{x}_2 x_4 \vee x_3 x_4.$$

Here, the consensus procedure stops. Note that the first two terms of ϕ''' actually have a consensus, but it is absorbed by the last term.   □

We have already observed that the operations of absorption and consensus transform DNFs, but do not change the Boolean functions that they represent. This is implied by the two lemmas below, which easily follow from the basic Boolean identities (see also Theorem 2.8).

Lemma 3.1. *For any two elementary conjunctions C and CD,*

$$C \vee CD = C.$$

Lemma 3.2. *For any two elementary conjunctions xC and $\overline{x}D$,*

$$xC \vee \overline{x}D = xC \vee \overline{x}D \vee CD.$$

The importance of the consensus procedure in the theory of Boolean functions derives from the following theorem, which asserts the correctness of Procedure CONSENSUS*.

Theorem 3.5. *Given an arbitrary DNF ϕ of a Boolean function f, the consensus procedure applied to ϕ produces the complete DNF of f, that is, the disjunction of all prime implicants of f.*

In view of its crucial role, we shall provide two alternative proofs of this theorem. In order to present the first proof, we start by establishing two technical lemmas.

Lemma 3.3. *Given an arbitrary DNF ϕ of a Boolean function f, a prime implicant of f can involve only those variables that are present in ϕ.*

Proof. If ϕ does not involve x, then the value of f does not change when only the value of x changes. Therefore, implicants of f involving x cannot be prime.   □

Lemma 3.4. *Given an arbitrary DNF ϕ of a Boolean function f, if C is an implicant of f that involves all variables present in ϕ, then C is absorbed by a term in ϕ.*

Proof. If C contains all the variables in ϕ, then the assignment that makes $C = 1$ assigns values to all variables in ϕ. Since C is an implicant of f, this assignment makes $\phi = 1$, and therefore at least one term in ϕ is 1. This term absorbs C.   □

We are now ready to proceed with a first proof of the theorem.

Proof of Theorem 3.5. We prove the statement by contradiction. Let us assume that there exists a Boolean function f and a DNF ϕ of f such that, when the consensus procedure is applied to ϕ, it returns a DNF ψ that does not contain some prime implicant C_0 of f. Lemma 3.3 implies that C_0 involves only variables present in ψ. Let us consider the set S of elementary conjunctions C satisfying the following three conditions:

1. C only involves variables present in ψ.
2. $C \le C_0$ (and therefore C is an implicant of f).
3. C is not absorbed by any term in ψ.

The set S is not empty, since C_0 satisfies all three conditions. Let C^m be a term of maximum degree in S. Since C^m is not absorbed by any term in ψ, by Lemma 3.4, C^m cannot involve all the variables present in ψ. Let x be a variable present in ψ and not present in C^m. The degree of the elementary conjunctions xC^m and $\overline{x}C^m$ exceeds that of C^m. Since the degree of C^m was assumed to be maximum, xC^m and $\overline{x}C^m$ do not belong to S, and therefore cannot satisfy all three conditions. Since they obviously satisfy the first two conditions, they must violate the last one, namely, there must exist terms C' and C'' in ψ such that $xC^m \le C'$ and $\overline{x}C^m \le C''$. Since C^m is not absorbed by either C' or C'', it follows that $C' = xD'$ and $C'' = \overline{x}D''$, where D' and D'' are elementary conjunctions that absorb C^m. This implies that D' and D'' do not conflict in any variable. Therefore, the consensus of C' and C'' exists: It is $D'D''$, and this term absorbs C^m. Since the consensus procedure stops on the DNF ψ, there must exist a term C''' in ψ that absorbs $D'D''$. Then C''' must also absorb C^m, contradicting the assumption that C^m belongs to S. $\square$

Before we proceed with the second proof of Theorem 3.5, we first establish a lemma. Recall from Section 1.9 that, if A and B are two disjoint subsets of $\{1,\dots,n\}$, then the set of all Boolean vectors in $\mathcal{B}^n$ whose coordinates in A are fixed at 1 and whose coordinates in B are fixed at 0, forms a subcube of $\mathcal{B}^n$. This subcube is denoted by $T_{A,B}$.

Lemma 3.5. *Let ϕ be a DNF closed under consensus and let $T_{A,B}$ be a subcube. The equation $\phi(X) = 0$ has a solution in $T_{A,B}$ if and only if no term of ϕ is identically 1 on $T_{A,B}$.*

Proof. The "only if" part of the statement is trivial. Let us prove the "if" part by contradiction. Assume that T_{A^*,B^*} is a subcube such that no term of ϕ is identically 1 on T_{A^*,B^*}, and such that no solution of $\phi(X) = 0$ exists in T_{A^*,B^*}; moreover, assume that $|A^* \cup B^*|$ has maximum cardinality among all subcubes satisfying these conditions. Clearly, $|A^* \cup B^*| \le n - 1$, since the statement trivially holds when $|A \cup B| = n$.

Let us select an arbitrary variable x_i such that $i \notin A^* \cup B^*$. Since each of the two subcubes $T_{A^* \cup \{i\},B^*}$ and $T_{A^*,B^* \cup \{i\}}$ is a subset of T_{A^*,B^*}, no solution of the equation $\phi(X) = 0$ exists either in $T_{A^* \cup \{i\},B^*}$ or in $T_{A^*,B^* \cup \{i\}}$. It follows, then, from the maximality of $|A^* \cup B^*|$ that, on each of the two subcubes $T_{A^* \cup \{i\},B^*}$ and $T_{A^*,B^* \cup \{i\}}$, at least one of the terms of ϕ is identically 1. Obviously, one of these terms must involve the literal x_i, while the other one must involve $\overline{x}_i$. Let the two terms in question be $x_i C$ and $\overline{x}_i D$. Clearly, C and D are elementary conjunctions that are both identically 1 on the subcube T_{A^*,B^*}. Therefore, C and D cannot conflict, and hence the consensus CD of $x_i C$ and $\overline{x}_i D$ exists. Since ϕ is closed under consensus, it must contain a term E that absorbs CD. Then this term E must be identically 1 on the subcube T_{A^*,B^*}, contradicting our assumption. $\qquad\square$

Lemma 3.6. *A DNF is closed under consensus if and only if it contains all prime implicants of the Boolean function it represents.*

Proof. We first prove the "if" part of the statement. It follows from Lemma 3.2 that the consensus of any two terms of a DNF ϕ is an implicant of the Boolean function f represented by ϕ. This implies in turn that, if ϕ contains all prime implicants of f, then it is closed under consensus.

To prove the "only if" part of the lemma, let us assume that a DNF ϕ representing the Boolean function f is closed under consensus, and that the conjunction $C = \bigwedge_{i \in P} x_i \bigwedge_{i \in N} \overline{x}_i$ is a prime implicant of f not contained in ϕ. Clearly, the partial assignment defined by

$$x_i = 1 \text{ for } i \in P \text{ and } x_i = 0 \text{ for } i \in N$$

makes no term in ϕ identically 1, since such a term would absorb C. Therefore, by Lemma 3.5, there exists a solution X^* to $\phi(X) = 0$ such that $X^* \in T_{P,N}$. Then $C(X^*) = 1$, while $\phi(X^*) = 0$, contradicting the assumption that C is an implicant of f. $\qquad\square$

We are now ready to present the next proof.

Proof of Theorem 3.5. Let ϕ' be the DNF produced by the consensus procedure applied to the given DNF ϕ. This DNF ϕ' is closed under absorption and consensus. By Lemma 3.6, ϕ' contains all the prime implicants of f. Since every implicant of f is absorbed by a prime implicant of f, it follows that ϕ' is the complete DNF of f. $\qquad\square$

Theorem 3.5 is equivalent to the following statement:

Corollary 3.2. *A DNF is closed under consensus and absorption if and only if it is the complete DNF of the function it represents.*

This statement is frequently used to check whether or not a DNF is complete. The following corollary shows that the completeness of a DNF can be verified in polynomial time. Recall that $\|\phi\|$ denotes the number of terms in a DNF ϕ.

Corollary 3.3. *Given a DNF ϕ of a Boolean function f, one can check in $O(n\|\phi\|^3)$ time whether ϕ is the complete DNF of f.*

Proof. Given two terms, one can check in $O(n)$ time whether one is absorbed by the other. Therefore, one can check in $O(n||\phi||^2)$ time whether the absorption operation can be applied to ϕ.

Given two terms, checking the existence of their consensus and producing it can be done in $O(n)$ time. Since there are $\binom{||\phi||}{2}$ pairs of terms of ϕ, it can be checked in $\binom{||\phi||}{2} O(n||\phi||) = O(n||\phi||^3)$ time whether every consensus of two terms of ϕ is absorbed by another term of ϕ.   □

The next corollary is essentially due to Robinson [787]. It shows how a solution of a consistent DNF equation can be efficiently computed once the prime implicants of the DNF are available.

Corollary 3.4. *If a DNF ϕ is closed under consensus, then one can find a solution of the equation $\phi(X) = 0$, or prove that the equation is inconsistent, in $O(n^2||\phi||)$ time.*

Proof. By Lemma 3.6, the equation $\phi(X) = 0$ is inconsistent if and only if 1 is one of the terms of ϕ. If this is not the case, then a solution of the equation is obtained by a simple "greedy" procedure: Fix successively the variables $x_1, x_2, \ldots, x_n$ to either 0 or 1, while avoiding making any term in the DNF identically equal to 1. Indeed, Lemma 3.5 implies that this procedure is correct, since any DNF that is closed under consensus will remain closed under consensus after substituting any Boolean values for any of the variables. The time bound follows from the fact that substituting a value for a variable in any of the DNFs obtained in the process of fixing variables can be done in $O(n||\phi||)$ time.   □

Variable depletion

A streamlined version of the consensus procedure called *variable depletion* was proposed by Blake [99] and later by Tison [864]. This method organizes the consensus procedure in the following way: First, a starting variable x_{i_1} is chosen, and all possible consensuses are formed using pairs of terms that conflict in x_{i_1}. After completing this stage and removing all absorbed terms, another variable x_{i_2} is chosen, and all consensuses on x_{i_2} are produced. The process is repeated in the same way until all variables have been exhausted.

The surprising fact, perhaps, is that after the stage based on an arbitrary variable x_i is completed, there is no need later on to apply again the consensus operation to any pair of terms conflicting in x_i. Before proving the correctness of this method, we first establish the following lemma (which extends Theorem 2.6).

Lemma 3.7. *Let f be a Boolean function depending on the variables $x_1, x_2, \ldots, x_n$, and let g, h, and l be Boolean functions depending on the variables $x_1, x_2, \ldots, x_{n-1}$ such that:*

$$f = x_n g \vee \overline{x}_n h \vee l.$$

A conjunction C not depending on x_n is an implicant (prime implicant) of f if and only if it is an implicant (prime implicant) of

$$f' = (g \vee l)(h \vee l) = gh \vee l.$$

Proof. Since $f' \leq f$, it follows that every implicant of f' is an implicant of f. Conversely, let us assume that C is an implicant of f that does not depend on x_n and let X^* be any point in $\mathcal{B}^{n-1}$. If $C(X^*) = 1$, then $f(X^*, 0) = f(X^*, 1) = 1$ (since C is an implicant of f), or equivalently $h(X^*) \vee l(X^*) = g(X^*) \vee l(X^*) = 1$, and hence $f'(X^*) = 1$. Thus, C is an implicant of f'.

Furthermore, if C is a prime implicant of f but not a prime implicant of f', then there exists another implicant C' of f' such that $C < C'$. Since every implicant of f' is an implicant of f, $C' > C$ is an implicant of f, contradicting the assumption that C is a prime implicant of f.

A similar reasoning shows that every prime implicant of f' is also a prime implicant of f.   □

Lemma 3.8. *Let f be a Boolean function depending on the variables $x_1, x_2, \ldots, x_n$, and let g, h, and l be Boolean functions depending on the variables $x_1, x_2, \ldots, x_{n-1}$ such that:*

$$f = x_n g \vee \overline{x}_n h \vee l.$$

A conjunction $x_n C$ is an implicant (prime implicant) of f if and only if it is an implicant (prime implicant) of

$$f'' = x_n g \vee l.$$

The proof of this statement is analogous to that of Lemma 3.7, and is therefore omitted.

We are now ready to formally state and prove the correctness of the method of variable depletion.

Theorem 3.6. *Given an arbitrary DNF ϕ of a Boolean function f, the method of variable depletion applied to ϕ produces the complete DNF of f.*

Proof. Let us call a variable x *non-unate* in a DNF ϕ if ϕ contains both a term xC' and a term $\overline{x}C''$. Let us prove the theorem by induction on the number of non-unate variables in the given DNF ϕ. If ϕ contains just one non-unate variable, then the variable depletion procedure stops after one step, and the resulting DNF is closed under consensus and absorption. Corollary 3.2 implies that this resulting DNF is the complete DNF of f, thus proving the basis of induction.

Let us assume now that the theorem holds if the number of non-unate variables is at most $n - 1$, and let ϕ contain n non-unate variables. Let x_n be the first variable used in the variable depletion procedure, and let us represent the DNF ϕ of f as

$$\phi = x_n \phi_1 \vee \overline{x}_n \phi_0 \vee \phi_2,$$

where the DNFs ϕ_0, ϕ_1, and ϕ_2 do not depend on x_n.

Note that the first step of the variable depletion procedure will generate all the terms of the conjunction $\phi_0\phi_1$. Therefore, the DNF ϕ' produced after the first step of variable depletion is

$$\phi' = x_n\phi_1 \vee \overline{x}_n\phi_0 \vee \phi_2 \vee \phi_0\phi_1.$$

Although some absorptions may be possible in ϕ', no term of $\phi_2 \vee \phi_0\phi_1$ can be absorbed by a term of $x_n\phi_1 \vee \overline{x}_n\phi_0$. It follows from Lemma 3.7 that every prime implicant of f that does not depend on x_n is a prime implicant of $\phi_2 \vee \phi_0\phi_1$. Note that the DNF $\phi_2 \vee \phi_0\phi_1$ has at most $n-1$ non-unate variables. By the inductive assumption, the variable depletion procedure applied to $\phi_2 \vee \phi_0\phi_1$ will generate all such prime implicants of f. All these prime implicants will also be generated by the variable depletion procedure applied to ϕ'. Indeed, since x_n was already "depleted", every consensus in this latter procedure, which involves a term depending on x_n, must result in a term depending on x_n. Additionally, a term not depending on x_n cannot be absorbed by a term depending on x_n. Thus, all prime implicants of f not depending on x_n will be generated by the variable depletion procedure.

Applying Lemma 3.8 to the DNF ϕ', one can see that a term x_nC is a prime implicant of f if and only if it is a prime implicant of $x_n\phi_1 \vee \phi_2 \vee \phi_0\phi_1$. Note that this DNF has at most $n-1$ non-unate variables. By the inductive assumption, the variable depletion procedure applied to this DNF will generate all those prime implicants of f that have the form x_nC. All these prime implicants will also be generated by the variable depletion procedure applied to ϕ'. Indeed, since x_n was already "depleted", every consensus in this latter procedure that involves a term containing $\overline{x}_n$ must result in a term containing $\overline{x}_n$. Additionally, a term not containing $\overline{x}_n$ cannot be absorbed by a term containing $\overline{x}_n$. Thus, all prime implicants of f having the form x_nC will be generated by the variable depletion procedure.

The case of prime implicants of f having the form $\overline{x}_nC$ is completely analogous to the above. □

Term disengagement

Another interesting variant of the consensus procedure based on *term disengagement* was introduced by Tison [864], who proved that it works for arbitrary DNFs; it was subsequently generalized by Pichat [747] to more abstract lattice-theoretic structures. The DISENGAGEMENT CONSENSUS procedure is described in Figure 3.2. It relies on the following principles:

Definition 3.2. *A consensus algorithm is said to be a* (term) disengagement *algorithm if it maintains a list $\mathcal{L}$ of implicants and proceeds in successive stages, where*

(i) *at each stage a term C in the current list $\mathcal{L}$ is selected and all possible consensuses of C with all other terms of $\mathcal{L}$ are generated;*

Procedure DISENGAGEMENT CONSENSUS(ϕ)

Input: A DNF expression $\phi = \bigvee_{k=1}^m C_k$ of a Boolean function f.
Output: The list $\mathcal{L}$ of all prime implicants of f.

begin
    $\mathcal{L} := (C_1, C_2, \ldots, C_m)$, the list of terms of ϕ;
    declare all terms C in $\mathcal{L}$ to be *engaged*;
    while $\mathcal{L}$ contains some *engaged* term **do**
        select an *engaged* term C;
        declare C to be *disengaged*;
        generate all possible consensuses of C and the other terms of $\mathcal{L}$;
        let $\mathcal{R}$ be the list of all such consensuses;
        $\mathcal{L}' := \mathcal{L} \cup \mathcal{R}$;
        for each C' in $\mathcal{R}$
            if C' is not absorbed by another term in $\mathcal{L}'$
            then add C' to $\mathcal{L}$ and declare C' to be *engaged*;
    end while
    return $\mathcal{L}$;
end

Figure 3.2. Procedure DISENGAGEMENT CONSENSUS

(ii) each of the newly generated terms is checked for absorption by any other (old or new) existing term; if it is not absorbed, then it is added to $\mathcal{L}$;

(iii) the term C can no longer be chosen as a parent (it is "disengaged") in any subsequent stages, although it can still absorb some new terms.

We refer to Tison [864] and Pichat [747] for a proof of correctness of the disengagement procedure.

3.2.3 Generation from a DNF representation: Complexity

In this subsection, we are going to discuss the computational complexity of generating the prime implicants of a Boolean function given in DNF. The most basic computational problem simply consists of checking whether a given elementary conjunction is an implicant of a Boolean function represented by a DNF:

IMPLICANT RECOGNITION
Instance: An elementary conjunction C and a Boolean function f in DNF.
Question: Is C an implicant of f?

Theorem 3.7. *The* IMPLICANT RECOGNITION *problem is co-NP-complete.*

Proof. Clearly, the problem belongs to the class co-NP, since one can easily check in polynomial time whether a Boolean point gives value 1 to the elementary conjunction C and value 0 to the function f.

A DNF equation $\phi = 0$ is inconsistent if and only if the empty conjunction $C = 1$ is an implicant of the function represented by ϕ. Since the DNF equation problem is NP-complete, it follows that the implicant recognition problem is co-NP-complete.   □

Theorem 3.7 already suggests that generating the prime implicants of a function given in DNF cannot be an easy task. Moreover, Theorem 3.17 will show that the number of prime implicants (that is, the length of the output) may be exponential in the length of the initial DNF (the input). Therefore, the computational complexity of prime implicant generation algorithms should be measured in terms of the sizes of their input and of their output (see Appendix B for a more detailed discussion of list-generation algorithms).

In fact, if it were possible to design a prime implicant generation algorithm that runs in polynomial total time (that is, polynomial in the combined sizes of the input and of the output), then this algorithm could be used to solve DNF equations in polynomial time, since the only prime implicant of a tautology is the constant 1 (see the proof of Theorem 3.7). This, of course, is not to be expected.

The next theorems will show that the computational complexity of the DNF equation problem actually is the main stumbling block on the way to the efficient recognition and generation of prime implicants. In order to state these results, let us recall that $|\phi|$ denotes the length (that is, the number of literals) of a DNF ϕ, and let $t(L)$ denote the computational complexity of solving a DNF equation of length at most L.

Theorem 3.8. *For any Boolean function f, any DNF ϕ representing f, and any elementary conjunction C, one can check in $O(|\phi|) + t(|\phi|)$ time whether C is an implicant of f.*

Proof. By definition, an elementary conjunction $C = \bigwedge_{i \in A} x_i \bigwedge_{j \in B} \overline{x}_j$ is an implicant of f if and only if the restriction of f to the subcube $T_{A,B}$ is a tautology. The latter property can be checked in $O(|\phi|) + t(|\phi|)$ time by fixing x_i to 1, for all $i \in A$, and x_j to 0, for all $j \in B$, in the DNF ϕ, and by solving the resulting DNF equation.   □

Corollary 3.5. *For any Boolean function f, any DNF ϕ representing f, and any implicant C of f, a prime implicant absorbing C can be constructed in $O(|C|(|\phi| + t(|\phi|)))$ time.*

Proof. If C is not prime, then there must exist a literal in C such that the elementary conjunction obtained from C by removing this literal remains an implicant of f. By Theorem 3.8, this process can be carried out in $O(|\phi|) + t(|\phi|)$ time for every literal in C.   □

Let us now denote by $\Pi(f)$ the set of prime implicants of a Boolean function f.

Theorem 3.9. *For any Boolean function f and any DNF ϕ representing f, the set $\Pi(f)$ can be generated by an algorithm that solves $O(n|\Pi(f)|^2)$ DNF equations of length at most $|\phi|$. If $t(L)$ is the computational complexity of solving DNF equations of length L, then the running time of this algorithm is $O(n|\Pi(f)|^2 (n|\Pi(f)| + |\phi| + t(|\phi|)))$.*

Proof. By Corollary 3.5, for every elementary conjunction C in ϕ, one can find a prime implicant of f absorbing C in $O(|C|(|\phi| + t(|\phi|)))$ time. In this way, ϕ can be reduced to a prime DNF (i.e., a disjunction of prime implicants) ϕ'. In order to obtain the complete DNF of f, we are going to add prime implicants to ϕ', as described below.

First, the terms of ϕ' are ordered arbitrarily, and the first term is marked. At each step of the algorithm,

- the first unmarked term is compared to every marked term, and their consensus – if any – is produced;
- for every consensus produced, a prime implicant of f absorbing it is found (as in Corollary 3.5), and this prime implicant is added to ϕ' if it is not already present.

After this, the term is marked, and the algorithm continues with the next unmarked term of ϕ'. The algorithm stops when all the terms of ϕ' are marked.

By construction, the resulting DNF ϕ' is closed under absorption and consensus. Therefore, by Corollary 3.2, the output DNF ϕ' is complete.

Since the number of terms of ϕ' never exceeds $|\Pi(f)|$, the number of steps of the algorithm does not exceed $|\Pi(f)|$. At each step, an unmarked term is compared to at most $|\Pi(f)|$ marked terms, and for every consensus produced, a prime implicant of f absorbing it can be found in $O(n(|\phi| + t(|\phi|)))$ time. Finally, it can be checked in $O(n|\Pi(f)|)$ time whether a prime implicant is already present in ϕ'. Therefore, the total running time of the algorithm is $O(n|\Pi(f)|^2 (n|\Pi(f)| + |\phi| + t(|\phi|)))$. $\square$

Note that the complexity of the algorithm in Theorem 3.9 depends, not only on the size of the output (namely, on $|\Pi(f)|$), but also on the complexity $t(\phi)$ of solving the DNF equation $\phi = 0$. For arbitrary DNFs, we expect $t(\phi)$ to be exponential in the input size.

It is natural, however, to consider the problem of generating the prime implicants of a Boolean function represented by a DNF in the special case where the associated DNF equation can be solved efficiently. More precisely, let us call a class C of DNFs *tractable* if the DNF equation $\phi = 0$ can be solved in polynomial time for every DNF ϕ in C and for every DNF obtained by fixing variables to either 0 or 1 in such a DNF. For instance, the class of quadratic DNFs the class of Horn DNFs are tractable (see Chapters 5 and 6).

Corollary 3.6. *For every tractable class C, there exists a polynomial $p(x, y)$ and an algorithm that, for every DNF $\phi \in C$, generates the set of prime implicants $\Pi(f)$ of the function represented by ϕ in polynomial total time $p(|\phi|, |\Pi(f)|)$.*

Proof. This statement follows immediately from the definition of a tractable class and from the proof of Theorem 3.9. $\square$

In the terminology of Johnson, Yannakakis, and Papadimitriou [538] and of Appendix B, the algorithm mentioned in Corollary 3.6 actually runs in polynomial incremental time. We leave this for the reader to verify.

Theorem 3.9 and Corollary 3.6 were originally established (in a slightly different form) by Boros, Crama, and Hammer [112].

3.2.4   Generation from a CNF representation

Finally, we briefly discuss the prime implicant generation problem PRIME IMPLI-CANTS when the function is given as a CNF. It will follow from Theorem 3.18 that no method can produce all prime implicants of a function f given by a CNF ϕ in time polynomial in $\|\phi\|$. Despite its computational intractability, this problem, which is a special case of the *dualization problem* investigated in Chapter 4, plays a major role in the theory of Boolean functions.

The term "dualization" is due to the fact that the dual expression ϕ^d is a DNF representing the dual Boolean function f^d, and therefore, the problem of generating all prime implicants of a function represented by a CNF is equivalent to the problem of generating all prime implicants of the dual of a Boolean function represented by a DNF. As a consequence, several dualization algorithms can be used to generate all the prime implicants of a Boolean function represented by a CNF; we refer to Chapter 4, in particular, Section 4.3, for a more thorough discussion of this topic.

In addition, the following fact is worth noticing. Suppose that $\mathcal{A}$ is a dualization algorithm that, when applied to a DNF representation of a function f, produces all the prime implicants of the dual function f^d. Then, the involution property of Boolean functions (namely, $(f^d)^d = f$, see Theorem 1.2) makes it possible to use $\mathcal{A}$ for generating all the prime implicants of f, by simply applying dualization twice; namely, given any DNF ϕ representing f, apply $\mathcal{A}$ to ϕ to produce all the prime implicants of f^d, and then apply again $\mathcal{A}$ to the complete DNF of f^d to produce all the prime implicants of f. This approach is sometimes known as the *double dualization method*, and is usually attributed to Nelson [705]. From the point of view of computational efficiency, it is clear that the advantages of the double dualization method over the consensus procedure, if any, must be confined to special situations.

3.3   Logic minimization

It was already observed in Chapter 1 that a Boolean function may have numerous DNF representations (see, e.g., Example 1.16). It was also mentioned there that in some applications a "short" DNF representation of a Boolean function is preferred over a longer one (see Section 1.13.1, describing how a system of implication rules in artificial intelligence can be replaced by a logically equivalent one, containing fewer and simpler rules).

The problem of constructing a short DNF representation of a Boolean function is usually referred to as the problem of *logic minimization*, or *two-level logic minimization*, or *Boolean function minimization*. This problem was originally studied within the context of electrical and computer engineering (see Section 1.13.2), where logic minimization is used to reduce the number of electronic components in a switching circuit that realizes a Boolean function. We refer, for instance, to Coudert [221]; Coudert and Sasao [222]; Czort [249]; Sasao [804]; Umans, Villa, and Sangiovanni-Vincentelli [877]; or Villa, Brayton, and Sangiovanni-Vincentelli [891] for surveys.

The complexity of a DNF ϕ can be measured in several ways. The two most popular measures used in logic minimization are $||\phi||$ (the number of terms) and $|\phi|$ (the number of literals) in ϕ. Note that in other areas of the theory of Boolean functions, different measures of DNF complexity can be more relevant, such as for instance the degree of ϕ, that is, the largest number of literals in a term of ϕ.

Let us remark that a $||\phi||$-minimizing DNF must be irredundant, while a $|\phi|$-minimizing DNF must be both irredundant and prime (see Definition 1.30, for the terminology). On the other hand, an arbitrary prime irredundant DNF of a Boolean function may be neither $||\phi||$-minimizing nor $|\phi|$-minimizing.

Example 3.4. *Consider the Boolean function $f(x_1, x_2, x_3)$ represented by the DNF*

$$\phi_1 = x_1\overline{x}_2 \vee \overline{x}_1 x_2 \vee x_1\overline{x}_3 \vee \overline{x}_1 x_3.$$

This DNF is neither $||\phi||$-minimizing nor $|\phi|$-minimizing, since f can also be represented by the DNF

$$\phi_2 = x_1\overline{x}_2 \vee \overline{x}_1 x_3 \vee x_2\overline{x}_3,$$

which has both fewer terms and fewer literals.   □

In our discussion of logic minimization, to avoid unnecessary technical complications, we shall focus on finding $||\phi||$-minimizing DNFs. Clearly, if a $||\phi||$-minimizing DNF is not prime, then this DNF can be simplified further by reducing each of its nonprime terms to a prime one. Therefore, in this section we limit our attention to those $||\phi||$-minimizing DNFs that are not only irredundant but also prime. (The reader should note at this point, however, that it may already be quite hard to recognize whether an arbitrary DNF is irredundant, or whether it is prime; see Exercises 8 and 9 at the end of this chapter.)

Finally, we shall also need to distinguish among different versions of the logic minimization problem, depending on the format of its input. Accordingly, we formally define the following algorithmic problems:

(T, F) $||\phi||$-MINIMIZATION
Instance: The complete truth table of a Boolean function f.
Output: A prime $||\phi||$-minimizing DNF of f.

T $||\phi||$-MINIMIZATION
Instance: The list of true points of a Boolean function f or, equivalently, the minterm DNF expression of f.
Output: A prime $||\phi||$-minimizing DNF of f.

$minT$ $||\phi||$-MINIMIZATION
Instance: The list of prime implicants of a Boolean function f or, equivalently, the complete DNF of f.
Output: A prime $||\phi||$-minimizing DNF of f.

$||\phi||$-MINIMIZATION
Instance: An arbitrary DNF expression of a Boolean function f.
Output: A prime $||\phi||$-minimizing DNF of f.

3.3.1 Quine-McCluskey approach: Logic minimization as set covering

We first present a fundamental result due to McCluskey [633] and to Quine [766], which will allow us to reformulate logic minimization problems as set covering problems.

Let us assume that a Boolean function $f(x_1, x_2, \ldots, x_n)$ is represented by the set $T(f)$ of its true points. As shown in Section 3.2.1 (Theorem 3.4), all the prime implicants $C_1, C_2, \ldots, C_k$ of $f(x_1, x_2, \ldots, x_n)$ can be generated in time polynomial in $n|T(f)|$.

Let us associate a $0, 1$-variable s_i with each of the prime implicants C_i, $i = 1, 2, \ldots, k$: The interpretation of these variables will be that $s_i = 1$ if C_i is retained in the construction of a DNF on the collection of terms $\{C_1, C_2, \ldots, C_k\}$. More formally, every Boolean point $S = (s_1, s_2, \ldots, s_k) \in \mathcal{B}^k$ defines a DNF

$$\phi_S(x_1, x_2, \ldots, x_n) = \bigvee_{i: s_i=1} C_i = \bigvee_{i=1}^{k} s_i C_i. \tag{3.2}$$

Since $\bigvee_{i=1}^{k} C_i$ is the complete DNF of the Boolean function f, for every Boolean vector S we have:

$$\phi_S(x_1, x_2, \ldots, x_n) \leq f(x_1, \ldots, x_n). \tag{3.3}$$

Clearly, every prime DNF of f corresponds to a vector S for which the inequality (3.3) holds as an equality and, conversely, every vector S for which (3.3) becomes an equality defines a prime DNF of f. It follows from (3.3) that to characterize those vectors S that correspond to the prime DNFs of f, it is sufficient to characterize those S for which the reverse inequality

$$\phi_S(x_1, x_2, \ldots, x_n) \geq f(x_1, \ldots, x_n) \tag{3.4}$$

also holds. Moreover, the inequality (3.4) can be reformulated as a system of $|T(f)|$ linear inequalities in the variables s_i, $i = 1, \ldots, k$:

$$\sum_{i=1}^{k} s_i C_i(X) \geq 1, \quad \text{for all } X \in T(f). \tag{3.5}$$

We now consider the Boolean function $\pi(s_1, s_2, \ldots, s_k)$ that takes value 1 exactly on those points $S = (s_1, s_2, \ldots, s_k)$ for which the system of inequalities (3.5) holds or, equivalently, on those points S for which (3.2) defines a prime DNF of f. This function is known in the literature as the *Petrick function* associated with f (see [744]). A CNF representation of the Petrick function follows directly from (3.5):

$$\pi(s_1, \ldots, s_k) = \bigwedge_{X \in T(f)} \left(\bigvee_{i=1}^{k} s_i C_i(X) \right). \tag{3.6}$$

This CNF representation clearly shows that the Petrick function is positive, a fact that can also be easily derived from its definition.

By definition of the Petrick function, there is a one-to-one correspondence between its positive implicants and the prime DNFs of the function f. Furthermore, one can easily see that there is a one-to-one correspondence between the prime implicants of the Petrick function and the prime irredundant DNFs of the function f. But of course, in general, the computational complexity of generating all the prime implicants of the Petrick function is prohibitively expensive.

In view of the preceding discussion, the problem of finding a $||\phi||$-minimizing DNF can be formulated as the problem of finding a minimum degree prime implicant of the Petrick function. Alternatively, the same problem can be formulated as the *set covering problem*

$$\text{minimize} \sum_{i=1}^{k} s_i \tag{3.7}$$

$$\text{subject to (3.5) and } (s_1, s_2, \ldots, s_k) \in \mathcal{B}^k. \tag{3.8}$$

Similarly, the problem of finding a $|\phi|$-minimizing DNF can be formulated as the *weighted set covering problem*

$$\text{minimize} \sum_{i=1}^{k} \deg(C_i) s_i \tag{3.9}$$

$$\text{subject to (3.5) and } (s_1, s_2, \ldots, s_k) \in \mathcal{B}^k \tag{3.10}$$

(where $\deg(C_i)$ denotes the degree of C_i, i.e., the number of literals in C_i).

Example 3.5. *Consider again the Boolean function* $f(x_1,x_2,x_3)$ *of Example 3.4, represented this time by its set of true points*

$$T(f) = \{(1,0,0),(0,1,0),(0,0,1),(1,1,0),(1,0,1),(0,1,1)\}.$$

Using the algorithm described in Section 3.2.1, we generate all the prime implicants of this function: $x_1\overline{x}_2,\ \overline{x}_1x_2,\ x_1\overline{x}_3,\ \overline{x}_1x_3,\ x_2\overline{x}_3,\ \overline{x}_2x_3$. *Associating with these prime implicants the binary variables* $s_1,s_2,\ldots,s_6$, *respectively, we can write the CNF (3.6) of the Petrick function as*

$$\pi(s_1,\ldots,s_6) = (s_1 \vee s_3)(s_2 \vee s_5)(s_4 \vee s_6)(s_3 \vee s_5)(s_1 \vee s_6)(s_2 \vee s_4).$$

The complete DNF of the Petrick function is obtained by dualization of this CNF (see Section 3.2.4):

$$\pi(s_1,\ldots,s_6) = s_1s_4s_5 \vee s_2s_3s_6 \vee s_1s_2s_3s_4 \vee s_1s_2s_5s_6 \vee s_3s_4s_5s_6.$$

From this DNF, we conclude that the function $f(x_1,x_2,x_3)$ *has five prime irredundant DNFs, two of them consisting of three prime implicants each, and three others consisting of four prime implicants each.*

   The problem of finding a $\|\phi\|$-*minimizing DNF of* f *(without necessarily listing all its prime irredundant DNFs) can be formulated as the following set covering problem:*

$$
\begin{array}{lllllll}
\text{minimize} & s_1+ & s_2+ & s_3+ & s_4+ & s_5+ & s_6 \\[2mm]
\text{subject to} & s_1+ & & s_3 & & & \geq 1 \\
& & s_2+ & & & s_5 & \geq 1 \\
& & & & s_4+ & & s_6 \geq 1 \\
& & & s_3+ & & s_5 & \geq 1 \\
& s_1+ & & & & s_6 & \geq 1 \\
& & s_2+ & & s_4 & & \geq 1
\end{array}
\qquad (3.11)
$$

$$s_i \in \{0,1\},\ i = 1,\ldots,6.$$

For this small example, it can be easily checked that the optimal solutions of the set covering problem (3.11) are $(1,0,0,1,1,0)$ *and* $(0,1,1,0,0,1)$, *corresponding to the two* $\|\phi\|$-*minimizing DNFs of* f:

$$\phi' = x_1\overline{x}_2 \vee \overline{x}_1x_3 \vee x_2\overline{x}_3,$$

and

$$\phi'' = \overline{x}_1x_2 \vee x_1\overline{x}_3 \vee \overline{x}_2x_3.$$

Since, in this example, all the prime implicants of f *have the same degree, its* $\|\phi\|$-*minimizing DNFs and* $|\phi|$-*minimizing DNFs coincide.* $\square$

   It follows from Section 3.2.1 that, given the set of true points of a Boolean function, the set covering formulation of the logic minimization problem can be constructed in polynomial time. However, it is well-known that the set covering problem is NP-hard; therefore, this approach to logic minimization does not necessarily provide a polynomial algorithm. In fact, it will be seen later in this chapter that the problem of logic minimization is intractable in general. Moreover, if a Boolean function is represented by an arbitrary DNF, or even by its complete DNF, then just the construction of the set covering formulation of the logic minimization problem can in itself be computationally difficult because of the possibly exponential number of true points.

3.3.2 Local simplifications of DNFs

A main challenge of the logic minimization problem stems from the fact that the same Boolean function can be represented by numerous DNFs of varying lengths, even if we restrict our attention only to prime and irredundant DNFs. While the set of prime implicants of a Boolean function is unique, the subsets of the prime implicants used in two distinct DNF representations of the same function can be quite different, and, as Example 3.5 shows, these subsets can even be disjoint. On the other hand, some of the prime implicants of a Boolean function can exhibit a consistent pattern of behavior regarding their participation in the prime and irredundant DNFs of the function, as illustrated in the following example.

Example 3.6. *Let us consider the Boolean function f whose set of prime implicants is*

$$\Pi(f) = \{x\overline{y}, \overline{x}y, xu, yu, \overline{u}w, xw, yw\}.$$

It can be verified that this function has exactly two prime and irredundant DNFs:

$$\phi_1 = x\overline{y} \vee \overline{x}y \vee xu \vee \overline{u}w$$

and

$$\phi_2 = x\overline{y} \vee \overline{x}y \vee yu \vee \overline{u}w.$$

Notice that the prime implicants $x\overline{y}$, $\overline{x}y$, and $\overline{u}w$ appear in all prime and irredundant DNFs of f, while the prime implicants xw and yw do not appear in any prime and irredundant DNFs of f.   □

In view of this example, let us introduce the following concepts (see Quine [769], Pyne and McCluskey [764]).

Definition 3.3. *A prime implicant of a Boolean function f is called* essential *if it appears in every prime DNF of f. A prime implicant of f is called* redundant *if it does not appear in any prime and irredundant DNF of f.*

In the foregoing example, $x\overline{y}$, $\overline{x}y$, and $\overline{u}w$ are essential prime implicants, while xw and yw are redundant prime implicants. The prime implicant xu (as well as yu) is neither essential nor redundant, since there exists a prime and irredundant DNF in which it appears, and another one in which it does not.

Clearly, the knowledge of essential and redundant prime implicants is very useful in solving logic minimization problems. As we will see, if a Boolean function is represented by the set of its true points, the detection of essential and redundant prime implicants can be carried out without major computational difficulties. To do so, we first return to the set covering formulation (3.5) of logic minimization.

With the set of linear inequalities (3.5), let us associate a $(0,1)$-matrix A with $|T(f)|$ rows and k columns. If $X^1, X^2, \ldots, X^{|T(f)|}$ are the true points of f, then the elements of A are defined as

$$a_{ji} = C_i(X^j) \text{ for } j = 1,2,\ldots,|T(f)| \text{ and } i = 1,2,\ldots,k. \qquad (3.12)$$

The rows of this matrix correspond to the true points of f and will be denoted by a_j, $j = 1,2,\ldots,|T(f)|$. The columns of the matrix correspond to the prime implicants of f and will be denoted by a^i, $i = 1,2,\ldots,k$.

Let us say that a $(0,1)$-point $S = (s_1, s_2, \ldots, s_k)$ satisfying the system of inequalities (3.5) is a *minimal solution* of (3.5) if no point obtained by changing any of the components of S from 1 to 0 also satisfies (3.5). We now discuss three computationally easy transformations which can be used to simplify the system of set covering inequalities (3.5), while preserving all its minimal solutions (the presentation is ours, but we refer to Gimpel [382], Pyne and McCluskey [764, 765], or Zhuravlev [937] for early references on this topic).

S1 If the matrix A contains a row a_{j*} with a single component, say i^*, equal to 1 (that is, $a_{j*i*} = 1$, and $a_{j*i} = 0$ for all $i \neq i^*$), then fix $s_{i*} = 1$ and remove from the matrix A the column a^{i^*} and all the rows a_j having $a_{ji*} = 1$.

S2 If the matrix A contains two comparable rows, say $a_{j'}$ and $a_{j''}$, such that $a_{j'} \leq a_{j''}$ (i.e., $a_{j'i} \leq a_{j''i}$ for every i), then remove the row $a_{j''}$ from A.

S3 If the matrix A contains a column a^{i^*} consisting only of 0 components, then fix $s_{i*} = 0$ and remove the column a^{i^*} from A.

It can be seen easily that the three simplifications S1, S2, and S3 preserve the set of minimal solutions of the set covering inequalities (3.5). Therefore, one can simplify (3.5) by repeatedly applying S1, S2, and S3 in an arbitrary order, for as long as possible. Let us denote the resulting matrix by $\widehat{A}$, the set of variables s that are fixed at 1 by $\widehat{S}_1$, and the set of variables s that are fixed at 0 by $\widehat{S}_0$. One would expect the matrix $\widehat{A}$ and the sets $\widehat{S}_1$ and $\widehat{S}_0$ to depend on the particular order in which the simplifications were applied. To avoid ambiguity, let us now specify an algorithm that first applies S1 for as long as possible, then applies S2 for as long as possible, and finally applies S3 for as long as possible. We shall call this algorithm the *essential reduction algorithm* (ERA). Let us denote the resulting matrix by A^*, and the set of variables which are fixed at 1 (respectively, 0) by S_1^* (respectively, S_0^*).

Theorem 3.10. *The end result of applying simplifications S1, S2, and S3 as long as possible does not depend on the order of their application: Every possible order always yields $\widehat{A} = A^*$, $\widehat{S}_1 = S_1^*$, and $\widehat{S}_0 = S_0^*$.*

Proof. The proof follows from three simple observations. First, let us observe that if an intermediate matrix A' contains a row with a single 1 component, then that row cannot contain more than one 1 in the original matrix A. Indeed, if a column was removed during the simplification process, then either this column had no 1's, and therefore its removal did not affect the number of 1's in the remaining rows, or all the rows in which this column had a 1 were also removed at the same step. Therefore, $\widehat{S}_1 = S_1^*$.

Second, an intermediate matrix A' contains two comparable rows if and only if these two rows are also comparable in the original matrix A. This is a direct consequence of the fact that none of the simplification steps S1, S2, or S3 affects the comparability of the remaining rows. It follows, then, from the foregoing two observations that the sets of rows of $\widehat{A}$ and A^* are exactly the same.

Third, the set of columns of $\widehat{A}$ and A^* consists exactly of those columns of the original matrix that have at least one 1 component in the remaining rows. Indeed, on the one hand, neither of the matrices contains a column consisting only of 0's. On the other hand, if a removed column did have some 1's, then all the rows, in which it had 1's, were also removed. In conclusion, the set of remaining columns is uniquely determined by the set of remaining rows. Since the sets of rows of $\widehat{A}$ and A^* coincide, we have $\widehat{A} = A^*$. It follows that exactly the same sets of variables were fixed in both procedures, and since we have already concluded that $\widehat{S}_1 = S_1^*$, we can now conclude that $\widehat{S}_0 = S_0^*$.   □

Lemma 3.9. *For every variable s of the system of set covering inequalities (3.5), s is not fixed by ERA if and only if there exists a minimal solution of (3.5) in which s = 1 and a minimal solution of (3.5) in which s = 0.*

Proof. The "if" part follows from the fact that the simplifications S1, S2, and S3 preserve all the minimal solutions.

We now prove the "only if" part. Let s_{i*} be a variable that is not fixed by ERA. On the one hand, since every row of A^* has at least two 1's, we can set $s_{i*} = 0$, and the problem will remain feasible, showing that there must exist a minimal solution of (3.5) in which $s_{i*} = 0$. On the other hand, since A^* has no columns consisting of all 0's, there must exist a row j^* in A^* such that $a_{j*i*} = 1$. Let us now set $s_i = 0$ for every $i \neq i^*$ such that $a_{j*i} = 1$. Since A^* has no comparable rows, the set covering system remains feasible, and in every solution of this reduced set covering system (including the minimal ones), s_{i*} must be equal to 1 because there is no other way to satisfy the inequality corresponding to row j^*.   □

Since prime and irredundant DNFs of a Boolean function are in one-to-one correspondence with the minimal solutions of the set covering inequalities (3.5), Lemma 3.9 implies the following characterizations of the essential and redundant prime implicants:

Theorem 3.11. *A prime implicant of a Boolean function is essential if and only if the corresponding variable s is fixed at 1 by ERA.*

Theorem 3.12. *A prime implicant of a Boolean function is redundant if and only if the corresponding variable s is fixed at 0 by ERA.*

It is important to observe that, from a computational point of view, ERA is relatively inexpensive. More specifically, given a $|T(f)| \times k$ set covering matrix A, the ERA simplifications can be carried out in $O(k|T(f)|^2)$ time. Indeed, all the simplifications S1 and S3 can be done in $O(k|T(f)|)$ time, since each of these two types of simplifications requires a single pass over the set covering matrix. Additionally, to carry out all the simplifications S2, one has to compare at most $\binom{|T(f)|}{2}$ pairs of rows, and each comparison can be done in $O(k)$ time.

Example 3.7. *Let us consider the set covering matrix A associated with the logic minimization problem for the Boolean function f given in Example 3.6:*

(x,y,u,w)	$x\overline{y}$	$\overline{x}y$	xu	yu	$\overline{u}w$	xw	yw
$(0,0,0,1)$	0	0	0	0	1	0	0
$(0,1,0,0)$	0	1	0	0	0	0	0
$(1,0,0,0)$	1	0	0	0	0	0	0
$(0,1,0,1)$	0	1	0	0	1	0	1
$(1,0,0,1)$	1	0	0	0	1	1	0
$(0,1,1,0)$	0	1	0	1	0	0	0
$(1,0,1,0)$	1	0	1	0	0	0	0
$(0,1,1,1)$	0	1	0	1	0	0	1
$(1,0,1,1)$	1	0	1	0	0	1	0
$(1,1,0,1)$	0	0	0	0	1	1	1
$(1,1,1,0)$	0	0	1	1	0	0	0
$(1,1,1,1)$	0	0	1	1	0	1	1

The twelve rows of this matrix correspond to the true points of the function, while the seven columns correspond to its prime implicants. Three applications of the simplification S1 show that $x\overline{y}$, $\overline{x}y$, and $\overline{u}w$ are essential prime implicants. The resulting simplified set covering matrix is

(x,y,u,w)	xu	yu	xw	yw
$(1,1,1,0)$	1	1	0	0
$(1,1,1,1)$	1	1	1	1

Applying now S2, the matrix reduces to

(x,y,u,w)	xu	yu	xw	yw
$(1,1,1,0)$	1	1	0	0

Finally, two applications of the simplification S3 show that xw and yw are redundant prime implicants. The set covering matrix of the remaining problem is

(x,y,u,w)	xu	yu
$(1,1,1,0)$	1	1

showing that every prime and irredundant DNF contains either xu or yu, but not both. These results confirm the statements made in Example 3.6. □

We have seen that the simplifications S1, S2, and S3, and therefore ERA, preserve the minimal solutions of the system of set covering inequalities (3.5); namely, they preserve the set of prime and irredundant DNFs. This property allows the application of S1, S2, and S3, and of ERA, to any type of logic minimization problem whose objective is to minimize the number of terms, the number of literals, or any monotonically increasing function of these two DNF complexity measures.

Let us now turn our attention to another type of simplifying transformation, which has a more limited scope of application, since it may not preserve all the minimal solutions of the set covering inequalities (3.5).

S4 If the matrix A contains two comparable columns, say, $a^{i'}$ and $a^{i''}$, such that $a^{i'} \geq a^{i''}$ (i.e., $a_{ji'} \geq a_{ji''}$ for every j), then fix $s_{i''} = 0$ and remove the column $a^{i''}$ from A.

Note that the simplification S3 introduced earlier is a special case of S4. The simplification S4 is guaranteed to preserve at least one minimum-cardinality solution of the system (3.5), namely, one optimal solution of the set covering problem (3.7)–(3.8). Indeed, a single application of S4 reduces the current set of minimal solutions in such a way that only those minimal solutions in which $s_{i''} = 0$ are preserved. Further, since $a^{i'} \geq a^{i''}$, if there is a current minimum-cardinality solution S with $s_{i''} = 1$, then the point S^*, which is equal to S in all components, except $s^*_{i''} = 0$ and $s^*_{i'} = 1$, is also a minimum-cardinality solution and is preserved by S4.

It is now clear that S4 can be applied to simplify those logic minimization problems whose objective is to find at least one minimum solution of the set covering problem (3.7)–(3.8), that is, to find a $||\phi||$-minimizing prime DNF.

Note that the simplification process can never start with S4 because, in our logic minimization problems, the initial set covering matrix A does not contain comparable columns (because no prime implicant is absorbed by another one). Nevertheless, S4 may become applicable after several applications of simplifications S1 or S2. On the other hand, the opposite phenomenon can also happen; namely, it is possible that neither S1 nor S2 is applicable but S4 is, and after several applications of S4 it may become possible to apply S1 or S2. Therefore, further simplifications can be achieved by alternatively applying ERA and S4 as long as possible.

3.3.3 Computational complexity of logic minimization

It was seen in the previous subsection that logic minimization problems can be reduced to set covering problems. This reduction makes it possible to solve logic minimization problems by generic set covering algorithms. Note, however, that the use of such generic algorithms may not be most appropriate for solving the resulting set covering problems if they appeared to possess some special properties that would allow the development of specialized, more efficient algorithms.

At first glance, set covering problems arising from logic minimization problems do display some special features. For example, since each column of the set covering matrix corresponds to a prime implicant (i.e., a subcube), the number of 1's in the column must be a power of 2. Similarly, the number of 1's in the intersection of any subset of columns must also be a power of 2.

In view of such special features, formally, not every set covering problem originates from logic minimization. Therefore, it comes as a surprise that every (nontrivial) set covering problem is, in fact, an S1-simplified version of a logic minimization problem. More precisely, given an arbitrary set covering problem without zero rows or columns, there exists a logic minimization problem, which – after several applications of the simplification S1 – reduces to it. This subsection is devoted to a proof of this result and its corollaries.

Let us consider the system of set covering inequalities

$$\sum_{i=1}^{m} a_{ji} s_i \geq 1, \quad j = 1, 2, \ldots, n \tag{3.13}$$

and the corresponding matrix $A = (a_{ji})_{j=1,\ldots,n}^{i=1,\ldots,m}$, which we assume to have no zero rows or columns. The construction of a logic minimization problem reducible to the given set covering system involves two steps. First, we construct a set of Boolean points and a set of terms in such a way that the given matrix A represents the associated set covering conditions. At this stage, the logic minimization problem is not completely defined because the constructed terms can also cover other Boolean points, besides the constructed ones. Then we extend the construction by adding some special terms and the corresponding true points. It will be shown that in the completely defined logic minimization problem constructed in this way, the terms added at the second stage are essential prime implicants. Moreover, the set covering inequalities associated with this logic minimization problem will be shown to be reducible (by using the simplification S1) to the originally given set covering system (3.13). Let us now describe the details of the construction.

As a first step, let us associate with each row j of A a Boolean point of dimension n, denoted $P^j = (p_1^j, p_2^j, \ldots, p_n^j)$, where, for $j, r = 1, 2, \ldots, n$,

$$p_r^j = \begin{cases} 1 & \text{if } r \neq j, \\ 0 & \text{if } r = j. \end{cases} \tag{3.14}$$

Also, with each column i of A, $i = 1, 2, \ldots, m$, let us associate an elementary conjunction C_i on variables from $\{x_1, x_2, \ldots, x_n\}$:

$$C_i = \bigwedge_{j: a_{ji}=0} x_j. \tag{3.15}$$

Example 3.8. *As a small example, let us consider the following set covering matrix:*

$$A = \begin{pmatrix} 1 & 0 & 1 \\ 1 & 1 & 0 \\ 0 & 0 & 1 \end{pmatrix}.$$

Then, the points associated to its rows are

$$\begin{pmatrix} P^1 \\ P^2 \\ P^3 \end{pmatrix} = \begin{pmatrix} 0 & 1 & 1 \\ 1 & 0 & 1 \\ 1 & 1 & 0 \end{pmatrix},$$

while the terms associated to its columns are

$$(C_1, C_2, C_3) = (x_3, x_1 x_3, x_2).$$

□

Lemma 3.10. *For every matrix $A \in \mathcal{B}^{n \times m}$ without zero rows, there holds*

(a) *for all $j = 1, 2, \ldots, n$ and $i = 1, 2, \ldots, m$, $a_{ji} = C_i(P^j)$;*
(b) *for all $j = 1, 2, \ldots, n$, P^j is a true point of the function represented by the DNF $\bigvee_{i=1}^m C_i$.*

Proof. To establish (a), notice that, by construction of C_i, $C_i(P^j) = 0$ if and only if there exists an index k such that $a_{ki} = 0$ and $P_k^j = 0$. But by definition of P^j, this is equivalent to $k = j$ and $a_{ji} = 0$.

To prove assertion (b), simply note that, for each P^j, there is at least one conjunction C_i such that $C_i(P^j) = 1$ (since A has no zero row).   □

Lemma 3.10 suggests that A comes close to being the matrix associated with a logic minimization problem, because it expresses the covering of the true points P^j by the terms C_i. However, as the Example 3.8 shows, absorption may possibly take place among the conjunctions C_i (indeed, x_3 absorbs $x_1 x_3$ in the example). Therefore, the construction has to be modified if we want the conjunctions C_i to represent the prime implicants of some Boolean function.

Let us call a column $a^{i'}$ of A *dominating* if there exists another column $a^{i''}$ in A such that $a^{i'} \geq a^{i''}$, and let us redefine the associated conjunctions C_i by

$$C_i := \begin{cases} C_i, & \text{if } a^i \text{ is not dominating,} \\ C_i y_i, & \text{if } a^i \text{ is dominating,} \end{cases} \tag{3.16}$$

where the y_i's represent additional Boolean variables. Obviously, after this transformation, there will be no absorption among the conjunctions C_i. In order to complete the construction, we shall extend the associated vectors P^j by adding additional components for each of the additional variables y_i, and defining the value of all these components to be 1. This modification preserves the property that A expresses the covering of the points P^j's by the conjunctions C_i's.

Example 3.9. *Returning to our Example 3.8, we find now:*

$$(C_1, C_2, C_3) = (x_3 y_1, x_1 x_3, x_2),$$

and

$$\begin{pmatrix} P^1 \\ P^2 \\ P^3 \end{pmatrix} = \begin{pmatrix} 0 & 1 & 1 & 1 \\ 1 & 0 & 1 & 1 \\ 1 & 1 & 0 & 1 \end{pmatrix}.$$

□

To define a logic minimization problem equivalent to the original set covering problem, we construct the DNF

$$\psi = \bigvee_{i=1}^{m} C_i, \qquad (3.17)$$

where the terms C_i are defined by (3.15) and (3.16). Note that ψ represents a positive Boolean function, say, f, and, since ψ is closed under absorption, it is the complete DNF of f. The true points of f include all the points P^j, $j = 1, 2, \ldots, n$ but can also include many additional points, say, Q^t, $t = 1, 2, \ldots, T$.

If we simply extend the set covering problem by adding to A all the rows corresponding to the additional true points Q^t, then the resulting matrix may not necessarily be reducible to A by using the simplifications S1, S2, and S3. To make this reduction possible, we introduce two additional variables, z_0 and z_1. For any Boolean point Q, let us denote by $[Q]$ the unique minterm (in the (x, y)-variables) covering Q, and let us say that Q is "even" (respectively, "odd") if it has an even (respectively, odd) number of components equal to 1. We can now define the DNF:

$$\psi^* = \bigvee_{i=1}^{m} z_0 z_1 C_i \vee \bigvee_{t:\, Q^t \text{ is even}} z_0 [Q^t] \vee \bigvee_{t:\, Q^t \text{ is odd}} z_1 [Q^t]. \qquad (3.18)$$

We let f^* be the Boolean function represented by ψ^*.

Example 3.10. *For the Example 3.9, there are eight additional true points Q^t:*

$$\begin{pmatrix} Q^1 \\ Q^2 \\ Q^3 \\ Q^4 \\ Q^5 \\ Q^6 \\ Q^7 \\ Q^8 \end{pmatrix} = \begin{pmatrix} 1 & 1 & 1 & 1 \\ 1 & 1 & 1 & 0 \\ 1 & 1 & 0 & 0 \\ 1 & 0 & 1 & 0 \\ 0 & 1 & 1 & 0 \\ 0 & 1 & 0 & 1 \\ 0 & 0 & 1 & 1 \\ 0 & 1 & 0 & 0 \end{pmatrix}.$$

The associated DNF is

$$\psi^* = x_3 y_1 z_0 z_1 \vee x_1 x_3 z_0 z_1 \vee x_2 z_0 z_1 \vee x_1 x_2 x_3 y_1 z_0 \vee x_1 x_2 \overline{x}_3 \overline{y}_1 z_0 \vee x_1 \overline{x}_2 x_3 \overline{y}_1 z_0 \vee$$

$$\overline{x}_1 x_2 x_3 \overline{y}_1 z_0 \vee \overline{x}_1 x_2 \overline{x}_3 y_1 z_0 \vee \overline{x}_1 \overline{x}_2 x_3 y_1 z_0 \vee x_1 x_2 x_3 \overline{y}_1 z_1 \vee \overline{x}_1 x_2 \overline{x}_3 \overline{y}_1 z_1.$$

□

Lemma 3.11. *The complete DNF of f^* is ψ^*.*

Proof. Let us write ψ^* as

$$\psi^* = z_0 z_1 \psi \vee z_0 \psi_0 \vee z_1 \psi_1.$$

By construction, no two terms of ψ absorb each other. The same holds for the terms of ψ_0 and ψ_1. Moreover, it is obvious that no term of $z_0 \psi_0$ can absorb a term of $z_1 \psi_1$, and vice versa. It is also obvious that no term of $z_0 z_1 \psi$ can absorb any term of $z_0 \psi_0$ or $z_1 \psi_1$. Since A has no zero columns, no term of ψ is a minterm on the (x, y)-variables. Then, since every term of ψ_0 and of ψ_1 is a minterm, no term of $z_0 \psi_0$ or $z_1 \psi_1$ can absorb a term of $z_0 z_1 \psi$. Thus, ψ^* is closed under absorption.

Let us now prove that ψ^* is closed under consensus. Obviously, no two terms of $z_0 z_1 \psi$ have a consensus because they are all positive. Moreover, any two terms of ψ_0 have at least two conflicting literals, and hence no two terms of $z_0 \psi_0$ have a consensus. For the same reason, no two terms of $z_1 \psi_1$ have a consensus.

Let us now assume that a term of $z_0 z_1 \psi$, say $z_0 z_1 C$, and a term of $z_0 \psi_0$, say $z_0[Q]$, have a consensus. This can only happen if there is a variable w in C such that $\overline{w}$ appears in $[Q]$. Since Q is a true point of f, there exists a prime implicant of f, say, C' that absorbs $[Q]$ and obviously does not contain $\overline{w}$. Then, $z_0 z_1 C'$ is a term of $z_0 z_1 \psi$ that absorbs the consensus of $z_0 z_1 C$ and $z_0[Q]$. Similarly, every consensus of a term in $z_0 z_1 \psi$ and a term in $z_1 \psi_1$ will be absorbed by a term in $z_0 z_1 \psi$.

Let us next assume that a term of $z_0 \psi_0$, say, $z_0[Q']$, and a term of $z_1 \psi_1$, say $z_1[Q'']$, have a consensus. Without loss of generality, let us assume that $[Q'] = wG$ and $[Q''] = \overline{w}H$. Again, there exists a prime implicant of f, say, C that absorbs $[Q'']$ and that does not contain $\overline{w}$. Then, $z_0 z_1 C$ is a term of $z_0 z_1 \psi$ that absorbs the consensus of $z_0[Q']$ and $z_1[Q'']$.

Thus, ψ^* is closed under consensus and, in view of Corollary 3.2, ψ^* is the complete DNF of f^*.   □

We now discuss the logic minimization problem for f^*. By Lemma 3.11, the columns of the set covering matrix A^* associated with f^* correspond to the terms of ψ^*. The rows of A^* correspond to the set of true points $T(f^*)$. These true points are derived from the points P^j, $j = 1, 2, \ldots, n$ and Q^t, $t = 1, 2, \ldots, T$ by extending them with two additional components, corresponding to z_0 and z_1, so that $T(f^*)$ consists of the following disjoint subsets:

- The set $\mathcal{P}$ of points $(P^j, 1, 1)$, $j = 1, 2, \ldots, n$.
- The set $\mathcal{Q}_{11}$ of points $(Q^t, 1, 1)$, $t = 1, 2, \ldots, T$.
- The set $\mathcal{Q}_{10}$ of points $(Q^t, 1, 0)$, where Q^t is even.
- The set $\mathcal{Q}_{01}$ of points $(Q^t, 0, 1)$, where Q^t is odd.

Let us see what happens when the simplification steps S1 are performed on A^*. Every true point of the form $(Q^t, \overline{\sigma}, \sigma)$ $(t = 1, \ldots, T; \sigma = 0, 1)$ is covered by a single prime implicant $z_\sigma[Q^t]$. Therefore, every prime implicant $z_\sigma[Q^t]$ is essential, and the application of the simplification S1 removes the corresponding columns and all the rows in $\mathcal{Q}_{10} \cup \mathcal{Q}_{01}$ from the set covering matrix A^*. Moreover, the rows in $\mathcal{Q}_{11}$ are also removed by S1, since every prime implicant of the form $z_\sigma[Q^t]$ covers $(Q^t, 1, 1)$.

So, the application of S1 only leaves in A^* the rows associated with the true points $(P^j, 1, 1)$, $j = 1, 2, \ldots, n$, and the columns associated with the prime implicants of the form $z_0 z_1 C_i$, $i = 1, 2, \ldots, m$. It now follows from Lemma 3.10 that this reduced set covering matrix coincides with the original matrix A. This completes the proof of the following result, due to Gimpel [381]:

Theorem 3.13. *Given an arbitrary set covering problem without zero rows or columns, there exists a logic minimization problem whose set covering formulation can be reduced to the given problem after several applications of the simplification S1.*

The foregoing arguments show how to transform an arbitrary set covering problem of size $n \times m$ to an equivalent logic minimization problem having at most $n + m + 1$ Boolean variables. Since it is well known that the set covering problem is NP-hard [371], one may be tempted to interpret this construction as an NP-hardness proof for the logic minimization problem. Unfortunately, this inference is incorrect, since the described reduction is not necessarily polynomial because the number of true points of f^* constructed above can be exponentially large in n and m. However, this difficulty is easy to overcome, and we can establish the following result (Gimpel [381]):

Theorem 3.14. *The logic minimization problem is NP-hard when its input is a Boolean function given by the set of its true points.*

Proof. It is known [371] that the set covering problem remains NP-hard in the special case in which every column of the set covering matrix contains at most three 1's, and the matrix does not contain any pair of comparable columns. In this case, because of the incomparability of the columns, no variable y is needed in the construction (3.17) of the DNF ψ. Moreover, the degree of every conjunction C_i is at least $n - 3$, hence the number T of the additional true points Q^t of f is at most $8m$. Therefore, the number of prime implicants of the Boolean function f^* is at most $9m$, and the DNF ψ^* can be constructed in polynomial time. It follows that, for this special case of set covering problems, the transformation to an equivalent logic minimization problem is polynomial, which completes the proof. $\square$

Theorem 3.14 describes the complexity of the logic minimization problem in the most commonly considered case, where the Boolean function is given by the set of its true points. However, there are many other ways to represent a Boolean function, for example, by an arbitrary DNF or CNF, by a complete truth table containing the value of the function in all the 2^n Boolean points, or by the set of its false points, and so on. It is important to note that the computational complexity of the logic minimization problem can depend on the representation of the input, since different representations of the same Boolean function are not polynomially equivalent; namely, the length of one representation may not necessarily be limited by a polynomial function of the length of another one.

Some representations of a Boolean function can be viewed as special cases of others. For example, the representation of a Boolean function by the set of its true points can be viewed as a special type of DNF representation. Thus, in particular, Theorem 3.14 implies that the logic minimization problem is NP-hard for Boolean functions expressed in DNF.

On the other hand, the representation of a Boolean function by the set of its true points can be exponentially shorter than its representation by a complete truth table. It is therefore surprising that the latter, possibly much larger, representation does not make the logic minimization problem significantly simpler. As a matter of fact, Masek [674] was able to prove that the logic minimization problem remains NP-hard when its input is a complete truth table. A more accessible proof (based on Gimpel's construction [381]) of the latter result was recently proposed by Allender et al. [16].

3.3.4 *Efficient approximation algorithms for logic minimization*

We saw in the previous subsection that the logic minimization problem is computationally equivalent to the set covering problem. Because of the NP-hardness of these problems, it is widely believed that solving them to optimality may require exponential time. This explains the importance of developing efficient approximation algorithms for their solution.

One of the most natural approaches to the solution of many optimization problems is to use a "greedy" procedure, that is, an iterative process of which each step is aimed at reaping the maximum immediate benefit, without the heavy computational expense required to analyze its global impact. The general philosophy of greedy procedures has found numerous implementations, often with excellent results.

We are going to describe in this subsection an efficient greedy procedure for solving logic minimization problems, derived from the associated set covering formulations with constraints (3.5).

The classical greedy procedure for a generic set covering problem of the form

$$\text{minimize} \quad \sum_{i=1}^{k} s_i$$

$$\text{subject to} \quad \sum_{i=1}^{k} a_{ji} s_i \geq 1, \quad j = 1, 2, \ldots, n,$$

$$(s_1, s_2, \ldots, s_k) \in \mathcal{B}^k,$$

is an iterative process at each step of which a variable s_i is chosen in such a way that setting this variable to 1 satisfies the largest possible number of yet unsatisfied constraints (3.5). When all the constraints are satisfied, the greedy procedure stops, and all those variables s_i, which have not been set to 1 in this process, are now set to 0.

Let us now describe this greedy procedure in terms of the set covering matrix $A = (a_{ji})_{j=1,\ldots,n}^{i=1,\ldots,k}$. Denote by $A(r)$ the reduced set covering matrix at the beginning of step r of the greedy procedure. Thus, $A(1)$ denotes the original set covering matrix A. At step r, the greedy procedure

1. calculates the number $|a(r)^i|$ of 1's in every column $a(r)^i$ of the matrix $A(r)$;
2. chooses a column $a(r)^{i_r}$ having the maximum number of 1's; and
3. reduces $A(r)$ to $A(r+1)$ by removing from $A(r)$ the chosen column $a(r)^{i_r}$ as well as all the rows $a(r)_j$ covered by it, namely, those with $a(r)_{j i_r} = 1$.

The process stops when all rows have been removed from the set covering matrix.

Let q be the number of steps of the greedy procedure. For simplicity, let us renumber the columns of the set covering matrix in such a way that the removed columns $i_1, i_2, \ldots, i_q$ become $1, 2, \ldots, q$; the remaining columns are numbered from $q+1$ to k. Let us denote by w_i^r the number $|a(r)^i|$ of 1's in the i-th column of $A(r)$. With this notation, w_r^r is the number of rows removed from the set covering matrix at step r of the greedy procedure. Note that w_1^1 is the maximum number of 1's in the columns of the original set covering matrix A (we assume, without loss of generality, that $w_1^1 \geq 1$).

Two important observations about the greedy procedure are in order. First, this procedure is very efficient. Indeed, if n is the number of rows and k is the number of columns in the set covering matrix, then the number of steps of the procedure does not exceed $\min\{n, k\}$, while each step takes $O(nk)$ time; therefore the computational complexity of the greedy procedure is $O(\min\{n, k\}nk)$. Second, despite its low computational cost, the greedy procedure produces very good solutions. To quantify this last statement, let us compare the size q of the greedy cover (that is, the number of variables fixed to 1 by the greedy procedure) with the size m of a minimum cover. Obviously, $m \leq q$. On the other hand, q cannot be "much worse" than the optimum, in view of the following surprising result:

Theorem 3.15. *For any set covering problem, if m is the size of a minimum cover, then the size q of the greedy cover is bounded by the relation*

$$q \leq H(w_1^1)m, \tag{3.19}$$

where w_1^1 is the maximum number of 1's in a column of the set covering matrix, and

$$H(d) = \sum_{i=1}^{d} \frac{1}{i} \quad \text{for all positive integers } d.$$

We refer to Chvátal [198], Johnson [536], or Lovász [623], for a proof of this classical result. It is easy to show (e.g., by induction) that $H(d) \leq 1 + \ln d$ for any positive integer d. Thus, Theorem 3.15 implies the following corollary (see also Slavík [837] for a slight improvement).

Corollary 3.7. *For any set covering problem, if m is the size of a minimum cover, then the size q of the greedy cover is bounded by the relation*

$$q \le (1 + \ln w_1^1)m \le (1 + \ln n)m, \tag{3.20}$$

where w_1^1 is the maximum number of 1's in a column of the set covering matrix, and n is the number of its rows.

Let us now consider the application of these approximation results to logic minimization. If a Boolean function is identically 1 and is represented by the set of its true points, then, obviously, the logic minimization problem is trivial. If a Boolean function is not identically 1, then each of its prime implicants covers at most 2^{n-1} points. Together with Corollary 3.7, this observation implies the following result.

Corollary 3.8. *Let f be a Boolean function of n variables, let q be the number of terms in its prime DNF constructed by the greedy procedure applied to the set covering formulation (3.7)–(3.8), and let m be the number of terms in a $\|\phi\|$-minimizing DNF of f. Then,*

$$q \le (1 - \ln 2 + n \ln 2)m. \tag{3.21}$$

We observed in Section 3.3.1 that, when the input is the set of true points of a Boolean function, the set covering formulation (3.7)–(3.8) of the logic minimization–problem can be constructed in polynomial time. Hence, the greedy procedure also runs in polynomial time on this input and provides a solution of the $\|\phi\|$-minimization problem that approximates its optimal value to within a factor $O(n)$.

A natural question to be asked now is whether there exists a polynomial time algorithm having a significantly better approximation ratio. In all likelihood, the answer to this question is negative. Indeed, Feldman [328] established the following result: *Even* when the input of the logic minimization problem consists of the complete truth table of a Boolean function f, there exists a constant $\gamma > 0$ such that it is NP-hard to approximate m to within a factor n^γ, where m is the number of terms in a $\|\phi\|$-minimizing DNF of f, and n is the number of variables. This result implies that the approximation factor achieved by the greedy algorithm is at most polynomially larger than the best ratio that can be achieved in polynomial time, unless P=NP. (When the input is an arbitrary DNF, Umans [876] proves stronger inapproximability results.)

Additionally, the following surprising fact was proved by Feige [324]: Under the assumption that NP-complete problems cannot be solved in $O(l^{O(\log \log l)})$ time (where l denotes the length of the input), it is shown in [324] that no polynomial time algorithm for the set covering problem can have an approximation ratio less than $(1 - o(1)) \ln \rho$. Since, by Corollary 3.7, the approximation ratio of the greedy procedure is $(1 + o(1)) \ln \rho$, the only remaining possibility is to improve the approximation ratio by a lower-order term $o(\ln \rho)$.

Chvátal [198] generalized Theorem 3.15 and Corollary 3.7 for the weighted version of the set covering problem in which nonnegative weights c_i are associated with the variables s_i, and the problem is the following:

$$\text{minimize} \quad \sum_{i=1}^{k} c_i s_i$$

$$\text{subject to} \quad \sum_{i=1}^{k} a_{ji} s_i \geq 1, \quad j = 1, 2, \ldots, n,$$

$$(s_1, s_2, \ldots, s_k) \in \mathcal{B}^k.$$

In this case, the generalized greedy procedure is defined in a similar way, the only difference being that at each iteration r a column $a(r)^i$ is chosen so as to maximize the ratio w_i^r / c_i of the number of 1's remaining in the column divided by its weight. The approximation results of Theorem 3.15 and Corollary 3.7 remain valid for this weighted set covering problem [198]. Therefore, if q_l is the number of literals in a prime DNF of an n-variable Boolean function f, constructed by the generalized greedy procedure applied to the set covering formulation (3.9)–(3.10), and if m_l is the number of literals in a $|\phi|$-minimizing DNF of f, then it follows, similarly to Corollary 3.8 that

$$q_l \leq (1 - \ln 2 + n \ln 2) m_l. \tag{3.22}$$

3.4 Extremal and typical parameter values

Several numerical parameters provide important information about Boolean functions and their DNFs. Typical examples of such parameters include the number of terms and the number of literals of a DNF, the degree of implicants, the number of irredundant and prime DNFs of a Boolean function, and so on. We discuss first several issues related to the number of prime implicants of a Boolean function.

3.4.1 Number of prime implicants

The number of different terms, or elementary conjunctions, in n Boolean variables equals 3^n, since each variable can be either present in uncomplemented form or present in complemented form, or can be absent in a term. We shall show that a Boolean function of n variables can have almost as many prime implicants as there are terms. We obtain this result by analyzing a special class of Boolean functions, called *symmetric*. The value of a symmetric function depends only on the number of 1's in the Boolean point where it is computed (see Exercise 5 in Chapter 1). An important subclass of symmetric functions consists of the so-called *belt* functions, denoted $b_n^{m,k}$ and defined by

$$b_n^{m,k}(x_1, \ldots, x_n) = \begin{cases} 1, & \text{if } m \leq \sum_{i=1}^{n} x_i \leq m+k, \\ 0, & \text{otherwise.} \end{cases}$$

Here m and k are nonnegative integers such that $m + k \leq n$.

Lemma 3.12. *A term $C = \bigwedge_{i \in P} x_i \bigwedge_{j \in N} \bar{x}_j$ is a prime implicant of a belt function $b_n^{m,k}$ if and only if $|P| = m$ and $|N| = n - m - k$.*

Proof. Clearly, every term with $|P| = m$ and $|N| = n - m - k$ is an implicant, since it covers only points whose number of 1's is between m and $m + k$. Every such implicant is prime, since removing any literal from the term will result in a term that covers a point with either fewer than m 1's or more than $m + k$ 1's.

On the other hand, an implicant of $b_n^{m,k}$ must have at least m positive and $n - m - k$ negative literals, and if an implicant has more than m positive or more than $n - m - k$ negative literals, then it is not prime, since the term that results from removing an extra literal will remain an implicant. □

Theorem 3.16. *There is a positive constant c such that, for every $n \geq 3$, there exists a Boolean function of n variables having at least $c \frac{3^n}{n}$ prime implicants.*

Proof. The statement holds for the belt function $b_n^{\frac{n}{3}, \frac{n}{3}}$. Indeed, it follows from Lemma 3.12 that the number of prime implicants of a belt function $b_n^{m,k}$ equals

$$\binom{n}{m}\binom{n-m}{n-m-k} = \frac{n!}{m!k!(n-m-k)!},$$

which, for $m = k = \frac{n}{3}$, equals

$$\frac{n!}{(\frac{n}{3}!)^3}. \tag{3.23}$$

Substituting into (3.23) the well-known Stirling formula (see, e.g., [314])

$$n! = \sqrt{2\pi n}(\frac{n}{e})^n(1 + o(1)),$$

one can see that there exists a positive constant c such that the number of prime implicants of $b_n^{\frac{n}{3}, \frac{n}{3}}$ is at least $c\frac{3^n}{n}$. □

The previous statement shows that the number of prime implicants of a Boolean function can be exponentially large in the number of Boolean variables. From the algorithmic point of view, it is also important to understand how large the number of prime implicants can be in terms of the length of an arbitrary DNF or CNF representation of a Boolean function. Interestingly, the number of prime implicants can be exponential in the length of a DNF, even for seemingly simple functions, as the following theorem shows.

Theorem 3.17. *For every integer $n \geq 1$, there exists a Boolean function f that has $2^n + 2n$ prime implicants and can be represented by a DNF having $2n + 1$ terms.*

Proof. Let f be the Boolean function represented by the DNF

$$\phi(x_1,\ldots,x_n,y_1,\ldots,y_n) = \left(\bigwedge_{i=1}^{n} x_i\right) \vee \bigvee_{i=1}^{n} (x_i \overline{y}_i \vee \overline{x}_i y_i),$$

which has $2n+1$ terms and can be easily seen to be prime.

If we apply Theorem 3.2 consecutively to every pair $\{x_i, y_i\}$, we can see that an elementary conjunction different from $x_i\overline{y}_i$ or $\overline{x}_i y_i$ is a prime implicant of f if and only if it has the form $\bigwedge_{i=1}^{n} u_i$, where each u_i is either x_i or y_i. Therefore, the number of prime implicants of f equals $2^n + 2n$. □

The argument in this proof can be easily modified (e.g., by adding to ϕ an additional linear term z) for the case of DNFs with an even number of terms.

Similarly, the number of prime implicants of a Boolean function can be exponentially large in the length of a CNF representation of the function.

Theorem 3.18. *For every integer $n \geq 1$, there exists a Boolean function f that has 3^n prime implicants and can be represented by a CNF having n clauses.*

Proof. Let us consider the positive function f of $3n$ variables represented by the CNF

$$\psi(x_1,\ldots,x_n,y_1,\ldots y_n,z_1,\ldots z_n) = \bigwedge_{i=1}^{n} (x_i \vee y_i \vee z_i).$$

This CNF has n clauses. It is clear from the CNF expression that, in each minimal true point of f, exactly one the variables x_i, y_i, and z_i equals 1, for every $i \in \{1, 2, \ldots, n\}$. Therefore, in view of Theorem 1.26, the complete DNF of f consists of elementary conjunctions of the form $\bigwedge_{i=1}^{n} u_i$, where each u_i is either x_i, y_i, or z_i. Hence, the function f has 3^n prime implicants. □

3.4.2 Extremal parameters of minimal DNFs

To better understand the nature of the logic minimization problem, we provide in this section some evaluations of the extremal values of a number of important DNF parameters.

We start our discussion with the analysis of the worst-case values of DNF parameters. Probably, the most interesting such parameter related to the logic minimization problem is the largest number of terms contained in a $\|\phi\|$-minimizing DNF of a Boolean function of n variables.

Theorem 3.19. *A $\|\phi\|$-minimizing DNF of a Boolean function of n variables cannot contain more than 2^{n-1} terms, and this number of terms can be attained.*

Proof. To establish the upper bound, we prove by induction on the number of variables that every Boolean function of n variables can be represented by a DNF containing at most 2^{n-1} terms. Clearly, the statement holds for $n = 1$. Assuming that the statement holds for all functions of up to $n - 1$ variables, let us consider an arbitrary function $f(x_1, x_2, \ldots, x_n)$ and, using the Shannon expansion, represent it as

$$f(x_1, x_2, \ldots, x_n) = \overline{x}_n f(x_1, x_2, \ldots, x_{n-1}, 0) \vee x_n f(x_1, x_2, \ldots, x_{n-1}, 1).$$

By the induction hypothesis, $f(x_1, x_2, \ldots, x_{n-1}, 0)$ and $f(x_1, x_2, \ldots, x_{n-1}, 1)$, being functions of $n - 1$ variables, have DNF representations ϕ_0 and ϕ_1 such that $||\phi_0|| \leq 2^{n-2}$ and $||\phi_1|| \leq 2^{n-2}$. Then, $f(x_1, x_2, \ldots, x_n)$ can be represented by the expression

$$\overline{x}_n \phi_0 \vee x_n \phi_1,$$

which immediately expands into a DNF ϕ such that $||\phi|| \leq ||\phi_0|| + ||\phi_1|| \leq 2^{n-1}$.

To show that the bound is attained, define the *parity function* of n variables to be the Boolean function whose value in the Boolean point $X = (x_1, x_2, \ldots, x_n)$ is 1 if and only if $\sum_{i=1}^{n} x_i$ is odd. Obviously, the number of true points of the parity function is 2^{n-1}. Since every two terms in the minterm DNF of the parity function have degree n and conflict in at least two variables, this DNF is closed under absorption and consensus, and is therefore the complete DNF of the parity function. Since the minterm DNF is obviously irredundant, it then follows that the parity function has a unique DNF representation, and that this representation has 2^{n-1} terms. $\qquad \square$

Another parameter of interest for logic minimization is the so-called *spread* of f: If ϕ_m is any $||\phi||$-minimizing DNF of f, the spread of f is

$$Y(f) = \max\left\{ \frac{||\phi||}{||\phi_m||} : \phi \text{ is a prime irredundant DNF of } f \right\}.$$

It was shown by Vasiliev [889] that the maximum value of $Y(f)$ over all Boolean functions of n variables is at least $2^{n-3\sqrt{n}}$, which clearly justifies the relevance of logic minimization.

Since among the $||\phi||$-minimizing DNFs there is always a prime irredundant one, it is also interesting to obtain some information about the number $\mathcal{I}(f)$ of different prime irredundant DNFs of a Boolean function f. It turns out (see [890]) that the maximum value of $\mathcal{I}(f)$ over all Boolean functions of n variables exceeds $2^{2^{n(1-o(1))}}$, where $o(1) \to 0$ when $n \to \infty$.

3.4.3 Typical parameters of Boolean functions and their DNFs

Since the number of distinct Boolean functions of n variables is 2^{2^n}, let us say that a certain property holds for *almost all* Boolean functions if the number of functions of n variables that have this property is $(1 - o(1))2^{2^n}$.

Theorem 3.20. *For almost all Boolean functions f of n variables, the number $|T(f)|$ of true points of f satisfies the inequalities:*

$$2^{n-1} - n2^{\frac{n}{2}} \le |T(f)| \le 2^{n-1} + n2^{\frac{n}{2}}. \tag{3.24}$$

Proof. The number of Boolean functions of n variables having exactly k true points is $\binom{2^n}{k}$, since every Boolean point is either a true point or a false point. Hence, the total number of Boolean functions with the property that their number of true points satisfies (3.24) is

$$\sum_{k=2^{n-1}-n2^{\frac{n}{2}}}^{2^{n-1}+n2^{\frac{n}{2}}} \binom{2^n}{k} = 2^{2^n} - 2\left(\sum_{k=0}^{2^{n-1}-n2^{\frac{n}{2}}-1} \binom{2^n}{k} \right).$$

The statement of the theorem follows from the fact that

$$\sum_{k=0}^{2^{n-1}-n2^{\frac{n}{2}}} \binom{2^n}{k} \le (2^{n-1} - n2^{\frac{n}{2}})\binom{2^n}{2^{n-1} - n2^{\frac{n}{2}}} = o(2^{2^n}),$$

where the last equality can be obtained by using the formula $\binom{m}{k} = \frac{m!}{k!(m-k)!}$, together with the following refined version of the Stirling formula (see, e.g., [314]):

$$\sqrt{2\pi n}(\frac{n}{e})^n e^{\frac{1}{12n+1}} < n! < \sqrt{2\pi n}(\frac{n}{e})^n e^{\frac{1}{12n}},$$

with the limits: $(1 + \frac{1}{m})^m \to e$ and $(1 - \frac{1}{m})^m \to \frac{1}{e}$ when $m \to \infty$. $\square$

A simple interpretation of Theorem 3.20 is that, for almost all Boolean functions, the number of true points is about the same as that of false points, namely, about 2^{n-1}. After establishing this fact, it is natural to ask in what way these two sets of true and false points are mixed in the Boolean hypercube. More specifically, one may wonder whether the set of true points of a typical Boolean function contains large subcubes. The next theorem states that a typical Boolean function has only "long" implicants, thus showing that the answer to the previous question is negative.

Theorem 3.21. *For almost all Boolean functions of n variables, the degree of every implicant is at least $n - \log_2(3n)$.*

Proof. Before proving the statement, we first calculate the average number of implicants of a fixed degree k over the set of all Boolean functions of n variables.

Note that the number of different terms of degree k is $\binom{n}{k}2^k$. Every such term takes the value 1 in exactly 2^{n-k} Boolean vectors. Therefore, every such term is an implicant of exactly $2^{2^n - 2^{n-k}}$ different Boolean functions of n variables.

Let us consider now a bipartite graph having two disjoint vertex sets A and B, where the nodes in A correspond to the terms of degree k over n variables, while the nodes in B correspond to the different Boolean functions of n variables; an edge (a, f) connects $a \in A$ to $f \in B$ if and only if a is an implicant of f. Clearly, the number of edges in this graph is

$$\binom{n}{k} 2^k 2^{2^n - 2^{n-k}}.$$

Since the total number of Boolean functions of n variables is 2^{2^n}, the average number of edges incident to a node in B is

$$\frac{1}{2^{2^n}} \binom{n}{k} 2^k 2^{2^n - 2^{n-k}} = \frac{\binom{n}{k} 2^k}{2^{2^{n-k}}}. \tag{3.25}$$

Obviously, this number is the average number of implicants of degree k over the set of all Boolean functions of n variables. It follows that at most $\frac{1}{n} 2^{2^n}$ Boolean functions of n variables can have

$$g(n,k) = \frac{n \binom{n}{k} 2^k}{2^{2^{n-k}}}$$

or more implicants of degree k, since, otherwise, the average number of implicants would exceed (3.25). Therefore, for almost all Boolean functions of n variables, the number of implicants of degree k is at most $g(n,k)$.

Obviously, if $g(n,k) < 1$, then it is true that almost all Boolean functions of n variables do not have implicants of degree k or less. Since neither $\binom{n}{k}$ nor 2^k can exceed 2^n, and since $n < 2^n$, the inequality

$$2^{3n} < 2^{2^{n-k}} \tag{3.26}$$

implies that $g(n,k) < 1$. Obviously, the inequality (3.26) is implied by the inequality

$$k < n - \log_2(3n). \tag{3.27}$$

This shows that if (3.27) holds, then $g(n,k) < 1$. Hence, for almost all Boolean functions of n variables, the degree of any implicant is at least $n - \log_2(3n)$. □

The next natural question concerns the number of terms (or literals) in a $||\phi||$-minimizing (or $|\phi|$-minimizing) DNF of a typical Boolean function of n variables. Several important results are known in this area. We shall not present the detailed proofs of these technical results here, and we give only a brief overview.

An interesting result obtained by Nigmatullin [712] shows that the number of terms (respectively, literals) in the $||\phi||$-minimizing (respectively, $|\phi|$-minimizing) DNFs of almost all Boolean functions of n variables is asymptotically the same. Let $t(n)$ and $l(n)$ represent "asymptotic estimates" of these two numbers. It follows from Theorem 3.21 that $l(n)$ behaves like $nt(n)$; thus, it is sufficient to estimate $t(n)$ only.

Glagolev [385] obtained the following lower bound on $t(n)$:

$$t(n) \geq \frac{2^{n-1}}{(\log_2 n)(\log_2 \log_2 n)}.$$

Moreover, an upper bound on $t(n)$ obtained by Sapozhenko [803] shows that

$$t(n) \leq \frac{2^n}{\log_2 n}.$$

Together with Theorem 3.19, these two bounds imply that the number of terms in the $||\phi||$-minimizing DNFs of almost all Boolean functions of n variables is asymptotically smaller than the worst possible one, but not by much.

To conclude, we stress that the results in this section are only intended to indicate the flavor of the research carried out in this area. A more complete presentation would substantially exceed the scope of this volume.

3.5 Exercises

1. Consider a set of ordered pairs $\Pi = \{(i,j)\}$, where $i,j \in \{1,2,\ldots,n\}$, and call a Boolean point $X = (x_1, x_2, \ldots, x_n)$ Π-*feasible* if, for every pair $(i,j) \in \Pi$, the implication "$x_i = 1$ implies $x_j = 1$" holds. Let f_Π be the Boolean function that takes the value 1 on Π-feasible Boolean points, and the value 0 on all the other Boolean points. Prove that $\overline{f}_\Pi$ has no prime implicants of degree 3.

2. Consider the linear inequality

$$\sum_{i=0}^{n} 2^i x_i \leq k,$$

where $x_i \in \{0,1\}$, $i = 0,1,\ldots,n$, and consider the Boolean function $f(x_0, x_1, \ldots, x_n)$ that takes the value 0 if and only if the Boolean point $(x_0, x_1, \ldots, x_n)$ satisfies the given inequality. Determine the maximum degree of a prime implicant of f if

   (a) $k = 2^m - 1$,
   (b) $k = 2^m - 2$,

   where m is a positive integer not exceeding n.

3. Prove that it is NP-hard to check whether all the prime implicants of a Boolean function given by a DNF are quadratic.

4. Prove Lemma 3.8.

5. Let $\psi = \bigvee_{k=1}^{m} C_k$ be a DNF of a Boolean function f. Prove that the following statements are equivalent:

   (a) The collection $\{C_k \mid k = 1,2,\ldots,m\}$ contains all prime implicants of f.
   (b) For every DNF ϕ, the implication $\phi \leq \psi$ holds if and only if each term of ϕ is absorbed by some term of ψ.

6. Prove that, if f and g are two Boolean functions on $\mathcal{B}^n$ such that $g \leq f$, then every implicant of g is absorbed by some prime implicant of f.

7. Prove that the following problem is co-NP-complete: Given an elementary conjunction C, and given a prime and irredundant DNF ψ, decide whether C is an implicant of the function represented by ψ. (Compare with Theorem 3.7.) *Hint:* Show that the DNF equation $\bigvee_{k=1}^{m} C_k = 0$ is consistent if and only if $C = y_1 y_2 \ldots y_m$ is not an implicant of the prime and irredundant DNF $\psi(X,Y) = \bigvee_{k=1}^{m} y_k C_k$.

8. Prove that it is NP-complete to decide whether a DNF ψ is irredundant. *Hint:* Show that the DNF equation $\bigvee_{k=1}^{m} C_k = 0$ is consistent if and only if the DNF $\psi(X, Y) = y_1 y_2 \ldots y_m \vee \bigvee_{k=1}^{m} y_k C_k$ is irredundant.

9. Prove that it is NP-complete to decide whether a DNF ψ is prime. (Compare with Corollary 3.3.) *Hint:* Show that the DNF equation $\bigvee_{k=1}^{m} C_k = 0$ is consistent if and only if the DNF $\psi(X, Y) = (\overline{y}_1 y_2 \ldots y_m) \vee (y_1 \overline{y}_2 y_3 \ldots y_m)$ $\vee \ldots \vee (y_1 \ldots y_{m-1} \overline{y}_m) \vee \bigvee_{k=1}^{m} y_k C_k$ is prime.

10. Let f be an arbitrary Boolean function on $\mathcal{B}^n$, let Π^+ be the set of its positive prime implicants, and let f_- be the largest positive minorant of f; namely, f_- is the largest positive function smaller than f. Prove that $f_- = \bigvee_{P \in \Pi^+} P$. (See Exercise 13 in Chapter 1, and Hammer, Johnson and Peled [443].)

11. If every term of a DNF of a Boolean function f contains at most one nonnegated variable, then show that the same property holds for each prime implicant of f.

12. Does the property described in Exercise 11 hold if "at most one" is replaced by "at least one"?

13. Does the property described in Exercise 11 hold if "at most one" is replaced by "at most two"?

14. Let us define a simplification algorithm ERA$^+$ consisting of the application of ERA followed by the repeated application of S4 as long as possible, and then the iteration of these two steps as long as possible. Let $\widehat{A}$ denote the final set covering matrix, let $\widehat{s}_1$ denote the number of the s variables fixed at 1, and let $\widehat{s}_0$ denote the number of the s variables fixed at 0. Let us also consider a procedure consisting of the repeated applications of S1, S2, S3, and S4 in any order and as long as possible. Let A^*, s_1^*, and s_0^* denote the final set covering matrix and the number of the s variables fixed at 1 and at 0, respectively. Prove that

   - $\widehat{s}_1 = s_1^*$,
   - $\widehat{s}_0 = s_0^*$,
   - the matrix $\widehat{A}$ can be obtained from A^* by a permutation of its rows and columns.

15. Prove that if a Boolean function f is not identically $\mathbf{1}$, then every linear implicant of f is an essential prime implicant.

16. Prove that if a Boolean function has linear prime implicants, then the DNF constructed by the greedy procedure will include all of them.

17. Use the result in the previous exercise to show that, if $n \geq 2$, then the approximation ratio of the greedy procedure is not greater than $n \ln 2 - 2 \ln 2 + 1$.

18. Construct an example showing that the DNF produced by the greedy procedure is not necessarily irredundant.

4

Duality theory

Yves Crama and Kazuhisa Makino

This chapter deals with yet another fundamental topic in the theory of Boolean functions, namely, duality theory. Some of the applications of duality were sketched in Chapter 1, and the concept has appeared at various occasions in Chapters 2 and 3. Here, we collect some of the basic properties of the dual of Boolean functions and, then characterize those functions that are comparable to (i.e., either imply, or are implied by) their dual. A large section of the chapter is then devoted to algorithmic aspects of dualization, especially for the special and most interesting case of positive functions expressed in disjunctive normal form. It turns out that the complexity of the latter problem remains incompletely understood, in spite of much recent progress on the question.

4.1 Basic properties and applications

4.1.1 Dual functions and expressions

Recall Definition 1.8 from Chapter 1, Section 1.3.

Definition 4.1. *The* dual *of a Boolean function f is the function f^d defined by*

$$f^d(X) = \overline{f(\overline{X})} \tag{4.1}$$

for all $X = (x_1, x_2, \ldots, x_n) \in \mathcal{B}^n$, where $\overline{X} = (\overline{x}_1, \overline{x}_2, \ldots, \overline{x}_n)$.

Example 4.1. *Let f be the 2-variable function defined by $f(0,0) = f(0,1) = f(1,1) = 1$ and $f(1,0) = 0$. Then the dual of f is defined by $f^d(0,0) = f^d(1,0) = f^d(1,1) = 0$ and $f^d(0,1) = 1$.* $\square$

The basic properties of dual functions are easily established.

Theorem 4.1. *If f and g are Boolean functions on $\mathcal{B}^n$, then*

(a) *$g = f^d$ if and only if, for all $X \in \mathcal{B}^n$, $f(X) \vee g(\overline{X}) = 1$ and $f(X) g(\overline{X}) = 0$;*
(b) *$(f^d)^d = f$ (involution: the dual of the dual is the function itself);*

167

(c) $(\overline{f})^d = \overline{(f^d)}$;
(d) $(f \vee g)^d = f^d g^d$;
(e) $(fg)^d = f^d \vee g^d$;
(f) $f \leq g$ if and only if $g^d \leq f^d$.

Proof. All these properties are trivial consequences of Definition 4.1 (properties (b)–(e) have already been verified in Theorem 1.2). $\square$

In view of the involution property (b), we sometimes say that two functions f, g are *mutually dual* when $g = f^d$ or, equivalently, when $f = g^d$.

Observe that the properties stated in Theorem 4.1 continue to hold when we replace dualization by complementation. As a matter of fact, investigating properties of the dual function f^d is tantamount to investigating properties of the function $\overline{f}$, namely, the complement of f, up to the "change of variables" $X \leftrightarrow \overline{X}$. It turns out, however, that the duality concept arises quite naturally in several applications. Therefore, we prefer to place our discussion in this framework.

If f is a function on $\mathcal{B}^n$ and P, N are disjoint subsets of $\{1, 2, \ldots, n\}$, then we denote by $f_{|P,N}$ the restriction of f obtained by fixing $x_i = 1$ for all $i \in P$ and $x_j = 0$ for all $j \in N$. The next property expresses in a formal way that "the dual of the restriction of a function is the restriction of the dual of the function to the complementary values."

Theorem 4.2. *Let f be a Boolean function on $\mathcal{B}^n$, and let $P, N \subseteq \{1, 2, \ldots, n\}$, with $P \cap N = \emptyset$. Then $(f_{|P,N})^d = (f^d)_{|N,P}$.*

Proof. This property follows from the definition of f^d. $\square$

Another easy, but useful, property is stated as follows:

Theorem 4.3. *Let f and g be two Boolean functions on $\mathcal{B}^n$. If f and g are mutually dual, then, for all $i \in \{1, 2, \ldots, n\}$, $f_{|x_i=0}$ and $g_{|x_i=1}$ are mutually dual, and $f_{|x_i=1}$ and $g_{|x_i=0}$ are mutually dual. Conversely, if for some $i \in \{1, 2, \ldots, n\}$, $f_{|x_i=0}$ and $g_{|x_i=1}$ are mutually dual, and $f_{|x_i=1}$ and $g_{|x_i=0}$ are mutually dual, then f and g are mutually dual.*

Proof. For every $i = 1, 2, \ldots, n$, we can write the Shannon expansions of g and f^d as

$$g = x_i g_{|x_i=1} \vee \overline{x}_i g_{|x_i=0}, \tag{4.2}$$

$$f^d = x_i (f^d)_{|x_i=1} \vee \overline{x}_i (f^d)_{|x_i=0}. \tag{4.3}$$

From (4.3) and Theorem 4.2,

$$f^d = x_i (f_{|x_i=0})^d \vee \overline{x}_i (f_{|x_i=1})^d. \tag{4.4}$$

The theorem follows by comparing (4.2) and (4.4). $\square$

We also recall that, by definition, the dual of a Boolean expression ϕ is the expression ϕ^d obtained by exchanging $\vee$ and $\wedge$ as well as the constants 0 and 1 in ϕ (Definition 1.9). We have shown that if the expression ϕ represents f, then ϕ^d represents f^d (Theorem 1.3). The latter property can be seen as a consequence of De Morgan's laws.

Example 4.2. *Consider again the function f in Example 4.1. Then $\varphi = \overline{x} \vee y$ represents f, and the dual expression $\varphi^d = \overline{x}\, y$ represents f^d.*   □

More generally, we mention the following fundamental *duality principle* of Boolean algebra:

Theorem 4.4. *Let I be a valid statement expressed in terms of the constants $0, 1$; the operations $\vee$, $\wedge$; the implication relations $\leq$, $\geq$; and Boolean functions. Then the "dual statement" I^d obtained from I by exchanging the symbols 0 and 1, $\vee$ and $\wedge$, $\leq$ and $\geq$, and by replacing every function by its dual, is also valid.*

Proof. We refer, for example, to Rudeanu [795] or Stoll [848].   □

4.1.2 Normal forms and implicants of dual functions

This subsection considers disjunctive and conjunctive normal forms, as well as (prime) implicants and implicates of dual functions. The following connection is an immediate consequence of the properties mentioned before; we record it explicitly because of its importance.

Theorem 4.5. *The DNF $\phi = \bigvee_{k=1}^{m} \left(\bigwedge_{j \in P_k} x_j \bigwedge_{j \in N_k} \overline{x}_j \right)$ represents the Boolean function f if and only if the CNF $\psi = \bigwedge_{k=1}^{m} \left(\bigvee_{j \in P_k} x_j \bigvee_{j \in N_k} \overline{x}_j \right)$ represents f^d.*

Proof. This holds because ψ is the dual expression of ϕ.   □

This theorem suggests a simple characterization of the (prime) implicants and implicates of dual functions.

Theorem 4.6. *For a Boolean function f,*

(i) *the elementary conjunction $C_{PN} = \bigwedge_{j \in P} x_j \bigwedge_{j \in N} \overline{x}_j$ is an impli-cant (respectively, a prime implicant) of f^d if and only if $D_{PN} = \bigvee_{j \in P} x_j \bigvee_{j \in N} \overline{x}_j$ is an implicate (respectively, a prime implicate) of f;*

(ii) *the elementary disjunction $D_{PN} = \bigvee_{j \in P} x_j \bigvee_{j \in N} \overline{x}_j$ is an implicate (respec-tively, a prime implicate) of f^d if and only if $C_{PN} = \bigwedge_{j \in P} x_j \bigwedge_{j \in N} \overline{x}_j$ is an implicant (respectively, a prime implicant) of f.*

Proof. Assertion (i) easily follows from the observation that $f \leq D_{PN}$ if and only if $D_{PN}^d \leq f^d$, and from the identity $D_{PN}^d = C_{PN}$. Assertion (ii) is obtained by interchanging the roles of f and f^d in the previous one.   □

The next result presents an alternative characterization of dual prime implicants, which is frequently useful. In words, this characterization expresses that,
when viewed as collections of literals, dual implicants and implicants always have
a nonempty intersection, and that dual prime implicants are minimal with this
property. (A similar characterization of prime implicates would be immediately
obtained by combining the statements of Theorems 4.6 and 4.7.)

Theorem 4.7. *Let* $\phi = \bigvee_{i=1}^{m} \left(\bigwedge_{j \in P_i} x_j \bigwedge_{j \in N_i} \overline{x}_j \right)$ *be an arbitrary DNF of
a Boolean function* f, *and let* $C_{PN} = \bigwedge_{j \in P} x_j \bigwedge_{j \in N} \overline{x}_j$ *be an elementary
conjunction. Then,*

(i) C_{PN} *is an implicant of* f^d *if and only if*

$$(P \cap P_i) \cup (N \cap N_i) \neq \emptyset \quad for\ i = 1, 2, \ldots, m; \tag{4.5}$$

(ii) C_{PN} *is a prime implicant of* f^d *if and only if* (4.5) *holds and, for every* $P' \subseteq P$
and $N' \subseteq N$ *with* $P' \cup N' \neq P \cup N$, *there exists an index* $i \in \{1, 2, \ldots, m\}$
such that

$$(P' \cap P_i) \cup (N' \cap N_i) = \emptyset.$$

Proof. By definition of dual functions (namely, $f^d(X) = \overline{f}(\overline{X})$), C_{PN} is an implicant of f^d if and only if $C_{NP} = \bigwedge_{j \in P} \overline{x}_j \bigwedge_{j \in N} x_j$ is an implicant of $\overline{f}$. Since
$f \wedge \overline{f} = 0$, the identity $C_{P_i N_i} \wedge C_{NP} = 0$ must hold for all implicants $C_{P_i N_i}$ of f,
which implies (4.5).

Conversely, if (4.5) holds, then $f \wedge C_{NP} = 0$ holds identically, meaning that
C_{NP} is an implicant of $\overline{f}$. This establishes assertion (i).

Assertion (ii) follows from the definition of prime implicants.   □

Observe that, in conditions (i) and (ii) of Theorem 4.7, the conjunctions $C_{P_i N_i}$
could be taken to be prime implicants, rather than arbitrary implicants of f.

4.1.3 Dual-comparable functions

Definition 4.2. *A Boolean function* f *is called* dual-minor *if* $f \leq f^d$, dual-major
if $f \geq f^d$, *and* self-dual *if* $f^d = f$. *A function is* dual-comparable *if it is either
dual-minor, dual-major, or self-dual.*

Example 4.3. *The function* $f = x_1 x_2 x_3$ *is dual-minor, since* $f^d = x_1 \vee x_2 \vee x_3$
satisfies $f \leq f^d$. *By Theorem 4.5, the dual of* $g = x_1 x_2 \overline{x}_3 \vee x_1 \overline{x}_2 x_3 \vee \overline{x}_1 x_2 x_3 \vee
\overline{x}_1 \overline{x}_2 \overline{x}_3$ *is*

$$g^d = (x_1 \vee x_2 \vee \overline{x}_3)(x_1 \vee \overline{x}_2 \vee x_3)(\overline{x}_1 \vee x_2 \vee x_3)(\overline{x}_1 \vee \overline{x}_2 \vee \overline{x}_3)$$

$$= x_1 x_2 \overline{x}_3 \vee x_1 \overline{x}_2 x_3 \vee \overline{x}_1 x_2 x_3 \vee \overline{x}_1 \overline{x}_2 \overline{x}_3,$$

and therefore g is self-dual.   □

The investigation of dual-comparable functions has proved useful in a variety of contexts (see, e.g., Muroga [698]). The next theorems present several characterizations of these functions. First, we observe that trivial examples of dual-comparable functions are easily provided.

Theorem 4.8. *Suppose that the function f has a prime implicant of degree 1. Then, f is dual-major. Moreover, f is dual-minor (and self-dual) if and only if f has no other prime implicant.*

Proof. Assume without loss of generality that $f(x_1, x_2, \ldots, x_n) = x_1 \vee g(x_2, x_3, \ldots, x_n)$. Then, $f^d = x_1 g^d$. Since $f = 0$ implies $x_1 = 0$, we see that f is dual-major. If f has no other prime implicant than x_1, then f is clearly self-dual. Conversely, if f has another prime implicant, then there exists a point $(x_1^*, x_2^*, \ldots, x_n^*) \in B^n$ such that $x_1^* = 0$ and $f(x_1^*, x_2^*, \ldots, x_n^*) = 1$. But $x_1^* = 0$ implies $f^d(x_1^*, x_2^*, \ldots, x_n^*) = 0$, and we conclude that f is not dual-minor. $\square$

Of course, the function g in Example 4.3 shows that there also exist nontrivial examples of dual-comparable functions. The next result is a simple restatement of Definition 4.2 (compare with Theorem 4.1(a)).

Theorem 4.9. *Let f be a Boolean function on B^n.*

(i)   *f is dual-minor if and only if the complement of every true point of f is a false point of f: For all $X \in B^n$, $f(X) = 1 \Rightarrow f(\overline{X}) = 0$ or, equivalently, $f(X) f(\overline{X}) = 0$.*

(ii) *f is dual-major if and only if the complement of every false point of f is a true point of f: For all $X \in B^n$, $f(X) = 0 \Rightarrow f(\overline{X}) = 1$ or, equivalently, $f(X) \vee f(\overline{X}) = 1$.*

(iii) *f is self-dual if and only if every pair of complementary points contains exactly one true point and one false point of f: For all $X \in B^n$, $f(X) = 1 \Leftrightarrow f(\overline{X}) = 0$.*

The next characterization of dual-minor functions is based on Theorem 4.7.

Theorem 4.10. *A function f is dual-minor if and only if*

$$(P \cap P') \cup (N \cap N') \neq \emptyset \tag{4.6}$$

for all pairs of (prime) implicants $C_{PN} = \bigwedge_{j \in P} x_j \bigwedge_{j \in N} \overline{x}_j$ and $C_{P'N'} = \bigwedge_{j \in P'} x_j \bigwedge_{j \in N'} \overline{x}_j$ of f.

Proof. If f is dual-minor, then every implicant C_{PN} of f is an implicant of f^d, since $C_{PN} \leq f \leq f^d$. In view of Theorem 4.7(i), this implies conditions (4.6).

On the other hand, if f is not dual-minor, then there exists a (prime) implicant C_{PN} of f such that $C_{PN} \not\leq f^d$, that is, such that C_{PN} is not an implicant of f^d. Hence, by Theorem 4.7(i), there exists a (prime) implicant $C_{P'N'}$ of f such that (4.6) does not hold. $\square$

We now give a necessary and sufficient condition for a function to be dual-major.

Theorem 4.11. *A function f is dual-major if and only if, for all $A \subseteq \{1,2,\ldots,n\}$, there exists a (prime) implicant $C_{PN} = \bigwedge_{j\in P} x_j \bigwedge_{j\in N} \overline{x}_j$ of f such that either $P \subseteq A$ and $N \cap A = \emptyset$, or $P \cap A = \emptyset$ and $N \subseteq A$.*

Proof. For any $X \in \mathcal{B}^n$, let $A = \{i \mid x_i = 1\}$. Then, the condition to be established is easily seen to be equivalent to condition (ii) in Theorem 4.9. $\square$

We now establish that self-dual functions are maximal among all dual-minor functions. More precisely, let us say that a dual-minor function f is *maximally dual-minor* if there exists no dual-minor function g such that $f \leq g$ and $f \neq g$.

Theorem 4.12. *A Boolean function is self-dual if and only if it is maximally dual-minor.*

Proof. If f is self-dual and g is a dual-minor function such that $f \leq g$, then we derive the sequence of inequalities

$$g^d \leq f^d = f \leq g \leq g^d.$$

Hence, $f = g$, implying that f is maximally dual-minor.

Conversely, assume that f is dual-minor, but not self-dual. Then, there exists a point X^* such that $f(X^*) = 0$ and $f^d(X^*) = 1$. Assume for instance that $x_1^* = 1$, and consider the function $g = f \vee f^d x_1$. Clearly, $f \leq g$, and $f \neq g$ since $g(X^*) = 1$. Moreover, g is dual-minor (actually, self-dual):

$$g^d = f^d (f \vee x_1) = f^d f \vee f^d x_1 = g$$

(the last equality holds because f is dual-minor). Therefore, f is not maximally dual-minor. $\square$

Of course, we would similarly show that:

Theorem 4.13. *A Boolean function is self-dual if and only if it is minimally dual-major.*

The construction used in the proof of Theorem 4.12 can be generalized to yield a simple, standard way of associating a self-dual function with an arbitrary Boolean function.

Definition 4.3. *For a Boolean function $f(x_1,x_2,\ldots,x_n)$, the self-dual extension of f is the function $f^{SD}(x_1,x_2,\ldots,x_n,x_{n+1})$, defined by*

$$f^{SD}(x_1,x_2,\ldots,x_n,x_{n+1}) = f(x_1,x_2,\ldots,x_n)\overline{x}_{n+1} \vee f^d(x_1,x_2,\ldots,x_n)x_{n+1}.$$
$$(4.7)$$

This terminology is well-justified.

Theorem 4.14. *For every Boolean function f, the function f^{SD} defined by (4.7) is self-dual. The mapping $SD: f \mapsto f^{SD}$ is a bijection between the set of Boolean functions of n variables and the set of self-dual functions of $n + 1$ variables.*

Proof. The dual of (4.7) is

$$(f^d(X) \vee \overline{x}_{n+1})(f(X) \vee x_{n+1}) = f^d(X) f(X) \vee f(X) \overline{x}_{n+1} \vee f^d(X) x_{n+1}$$

$$= f(X) \overline{x}_{n+1} \vee f^d(X) x_{n+1}.$$

Hence, f^{SD} is self-dual.

The mapping SD is injective, since the restriction of f^{SD} to $x_{n+1} = 0$ is exactly f. Moreover, SD has an inverse, defined by $g \mapsto g_{|x_{n+1}=0}$ for every self-dual function $g \in \mathcal{B}^{n+1}$. Indeed, if g is self-dual, then

$$(g_{|x_{n+1}=0})^{SD} = g_{|x_{n+1}=0} \overline{x}_{n+1} \vee (g_{|x_{n+1}=0})^d x_{n+1}$$

$$= g_{|x_{n+1}=0} \overline{x}_{n+1} \vee (g^d)_{|x_{n+1}=1} x_{n+1}$$

$$= g_{|x_{n+1}=0} \overline{x}_{n+1} \vee g_{|x_{n+1}=1} x_{n+1},$$

and this last expression is exactly the Shannon expansion of g. $\square$

Note that, when applied to dual-minor functions, the definition of self-dual extensions assumes a simpler form.

Theorem 4.15. *If f is a dual-minor function on $\mathcal{B}^n$, then $f^{SD} = f \vee f^d x_{n+1}$.*

Proof. This holds because, for all $a, b, x \in \mathcal{B}$, $a \leq b$ implies $a\overline{x} \vee bx = a \vee bx$. $\square$

Theorem 4.14 implies, in particular, that there are $2^{2^{n-1}}$ self-dual functions of n variables, as compared to 2^{2^n} Boolean functions of n variables (this could have been deduced from Theorem 4.9(iii) as well).

Another corollary of Theorem 4.14 is that dual comparability is not preserved under fixation of variables, a fact which also follows directly from the observation that the constant function $\mathbf{1}_n$ is not dual-minor, and that the constant function $\mathbf{0}_n$ is not dual-major. Interestingly, however, self-duality is preserved under composition of Boolean functions.

Theorem 4.16. *If $f_1(x_1, x_2, \ldots, x_n, x_{n+1})$ and $f_2(y_1, y_2, \ldots, y_m)$ are self-dual functions (where f_1 and f_2 may depend on common variables), then the function*

$$g(x_1, x_2, \ldots, x_n, y_1, y_2, \ldots, y_m) = f_1(x_1, x_2, \ldots, x_n, f_2(y_1, y_2, \ldots, y_m))$$

is self-dual.

Proof. Let $X = (x_1, x_2, \ldots, x_n) \in \mathcal{B}^n$ and $Y = (y_1, y_2, \ldots, y_m) \in \mathcal{B}^m$. Then,

$$
\begin{aligned}
g(\overline{X}, \overline{Y}) &= f_1(\overline{x}_1, \overline{x}_2, \ldots, \overline{x}_n, f_2(\overline{Y})) \\
&= f_1(\overline{x}_1, \overline{x}_2, \ldots, \overline{x}_n, \overline{f_2(Y)}) \\
&= \overline{f_1}(x_1, x_2, \ldots, x_n, f_2(Y)) \\
&= \overline{g}(X, Y),
\end{aligned}
$$

which shows that g is self-dual.   □

4.1.4 Applications

Duality plays a central role in various applications arising in artificial intelligence, computational logic, data mining, reliability theory, game theory, integer programming, and so on. Some of these applications have already been mentioned in previous chapters. We have observed several times, for instance, that one way of solving the Boolean equation $\phi = 0$ is to compute a CNF representation of ϕ or, equivalently, a DNF representation of ϕ^d. Actually, if a DNF expression of ϕ^d is at hand, then *all* solutions of the equation $\phi = 0$ are readily available (see Section 2.11.2).

Example 4.4. *Let* $f(x, y, z, u) = \overline{x}y \vee xz\overline{u} \vee x\overline{y}z \vee \overline{y}zu$. *It can be checked that* $f^d = \overline{x}z \vee yz \vee xyu \vee \overline{x}\,\overline{y}\,\overline{u}$. *Hence, the solutions of* $f = 0$ *have the form* $(1, *, 0, *)$, $(*, 0, 0, *)$, $(0, 0, *, 0)$, *or* $(1, 1, *, 1)$, *where* $*$ *denotes an arbitrary 0–1 value.*   □

We now present a few additional models involving dual functions. Other applications will be presented in Section 4.2, when we concentrate more specifically on positive functions.

Application 4.1. (Artificial intelligence, electrical engineering.) *Reiter [783] proposes a logic-based framework for the analysis of diagnosis problems and presents an application to fault diagnosis in combinational circuits. If we restrict ourselves to propositional logic, then Reiter's approach can be sketched as follows (even in propositional logic, Reiter's model is actually more general than the one below; but our formulation is already sufficiently general, for instance, to encompass the circuit fault diagnosis problem):*

Consider a complex system Σ *consisting of* m *interrelated components. The intended operation of* Σ *is modeled by a collection of Boolean equations* $\phi_k(X, Y) = 0$ ($k = 1, 2, \ldots, m$), *with the following interpretation: For every point* $X^* \in \mathcal{B}^n$, *there exists a unique point* $Y^* \in \mathcal{B}^t$ *such that* $\bigvee_{k=1}^{m} \phi_k(X^*, Y^*) = 0$ *(see Section 1.13.2). Each point* $X^* \in \mathcal{B}^n$ *is called an* observation. *Assume now that a particular observation* X^* *is such that the equation* $\bigvee_{k=1}^{m} \phi_k(X^*, Y) = 0$ *is inconsistent. This means that the behavior of the system* Σ *deviates from its specification, that is,* Σ *is faulty. The diagnosis issue is now, intuitively, to understand*

what went wrong with the system. More precisely, Reiter [783] defines a diagnosis
as a minimal subset $\Delta \subseteq \{1,2,\ldots,m\}$ *such that*

$$\bigvee_{\substack{k=1 \\ k \notin \Delta}}^{m} \phi_k(X^*,Y) = 0$$

is consistent. The idea is that, were it not for the components in Δ, *then* Σ *would
have been functioning properly. The minimality of* Δ *translates what Reiter calls
the "Principle of Parsimony."*

 *The task of the analyst is now to produce all diagnoses associated with a given
observation* X^*. *Let* $p_1, p_2, \ldots, p_m$ *be m new Boolean variables, and define the
function*

$$f(Y,P) = \bigvee_{k=1}^{m} p_k \phi_k(X^*,Y).$$

Let also

$$\bigwedge_{i \in \Delta_k} p_i \bigwedge_{j \in A_k} y_j \bigwedge_{j \in B_k} \overline{y}_j, \qquad k = 1,2,\ldots,r,$$

denote the prime implicants of the dual function f^d. *We leave it to the reader to
check that diagnoses are exactly the minimal members of the collection of sets*
$\{\Delta_1, \Delta_2, \ldots, \Delta_r\}$. *Reiter proposes an ad hoc algorithm that produces all the diagnoses and that uses as a subroutine a simple dualization algorithm for positive
functions (see also the exercises at the end of this chapter).*   □

Application 4.2. (Complexity theory.) *Theoretical computer scientists have introduced several measures of complexity reflecting the difficulty to compute Boolean
functions, and they have analyzed the relation between them. One such measure,
albeit a rather primitive one, is the degree* $deg(f)$ *of the Boolean function* f, *that
is, the minimum degree of a DNF representing* f. *A more elaborate measure is the
decision tree complexity of* f. *Remember that decision trees were introduced in
Section 1.12.3. The depth* $\delta(T)$ *of a decision tree* T *is the length (that is, the number
of arcs) of a longest path from the root to a leaf of* T. *The decision tree complexity
$DT(f)$ of a Boolean function* f *is the minimum of* $\delta(T)$ *over all decision trees
computing* f. *This measure of complexity has been extensively investigated; see,
for example, [902, 903]. Now, assume that* f *is computed by a tree of depth* δ.
Then, as we noted at the end of Section 1.12.3, f *and* $\overline{f}$ *can both be represented
by (orthogonal) DNFs of degree at most* δ. *Therefore, we obtain the relation*

$$\max(deg(f), deg(f^d)) \leq DT(f).$$

However, a more subtle relation also holds, namely,

$$DT(f) \leq deg(f) deg(f^d).$$

*We can prove this inequality by induction on the number of variables ([626, 903]).
Let* $f = \phi(x_1, x_2, \ldots, x_n)$, *and* $f^d = \psi(x_1, x_2, \ldots, x_n)$, *where* ϕ *and* ψ *are DNFs*

of degrees $deg(f)$ and $deg(f^d)$, respectively; let C be any term of ϕ, and let (without loss of generality) $\{x_1, x_2, \ldots, x_k\}$ be the set of variables occurring in C. Thus, $k \leq deg(f)$. We construct a decision tree for f as shown in Figure 1.5, branching on the variables in the natural order $(x_1, x_2, \ldots, x_n)$. Thus, the root of the tree is labeled by x_1. More generally, if u is an internal vertex at depth i from the root $(0 \leq i < k)$, then u is labeled by x_{i+1}. Now, consider any internal vertex at depth $k - 1$ (if there is no such vertex, then the tree has depth at most $k - 1$ and the required inequality holds). Let v and w be the children of this vertex. Then, the subtree hanging from v (respectively, from w) is a decision tree for a function of the form $g = f_{|P,N}$ (resp., $h = f_{|P \setminus \{k\}, N \cup \{k\}}$), where (P, N) is a partition of $\{1, 2, \ldots, k\}$ and $k \in P$.

We can assume that the subtrees representing g and h both have optimal depth. In this way, we obtain for f a decision tree with depth $\max(DT(g), DT(h)) + k$. Assume that $\max(DT(g), DT(h)) = DT(g)$ (the other case is similar). By induction, we can assume that $DT(g) \leq deg(g) deg(g^d)$. Note that $deg(g) \leq deg(f)$, since g is a restriction of f. Moreover, by Theorem 4.2,

$$g^d = (f_{|P,N})^d = (f^d)_{|N,P}. \tag{4.8}$$

Since $P \cup N = \{1, 2, \ldots, k\}$ is the set of indices of the variables in C, Theorem 4.7, together with (4.8), implies that a DNF of g^d is obtained by fixing at least one variable to either 0 or 1 in each term of ψ. Therefore, $deg(g^d) \leq deg(f^d) - 1$.

So, we have represented f by a decision tree of depth

$$DT(g) + k \leq deg(g) deg(g^d) + k \leq desg(f)(deg(f^d) - 1) + deg(f)$$
$$= deg(f) deg(f^d),$$

which proves the required inequality. $\qquad\qquad\qquad\qquad\qquad\qquad\qquad \square$

4.2 Duality properties of positive functions

Much of the literature on Boolean duality has focused on the special case of positive functions, where the results usually have simple combinatorial interpretations. In this section, we reexamine some of the results of Section 4.1 within this framework and discuss their meaning within various fields of application. Most of these results have actually been independently discovered in several areas; see, for example, Benzaken [64], Berge [72], and Muroga [698].

4.2.1 Normal forms and implicants of dual functions

Recall from Section 1.10 that a Boolean function f is *positive* if and only if $X \leq Y$ implies $f(X) \leq f(Y)$ for all $X, Y \in \mathcal{B}^n$, and that f is positive if and only if it can be represented by a positive expression (namely, a Boolean expression which contains only positive literals). In fact, the complete DNF of a positive Boolean function is positive and is its unique prime irredundant DNF (see Theorem 1.23).

Thus, in view of Theorem 4.5, positivity is preserved under dualization.

Theorem 4.17. *A function* f *is positive if and only if its dual* f^d *is positive.*

For a positive function f, we denote by $minT(f)$ the set of minimal true points of f, and by $maxF(f)$ the set of its maximal false points. Theorem 1.26 describes a simple one-to-one correspondence between the prime implicants of f and its minimal true points: Namely, $X^* \in minT(f)$ if and only if $C = \bigwedge_{i \in supp(X^*)} x_i$ is a prime implicant of f, where $supp(X^*) = \{i \in \{1,2,\dots,n\} \,|\, x_i^* = 1\}$.

Theorem 1.27 establishes a similar relationship between the prime implicates of f and its maximal false points. In duality terms, this result translates as follows:

Theorem 4.18. *Let* f *be a positive Boolean function on* $\mathcal{B}^n$. *The point* $X^* \in \mathcal{B}^n$ *is a maximal false point of* f *if and only if the elementary conjunction* $C = \bigwedge_{i \in supp(\overline{X^*})} x_i$ *is a prime implicant of* f^d, *where* $supp(\overline{X^*}) = \{i \in \{1,2,\dots,n\} \,|\, x_i^* = 0\}$.

Example 4.5. *Let* $f = x_1 \vee x_2 x_3$. *Its dual is* $f^d = x_1 x_2 \vee x_1 x_3$. *One can check that the maximal false points of* f *are* $(0,0,1)$ *(corresponding to the prime implicant* $x_1 x_2$ *of* f^d*) and* $(0,1,0)$ *(corresponding to the prime implicant* $x_1 x_3$ *of* f^d*).*   □

Other useful characterizations of dual prime implicants of positive functions are best stated in hypergraph terminology. Recall from Chapter 1 and Appendix A that, if $\mathcal{H} = (N, \mathcal{E})$ is a hypergraph and $S \subseteq N$ is a subset of vertices, then S is called a *transversal* of $\mathcal{H}$ (or of $\mathcal{E}$) if $S \cap E \neq \emptyset$ holds for all edges $E \in \mathcal{E}$, and S is called *stable* if its complement is a transversal (namely, if S does not include any edge of $\mathcal{H}$). A transversal S is *minimal* if it does not (properly) include any other transversal.

Now, for a positive Boolean function f on $\mathcal{B}^n$, let $\mathcal{H}_f = (N, \mathcal{P})$, where $N = \{1,2,\dots,n\}$ and $\mathcal{P}$ denotes the family of all subsets $P \subseteq \{1,2,\dots,n\}$ such that $\bigwedge_{i \in P} x_i$ is a prime implicant of f. We know that $\mathcal{H}_f$ is a *clutter* (or a *Sperner hypergraph*), meaning that no set in $\mathcal{P}$ contains another set in $\mathcal{P}$ (sometimes, we may say that $\mathcal{P}$ itself is the clutter).

Theorem 4.19. *Let* $f = \bigvee_{P \in \mathcal{P}} \bigwedge_{i \in P} x_i$ *and* $g = \bigvee_{T \in \mathcal{T}} \bigwedge_{i \in T} x_i$ *be the complete DNFs of two positive functions on* $\mathcal{B}^n$. *The following statements are equivalent:*

(a) *$g = f^d$.*

(b) *For every partition of* $N = \{1,2,\dots,n\}$ *into two sets* A *and* $\overline{A}$, *there is either a member of* $\mathcal{P}$ *contained in* A *or a member of* $\mathcal{T}$ *contained in* $\overline{A}$, *but not both.*

(c) *$\mathcal{T}$ is exactly the family of minimal transversals of* $\mathcal{P}$.

Proof. The equivalence of (a) and (b) is a restatement of Theorem 4.1(a). Statement (c) is a corollary of Theorem 4.7.   □

Example 4.6. *As in Example 4.5, consider the function* $f = x_1 \vee x_2 x_3$ *and its dual* $f^d = x_1 x_2 \vee x_1 x_3$. *The hypergraph* $\mathcal{H}_f$ *has the edge-set* $\mathcal{E} = \{\{1\},\{2,3\}\}$. *One easily checks that* $\{1,2\}$ *and* $\{1,3\}$ *are exactly the minimal transversals of* $\mathcal{H}_f$. $\square$

The previous results provide an efficient characterization of the dual (prime) implicants of a positive function: Namely, given any reasonable description of a positive function f (e.g., its complete DNF) and given an elementary conjunction $C = \bigwedge_{i \in T} x_i$, Theorem 4.18 allows us to verify efficiently whether C is a prime implicant or an implicant of f^d (see also Theorem 1.27).

It turns out to be more difficult to decide whether C is a *dual subimplicant* of f, that is, to determine whether there exists a set of indices S such that $T \subseteq S$ and $\bigwedge_{i \in S} x_i$ is a prime implicant of f^d. Boros, Gurvich, and Hammer [121] proved that this question is NP-complete when f is given in DNF, but they also gave a characterization of dual subimplicants which can be efficiently tested when $|T|$ is bounded. We defer a presentation of this result to Chapter 10 (see Theorem 10.4), where it will constitute a main tool for the recognition of read-once Boolean functions.

4.2.2 Dual-comparable functions

Let us now turn to the characterization of dual-comparable functions. We say that a hypergraph $\mathcal{H} = (N, \mathcal{E})$ (or the family of sets $\mathcal{E}$) is *intersecting* if $E \cap E' \neq \emptyset$ for all $E, E' \in \mathcal{E}$.

Theorem 4.20. *A positive function* f *is dual-minor if and only if* $\mathcal{H}_f$ *is intersecting.*

Proof. This follows from Theorem 4.10. $\square$

Let $\mathcal{H} = (N, \mathcal{E})$ be an arbitrary hypergraph, and let $k \geq 1$ be an integer. A *k-coloring* of $\mathcal{H}$ is a partition of N into k stable sets $N_1, N_2, \ldots, N_k$. We say that $\mathcal{H}$ is *k-colorable* if it admits a k-coloring, and we denote by $\chi(\mathcal{H})$ the *chromatic number* of $\mathcal{H}$, that is, the smallest integer k such that $\mathcal{H}$ is k-colorable. Note that $\chi(\mathcal{H})$ is finite, except when $\mathcal{H}$ has either an empty edge or an edge of cardinality 1. We let $\chi(\mathcal{H}) = +\infty$ in either of these two cases. On the other hand, $\chi(\mathcal{H}) = 1$ exactly when $\mathcal{H}$ has no edge.

For a positive Boolean function f, the hypergraph $\mathcal{H}_f$ has an empty edge only if $f = \mathbf{1}$, and it has an edge of cardinality 1 only if f has a linear prime implicant. In view of Theorem 4.8, we do not lose much if we disregard linear prime implicants in the next statement.

Theorem 4.21. *If* f *is a dual-minor positive function without prime implicants of degree 1, then* $\chi(\mathcal{H}_f) \leq 3$.

Proof. If f is dual-minor, then $f \neq \mathbf{1}$, and hence its chromatic number is finite. Consider an arbitrary coloring of $\mathcal{H}_f = (N, \mathcal{P})$ into k stable sets $N_1, N_2, \ldots, N_k$, and assume that $k \geq 4$. One of the sets $A = N_1 \cup N_2$ or $\overline{A} = N_3 \cup \ldots \cup N_k$ is stable:

Otherwise, there are two sets $P, P' \in \mathcal{P}$ such that $P \subseteq A$ and $P' \subseteq \overline{A}$, and thus $P \cap P' = \emptyset$, in contradiction with Theorem 4.20. Therefore, either $(N_1, N_2, \overline{A})$ or $(A, N_3, \ldots, N_k)$ is a coloring of $\mathcal{H}_f$ involving fewer than k classes. $\square$

Theorem 4.22. *A positive function* f *is dual-major if and only if* $\chi(\mathcal{H}_f) \geq 3$.

Proof. The clutter $\mathcal{H}_f$ is 2-colorable if and only if there exists a partition $(A, \overline{A})$ of $\{1, 2, \ldots, n\}$ such that $P \cap A \neq \emptyset$ and $P \cap \overline{A} \neq \emptyset$ for all $P \in \mathcal{P}$. In view of Theorem 4.11, this means that f is not dual-major.

The case $\chi(\mathcal{H}_f) = 1$ corresponds to the constant function $f = \mathbf{0}$ which is not dual-major. $\square$

From these results, we derive a characterization of self-dual positive functions.

Theorem 4.23. *A positive function* f *without prime implicants of degree* 1 *is self-dual if and only if* $\mathcal{H}_f$ *is intersecting and* $\chi(\mathcal{H}_f) = 3$.

Proof. This follows directly from Theorems 4.20, 4.21, and 4.22. $\square$

Finally, let us note that the proof of Theorem 4.12 is easily adapted to establish the next result.

Theorem 4.24. *A positive Boolean function* f *on* $\mathcal{B}^n$ *is self-dual if and only if it is maximal among all positive dual-minor functions or, equivalently, if and only if* $\{supp(X): f(X) = 1\}$ *is a maximal intersecting family of subsets of* $\{1, 2, \ldots, n\}$.

The number of positive self-dual functions on $\mathcal{B}^n$ is not as easily determined as the total number of self-dual functions, but asymptotic formulas have been derived by Korshunov [579] (see also Bioch and Ibaraki [88]; Loeb and Conway [621]).

4.2.3 Applications

Application 4.3. (Combinatorics.) *We saw in Section 1.13.5 that positive functions are in one-to-one correspondence with clutters, by way of the mapping*

$$f(x_1, x_2, \ldots, x_n) = \bigvee_{A \in \mathcal{P}} \bigwedge_{j \in A} x_j \mapsto \mathcal{P}.$$

Let $\phi = \bigvee_{T \in \mathcal{T}} (\bigwedge_{j \in T} x_j)$ *be the complete DNF of* f^d. *By Theorem 4.19, every set* T *in* $\mathcal{T}$ *is a minimal transversal of* $\mathcal{H}$. *In hypergraph terminology,* $\mathcal{T}$ *is the transversal clutter or blocker of* $\mathcal{H}$ *(see, e.g., Berge [72]; Eiter and Gottlob [295]; the terminology blocker is due to Edmonds and Fulkerson [288, 353]).*

Let $\mathcal{T}(H)$ *denote the blocker of an arbitrary clutter* $\mathcal{H}$. *Many elementary properties of blockers are probably best viewed in a Boolean context (and, in this context, can be extended to nonpositive functions). For instance, Lawler [603] and Edmonds and Fulkerson [288] observed that* $\mathcal{T}(\mathcal{T}(H)) = H$, *a property that is equivalent to the Boolean identity* $(f^d)^d = f$. *Similarly, we can deduce from Theorem 4.2 the following property mentioned in Seymour [821]: For all* $S \subseteq \{1, 2, \ldots, n\}$,

$$T(\mathcal{H}) \setminus S = T(\mathcal{H} / S) \text{ and } T(\mathcal{H}) / S = T(\mathcal{H} \setminus S),$$

where the deletion (\\) and contraction (/) operations have been introduced in Section 1.13.5.

Properties of intersecting clutters (that is, dual-minor functions), (non) 2-colorable clutters (that is, dual-major functions), maximal intersecting hypergraphs (corresponding to self-dual functions), and k-colorable hypergraphs have been extensively studied in the literature; see, for instance, Berge [72] or Schrijver [814]. Their connections with Boolean duality have been stressed in a series of papers by Benzaken [62, 63, 64, etc.].   □

Application 4.4. *(Integer programming and combinatorial optimization.) Consider a set covering problem* SCP, *as introduced in Section 1.13.6:*

$$\text{minimize} \quad z(y_1, y_2, \ldots, y_n) = \sum_{i=1}^{n} c_i y_i \tag{4.9}$$

$$\text{subject to} \quad \sum_{i \in A_k} y_i \geq 1, \quad k = 1, 2, \ldots, m \tag{4.10}$$

$$(y_1, y_2, \ldots, y_n) \in \mathcal{B}^n, \tag{4.11}$$

and let $\mathcal{P} = \{A_1, A_2, \ldots, A_m\}$. *Clearly, the (minimal) feasible solutions of* SCP *are the characteristic vectors of the (minimal) transversals of* $\mathcal{P}$. *Therefore, if we define a Boolean function* f *by*

$$f = \bigvee_{k=1}^{m} \bigwedge_{i \in A_k} x_i = \bigvee_{P \in \mathcal{P}} \bigwedge_{i \in P} x_i,$$

then the (minimal) feasible solutions of SCP *are exactly the (minimal) true points of* f^d. *In particular, any algorithm that computes the dual of* f *could be used, in principle, to solve the set covering problem (see, e.g., Lawler [603] for early work based on these observations).*

More generally, dual blocking pairs $(\mathcal{P}, T)$, *where* T *is the blocker of* $\mathcal{P}$, *play a very important role in the theory of combinatorial optimization. A paradigmatic example of such a pair is provided by the set* $\mathcal{P}$ *of elementary paths joining two vertices s and t in a directed graph, and by the set* T *of minimal cuts separating s from t. Another example consists of the set* $\mathcal{P}$ *of all chains in a partially ordered set and the set* T *of all antichains.*

We have just seen that the set covering problem SCP *is equivalent to the minimization problem:* $\min_{T \in T} \sum_{i \in T} c_i$. *If we replace the sum by a max-operator in the objective function, then we obtain a class of bottleneck optimization problems, expressed as*

$$\min_{T \in T} \max_{i \in T} c_i.$$

Edmonds and Fulkerson [288] have established that this class of problems displays a very strong property which, in fact, provides a rather unexpected characterization of duality for positive Boolean functions.

Theorem 4.25. *Let $\mathcal{P}$ and $\mathcal{T}$ be two nonempty clutters on $\{1, 2, \ldots, n\}$. Then, the equality*

$$\max_{P \in \mathcal{P}} \min_{i \in P} c_i = \min_{T \in \mathcal{T}} \max_{i \in T} c_i \qquad (4.12)$$

holds for all choices of real coefficients $c_1, c_2, \ldots, c_n$ if and only if $\mathcal{T}$ is the blocker of $\mathcal{P}$.

Proof. Assume first that $\mathcal{T}$ is the blocker of $\mathcal{P}$ and fix the coefficients $c_1, c_2, \ldots, c_n$. Consider any $P \in \mathcal{P}$ and $T \in \mathcal{T}$. Since $P \cap T \neq \emptyset$, $\min_{i \in P} c_i \leq \max_{i \in T} c_i$. Therefore, the left-hand side of (4.12) is no larger than its right-hand side.

Now, assume without loss of generality that $c_1 \geq c_2 \geq \ldots \geq c_n$, and consider the smallest index j such that $\{1, 2, \ldots, j\}$ contains a member of $\mathcal{P}$; say $P^* \subseteq \{1, 2, \ldots, j\}$ and $P^* \in \mathcal{P}$. Then, $\min_{i \in P^*} c_i = c_j$. Note that $\{j + 1, j + 2, \ldots, n\}$ does not contain any set $T \in \mathcal{T}$ because such a set T would not intersect P^*. On the other hand, $\{j, j + 1, \ldots, n\}$ is a transversal of $\mathcal{P}$ (since its complement is stable in $\mathcal{P}$, by choice of j), and hence it contains some set $T^* \in \mathcal{T}$. Therefore, $\max_{i \in T^*} c_i = c_j$, and equality holds in (4.12).

For the converse implication, let us assume that (4.12) holds for all choices of $c_1, c_2, \ldots, c_n$, and let us establish condition (b) in Theorem 4.19. Let $(A, \overline{A})$ be a partition of $\{1, 2, \ldots, n\}$ into two sets, and let $c_i = 1$ if $i \in A$, $c_i = 0$ if $i \in \overline{A}$. By assumption, (4.12) holds for this choice of $c_1, c_2, \ldots, c_n$. If both sides of the equation are equal to 1, this means that there is a set $P^* \in \mathcal{P}$ such that $P^* \subseteq A$, and that no set in $\mathcal{T}$ is entirely contained in $\overline{A}$. On the other hand, if both sides of (4.12) are equal to 0, then the reverse conclusion holds. Hence, by Theorem 4.19(b), $\mathcal{T}$ is the blocker of $\mathcal{P}$.   □

Gurvich [421] generalized Theorem 4.25 in order to characterize Nash-solvable game forms.   □

Application 4.5. (Reliability theory.) *As in Section 1.13.4, let f_S be the (positive) structure function of a coherent system S. We have already seen that each prime implicant $\bigwedge_{i \in P} x_i$ corresponds to a minimal pathset of S, namely, a minimal set of components P with the property that the whole system S works whenever the components in P work.*

Similarly, every (prime) implicant $\bigwedge_{i \in T} x_i$ of f_S^d is associated with a subset T of components called a (minimal) cutset of S. A cutset T has the distinguishing property that, if X^* describes a state of the components such that $x_i^* = 0$ for all $i \in T$, then $f_S^d(\overline{X^*}) = 1$, and hence $f_S(X^*) = 0$. In other words, the system S fails whenever all components in the cutset fail, irrespectively of the operating state of the other components. Therefore, the dual function f_S^d describes the system S in terms of failing states.

This duality relationship between minimal pathsets and minimal cutsets is well-known in the context of reliability theory, as stressed by Ramamurthy [777]. Several authors have actually investigated the use of Boolean dualization techniques to generate the list of minimal cutsets of a system from a list of its minimal pathsets; see, for instance, Locks [616] or Shier and Whited [830].   □

Application 4.6. (Game theory.) *Let v be a simple game on the player set N, and let f_v be the positive Boolean function associated with v, as explained in Section 1.13.3. Then, the prime implicants of f_v correspond to the minimal winning coalitions of the game, namely, to those minimal subsets P of players such that $v = 1$ whenever all players in P vote "Yes."*

If $\bigwedge_{i \in T} x_i$ is a prime implicant of f_v^d, then, in view of Theorem 4.18, T is the complement of a maximal losing coalition. In other words, T is a blocking coalition, that is, a minimal subset of players such that $v = 0$ if all players in T vote "No."

When modeling real-world voting bodies, it often makes sense to consider certain special classes of games (see, e.g., Ramamurthy [777] or Shapley [828]). A game v is called proper if two complementary coalitions S and $N \setminus S$ cannot be simultaneously winning. It follows from Theorem 4.9(i) (or from Theorem 4.20) that the game v is proper if and only if the function f_v is dual-minor. On the other hand, in a strong game, two complementary coalitions cannot be simultaneously losing. By Theorem 4.9(ii) (or Theorem 4.22), a game v is strong if and only if f_v is dual-major. Finally, v is called decisive (or constant-sum) if exactly one of any two complementary coalitions is winning. So, v is decisive if and only if f_v is self-dual.

For obvious reasons, most practical voting rules are proper. For instance, when all the players carry one vote and decisions are made based on the majority rule with threshold $q > \frac{n}{2}$, then the resulting game is proper. If the number of players is odd and $q = \frac{n+1}{2}$, then the game is also decisive.

The concept of self-dual extension has been studied in the game-theoretic literature under the name of constant-sum extension.

Unexpectedly, perhaps, Boolean duality also plays an important role in the investigation of solution concepts for nonsimple games, such as 2-person (or n-person) positional games; we refer to Gurvich [421, 423, 424, etc.] and to Chapter 10 for illustrations.   □

Application 4.7. (Distributed computing systems.) *Dual-comparable Boolean functions have also found applications in several areas of computer science. Lamport [593], for instance, has proposed to use them in order to achieve mutual exclusion in distributed computing systems (see also Davidson, Garcia-Molina, and Skeen [258]; Garcia-Molina and Barbara [370]; Bioch and Ibaraki [88]; Ibaraki and Kameda [516]; etc.). In this context, intersecting clutters (corresponding to the prime implicants of positive dual-minor functions) are usually called coteries, and each member of a coterie is called a quorum.*

More precisely, let $N = \{1, 2, \ldots, n\}$ represent the sites in a distributed system and let C be a coterie on N. Lamport [593] proposed that a task (e.g., updating data in a replicated database) should be allowed to enter a critical section only *if it can get permission from all the members of a quorum $T \in C$, where each site is allowed to issue at most one permission at a time. The intersecting property of coteries guarantees that at most one task can enter the critical section at any time (meaning, e.g., that conflicting updates cannot be performed concurrently in the database).*

A coterie C is said to dominate another coterie D if, for each quorum $T_1 \in D$, there is a quorum $T_2 \in C$ satisfying $T_2 \subseteq T_1$ (see Garcia-Molina and Barbara [370]). Non-dominated coteries have maximal "efficiency" and are therefore important in practical applications. Theorem 4.24 shows that nondominated coteries are nothing but self-dual positive functions in disguise. Theorems 4.22 and 4.23 have also been rediscovered in this context (see [370]). □

Further discussions of the use of duality concepts in applications can be found, for instance, in papers by Domingo, Mishra, and Pitt [275] or Eiter and Gottlob [295].

4.3 Algorithmic aspects: The general case

4.3.1 Definitions and complexity results

The applications presented in the previous sections have established the need for an algorithm that computes an expression of f^d from an expression of f. Since we know that an expression of f^d can be obtained by exchanging $\vee$ and $\wedge$, as well as the constants 0 and 1, in any given expression of f, the problem has to be stated more precisely in order to avoid trivialities. A closer look at the applications shows that, in many cases, we are more specifically interested in one of the following algorithmic problems.

DUAL RECOGNITION
Instance: DNF representations of two Boolean functions f and g.
Question: Is $g = f^d$?

DUALIZATION
Instance: An arbitrary expression of a Boolean function f.
Output: The complete DNF of f^d or, equivalently, a list of all prime implicants of f^d.

DNF DUALIZATION
Instance: A DNF representation of a Boolean function f.
Output: The complete DNF of f^d or, equivalently, a list of all prime implicants of f^d.

In this section, we examine more closely the algorithmic complexity of these dualization problems, as well as their relationship with the solution of Boolean equations and the generation of prime implicants. We start with an easy result.

Theorem 4.26. DUAL RECOGNITION *is co-NP-complete, even if f is a positive function represented by its complete DNF.*

Proof. Consider an arbitrary DNF equation $\phi = 0$. Let $f = \mathbf{0}$, and let g be the function represented by ϕ. Then, $g = f^d = \mathbf{1}$ if and only if the equation $\phi = 0$ is inconsistent.   □

Theorem 4.26 already underlines the intrinsic complexity of the dualization problem in its decision version (which simply requires a Yes or No answer). When we turn to the list-generation problems DUALIZATION and DNF DUALIZATION, another difficulty arises. Indeed, we have observed in Theorem 3.18 that the number of prime implicants of f^d can be exponentially larger than the number of terms in a DNF of f, even when f is a positive function. Thus, the size of the output of the DUALIZATION and DNF DUALIZATION problems is generally not polynomially bounded in the size of their input. In view of this unavoidable difficulty, the complexity of dualization algorithms is most meaningfully expressed as a function of the combined size of their input *and* of their output (we refer to [538, 605] and to Appendix B for a discussion of the complexity of list-generation algorithms).

Remark. The reader should note that on some occasions, it may be easier to generate a shortest DNF of f^d, or even an arbitrary DNF of f^d, rather than its complete DNF. Indeed, Theorem 3.17 shows that for some Boolean functions, the size of the complete DNF may be exponentially larger than the size of certain appropriately selected DNF representations. (This can only hold for nonmonotone functions. Indeed, for monotone functions, the complete DNF is necessarily shorter than any other DNF; see Theorem 1.24.)

It turns out, however, that practically all dualization algorithms generate the complete DNF of f^d, rather than an arbitrary DNF. Moreover, analyzing the complexity of the "incomplete" version of the problem requires special care, since the output of the problem is not univocally defined, or may not have an efficient characterization (e.g., when the objective is to generate a shortest DNF of f^d). These reasons explain why we mostly concentrate here on generating the complete DNF of f^d. Exceptions will be found in Theorem 4.29 and, indirectly, in the proof of Theorem 4.28.   □

As mentioned in Sections 2.11.2 and 4.1.4, and as expressed by the proof of Theorem 4.26, DUALIZATION and DNF DUALIZATION can be seen as generalizations of the problem of solving (DNF) equations. Therefore, both problems are certainly hard. More precisely:

Theorem 4.27. *Unless* P = NP, *there is no polynomial total time algorithm for* DUALIZATION *or* DNF DUALIZATION, *even if their input is restricted to cubic DNFs.*

Proof. Assume that there is a polynomial total time algorithm $\mathcal{A}$ for either problem. Denote by $r(L,U)$ the running time of $\mathcal{A}$, where $r(x,y)$ is a bivariate polynomial, L is the input length and U is the output length.

Let ϕ be a cubic DNF. From Theorem 2.1, we know that, unless P = NP, there is no polynomial time algorithm for deciding whether the equation $\phi(X) = 0$ is consistent. Note that $\phi = 0$ is inconsistent exactly when ϕ^d is identically 0, that is when ϕ^d has no implicant.

We now consider any of the two dualization problems with the input ϕ. Run the algorithm $\mathcal{A}$ on ϕ until either (i) it halts or (ii) the time limit $r(|\phi|,0)$ is exceeded. In case (i), if $\mathcal{A}$ outputs some implicant of f^d, then the equation $\phi(X) = 0$ is consistent; otherwise, it is inconsistent. In case (ii), the equation $\phi(X) = 0$ is consistent. Therefore, in both cases, the equation has been solved in time polynomial in $|\phi|$, which can only happen if P = NP. $\square$

A converse of Theorem 4.27 holds. Indeed, if P = NP, then the following result implies the existence of a polynomial total time algorithm for DUALIZATION (and hence, for DNF DUALIZATION):

Theorem 4.28. *There is an algorithm for* DUALIZATION *which, given an arbitrary Boolean expression* $\phi(x_1, x_2, \ldots, x_n)$ *of a function* f, *produces the complete DNF* ψ *of* f^d *by solving* $O(np)$ *Boolean equations of size at most* $|\phi| + |\psi|$, *where* p *is the number of prime implicants of* f^d. *If* $t(L)$ *is the complexity of solving a Boolean equation with input length at most* L, *then the running time of this algorithm is polynomial in* $|\phi|$, p *and* $t(|\phi| + |\psi|)$.

Proof. The algorithm combines the arguments developed in Theorem 2.20, Corollary 3.5, and Theorem 3.9. It consists of two phases.

In Phase 1, as in the proof of Theorem 2.20, assume that the prime implicants $C_1, C_2, \ldots, C_k$ of f^d have already been produced ($k \leq p$). In the next iteration, the algorithm solves the equation

$$\phi(X) \vee \bigvee_{i=1}^{k} C_i(\overline{X}) = 0. \tag{4.13}$$

If $X^* \in \mathcal{B}^n$ is a solution of (4.13), then X^* is a false point of ϕ, that is, $\overline{X^*}$ is a true point of f^d, and $\overline{X^*}$ is not covered by any of $C_1, C_2, \ldots, C_k$. In other words, the minterm $C^* = \bigwedge_{j \notin supp(X^*)} x_j \bigwedge_{j \in supp(X^*)} \overline{x}_j$, is an implicant of f^d that is not absorbed by any of the known prime implicants. Therefore, as in Corollary 3.5, solving n Boolean equations allows us to produce a new prime implicant of f^d that absorbs C^*. The algorithm adds this new prime implicant to the list and proceeds with the next iteration of Phase 1.

Conversely, if (4.13) is inconsistent, then it means that every true point of f^d is covered by one of $C_1, C_2, \ldots, C_k$. So, at this point, the DNF $\phi'(X) = \bigvee_{i=1}^{k} C_i(X)$ represents f^d, although some prime implicants of f^d may still be missing. Then, Phase 1 terminates and the algorithm enters a second phase where the consensus procedure is applied to the DNF ϕ' (with the same modifications as in the proof of Theorem 3.9), until the complete DNF of f^d has been obtained.

Clearly, the whole algorithm runs in time polynomial in $|\phi|$, p and $t(|\phi| + |\psi|)$. $\square$

A result similar to Theorem 4.28 holds for generating the minterm expression of the dual (remember that the minterm expression of a function is a special type of DNF representation; see Definition 1.11).

Theorem 4.29. *There is an algorithm which, given an arbitrary Boolean expression $\phi(x_1, x_2, \ldots, x_n)$ of a function f, produces the minterm expression of f^d by solving $q + 1$ Boolean equations of size at most $|\phi| + nq$, where q is the number of minterms of f^d. If $t(L)$ is the complexity of solving a Boolean equation with input length at most L, then the running time of this algorithm is polynomial in $|\phi|$, q and $t(|\phi| + nq)$.*

Proof. This is an immediate corollary of Theorem 2.20 and of the fact that X is a false point of f if and only if $\overline{X}$ is a true point of f^d. $\square$

Together with the results obtained in previous chapters, Theorems 4.28 and 4.29 stress once again the close connection among three fundamental problems on Boolean functions, namely, the solution of Boolean equations, the generation of prime implicants, and the dualization problem. Essentially, these results show that an algorithm for any of these three problems can be used as a black box for the solution of the other two problems. Indeed, assume that $\mathcal{A}$ is an algorithm taking as input an arbitrary Boolean expression ϕ, and let f be the function represented by ϕ:

(i) If $\mathcal{A}$ is a dualization algorithm or an algorithm that generates all prime implicants of f, then $\mathcal{A}$ trivially solves the equation $\phi = 0$.

(ii) Conversely, if $\mathcal{A}$ is an algorithm for the solution of Boolean equations, then $\mathcal{A}$ can be used to produce all prime implicants of f (see Theorem 3.9) as well as all prime implicants of f^d (see Theorem 4.28).

4.3.2 Dualization by sequential distributivity

The algorithms sketched in Theorems 4.29 and 4.28 are valid when ϕ is not in disjunctive normal form, but they require subroutines (i.e., NP-oracles) for the solution of Boolean equations. In this section, we present a simple dualization algorithm for the most important case, namely, the DNF DUALIZATION problem. It is based on Theorem 4.5, which shows that a CNF of f^d can be immediately deduced from any DNF of f. Then, by repeated use of the distributivity law and of absorption, the available CNF can easily be transformed into a DNF of f^d.

More formally, for the input DNF $\phi = \bigvee_{i=1}^{m} C_i$, let $\phi_k = \bigvee_{i=1}^{k} C_i$ and let f_k denote the function represented by ϕ_k ($k = 1, 2, \ldots, m$). The k-th iteration of the algorithm computes all prime implicants of f_k^d, so that the task is complete after the m-th iteration.

For $i = 1, 2, \ldots, m$, let $C_i = \left(\bigwedge_{j \in L_i} \ell_j \right)$, where $\ell_1, \ell_2, \ldots$ are literals. The prime implicants of $f_1^d = C_1^d$ are exactly the literals ℓ_j ($j \in L_1$). For $k > 1$, suppose that f_{k-1}^d is expressed by its complete DNF $\bigvee_{T \in \mathcal{T}} P_T$. Then, by Theorem 4.5,

$$f_k^d = \bigwedge_{i=1}^{k} \left(\bigvee_{j \in L_i} \ell_j \right) = \left(\bigvee_{T \in \mathcal{T}} P_T \right) \wedge \left(\bigvee_{j \in L_k} \ell_j \right),$$

and, by distributivity,

$$f_k^d = \bigvee_{T \in \mathcal{T}} \bigvee_{j \in L_k} P_T \ell_j.$$

So, we obtain all prime implicants of f_k^d from those of f_{k-1}^d by first generating all terms $P_T \ell_j$ ($j \in L_k$), for each prime implicant P_T of f_{k-1}^d, and then removing the terms that are absorbed.

Example 4.7. *Let $\phi = \overline{x}y \vee xz\overline{u} \vee x\overline{y}z$. Then, $\phi_1 = \overline{x}y$ and f_1^d has two prime implicants, namely, $\overline{x}$ and y. Consider now $\phi_2 = \overline{x}y \vee xz\overline{u}$. Applying the distributivity law to the dual expression $\phi_2^d = (\overline{x} \vee y)(x \vee z \vee \overline{u})$, we generate the terms $\overline{x}x$, $\overline{x}z$, $\overline{x}\,\overline{u}$, xy, yz, and $y\overline{u}$. The first term is absorbed by the other ones, so that ϕ_2^d has 5 prime implicants: $\overline{x}z$, $\overline{x}\,\overline{u}$, xy, yz, and $y\overline{u}$. Finally, we obtain that*

$$\phi_3^d = (\overline{x}z \vee \overline{x}\,\overline{u} \vee xy \vee yz \vee y\overline{u})(x \vee \overline{y} \vee z),$$

and we generate the terms $\overline{x}\,\overline{y}z$, $\overline{x}z$, $\overline{x}\,\overline{y}\,\overline{u}$, $\overline{x}z\overline{u}$, xy, xyz, yz, $xy\overline{u}$, and $yz\overline{u}$. Since $\overline{x}\,\overline{y}z$, $\overline{x}z\overline{u}$, xyz, $xy\overline{u}$, and $yz\overline{u}$ are absorbed, we conclude that $\phi^d (= \phi_3^d)$ has 4 prime implicants, namely, $\overline{x}z$, $\overline{x}\,\overline{y}\,\overline{u}$, xy, and yz. ☐

The resulting procedure is called SD-DUALIZATION (for "sequential-distributive dualization") and is stated more formally in Figure 4.1.

Theorem 4.30. *Procedure SD-DUALIZATION outputs all the prime implicants of f^d.*

Proof. The statement follows easily from Theorem 4.7. ☐

Procedure SD-DUALIZATION is part of the folklore of the field and has been repeatedly proposed by numerous authors, often in the context of the dualization of positive DNFs; see Fortet [342], Maghout [643], Pyne and McCluskey [765], Kuntzmann [589], Benzaken [61], Lawler [603], and so on. (Some authors [119, 432] recently called it "Berge multiplication," in reference to its description in [71, 72].) Nelson [705] proposed using it as a subroutine in his so-called *double*

```
Procedure SD-DUALIZATION
Input: A DNF φ = ⋁ᵢ₌₁ᵐ (⋀ⱼ∈Lᵢ ℓⱼ) of a Boolean function f.
Output: The set of prime implicants of fᵈ.
begin
     T* := {ℓⱼ | j ∈ L₁};
     for k = 2 to m do
     begin
          T := ∅;
          for all P ∈ T* and for all j ∈ Lₖ do T := T ∪ {P ℓⱼ};
          remove from T every term which is absorbed by another term in T;
          T* := T;
     end
     return T*;
end
```

Figure 4.1. Procedure SD-DUALIZATION.

dualization method for generating all prime implicants of a function f represented by a DNF ϕ: Indeed, all prime implicants of f can be obtained by applying SD-DUALIZATION twice in succession, first on ϕ, then on the complete DNF of f^d obtained after this first step (see Section 3.2.4).

From a practical viewpoint, this simple algorithm is reasonably efficient for small problem sizes and can easily be accelerated by various procedural shortcuts, such as those based on the following result, found in Benzaken [61]:

Theorem 4.31. *Let $P_1, P_2, \ldots, P_k$ be the prime implicants of a Boolean function f, and let $C = \bigwedge_{j=1}^{p} \ell_j$ be an elementary conjunction, where $\ell_1, \ell_2, \ldots, \ell_p$ are literals. Assume that $P_1, P_2, \ldots, P_k$ are sorted into $p+2$ classes $T(1), T(2), \ldots, T(p+2)$, where*

- *for $i = 1, 2, \ldots, p$, each conjunction in $T(i)$ involves the literal ℓ_i and no other literal from C;*
- *each conjunction in $T(p+1)$ involves at least 2 literals from C; and*
- *the conjunctions in $T(p+2)$ do not involve any literal from C.*

Then, the prime implicants of $f \wedge (\bigvee_{j=1}^{p} \ell_j)$ are exactly

   (i) *the conjunctions in $T(1), T(2), \ldots, T(p+1)$; and*
   (ii) *the conjunctions of the form $P_i \ell_j$, where $P_i \in T(p+2)$, $j \in \{1, 2, \ldots, p\}$, and $P_i \ell_j$ is not absorbed by any conjunction in $T(j)$.*

Proof. Left as exercise for the reader.   □

Additional shortcuts and other improvements of SD-DUALIZATION have been proposed and implemented by several researchers, such as Benzaken [61], Locks [616, 617], Shier and Whited [830], etc. More recently, the application of this algorithm to positive DNFs has received special attention, and its efficiency has been improved in various ways, for instance, by Bailey, Manoukian, and Ramamohanarao [40]; Dong and Li [276]; or Kavvadias and Stavropoulos [559].

SD-DUALIZATION does not run in polynomial total time (even on positive DNFs), namely, its running time may be exponentially large in the combined input *and* output size of the problem. In fact, it tends to generate many useless terms in its intermediate iterations (for $k = 2, \ldots, m - 1$), and it only generates the prime implicant of f^d in its very last iteration (when $k = m$), after exponentially many operations may already have been performed. This behavior was described more accurately by Takata [856], who showed that on some examples, SD-DUALIZATION may produce a superpolynomial blowup for *every* possible ordering of the terms of the input DNF (see also Hagen [432]). By contrast however, Boros, Elbassioni and Makino [119] proved that SD-DUALIZATION can be implemented to run in output-subexponential time on positive DNFs, and to run in polynomial total time on certain special classes of positive DNFs, such as bounded-degree DNFs or read-once DNFs (see also Exercise 7).

4.4 Algorithmic aspects: Positive functions

4.4.1 Some complexity results

This section focuses on the dualization problem for positive Boolean functions. Just as in the general case, this problem appears to be intractable (Lawler, Lenstra, and Rinnooy Kan [605]).

Theorem 4.32. *Unless* P = NP, *there exists no polynomial total time algorithm for* DUALIZATION, *even if its input represents a positive function.*

Proof. Consider a DNF equation $\psi(x_1, x_2, \ldots, x_n) = 0$, and assume that each of the literals x_i and $\overline{x}_i$ appears at least once in ψ, for $i = 1, 2, \ldots, n$. Clearly, solving this type of DNF equation is NP-complete.

Now, let $\psi^*(x_1, x_2, \ldots, x_n, x_{n+1}, x_{n+2}, \ldots, x_{2n})$ be the positive DNF obtained after replacing each negative literal $\overline{x}_i$ by a new variable x_{n+i} in ψ ($i = 1, 2, \ldots, n$). Notice that $\psi(X) = 0$ if and only if $\psi^*(X, \overline{X}) = 0$. Define further the positive expression:

$$\phi(x_1, x_2, \ldots, x_{2n}) = \psi^* \wedge \bigwedge_{i=1}^{n} (x_i \vee x_{n+i}). \tag{4.14}$$

Let $X^{(i)}$, $i = 1, 2, \ldots, n$, be the point of $\mathcal{B}^{2n}$ having all its components equal to 1 except for the i-th and $(n+i)$-th components. Clearly, $X^{(i)}$ is a maximal false point of ϕ. We now claim that the maximal false points of ϕ are exactly the points $X^{(1)}, X^{(2)}, \ldots, X^{(n)}$ if and only if the equation $\psi = 0$ has no solution.

Let us first assume that ϕ has a maximal false point $Y \in \mathcal{B}^{2n}$ other than $X^{(1)}, X^{(2)}, \ldots, X^{(n)}$. If $y_i = y_{n+i} = 0$ holds for some index i, then $Y \leq X^{(i)}$, a contradiction. So, $\bigwedge_{i=1}^{n}(y_i \vee y_{n+i}) = 1$, and there follows that $\psi^*(Y) = 0$.

Let $U = (y_1, y_2, \ldots, y_n)$ and note that $(U, \overline{U}) \leq Y$. Hence (by positivity of ψ^*) $\psi^*(U, \overline{U}) = 0$ and $\psi(U) = 0$, thus proving the "if" part of the claim.

Conversely, if $\psi = 0$ has a solution, say, $\psi(U) = 0$, then $\psi^*(U, \overline{U}) = \phi(U, \overline{U}) = 0$. Thus, there exists a maximal false point of ϕ, say, $Y \in \mathcal{B}^{2n}$, such that $(U, \overline{U}) \leq Y$. Note that $y_i = y_{n+i} = 0$ cannot hold for any index i, and hence Y is distinct from $X^{(1)}, X^{(2)}, \ldots, X^{(n)}$, proving the "only if" part of the claim

Using Theorem 4.18, we obtain that the dual of ϕ has exactly n prime implicants if and only if the equation $\psi = 0$ has no solution. Now, the proof is easily completed by the same type of argument as in the proof of Theorem 4.27. $\square$

Observe that the expression (4.14) is not a disjunctive normal form, so that Theorem 4.32 does not settle the complexity of DUALIZATION when its input ϕ is restricted to positive DNFs: Let us call this problem POSITIVE DNF DUALIZATION. Clearly, we can assume without loss of generality that the input of POSITIVE DNF DUALIZATION is the complete DNF of a positive function f, that is, a positive DNF consisting of all prime implicants of f. Thus, formally, we define POSITIVE DNF DUALIZATION as follows:

POSITIVE DNF DUALIZATION
Instance: The complete DNF of a positive Boolean function f.
Output: The complete DNF of f^d.

For simplicity, and when no confusion can arise, we often use the same notation for a positive function f and for its complete DNF ϕ in the sequel. For instance, we denote by $|f|$ the size of the complete DNF of f, that is, we let $|f| \overset{\text{def}}{=} |\phi|$.

POSITIVE DNF DUALIZATION is known to be equivalent to many interesting problems encountered in various fields (see Section 4.2 and [295]). Within Boolean theory alone, several authors – in particular, Bioch and Ibaraki [89]; Eiter and Gottlob [295]; Fredman and Khachiyan [347]; Johnson, Yannakakis, and Papadimitriou [538] – have observed that this problem is polynomially equivalent to the fundamental problem of recognizing whether two positive functions f and g are *mutually dual*, namely, whether $f = g^d$ (note that this is just the positive version of the DUAL RECOGNITION problem introduced in Section 4.3.1):

POSITIVE DUAL RECOGNITION
Instance: The complete DNFs of two positive Boolean functions f and g.
Question: Is $g = f^d$?

If f and g are not mutually dual, then by definition of duality, there exists a point $X^* \in \mathcal{B}^n$ such that $f(X^*) = g(\overline{X^*})$. Let us now establish that solving POSITIVE DUAL RECOGNITION indirectly allows us to determine such a point X^*. (It is interesting to observe that a similar result holds without the positivity assumptions.)

Theorem 4.33. *If f and g are two positive functions on $\mathcal{B}^n$ expressed by their complete DNFs, and if f and g are not mutually dual, then a point $X^* \in \mathcal{B}^n$*

such that $f(X^*) = g(\overline{X^*})$ *can be found by solving n instances of* POSITIVE DUAL RECOGNITION *with size at most* $|f| + |g|$.

Proof. The proof is by induction on n. Let $\mathcal{A}$ be an algorithm for POSITIVE DUAL RECOGNITION, and assume that $\mathcal{A}$ returns the output No on the instance (f, g). Theorem 4.3 implies that either $f_{|x_n=0}$ and $g_{|x_n=1}$ are not mutually dual or $f_{|x_n=1}$ and $g_{|x_n=0}$ are not mutually dual. Let us assume, without loss of generality, that $f_{|x_n=0}$ and $g_{|x_n=1}$ are not mutually dual (one call on the algorithm $\mathcal{A}$ suffices to find out). By induction on the number of variables, $n-1$ additional calls on $\mathcal{A}$ can be used to compute a point $Y^* \in \mathcal{B}^{n-1}$ such that $f_{|x_n=0}(Y^*) = g_{|x_n=1}(\overline{Y^*})$. Then, $f(Y^*, 0) = g(\overline{Y^*}, 1)$, and the point $X^* = (Y^*, 0)$ is as required. $\square$

We are now in a position to establish the equivalence of POSITIVE DNF DUALIZATION and POSITIVE DUAL RECOGNITION.

Theorem 4.34. POSITIVE DNF DUALIZATION *and* POSITIVE DUAL RECOGNITION *are polynomially equivalent. More precisely:*

(i) *There is an algorithm for* POSITIVE DUAL RECOGNITION *which, given the complete DNFs of two positive functions f and g, decides whether f and g are mutually dual by solving one instance of* POSITIVE DNF DUALIZATION. *If $r(|f|, |f^d|)$ is the complexity of solving* POSITIVE DNF DUALIZATION *on the input f, then the running time of this algorithm is polynomial $|f|$, $|g|$ and $r(|f|, |g|)$.*

(ii) *Conversely, there is an algorithm for* POSITIVE DNF DUALIZATION *which, given the complete DNF of a positive function f, produces the complete DNF of f^d by solving $O(np)$ instances of* POSITIVE DUAL RECOGNITION *of size at most $|f| + |f^d|$, where p is the number of prime implicants of f^d. If $t(f_1, f_2)$ is the complexity of solving* POSITIVE DUAL RECOGNITION *on the input (f_1, f_2), then the running time of this algorithm is polynomial in $|f|$, p and $t(|f|, |f^d|)$.*

Proof. (i) If $\mathcal{A}$ is a dualization algorithm with running time $r(|f|, |f^d|)$, and (f, g) is the input to POSITIVE DUAL RECOGNITION, then we run $\mathcal{A}$ on the input f. If $\mathcal{A}$ does not stop at time $r(|f|, |g|)$, then it means that $g \neq f^d$. Otherwise, the output of $\mathcal{A}$ can be used to determine whether $g = f^d$ and to answer POSITIVE DUAL RECOGNITION.

(ii) Assume that $\mathcal{A}$ is an algorithm for POSITIVE DUAL RECOGNITION and assume that, at some stage, the prime implicants P_J ($J \in \mathcal{G}$) of f^d have already been produced, where $|\mathcal{G}| \leq p$. In the next iteration, the algorithm considers the positive function

$$g = \bigvee_{J \in \mathcal{G}} P_J(X). \qquad (4.15)$$

The algorithm $\mathcal{A}$ can be used to decide whether $g = f^d$. In the affirmative, we can stop. Otherwise, $\mathcal{A}$ can again be used (as in Theorem 4.33) to compute a point $X^* \in \mathcal{B}^n$ such that $f(X^*) = g(\overline{X^*})$. Since $g \leq f^d$, it must be the case that $f(X^*) = g(\overline{X^*}) = 0$. Then, we can find (in polynomial time) a maximal false point of f, say, Y^*, such that $X^* \leq Y^*$. By Theorem 4.18, the term $P = \bigwedge_{j \in supp(\overline{Y^*})} x_j$ is a prime implicant of f^d and $P(\overline{Y^*}) = 1$. On the other hand, by positivity of g, $g(\overline{Y^*}) = 0$, which implies that the prime implicant P is not in the current list $(P_J, J \in \mathcal{G})$.

This process can be repeated $p + 1$ times in order to produce all the prime implicants of f^d. $\qquad\square$

Lawler, Lenstra, and Rinnooy Kan [605] and several other researchers (see Garcia-Molina, and Barbara [370]; Johnson, Yannakakis, and Papadimitriou [538]; Bioch and Ibaraki [89]; Eiter and Gottlob [295]) have asked whether POSITIVE DNF DUALIZATION can be solved in polynomial total time or, equivalently, whether POSITIVE DUAL RECOGNITION can be solved in polynomial time. This central question of duality theory remains open to this day. A breakthrough result by Fredman and Khachiyan [347], however, has established the existence of *quasi-polynomial* time algorithms for POSITIVE DUAL RECOGNITION and for POSITIVE DNF DUALIZATION. This is in stark contrast with the NP-hardness results obtained for the general DUAL RECOGNITION (Theorem 4.26) and DNF DUALIZATION problems (Theorem 4.27), since it is widely believed that NP-hard problems have no quasi-polynomial time algorithm.

4.4.2 A quasi-polynomial dualization algorithm

We now describe a simplest version of the dualization algorithm proposed by Fredman and Khachiyan [347], which builds on the approach developed in Theorems 4.33 and 4.34. Consider the complete DNFs of a positive function f on $\mathcal{B}^n$ and of its dual f^d, say,

$$f = \bigvee_{I \in \mathcal{F}} \left(\bigwedge_{i \in I} x_i \right) \tag{4.16}$$

and

$$f^d = \bigvee_{J \in \mathcal{F}^d} \left(\bigwedge_{j \in J} x_j \right). \tag{4.17}$$

As in Theorem 4.34, let $\mathcal{G} \subseteq \mathcal{F}^d$ represent the collection of prime implicants of f^d which are currently known, and let

$$g = \bigvee_{J \in \mathcal{G}} \left(\bigwedge_{j \in J} x_j \right). \tag{4.18}$$

The algorithm proceeds to determine whether f and g are mutually dual and, in the negative, to find a point point $X^* \in \mathcal{B}^n$ such that

$$f(X^*) = g(\overline{X^*}) = 0. \tag{4.19}$$

However, since an efficient procedure is not immediately available for deciding whether f and g are mutually dual (i.e., for solving POSITIVE DUAL RECOGNITION), we cannot apply the same recursive approach used in the proof of Theorem 4.33. Therefore, we introduce here two crucial modifications. First, instead of exactly solving an instance of POSITIVE DUAL RECOGNITION at every step of the recursion (as in the proof of Theorem 4.33), we rely on an incomplete test based on examining the quantity

$$E(f,g) \stackrel{\text{def}}{=} \sum_{I \in \mathcal{F}} 2^{-|I|} + \sum_{J \in \mathcal{G}} 2^{-|J|}. \tag{4.20}$$

Theorem 4.35. *Let f and g be two positive functions defined by (4.16) and (4.18). If $E(f,g) < 1$, then f and g are not mutually dual, and a point X^* satisfying (4.19) can be computed in polynomial time.*

Proof. We use the same approach in the proofs of Theorems 2.26 and 2.27. Namely, consider the polynomial

$$F(X) = \sum_{I \in \mathcal{F}} \prod_{i \in I} x_i + \sum_{J \in \mathcal{G}} \prod_{j \in J} (1 - x_j).$$

It defines a real-valued function on $[0,1]^n$. Let $X^H = (\frac{1}{2}, \frac{1}{2}, \ldots, \frac{1}{2})$ denote the center of the unit hypercube. There holds $F(X^H) = E(f,g)$, and (using Rosenberg's results [789], as in the proof of Theorem 2.26), one can compute in polynomial time a point $X^* \in \{0,1\}^n$ such that $F(X^*) \leq F(X^H)$. In particular, if $E(f,g) < 1$, then $F(X^*) = 0$, which implies that $f(X^*) = g(\overline{X^*}) = 0$, and f and g are not mutually dual.   □

Thus, when $E(f,g) < 1$, Theorem 4.35 can be used as a substitute for Theorem 4.33. When $E(f,g) \geq 1$, however, we cannot draw any immediate conclusion, and we turn instead to a recursive divide-and-conquer procedure based on Theorem 4.3. But rather than decomposing f and g on an arbitrary variable, we are going to show how to choose a "good" variable x_i, so that the size of the resulting subproblems is relatively small. Observe first that when $E(f,g) \geq 1$, either f or g contains a prime implicant of only logarithmic length.

Lemma 4.1. *Let f and g be two positive functions defined by (4.16) and (4.18). If $E(f,g) \geq 1$, then either f or g has a prime implicant with degree at most $\log(|\mathcal{F}| + |\mathcal{G}|)$.*

Proof. Let $\delta = \min\{|I| \mid I \in \mathcal{F} \cup \mathcal{G}\}$ be the degree of a shortest prime implicant of either f or g. By definition (4.20), $(|\mathcal{F}| + |\mathcal{G}|)2^{-\delta} \geq E(f,g) \geq 1$.   □

For $\epsilon \in [0,1]$ and $i \in \{1,2,\ldots,n\}$, we say that variable x_i *occurs in f with frequency at least ϵ* if

$$\frac{|\{I \mid i \in I, I \in \mathcal{F}\}|}{|\mathcal{F}|} \geq \epsilon.$$

We say that x_i is a *frequent variable* for the pair (f,g) if x_i occurs with frequency at least $1/\log(|\mathcal{F}| + |\mathcal{G}|)$ either in f or in g.

Procedure RECOGNIZE DUAL

Input: Two positive Boolean functions f and g on $\mathcal{B}^n$ expressed by their complete DNFs (4.16) and (4.18), with $I \cap J \neq \emptyset$ for all $I \in \mathcal{F}$ and all $J \in \mathcal{G}$.

Output: Yes if f and g are mutually dual. Otherwise, a point $X^* \in \mathcal{B}^n$ such that $f(X^*) = g(\overline{X^*}) = 0$.

begin

Step 1: **if** $E \geq 1$ **then** go to Step 2

         else return a vector $X^* \in \mathcal{B}^n$ such that $f(X^*) = g(\overline{X^*})$;

Step 2: **if** $|\mathcal{F}||\mathcal{G}| \leq 1$ **then** check directly whether $g = f^d$;

         if $g \neq f^d$ **then** return $X^* \in \mathcal{B}^n$ such that $f(X^*) = g(\overline{X^*})$

         else return "Yes";

Step 3: select a frequent variable x_i for the pair (f, g);

         call RECOGNIZE DUAL$(f|_{x_i=0}, g|_{x_i=1})$;

         if the returned value is $Y^* \in \mathcal{B}^{n-1}$

           then return $X^* \in \mathcal{B}^n$, where $x_j^* := y_j^*$ for all $j \neq i$ and $x_i^* := 0$

           else begin

               call RECOGNIZE DUAL$(f|_{x_i=1}, g|_{x_i=0})$;

               if the returned value is $Y^* \in \mathcal{B}^{n-1}$

                  then return $X^* \in \mathcal{B}^n$, where $x_j^* := y_j^*$ for all $j \neq i$ and $x_i^* := 1$

                  else return "Yes";

           end

end;

Figure 4.2. Procedure RECOGNIZE DUAL.

Theorem 4.36. *Let f and g be two positive functions defined by* (4.16) *and* (4.18), *and assume that*

$$I \cap J \neq \emptyset \quad \text{for all } I \in \mathcal{F} \text{ and } J \in \mathcal{G}. \tag{4.21}$$

If $E(f, g) \geq 1$ and $|\mathcal{F}||\mathcal{G}| \geq 1$, then there exists a frequent variable for the pair (f, g).

Proof. By Lemma 4.1, either f or g has a prime implicant with degree at most $\log(|\mathcal{F}| + |\mathcal{G}|)$. Let us assume without loss of generality that $J \in \mathcal{G}$ defines such a short implicant. Then, in view of (4.21), some variable x_i, $i \in J$, must occur in f with frequency $1/|J| \geq 1/\log(|\mathcal{F}| + |\mathcal{G}|)$. $\square$

We now have all the necessary ingredients to present the important quasi-polynomial time algorithm proposed by Fredman and Khachiyan [347] for the solution of POSITIVE DUAL RECOGNITION. A formal description of the algorithm is given in Figure 4.2.

Theorem 4.37. *Procedure* RECOGNIZE DUAL *is correct and runs in time* $m^{4\log^2 m + O(1)}$, *where $m = |\mathcal{F}| + |\mathcal{G}|$.*

Proof. The correctness of the procedure follows from the above discussion. Theorem 4.35 implies that Step 1 can be executed in time polynomial in the input size $|f| + |g|$. It can be checked that, if $g = f^d$, then $n \leq |\mathcal{F}||\mathcal{G}| \leq m^2$ (see Exercise 5), and hence $|f| + |g| = O(nm) = O(m^3)$. Step 2 is easily done in $O(1)$ time. Therefore, up to a polynomial factor $m^{O(1)}$, the running time of the procedure is bounded by the number of recursive calls.

Fix m, and let $\epsilon = 1/\log m$. Let $v = |\mathcal{F}||\mathcal{G}|$ be the *volume* of the pair (f, g), and let $a(v)$ be the maximum number of recursive calls of the procedure when running on a pair with size at most m and volume at most v. We are going to show that

$$a(v) \leq m^{4 \log^2 m}. \tag{4.22}$$

Note that the size of each pair involved in a recursive call is smaller than m. So, the frequent variable x_i selected in Step 3 always has frequency at least ϵ either in f or in g. Suppose, without loss of generality, that x_i occurs with frequency ϵ in f.

Then, the number of terms of $f_{|x_i=0}$ is at most $(1 - \epsilon)|\mathcal{F}|$, and the number of terms of $g_{|x_i=1}$ is at most $|\mathcal{G}|$, so that the volume of the pair $(f_{|x_i=0}, g_{|x_i=1})$ is at most $(1 - \epsilon)v$. Also, the number of terms of $f_{|x_i=1}$ is at most $|\mathcal{F}|$ and the number of terms of $g_{|x_i=0}$ is at most $|\mathcal{G}| - 1$, so that the volume of the pair $(f_{|x_i=1}, g_{|x_i=0})$ is at most $v - 1$.

We thus obtain the following recurrence:

$$a(v) \leq 1 + a((1 - \epsilon)v) + a(v - 1) \quad \text{and} \quad a(1) = 1.$$

From this recurrence, we obtain $a(v) \leq k + ka((1 - \epsilon)v) + a(v - k)$ for all $k \leq v$. Letting $k = \lceil v\epsilon \rceil$ yields $a(v) \leq (3 + 2v\epsilon)a((1 - \epsilon)v)$, and hence

$$a(v) \leq (3 + 2v\epsilon)^{(\log v)/\epsilon}.$$

The bound (4.22) on $a(v)$ follows from $v = |\mathcal{F}||\mathcal{G}| \leq (|\mathcal{F}| + |\mathcal{G}|)^2/4 \leq m^2/4$ and $\epsilon = 1/\log m$. $\square$

A dualization algorithm for positive functions in DNF can be obtained as a by-product of RECOGNIZE DUAL, just as in Theorem 4.34. The procedure is described in Figure 4.3.

As an immediate consequence of the above results, we obtain:

Theorem 4.38. *Procedure* FK-DUALIZATION *is correct and runs in time* $m^{4 \log^2 m + O(1)}$, *where* $m = |\mathcal{F}| + |\mathcal{F}^d|$.

Fredman and Khachiyan [347] have improved the time complexity of RECOGNIZE DUAL (or FK-DUALIZATION) to $m^{o(\log m)}$ (see also Elbassioni [309]). But, as already mentioned, it remains an important open question to determine whether the dual recognition problem can be solved in polynomial time or, equivalently, whether POSITIVE DNF DUALIZATION can be solved in polynomial total time.

The results presented in this section have been a source of inspiration for much subsequent research and have been generalized in many ways. For instance, Boros et al. [117, 123, 124, 562, etc.] considered natural generalizations of positive dualization problems that allow them to model numerous interesting applications

---

Procedure FK-DUALIZATION
Input: A positive Boolean function f on $\mathcal{B}^n$ expressed by its complete DNF.
Output: The complete DNF of f^d.

Step 0: $g := 0$;

Step 1: Call RECOGNIZE DUAL on the pair (f, g);
       if the returned value is "Yes" **then** halt;
       else let $X^* \in \mathcal{B}^n$ be the point returned by RECOGNIZE DUAL;
          compute a maximal false point of f, say Y^*, such that $X^* \le Y^*$;
          $g := g \vee \bigwedge_{j \in supp(\overline{Y^*})} x_j$;
          return to Step 1.

Figure 4.3. Procedure FK-DUALIZATION.

[9, 651, 654, 839]. We refer to Eiter, Makino, and Gottlob [302] and to Boros, Elbassioni, Gurvich, and Makino [118] for surveys of related results. It is also worth recalling at this point that the sequential-distributive algorithm SD-DUALIZATION has been recently shown to run in subexponential time on positive DNFs ([119]; see Section 4.3.2).

4.4.3 Additional results

Bioch and Ibaraki [89] and Eiter and Gottlob [295] have systematically investigated several algorithmic problems that turn out to be polynomially equivalent to dualization. We have already mentioned the equivalence of POSITIVE DNF DUALIZATION and POSITIVE DUAL RECOGNITION. It can also be shown that POSITIVE DUAL RECOGNITION is equivalent to the (apparently more restrictive) problem of deciding whether a positive function given in complete disjunctive normal form is self-dual or not, and to the following (apparently more general) *identification problem*:

IDENTIFICATION
Instance: A black-box oracle to evaluate a positive Boolean function f at any given point.
Output: All prime implicants of f and of f^d.

The importance of this problem, where knowledge of f can only be gained through queries of the form: "What is the value of f at the point X?" has been underlined by Bioch and Ibaraki [89] and has been investigated especially in the machine learning literature in relation to various other models of "exact learning by membership queries"; see [21, 22, 29, 275, 429, 651, 652, 653, 654, 838, 884, etc.] and [233] or Chapter 12 for related considerations. Incremental approaches of the type used in Theorems 4.29, 4.28, 4.34, in particular, have proved useful in the oracle context (see, for instance, Lawler, Lenstra, and Rinnooy Kan [605] or Angluin [21]).

Many researchers have investigated natural special cases of POSITIVE DNF DUALIZATION [74, 129, 225, 275, 295, 538, 652, 653, 735, 736]. If ϕ is a positive quadratic DNF, then the dualization problem is equivalent to the problem of generating all maximal stable sets of a graph and can be solved with polynomial delay [873, 605, 538]. More generally, the dualization problem has a polynomial total time algorithm when its input is restricted to positive DNFs of degree at most k, where k is viewed as a constant [119, 121, 295]. Many other subcases can also be solved in polynomial total time; this is the case when f is regular, threshold, matroidal, read-once, acyclic, and so on (see [74, 119, 129, 225, 275, 295, 429, 605, 652, 653, 735, 736]). We refer to a survey by Eiter, Makino, and Gottlob [302] for more details.

Finally, Lawler, Lenstra, and Rinnooy Kan [605] observed that a general approach (inspired from previous work by Paull and Unger [733]) can be used to derive polynomial dualization algorithms for certain special classes of positive functions. This approach is quite different from those described so far: Instead of producing the prime implicants of f^d one by one, as in Theorem 4.34 or procedure FK-DUALIZATION, it recursively dualizes $f_{|x_1=\ldots=x_n=0}$, then $f_{|x_2=\ldots=x_n=0}$, $\ldots$, $f_{|x_n=0}$, and finally f. For $j = 1, 2, \ldots, n$, consider the following subproblem:

ADD-j
Instance: A prime implicant P of $(f_{|x_j=\ldots=x_n=0})^d$, where f is a positive Boolean function on $\mathcal{B}^n$ expressed in DNF.
Output: All prime implicants of $(f_{|x_{j+1}=\ldots=x_n=0})^d$ that are absorbed by P.

Theorem 4.39. *If $\mathcal{C}$ is a class of positive functions such that ADD-j can be solved in polynomial total time on $\mathcal{C}$ for all $j = 1, 2, \ldots, n$, then POSITIVE DNF DUALIZATION can be solved in polynomial total time on $\mathcal{C}$.*

Proof. We only sketch the proof. For every positive function f,

$$
\begin{aligned}
f^d &= \left(x_n f_{|x_n=1} \vee f_{|x_n=0} \right)^d \\
&= \left(x_n \vee (f_{|x_n=1})^d \right) (f_{|x_n=0})^d \\
&= x_n (f_{|x_n=0})^d \vee (f_{|x_n=1})^d (f_{|x_n=0})^d.
\end{aligned}
$$

It follows that every prime implicant of f^d is absorbed by some prime implicant of $(f_{|x_n=0})^d$ and, therefore, f^d can be computed by repeatedly solving ADD-n for all prime implicants of $(f_{|x_n=0})^d$.

Similarly, $(f_{|x_{j+1}=\ldots=x_n=0})^d$ can be computed for all j by repeatedly solving instances of ADD-j. Details, and ways to accelerate the algorithm, can be found in Lawler, Lenstra, and Rinnooy Kan [605]. $\square$

Despite its apparent simplicity, the approach sketched in Theorem 4.39 has a surprisingly broad range of applicability. Several related approaches are mentioned by Eiter, Makino, and Gottlob [302]; see also Grossi [413].

4.5 Exercises

1. Consider Reiter's analysis of the diagnosis problem (Application 4.1).
   (a) Prove that the characterization of diagnoses is correct.
   (b) With the same notations as in Application 4.1, define a *conflict set* to be a minimal subset $\Gamma \subseteq \{1,2,\ldots,m\}$ such that

$$\bigvee_{\substack{k=1 \\ k \in \Gamma}}^{m} \phi_k(X^*,Y) = 0$$

is inconsistent. Show that $\Gamma \subseteq \{1,2,\ldots,m\}$ is a conflict set if and only if $\bigwedge_{k \in \Gamma} p_k$ is a prime implicant of f.
   (c) Prove that the diagnoses are exactly the transversals of the conflict sets.

2. Prove that the composition of dual-minor positive functions is dual-minor, and the composition of dual-major positive functions is dual-major. Show that these results do not hold without the positivity assumption.

3. Show that, if $f(x_1,x_2,\ldots,x_n)$ is a Boolean function, then $g(x_1,x_2,\ldots,x_n,x_{n+1},$ $x_{n+2}) = x_{n+1}x_{n+2} \vee x_{n+1}f \vee x_{n+2}f^d$ is self-dual.

4. Show that there exists a positive function f such that $\chi(\mathcal{H}_f) \leq 3$, but f is not dual-minor (compare with Theorem 4.21).

5. Prove that, if f is a positive Boolean function on n variables, then $n \leq pq$, where p (respectively, q) is the number of prime implicants of f (respectively, f^d).

6. Show that the procedure SD-DUALIZATION presented in Section 4.3.1 does not run in polynomial total time.

7. Consider a variant of SD-DUALIZATION where the prime implicants of f are sorted in such a way that, for $j = 1,2,\ldots,n$, the prime implicants on $\{x_1,x_2,\ldots,x_j\}$ precede any prime implicant containing x_{j+1}. Prove that this variant can be implemented to run in polynomial total time on quadratic positive functions. (Note: this implies that all maximal stable sets of a graph can be generated in polynomial total time).

8. Prove Theorem 4.31.

9. Let ψ be a DNF of the Boolean function $f(x_1,x_2,\ldots,x_n)$. Show that the complete DNF of f^d can be generated by the following procedure: (a) In ψ, replace every occurence of the literal $\overline{x}_i$ by a new variable y_i ($i = 1,2,\ldots,n$), thus producing a positive DNF $\phi(x_1,x_2,\ldots,x_n,y_1,y_2,\ldots,y_n)$; (b) Generate the complete DNF of ϕ^d, say $\eta(x_1,x_2,\ldots,x_n,y_1,y_2,\ldots,y_n)$; (c) In η, replace every occurence of y_i by $\overline{x}_i$, and remove the terms which are identically zero. Is this sufficient to conclude that the problem DNF DUALIZATION is no more difficult than POSITIVE DNF DUALIZATION?

10. Show that the bounds in Lemma 4.1 and Theorem 4.36 are tight up to a factor of 2. (Fredman and Khachiyan [347].)

11. Show that Theorem 4.35, Lemma 4.1, and Theorem 4.36 hold for arbitrary, not necessarily positive functions. (Fredman and Khachiyan [347].)

12. Prove that POSITIVE DUAL RECOGNITION is polynomially equivalent to deciding whether a positive function given in complete disjunctive normal form is self-dual.

13. Prove that POSITIVE DNF DUALIZATION is polynomially equivalent to the IDENTIFICATION problem.

14. Complete the proof of Theorem 4.39. Show that ADD-j can be solved in polynomial time on the class C of quadratic positive functions. (Compare with Exercise 7.)

Question for thought

15. What is the complexity of the following problem: Given the complete DNFs of two Boolean functions f and g, decide whether $g = f^d$?

Part II

Special Classes

5

Quadratic functions

Bruno Simeone

This chapter is devoted to an important class of Boolean functions, namely, quadratic Boolean functions, or Boolean functions that can be represented by DNFs of degree at most two. Since linear functions are trivial in many respects, quadratic functions are in a sense the simplest interesting Boolean functions: Most of the fundamental problems introduced in the first part of this monograph – solving Boolean equations, generating prime implicants, dualization – turn out to be efficiently solvable for quadratic functions expressed in DNF. Their solution, however, requires a good understanding of structural properties of quadratic functions, as well as clever algorithms. Graph-theoretical models play a central role in these developments, and we will see that, conversely, many questions about graphs can also be fruitfully rephrased as questions involving quadratic Boolean functions.

5.1 Basic definitions and properties

We start with basic definitions and properties.

Definition 5.1. *We call a DNF*

$$\phi(x_1,\ldots,x_n) = \bigvee_{i=1}^{m} \left(\bigwedge_{j \in P_i} x_j \bigwedge_{j \in N_i} \overline{x}_j \right)$$

quadratic *if all its terms are quadratic, that is, if they are conjunctions of at most two literals:* $|P_i \cup N_i| \le 2$ *for all* $i \in \{1,\ldots,m\}$. *A term is called* linear *or* purely quadratic *according to whether it consists of exactly one or exactly two literals.*

Definition 5.2. *A Boolean function* f *is called* quadratic *if it admits a quadratic DNF.*

In a similar fashion, we call a CNF *quadratic* if all its clauses are disjunctions of at most two literals.

Definition 5.3. *A Boolean function f is called* dually quadratic *if it admits a quadratic CNF.*

This definition is equivalent to the property that the dual function f^d is quadratic.

Recall from Chapter 2 (Definition 2.5) that the *consensus* of two terms xC and $\overline{x}D$ is the term CD (provided it is not identically 0). Note that if both xC and $\overline{x}D$ are quadratic, then their consensus CD is quadratic, too.

An important consequence of this observation is the following:

Theorem 5.1. *All prime implicants of a quadratic Boolean function are quadratic.*

Proof. Let f be a quadratic Boolean function, and let ϕ be an arbitrary quadratic DNF representing f. By Theorem 3.5, all prime implicants of f can be obtained by applying the consensus procedure to ϕ. By the above observation, all terms obtained by this procedure, and, in particular, all prime implicants of f, must be quadratic.   □

Definition 5.4. *A quadratic Boolean function f is called* purely quadratic *if it is not constant and if it has no linear prime implicant or, equivalently, if no linear term appears in any DNF of f.*

The following statement follows immediately from the definitions:

Lemma 5.1. *If f is purely quadratic, then in every quadratic DNF of f every term is a prime implicant.*

Let us remark that a function might be quadratic even though at first sight it does not appear as such. In other words, it is quite possible for a quadratic function to be represented by a DNF of higher degree.

Example 5.1. *The function*

$$f = x_1 x_2 \overline{x}_3 \overline{x}_4 \vee x_1 x_2 \overline{x}_3 x_4 \vee \overline{x}_1 x_2 \overline{x}_3 x_4 \vee x_1 x_2 x_3 \vee \overline{x}_2 \overline{x}_3 x_4 \vee \overline{x}_1 x_3 \vee \overline{x}_2 \overline{x}_4$$

is quadratic, since it also admits the DNF

$$f = x_1 x_2 \vee \overline{x}_1 x_3 \vee \overline{x}_2 \overline{x}_4 \vee \overline{x}_3 x_4.$$ □

As noted in Chapter 1, Theorem 1.31, the problem of recognizing whether a given DNF represents a quadratic Boolean function is co-NP-complete. Therefore, we often assume that a quadratic Boolean function is given by a quadratic DNF. In particular, this is the case in Definition 5.5, which introduces one of the most important notions of this chapter.

Definition 5.5. *A quadratic Boolean equation is a DNF equation of the form*

$$\varphi(X) = 0,$$

where φ is a quadratic DNF.

Many authors prefer to concentrate on quadratic CNF equations of the form

$$\psi(X) = 1,$$

where ψ is a quadratic CNF. The problem of deciding whether an equation of the latter form has solutions is known under the name 2-SATISFIABILITY (2-SAT for short). As follows from the discussion in Section 1.4 and Section 2.2, however, the DNF and CNF forms of quadratic equations are strictly equivalent.

5.2   Why are quadratic Boolean functions important?

Quadratic Boolean functions are interesting, and worthy of investigation, for many reasons. Here, we list the main ones:

(1) Quadratic Boolean functions are "abundant in nature."
(2) There are strong connections in both directions between quadratic Boolean functions and graphs.
(3) Many significant combinatorial problems can be reduced to 2-SAT.
(4) Low complexity algorithms are available for solving 2-SAT, as well as for finding all prime implicants and irredundant normal forms of quadratic Boolean functions.

We now briefly comment on these points. Item (2) is discussed at length in Section 5.4, item (3) in Section 5.5, and item (4) in Sections 5.6 and 5.8.

(1) *Quadratic Boolean functions are "abundant in nature."*
     The most common types of logical relations, like

     "P implies Q."
     "Either P or Q is true."
     "Either P or Q is false."
     "Exactly one of P or Q is true."
     "P is true if and only if Q is true."

     can be represented by quadratic equations, such as

$$p\overline{q} = 0,$$
$$\overline{p}q = 0,$$
$$pq = 0,$$
$$pq \vee \overline{p}\,\overline{q} = 0,$$
$$p\overline{q} \vee \overline{p}q = 0.$$

     In fact, it has been estimated that about 95% of the production rules in expert systems are of the foregoing types and, hence, can be represented by quadratic equations (see Jaumard, Simeone, and Ow [531]).

(2) *Quadratic Boolean functions and graphs.*

The theory of quadratic Boolean functions has a strong combinatorial appeal. This is mainly due to the fact that, with any given quadratic Boolean function f, one can associate in many ways a graph that "represents" f, and vice versa. Depending on f, the graph is either undirected or directed, or bidirected. (A bidirected graph is a graph in which a label from the set $\{-1, 1\}$ is independently assigned to each endpoint of every edge. The arc associated with edge (i, j), according to the labels of its two endpoints, can be viewed as going either from i to j or from j to i, or as being directed into both i and j or out of both i and j; see Figure 5.1.)

This two-way correspondence between quadratic functions and graphs is very useful. For several important subclasses of quadratic Boolean functions (discussed in detail in Section 5.3), the recognition problem can be formulated as a problem in graph theory. The most efficient procedures for solving quadratic Boolean equations known so far are graph algorithms (see Section 5.6). In the opposite direction, many graph-theoretic properties, such as bipartiteness or the Kőnig-Egerváry property, can be naturally expressed as quadratic Boolean equations.

(3) *Many significant combinatorial problems can be reduced to quadratic Boolean equations.*

Another good reason for studying quadratic Boolean functions is that a host of significant combinatorial decision problems can be formulated as quadratic equations. Early examples (recognition of bipartiteness and signed graph balance) already appear in Maghout [644] and Hammer [436].

In Section 5.5, we present a collection of problems that are reducible to quadratic equations. For each of these problems, the reduction can be obtained in polynomial time, and for some, even in linear time. Since, as we show in Section 5.6, there are quite fast (indeed, linear) algorithms for quadratic equations, each of the above problems turns out to be efficiently solvable in polynomial, or even in linear time.

Just as 3-SAT problems (or cubic equations) are a "template" for a broad class of "hard" combinatorial decision problems (including maximum clique, vertex cover, chromatic number, subset sum, set covering, traveling salesman, etc.) that can all be reduced to 3-SAT in polynomial time, 2-SAT problems (or quadratic Boolean equations) can be taken to be the "template" of a rich class of "easy," although nontrivial, combinatorial problems (including the above-mentioned collection of problems and many others), all of which are efficiently reducible to 2-SAT.

(4) *Low-complexity algorithms are available for quadratic equations as well as for finding all prime implicants and irredundant normal forms of quadratic Boolean functions.*

As we show in Section 5.5, the recognition of the Kőnig-Egerváry property in graphs and quadratic equations are mutually reducible to each other. On this ground, an efficient algorithm of Gavril [374] for testing the Kőnig-Egerváry property can be easily translated into a linear-time algorithm for quadratic equations, which is actually the fastest currently available algorithm for quadratic equations.

Another nice feature of quadratic Boolean functions, which is not enjoyed by those of higher degree, is that, starting from an arbitrary quadratic DNF, one can produce in polynomial time all the prime implicants of the function, as well as an irredundant DNF of it (for a general definition of these notions, see Section 1.7). Efficient algorithms for these problems are presented in Section 5.8.

5.3 Special classes of quadratic functions

5.3.1 Classes

In this section, we introduce several classes of quadratic Boolean functions and then – starting from the class of all quadratic Boolean functions – we point out characterizations of some of these classes by functional inequalities.

In any DNF, a quadratic term may take one of the three forms

$$xy, \ \overline{x}\overline{y}, \ x\overline{y},$$

where x and y are variables. By forbidding all terms having one or more of these three forms, one can naturally define meaningful special subclasses of quadratic DNFs and, accordingly, of quadratic Boolean functions.

Let us now introduce some special classes of general, not necessarily quadratic, DNFs. We start with the definitions of Horn, co-Horn and polar DNFs, which are thoroughly studied in Chapters 6 and 11.

Definition 5.6. *A* Horn *DNF is a DNF in which every term contains at most one complemented variable.*

Definition 5.7. *A* co-Horn *DNF is a DNF in which every term contains at most one uncomplemented variable.*

Definition 5.8. *A* polar *DNF is a DNF in which no term contains both a complemented and an uncomplemented variable.*

In Section 5.4, we extensively refer to those quadratic DNFs in which every quadratic term consists of one complemented and one uncomplemented variable.

Definition 5.9. *A* mixed *DNF is a DNF that is both Horn and co-Horn.*

As mentioned above, these important subclasses of DNFs, when restricted to quadratic DNFs, can be simply characterized by means of forbidden terms (see Table 5.1).

Any of these types of DNFs defines in a natural way a corresponding class of Boolean functions. For example, we say that a Boolean function is *Horn* if it is representable by a Horn DNF.

Table 5.1. Subclasses of quadratic DNFs and their
forbidden terms

Class	Forbidden terms
Horn	$\overline{x}\,\overline{y}$
co-Horn	xy
polar	$x\overline{y}$
mixed	$xy, \overline{x}\,\overline{y}$
positive (purely quadratic)	$x\overline{y}, \overline{x}\,\overline{y}$

Before we proceed with functional characterizations of these subclasses, let us mention a result on DNF representations of purely quadratic Boolean functions. Recall from Section 1.10 that by a *positive DNF*, we mean a DNF containing no complemented variables.

Lemma 5.2. *Let φ be a quadratic DNF of a purely quadratic Boolean function f. If f is a positive, Horn, co-Horn, or mixed Boolean function, then φ is a positive, Horn, co-Horn, or mixed DNF, respectively.*

Proof. Let f be a purely quadratic Boolean function, and let φ be any quadratic DNF of f. By Lemma 5.1, every term of φ is a prime implicant of f. If f is positive, then every prime implicant of f is positive; hence, φ is positive.

Since every term of φ is a prime implicant of f, it can be generated by the consensus algorithm of Chapter 2, executed on an arbitrary quadratic DNF of f. On the other hand, each consensus operation, when performed on a pair of quadratic terms, preserves the Horn, co-Horn, and mixed types.   □

Note that Lemma 5.2 does not extend to polar DNFs.

5.3.2 Characterizations by functional relations

Ekin, Foldes, Hammer, and Hellerstein [305] obtained, for every class of Boolean functions in Table 5.2 (and for others), a characterization in terms of functional inequalities satisfied by every function in the class.

In Table 5.2, the inequalities are understood to be universally quantified over all vectors X, Y, Z in $\mathcal{B}^n$; XY and $X \vee Y$ are the vectors in $\mathcal{B}^n$ whose ith component is given by $x_i y_i$ and by $x_i \vee y_i$, respectively, for $i = 1, \ldots, n$. The functional characterization of quadratic Boolean functions on the first line of the table was obtained by Schaefer [807]. A proof of this result, due to Ekin, Foldes, Hammer, and Hellerstein [305], will be presented in Chapter 11 together with proofs of the other functional characterizations in Table 5.2.

In view of their functional characterization, polar functions are sometimes called *supermodular* and mixed ones *submodular*. (Notice the formal analogy with the supermodular and submodular (real-valued) set functions defined in Chapter 13.)

Table 5.2. Characterizations of classes of Boolean functions

Class	Functional relations
quadratic	$f(XY \vee XZ \vee YZ) \le f(X) \vee f(Y) \vee f(Z)$
dually quadratic	$f(X)f(Y)f(Z) \le f((X \vee Y)(X \vee Z)(Y \vee Z))$
Horn	$f(XY) \le f(X) \vee f(Y)$
co-Horn	$f(X \vee Y) \le f(X)f(Y)$
polar	$f(X) \vee f(Y) \le f(XY) \vee f(X \vee Y)$
mixed	$f(XY) \vee f(X \vee Y) \le f(X) \vee f(Y)$

Table 5.3. Quadratic Boolean functions
and graphs

Quadratic Boolean functions	Graphs
positive	undirected
mixed	directed
arbitrary	bidirected

Further properties of sub- and supermodular Boolean functions are discussed in Chapter 6 and Chapter 11.

5.4   Quadratic Boolean functions and graphs

5.4.1   Graph models of quadratic functions

There is a quite natural correspondence between certain classes of quadratic Boolean functions on one side, and graphs, digraphs, and bidirected graphs on the other side, as shown in Table 5.3.

In fact, as explained in Section 1.13.5, one can associate with any undirected graph $G = (V, E)$ its *stability function*, namely, the positive quadratic Boolean function given by

$$f = \bigvee_{(i,j) \in E} x_i x_j. \qquad (5.1)$$

Note that the prime implicants of f are precisely the terms $x_i x_j$ of this DNF, which is also the unique irredundant DNF of f. It follows that the correspondence between positive purely quadratic Boolean functions and undirected graphs is one-to-one.

Let now $D = (N, A)$ be a directed graph, with $N = \{1, 2, \ldots, n\}$. We can associate with D a quadratic mixed DNF $\varphi \equiv \varphi(D)$ as follows: We associate with every vertex $i \in N$ a variable x_i of φ, and with every arc $(i, j) \in A$ a quadratic term $x_i \overline{x}_j$ of φ. Conversely, given any mixed quadratic DNF φ (without linear terms), one

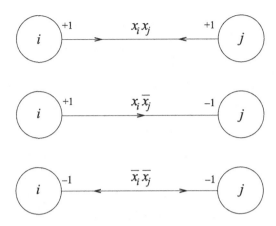

Figure 5.1. Terms associated with bidirected arcs.

can uniquely reconstruct the directed graph $D \equiv D(\varphi)$ whose associated DNF is φ.

However, this time the correspondence between digraphs and quadratic mixed Boolean functions is not one-to-one: Indeed, a purely quadratic mixed Boolean *function f* may be represented by many irredundant quadratic mixed *DNFs*. In order to state this relation more precisely, we need the notion of transitive closure of a digraph (see also Appendix A): Given a digraph $D = (N, A)$, its *transitive closure* is the digraph obtained from D by adding to A all the arcs (u, v) such that there is a directed path from u to v in D.

Theorem 5.2. *Two digraphs correspond to the same quadratic mixed Boolean function if and only if their transitive closures are identical.*

Proof. Two mixed DNFs represent the same quadratic Boolean function if and only if the two sets of prime implicants that one can obtain from them by the consensus algorithm are the same. It is easy to see that these implicants are quadratic mixed terms, and that the digraph associated with their disjunction is transitively closed. $\square$

Finally, if $B = (N, H)$ is a bidirected graph, one introduces again the variables $\{x_1, x_2, \ldots, x_n\}$ associated with its n vertices as above. Quadratic terms are associated with the arcs of B as indicated in Figure 5.1. Then, φ is the DNF consisting of the disjunction of all such quadratic terms. Conversely, B can be reconstructed from φ.

5.4.2 The matched graph

Another graph that can be conveniently associated with a quadratic DNF φ is the *matched graph* G_φ, introduced by Simeone [834]. This undirected graph has $2n$

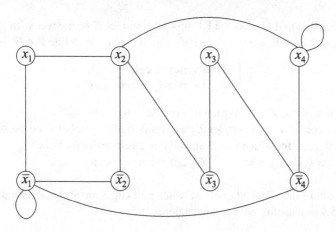

Figure 5.2. A matched graph.

vertices corresponding to the $2n$ literals $\{x_1,\ldots,x_n,\overline{x}_1,\ldots,\overline{x}_n\}$. Its set of edges is

$$\{(x_i,\overline{x}_i) : i \in \{1,\ldots,n\}\} \cup \{(\xi,\eta) : \xi\eta \text{ is a term of } \varphi\}.$$

If φ contains linear terms, a loop (ξ,ξ) is introduced for every such term ξ.

Example 5.2. *The matched graph associated with the DNF*

$$\varphi = \overline{x}_1 \vee x_4 \vee x_1 x_2 \vee \overline{x}_1 \overline{x}_2 \vee \overline{x}_1 \overline{x}_4 \vee x_2 \overline{x}_3 \vee x_2 x_4 \vee x_3 \overline{x}_4 \qquad (5.2)$$

is shown in Figure 5.2.   □

   The edges of G_φ are classified as *positive, negative, mixed,* or *null* edges according to whether they have the form (x_i,x_j), $(\overline{x}_i,\overline{x}_j)$, $(x_i,\overline{x}_j)$, or $(x_i,\overline{x}_i)$, respectively.

   The consistency of the quadratic Boolean equation $\varphi = 0$ has a nice graph-theoretic counterpart for G_φ. In order to state this property, we need some terminology.

   If $\mu(G)$ and $\tau(G)$ respectively denote the maximum cardinality of a matching and the minimum cardinality of a (vertex) cover of an arbitrary graph G (see definitions in Appendix A), then the following relation always holds:

$$\mu(G) \le \tau(G). \qquad (5.3)$$

The graph G is said to have the *Kőnig-Egerváry (KE) property* if equality holds in (5.3).

Theorem 5.3. *The quadratic Boolean equation $\varphi = 0$ is consistent if and only if the matched graph G_φ has the Kőnig-Egerváry property.*

Proof. The n null edges form a maximum matching of G_φ. Therefore, G_φ has the KE property if and only if there is a cover C in G_φ with $|C| = n$.

Assume first that G_φ has the KE property, and let C be a cover with $|C| = n$. As every null edge has exactly one endpoint in C, we can define $Z \in \mathcal{B}^n$ by

$$z_i = \begin{cases} 0 & \text{if vertex } x_i \text{ belongs to } C \\ 1 & \text{if vertex } \overline{x}_i \text{ belongs to } C. \end{cases}$$

Since C is a cover, Z is a solution of the equation $\varphi = 0$.

For the converse direction, let Z be a solution of $\varphi = 0$. Let C be the set of all those vertices x_i for which $z_i = 0$ and all those vertices $\overline{x}_i$ for which $z_i = 1$. Then C is a cover with $|C| = n$, and so G_φ has the KE property. $\square$

A variant of the matched graph in which null edges are absent is introduced in Section 5.9 as a useful tool for dualization.

5.4.3 The implication graph

As an alternative to the matched graph G_φ, one can associate with the quadratic DNF φ a directed graph D_φ, called the *implication (di)graph* of φ, and again characterize the consistency of $\varphi = 0$ in terms of a simple property of D_φ. As we shall see in Section 5.8, the implication graph also turns out to be a convenient tool for the efficient solution of two other fundamental problems, namely, finding all prime implicants or computing an irredundant DNF of a quadratic Boolean function. Moreover, the implication graph will prove useful in obtaining a concise parametric product form of the solutions of a quadratic Boolean equation and in getting a fast on-line 2-SAT algorithm (Section 5.7).

The definition of an implication graph naturally arises from the observation that the relation

$$\xi\eta = 0$$

is equivalent to the implication

$$\xi \Rightarrow \overline{\eta}, \tag{5.4}$$

as well as to the implication

$$\eta \Rightarrow \overline{\xi}. \tag{5.5}$$

As in the matched graph G_φ, the vertices of the implication graph D_φ correspond to the $2n$ literals $\{x_1, \ldots, x_n, \overline{x}_1, \ldots, \overline{x}_n\}$. For each quadratic term $\xi\eta$, in view of (5.4) and (5.5), there are in D_φ two arcs $(\xi, \overline{\eta})$ and $(\eta, \overline{\xi})$ (either arc will be called the *mirror arc* of the other one, and the simultaneous presence of these two arcs will be referred to as the *Mirror Property*). For each linear term ξ, there is an arc $(\xi, \overline{\xi})$ in D_φ.

Example 5.3. *The implication graph associated with the DNF (5.2) is shown in Figure 5.3.* $\square$

The notion of implication graph was introduced by Aspvall, Plass, and Tarjan [34]. Their representation of linear terms, however, is different from ours:

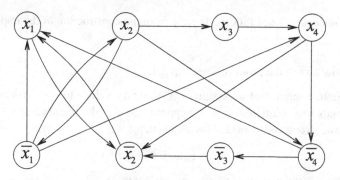

Figure 5.3. An implication graph.

They add two dummy vertices x_0 (representing the constant 0) and $\overline{x}_0$ (representing the constant 1) and, for each linear term ξ, two arcs (x_0, ξ) and $(\overline{\xi}, \overline{x}_0)$, again mirroring each other. The advantages of our representation will become apparent in Section 5.8, when we discuss the relationship between prime implicants of a quadratic DNF and transitive closures.

One should notice that, through the implication graph, a quadratic Boolean equation is represented by an equivalent system of logical implications – a *deductive knowledge base* in the terminology of artificial intelligence (see Nilsson [713]).

The most important property of the implication graph D_φ relates its strong components to the consistency of the quadratic Boolean equation $\varphi = 0$. According to definitions in Appendix A, a *strongly connected component* (or, briefly, a *strong component*) of $D_\varphi = (N, A)$ is any maximal subset C of vertices with the property that any two vertices of C lie on some closed directed walk consisting only of vertices of C. The strong components of D_φ form a partition of its vertex-set N, and they can be computed in $O(m)$ time, where $m = |A|$ (Tarjan [858]). By shrinking each strong component into a single vertex, one obtains an acyclic digraph $\hat{D}_\varphi$, the *condensed implication graph* of φ. Notice that, in view of the Mirror Property, the strong components of D_φ come in pairs: If C is a strong component, then the set $\overline{C}$ consisting of the negations of all literals in C also is a strong component.

Aspvall, Plass, and Tarjan [34] proved:

Theorem 5.4. *The quadratic Boolean equation $\varphi = 0$ is consistent if and only if no strong component of D_φ contains both a literal ξ and its complement $\overline{\xi}$.*

To prove this theorem, let us state a simple, but useful, result.

Lemma 5.3. *An assignment of binary values to the vertices of D_φ corresponds to a solution of the equation $\varphi = 0$ if and only if*

   (i) *for all i, vertices x_i and $\overline{x}_i$ receive complementary values, and*
   (ii) *no arc (and hence no directed path) goes from a 1-vertex (that is, a vertex with value 1) to a 0-vertex (that is, a vertex with value 0).*

Proof. This equivalence follows directly from the construction of the implication graph. □

We now turn to the proof of Theorem 5.4.

Proof. First, assume that an assignment of binary values to the vertices of D_φ corresponds to a solution of $\varphi = 0$. Suppose also that the literals ξ and $\bar{\xi}$ belong to the same strong component. This means that

$$\xi \Rightarrow \bar{\xi} \text{ and } \bar{\xi} \Rightarrow \xi.$$

Therefore, the literal ξ must take the values 0 and 1 at the same time, a contradiction. Hence $\varphi = 0$ has no solution.

For the converse direction, let us show that if no strong component of D_φ contains both a literal and its complement, then $\varphi = 0$ has a solution. The proof is by induction on the number s of strong components of D_φ (which is always even).

If $s = 2$ and C is a strong component, then the other strong component is $\overline{C}$. Since C and $\overline{C}$ are different strong components, we may assume that all the arcs between C and $\overline{C}$ (if any) go from a vertex of C to a vertex of $\overline{C}$. Now, assign the value 0 to all literals in C and the value 1 to those in $\overline{C}$. Properties (i) and (ii) of Lemma 5.3 are satisfied and thus the assignment defines a solution of $\varphi = 0$.

Assume now that the statement is true whenever the implication graph has at most $s - 2$ strong components ($s \geq 4$), and let D_φ have s strong components. Consider the acyclic condensed digraph $\hat{D}_\varphi$ obtained from D_φ upon contraction of the strong components of D_φ.

Let C be the strong component of D_φ corresponding to a source in $\hat{D}_\varphi$. Then, by the Mirror Property, $\overline{C}$ is a sink of $\hat{D}_\varphi$. By the definitions of source and sink, no arc of D_φ goes into C and no arc leaves $\overline{C}$. Remove both C and $\overline{C}$ from D_φ. Let D' be the resulting subdigraph of D_φ. The digraph D' has $s - 2$ strong components. Hence the statement of Theorem 5.4 holds for D' by the inductive hypothesis, and therefore there is an assignment of binary values to the vertices of D' satisfying (i) and (ii) of Lemma 5.3. Such an assignment can be extended to D_φ by assigning the value 0 to all literals in C and the value 1 to all literals in $\overline{C}$. It is immediate to verify that the extended assignment still satisfies (i) and (ii) of Lemma 5.3 in the digraph D_φ. Hence, it yields a solution of $\varphi = 0$. □

The implication graph enables us not only to determine the consistency of the corresponding quadratic Boolean equation but also, in case of consistency, to infer further properties of its solutions.

We say that a literal ξ is *forced to the value* α (for $\alpha \in \{0, 1\}$) if either the quadratic Boolean equation $\varphi = 0$ is inconsistent, or if ξ takes the value α in all its solutions.

Theorem 5.5. *Suppose that the equation $\varphi = 0$ is consistent. Then, the literal ξ is forced to 0 if and only if there exists a directed path from ξ to $\bar{\xi}$ in D_φ.*

Proof. If there is a directed path from ξ to $\overline{\xi}$ and $\xi = 1$ in some solution, then this contradicts part (ii) of Lemma 5.3.

For the converse direction, suppose that there is no directed path from ξ to $\overline{\xi}$, and let X be any solution of $\varphi = 0$. If $\xi = 1$ in X, then we are done. Else, let us modify X as follows: Assign to ξ and to all its successors the value 1; assign to $\overline{\xi}$ and to all its ancestors the value 0. Let X' be the resulting assignment. First of all, X' is well defined: No conflicting values may arise, since no ancestor of $\overline{\xi}$ can be a successor of ξ (as this would yield a directed path from ξ to $\overline{\xi}$).

Let us show that X' is a solution. If not, by Lemma 5.3 (ii), there is a path from a 1-vertex α to a 0-vertex β. Since this path did not exist for X, either α is a successor of ξ or β is an ancestor of $\overline{\xi}$. By symmetry, it is enough to consider the former case. But, if α is a successor of ξ, so is β, and hence β should take the value 1 in X', which is a contradiction. $\square$

Theorem 5.6. *Let ξ be a literal not forced to 0, and let η be a literal not forced to 1. The relation $\xi \le \eta$ holds in all solutions of the quadratic Boolean equation $\varphi = 0$ if and only if there is a directed path from ξ to η in D_φ.*

Proof. The "if" part is obvious after part (ii) of Lemma 5.3. Let us prove the "only if" part. Assume there is no directed path from ξ to η, and let us prove that, if there is a solution at all, then there is also a solution in which $\xi = 1$ and $\eta = 0$.

Consider an arbitrary solution X. By part (ii) of Lemma 5.3 there is no directed path from any 1-vertex to a 0-vertex. If in X we have $\xi = 1$ and $\eta = 0$, we are done. Otherwise, let us modify X as follows: Assign the value 1 to ξ and the value 0 to η. Also, assign the value 1 to all successors of ξ and 0 to all ancestors of η. Taking into account the Mirror Property, assign the value 0 to all ancestors of $\overline{\xi}$ and the value 1 to all successors of $\overline{\eta}$. Leave the remaining values unchanged. We claim that the assignment of values X' obtained in this way is also a solution of $\varphi = 0$.

First of all, X' is well-defined: No conflicting values may arise, since no successor of ξ may be an ancestor of η (as this would yield a directed path from ξ to η, against our assumption).

Furthermore, no successor of ξ can be also an ancestor of $\overline{\xi}$, else there would be a directed path from ξ to $\overline{\xi}$, and ξ would be forced to 0. Similarly, no ancestor of η can be a successor of $\overline{\eta}$. Suppose that in X' there is a directed path from a 1-vertex α to a 0-vertex β. Then α is a successor either of ξ or $\overline{\eta}$. But then, so is β; hence β should take the value 1 in X', a contradiction. $\square$

Two literals ξ and η are said to be *twins* if $\xi = \eta$ in every solution of the quadratic Boolean equation $\varphi = 0$.

Corollary 5.1. *Suppose that the two literals ξ and η are not forced. Then, they are twins if and only if they are in the same strong component of the implication graph D_φ.*

Proof. The equality $\xi = \eta$ is equivalent to the pair of relations $\xi \leq \eta, \eta \leq \xi$. The statement then follows from Theorem 5.6. $\qquad\square$

5.4.4 Conflict codes and quadratic graphs

In this section, we describe yet another way of associating a graph with a DNF. Let us say that two elementary conjunctions *conflict* if there is a variable that appears complemented in one of them and uncomplemented in the other one. Given an arbitrary DNF φ, the *conflict graph* C_φ of φ is the undirected graph whose vertices are the terms of φ, and whose edges are the pairs of conflicting terms (see Hammer [437, 465]).

Conversely, given a graph G, a *(conflict) code* of G is an assignment of elementary Boolean conjunctions to the vertices of G such that, if φ is the disjunction of these conjunctions, then $G = C_\varphi$.

Example 5.4. *Figure 5.4 shows a graph G and two of its conflict codes.* $\qquad\square$

Let us introduce some additional terminology. Consider an arbitrary DNF φ and its conflict graph $C_\varphi = (V, E)$. For a variable x of φ, we call *color of x* the set of all edges $(T, T') \in E$ such that x is complemented in T and uncomplemented in T', or vice-versa. Clearly, each color spans a (possibly empty, and not necessarily induced) complete bipartite subgraph of C_φ. Moreover, the union of all colors corresponding to the variables of φ covers the edge-set of C_φ.

Conversely, for an arbitrary graph $G = (V, E)$, any collection of complete bipartite subgraphs that covers E defines a conflict code of G. It easily follows from this observation that every graph has at least one, and generally many distinct conflict codes, as illustrated by Example 5.4.

As pointed out by Hammer [438, 465]; Benzaken, Hammer, and Simeone [68, 69]; and Hammer and Simeone [463], the non-uniqueness of a conflict code of a graph can be exploited in order to preprocess and simplify weighted maximum stable set problems in graphs, weighted maximum satisfiability problems (MAX SAT), and unconstrained nonlinear binary optimization problems; see Section 13.4.4.

Notice that the DNF corresponding to the conflict code of the graph in Figure 5.4(b) is quadratic, whereas the one corresponding to Figure 5.4(a) is not. Naturally, one may ask which graphs admit a quadratic code. Such graphs are called *quadratic* by Benzaken, Hammer, and Simeone [68, 69]; an equivalent graph-theoretic definition is that a graph is quadratic if and only if its edge-set can be covered by complete bipartite graphs (corresponding to colors) so that at most two different colors meet at each vertex. If, furthermore, the colors can be chosen to be stars, then the graph is called *bistellar* (Hammer and Simeone [461]).

Since two terms may have more than one conflicting variable, the colors generally form a covering, but not necessarily a partition, of the edge-set of C_φ. However,

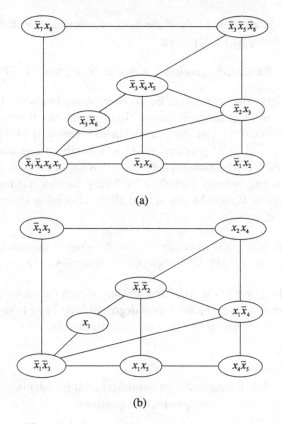

Figure 5.4. Two conflict codes of the same graph.

they do form a partition when the DNF φ is both quadratic and *primitive*, that is, when two different terms of φ do not involve exactly the same set of variables.

A quadratic graph is called *primitive*, *Horn*, or *mixed* if it admits a primitive, Horn, or mixed quadratic code, respectively.

The complexity of recognizing quadratic graphs appears to be still an open question. However, the following negative result was established by Crama and Hammer [230].

Theorem 5.7. *Recognizing quadratic primitive graphs is NP-complete.*

Actually, they proved the following stronger result.

Theorem 5.8. *Recognizing whether the edge-set of a bipartite graph can be partitioned into colors, so that all colors are either stars or squares (that is, C_4's), and at most two colors meet at each vertex, is an NP-complete problem.*

Benzaken, Hammer, and Simeone [69] remarked that quadratic primitive mixed graphs are precisely the adjoints of directed graphs (where the *adjoint* of a digraph D is the undirected graph whose vertices are the arcs of D, and where two vertices

u and v are adjacent if and only if the head of v coincides with the tail of u).
Chvátal and Ebenegger [200] proved:

Theorem 5.9. *Recognizing quadratic primitive mixed graphs is NP-complete.*

On the positive side, Benzaken, Boyd, Hammer, and Simeone [65] obtained
a characterization of quadratic primitive Horn graphs, and Hammer and Simeone [461] characterized bistellar graphs. In the statement of Theorem 5.10
hereunder, the word "configuration" refers to a family of digraphs on a given
set S of vertices. A configuration is defined by two disjoint subsets $A, B \subseteq S \times S$.
The meaning is that, in every digraph of the family, the arcs in A must always be
present, the arcs in B must be absent, and all the remaining arcs may be either
present or absent.

Theorem 5.10. *A graph G is quadratic primitive Horn if and only if it admits an
edge-orientation that avoids the ten special configurations of Figure 5.5.*

Theorem 5.11. *A graph G is bistellar if and only if each connected component of
the subgraph of G induced by vertices of degree at least 3 is a 1-tree, that is, it is
either a tree or it becomes a tree after deletion of one edge.*

5.5 Reducibility of combinatorial problems
to quadratic equations

5.5.1 Introduction

As noted earlier, the importance of quadratic Boolean functions is substantiated by
the fact that many combinatorial decision problems can be efficiently reduced to
quadratic equations. A partial list, to be further discussed in this section, includes
checking bipartiteness of a graph, balance in signed graphs, recognition of split
graphs, recognition of the Kőnig-Egerváry property, and single-bend drawings
of electronic circuits. For some of these problems, the reduction can even be
performed in linear time. Conversely, some of them also admit a linear time reduction *from* quadratic Boolean equations, which makes the former equivalent, in a
well-defined sense, to the latter.

Additional applications of quadratic Boolean functions and equations can be
found in papers by Waltz [895] (computer vision); Even, Itai, and Shamir [318]
(timetabling); Hansen and Jaumard [467] (minimum sum-of-diameters clustering); Boros, Hammer, Minoux, and Rader [132] (VLSI design); Eskin, Halperin,
and Karp [316] (phylogenetic trees) Miyashiro and Matsui [688] (selection of
home and away games in round-robin tournaments), Wang et al. [898] (routing
on the internet), and so forth. In Section 6.10.1, we present yet another application of quadratic Boolean equations: Namely, the recognition of renamable Horn
functions.

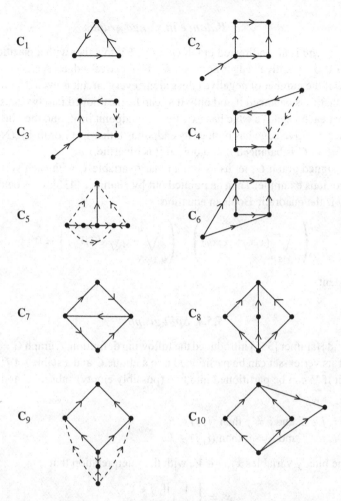

Figure 5.5. The ten forbidden configurations for quadratic primitive Horn graphs. Continuous arcs must be present; dashed ones must be absent.

5.5.2 Bipartite graphs

Recall that an undirected graph $G = (V, E)$ is *bipartite* if its vertex-set V can be partitioned into two subsets V_1 and V_2 such that every edge of G has exactly one endpoint in V_1 and the other endpoint in V_2. Introduce binary variables x_i, $i \in V$, where $x_i = 1$ or 0 according to whether vertex i belongs to V_1 or to V_2. Then, the graph G is bipartite if and only if the quadratic Boolean equation

$$\bigvee_{(i,j)\in E} (x_i x_j \vee \overline{x}_i \overline{x}_j) = 0 \tag{5.6}$$

is consistent.

5.5.3 Balance in signed graphs

A *signed graph* is an undirected graph $G = (V, E)$, together with a partition of E into a set P of "positive" edges and a set N of "negative" edges. A signed graph is *balanced* if the number of negative edges along every circuit is even. Harary [475] showed that G is balanced if and only if V can be partitioned into two sets V_1 and V_2, so that each negative edge has exactly one endpoint in V_1 and the other in V_2, while each positive edge has both of its endpoints either in V_1 or in V_2. (Note that when $E = N$, G is balanced if and only if it is bipartite.)

For a signed graph G, let us assign a binary variable x_i with each vertex i, as in the previous example. Then as pointed out by Hammer [436], G is balanced if and only if the quadratic Boolean equation

$$\left(\bigvee_{(i,j) \in P} (x_i \overline{x}_j \vee \overline{x}_i x_j) \right) \vee \left(\bigvee_{(i,j) \in N} (x_i x_j \vee \overline{x}_i \overline{x}_j) \right) = 0 \qquad (5.7)$$

is consistent.

5.5.4 Split graphs

Foldes and Hammer [335] introduced the following definition: A graph $G = (V, E)$ is *split* if its vertex-set can be partitioned into a clique C and a stable set I; that is, G is split if V can be partitioned into two (possibly empty) subsets C and I such that

(i) if $i, j \in C$ and $i \neq j$ then $(i, j) \in E$;
(ii) if $i, j \in I$ and $i \neq j$, then $(i, j) \notin E$.

Define binary variables x_j, $j \in V$, with the interpretation that

$$x_j = \begin{cases} 1 & \text{if } j \in C, \\ 0 & \text{if } j \in I. \end{cases}$$

Then, conditions (i) and (ii) hold if and only if the quadratic Boolean equation

$$\left(\bigvee_{(i,j) \in E} \overline{x}_i \overline{x}_j \right) \vee \left(\bigvee_{(i,j) \notin E} x_i x_j \right) = 0$$

is consistent.

The related class of *bisplit* graphs has been investigated by Brandstädt, Hammer, Le and Lozin [151]. Their recognition turns out again to be reducible to a quadratic Boolean equation.

5.5.5 Forbidden-color graph bipartition

Gavril [375] has studied the following decision problem in graph theory (presented here in a slightly different, but equivalent, form).

FORBIDDEN-COLOR GRAPH BIPARTITION
Instance: A graph $G = (V, E)$, together with an edge-coloring of G (that is, a partition of E) consisting of at least two colors, say "red" and "blue," and possibly other colors.
Question: Is there a partition of V into two (possibly empty) subsets U and W such that

(i) no red edge is entirely contained in U;
(ii) no blue edge is entirely contained in W?

We use the shorthand FCGB to denote the foregoing problem. Gavril [375] showed that several combinatorial decision problems are polynomial-time reducible (and, in fact, log-space reducible) to FCGB. For example, the recognition of split graphs is a special case of FCGB on the complete graph K_n ($n = |V|$): Just color "red" the edges of G, and "blue" those of the complement $\overline{G}$.

Furthermore, Gavril showed that quadratic equations and FCGB are mutually reducible in linear time. Here we show that FCGB is reducible to quadratic equations. In fact, let R and B be the sets of red and blue edges of G, respectively. Introduce binary variables x_j, $j \in V$, such that

$$x_j = \begin{cases} 1 & \text{if } j \in U, \\ 0 & \text{if } j \in W. \end{cases}$$

Then, the answer to FCGB is Yes if and only if the quadratic Boolean equation

$$\left(\bigvee_{(i,j) \in R} x_i x_j \right) \vee \left(\bigvee_{(i,j) \in B} \overline{x}_i \overline{x}_j \right) = 0$$

is consistent.

5.5.6 Totally unimodular matrices with two nonzero entries per column

Definition 5.10. *A matrix is totally unimodular (TU) if all its square submatrices have determinant 0, 1 or −1.*

Clearly, all entries of a TU matrix must be 0, 1, or −1. TU matrices are very important in integer programming in view of the following classical result of Hoffman and Kruskal [495].

Theorem 5.12. *Let A be an $m \times n$ TU matrix, and let $b \in \mathbb{Z}^m$ be an arbitrary integral m-vector. Then, each extreme point of the polyhedron*

$$P = \{x \in \mathbb{R}^n : Ax \leq b\}$$

is integral.

Proof. See Hoffman and Kruskal [495]. $\square$

Theorem 5.12 and the Fundamental Theorem of Linear Programming (see e.g., [199, 812]) imply the following corollary:

Corollary 5.2. *Let A be an $m \times n$ TU matrix, let $c \in \mathbb{R}^n$, and let $b \in \mathbb{Z}^m$ be an integral m-vector. If the linear program*

$$\begin{array}{c} \text{maximize } cx \\ \text{subject to } Ax \leq b, x \in \mathbb{R}^n \end{array} \tag{5.8}$$

has a finite optimum, then it has an integral optimal solution.

Hence the integer linear program obtained from (5.8) by the addition of integrality constraints on x can be solved by ordinary linear programming.

A complete characterization of TU matrices was obtained by Seymour [823]; a polynomial-time recognition algorithm based on this result can be found in Schrijver [812].

For the special case of matrices with two nonzero entries per column, however, Heller and Tompkins [483] gave more efficient characterizations of totally unimodular matrices.

Theorem 5.13. *A necessary and sufficient condition for a $(-1,0,1)$-matrix A with two nonzero entries per column to be totally unimodular is that its set of rows can be partitioned into two (possibly empty) subsets R_1 and R_2 such that, for each column a^j:*

   (i) *if the two nonzero entries of a^j are different, then they both belong to R_1, or they both belong to R_2;*

   (ii) *if the two nonzero entries of a^j are equal, then one of them belongs to R_1, and the other one belongs to R_2.*

Clearly, these conditions can be expressed in terms of consistency of a quadratic Boolean equation with two quadratic terms per column.

Example 5.5. *The matrix*

$$A = \begin{bmatrix} 0 & 1 & 1 & -1 \\ -1 & 0 & -1 & 0 \\ 1 & 1 & 0 & -1 \end{bmatrix}$$

is not TU, since the associated quadratic Boolean equation

$$\overline{x}_2 x_3 \vee x_2 \overline{x}_3 \vee x_1 x_3 \vee \overline{x}_1 \overline{x}_3 \vee x_1 \overline{x}_2 \vee \overline{x}_1 x_2 \vee x_1 x_3 \vee \overline{x}_1 \overline{x}_3 = 0$$

has no solution (the submatrix formed by the first three columns has determinant -2).　　　　□

5.5.7 The Kőnig-Egerváry property for graphs

In Section 5.4.2, we have proved that a quadratic Boolean equation $\varphi = 0$ is consistent if and only if the matched graph G_φ associated with φ has the Kőnig-Egerváry property. Here we show that, conversely, the validity of the Kőnig-Egerváry property for graphs can be reduced to the satisfiability of a quadratic Boolean function.

Let $G = (V, E)$ be an arbitrary graph, and let M be a maximum matching of G. Note that M can be found in $O(|V|^{2.5})$ time (see, e.g., Papadimitriou and Steiglitz [726]). Let F be the set of all free vertices, that is, the set of vertices that are not endpoints of any edge in M. For each edge $e_i \in M$, let us associate the literal x_i with one of the endpoints of e_i, and the literal $\overline{x}_i$ with the other endpoint; moreover, we associate a literal x_j with each $j \in F$.

Finally, denoting by $\xi(v)$ the literal associated with vertex $v \in V$, we set

$$\varphi = \left(\bigvee_{(u,v) \in E \setminus M} \xi(u)\xi(v) \right) \vee \left(\bigvee_{w \in F} \overline{\xi}(w) \right). \tag{5.9}$$

Simeone [834] proved:

Theorem 5.14. *The graph G has the Kőnig-Egerváry property if and only if the quadratic Boolean equation $\varphi = 0$ is consistent, where φ is defined by (5.9).*

Before proving the theorem, let us introduce the notion of "rake," and let us state two related results. A pair (C, M), where $C \subseteq V$ is a cover and M is a matching, is called a *rake* if every $v \in C$ is an endpoint of exactly one edge of M, and every $e \in M$ has exactly one endpoint in C. Note that if (C, M) is a rake, then C necessarily is a minimum cover and M necessarily is a maximum matching. The next two results are due to Klee (as reported in [604]) and to Gavril [374], respectively.

Theorem 5.15. *A graph has the Kőnig-Egerváry property if and only if it has a rake.*

Theorem 5.16. *A graph has the Kőnig-Egerváry property if and only if, for every minimum cover C and every maximum matching M, the pair (C, M) is a rake.*

Now we can prove Theorem 5.14.

Proof. Assume that the Boolean equation $\varphi = 0$ is consistent, and let X^* be a solution. Let I be the set of all vertices $v \in V$ such that the associated literal $\xi(v)$ takes value 1 in X^*. The set I must be stable; hence, $C = V \setminus I$ is a cover. Moreover, all vertices in C must be matched because $F \subseteq I$. On the other hand, every edge of M must have exactly one endpoint in C and one in I. Hence, (C, M) is a rake, and by the "if" part of Theorem 5.15, G has the KE property.

Conversely, assume that the KE property holds for G. If C is an arbitrary minimum cover of G, then (C, M) must be a rake by the "only if" part of Theorem 5.16.

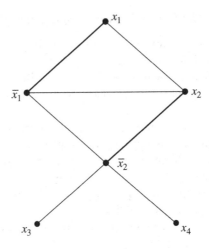

Figure 5.6. Graph for Example 5.6.

The set $I = V \setminus C$ is stable and must include F because all vertices in C are matched. Hence, if we assign the value 0 or 1 to $\xi(v)$ according to whether $v \in C$ or $v \in I$, we obtain a solution of the equation $\varphi = 0$.   □

Example 5.6. *Consider the graph of Figure 5.6, where the matching is represented by thick edges. The associated Boolean equation is*

$$\varphi \equiv x_1 x_2 \vee \overline{x}_1 x_2 \vee \overline{x}_1 \overline{x}_2 \vee \overline{x}_2 x_3 \vee \overline{x}_2 x_4 \vee \overline{x}_3 \vee \overline{x}_4 = 0.$$

It is easy to see that this equation is inconsistent, and that the graph does not have the KE property.   □

5.5.8 Single-bend wiring

In the design of microwave integral circuits, some prescribed pairs of pins with known locations on a rectangular board are to be connected. When the conductors are microstrip lines, it is desirable that each connection consist only of a horizontal segment and of a vertical one; due to this *single-bend wiring* requirement, only two connections, called *upper* and *lower*, respectively, are allowed for any given pair of pins (see Figure 5.7).

For technological reasons, we want to find, if there is one, a set of pairwise non-crossing connections for the prescribed pairs. (We may assume that the pins are in "general position," that is, no two of them are aligned along the same horizontal or vertical line. This assumption simplifies the discussion.)

Let us associate a Boolean variable x with every pair of pins to be connected, where $x = 1$ or 0, respectively, depending on whether an upper or a lower connection is chosen for the pair.

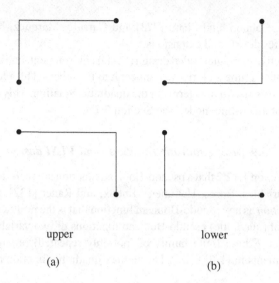

upper   lower

(a)   (b)

Figure 5.7. Single bend wiring.

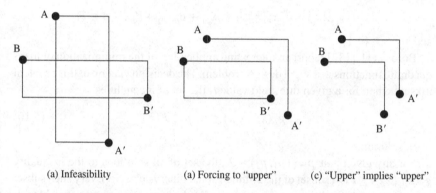

(a) Infeasibility   (a) Forcing to "upper"   (c) "Upper" implies "upper"

Figure 5.8. Some basic patterns in the single bend wiring problem.

It is easy to see that the relative positions of two given pairs of pins may induce some constraints on the connections between each pair, and hence, on the corresponding Boolean variables. Figure 5.8 shows some of the patterns that may occur. In case (a), no matter whether connections AA' and BB' are upper or lower, they must cross each other, giving rise to an infeasible situation. In case (b), regardless of whether connection BB' is upper or lower, connection AA' is forced to be upper, else it would cross BB'. In case (c), if connection BB' is upper, then also AA' must be upper, else it would cross BB'. In every case, each constraint involves only two connections; hence it can be represented by quadratic conditions on the corresponding Boolean variables. Therefore, checking the existence of a feasible noncrossing wiring can be reduced to the solution of a quadratic Boolean equation;

see Raghavan, Cohoon, and Shani [773] and Garrido, Márquez, Morgana, and
Portillo [373] for details and extensions.

In an interactive computer-aided design (CAD) environment, one usually places
one component at a time and tries to connect it to the others. The addition of such
a component gives rise to new terms in the quadratic equation. This motivates the
investigation of an on-line model; see Section 5.7.4.

5.5.9 Max-quadratic functions and VLSI design

Recall from Section 1.12.2 that a pseudo-Boolean function is a real-valued function
of Boolean variables. Boros, Hammer, Minoux, and Rader [132] define a *max-
quadratic function* as any pseudo-Boolean function that is the pointwise maximum
of a finite set of (quadratic) pseudo-Boolean functions of two variables.

Formally, let $\mathcal{F}$ be a finite family of (possibly repeated) ordered pairs $p = (p_1, p_2)$ of elements in $\{1, 2, \ldots, n\}$. Then a max-quadratic function has the form

$$g(x_1, x_2, \ldots, x_n) = \max_{p \in \mathcal{F}} g_p(x_{p_1}, x_{p_2}),$$

where

$$g_p(x_{p_1}, x_{p_2}) = a_p x_{p_1} x_{p_2} + b_p x_{p_1} + c_p x_{p_2} + d_p, \quad p \in \mathcal{F}$$

and all $x_j \in \{0, 1\}$.

Boros et al. [132] report an interesting application of the minimization of max-
quadratic functions to a VLSI design problem. The decision version of this problem
asks whether, for a given threshold value t, the set of inequalities

$$g_p(x_{p_1}, x_{p_2}) \leq t, \quad p \in \mathcal{F} \tag{5.10}$$

has a solution.

For any given pair $p = (p_1, p_2) \in \mathcal{F}$, the set of all solutions to the inequality
$g_p(x_{p_1}, x_{p_2}) \leq t$ is a subset of the 2-dimensional binary cube $\mathcal{B}^2$. Every such subset
is itself the set of solutions of a quadratic Boolean equation in two variables. It
follows that the set of solutions of the system of inequalities (5.10) is also the set
of solutions of a quadratic Boolean equation.

Example 5.7. *Let*

$$g_{(2,5)}(x_2, x_5) = 7 - 3x_2 - 2x_5 + 4x_2 x_5$$

*and let $t = 5$. Then the set of solutions of the inequality $g_{(2,5)}(x_2, x_5) \leq t$ consists
of the points $(x_2, x_5) = (0, 1)$ and $(x_2, x_5) = (1, 0)$. Hence, the set of solutions of
the inequality $g_{(2,5)}(x_2, x_5) \leq t$ coincides with the set of solutions of the quadratic
Boolean equation*

$$x_2 x_5 \vee \overline{x}_2 \overline{x}_5 = 0.$$

$\square$

5.5.10 A level graph drawing problem

A *level graph* is a directed acyclic graph (DAG) $G = (V, A)$ together with a *level function*, that is, a function l from V onto $J_r \equiv \{1, \ldots, r\}$ (r being a positive integer) such that

$$(u, v) \in A \Rightarrow l(v) > l(u).$$

In G, *level* h ($h = 1, \ldots, r$) is defined to be the set

$$L_h = \{v \in V : l(v) = h\},$$

which is certainly nonempty by our assumption that l is surjective.

A level graph is *proper* if the stronger condition

$$(u, v) \in A \Rightarrow l(v) = l(u) + 1$$

holds. A *level-planar embedding* of the level graph G is an embedding of G in the plane such that

(i) the vertices of each level L_h are aligned along a straight vertical line which differs from level to level;

(ii) all arcs are represented by straight line segments whose endpoints must lie on two consecutive vertical lines;

(iii) any two such straight line segments, if different, may intersect only in a common endpoint.

Checking whether a given level graph admits a level-planar embedding is a question of practical importance in the area of graph drawing, in view of its applications to software engineering, database design, and project management.

The essence of the problem lies in finding suitable linear orders of each level L_h such that, if the vertices in L_h are placed along a vertical line from top to bottom according to the linear order in L_h, no arc-crossing arises.

Thus, checking the existence of a level-planar embedding of a proper level graph can be rephrased in order-theoretic terms as follows. Let $(R, \leq)$ and $(S, \preceq)$ be two finite linearly ordered sets. Let φ be a one-to-many mapping of R into S. The mapping φ is said to be *isotonic* if

$$(x, y \in R \text{ and } x < y) \Rightarrow (\xi \preceq \eta \text{ for all } \xi \in \varphi(x) \text{ and } \eta \in \varphi(y)).$$

Consider the following decision problem:

ISOTONY

Instance: r mutually disjoint finite sets $L_1, \ldots, L_r$; for each $h = 1, \ldots, r - 1$, a one-to-many mapping φ_h from L_h to L_{h+1}.

Question: Are there r linear orders $\preceq_1, \ldots, \preceq_r$ on $L_1, \ldots, L_r$, respectively, such that φ_h is an isotonic mapping from $(L_h, \preceq_h)$ into $(L_{h+1}, \preceq_{h+1})$, for $h = 1, \ldots, r - 1$?

Clearly, the above embedding problem is reducible to ISOTONY.

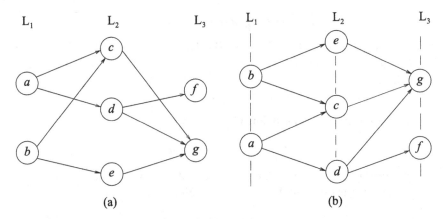

Figure 5.9. The level graph for Example 5.8.

Example 5.8. *Consider the proper level graph G shown in Figure 5.9(a). The levels of G are* $L_1 = \{a,b\}, L_2 = \{c,d,e\}, L_3 = \{f,g\}$. *A level-planar embedding of G is shown in Figure 5.9(b). With reference to the corresponding* ISOTONY *formulation, the mappings* φ_1 *and* φ_2 *are given by*

$$\varphi_1(a) = \{c,d\}, \quad \varphi_1(b) = \{c,e\};$$

$$\varphi_2(c) = \{g\}, \quad \varphi_2(d) = \{f,g\}, \quad \varphi_2(e) = \{g\}.$$

The answer to ISOTONY *is Yes and the required linear orders* $\preceq_1, \preceq_2, \preceq_3$ *are given by*

$$b \prec_1 a; \; e \prec_2 c \prec_2 d; \; g \prec_3 f. \qquad \square$$

Randerath, Speckenmeyer, Boros, Čepek, Hammer, Kogan, Makino, and Simeone [778] pointed out a simple reduction of ISOTONY to a cubic Boolean equation (or equivalently, to 3-SAT). In order to describe it, let us introduce binary variables

$$z^h_{ij} = \begin{cases} 1 & \text{if } i,j \in L_h, i \neq j, \text{ and } i \prec_h j, \\ 0 & \text{otherwise.} \end{cases}$$

In other words, $Z^h = [z^h_{ij}]$ is the incidence matrix of the (unknown) linear order $\preceq_h, h = 1,\dots,r$. Then, the following constraints must be satisfied:

(i) For each $h = 1,\dots,r-1; i,p \in L_h, i \neq p; j \in \varphi_h(i), q \in \varphi_h(p)$:

$$i \prec_h p \Rightarrow j \preceq_{h+1} q \qquad \text{(isotony)}$$

or, equivalently,

$$z^h_{ip} \overline{z}^{h+1}_{jq} = 0 \qquad\qquad (5.11)$$

since each $\preceq_h$ is a linear order.

(ii) For each $h = 1,\dots,r; i,p \in L_h, i \neq p$:

$$i \preceq_h p \Longleftrightarrow p \npreceq_h i \qquad \text{(asymmetry and completeness)}$$

Table 5.4. Complexity of reductions to quadratic equations

Problem	Complexity of the reduction
Bipartiteness	$O(m)$
Balance in signed graphs	$O(m)$
Recognition of split graphs	$O(n^2)$
Forbidden-color graph bipartition	$O(m)$
Totally unimodular matrices	linear
Kőnig-Egerváry property	$O(n^{2.5})$
Single bend wiring	quadratic
Max-quadratic functions	linear
Level graph drawing	quadratic

or, equivalently,

$$z_{ip}^h z_{pi}^h \vee \overline{z}_{ip}^h \overline{z}_{pi}^h = 0. \tag{5.12}$$

(iii) For each $h = 1, \ldots, r; i, k, p \in L_h$:

$$i \preceq_h k \text{ and } k \preceq_h p \Rightarrow i \preceq_h p \qquad \text{(transitivity)}$$

or, equivalently,

$$z_{ik}^h z_{kp}^h \overline{z}_{ip}^h = 0. \tag{5.13}$$

Summing up, the answer to ISOTONY is Yes if and only if the cubic Boolean equation

$$F(Z) = 0$$

is consistent, where F is the disjunction of all the left-hand sides of (5.11), (5.12), and (5.13). Randerath et al. [778] proved the following surprising result.

Theorem 5.17. *The cubic constraints (5.13) are redundant. Therefore,* ISOTONY *is polynomially reducible to a quadratic Boolean equation, and it can be answered in polynomial time.*

Proof. The proof is lengthy and must be omitted here. The reader may consult the paper by Randerath et al. [778].   □

5.5.11 A final look into complexity

Most of the reductions to quadratic Boolean equations discussed in Sections 5.5.2 to 5.5.10 can be performed in linear time, and all of them in polynomial time. The complexity of these reductions is summarized in Table 5.4. In this table, n and m stand for the number of vertices and edges of the input graph, while "linear" and "quadratic" are meant with respect to the input size.

In conclusion, we see that quadratic Boolean equations, or 2-SAT problems, play, within a wide class of "tractable" problems, an analogous role to that of cubic equations, or 3-SAT problems, for the class of "untractable" NP-complete problems. Formally, it can be proved (see Papadimitriou [725]) that 2-SAT is NL-complete, where NL denotes the class of those problems that can be solved by a nondeterministic Turing machine using a logarithmic amount of memory space. This result supports the view that, in a sense, quadratic Boolean equations are among the "hardest easy" discrete problems.

5.6 Efficient graph-theoretic algorithms for quadratic equations

5.6.1 Introduction

As discussed in Chapter 2, solving Boolean equations is one of the most fundamental and important problems on Boolean functions. Although intractable in general, this problem admits efficient algorithms when the input is restricted to quadratic DNFs. Indeed, the polynomial-time solvability of quadratic equations was already pointed out by Cook [208] in his seminal paper on the NP-completeness of SAT-ISFIABILITY. Here, we provide a simple argument to establish this fact (in DNF formulation).

Given a quadratic DNF equation in n variables, apply the classical variable elimination method presented in Section 2.6, maintaining at each iteration a current list of terms. Eliminating an arbitrary variable requires computing the conjunction (product) of two linear expressions, which results in $O(n^2)$ quadratic terms, and checking whether each of the generated terms is absorbed by some term in the current list, which takes $O(n^4)$ time. The backward step for retrieving a solution costs only $O(n^2)$ time. In conclusion, an $O(n^5)$ algorithm ensues.

The key property that allows this procedure to run in polynomial time is that the equation obtained after eliminating a variable remains quadratic. As a consequence, no exponential blowup can occur in the course of the algorithm. A similar reasoning would apply to the consensus procedure described in Section 2.7, since the consensus of any two quadratic terms is again quadratic.

However, one can do much better in terms of complexity. In the rest of this section, we describe four fast algorithms for the solution of quadratic Boolean equations:

- The Labeling algorithm of Gavril [374].
- The Alternative Labeling algorithm of Even, Itai, and Shamir [318] (this paper contains only an outline of the algorithm; more detailed descriptions can be found in Gavril [374] and Simeone [834]).
- The Switching algorithm of Petreschi and Simeone [741].
- The Strong Components algorithm of Aspvall, Plass, and Tarjan [34].

All four algorithms above are graph theoretic: In the first three algorithms, the quadratic Boolean expression φ is represented by an undirected graph (namely, the matched graph introduced in Section 5.4.2), whereas the fourth algorithm exploits a digraph model (namely, the implication graph introduced in Section 5.4.3).

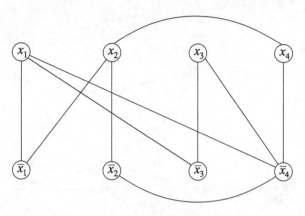

Figure 5.10. The matched graph for Example 5.9.

Consider a quadratic Boolean equation in n variables $x_1, \ldots, x_n$ and in m terms, say,

$$\varphi \equiv T_1 \vee \cdots \vee T_m = 0, \tag{5.14}$$

where, without loss of generality, we may assume that each term is the conjunction of exactly two literals and that no term appears more than once in the expression φ.

Example 5.9. *The four algorithms to be described will be demonstrated on the quadratic Boolean equation*

$$\varphi = x_1 \overline{x}_3 \vee x_1 \overline{x}_4 \vee \overline{x}_1 x_2 \vee x_2 x_4 \vee \overline{x}_2 \overline{x}_4 \vee x_3 \overline{x}_4 = 0. \tag{5.15}$$

The corresponding matched graph G_φ and implication graph D_φ are shown in Figures 5.10 and 5.11, respectively.   □

5.6.2 Labeling algorithm (L)

The basic principle of the Labeling algorithm for quadratic Boolean equations can be traced back to the algorithm proposed by Gavril [374] for the recognition of the Kőnig-Egerváry property in graphs, see Petreschi and Simeone [742]. The idea is to *guess* the value of an arbitrary literal ξ and to *deduce* — essentially, by the unit literal rules of Section 2.5.2 – the possible consequences of this guess on other variables. One keeps track of these consequences by a 0–1 labeling of the literals occurring in the input DNF φ.

Initially all terms are declared to be "unscanned." An arbitrary literal ξ is selected and is given the label 1; at the same time $\overline{\xi}$ is given the label 0. Then the labeling is propagated to as many literals as possible through repeated execution of the following STEP:

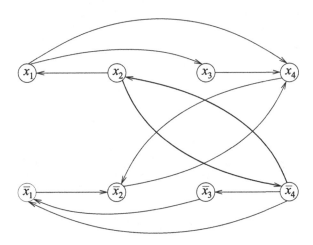

Figure 5.11. The implication graph for Example 5.9.

STEP:
Pick an arbitrary unscanned term $\eta\zeta$ such that η has the label 1, and assign to ζ and to $\bar{\zeta}$ the labels 0 and 1, respectively, making sure that ζ did not previously receive the label 1. Declare the term $\eta\zeta$ "scanned."

If a conflict arises because ζ was previously assigned the label 1, and the algorithm now tries to assign the label 0 to ζ, then the labeling stops, all labels are erased, and the alternative guess $\xi = 0$ is made. The labeling procedure starts again, and if a new conflict occurs at a later stage, then the algorithm terminates with the conclusion that the equation has no solution. On the other hand, if all literals are successfully labeled, then the algorithm concludes that the equation is consistent and the labeling directly yields a solution. However, a third possibility may occur: The labeling "gets stuck," in the sense that no conflict has occurred, but some literals are still unlabeled. This may happen only when no unscanned term contains a literal labeled 1, in other words, when each literal appearing in an unscanned term is either unlabeled or has the label 0. If this situation occurs, then the labeled variables are fixed according to the current labels, and the labeling restarts with a new guess on the reduced expression involving only the unlabeled literals.

Theorem 5.18. *The Labeling algorithm is correct and runs in $O(mn)$ time.*

Proof. The algorithm makes a guess on the value of some literal, and then it deduces the values of as many literals as possible. Label propagation (that is, value assignment) stops in three cases:

Case 1: All literals have been labeled without conflicts.

In this case the labels assigned to the literals define a solution X^* of the quadratic Boolean equation $\varphi = 0$. Indeed, the algorithm is such that

(i) the labels assigned to variable x_i and to its complement $\bar{x}_i$ are different for $i = 1, \ldots, n$;

(ii) the label 1 is never simultaneously assigned to two literals appearing in a same quadratic term.

Case 2: A conflict occurs.

In this case the initial guess $\xi = 1$ was wrong. This means that one must have $\xi = 0$ in every solution (if any) of the Boolean equation $\varphi = 0$; equivalently, ξ is a linear implicant of the quadratic Boolean function f associated with φ. If the label propagation consequent to the alternative guess $\xi = 0$ also ends in a conflict, then $\bar{\xi}$ must be a linear implicant of f, too, meaning that the equation $\varphi = 0$ has no solution.

Case 3: The algorithm "gets stuck," that is, a proper subset of literals is labeled and the labeling cannot be extended further.

In this case, let L and U be the sets of labeled and unlabeled literals, respectively. Let φ_U be the subexpression of φ obtained after fixing all the labeled variables to their current labels; thus, φ_U involves only the unlabeled literals. We claim that $\varphi = 0$ is consistent if and only if $\varphi_U = 0$ is consistent.

Observe that each term of T of φ_U is among the terms of φ: Indeed, if this is not the case, then T must result from some term $T_i = \eta\zeta$ of φ by fixation of one of its literals to 1. But then, the propagation step implies that the other literal of T_i should have been labeled 0, so that T should not appear in φ_U.

Now, assume that $\varphi = 0$ is consistent and that $\varphi(X^*) = 0$. Then, in view of the previous observation, the restriction of X^* to the variables associated with unlabeled literals defines a solution of $\varphi_U = 0$.

Conversely, assume that $\varphi_U = 0$ is consistent, and consider the labeling corresponding to an arbitrary solution. Such labeling, together with the one already obtained for L, defines a complete labeling of the literals of φ having the above properties (i) and (ii), and hence, a solution of $\varphi = 0$.

It follows that the labels of L can be made permanent, and that the labeling can restart from U after the process gets stuck. Hence, the algorithm is correct.

The total number of initial guesses made by the algorithm is at most $2n$. After each guess, the corresponding label propagation stage explores at most m terms. Hence, the worst-case complexity of the Labeling algorithm is $O(mn)$.   □

Example 5.9 (continued). *The history of the execution of the Labeling algorithm on the quadratic equation (5.15) is shown in Table 5.5. After Step 10 all literals have been labeled without conflicts. Hence, the equation $\varphi = 0$ is consistent, and a solution is $x_1^* = 0$, $x_2^* = 0$, $x_3^* = 1$, $x_4^* = 1$.*   □

Table 5.5. Execution of the Labeling algorithm on
the equation (5.15)

Step	Term	Labels	State
0	/	$x_2 = 1, \bar{x}_2 = 0$	guess
1	$\bar{x}_1 x_2$	$\bar{x}_1 = 0, x_1 = 1$	
2	$x_2 x_4$	$x_4 = 0, \bar{x}_4 = 1$	
3	$x_1 \bar{x}_3$	$\bar{x}_3 = 0, x_3 = 1$	
4	$x_1 \bar{x}_4$	$\bar{x}_4 = 0, x_4 = 1$	conflict
5	/	$x_2 = 0, \bar{x}_2 = 1$	alternative guess
6	$\bar{x}_2 \bar{x}_4$	$\bar{x}_4 = 0, x_4 = 1$	stuck
7	/	$x_1 = 0, \bar{x}_1 = 1$	guess
8	/		stuck
9	/	$x_3 = 1, \bar{x}_3 = 0$	guess
10	/		end

Example 5.10. *Consider the DNF*

$$\varphi = \bar{x}_1 x_2 \vee \bar{x}_2 x_3 \vee \ldots \vee \bar{x}_{n-1} x_n \vee \bar{x}_{n-1} \bar{x}_n$$

with $m = n$. If the initial guess $x_1 = 0$ is made, then the ensuing label propagation stage discovers a conflict very late, that is, after n steps, when x_n must successively receive the labels 0 and 1. For the alternative guess $x_1 = 1$, label propagation immediately gets stuck, and no further variable may be labeled. Afterwards, the wrong guess $x_2 = 0$ can be made, and so on. So, in the worst case, $n + (n-1) + \cdots + 2 + 1 = \frac{1}{2}(n+1)n = \frac{1}{2}(n+1)m$ terms are scanned (with repetitions) by the algorithm, and the total number of operations performed is of the order of $\Omega(mn)$. $\square$

5.6.3 Alternative Labeling algorithm (AL)

The idea of the Alternative Labeling algorithm is again to *guess* the value of an arbitrary literal ξ in some solution and to *deduce* the possible consequences of this guess on other variables appearing in the expression. Since ξ can take either the value 0 or the value 1, the algorithm analyzes in parallel the consequences of these two alternative guesses on ξ. It keeps track of these consequences by a "red" labeling (corresponding to the guess $\xi = 1$) and by a "green" labeling (corresponding to the guess $\xi = 0$). The purpose of propagating the two labelings in parallel is to avoid wasting time on the green labeling, say, as soon as the red one either detects an early conflict or gets stuck.

Initially, all terms are declared to be "red-unscanned" and "green-unscanned." Then, the algorithm selects an arbitrary literal ξ and assigns to it both the red label 1 and the green label 0, while the complementary literal $\bar{\xi}$ receives the red label 0 and the green label 1.

Table 5.6. Execution of the Alternative Labeling algorithm on the
equation (5.15)

		RED LABELING			GREEN LABELING	
Step	Term	Labels	State	Term	Labels	State
0	/	$x_2 = 0, \bar{x}_2 = 1$	guess	/	$x_2 = 1, \bar{x}_2 = 0$	guess
1	$\bar{x}_2\bar{x}_4$	$\bar{x}_4 = 0, x_4 = 1$		$\bar{x}_1 x_2$	$\bar{x}_1 = 0, x_1 = 1$	
2	$x_2 x_4$	$x_2 = 0, \bar{x}_2 = 1$	stuck			
3	/	$x_3 = 0, \bar{x}_3 = 1$	guess	/	$\bar{x}_3 = 0, x_3 = 1$	guess
4	$x_1\bar{x}_3$	$x_1 = 0, \bar{x}_1 = 1$	end			

The two labelings are then extended to as many literals as possible through the
alternate execution of the following STEP for the red labeling and for the green
one:

STEP:
Pick an arbitrary unscanned term $\eta\zeta$ such that η has the label 1, and assign to ζ
and to $\bar{\zeta}$ the labels 0 and 1, respectively, making sure that ζ did not previously
receive the label 1. Declare the term $\eta\zeta$ "scanned." (Here, terms like "label,"
"unscanned," "scanned" are relative to the color currently under consideration.)

If a conflict arises, say, for the red labeling (i.e., some literal that was previously
red-labeled 1 is forced to get the red label 0, or vice versa), the red labeling stops
and the red labels are erased. If, at a later stage, a conflict occurs also for the green
labeling, the algorithm stops and the equation has no solution. It may happen that
one of the labelings, say, the red one, "gets stuck," meaning that no conflict has
occurred, but that there are still literals having no red label. This is possible only
when, for each red-unscanned term, the literals appearing in that term are either
red-unlabeled or have red label 0. If this situation occurs, then the red labels are
made permanent, and both the red and the green labeling are restarted on the
reduced expression involving only the red-unlabeled literals.

The algorithm can be shown to run in $O(m)$ time (see Gavril [374]).

Example 5.9 (continued). *Table 5.6 summarizes a run of the algorithm on the
equation (5.15). After step 4, all literals have been (red-)labeled without conflicts.
Hence the equation is consistent and a solution is given by $x_1^* = 0, x_2^* = 0, x_3^* =
0, x_4^* = 1$.* □

5.6.4 Switching algorithm (S)

The Switching algorithm relies on the idea of Horn-renamability. Lewis [612]
introduced the class of Horn-renamable DNFs, consisting of those DNFs that

can be written as Horn DNFs after *switching* a subset of variables, that is, after performing the change of variables that replaces some of the original variables x_i by new variables $y_i = \overline{x}_i$. He provided a 2-SAT characterization of Horn-renamability (see Section 6.10.1 for details). For quadratic DNFs, a sort of converse relation holds.

Theorem 5.19. *Given a pure quadratic Boolean DNF φ, the equation $\varphi = 0$ is consistent if and only φ is Horn-renamable.*

Proof. The proof is left as an easy exercise. $\square$

On the basis of Theorem 5.19, the Switching algorithm tries to transform the given expression φ into a Horn expression, if possible, through a sequence of switches of variables. The algorithm first identifies an arbitrary negative term, say $\overline{x}_i \overline{x}_r$; if this term is to be transformed into a Horn term, then at least one of the variables x_i, x_r needs to switched. The algorithm accordingly picks one of the variables, say x_i, and tries to deduce the consequences of this choice.

In order to describe more formally the algorithm, it is convenient to introduce some preliminary definitions. (We use the tree terminology of Appendix A.) An *alternating tree rooted at* $\overline{x}_i$ is a subgraph $T(\overline{x}_i)$ of the matched graph G_φ with the following properties:

(1) $T(\overline{x}_i)$ is a tree, and $\overline{x}_i$ is its root.
(2) If x_j is a vertex of $T(\overline{x}_i)$, then its father in $T(\overline{x}_i)$ is $\overline{x}_j$.
(3) If $\overline{x}_j$ is a vertex of $T(\overline{x}_i)$ and $j \neq i$, then its father is a vertex x_r of $T(\overline{x}_i)$ such that $(x_r, \overline{x}_j)$ is a mixed edge of G_φ.
(4) If x_r is a vertex of $T(\overline{x}_i)$ and $(x_r, \overline{x}_j)$ is a mixed edge of G_φ, then $\overline{x}_j$ is a vertex of $T(\overline{x}_i)$.

Note that it is easy to "grow" a maximal alternating tree $T(\overline{x}_i)$ rooted at a vertex $\overline{x}_i$ of a matched graph. Indeed, suppose that T is any tree T which satisfies conditions (1)–(3) (initially, T may contain the isolated vertex $\overline{x}_i$ only), and perform the following steps as long as possible:

(i) If T has a leaf of the form $\overline{x}_j$, then add vertex x_j and edge $(\overline{x}_j, x_j)$ to T.
(ii) If T has a leaf x_r, then add to T all vertices $\overline{x}_j$ and edges $(x_r, \overline{x}_j)$ such that $(x_r, \overline{x}_j)$ is a mixed edge of G_φ and $\overline{x}_j$ is not already in T.

It is clear that conditions (1)–(3) are maintained by both steps (i) and (ii). Moreover, when step (ii) no longer applies, then condition (4) is also satisfied; hence, T is an alternating tree rooted at x_i.

Let us now record two useful properties of alternating trees.

Lemma 5.1. *Let $T(\overline{x}_i)$ be an alternating tree of G_φ rooted at $\overline{x}_i$, let x_j be any vertex of $T(\overline{x}_i)$, and let $P(i, j)$ be the unique path from x_i to x_j in $T(\overline{x}_i)$. If X^* is a solution of the equation $\varphi(X) = 0$ such that $x_j^* = 0$, then $x_k^* = 0$ for all vertices x_k lying on $P(i, j)$.*

Proof. The proof is by induction on the length of the path $P(i, j)$. If $P(i, j)$ has length 0, then $i = j$ and the statement is trivial. Otherwise, observe that $\overline{x}_j$ is the father of x_j in $T(\overline{x}_i)$, and consider the father of $\overline{x}_j$; in view of condition (3), this is a vertex x_r such that $x_r\overline{x}_j$ is a term of of φ. Since $\varphi(X^*) = 0$ and $x_j^* = 0$, we obtain that $x_r^* = 0$. Now, the conclusion follows by induction, since the path from x_i to x_r is shorter than $P(i, j)$. $\square$

To state the next property, we define the *join* of two vertices of $T(\overline{x}_i)$ to be their common ancestor that is farthest away from the root $\overline{x}_i$. Note that the join of any two vertices necessarily corresponds to an uncomplemented variable.

Lemma 5.2. *If x_h and x_k are two vertices of an alternating tree $T(\overline{x}_i)$, and $x_h x_k$ is a positive term of φ, then the variable x_j associated with the join of x_h and x_k is forced to 0 in all solutions of $\varphi = 0$.*

Proof. In every solution (if any) of $\varphi = 0$, either x_h or x_k must take value 0. Since x_j is on the path from x_i to x_h and on the path from x_i to x_k, the conclusion follows from Lemma 5.1. $\square$

We are now ready to describe the Switching algorithm. The algorithm works on the matched graph G_φ. An endpoint $\overline{x}_i$ of a negative edge $(\overline{x}_i, \overline{x}_r)$ is selected, and an alternating tree $T(\overline{x}_i)$ is grown, as explained above. As soon as a new vertex x_h of $T(\overline{x}_i)$ is generated, one checks whether G_φ has a positive edge (x_h, x_k) linking x_h to a previously generated vertex x_k of $T(\overline{x}_i)$. If this is the case, the variable x_j corresponding to the join of x_h and x_k must be forced to 0 by Lemma 5.2.

As a consequence, other variables are forced in cascade according to the following rules:

- If ξ is forced to 0, then $\overline{\xi}$ is forced to 1.
- If ξ is forced to 1 and (ξ, η) is an edge of G_φ, then η is forced to 0.

If a conflict occurs during this process (that is, if some variable is forced both to 0 and to 1), then the algorithm stops and concludes that the equation is inconsistent. Otherwise, we obtain a reduced equation involving fewer variables, and a new iteration begins. If the construction of $T(\overline{x}_i)$ has been completed and no positive edge between two vertices of $T(\overline{x}_i)$ has been detected, then a switch is performed on all the variables corresponding to the vertices of $T(\overline{x}_i)$. In this way, we produce an equivalent expression, and a new iteration begins. The procedure is iterated until either a Horn equation is obtained or all variables are forced. In both cases, a solution of the original equation $\varphi = 0$ can be found by inspection of the lists of the forced variables and of the switched ones.

Example 5.9 (continued). *The matched graph G_φ of Figure 5.10 has a negative edge $(\overline{x}_2, \overline{x}_4)$. Hence, the alternating tree $T(\overline{x}_2)$ shown in Figure 5.12 is grown, until the positive edge (x_2, x_4) is detected.*

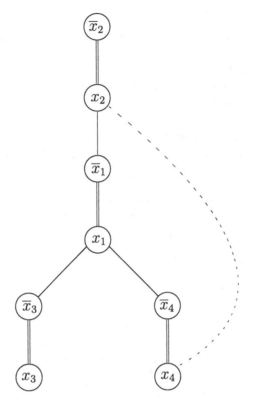

Figure 5.12. Alternating tree rooted at $\overline{x}_2$ for the matched graph of Figure 5.10.

Since the join of x_2 and x_4 is x_2 itself, the variable x_2 is forced to 0. Because of the term $\overline{x}_2\overline{x}_4$, variable x_4 is forced to 1. The subgraph of G_φ induced by x_1, x_3 and their complements has no negative edge, so the associated DNF ψ' is Horn, and the equation $\psi' = 0$ has the trivial solution $x_1 = x_3 = 0$. It follows that the equation $\varphi = 0$ is consistent, and that it admits the solution $x_1 = x_2 = x_3 = 0$, $x_4 = 1$. $\qquad\square$

Theorem 5.20. *The Switching algorithm is correct and can be implemented to run in $O(mn)$ time.*

Proof. In view of Lemma 5.2, the consistency of the original equation is not affected when we fix variables as explained in the algorithm. Also, switching a set of variables does not affect consistency. Therefore, if the algorithm terminates (either because the equation is proved to be inconsistent or because a solution has been produced), then it necessarily returns the correct answer. Thus, we only need to prove that the algorithm always terminates. To see this, let us show that each vertex $\overline{x}_i$ can occur at most once as the root of an alternating tree during the execution of the algorithm. Consider what can happen when the tree $T(\overline{x}_i)$ is generated.

- If the equation is declared inconsistent, then the algorithm stops.
- If a positive edge (x_h, x_k) is encountered and the join of x_h, x_k is forced to 0, then as a consequence of Lemma 5.1, x_i is subsequently fixed to 0 as well, and this variable disappears from the remaining equation.
- If all variables occurring in $T(\overline{x}_i)$ are switched when $T(\overline{x}_i)$ has been completely generated, then we claim that no new negative edges arise in the process (in other words, a positive edge or a mixed edge is never transformed into a negative edge in the course of the algorithm). This implies, in particular, that $\overline{x}_i$ will never appear in a negative edge in any subsequent iteration of the algorithm. To prove the claim,
  - consider any positive edge (x_h, x_k) of G_φ; at most one of x_h and x_k can belong to $T(\overline{x}_i)$; otherwise, the positive edge (x_h, x_k) would have been detected and handled earlier by the algorithm; hence, this edge either remains positive or becomes mixed after switching;
  - consider a mixed edge $(x_h, \overline{x}_k)$ of G_φ; in view of condition (4) in the definition of alternating trees, it cannot be the case that x_h is a vertex of $T(\overline{x}_i)$ but x_k is not; hence, $(x_h, \overline{x}_k)$ cannot be transformed into a negative edge.

Petreschi and Simeone [741] describe an implementation of the Switching algorithm with complexity $O(mn)$.   □

5.6.5 Strong Components algorithm (SC)

This algorithm is based on Theorem 5.4 and Lemma 5.3. It works on the implication graph $D = D_\varphi$ and preliminarily finds the strong components of D in reverse topological order (see Appendix A and Tarjan [858]). The Mirror Property of D (see Section 5.4.3) implies that for every strong component C of D, there exists a "mirror" component $\overline{C}$, the *complement* of C, induced by the complements of the vertices in C. Hence, Theorem 5.4 can be restated as follows: "φ is satisfiable if and only if no strong component of D coincides with its complement."

The general step of the Strong Components algorithm implements the procedure described in the proof of Theorem 5.4. Namely, it processes the strong components of D (or equivalently, the vertices of the condensed implication graph $\hat{D}$) in reverse topological order, starting from a sink, and it labels them in the following way. For each strong component C, one of the following cases must occur:

(a) C is already labeled. Then, the algorithm processes the next strong component.

(b) $C = \overline{C}$. Then, the algorithm stops. In view of Theorem 5.4, the equation $\varphi = 0$ is inconsistent.

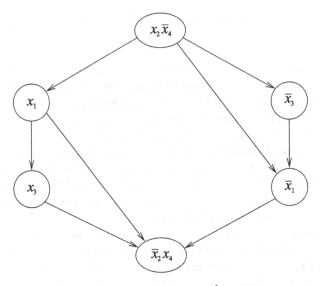

Figure 5.13. The condensed implication graph $\hat{D}_\varphi$ for the equation (5.15)

(c) C is unlabeled. Then, the algorithm assigns the label 1 to C and the label 0 to $\overline{C}$.

It is easy to see that, if C_1 and C_2 are two strong components, if there exists an arc from some vertex of C_1 to some vertex of C_2 in D, and if C_1 is labeled 1, then C_2 is necessarily labeled 1 as well. Thus, if we assign to each vertex ξ the label of the component containing ξ, we get a solution to the equation $\varphi = 0$ (by virtue of Lemma 5.3).

Example 5.9 (continued). *We consider again the equation $\varphi = 0$ given in (5.15) and the associated implication graph D_φ. As can be seen from Figure 5.11, the strong components of D_φ are $\{\overline{x}_2, x_4\}$, $\{x_1\}$, $\{x_3\}$ and their mirror components. The condensed implication graph $\hat{D}_\varphi$ is shown in Figure 5.13.*

Since no pair $x_i, \overline{x}_i$ belongs to the same strong component for any i, $\varphi = 0$ is consistent. The strong components of D_φ are labeled in the order shown in Table 5.7. Hence, a solution of the quadratic Boolean equation $\varphi = 0$ is given by $x_1^* = 1, x_2^* = 0, x_3^* = 1, x_4^* = 1$.   □

Aspvall, Plass and Tarjan [34] show that the Strong Components algorithm has complexity $O(m)$. A randomized version of the algorithm, with expected $O(n)$ time complexity, has been described by Hansen, Jaumard, and Minoux [470].

Table 5.7. Labeling of the strong
components of D_φ for the equation (5.15)

Strong component	Label
$\{\overline{x}_2, x_4\}$	1
$\{x_2, \overline{x}_4\}$	0
$\{x_3\}$	1
$\{\overline{x}_3\}$	0
$\{x_1\}$	1
$\{\overline{x}_1\}$	0

5.6.6 An experimental comparison of algorithms for quadratic equations

Petreschi and Simeone [742] report on the results of an experimental study in which the performance of the four algorithms for quadratic equations described in Sections 5.6.2–5.6.5 has been compared on 400 randomly generated test problems with up to 2000 variables and 8000 terms.

In all test problems, the *density* $\frac{m}{n}$ was nearly constant and equal to 4. With such density, almost all random quadratic equations instances are unsatisfiable under mild assumptions on the probability distribution of their terms (see Theorem 2.16 in Chapter 2 and Exercises 12–13 at the end of the current chapter). Therefore, 200 random instances were generated, and *all* of them proved to be unsatisfiable. The remaining 200 instances were randomly generated so as to be renamable Horn and thus *provably* satisfiable. One shortcoming of the uniform probability model was that almost all the strong components of the implication graph were singletons, except for one (in the unsatisfiable case) or two (in the satisfiable case) "megacomponents": This is in agreement with the theoretical probabilistic results in Hansen, Jaumard, and Minoux [470]. In order to eliminate these and other related anomalies, another instance generator was built, which produced strong components with binomially distributed sizes. Then random instances were generated by this "binomial" generator.

An analysis of the results led to the following main conclusions:

1) The first, and perhaps most important, observation is that quadratic Boolean equations are indeed easy to solve: Even the slowest algorithm took only 44 milliseconds (on an IBM 3090 – nowadays an archaic computer!) to solve the largest problem (2000 variables and 8000 terms).

2) In the satisfiable case, the foregoing experiments show a clear-cut ranking of the four algorithms with respect to running times: L is unquestionably the fastest one, followed by AL, S, and SC (see Figure 5.14).

3) In the unsatisfiable case, the running times of L, AL, and S are roughly comparable, whereas the running time of SC is by far larger; except for SC, the running times were much smaller in the unsatisfiable case than in

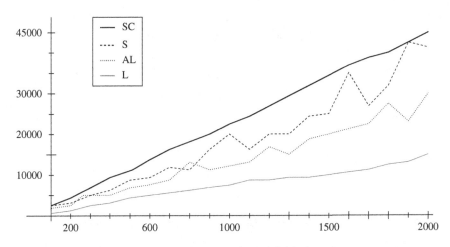

Figure 5.14. Running times for satisfiable formulas.

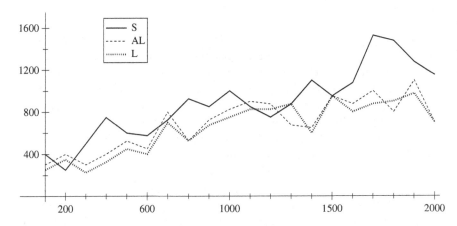

Figure 5.15. Running times for unsatisfiable formulas.

the satisfiable one (see Figure 5.15, where the SC-graph is oversized and hence, is not shown).

4) In the satisfiable case, the running times of SC and L grow quite regularly with the problem size. In fact, they are very well fitted by a straight line: The authors found that $TIME_L = 5.94n$ and $TIME_{SC} = 21.99n$, the squared correlation coefficients being $R_L^2 = 0.999$ and $R_{SL}^2 = 1$, respectively. On the other hand, the graph of the running times of AL and S as a function of n is less regular, but it lies between two straight lines corresponding to L and SC (see Figure 5.14).

In the unsatisfiable case, the behavior of SC is as regular as it is in the satisfiable case. The other three algorithms, however, behave very irregularly and exhibit frequent nonmonotonicities (see Figure 5.15). At any rate, their complexity turns out to be sublinear: Roughly speaking, the running times are proportional to the square root of n. Furthermore, the running times of L and AL are seen to be highly correlated.

In conclusion, the experimental average complexity of both L and S is lower than their worst-case complexity, and is, in any case, bounded above by a linear function of n. Similar conclusions are reached with the binomial generator.

5) In the satisfiable case, the vast majority of the variables turned out to be forced.

6) A direct comparison between L and AL shows that the latter algorithm, despite its $O(m)$ worst-case complexity, is more than twice slower than the former one, whose worst-case complexity is $O(mn)$.

The main point is that L "capitalizes on luck," whereas AL follows a more "pessimistic" approach, and L is less affected by random factors, which may increase its running time in the worst-case but may also decrease it on average. Actually, for L to reach its $O(mn)$ worst-case complexity, the following events must take place:

- Every time a guess is made, it is always the wrong one.
- Every time a wrong guess is made, the resulting conflict is detected very late.
- Every time a conflict takes place, the alternative guess results in very early blocking.

However, under both probability models, things do not go that way:

- A guess is successful in about 50% of the cases.
- Every time a wrong guess is made, the resulting conflict is detected rather early because conflicts are due to "local obstructions" (Simeone [834]).
- Every time a conflict takes place, a certain literal ζ is recognized as being forced; as a consequence, a large set $C(\zeta)$ of literals is then forced.

5.7 Quadratic equations: Special topics

5.7.1 The set of solutions of a quadratic equation

There is a nice connection between the set of solutions of a quadratic Boolean equation and median graphs. A *median graph* is an undirected graph having the property that, for any three vertices x, y, z, there exists a unique vertex w (called the *median* of x, y, z) that at the same time lies on some shortest path between x and y, on some shortest path between x and z, and on some shortest path between y and z. Median graphs display many interesting properties and remarkable connections with other branches of mathematics, computer science, natural sciences, and social sciences; see Bandelt and Chepoi [50]; Chung, Graham, and Saks [195]; Mulder [694]; Mulder and Schrijver [695], and so on.

Given a quadratic DNF φ and its implication graph D_φ, let us introduce an undirected graph H whose vertices are all the solutions of the quadratic Boolean equation $\varphi = 0$, where two solutions X^* and Y^* are adjacent if there exists some strong component C of D_φ having the property that Y^* is obtained from X^* (and vice versa) by switching the values of some variables x_i such that either x_i or $\overline{x}_i$ belongs to C. For instance, if $X^* = (0,1,0,0,1,0)$, $Y^* = (0,0,1,0,0,0)$, and there exists a strong component $\{x_2,\overline{x}_3,\overline{x}_5,x_6\}$, then X^* and Y^* are adjacent in H, since one obtains Y^* from X^* by switching the values of the second, third, and fifth variables, and the literals $x_2,\overline{x}_3,\overline{x}_5$ all belong to the strong component.

Theorem 5.21. *The foregoing construction always produces a median graph, and all median graphs can be obtained in this way.*

This result follows from work of Schaefer [807]; see also Bandelt and Chepoi [50] and Feder [323]. An interesting "closure" property can be derived from it. (This property is in fact a restatement of the characterization of quadratic functions given in Section 5.3.2.)

Corollary 5.3. *Let X, Y, Z be any three solutions of a quadratic Boolean equation in n variables. Let W be the point of $\mathcal{B}^n$ defined as follows: for each $i = 1, 2, \ldots, n$, the i-th component of W takes the value 1 if and only if the i-th components of at least two out of the three vectors X, Y, Z take the value 1; that is, W is obtained from these three vectors according to the majority rule (componentwise). Then W also is a solution of the quadratic Boolean equation.*

The number of solutions of a quadratic Boolean equation

$$\varphi(x_1, x_2, \ldots, x_n) = 0 \tag{5.16}$$

may be exponentially larger than the number of its variables, and generating them all is generally a prohibitive task. In fact, Valiant [883] proved that even determining the number of such solutions is #P-complete, and hence, probably very difficult. It is perhaps worth mentioning here that merely *counting* the solutions is somewhat "easier" than *generating* them; see Dahlöf, Jonsson, and Wahlström [252]; Fürer and Kasiviswanathan [354].

Feder [322, 323] proposed a generating algorithm, which we now sketch. For ease of presentation, we assume that the quadratic equation given by (5.16) is pure and Horn, that is, all its terms are quadratic and either positive (they involve only uncomplemented variables) or mixed (they involve exactly one complemented and one uncomplemented variable). This assumption is not restrictive, since, in view of Theorem 5.19, every consistent purely quadratic Boolean equation can always be cast into a Horn equation after some of its variables are renamed.

For every pair of Boolean variables x_k, x_j, the following equivalences hold:

$$x_k \overline{x}_j = 0 \quad \text{if and only if} \quad x_k \le x_j,$$
$$x_k x_j = 0 \quad \text{if and only if} \quad x_k \le \overline{x}_j.$$

Therefore, (5.16) can be rewritten (in more than one way) as a system of Boolean implications of the form

$$x_k \le x_j \quad \text{for all } x_j \in D_k, \tag{5.17}$$

$$x_k \le \overline{x}_j \quad \text{for all } \overline{x}_j \in D_k, \tag{5.18}$$

where $D_k \subseteq X \cup \overline{X}$ for $k = 1, 2, \ldots, n$.

We also assume, without loss of generality, that there are no forced variables and no twin literals in the equation, since these can easily be detected and handled in a preprocessing phase. As a consequence of our assumptions, the implications (5.17)–(5.18) can be written in such a way that $k < j$ when either $x_j \in D_k$ or $\overline{x}_j \in D_k$.

Feder [322, 323] observed:

Theorem 5.22. *Let $X^* \in \mathcal{B}^n$ be a nonzero solution of (5.17)–(5.18), and let $\ell \le n$ be such that $x_\ell^* = 1$ and $x_i^* = 0$ for $1 \le i < \ell$. Then, the point Y^* obtained after replacing x_ℓ^* by 0 is again a solution of (5.17)–(5.18).*

Proof. Because $y_\ell^* = 0$, the point Y^* clearly satisfies all implications of the form (5.17) for $x_j \in D_\ell$, as well as all implications of the form (5.18) for $\overline{x}_j \in D_\ell$ and for $\overline{x}_\ell \in D_k$. Moreover, when $x_\ell \in D_k$, the implication (5.17) is necessarily satisfied by Y^* because $k < \ell$, and hence, $y_k^* = 0$. $\qquad\square$

Example 5.11. *Consider the quadratic Boolean equation*

$$x_1x_2 \vee x_1x_3 \vee x_1\overline{x}_5 \vee x_2\overline{x}_7 \vee x_3\overline{x}_4 \vee x_3\overline{x}_8 \vee x_4x_5 \vee x_4\overline{x}_7 \vee x_5x_6 \vee x_6x_7 \vee x_6\overline{x}_8 \vee x_7x_8 = 0. \tag{5.19}$$

This equation is equivalent to the system of inequalities

$$x_1 \le \overline{x}_2, \; x_1 \le \overline{x}_3, \; x_1 \le x_5, \; x_2 \le x_7, \ldots, \; x_7 \le \overline{x}_8. \tag{5.20}$$

Because $X^* = (0, 1, 0, 0, 1, 0, 1, 0)$ is a solution of the equation, we can deduce that $Y^* = (0, 0, 0, 0, 1, 0, 1, 0)$ (obtained after replacing x_2^* by 0) also is a solution. $\qquad\square$

We say that Y^* is the *father* of the solution X^* if Y^* and X^* are in the relation described by Theorem 5.22. Note that every nonzero solution has exactly one father. Consider now the digraph $T = (S, A)$, where S is the set of solutions of (5.17)–(5.18) (or, equivalently, of the quadratic equation $\varphi = 0$), and where an arc (Y^*, X^*) is in A if and only if Y^* is the father of X^*. Then, T defines an arborescence rooted at the all-zero solution. Given any solution $Y^* \in S$, the children of Y^* in T can easily be generated: If y_j^* is the first nonzero component of Y^*, then the children of Y^* are exactly the points of the form $X^* = Y^* \vee e_i, i < j$, such that $X^* \in S$. It follows that the arborescence T can be generated and traversed efficiently (in fact, with polynomial delay; see Appendix B.8).

Feder [322, 323] describes a low-complexity implementation of this procedure.

Theorem 5.23. *The solutions of a quadratic equation with n variables and m terms can be generated after $O(m)$ preprocessing time in $O(n)$ time per solution, using $O(m)$ space.*

Proof. We refer the reader to Feder [322, 323] for details of the analysis.   □

5.7.2 Parametric solutions

In spite of the high complexity of generating the solutions of a quadratic Boolean equation, Crama, Hammer, Jaumard, and Simeone [234] showed that one can obtain a concise *product-form parametric representation* for all such solutions. The representation uses no more than n free Boolean parameters for an equation in n variables. Each variable (or its complement) is expressed as a product of these parameters or their complements, and these expressions provide a complete description of the solution set of the equation. Furthermore, the representation can be computed in $O(n^3)$ time.

In fact, algebraic methods for determining parametric representations in the case of general Boolean equations have been known for a long time (see Löwenheim [628, 629] and Section 2.11.3). When specialized to quadratic equations, Löwenheim's method produces (in polynomial time) a parametric representation of the solution set, each variable being associated with some Boolean expression of the parameters. The resulting expressions are generally in neither disjunctive nor conjunctive normal form, and reducing them to such a convenient format can be computationally expensive. This is to be contrasted with the very simple form of the representation proposed by Crama et al. [234].

Let us sketch the basic ideas leading to this parametric representation. As in the previous section, we assume that the quadratic equation is represented by the system of Boolean implications (5.17)–(5.18), where $x_k \notin D_k, \bar{x}_k \notin D_k, D_k$ does not contain both a variable and its complement, and $x_k \notin D_j$ when $x_j \in D_k$, for $k, j \in \{1,\ldots,n\}$ (otherwise, the equation can be simplified). The system (5.17)–(5.18) is in turn equivalent to the following one:

$$x_k \le \left(\bigwedge_{j:x_j \in D_k} x_j \right) \left(\bigwedge_{j:\bar{x}_j \in D_k} \bar{x}_j \right) \quad (k = 1, 2, \ldots, n), \qquad (5.21)$$

and hence, also to the system of equations:

$$x_k = x_k \left(\bigwedge_{j:x_j \in D_k} x_j \right) \left(\bigwedge_{j:\bar{x}_j \in D_k} \bar{x}_j \right) \quad (k = 1, 2, \ldots, n). \qquad (5.22)$$

In the remainder of this section, we focus on the equivalent expression (5.22) of the original quadratic equation.

The expression (5.22) suggests the following construction. Let $P = (p_1, p_2, \ldots, p_n)$ denote a vector of free Boolean parameters, and define the functions

$$g_k(P) = g_k(p_1, \ldots, p_n) = p_k \left(\bigwedge_{j:x_j \in D_k} p_j \right) \left(\bigwedge_{j:\overline{x}_j \in D_k} \overline{p}_j \right) \qquad (5.23)$$

for $k = 1, 2, \ldots, n$. Let

$$Q = \{(g_1(P), \ldots, g_n(P)) : P \in \{0,1\}^n\}. \qquad (5.24)$$

Then, we can prove:

Lemma 5.4. *If S is the set of solutions of the system (5.22), and if Q is defined by (5.23)–(5.24), then $S \subseteq Q$.*

Proof. If $(x_1^*, \ldots, x_n^*) \in S$, then $x_k^* = g_k(x_1^*, \ldots, x_n^*)$ for $k = 1, 2, \ldots, n$ in view of (5.22). Hence, $(x_1^*, \ldots, x_n^*) \in Q$. $\qquad \square$

The next proposition states a necessary and sufficient condition under which equality holds between S and Q. We first introduce some additional notation. With the system (5.22), we associate the directed graph $H = (X \cup \overline{X}, A)$, defined as follows: For all x_k in X and μ in $X \cup \overline{X}$, the arc (x_k, μ) is in A if and only if $\mu \in D_k$. (H is in general a subgraph of the implication graph of the original equation (5.16).)

Theorem 5.24. *If S is the set of solutions of the system (5.22), and if Q is defined by (5.23)–(5.24), then $S = Q$ if and only if the digraph H is transitive.*

Proof. Assume in the first place that H is transitive. By Lemma 5.4, we only have to prove that every point $(g_1(P), \ldots, g_n(P))$ in Q is a solution of (5.17)–(5.18).

Let $\overline{x}_j \in D_k$. If $g_k(P) = 1$, then $p_j = 0$, and hence, $g_j(P) = 0$. This shows that the implications (5.18) are satisfied by $(g_1(P), \ldots, g_n(P))$.

Let $x_j \in D_k$. If $g_j(P) = 0$, then either (i) $p_j = 0$, or (ii) $p_i = 0$ for some i such that $x_i \in D_j$, or (iii) $p_i = 1$ for some i such that $\overline{x}_i \in D_j$. In case (ii), $x_i \in D_k$ by transitivity of H. Similarly, in case (iii), $\overline{x}_i \in D_k$. Hence, in all cases, $g_k(P) = 0$, and the implications (5.17) are satisfied by $(g_1(P), \ldots, g_n(P))$.

Conversely, assume that H is not transitive. This means that, for some $x_k, x_j \in X$ and $\mu \in X \cup \overline{X}$, (x_k, x_j) and (x_j, μ) are in A, but (x_k, μ) is not in A. Assume for instance that $\mu \in X$, that is, $\mu = x_i$ for some $i \in \{1, \ldots, n\}$ (the proof is similar if $\mu \in \overline{X}$). So, $x_i \in D_j$, but $x_i \notin D_k$. Notice that $i \neq k$, by our assumptions on the system (5.17)–(5.18).

Let $P = (p_1, \ldots, p_n)$, where $p_k = 1, p_i = 0, p_l = 1$ if $x_l \in D_k$ and $p_l = 0$ otherwise (this is a valid assignment of values to the parameters). Then, $g_k(P) = 1$ and $g_j(P) = 0$. So $(g_1(P), \ldots, g_n(P))$ is not a solution of (5.17)–(5.18) and $S \neq Q$. $\qquad \square$

So, when H is transitive, the expressions $g_k(P)$ $(k = 1, 2, \ldots, n)$ defined by (5.23) yield a simple, *product-form parametric representation* of the solutions of (5.22), and hence of the original equation (5.16). Notice that, even if H is not transitive, (5.22) can always be transformed into an equivalent system for which the associated graph is transitive, by adding to it the necessary missing terms. More precisely, if $x_k \leq x_j$ and $x_j \leq \mu$ are two inequalities in the system (5.17)–(5.18), then the inequality $x_k \leq \mu$ is redundant, and it can always be added to the system. Iterating this operation until the resulting graph is transitive amounts to computing the transitive closure of H (see Section 5.8).

Crama et al. [234] rely on these ideas and on the properties of implication graphs to derive an efficient algorithm with complexity $O(\max\{m, n^3\})$ that computes a product-form parametric representation for an arbitrary quadratic equation. We refer to their paper for details.

Example 5.12. *Consider again the quadratic Boolean equation (5.19), which is equivalent to the system of inequalities (5.20). The digraph H associated with the system (5.20) is represented in Figure 5.16. The transitive closure H^* of H is displayed in Figure 5.17. (At this point, we can notice that x_3 must be equal to zero in all solutions of (5.19), because x_8 and $\overline{x}_8$ are successors of x_3 in H^*.)*

Using Theorem 5.24, we derive the following product-form parametric representation of the solutions of (5.19):

$$x_1 = p_1 \overline{P}_2 \overline{P}_3 p_5 \overline{P}_6,$$

$$x_2 = p_2 p_7 \overline{P}_8,$$

$$x_3 = 0,$$

$$x_4 = p_4 \overline{P}_5 p_7 \overline{P}_8,$$

$$x_5 = p_5 \overline{P}_6,$$

$$x_6 = p_6 \overline{P}_7 p_8,$$

$$x_7 = p_7 \overline{P}_8,$$

$$x_8 = p_8.$$

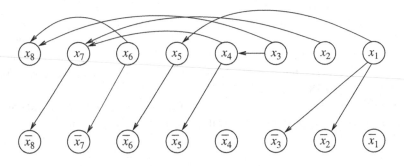

Figure 5.16. The digraph H associated with (5.20) in Example 5.12.

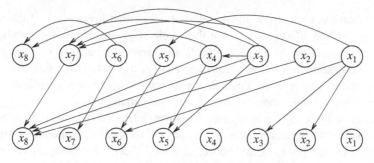

Figure 5.17. The transitive closure of H in Example 5.12.

The reader can check that all solutions of (5.19) are generated by giving all possible 0–1 values to the parameters $p_1, p_2, \ldots, p_8$.

Note also that, since the system of inequalities (5.20) is not uniquely defined by (5.19), it is possible to derive from Theorem 5.24 several product-form parametric representations of the solutions of (5.19).   □

5.7.3 Maximum 2-satisfiability

For a quadratic DNF φ on $\mathcal{B}^n$, the maximum 2-satisfiability problem, or MAX 2-SAT, consists in finding a point $X^* \in \mathcal{B}^n$ that cancels the maximum number of terms of φ. Of course, if the quadratic equation $\varphi = 0$ is consistent, then any false point of φ is a solution of MAX 2-SAT. In contrast with quadratic Boolean equations, however, MAX 2-SAT is an NP-hard problem, and remains hard even if we are only interested in finding a "provably good" approximate solution of the problem. This optimization problem was discussed extensively in Section 2.11.4. We refer to this section for more information and references on MAX 2-SAT; see also Chapter 13 for a brief discussion of quadratic binary optimization problems placed in the broader framework of pseudo-Boolean optimization problems.

5.7.4 On-line quadratic equations

In some applications, rather than a single quadratic Boolean equation (or 2-SAT problem), one is required to solve a nested sequence of m equations, where each formula is the disjunction of the previous one with an additional term of degree 2 (the initial formula is void and represents the constant 0). This problem is called an *on-line quadratic equation*, or *on-line 2-SAT*. Note, in particular, the following:

• The on-line model is quite natural in interactive environments.
• It leads to an early detection of inconsistency at its very onset.
• As soon as the equation becomes inconsistent, the removal of the last term immediately restores consistency.

Clearly, on-line quadratic equations can be solved in $O(1 + 2 + \ldots + m) = O(m^2)$ time by a naive approach. The main idea of an on-line algorithm, however, is, to update at each step a suitable data structure that keeps track of the work done so far and allows us to solve the whole sequence of problems with less computational effort. In this case, the classical worst-case analysis of the cost of a single operation may not be adequate to analyze the cost of the whole sequence of operations, and amortized complexity arguments are more appropriate. For a general discussion of amortized complexity, see Tarjan [859].

For an on-line equation involving n variables and m terms, Jaumard, Marchioro, Morgana, Petreschi, and Simeone [528] present an algorithm running in (amortized) $O(n)$ time per term, and hence, in overall $O(mn)$ time. For each formula in the nested sequence, not only does the algorithm check whether the formula is consistent or not, but it also yields an explicit solution, if any, and detects the sets of forced and twin (or identical) variables.

One can hardly conceive on-line algorithms with lower complexity, since simply writing out the solutions to m equations already requires $O(mn)$ time. For details, we refer to the paper by Jaumard et al. [528].

5.8 Prime implicants and irredundant forms

5.8.1 Introduction

In this section, we consider the following two problems (recall the definitions of prime implicants and irredundant DNF from Section 1.7):

(1) Given a quadratic DNF φ of a quadratic Boolean function f, find all prime implicants of f.
(2) Given a quadratic DNF φ of a quadratic Boolean function f, find an irredundant DNF of f.

Because all the prime implicants of a quadratic Boolean function in n variables are quadratic, their number is $O(n^2)$; moreover, as we mentioned in Section 5.6.1, the consensus method, starting from φ, generates all of them in polynomial time (actually, in time $O(n^6)$). Similar conclusions follow from Theorem 3.9 and Corollary 3.6 in Chapter 3.

However, much faster algorithms can be obtained on the basis of the close relationship that exists between the generation of all prime implicants of f and the generation of the transitive closure of a digraph. As we show in Section 5.8.2, the prime implicants of f can be easily obtained from the transitive closure of the implication graph of φ.

The disjunction of all the prime implicants of a Boolean function f is, in a sense, the most detailed and explicit DNF of f: Along with *each* pair of terms it explicitly features their consensus (or some term absorbing it); so, all logical implications derivable from those appearing in the DNF are themselves featured in the DNF. At the opposite extreme, irredundant DNFs are the most succinct and implicit DNFs of f: *No* consensus of pairs of terms appearing in any such DNF is also present in it, and the logical implications derivable from those appearing in the DNF are implicitly, rather than explicitly, present.

A polynomial bound can be derived for the complexity of finding an irredundant DNF of a quadratic Boolean function f, starting from an arbitrary quadratic DNF of f. This bound can be estimated as follows: Generate in $O(n^6)$ time, as earlier, the disjunction ψ of all prime implicants of f. Choose any term T of ψ and check in $O(n^2)$ time whether T is an implicant of the DNF ψ' resulting from the deletion of T in ψ (as e.g., in Theorem 3.8 of Chapter 3). If so, then T is redundant and ψ can be replaced by ψ'; otherwise, ψ remains unchanged. At this point, choose another term T' and repeat. The process ends when all terms have been checked for redundancy, and possibly deleted. Since the number of terms in ψ is $O(n^2)$, the overall complexity of the foregoing procedure is $O(n^6)$ – again a polynomial bound.

However, much faster algorithms can be obtained for this problem, too. As mentioned above, the graph-theoretic tool of choice for the generation of all prime implicants of a quadratic Boolean function f is the transitive closure of the implication digraph. On the other hand, as we show in Section 5.8.4, the appropriate notion for the generation of an irredundant quadratic DNF of f is that of transitive reduction of a digraph – just the converse of the transitive closure.

5.8.2 A transitive closure algorithm for finding all prime implicants

Let D_φ be the implication graph associated with the quadratic DNF φ. An elementary, but important, property of D_φ is that if $\xi\eta$ and $\overline{\eta}\zeta$ are any two terms for which there is a consensus $\xi\zeta$, then the corresponding arcs $(\xi,\overline{\eta})$, $(\overline{\eta},\overline{\zeta})$, and $(\xi,\overline{\zeta})$ form a transitive triplet, as shown in Figure 5.18. The arc $(\xi,\overline{\zeta})$ is present in D_φ if and only if the consensus $\xi\zeta$ appears in φ. Analogous statements hold for the "mirror" arcs (ζ,η), $(\eta,\overline{\xi})$, and $(\zeta,\overline{\xi})$.

Recall from Appendix A that the transitive closure of a digraph $D = (V,A)$, is the digraph $D^* = (V,A^*)$, where $A^* = A \cup \{(u,v)$: there is a directed path from u to v in $D\}$. Each consensus operation can be interpreted on D_φ as the addition of two mirror transitive arcs, and vice versa. Hence, in the transitive closure D_φ^* of D_φ, each pair $(\alpha,\overline{\beta})$ and $(\beta,\overline{\alpha})$ of mirror arcs corresponds to a quadratic implicant $\alpha\beta$ if $\alpha \neq \beta$, and to a linear implicant α if $\alpha = \beta$. Some of the quadratic implicants associated with arcs of D_φ^* may not be prime, since they might be absorbed by linear ones. However, it follows from Theorem 5.6 that all *prime implicants* must correspond to some pair of arcs of D_φ^*.

The obvious idea for generating all prime implicants of the quadratic DNF φ, then, is to compute the transitive closure D_φ^* and to efficiently perform absorption in order to remove nonprime quadratic implicants. The operation of absorption also has a simple interpretation on D_φ^*. Suppose that the linear term ξ absorbs the quadratic term $\xi\eta$. Then, the arcs $(\xi,\overline{\xi})$, $(\xi,\overline{\eta})$, and $(\eta,\overline{\xi})$ have to be present in D_φ^* (see Figure 5.19).

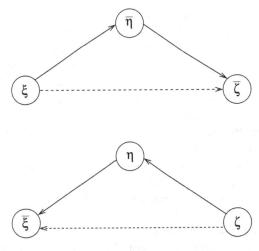

Figure 5.18. Transitive arcs in the implication graph.

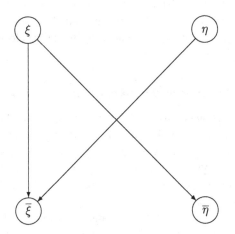

Figure 5.19. Absorption in the implication graph.

Therefore, absorption can be performed directly on D_φ^* by application of the following rule: Whenever an arc $(\xi, \bar{\xi})$ is present, remove all arcs leaving ξ (except for $(\xi, \bar{\xi})$), as well as all arcs entering $\bar{\xi}$ (again, except for $(\xi, \bar{\xi})$).

A survey on transitive closures is given in van Leeuwen [887]. Most of the known transitive closure algorithms are of two kinds.

1) *Algorithms that perform a sequence of transitive arc additions.*
   The $O(mn)$ algorithms of Goralcikova and Koubek [404], Ebert [286], Schmitz [810], Jaumard and Minoux [529], Chen and Cooke [191] belong to this class (for some of these algorithms, stronger complexity bounds hold, depending also on size parameters other than n and m.)

2) *Algorithms based on Boolean matrix multiplication.*
  A straightforward implementation results in an $O(n^3)$ transitive closure algorithm (Warshall [899]). Strassen-like matrix multiplication methods typically achieve complexities of $O(n^{2+\alpha} \log n)$, where $0 < \alpha < 1$; see Furman [355], Fischer and Meyer [331], Munro [697], Booth [104], Coppersmith and Winograd [212].

Munro [697] was apparently first to point out that, when computing the transitive closure of a digraph D, one may assume, without loss of generality, that D is connected (in the sense that its underlying undirected graph is connected) and acyclic. As a matter of fact, if D is disconnected, its transitive closure D^* is the union of the transitive closures of the connected components of D. If D has cycles, then one can preliminarily find the strong components of D by the $O(m)$ algorithm of Tarjan [858], and subsequently generate the acyclic condensation $\hat{D}$ of D by shrinking each strong component into a single supervertex. Once $\hat{D}^*$ has been computed, D^* can be obtained as follows:

Let A^* and $\hat{A}^*$ be the arc sets of D^* and $\hat{D}^*$, respectively. Then,

$$(x,y) \in A^* \Leftrightarrow \begin{cases} \text{there exists } (u,v) \in \hat{A}^* \text{ such that} \\ x \text{ belongs to the strong component of } D \\ \text{represented by } u, \text{ and } y \text{ belongs to} \\ \text{the strong component of } D \text{ represented by } v. \end{cases} \qquad (5.25)$$

Let n_k be the number of vertices in the kth strong component of $D, k = 1, \ldots, r$; let $\hat{n}$ and $\hat{m}$ be the number of vertices and arcs of $\hat{D}$, respectively. Besides the $O(\hat{m}\hat{n})$ operations required to generate $\hat{D}^*$, one needs

$$\sum_{(i,j) \in \hat{A}^*} n_i n_j \le (n_1 + \cdots + n_r)^2 = n^2$$

elementary operations to compute A^*, according to (5.25). But $\hat{n} \le n$, $\hat{m} \le n$ and $m \ge n - 1$ under the assumption that D is connected. It follows that D^* can be computed in $O(\hat{m}\hat{n} + n^2)$ time and thus in $O(mn)$ time. We state in Figure 5.20 a formal description of a transitive closure algorithm for the generation of all the prime implicants of a quadratic Boolean function f.

  Clearly Step 2 can be implemented in $O(mn)$ time, and this is also the overall complexity of the algorithm. From the discussion at the beginning of this section, we obtain the following results:

Theorem 5.25. *The algorithm* QUADRATIC PRIME IMPLICANTS *is correct, that is, it produces all prime implicants of the quadratic Boolean function f represented by the input DNF φ.*

Example 5.13. *Let f be the quadratic Boolean function represented by the DNF*

$$\varphi = x_1 \overline{x}_2 \vee x_1 \overline{x}_3 \vee x_2 x_3 \vee \overline{x}_3 x_4.$$

Procedure QUADRATIC PRIME IMPLICANTS(φ)
Input: A quadratic DNF φ.
Output: All prime implicants of the quadratic Boolean function f represented by φ.

begin
Step 1: construct the implication graph D_φ;
Step 2: run a transitive closure algorithm on the input D_φ;
       let $H = D_\varphi^*$ be the (transitive) graph obtained at the end of this step;
Step 3: for each arc $(\xi, \overline{\xi})$ in H, remove all arcs leaving ξ (except $(\xi, \overline{\xi})$)
       and all arcs entering $\overline{\xi}$ (except $(\xi, \overline{\xi})$); let Q be the resulting digraph;
Step 4: if there is a pair of arcs $(\xi, \overline{\xi})$, $(\overline{\xi}, \xi)$ in Q, **then** the Boolean constant $\mathbb{1}_n$
       is the only prime implicant of φ;
       else
       for each arc $(\xi, \overline{\xi})$ in Q, the linear term ξ is a prime implicant of φ;
       for each pair of mirror arcs $(\xi, \overline{\eta})$ and $(\eta, \overline{\xi})$, the quadratic term $\xi\eta$
       is a prime implicant of φ;
Step 5: return the list of prime implicants constructed in Step 4.
end

Figure 5.20. Procedure QUADRATIC PRIME IMPLICANTS.

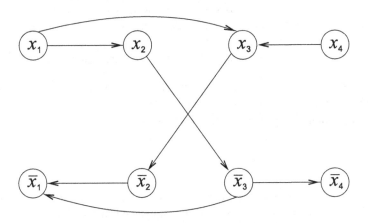

Figure 5.21. The implication graph D_φ.

The implication graph D_φ is shown in Figure 5.21; the graphs H and Q are shown in Figures 5.22 and 5.23, respectively. It follows that the disjunction of all the prime implicants of f is given by

$$x_1 \vee x_2 x_3 \vee x_2 x_4 \vee \overline{x}_3 x_4.$$

This can also be checked by the consensus method. $\qquad\qquad\square$

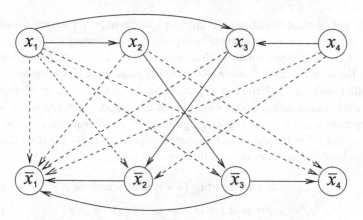

Figure 5.22. The graph H; dashed lines represent arcs added to D_φ.

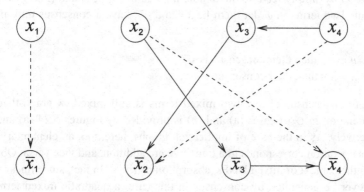

Figure 5.23. The graph Q.

5.8.3 A restricted consensus method and its application to computing the transitive closure of a digraph

In the present subsection, following a direction opposite to the previous one, we show how to obtain a fast and simple $O(mn)$ algorithm for the transitive closure of a digraph G through the execution of a very restricted form of consensus algorithm on a (quadratic) mixed DNF naturally associated with G. Unlike other transitive closure algorithms with the same complexity, this one has a very simple implementation and does not require complex data structures. The material in this subsection is drawn from recent work of Boros, Foldes, Hammer, and Simeone [120]. We refer to Section 3.2.2 and Section 6.5 for the general notions of disengagement consensus and input consensus, respectively.

Definition 5.1. *A consensus algorithm is said to be an input disengagement algorithm if it is both an input algorithm and a disengagement algorithm.*

Whether an input disengagement algorithm works or not for a given (quadratic) mixed Boolean function may actually depend on the disengagement order of the terms of its DNF representation. However, we will prove that, for an arbitrary mixed Boolean function f, there always exists *some* input disengagement algorithm that works for f. Before giving examples, let us work out a graph-theoretic framework which makes things easier to visualize and, as an additional bonus, leads to an efficient transitive closure algorithm. We recall from Section 5.4.1 that, for a mixed DNF φ, one can define a directed graph $G \equiv G(\varphi)$ – not to be confused with the implication graph D_φ – as follows:

$$x_i \text{ is a vertex of } G \Longleftrightarrow x_i \text{ is a variable of } \varphi, \tag{5.26}$$

$$(x_i, x_j) \text{ is an arc of } G \Longleftrightarrow x_i \overline{x}_j \text{ is a term of } \varphi. \tag{5.27}$$

Conversely, given an arbitrary digraph G, one can associate with G a mixed DNF $\varphi \equiv \varphi(G)$ by simply reading the double implications (5.26) and (5.27) from left to right. Two terms in φ, let them be $x\overline{y}$ and $u\overline{v}$, have a consensus only in two cases:

(a) $u = y$: then their consensus is $x\overline{v}$;
(b) $x = v$: then their consensus is $u\overline{y}$.

Thus, the consensus of any two mixed terms is still mixed. A graph-theoretic interpretation in G of cases (a) and (b) is provided by Figures 5.24 (a) and (b), respectively. As in the case of implication graphs, here, too, an elementary consensus operation corresponds to a transitive arc addition, and vice versa. Observe that in the context of mixed DNFs, absorption is trivial. In fact, since linear terms can never be generated by consensus in this case, a quadratic mixed term can be absorbed only by itself; that is, it is absorbed only if it is already present in the current list of terms. Accordingly, any consensus algorithm whose input is a mixed DNF φ can be interpreted as a transitive closure algorithm on the associated digraph G, and vice versa (recall also Theorem 5.2).

Now we are ready to give a graph-theoretic description of a generic input disengagement consensus algorithm. We assume that the algorithm directly takes as input, instead of a mixed DNF, a digraph $G = (V, E)$. As in Section 5.8.2, we may assume, without loss of generality, that G is a connected directed acyclic graph or a connected DAG.

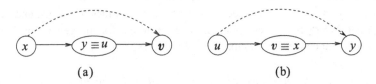

(a)   (b)

Figure 5.24. Transitive arc additions.

Procedure INPUT DISENGAGEMENT CONSENSUS $(G, \prec)$

Input: A connected DAG $G = (V, E)$ and a disengagement order $\prec$ on E.
Output: A DAG $H = (V, F)$, $F \supseteq E$.

begin
    let $F := E$;
    declare all arcs of E to be *engaged*;
    while there is some engaged arc **do** { process arc a }
        select the first (with respect to $\prec$) engaged arc a;
        declare arc a to be *disengaged*;
        let $a = (h, k)$;
        for each arc $(p, h) \in F$ **do** add arc (p, k) to F (if missing);
        for each arc $(k, q) \in F$ **do** add arc (h, q) to F (if missing);
    end while
    return $H = (V, F)$;
end

Figure 5.25. Procedure INPUT DISENGAGEMENT CONSENSUS.

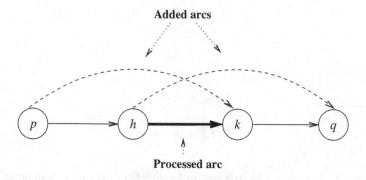

Figure 5.26. Processing an arc in the INPUT DISENGAGEMENT CONSENSUS procedure.

Definition 5.2. *A disengagement order $\prec$ is any strict linear order on the arc set of G.*

A disengagement order is meant to represent the order in which the arcs are disengaged in the INPUT DISENGAGEMENT CONSENSUS algorithm described in Figure 5.25 (compare with the DISENGAGEMENT CONSENSUS procedure in Figure 3.2 of Section 3.2.2, and see also Figure 5.26).

Does the digraph $H = (V, F)$ output by the INPUT DISENGAGEMENT CONSENSUS procedure coincide with the transitive closure G^* of G? The answer may depend on the chosen disengagement order $\prec$, as illustrated by the following example:

Example 5.14. *Consider the directed path $G = P_5$, and label its four arcs as shown in Figure 5.27. If the disengagement order is $1 \prec 4 \prec 3 \prec 2$, then H is a proper subgraph of G^*, since arc (v_1, v_5) is missing (see Figure 5.28; here and*

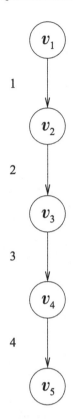

Figure 5.27. The dipath P_5.

in Figure 5.29, all arcs are assumed to be directed from top to bottom; at each iteration, the thick arc is the one that is being processed, and the dashed arcs are the ones that are being added).

On the other hand, if the disengagement order is $1 \prec 3 \prec 4 \prec 2$, then $H = G^*$ (see Figure 5.29). Interestingly, in this case, only three iterations are needed in order to generate G^*.   □

Definition 5.3. *A disengagement order $\prec$ is successful (for the digraph G) if the input disengagement algorithm outputs G^* when it runs on the input $(G, \prec)$.*

Can successful disengagement orders be characterized? Theorem 5.26 yields some insights into this question, providing a full characterization in the case of dipaths; this characterization proves useful in establishing our main Theorem 5.28.

Let us first introduce some preliminary definitions and notation. We denote by P_n the standard dipath whose vertices are $v_1, \ldots, v_n$ and whose arcs are $(v_1, v_2), (v_2, v_3), \ldots, (v_{n-1}, v_n)$. We let $m = n - 1$ and label arc (v_i, v_{i+1}) as i, for $i = 1, 2, \ldots, m$.

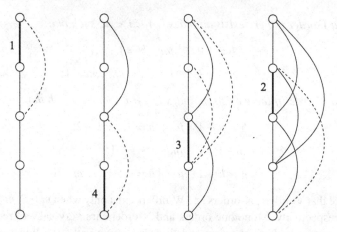

Figure 5.28.   *H does not coincide with G^*.*

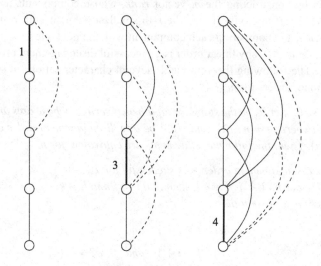

Figure 5.29.   *H = G^*.*

Definition 5.4. *A disengagement order $\prec$ on the arc set $\{1, 2, \ldots, m\}$ of P_n is said to be*

(i) *monotone if*

$$either \quad 1 \prec 2 \prec \cdots \prec m,$$

$$or \quad 1 \succ 2 \succ \cdots \succ m;$$

(ii) *an N-order if*

$$either \quad 1 \prec 2 \ and \ 2 \succ 3 \ and \ 3 \prec 4 \prec \cdots \prec m,$$

$$or \ (symmetrically) \ 1 \succ 2 \succ \cdots \succ m-2 \ and \ m-2 \prec m-1$$

$$and \ m-1 \succ m;$$

(iii) *a V-order if there exists an index i, $1 < i < m$, such that:*

$$h \succ h+1 \quad for \quad h = 1,\ldots,i-1;$$
$$h \prec h+1 \quad for \quad h = i,\ldots,m-1;$$

(iv) *a W-order if there exists an index i, $2 < i < m-1$, such that:*

$$h \succ h+1 \quad for \quad h = 1,\ldots,i-2;$$
$$i-1 \prec i \quad and \quad i \succ i+1;$$
$$h \prec h+1 \quad for \quad h = i+1,\ldots,m-1.$$

Notice that V-orders, N-orders and W-orders exist only when $m \geq 3$, $m \geq 4$, and $m \geq 5$, respectively. Monotone orders and N-orders are very easy to recognize. One can recognize both V-orders and W-orders among all strict linear orders in $O(m)$ time by constructing the m-vector *rank*, whose components are defined by $rank(h) = r$ if and only if h is the r-th smallest element with respect to $\prec$ $(h = 1,\ldots,m)$, and comparing each component with the next one.

Clearly, for $m \leq 3$, any linear order is a successful disengagement order for the path P_{m+1}. The following theorem yields several characterizations of successful disengagement orders for $m \geq 4$:

Theorem 5.26. *Let P_n be the standard dipath on n vertices whose arcs are labeled $1,2,\ldots,m$ (where $m = n-1$), and let $\prec$ be any disengagement order on the set $\{1,2,\ldots,m\}$. Then, the following statements are equivalent for $m \geq 4$:*

(a) *The disengagement order $\prec$ is successful for P_n.*
(b) *There are no $i < j < h < k$ such that $i \prec j$ and $h \succ k$.*
(c) *There is no i such that*

$$either \quad i \prec i+1 \quad and \quad i+2 \succ i+3,$$
$$or \quad i \prec i+1 \quad and \quad i+3 \succ i+4.$$

(d) *The disengagement order $\prec$ is either monotone, or an N-order, or a V-order, or a W-order.*
(e) *There is an arc a such that, for each $t = 1,2,\ldots,m-1$, $A_t \cup \{a\}$ induces a subpath of P_n, where A_t consists exactly of the first t arcs of P_n with respect to $\prec$.*

Proof. See Boros et al. [120]. $\square$

The following result is also worth mentioning:

Theorem 5.27. *The minimum cardinality of a set of arcs to be disengaged in order to generate the transitive closure of P_n is $n-2$. For any successful disengagement order and any arc a as in Theorem 5.26(e), one can obtain one such minimum cardinality set by moving a to the last rank in the order.*

Proof. See Boros et al. [120]. For instance, in the second case of Example 5.14, it is enough to disengage the arcs 1, 3, 4 in order to generate the transitive closure of P_5 (see Figure 5.29), but this is impossible with one or two arcs.   □

The main result of this section can now be stated.

Theorem 5.28. *For an arbitrary DAG G, there always exists a disengagement order that is successful for G. Such a disengagement order can be found in time $O(m)$, where m is the number of arcs of G.*

Proof. We label the arcs of G from 1 to m, as follows: The vertices of G are visited in reverse topological order and for each vertex i, the arcs going into i are assigned the highest previously unassigned labels (ties can be broken arbitrarily). This can be done in time $O(m)$ as in Tarjan [858].

Now, let $\prec$ be the disengagement order, defined by

$$i \prec j \Longleftrightarrow \text{label}(i) < \text{label}(j).$$

Since the arc labels are strictly increasing along each dipath P of G, $\prec$ induces a monotone strict linear order on the arcs of each dipath. By Theorem 5.26, the input disengagement algorithm running on the instance $(G, \prec)$ must generate the transitive closure of each maximal dipath of G. Since a DAG is transitively closed if and only if each of its maximal dipaths is such, it follows that the DAG H produced by the input disengagement algorithm must coincide with G^*.   □

The final result of this subsection concerns the complexity of the input disengagement algorithm.

Theorem 5.29. *The complexity of the* INPUT DISENGAGEMENT CONSENSUS *algorithm is $O(mn)$.*

Proof. Since the algorithm is an input consensus one and since all arcs of G are disengaged after processing, the algorithm consists of m stages, one for each arc of G. At each stage, an arc (h,k) of G is processed: All its predecessors (p,h) and all its successors (k,q) are examined and the arcs (p,k) and (h,q) added to the current set F, provided that they are not already present. Since the initial digraph G is acyclic and each transitive arc addition transforms a DAG again into a DAG, no predecessor p of h can coincide with a successor q of k. Hence, for a fixed arc (h,k), the number of all such vertices p and q is at most $n - 2$. Therefore, there are at most $m(n - 2)$ transitive arc additions, and the thesis follows.   □

5.8.4   Irredundant normal forms and transitive reductions

We turn our attention to the second problem stated in Section 5.8.1: Given a quadratic DNF φ of a quadratic Boolean function f, find an irredundant DNF of f.

We restrict ourselves to finding prime irredundant DNFs. Recall from Section 1.7 that a prime irredundant DNF of f has the following two properties:

- It is a disjunction of *prime* implicants of f.
- It does not have any redundant terms, that is, terms whose deletion results in a shorter DNF representation of f.

Therefore, a natural algorithmic strategy for finding a prime irredundant DNF is the following:

1) Generate all linear implicants of f from the input DNF φ.
2) If there are two linear implicants of the form ξ and $\bar{\xi}$, then the constant $\mathbf{1}_n$ is the only prime implicant, and hence, also the only prime irredundant form of f; stop.
3) Otherwise, perform all possible absorptions of quadratic terms by linear ones. The resulting DNF χ is prime.
4) Check whether any term of χ is redundant.

Step 1 can be efficiently implemented as follows: To check whether the linear term ξ is an implicant of f, assign to ξ the value 1 and deduce the values of as many literals as possible, exactly as in the Labeling algorithm of Section 5.6.2. Then, ξ is an implicant of f if and only if a conflict arises. Another efficient alternative is to work on the implication graph D_φ using Theorem 5.5 to check whether ξ is forced to 0. Each of these two approaches takes $O(mn)$ time.

Steps 2 and 3 are easy to implement.

An efficient implementation of Step 4 relies on the notion of transitive reduction. A *transitive reduction* of a digraph $D = (V, A)$ is any digraph $D' = (V, A')$ such that the transitive closure of D is equal to the transitive closure of D', and such that the cardinality of A' is minimum with this property. In the case of acyclic digraphs, the transitive reduction is unique and can be computed in polynomial time (see Aho, Garey, and Ullman [10]).

Let D_χ be the implication graph of the DNF χ found in Step 3. At this point, we may assume that no linear term is redundant. Also, we may assume, without loss of generality, that D_χ is an acyclic digraph.

Lemma 5.5. *A quadratic term $\xi\eta$ is redundant in χ if and only if, in the transitive reduction of D_χ, the arcs $(\xi, \bar{\eta})$ and $(\eta, \bar{\xi})$ are both missing.*

Proof. A term $\xi\eta$ of χ is redundant if and only if it can be obtained from the remaining terms of χ through a sequence of consensus operations. In view of the interpretation of consensus as a transitive arc addition, $\xi\eta$ is redundant if and only if in D_χ there is a directed path from ξ to $\bar{\eta}$ (and hence, also from η to $\bar{\xi}$), that is, if and only if both arcs $(\xi, \bar{\eta})$ and $(\eta, \bar{\xi})$ are missing in the transitive reduction of D_χ. $\qquad\square$

As a consequence of Lemma 5.5, one gets the following simple implementation of Step 4: Build the implication graph D_χ and its transitive reduction D_χ^r. Delete from χ all the quadratic implicants $\xi\eta$ such that both arcs $(\xi,\overline{\eta})$ and $(\eta,\overline{\xi})$ are missing in D_χ^r.

The resulting DNF Ψ is a prime irredundant DNF of f.

Aho, Garey, and Ullman [10] have shown that the transitive reduction of an arbitrary DAG can be generated with the same order of complexity as its transitive closure. In particular, $O(mn)$ algorithms are available. Hence, an irredundant DNF of f can be obtained within the same complexity.

5.9   Dualization of quadratic functions
(Contributed by Oya Ekin Karaşan)

5.9.1 Introduction

Several algorithmic problems related to dualization of Boolean functions were introduced in Section 4.3. We now consider the following special case (recall that the complete DNF of a Boolean function consists of the disjunction of all its prime implicants):

QUADRATIC DNF DUALIZATION
Instance: The complete DNF of a quadratic Boolean function f.
Output: The complete DNF of f^d.

We observe that for a quadratic function f, there is no serious loss of generality from assuming that f is given by its complete DNF, rather than by an arbitrary DNF representation. Indeed, fast algorithms can be used to generate all prime implicants of f from any DNF, as explained in Section 5.8.

We should note, however, that there are quadratic Boolean functions f whose dual f^d has exponentially more prime implicants than f. An example of such a function is given (for even n) by the DNF

$$\bigvee_{i=1}^{n/2} x_{2i-1} x_{2i},$$

whose dual has $2^{n/2}$ prime implicants.

Since the output may be large, the question of interest again becomes designing algorithms that run either with polynomial delay, in polynomial incremental time, or in polynomial total time (see Appendix B).

Recall that the problem of dualizing a *positive* quadratic Boolean function was mentioned in Chapter 4. There, it has been noted that, due to its relationship with the problem of generating all maximal stable sets of a graph, the problem can be solved with polynomial delay (cf. also Exercise 7 of Chapter 4).

In fact, as showed by Ekin [303, 304], this relation can be further exploited in order to develop a polynomial-delay algorithm for the dualization of general, not necessarily positive, quadratic DNFs. We discuss this in more detail in the following subsection.

5.9.2 The dualization algorithm

Let f be a quadratic Boolean function, and consider the complete DNF φ or, equivalently, the list of prime implicants of f. To solve the problem QUADRATIC DNF DUALIZATION, we may assume, without loss of generality, that φ is purely quadratic, meaning that all prime implicants of f are quadratic. Indeed,

- if $\varphi \equiv \mathbf{1}_n$, then $f^d = \mathbf{0}_n$;
- if ξ is a linear prime implicant of f, then all prime implicants of f^d contain ξ.

Let $G_f = (V, E)$ be the matched graph associated with φ, from which the null edges have been deleted; thus, the edges in E are in one to one correspondence with the prime implicants of f. It immediately follows from the definition of the dual that the prime implicants of f^d are in one-to-one correspondence with those minimal vertex covers of G_f that do not contain both a vertex ξ and its negation $\overline{\xi}$.

As in Section 5.4.3, we say that literals ξ and η are *twins* if both $\xi \overline{\eta}$ and $\overline{\xi} \eta$ are prime implicants of f, that is, if $\xi = \eta$ for all false points of f.

Let C be a minimal vertex cover of G_f that does not contain both a vertex and its negation. It is easy to see that if ξ and η are twin literals, then C contains either both ξ and η or neither of them. Indeed, if $\xi \in C$, and ξ and η are twins, then $(\eta, \overline{\xi}) \in E$. Therefore, as $\overline{\xi} \notin C$ by assumption, C must contain η in order to intersect the edge $(\eta, \overline{\xi})$.

Let us construct a graph $G_f^* = (V^*, E^*)$ from G_f as follows: The vertex set V^* consists of all equivalence classes induced by the "twin-relation" on the set V of literals; that is, ξ and η belong to the same equivalence class if and only if they are twins. Note that the negations of all vertices in an equivalence class I also form an equivalence class; we denote it by $\overline{I}$ and call it the *negation of the equivalence class I*. There is an edge in G_f^* between two equivalence classes I and J if and only if (ξ, η) is an edge of G_f for some $\xi \in I$ and $\eta \in J$.

Observe that no edge of G_f joins two twins, for this would violate the primality of the term in φ corresponding to this edge. Additionally, if $(I, J) \in E^*$, then $\xi \eta$ is a prime implicant of f for every $\xi \in I$ and $\eta \in J$, which is simply a consequence of the consensus operation. Hence, we can conclude that no information is lost in the process of identifying a set of twins, and the graph G_f^* summarizes all the information present in G_f.

As mentioned above, a minimal vertex cover of G_f that does not contain both a vertex and its negation contains only entire equivalence classes. Hence, a minimal vertex cover of G_f that does not contain both a vertex and its negation corresponds to a minimal vertex cover of G_f^* that does not contain both a vertex and its negation. In fact, the following stronger statement is valid:

Lemma 5.6. *No minimal vertex cover of G_f^* contains both a vertex and its negation.*

Proof. Assume by contradiction that C is a minimal vertex cover of G_f^* containing both I and $\overline{I}$. Since C is minimal, there is an edge $(I, J) \in E^*$ such that $J \notin C$.

Similarly, there is an edge $(\overline{I}, K) \in E^*$ such that $K \not\subseteq C$. Note that it is not possible to have $J = K$ as this would contradict primality. Moreover, $J = \overline{K}$ is not possible either, since it would mean that $I \cup K$ is an equivalence class.

Because $(I, J) \in E^*$ and $(\overline{I}, K) \in E^*$, there exist $\xi \in I$, $\eta \in J$, $\gamma \in K$ such that both $\xi\eta$ and $\overline{\xi}\gamma$ are prime implicants of f. It follows that their consensus $\eta\gamma$ gives rise to the edge (J, K) in E^*. But this edge is not covered by C, a contradiction. $\qquad\square$

We conclude that the minimal vertex covers of G_f that do not contain both a vertex and its negation (namely, the prime implicants of f^d) correspond precisely to the minimal vertex covers of G_f^*.

Several algorithms are available in the literature for generating all maximal stable sets of a graph with polynomial delay (and even in linear space) [538, 605, 873]. Since maximal stable sets are precisely the complements of minimal vertex covers, we obtain the following result due to Ekin [303, 304].

Theorem 5.30. *The problem* QUADRATIC DNF DUALIZATION *can be solved with polynomial delay.*

Let us illustrate the dualization algorithm with an example.

Example 5.15. *Let the Boolean function f be given by the quadratic DNF*

$$x_1 \vee \overline{x}_1 x_2 \vee \overline{x}_3\overline{x}_4 \vee x_4 x_5 \vee x_3\overline{x}_5 \vee x_4\overline{x}_6 \vee x_5\overline{x}_7 \vee x_6 x_8.$$

- **Step 1:** *Find the complete DNF representation φ of f. We obtain*

$$\varphi = x_1 \vee x_2 \vee \overline{x}_3\overline{x}_4 \vee \overline{x}_3 x_5 \vee x_3 x_4 \vee x_4 x_5 \vee x_3\overline{x}_5 \vee \overline{x}_4\overline{x}_5 \vee \overline{x}_3\overline{x}_6 \vee x_4\overline{x}_6$$
$$\vee \overline{x}_5\overline{x}_6 \vee x_3\overline{x}_7 \vee \overline{x}_4\overline{x}_7 \vee x_5\overline{x}_7 \vee \overline{x}_6\overline{x}_7 \vee x_6 x_8 \vee \overline{x}_7 x_8$$
$$\vee \overline{x}_3 x_8 \vee x_4 x_8 \vee \overline{x}_5 x_8.$$

We observe that $f \not\equiv \mathbb{1}$, and that the variables x_1, x_2 can be removed from further consideration (they appear in every dual prime implicant).
- **Step 2:** *Identify the equivalence classes. In this example, literals $\overline{x}_3, x_4$, and $\overline{x}_5$ are equivalent, and so are $x_3, \overline{x}_4$, and x_5.*
- **Step 3:** *Construct G_f^*; see Figure 5.30.*
- **Step 4:** *Find all maximal stable sets of G_f^*. There are three such sets, and each of them yields a prime implicant of f^d.*

Maximal stable sets	Corresponding prime implicants of f^d
$\{\overline{x}_3 x_4\overline{x}_5, x_6, \overline{x}_7\}$	$x_1 x_2 x_3\overline{x}_4 x_5\overline{x}_6 x_8$
$\{x_3\overline{x}_4 x_5, \overline{x}_6, x_6\}$	$x_1 x_2\overline{x}_3 x_4\overline{x}_5\overline{x}_7 x_8$
$\{x_3\overline{x}_4 x_5, \overline{x}_6, x_8\}$	$x_1 x_2\overline{x}_3 x_4\overline{x}_5 x_6\overline{x}_7$

$\qquad\square$

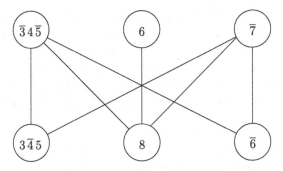

Figure 5.30. The graph G_f^*.

5.10 Exercises

1. In a plant, two machines are available for processing n jobs. Each job i has a fixed start time s_i and a fixed end time t_i, and it must be processed without interruption by either machine. No job can be processed by both machines, and neither machine can process more than one job at a time. When a job ends, the next one can start instantaneously on the same machine. Set up a quadratic Boolean equation that is consistent if and only if a feasible schedule exists for the n jobs.

2. Solve the quadratic Boolean equation $\varphi = 0$, with

$$\varphi = x_1\overline{x}_2 \vee x_1 x_6 \vee \overline{x}_1 x_2 \vee \overline{x}_1 \overline{x}_5 \vee \overline{x}_1 \overline{x}_6 \vee \overline{x}_1 \overline{x}_9 \vee x_2 x_6 \vee \overline{x}_2 x_3 \vee \overline{x}_2 \overline{x}_6 \vee x_3 x_4$$
$$\vee \overline{x}_3 \overline{x}_4 \vee \overline{x}_3 x_7 \vee \overline{x}_3 \overline{x}_8 \vee x_4 x_5 \vee x_4 x_6 \vee \overline{x}_4 x_7 \vee x_5 \overline{x}_8 \vee \overline{x}_5 x_8 \vee x_6 \overline{x}_7 \vee x_6 \overline{x}_8$$
$$\vee \overline{x}_7 x_9 \vee x_8 x_9,$$

by the four algorithms of Section 5.6.

3. Show that the quadratic Boolean equation

$$x_1\overline{x}_3 \vee x_1 x_5 \vee \overline{x}_1 x_2 \vee \overline{x}_2 x_4 \vee \overline{x}_2 x_7 \vee x_3 \overline{x}_6 \vee x_3 \overline{x}_8 \vee \overline{x}_3 x_5 \vee \overline{x}_4 x_6 \vee \overline{x}_4 x_8$$
$$\vee \overline{x}_5 \overline{x}_7 \vee \overline{x}_5 x_8 \vee \overline{x}_7 x_8 = 0$$

has no solution by pinpointing a strong component containing both a variable and its complement in the implication graph.

4. Given a pure quadratic Boolean DNF φ, show that the equation $\varphi = 0$ is consistent if and only φ is Horn-renamable.

5. Show that the Alternative Labeling algorithm can be implemented to run in $O(m)$ time.

6. Exhibit an example showing that the Switching algorithm in Section 5.6.4 can attain its $O(mn)$ worst-case complexity bound.

7. Prove that, for every $n \geq 2$, the number of solutions of the quadratic Boolean equation

$$x_1 x_2 \vee x_2 x_3 \vee \ldots \vee x_{n-1} x_n = 0$$

is given by the Fibonacci number F_{n+1} and thus grows exponentially with n.

8. Find all prime implicants of the quadratic Boolean function

$$f(x_1,\ldots,x_7) = x_1x_2 \vee x_1\overline{x}_7 \vee \overline{x}_2x_3 \vee \overline{x}_2x_4 \vee \overline{x}_3x_4 \vee \overline{x}_4\overline{x}_5 \vee \overline{x}_4x_6 \vee x_5\overline{x}_6 \vee \overline{x}_6x_7$$

by the algorithm in Section 5.8.2.

9. Find an irredundant DNF of the quadratic DNF

$$x_1x_2 \vee x_1\overline{x}_4 \vee x_1x_6 \vee x_2\overline{x}_3 \vee x_2x_5 \vee \overline{x}_2\overline{x}_4 \vee x_3x_5 \vee x_3\overline{x}_6 \vee \overline{x}_3\overline{x}_4 \vee x_4\overline{x}_5$$

$$\vee \overline{x}_4\overline{x}_6 \vee \overline{x}_5\overline{x}_6$$

by the algorithm in Section 5.8.4.

10. A posiform is a multilinear polynomial in the $2n$ variables $x_1, x_2, \ldots, x_n, \overline{x}_1,$ $\overline{x}_2, \ldots, \overline{x}_n$ with nonnegative real coefficients.

(i) Show that for every quadratic pseudo-Boolean function $f(X)$ on $\mathcal{B}^n$, there exist a constant c and a quadratic posiform $\phi(X, \overline{X})$ such that $f(X) = c + \phi(X, \overline{X})$ for all $X \in \mathcal{B}^n$.

(ii) Clearly, c is a lower bound on the minimum of f in $\mathcal{B}^n$. Show that this lower bound is tight if and only if a certain quadratic Boolean equation is consistent.

(See Hammer, Hansen, and Simeone [440] and Chapter 13.)

11. Let $\phi(x_1, x_2, \ldots, x_n) = \bigvee_{(i,j) \in E} x_i x_j$ be a positive quadratic DNF of n variables, let $V = \{1, 2, \ldots, n\}$, and let the graph $G = (V, E)$ be connected. Assume we know that, for some reason, the condition $x_i \leq x_j$ must hold between two variables x_i and x_j. Because this condition is equivalent to $x_i x_j = x_i$, it follows that the term $x_i x_j$ of ϕ can be "linearized" when the constraint $x_i \leq x_j$ holds. The question arises: What is the minimum number of binary order constraints that need to be imposed in order to make ϕ linear? (We only count the order constraints that are explicitly imposed, not those that are implied by the transitivity of the order relation $\leq$.)

(i) Show that, in order to linearize ϕ, at least $n - 1$ order constraints need to be imposed.

(ii) Show that a set of $n - 1$ order constraints linearizing ϕ is given by the set $\{x_i \leq x_j : (i, j) \in A\}$, where A is the set of arcs of a depth-first search tree T of G with the following property: If vertices i and j are adjacent in G, then i is an ancestor of j in T, or vice versa.

(See Tarjan [858]; Hammer and Simeone [462].)

12. Consider the probability model in which all the quadratic Boolean equations with n variables and m terms are equally likely. Show that a random quadratic equation is almost surely satisfiable when $m < n$ and almost surely unsatisfiable when $m > n$.

(See Chvátal and Reed [202].)

13. Show that if a quadratic Boolean equation with n variables and m terms is generated at random in the preceding probability model, then one can solve

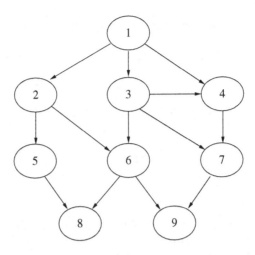

Figure 5.31. A directed acyclic graph.

it in expected $O(n)$ time.

(Hint: Randomly select $4n$ terms and solve the corresponding equation in $O(n)$ time. With high probability, the equation is inconsistent. If not, solve the full equation in $O(m)$ time. See Hansen, Jaumard, and Minoux [470].)

14. Find the transitive closure of the DAG in Figure 5.31 by the input disengagement method of Section 5.8.3.

15. Show that Lemma 5.6 does not hold in general for the graph G_f defined in Section 5.9.2.

6

Horn functions

Endre Boros

In this chapter, we study the class of Horn functions. The importance of Horn functions is supported by their basic role in complexity theory (see, e.g., Schaefer [807]), by the number of applications involving these functions, and, last but not least, by the beautiful mathematical properties that they exhibit.

Horn expressions and Horn logic were introduced first in formal logic by McKinsey [638] and Horn [509] and were later recognized as providing a proper setting for universal algebra by Galvin [367], Malcev [657], and McNulty [640]. Horn logic proved particularly useful and gained prominence in logic programming [19, 185, 488, 489, 494, 521, 552, 582, 648, 656, 721, 855, 816], artificial intelligence [186, 277, 318, 612, 853], and in database theory through its proximity to functional dependencies in relational databases [179, 267, 319, 320, 646, 647, 797]. The basic principles of Horn logic have been implemented in several widely used software products, including the programming language PROLOG and the query language DATALOG for relational databases [494, 648]. Though many of the cited papers are about first-order logic, the simplicity, expressive power, and algorithmic tractability of propositional Horn formulae are at the heart of these applications.

6.1 Basic definitions and properties

Horn functions, just like monotone and quadratic functions, are customarily defined by the syntax of their DNF (or CNF) expressions. It is important to note, however, that this syntactical property of a particular representation of a Horn function propagates, in fact, to all its prime representations. In this sense, the customarily used syntactical description of Horn functions does indeed define a class of functions, and not merely a family of expressions.

To see this, let us start with some basic definitions (see also Section 1.13 in Chapter 1).

Definition 6.1. *An elementary conjunction*

$$T(x_1,\ldots,x_n) = \bigwedge_{j \in P} x_j \bigwedge_{k \in N} \overline{x}_k \qquad (6.1)$$

is called a Horn *term if* $|N| \leq 1$, *that is, if* T *contains at most one complemented variable. The term* T *is called* pure Horn *if* $|N| = 1$, *and* positive *if* $N = \emptyset$.

Definition 6.2. *A DNF*

$$\eta(x_1,\ldots,x_n) = \bigvee_{i=1}^{m} \left(\bigwedge_{j \in P_i} x_j \bigwedge_{k \in N_i} \overline{x}_k \right) \qquad (6.2)$$

is called Horn (pure Horn) *if all of its terms are Horn (pure Horn).*

Note that the same function may have both Horn and non-Horn DNF representations.

Example 6.1. *The DNF*

$$\eta_1(x_1,x_2,x_3) = x_1\overline{x}_2 \vee x_1\overline{x}_3 \vee x_2x_3$$

is Horn because its first two terms are pure Horn and its last term is positive, whereas the following DNF of the same (monotone) Boolean function,

$$\eta_2(x_1,x_2,x_3) = x_1x_2 \vee x_1x_3 \vee x_1\overline{x}_2\overline{x}_3 \vee x_2x_3,$$

is not Horn because its third term contains two complemented variables. □

Definition 6.3. *For a pure Horn term* $T = \overline{x}_k \wedge \left(\bigwedge_{j \in P} x_j \right)$, *variable* x_k *is called the* head *of* T, *while variables* x_j, $j \in P$, *are called the* subgoals *of* T.

To simplify subsequent discussions, we further introduce the following notations. Given a subset $P \subseteq \{1,2,\ldots,n\}$, we also use the letter P to denote the corresponding elementary conjunction as well as the Boolean function defined by that conjunction:

$$P = P(x_1,\ldots,x_n) = \bigwedge_{j \in P} x_j,$$

whenever this notation does not cause any confusion. Thus, a Horn DNF can be written as

$$\eta = \left(\bigvee_{P \in \mathcal{P}_0} P \right) \vee \left(\bigvee_{i=1}^{n} \bigvee_{P \in \mathcal{P}_i} P \overline{x}_i \right), \qquad (6.3)$$

where $\mathcal{P}_0$ denotes the set of positive terms, while $\mathcal{P}_i$ denotes the family of subgoals of the terms with head x_i, for $i = 1,\ldots,n$. We interpret the families $\mathcal{P}_i$, $i = 0,\ldots,n$, as hypergraphs over the base set $\{1,2,\ldots,n\}$.

Example 6.2. *Consider the Boolean expression*

$$\eta = \overline{x}_1 \vee x_1\overline{x}_2 \vee x_1x_2\overline{x}_3 \vee x_2x_3\overline{x}_1. \qquad (6.4)$$

This is is a Horn expression, for which $\mathcal{P}_0 = \emptyset$, $\mathcal{P}_1 = \{\emptyset, \{2,3\}\}$, $\mathcal{P}_2 = \{\{1\}\}$, *and* $\mathcal{P}_3 = \{\{1,2\}\}$. *Since* $\mathcal{P}_0 = \emptyset$, η *is in fact a pure Horn formula.*   □

Recall Definition 2.5: If $A\overline{x}$ and Bx are two terms such that $AB \neq \mathbf{0}$, then AB is called their *consensus*. The term AB is the largest elementary conjunction satisfying $AB \leq A\overline{x} \vee Bx$, and thus, whenever both $A\overline{x}$ and Bx are implicants of a same Boolean function f, then $AB \neq \mathbf{0}$ is an implicant of f, too.

Theorem 6.1. *The consensus of two Horn terms is Horn. More precisely, the consensus of two pure Horn terms is pure Horn, while the consensus of a positive and a pure Horn term is positive.*

Proof. Assume, without any loss of generality that $A\overline{x}$ and Bx are two Horn terms that have a consensus (at least one of the terms must be pure Horn for their consensus to exist). Then, A must contain only positive literals, and B can contain at most one negated variable (which cannot belong to A). Hence, their consensus AB contains at most one negative literal; thus it is Horn. More precisely, AB is positive (respectively, pure Horn) if Bx is positive (respectively, pure Horn).   □

An important consequence of this observation is the following:

Theorem 6.2. *If h is a Boolean function which can be represented by a (pure) Horn DNF, then all prime implicants of h are (pure) Horn.*

Proof. Consider a (pure) Horn DNF η representing h. According to Theorem 3.5, all prime implicants of h can be obtained by applying the consensus method to η. Thus all prime implicants of h can be obtained by a sequence of consensus operations starting with terms present in η, that is, with (pure) Horn terms. Thus, by Theorem 6.1 all terms obtained by that procedure must also be (pure) Horn, and in particular, all prime implicants of h must be (pure) Horn.   □

Example 6.3. *Returning to Example 6.2, we observe that among the terms of η, only $\overline{x}_1$ is a prime implicant of the function h represented by η. All other terms in (6.4) are nonprime. In fact, h has three prime implicants, namely, $\overline{x}_1$, $\overline{x}_2$ and $\overline{x}_3$, the disjunction of which is another representation of h.*   □

Theorem 6.2 implies the following statement:

Corollary 6.1. *If h is a Boolean function that can be represented by a (pure) Horn DNF, then all prime DNF representations of h are (pure) Horn.*

This fact provides our motivation for the following definition:

Definition 6.4. *A Boolean function h is called a (pure) Horn function if it can be represented by a (pure) Horn DNF.*

Remark that the constant function $h = \mathbf{1}_n$ is Horn since, for instance, $\mathbf{1}_n = x_1 \vee \overline{x}_1$ is a Horn DNF, but it is not pure Horn, since $h(1,1,\ldots,1) = 0$ must hold for all pure Horn functions. Let us further add that we can consider $\mathbf{0}_n$ to be both Horn and pure Horn, by definition, since its only DNF representation is the empty DNF.

Although pure Horn functions play an important role in parts of this chapter, they are not fundamentally different from Horn ones. Indeed:

Theorem 6.3. *A function h on B^n is pure Horn if and only if*

(a) *h is Horn, and*
(b) $h(1,1,\ldots,1) = 0$.

Proof. Necessity of (a)–(b) is obvious from the definition of pure Horn functions. Conversely, if (a) holds then h can be represented by a Horn DNF, and if (b) holds then this DNF cannot contain any positive term. $\square$

Another easy relation is established by considering the pure Horn function p_n on B^n represented by $\pi = \bar{x}_1 \vee \bar{x}_2 \vee \ldots \vee \bar{x}_n$. Note that p_n is 1 everywhere except at $(1,1,\ldots,1)$.

Theorem 6.4. *A function h on B^n is Horn if and only if hp_n is pure Horn.*

Proof. Assume that h is represented by a Horn DNF ϕ, and let T be a term of ϕ. If T is pure Horn, then $T\pi = T$. If T is positive, then $T\pi$ is a pure Horn DNF (possibly 0). Thus, $hp_n = \phi\pi$ is pure Horn.

Conversely, assume that hp_n is pure Horn. If $h(1,1,\ldots,1) = 0$, then $hp_n = h$ and h is Horn. If $h(1,1,\ldots,1) = 1$, then $h = hp_n \vee x_1 x_2 \ldots x_n$, and h is Horn, too. $\square$

When dealing with Horn functions, we usually assume that the function is represented by one of its Horn DNFs. As a matter of fact, recognizing Horn functions expressed by arbitrary DNFs turns out to be hard.

Theorem 6.5. (a) *It is co-NP-complete to recognize whether an arbitrary DNF represents a Horn function.*
(b) *It is co-NP-complete to recognize whether an arbitrary DNF represents a pure Horn function.*

Proof. In statement (a), NP-hardness follows from Theorem 1.30, and membership in co-NP is an easy consequence of Corollary 6.2 to be proved in Section 6.3.

In statement (b), NP-hardness is implied by statement (a) and Theorem 6.4, whereas membership in co-NP is implied by Theorem 6.3 and Corollary 6.2. $\square$

Finally, note that the number of prime implicants of a Horn function can be much larger than the number of terms in an arbitrary defining Horn expression of it.

Example 6.4. *The expression given in the proof of Theorem 3.17 is such a Horn DNF. We can also consider the following, somewhat simpler expression:*

$$\eta_2 = \left(\bigvee_{i=1}^{k} x_i \overline{y}_i \right) \vee \bigwedge_{i=1}^{k} y_i. \tag{6.5}$$

Clearly, η_2 is a Horn expression in $2k$ variables and $k+1$ terms, and it has more than 2^k prime implicants. For instance, all terms of the form

$$\bigwedge_{i \in S} x_i \bigwedge_{i \notin S} y_i$$

for any subset $S \subseteq \{1, 2, \ldots, k\}$, are prime implicants of η_2.   □

6.2 Applications of Horn functions

Horn functions appear in many different disciplines, though sometimes in disguised form. We now describe a few examples of such applications.

6.2.1 Propositional rule bases

Expert systems, in particular, propositional production rule-based systems, are widely used for decision support (see, e.g., Ignizio [519] and Section 1.13.1). Boolean variables (propositions) are used in such systems to represent simple statements about the state of the world. To use statements about a sick person as examples, we may consider propositions like: $x_1 =$ "has a headache," $x_2 =$ "must take aspirin," $x_3 =$ "coughs," $x_4 =$ "must go to doctor," and so on. In a rule base, we can include simple implications, corresponding to statements which are known (or required) to be true:

$$R = \{x_1 \wedge x_3 \Longrightarrow x_2, \ x_1 \wedge x_3 \Longrightarrow x_4, \ \cdots\}.$$

In certain situations, some of the values of these propositional variables are known, and the rule base R is used to derive the values of the other variables (e.g., to choose which actions to take) so that all rules remain valid. In other cases, we might just want to check whether a certain chain of events (assignments of truth values to the propositional variables) obeys all the rules, or not.

We can easily see that such a rule base can equivalently be represented by a Horn DNF

$$h = x_1 x_3 \overline{x}_2 \vee x_1 x_3 \overline{x}_4 \vee \cdots.$$

More precisely, a binary assignment X to the propositional variables satisfies all the rules of R exactly when $h(X) = 0$ (such an assignment X is called a *model* of R). In other words, the models of R are exactly the false points of h.

Important problems arising in this context include deciding the consistency of a given rule base (namely, finding a solution to the Horn equation $h = 0$, see Section

6.4), deriving all consequences of a partial assignment in a system in which all rules of R must remain valid (namely, computing the *forward chaining closure* with respect to h, see Section 6.4), finding a simpler equivalent expression of a given rule base (namely, finding a "shorter" DNF of the Horn function representation, see Section 6.7), etc. (see for instance [108, 112, 172, 173, 297, 298, 299, 300, 308, 391, 446, 447, 449, 450, 564]).

6.2.2 Functional dependencies in databases

For simplicity, we can imagine a database as a large array in which every row corresponds to a particular item, usually called *record*, and in which the columns correspond to the various *attributes* those records may have. The entries are strings of text, numbers, dates, or more complex data structures themselves, and not every attribute value is necessarily defined for a particular record. As a typical example, we can think of each record as corresponding to a transaction with a customer (such as bill sent, payment received, reminder sent, etc.) of a large company, which may have thousands of customers, and many transactions with each of these customers. Such a large database is typically highly redundant; for example, in each transaction the customer may be identified by name, address, phone, and account number, implying that all these attributes appear repeatedly in many of the records. To handle such large amount of data, to produce various reports quickly, to check consistency efficiently, and for many other typical operations, it is crucial to store and access the database in an efficient way.

Functional dependencies provide one of the most important and most widely used theoretical tool to model these issues (see e.g. [30, 176, 264, 319, 320, 517, 565, 646, 647, 663, 664, 874, 875]). For subsets X and Y of the attributes, we say that X *determines* Y, and we write $X \to Y$, if in every record of the database, the values of the attributes in X determine uniquely the values of those in Y.

For instance, consider the following small database containing 4 records with attributes A, B, C, and D:

A	B	C	D
a	b	c	d
a	bb	c	dd
aa	b	cc	d
aa	bb	cc	dd

We can observe that $\{A\} \to \{C\}$, $\{B,C\} \to \{A,D\}$, and $\{D\} \to \{B\}$ are a few of the many functional dependencies in this database. For instance, $\{D\} \to \{B\}$ means that, whenever we know the value of attribute D, we "know" the value of B as well: Whenever $D = d$ in the above database, we also have $B = b$. In fact, using $\{A\} \to \{C\}$ and $\{D\} \to \{B\}$, we can uniquely recompute all records of the given database from the following two small tables:

A	C
a	c
aa	cc

and

D	B
d	b
dd	bb

Furthermore, it is obvious that the functional dependency $X \to Y$ is equivalent to the set of functional dependencies $X \to \{y\}$ for $y \in Y$, and that a set of functional dependencies $X_i \to \{y_i\}, i = 1, \ldots, m$ can equivalently be represented by the Horn system $\eta = \bigvee_{i=1}^{m} \left(\bigwedge_{x \in X_i} x \right) \wedge \overline{y}_i$ (see, e.g., Sagiv et al. [797]; Ibaraki, Kogan, and Makino [517]). This connection faithfully preserves all logical inferences, too, namely, all implicants of η correspond to valid functional dependencies of the same database, and vice versa (cf. so-called Armstrong's axioms [30, 647]).

6.2.3 Directed graphs, hypergraphs, and Petri nets

Simple examples of (pure) Horn systems arise from the following correspondence with directed graphs: Given a directed graph $G = (V, A)$ on the vertex set $V = \{1, 2, \ldots, n\}$, we can associate with it a quadratic pure Horn DNF by defining

$$\eta_G = \bigvee_{(i,j) \in A} x_i \overline{x}_j.$$

It is easy to see that all prime implicants of η_G are also quadratic and pure Horn, and that they are in a one-to-one correspondence with the directed paths in G, namely, $x_i \overline{x}_j$ is a prime implicant of η_G if and only if there exists a directed path from i to j in G. Algorithms and graph properties for directed graphs naturally correspond to operations and properties of Horn functions. For instance, strong components of G correspond in a one-to-one way to logically equivalent variables of η_G, the transitive closure of G corresponds to the set of prime implicants of η_G, etc. (see Chapter 5 for more details, in particular, Sections 5.4 and 5.8).

Directed hypergraphs $(V, \mathcal{A})$ provide a natural generalization of directed graphs. They consist of *hyperarcs* of the form $T \to h$, where $T \subseteq V$ and $h \in V$. The set T is called the *tail* (or *source set*) of the hyperarc $T \to h$, while h is called its *head* (see Ausiello, D'Atri, and Sacca [37] or Gallo et al. [361]). The connection with Horn expressions is quite obvious, and several algorithmic problems and procedures of logical inference on Horn systems can naturally be reformulated on directed hypergraphs (see, e.g., [168, 360, 363, 756]).

The more general notion of *Petri nets* was introduced for modeling and analyzing finite state dynamic systems (see Petri [743]). Many important aspects of Petri nets can equivalently be modeled by associated Horn expressions, providing efficient algorithmic solutions to some of the basic problems of system design and analysis (see, e.g., Barkaoui and Minoux [53, 683]).

6.2.4 Integer programming and polyhedral combinatorics

Just as monotone Boolean functions correspond naturally to set covering problems (see Chapter 1), many examples of Horn systems also arise in integer programming. Conditional covering problems involve binary variables and inequalities of the form

$$\sum_{i \in P} x_i \geq 1 \qquad \text{for} \quad P \in \mathcal{P},$$

$$\sum_{i \in H} x_i \geq x_j \qquad \text{for} \quad (H, j) \in \mathcal{H},$$

and are used to model certain facility location problems (see Moon and Chaudhry [691] or Chaudhry, Moon, and McCormick [189]). A similar model is used by Salvemini, Simeone, and Succi [800] to model shareholders' networks and to determine optimal ownership control.

For another type of connection, let us consider a Horn DNF, say, for instance,

$$\eta = x_1 x_2 x_3 \overline{x}_4 \vee x_3 \overline{x}_2 \vee x_1 x_4 \vee \cdots$$

and observe that a binary assignment X is a false point of η if and only if the corresponding system of linear inequalities

$$
\begin{array}{rrrrcr}
-x_1 & -x_2 & -x_3 & +x_4 & \geq & -2 \\
     & x_2 & -x_3 &   & \geq & 0 \\
-x_1 &   &   & -x_4 & \geq & -1 \\
     &   &   &   & \vdots &
\end{array}
$$

is satisfied by X. One characteristic of this system of inequalities is that each row has at most one positive coefficient. This feature turns out to imply interesting properties of the set of feasible solutions. Namely, it was proved by Cottle and Veinott [216] that a nonempty convex polyhedron of the form

$$P = \{x \mid A^T x \geq b, \ x \geq 0\} \tag{6.6}$$

has a least element if each row of the integral matrix A has at most one positive element. Furthermore, as was shown by Chandrasekaran [180], the polyhedron P has an integral least element for every integral right-hand side vector b if A has at most one positive element in each row, and all positive elements in A are equal to 1. For the special case of $0, \pm 1$ matrices, this property was also observed and utilized by Jeroslow and Wang [534] and Chandru and Hooker [182].

The property that P has a least element is perfectly analogous to the fact that Horn functions have a unique minimal false point, and it can in fact be established analogously to Theorem 6.6. This very useful property implies that for a linear integer minimization problem over a polytope of the form (6.6), a simple rounding procedure provides the optimal solution.

For further connections between cutting planes in binary integer programming and prime implicant generation techniques for Boolean functions and, in particular, those specialized for Horn DNFs, we refer the reader to the book by Chandru and Hooker [184] and to the survey by Hooker [503].

The next interesting connection is between $(0, \pm 1)$ matrices, certain associated polyhedra, and Horn functions. It is quite natural to associate with an $m \times n$, $(0, \pm 1)$ matrix A the DNF

$$\phi_A = \bigvee_{i=1}^{m} \left(\bigwedge_{j:a_{ij}=1} x_j \right) \wedge \left(\bigwedge_{k:a_{ik}=-1} \overline{x}_k \right).$$

The association $A \longleftrightarrow \phi_A$ is one-to-one between $(0, \pm 1)$ matrices and Boolean DNFs. Though this association is merely syntactical, in some cases, it covers a much deeper connection. Perfect $(0, \pm 1)$ matrices, introduced by Conforti, Cornuéjols, and de Francesco [207], constitute just such an interesting case. This family of matrices generalize perfect $(0, 1)$ matrices (i.e., matrices that are the clique vertex incidence matrices of maximal cliques of perfect graphs; see Lovász [622] and Padberg [722]), totally unimodular matrices, and balanced $(0, \pm 1)$ matrices (see, e.g., the books by Truemper [871] or Cornuejols [215]). A $(0, \pm 1)$ matrix A is called *perfect* if the polyhedron

$$P_A = \{x \mid Ax \leq 1 - n(A), \ 0 \leq x \leq 1\}$$

has integral vertices, where $n(A)$ is an integer vector, the ith component of which is the number of negative entries in row i of A, for $i = 1, \dots, m$, and where m is the number of rows in A. Perfect $(0, \pm 1)$ matrices have several characterizations in terms of the perfection of associated graphs (see [107, 207, 419]), and are also connected to the family of Horn functions. Namely, it was shown by Boros and Čepek [107] that a $(0, \pm 1)$ matrix A satisfying $P_A \neq \emptyset$, is perfect only if ϕ_A belongs to a subclass of *acyclic renamable Horn* functions (see Section 6.9 for definitions).

6.3 False points of Horn functions

Given a Boolean function f on $\mathcal{B}^n$, let us recall from Chapter 1 that $T(f)$ and $F(f)$ denote, respectively, the sets of its true points and false points. Given binary vectors $X, Y \in \mathcal{B}^n$, we call $Z = X \wedge Y$ their *conjunction*, defined by componentwise conjunction. In other words, if $X = (x_1, \dots, x_n)$, $Y = (y_1, \dots, y_n)$, and $Z = (z_1, \dots, z_n)$, then $z_j = x_j \wedge y_j$ for $j = 1, \dots, n$.

Definition 6.5. *For a nonempty subset $S \subseteq \mathcal{B}^n$, let us define*

$$S^\wedge = \left\{ \bigwedge_{X \in R} X \ \middle| \ \emptyset \neq R \subseteq S \right\},$$

and define $\emptyset^\wedge = \emptyset$. We call $S^\wedge$ the conjunction closure of S. Finally, we say that a subset $S \subseteq \mathcal{B}^n$ is closed under conjunction, or conjunction-closed, if $S = S^\wedge$.

Example 6.5. *If $S = \{(0, 1, 0, 1), (0, 1, 1, 0), (1, 1, 1, 1)\}$, then*

$$S^\wedge = \{(0, 1, 0, 1), (0, 1, 1, 0), (1, 1, 1, 1), (0, 1, 0, 0)\}. \qquad \square$$

Note that the mapping $S \to S^\wedge$ satisfies the usual properties of closure operations, justifying its name:

Lemma 6.1. *For all subsets $A \subseteq B \subseteq \mathcal{B}^n$, we have $A \subseteq A^\wedge$, $A^\wedge \subseteq B^\wedge$, and $(A^\wedge)^\wedge = A^\wedge$.*

Proof. Immediate by the definition. $\Box$

Since Boolean functions can also be defined by their sets of true and/or false points, and since Horn functions constitute a proper subfamily of all Boolean functions, not all subsets of $\mathcal{B}^n$ can appear as sets of false points of Horn functions. Indeed, the set of false points of a Horn function has a very special property, observed first by McKinsey [638] and also by Horn [507; Lemma 7].

Theorem 6.6. *A Boolean function is Horn if and only if its set of false points is closed under conjunction.*

Proof. Let us consider a Boolean function h on $\mathcal{B}^n$, and let $T_1, \ldots, T_p$ denote its prime implicants.

Assume first that h is Horn or, equivalently, by Theorem 6.2, that all its prime implicants are Horn. Let us note that $F(h) = \cap_{k=1}^p F(T_k)$, and that the intersection of conjunction-closed sets is conjunction-closed again. Hence, to prove the first half of the statement, it is enough to show that the set of false points $F(T)$ of a Horn term T is closed under conjunction. Since this is obvious for a positive term, let us assume that $T = \bigwedge_{j \in P} x_j \bar{x}_i$, and let us consider binary vectors X, Y, and Z for which $Z = X \wedge Y$ and $T(Z) = 1$. Then, we must have $z_i = 0$ and $z_j = 1$ for all $j \in P$, implying by the definition of conjunction that $x_j = y_j = 1$ for all $j \in P$ and $x_i \wedge y_i = 0$. Thus, at least one of x_i and y_i must be equal to 0, say, $x_i = 0$, and therefore $T(X) = 1$ follows. This implies that $F(T)$ is closed under conjunction.

For the reverse direction, let us assume that h is not Horn, and let us consider a non-Horn prime implicant $T = \left(\bigwedge_{j \in P} x_j \right) \wedge \left(\bigwedge_{k \in N} \bar{x}_k \right)$ of h, where $|N| \geq 2$. According to the Definition 1.18 of prime implicants, deleting any literal from T yields a non-implicant of h. Thus in particular, for every index $i \in N$ there exists a binary vector $X^i \in \mathcal{B}^n$ such that $x_j^i = 1$ for $j \in P$, $x_k^i = 0$ for $k \in N \setminus \{i\}$, and $x_i^i = 1$, and for which $h(X^i) = 0$ holds. Therefore, $T(X^i \wedge X^{i'}) = 1$ follows for any two distinct indices $i \neq i'$, $i, i' \in N$, implying $h(X^i \wedge X^{i'}) = 1$, and thus proving that $F(h)$ is not closed under conjunction. $\Box$

This result has several interesting consequences. First, it implies a simple characterization of Horn functions, which can serve as the basis for learning Horn theories (see, e.g., [139, 264, 650]), and which was generalized to several other classes of Boolean functions (see, e.g., [305, 303] and Chapter 11 in this book).

Corollary 6.2. *A Boolean function f on $\mathcal{B}^n$ is Horn if and only if*

$$f(X \wedge Y) \leq f(X) \vee f(Y) \tag{6.7}$$

holds for every $X, Y \in \mathcal{B}^n$.

Proof. Indeed, (6.7) implies that $F(f)$ is closed under conjunction, namely, f is Horn, by Theorem 6.6. Conversely, if f is Horn, then the left-hand side of (6.7) is zero whenever the right-hand side is zero, again by Theorem 6.6. $\square$

Another implication of Theorem 6.6 is the following statement:

Corollary 6.3. *For every Horn function h on B^n, $h \neq \mathbf{1}_n$, there exists a unique minimal false vector $X^h \in F(h) \subseteq B^n$.*

Proof. According to Theorem 6.6, the false vector

$$X^h = \bigwedge_{Y \in F(h)} Y$$

is well-defined, unique and satisfies the inequalities $X^h \leq Y$ for all $Y \in F(h) \neq \emptyset$. $\square$

Theorem 6.6 also implies that every Boolean function f has a unique maximal Horn minorant h, that is, a Horn function h such that $h \leq f$ and the inequalities $h \leq h' \leq f$ hold for no other Horn function $h' \neq h$.

Theorem 6.7. *Given a Boolean function f, let h be the function defined by $F(h) = F(f)^\wedge$. Then h is the unique maximal Horn minorant of f.*

Proof. Clearly, h is well defined, and since $F(h)^\wedge = (F(f)^\wedge)^\wedge = F(f)^\wedge = F(h)$, it is also Horn by Theorem 6.6. It is also clear that $F(h) = F(f)^\wedge \supseteq F(f)$, and hence, $h \leq f$. Furthermore, for any Horn minorant $h' \leq f$ we have $F(h') \supseteq F(f)$, and thus, by Theorem 6.6, $F(h') = F(h')^\wedge \supseteq F(f)^\wedge = F(h)$, which implies $h \geq h'$. $\square$

6.3.1 Deduction in AI

The false points of Horn functions play a role in artificial intelligence in a slightly different context, though the characterization by Theorem 6.6 remains essential. In the artificial intelligence literature, typically, Horn CNFs instead of Horn DNFs are considered. A Horn CNF is a conjunction of elementary disjunctions, called *clauses*, in which at most one literal is positive. Due to De Morgan's laws, $\eta = \bigvee_{i=1}^m \left(\bigwedge_{j \in P_i} x_j \bigwedge_{k \in N_i} \overline{x}_k \right)$ is a Horn DNF if and only if $\overline{\eta} = \bigwedge_{i=1}^m \left(\bigvee_{j \in P_i} \overline{x}_j \bigvee_{k \in N_i} x_k \right)$ is a Horn CNF. Accordingly, the solutions of the Boolean equation $\eta = 0$, that is, the false points of the Boolean mapping η are referred to as the *models* of $\overline{\eta}$ or, more precisely, as the models of the Boolean function h represented by the DNF η.

One of the frequently arising tasks in this context is *deduction*, that is, the problem of recognizing whether another logical expression η' is consistent with the given *knowledge base* h represented by η. Here, consistency means that all models of η are also models of η'. Thus, deduction is equivalent with recognizing whether $\eta' \leq \eta$. Such a task is solved customarily by algebraic manipulations of the expressions η and η' (e.g., consensus operations). As a new approach, *model-based reasoning* was introduced recently by a number of authors (see, e.g., [554, 566]). In this approach, based on the equivalence $\eta' \leq \eta \iff F(\eta') \supseteq F(\eta)$, the relation $\eta' \leq \eta$ is tested by checking the values of η' on the set $F(\eta)$. Though, this approach may be inefficient for general Boolean functions, a more efficient variant of it was introduced by Khardon and Roth [566] for Horn knowledge bases. The following observation serves as the basis for the improvement:

Theorem 6.8. *Let A, B, S be subsets of $\mathcal{B}^n$ such that $A \subseteq S \subseteq A^\wedge$ and $B \subseteq S \subseteq B^\wedge$. Then, $S \subseteq (A \cap B)^\wedge$.*

Proof. Assume indirectly that $S \not\subseteq (A \cap B)^\wedge$. It follows from Lemma 6.1 that $A \not\subseteq (A \cap B)^\wedge$ and $B \not\subseteq (A \cap B)^\wedge$. Let us choose a maximal point $X \in (A \cup B) \setminus (A \cap B)^\wedge$ with respect to the usual componentwise comparison. We can assume, without any loss of generality, that $X \in A$ (and hence, $X \notin B$). Since $A \subseteq S \subseteq B^\wedge$, there exist k binary vectors $Y_1, \ldots, Y_k \in B$ such that $X = Y_1 \wedge Y_2 \wedge \cdots \wedge Y_k$. Furthermore, since $X \notin B$, we have $k \geq 2$ and $Y_j \neq X$ for $j = 1, \ldots, k$. By the maximality of X in $(A \cup B) \setminus (A \cap B)^\wedge$, we must have $Y_j \in (A \cap B)^\wedge$ for all $j = 1, \ldots, k$, implying $X \in (A \cap B)^\wedge$ by Lemma 6.1, and hence, contradicting the choice of X. $\qquad \square$

Corollary 6.4. *For every nonempty subset $S \subseteq \mathcal{B}^n$, there exists a unique minimal subset $Q(S) \subseteq S$ such that $Q(S)^\wedge = S^\wedge \supseteq S$.*

Proof. Define

$$Q(S) = \bigcap_{Q \subseteq S \subseteq Q^\wedge} Q. \qquad (6.8)$$

Clearly, $Q(S) \subseteq S$, and by Theorem 6.8, $S \subseteq Q(S)^\wedge$. It follows by Lemma 6.1 that $Q(S)^\wedge = S^\wedge$. $\qquad \square$

In particular, for a Horn function h, we have $F(h) = Q(F(h))^\wedge$ by Theorem 6.6 and by Corollary 6.4. The elements of $Q(F(h))$ are called the *characteristic models* of h by Khardon and Roth [566], who argue that these points are enough for model-based deduction.

Theorem 6.9. ([566]). *Given a Horn function h and a Horn DNF η, we have $\eta \leq h$ if and only if $\eta(X) = 0$ holds for all $X \in Q(F(h))$.*

Proof. Assume first that $\eta \leq h$ holds. Then we have $\eta(X) = h(X) = 0$ for all $X \in F(h)$; and the claim follows by $F(h) \supseteq Q(F(h))$.

Assume next that $\eta(X) = 0$ for all $X \in Q(F(h))$. This means that $F(\eta) \supseteq Q(F(h))$; hence $F(\eta) \supseteq Q(F(h))^\wedge = F(h)$ by Theorem 6.6 and Corollary 6.4. $\qquad \square$

Further characterizations and properties of characteristic models are stated as exercises at the end of this chapter (see also [566]).

6.4 Horn equations

One of the main reasons Horn expressions appear in applications is that the tautology problem for Horn DNFs (or, equivalently, the satisfiability problem for Horn CNFs) can be solved efficiently (see Even, Itai, and Shamir [318] or Dowling and Gallier [277]). In this section we recall this result as well as several related algorithmic ideas.

6.4.1 Horn equations and the unit literal rule

Let us first observe, as a further important implication of Theorem 6.6, that the unique minimal false point X^h of a Horn function h, as defined in Corollary 6.3, provides us with a characterization of the negative linear prime implicants of h.

Corollary 6.5. *Given a Horn function h, a negative linear term $T = \overline{x}_j$ is an implicant of h if and only if $x_j^h = 1$.*

Proof. Assume first that $T = \overline{x}_j$ is an implicant of h, that is, $\overline{x}_j \leq h$. This implies that $y_j = 1$ for all vectors $Y \in F(h)$, and hence, $x_j^h = 1$, by the characterization of X^h in the proof of Corollary 6.3.

For the converse direction, consider an index j for which $x_j^h = 1$. Then $y_j = 1$ for all vectors $Y \in F(h)$ by the definition of X^h, and hence, $\overline{x}_j \leq h$, that is, the term $T = \overline{x}_j$ is indeed an implicant of h. $\square$

To prove that the tautology problem for Horn DNFs can be solved efficiently, we shall show below that given a Horn DNF η representing the Horn function h, the unique minimal false point $X^h \in F(h)$ can be found in linear time in the size $|\eta|$ of the DNF η. (As before, $|\eta|$ denotes the number of literals occurring in the DNF η.) Furthermore, $h = \mathbf{1}$ can also be recognized with the same effort whenever $F(h) = \emptyset$.

Consider a Horn DNF η of the form (6.2), and denote by h the Horn function represented by η. We assume, without loss of generality, that $|P_i \cup N_i| > 0$ for all $i = 1, \ldots, m$.

Note first that if $P_i \neq \emptyset$ for all terms $i = 1, \ldots, m$, then the vector $\mathbf{0} = (0, 0, \ldots, 0) \in \mathcal{B}^n$ is a solution of the equation $\eta(X) = 0$, and clearly, $X^h = \mathbf{0}$ is the unique minimal false point in this case.

Consider next the case in which $P_i = \emptyset$ for some term T_i of η. In this case, T_i must be a negative linear term of the form $T_i = \overline{x}_j$ for some index j. Clearly, for all solutions of the equation $\eta(X) = 0$ (i.e., for all false points $X \in F(h)$), we have $x_j = 1$, and thus $x_j^h = 1$ is implied, too.

Based on these observations, a naïve approach to solving the equation $\eta(X) = 0$ could proceed as shown in Figure 6.1. We can observe that this procedure is a restricted version of the so-called *Unit Literal Rule* employed by most satisfiability algorithms (see Chapter 2). In this version only negative linear terms are used, and hence, we call it the NEGATIVE UNIT LITERAL RULE procedure (NULR).

Procedure NULR(η)

Input: A Horn DNF η representing the Horn function h.
Output: A false point of h or a proof that $h = \mathbf{1}$.

set $\eta^0 := \eta$ and $k := 0$.
repeat
    if there is an empty term in η^k
    then stop {comment: no solution, $h = \mathbf{1}$}
    else find j such that $\overline{x}_j$ is a negative linear term of η^k;
        if there is no such index j
        then set all remaining variables to 0,
            return $X = X^h$ and **stop** {comment: solution found}
        else
            set $x_j := 1$, $\eta^{k+1} := \eta^k|_{x_j=1}$, and $k := k+1$.

Figure 6.1. Procedure NULR.

Theorem 6.10. *Let η be a Horn DNF of the Horn function h on $\mathcal{B}^n$. Then, algorithm NULR(η) runs in $O(n|\eta|)$ time, and either it detects that $h \equiv \mathbf{1}$ or it finds the vector $X^h \in F(h)$.*

Proof. Let us denote by l the value of index k at termination, and let j_k denote the index of the variable fixed at 1 in step $k-1$. Observe that for every $k = 1,\ldots,l$ we have $\eta^k = \eta^{k-1}|_{x_{j_k}=1}$, and hence,

$$\eta^k = \eta|_{x_{j_1}=1, x_{j_2}=1,\ldots,x_{j_k}=1}. \tag{6.9}$$

We claim that $\overline{x}_{j_k}, k = 1,\ldots,l$, are negative linear implicants of h. This is clearly true for $k = 1$, since $\overline{x}_{j_1}$ is a term of η by the choice of j_1. Let us prove the claim by induction on k, and let us assume that it is true for $k < i \le l$. Then $\overline{x}_{j_i}$ is a term of η^{i-1} by the choice of j_i; hence, by (6.9), η must have a term T of the form $T = (\bigwedge_{j \in S} x_j)\overline{x}_{j_i}$ for some subset $S \subseteq \{j_1,\ldots,j_{i-1}\}$. Since the terms $\overline{x}_j$ for $j \in S$ are linear implicants of h by our assumption, $\overline{x}_{j_i} \le T \vee \bigvee_{j \in S} \overline{x}_j \le h$ follows, proving that $\overline{x}_{j_i}$ is an implicant of h and concluding the inductive proof of the claim.

If algorithm NULR terminates with finding an empty term in η^l, then it follows by (6.9) that η must contain a term T of the form

$$T = \bigwedge_{i \in S} x_i$$

for some subset $S \subseteq \{j_1, j_2, \ldots, j_{l-1}\}$. Therefore,

$$1 = \left(\bigwedge_{i \in S} x_i\right) \vee \bigvee_{i \in S} \overline{x}_i \le h,$$

follows, implying $h \equiv \mathbf{1}_n$.

On the other hand, if NULR terminates with finding a solution, let us denote this solution by X^*; thus, X^* is the point defined as

$$x_i^* = \begin{cases} 1 & \text{if } i \in \{j_1, \ldots, j_l\}, \\ 0 & \text{otherwise.} \end{cases}$$

Then, since η^l has neither an empty term, nor a negative linear term, $\eta^l(0, 0, \ldots, 0) = 0$ follows. By (6.9), we have $0 = \eta^l(0, 0, \ldots, 0) = \eta(X^*)$, implying $X^* \in F(h)$. Then $X^* \geq X^h$ follows by Corollary 6.3. Since we have shown that all the terms $\overline{x}_{j_k}$ for $k = 1, \ldots, l$ are negative linear implicants of h, and since $h \not\equiv \mathbf{1}_n$, all these terms must be negative linear prime implicants of h, implying thus $X^* \leq X^h$ by the definition of X^* and by Corollary 6.5. Hence, $X^* = X^h$ follows, concluding the proof of correctness.

Finally, we note that all operations of the repeat loop can obviously be carried out in linear time in the size of the input DNF η, hence the total running time of NULR can be bounded by $O(n|\eta|)$. $\square$

The procedure NULR can actually be implemented to run in linear time, by representing the input DNF in an appropriate data structure:

Theorem 6.11. *Procedure NULR can be implemented to run in $O(n + |\eta|)$ time.*

Proof. We leave the proof as an exercise to the reader (see, e.g., Exercise 6 at the end of this chapter). $\square$

A first important consequence of the previous results is that, unlike in the case of general Boolean functions, we can decide in polynomial time whether or not a given term is an implicant of a Horn function.

Corollary 6.6. *Given a Horn DNF η of a Horn function h, one can decide in $O(n + |\eta|)$ time whether a given term T is an implicant of h.*

Proof. Follows readily by Theorems 3.8 and 6.11, since the restriction of η for $T = 1$ is, again, a Horn DNF. $\square$

Recall from Chapter 1 that a DNF η of the Boolean function h is called *prime* if all terms of η are prime implicants of h and called *irredundant* if no terms can be deleted from η without changing the Boolean function it represents.

Theorem 6.12. *Given a Horn DNF, η of a Horn function h, one can construct in $O(|\eta|(n + |\eta|))$ time an irredundant and prime Horn DNF of h.*

Proof. For a term T of η, let η' denote the DNF obtained from η by deleting the term T. Clearly, $\eta = \eta'$ if and only if T is an implicant of η', which we can test whether $O(n + |\eta'|)$ time in view of Corollary 6.6. Repeating this for all terms of η one by one, and deleting redundant terms, we can produce in $O(m(n + |\eta|))$ time an irredundant DNF of h.

To achieve primality, let us take a term T of the current Horn DNF, and let T' denote the term obtained from T by deleting a literal u of T. By definition, if T' is an implicant of h, then we can replace T by T'. According to Corollary 6.6, we can test whether T' is an implicant in $O(n + |\eta|)$ time. Thus, by repeating this procedure for all literals of T, replacing T by T' whenever T' is proved to be an implicant, and repeating for all terms of η, we can derive in $O(|\eta|(n + |\eta|))$ time a prime DNF of h.

Since $|\eta| \geq m$, the claim follows. $\square$

6.4.2 Pure Horn equations and forward chaining

When dealing with *pure* Horn DNFs, the tautology problem is trivial in view of the following observation:

Remark 6.1. If η is a pure Horn DNF, then $\eta(1,1,\ldots,1) = 0$. $\square$

Thus, in order to solve a pure Horn DNF equation, it is enough to read the input to confirm that it is indeed pure Horn; no additional computations are needed. However, to find the unique *minimal* solution of a pure Horn equation, one needs to employ NULR or a similar procedure. Such a variant of NULR, applied to pure Horn expressions, is widely used in artificial intelligence, where it is known as the *forward chaining* procedure.

To derive this procedure, we first consider a slightly more general inference problem that frequently arises in the AI literature. Given a DNF η and a subset $S \subseteq \{1,2,\ldots,n\}$ of indices of its variables, let us denote by $\eta|_S$ the DNF $\eta|_{x_i=1, i \in S}$ obtained by fixing all variables x_j to 1 in η, for $j \in S$. With these notations, we are interested in the inference problem: Given a pure Horn DNF η and a subset S of indices, find all other indices $j \notin S$ such that $x_j = 1$ is implied by the equation $\eta|_S = 0$. Clearly, a pure Horn term of the form $\left(\bigwedge_{i \in S} x_i\right) \overline{x}_j$ is an implicant of η if and only if $x_j = 1$ is such an implied assignment. Thus, in other words, we would like to determine the set of all negative linear terms of the pure Horn expression $\eta|_S$. Our previous results show that this can be done in linear time by using NULR. However, the computation can be organized in a somewhat simpler way in this special case, as described in Figure 6.2.

It is easy to see that this forward chaining procedure can be implemented to run in linear $O(n + |\eta|)$ time, just like NULR (see Theorem 6.11). However, there are two major differences between forward chaining and NULR: Namely, forward chaining starts fixing the variables x_j for $j \in S$ in the first step of the procedure, and it does not check for inconsistency of the input expression. This is well justified by Remark 6.1, since forward chaining is defined for pure Horn expressions only. NULR can in fact be viewed as a natural generalization of forward chaining for general Horn expressions, starting with $S = \emptyset$.

For a pure Horn DNF η, we look at the set S^η as the *logical closure* of S; the mapping $S \longmapsto S^\eta$ satisfies the usual properties of a closure operator.

Procedure FORWARD CHAINING(η, S)

Input: A pure Horn DNF η on $\mathcal{B}^n$, and a subset $S \subseteq \{1, 2, \ldots, n\}$.
Output: A superset S^η of S.

set $S^0 := S$, $\eta^0 := \eta|_S$, and $k := 0$;
repeat while there is a negative linear term $\overline{x}_j$ of η^k:
    set $S^{k+1} := S^k \cup \{j\}$, $\eta^{k+1} := \eta^k|_{x_j=1}$, and $k := k+1$;
return $S^\eta := S^k$

Figure 6.2. Procedure FORWARD CHAINING.

Lemma 6.2. *If η is a pure Horn DNF, then $S^\eta \supseteq S$, and $(S^\eta)^\eta = S^\eta$ for all subsets $S \subseteq \{1, 2, \ldots, n\}$.*

Proof. Follows directly by the definition. $\qquad\square$

Note that, in fact, the set S^η depends only on the pure Horn function h represented by η, and not on the particular representation η. Hence, we often prefer to use the notation S^h rather than S^η. Still, although the set S^h does not depend on the given representation of h, its computation may; hence, the notation S^η will also be used when necessary to avoid computational ambiguity.

The forward chaining procedure can also be viewed as producing the unique minimal false point of h within a subcube of $\mathcal{B}^n$. Recall from Chapter 1 that for a subset $S \subseteq \{1, 2, \ldots, n\}$ we denote by e_S the characteristic vector of S, and by $T|_{S, \emptyset}$ the sub-cube of vectors $X \geq e_S$: $T|_{S, \emptyset} = \{X \in \mathcal{B}^n \mid x_i = 1 \text{ for all } i \in S\}$. With these notations the following statement follows directly from the forward chaining procedure:

Remark 6.2. Given a pure Horn function h and a subset $S \subseteq \{1, 2, \ldots, n\}$ the point e_{S^h} is the unique minimal point in $T|_{S, \emptyset} \cap F(h)$. $\qquad\square$

Let us add that the simple linear-time forward chaining procedure is also instrumental in testing if a given term is an implicant of a Horn function.

Lemma 6.3. *Given a pure Horn DNF η of the pure Horn function h, a term $T = \bigwedge_{j \in P} x_j \bigwedge_{j \in N} \overline{x}_j$ is an implicant of h if and only if $N \cap P^\eta \neq \emptyset$.*

Proof. If T is not an implicant of h, then there must exist a vector $X^* \in \mathcal{B}^n$ such that $h(X^*) = 0$ and $T(X^*) = 1$, implying $x_j^* = 0$ for all $j \in N$. Moreover, for all indices $j \in P^\eta$ we must have $x_j^* = 1$ by the definition of P^η. Thus, $N \cap P^\eta = \emptyset$ follows. Conversely, if $i \in N \cap P^\eta$, then the term $T' = \left(\bigwedge_{j \in P} x_j \right) \overline{x}_i$ is an implicant of h, as we observed earlier, and thus $T \leq T'$ is an implicant, too. $\qquad\square$

6.4.3   More on Horn equations

We conclude this section with a few remarks about related results and techniques.

First, we note that the polynomial solvability of Horn equations was probably well-known "folklore," and some implementations were made independently in AI (see, e.g., the development of the programming language PROLOG [648]) and in database theory (see, e.g., [19, 185, 267, 489, 539, 582]), well before linear time solvability was formally proved. Several linear time algorithms have been proposed for Horn equations, using mainly graph or directed hypergraph models (see, e.g., [277, 318, 815]).

A special variant of this problem, the so-called *unique Horn satisfiability* problem also gained some popularity in the literature. In the DNF variant of this problem, a Horn DNF η is given, and the problem is to decide whether the Boolean equation $\eta = 0$ has a unique solution or, in other words, to decide whether $|F(h)| = 1$ for the function h represented by η. The difficulty of this problem comes from the fact that, while it is relatively easy to generate the negative linear prime implicants of a Horn function, one has to employ a more complicated algorithm to efficiently generate the positive linear prime implicants. Minoux [682] presented an $O(|\eta| + n \log n)$ algorithm, improved later by Schlipf et al. [809] to $O(|\eta| + m\alpha(m + n, n))$, where α denotes the inverse Ackerman function, and m is the number of terms in η. The existence of a truly linear time algorithm for unique Horn satisfiability is an open problem, as of now. Using the somewhat different computational model of random access machines, Pretolani [757] provides a linear time algorithm (based on a result of Gabow and Tarjan [357]).

The unit literal rule is widely used and is one of the basic procedures in most satisfiability solvers. In fact, it was shown to provide polynomial time solution, not only to Horn, but also to a much larger class of Boolean equations (see, e.g., Schlipf et al. [809])

In order to attack hard general (non-Horn) equations and satisfiability problems, several heuristics and approximations rely on Horn approximations of Boolean expressions (see, e.g., [106, 365, 554, 555, 596] and Section 2.5.2). We pose as exercises the related problems of finding tight Horn minorants and/or majorants of Boolean expressions.

6.5   Prime implicants of Horn functions

Logical inference is a central problem in various areas, including theorem proving, logic programming, databases, and so on. We can formulate logical inference as the problem of generating the prime implicants of a given DNF.

As we saw in Chapter 3, the consensus method is one of the general methods used to obtain all prime implicants of a Boolean function. Let us recall (see Theorem 3.9 and Corollary 3.6) that, for function classes for which the corresponding Boolean equation is tractable, prime implicants can efficiently be generated in total time (see also Appendix B). This certainly implies, according to Theorems 6.10 and

6.11, that prime implicants of a Horn function can be generated in polynomial total time. For the sake of completeness, let us briefly repeat here the heart of the argument.

Going into the details of the general consensus method, we can see that a great part of its computational redundancy is caused by the fact that for every prime implicant T of the input function f, the algorithm may generate many (maybe exponentially many) implicants, all of which will eventually be absorbed by T. Hence, if we could repeatedly simplify the list of terms representing f, allowing only prime implicants in it, we would cut drastically the length of the computations. For Horn DNFs this can be done efficiently, by Theorem 6.12. Using this idea Boros, Crama, and Hammer [112] proved that prime implicants of a Horn DNF can be generated in *polynomial incremental* time. More precisely, it was shown in [112] that given two Horn DNFs, ϕ and ψ, one can decide whether ψ contains all prime implicants of ϕ, and if not, one can find a new prime implicant of ϕ in $poly(|\phi|, |\psi|)$ time. Let us note that the same decision problem for general DNFs is a hard problem, since the testing of whether the terms of ψ are indeed (prime) implicants of ϕ is already a co-NP-complete problem by Theorem 3.7.

In this section, we present some further specialized versions of the consensus method that for Horn functions provide incrementally efficient ways to generate all prime implicants.

The first variant is the PRIME IMPLICANT DEPLETION procedure described in Figure 6.3. It runs on a prime DNF, and it is based on applying the consensus operation with the prime implicants of the input Horn DNF one-by-one, in any order, without ever returning to the same prime implicant. (This is akin to the *variable* depletion procedure in Section 3.2.2.)

To see the correctness of the algorithm in Figure 6.3, let us first prove the following lemma.

Procedure PRIME IMPLICANT DEPLETION(η)

Input: A prime DNF $\eta = T_1 \vee T_2 \vee \ldots \vee T_m$.
Output: All prime implicants of the function f represented by η.

initialize $\mathcal{P} := \emptyset$, $\mathcal{L} = \{T_1, \ldots, T_m\}$;
repeat while $\mathcal{L} \neq \emptyset$
    select a term $T \in \mathcal{L}$ and set $\mathcal{L} := \mathcal{L} \setminus \{T\}$ and $\mathcal{P} := \mathcal{P} \cup \{T\}$;
    generate all consensuses of T with terms $T' \in \mathcal{L}$, and add the
        produced terms to $\mathcal{L}$;
    substitute each term in $\mathcal{L}$ by a corresponding prime implicant
        of f absorbing this term;
    eliminate duplicates from $\mathcal{L}$, as well as those terms which also
        appear in $\mathcal{P}$;
end while
return the list of terms in $\mathcal{P}$

Figure 6.3. Procedure PRIME IMPLICANT DEPLETION.

Lemma 6.4. *Let f be a Boolean function represented by the DNF $\varphi = A \vee \psi$, where A is an elementary conjunction, and let us denote by η the disjunction of all the terms obtained by consensus of A with the terms of ψ. Then, for every implicant T of f, we have either $T \leq A$ or $T \leq \psi \vee \eta$.*

Proof. Let us assume that T is an implicant of f for which $T \leq A$ does not hold; hence, A contains a literal u that is not in T. Let us show that $T \leq \psi \vee \eta$.

Consider any binary point $X \in \mathcal{B}^n$ for which $T(X) = 1$. Since $T \leq f$, we must have $A(X) \vee \psi(X) = 1$. If $\psi(X) = 1$, we are done. Otherwise, $A(X) = 1$, and in particular, $u(X) = 1$. Let us denote by Y the binary point obtained from X by switching the value of the literal u. Since $u \notin T$, we have $T(Y) = 1$, hence $f(Y) = 1$. On the other hand, since $u \in A$, we have $A(Y) = 0$, thus implying $\psi(Y) = 1$. This, together with $\psi(X) = 0$, implies that there is a term B in ψ involving the literal $\bar{u}$, for which $B(Y) = 1$. Hence, the terms A and B have exactly one conflict, and thus their consensus $C = (A \cup B) \setminus \{u, \bar{u}\}$ must be a term of η, implying $C(X) = 1 \leq \eta(X)$. This proves the lemma. $\square$

Theorem 6.13. *The PRIME IMPLICANT DEPLETION procedure generates the complete list of prime implicants of the Boolean function f represented by the prime DNF η. Furthermore, when η is a Horn DNF, the procedure runs in polynomial incremental time and each main* **while** *loop takes $O(n(n + |\eta|)|\mathcal{L}|)$ time, where $\mathcal{L}$ is the current list of prime implicants at the beginning of the loop.*

Proof. Let $\Psi_{\mathcal{L}}$ denote the disjunction of prime implicants in the current list $\mathcal{L}$. We argue by induction on the size of $\mathcal{P}$ that at any moment during the procedure, every prime implicant of f is either explicitly listed in $\mathcal{P}$, or is an implicant of $\Psi_{\mathcal{L}}$. This is clearly the case at the very beginning of the procedure. According to Lemma 6.4, this property is not changed when we move a term T (a prime implicant of f) from $\mathcal{L}$ to $\mathcal{P}$, and then increment $\mathcal{L}$ with the consensuses obtained with T. The property also remains unchanged when we substitute the terms in $\mathcal{L}$ by some absorbing prime implicants, since such a substitution does not change the function represented by $\Psi_{\mathcal{L}}$. Similarly, the property remains valid when we eliminate duplicates from $\mathcal{L}$.

Now, when the algorithm stops $\mathcal{L}$ is empty; hence, $\mathcal{P}$ contains all prime implicants of f.

To see the complexity claim, let us observe that the consensus of two terms can be carried out in $O(n)$ steps, where n is the number of variables in η; hence all consensuses in a main iteration take $O(n|\mathcal{L}|)$ time. This step introduces at most $|\mathcal{L}|$ new terms. For each term, we need to find a prime implicant of f that absorbs it, which can be done, for instance, by forward chaining in $O(n(n + |\eta|))$ time. Hence, a prime list can be obtained in $O(n(n+|\eta|)|\mathcal{L}|)$ time. Finally, by keeping $\mathcal{L}$ and $\mathcal{P}$ in a hash, the elimination of duplicates can be accomplished in $O(|\mathcal{L}| \log n)$ time, proving the claim. $\square$

A further improvement can be achieved by introducing the restricted version of the consensus method in which only those consensuses are considered where at least one of the terms belongs to the original input DNF. More precisely, given a DNF η of the function f, let us call a consensus between two implicants of f an *input consensus* if at least one of these implicants is present in η.

Let us remark that input consensus is not necessarily complete for an arbitrary input DNF, in the sense that not all prime implicants can be generated in this way.

Example 6.6. *Consider the DNF*

$$\phi = x_1\overline{x}_3x_7 \vee x_2x_3x_6 \vee x_1\overline{x}_2x_7 \vee \overline{x}_1\overline{x}_4x_7 \vee x_4x_5x_8 \vee \overline{x}_1\overline{x}_5x_7.$$

It is easy to check that the term $T = x_6x_7x_8$ is a prime implicant of ϕ. However, it cannot be obtained from ϕ by input consensus, since all terms in ϕ have at least two variables not present in T. $\qquad\qquad\square$

Furthermore, even when $\phi = \mathbb{1}_n$, the unique prime implicant of ϕ can be generated by input consensus if and only if it can also be generated by unit consensuses, that is, by the NULR procedure (see Exercise 10 in Chapter 2).

For Horn DNFs, however, this restricted variant of the consensus method works well. It is described more precisely in Figure 6.4.

Before we prove the correctness of this procedure, we need to establish a result shown originally by Chang and Lee [186] and by Jones and Laaser [539].

Lemma 6.5. *Let us assume that η is a Horn DNF representing the function $\mathbb{1}$. An empty term can then be derived from η by a sequence of input consensuses such that each term of η is used at most once in the sequence.*

Proof. Let us consider the procedure NULR(η). Since $\eta = \mathbb{1}$, NULR terminates by finding an empty term in $\eta^k = \eta|_{x_{j_1}=1,\dots,x_{j_k}=1}$, and we can conclude, as in the proof of Theorem 6.10, that there is a corresponding positive term T_0 in η such that $T_0 \subseteq \{j_1,\dots,j_k\}$.

Procedure INPUT CONSENSUS(η)

Input: A DNF $\eta = T_1 \vee T_2 \vee \dots \vee T_m$.

Output: All prime implicants of the function f represented by η.

initialize $\mathcal{P} := \emptyset$, $\mathcal{L} = \{T_1,\dots,T_m\}$;

repeat while $\mathcal{L} \neq \emptyset$

    select a term $T \in \mathcal{L}$ and set $\mathcal{L} := \mathcal{L} \setminus \{T\}$ and $\mathcal{P} := \mathcal{P} \cup \{T\}$;

    generate all consensuses of T with the input terms T_1, ..., T_m, and

           add the obtained new terms to $\mathcal{L}$;

    absorption: delete from $\mathcal{P} \cup \mathcal{L}$ all terms which are absorbed

           by some other terms of $\mathcal{P} \cup \mathcal{L}$;

end while

return the list of terms in $\mathcal{P}$

Figure 6.4.   Procedure INPUT CONSENSUS.

For all $i = 1, \ldots, k$, we can also observe that, since $\overline{x}_{j_i}$ is a negative linear term of η^{i-1}, η must contain a corresponding term of the form

$$T_i = \left(\bigwedge_{j \in S_i} x_j \right) \wedge \overline{x}_{j_i}$$

where $S_i \subseteq \{j_1, \ldots, j_{i-1}\}$ (e.g., $S_1 = \emptyset$).

Let us then define $C_0 = T_0$, and let C_i be the consensus of C_{i-1} and T_{k-i+1}, for $i = 1, \ldots, k$. It is easy to verify by induction on i that these terms indeed have a consensus (since otherwise NULR(η) would have stopped earlier), and that C_i is a positive term with $C_i \subseteq \{j_1, \ldots, j_{k-i}\}$ for $i = 1, \ldots, k-1$. Therefore, C_k is the empty term.

Since T_0 and T_i for $i = 1, \ldots, k$ are all different terms of η, this chain of consensuses provides an input consensus derivation of the empty term with no repetitions.   □

Using the preceding lemma, we can now prove that the input consensus algorithm indeed works for Horn DNFs (see Hooker [498]).

Lemma 6.6. *Let η be a Horn DNF of the Horn function h, and let T be a prime implicant of h. Then, T can be obtained from η by a sequence of input consensuses such that each term of η is used at most once in the sequence.*

Proof. Let us consider the DNF $\eta' = \eta|_{T=1}$ obtained from η by substituting the value 1 for all literals in T. Then $\eta' \equiv \mathbf{1}$, and hence, there is a subset of its terms, say, $D_1, \ldots, D_l$, such that the empty term can be obtained from these by a sequence of consensuses, without repetitions. Since $D_1, \ldots, D_l$ are terms of η', each of them corresponds to a term T_i of η, for $i = 1, \ldots, l$. Performing exactly the same sequence of consensuses on $T_1, \ldots, T_l$, yields T.   □

It follows immediately from Lemma 6.6 that:

Corollary 6.7. *The* INPUT CONSENSUS *procedure correctly generates all prime implicants of any Horn DNF.*

The complexity of the INPUT CONSENSUS algorithm, however, may not be polynomial in the number of prime implicants of the input DNF: To achieve polynomiality, we have to perform again the same "prime substitution" step as in the PRIME IMPLICANT DEPLETION procedure; that is, whenever a new term T is generated and added to the list $\mathcal{L}$, we should subsequently substitute T by a prime implicant absorbing it. This leads us to the INPUT PRIME CONSENSUS procedure displayed in Figure 6.5.

We next prove that this modification is acceptable, and that the INPUT PRIME CONSENSUS method correctly generates all prime implicants of the input function. We first state an easy technical lemma.

Procedure INPUT PRIME CONSENSUS(η)

Input: A DNF $\eta = T_1 \vee T_2 \vee \ldots \vee T_m$.
Output: All prime implicants of the function f represented by η.

initialize $\mathcal{P} := \emptyset$, $\mathcal{L} = \{T_1, \ldots, T_m\}$;
repeat while $\mathcal{L} \neq \emptyset$
    select a term $T \in \mathcal{L}$ and set $\mathcal{L} := \mathcal{L} \setminus \{T\}$ and $\mathcal{P} := \mathcal{P} \cup \{T\}$;
    generate all consensuses of T with the input terms $T_1, \ldots, T_m$;
    replace each such consensus by a prime implicant of f absorbing it;
    check if each of these prime implicants is in $\mathcal{P} \cup \mathcal{L}$, and if not,
    add the new ones to $\mathcal{L}$.
end while
return the list of terms in $\mathcal{P}$

Figure 6.5. Procedure INPUT PRIME CONSENSUS.

Lemma 6.7. *Let us assume that P, Q, R are implicants of a function f and that P is the consensus of Q and R. Let us assume further that R' is a prime implicant of f absorbing R. Then, P is absorbed either by R' or by the consensus of Q and R'.*

Proof. Assume first that Q and R' do not have a consensus. Since $R \leq R'$, this implies that $P \leq R'$. Assume next that Q and R' have a consensus, say T. Then, Q and R' must have the same conflicting variable as Q and R, and thus, $P \leq T$ is implied. $\square$

Theorem 6.14. *When η is a Horn DNF, the INPUT PRIME CONSENSUS procedure correctly generates all prime implicants of the function represented by η.*

Proof. Consider an arbitrary prime implicant P. In view of Lemma 6.6, P can be generated by input consensus from η. Let T_{i_j}, $j = 0, 1, \ldots, k$, be the input terms used in this consensus derivation of P, and let R_j, $j = 1, \ldots, k$, be the implicants generated by these consensuses; more precisely, R_1 is the consensus of T_{i_0} and T_{i_1}, and R_j is the consensus of R_{j-1} and T_{i_j} for $j = 2, \ldots, k$. Finally, $P = R_k$.

We claim that, for all $j = 1, \ldots, k$, the list $\mathcal{P}$ contains a prime implicant $P_j \in \mathcal{P}$ absorbing R_j. Since $P = R_k$ is a prime implicant of f, this implies that $P = P_k \in \mathcal{P}$, which completes the proof of the theorem.

Let us establish the claim. Clearly, the consensus of T_{i_0} and T_{i_1} is executed by procedure INPUT PRIME CONSENSUS(η); thus, we must have a prime implicant $P_1 \in \mathcal{P}$ absorbing R_1. Assume now, for $j < k$, that there is a prime implicant $P_{j-1} \in \mathcal{P}$ absorbing R_{j-1}, and consider R_j. By Lemma 6.7, either P_{j-1} absorbs R_j, or the consensus C of T_{i_j} and P_{j-1} absorbs R_j. In the latter case the consensus C must have been generated by procedure INPUT PRIME CONSENSUS(η), since $P_{j-1} \in \mathcal{P}$ and T_{i_j} is an input term; therefore, there is a prime implicant $P_j \in \mathcal{P}$ absorbing C, and hence P_j absorbs R_j. $\square$

Corollary 6.8. *The complete list of prime implicants of the Boolean function represented by a Horn DNF η can be generated with polynomial delay using procedure* INPUT PRIME CONSENSUS(η).

Proof. Let us remark first that the incremental complexity of the previously described methods for prime implicant generation (namely, PRIME IMPLICANT DEPLETION and INPUT CONSENSUS) resulted from the fact that, in each main cycle, we had to check for absorption, a task requiring time proportional to the length of the lists $\mathcal{P}$ and $\mathcal{L}$. The speedup of INPUT PRIME CONSENSUS is due to the fact that, instead of *absorption*, we have to check now for *membership* in $\mathcal{P}$ and $\mathcal{L}$; this can be done in $O(n)$ time with an appropriate data structure, independently of the length of those lists.

More precisely, let us assume that we keep both $\mathcal{P}$ and $\mathcal{L}$ in a hash table. Then inserting a new member, deleting a member, or checking membership can all be done in $O(n)$ time. Now, it is easy to see that with every execution of the main **while** loop, we add exactly one new element to the output list $\mathcal{P}$. In the **while** loop, selecting a term T, deleting it from $\mathcal{L}$ and adding it to $\mathcal{P}$ takes $O(n)$ time; generating the consensus of T with $T_1, ..., T_m$ can be done in $O(nm)$ time; replacing the (at most m) consensuses by prime implicants can be done in $O(nm|\eta|)$ time; checking membership in $\mathcal{P}$ and $\mathcal{L}$ takes $O(nm)$ time; and adding the new terms to $\mathcal{L}$ can be done in $O(nm)$ time. It follows that all prime implicants can be generated with polynomial delay $O(nm|\eta|)$ between two successive prime implicants. $\square$

6.6 Properties of the set of prime implicants

Horn and pure Horn functions appear in many areas of applications, primarily because several of the tasks arising in those applications can be reduced to solving Boolean equations and thus, as we saw in Section 6.4, can be handled efficiently for Horn systems. The actual complexity of these procedures depends, however, on the representation of the underlying Horn function. Since a Horn function can typically be represented by many different DNFs (and/or CNFs, etc.) of widely varying sizes, it is a natural problem to find a "most efficient" representation of a given Horn function. This is a very important practical problem, frequently considered in the literature: Executing queries, or checking consistency in Horn rule bases, is faster on "shorter" DNFs; the storage efficiency of relational databases is improved if the Horn system of relations is represented in a most condensed form, and so on.

The basic problem of finding a "shortest," or "most economical" Horn DNF of a given Horn function is, in principle, a special case of logic minimization, a topic that we considered in Chapter 3; see, in particular, Section 3.3. In this special case, however, we are not only able to state more precise results, but we can also introduce specific measures expressing what "shortest" should really mean in different contexts.

While logic minimization is a hard problem in general, it becomes tractable for certain measures of size in the special case of Horn functions. To be able to present some of these positive results about Horn minimization, we need to establish further results about the structure of the family of implicants and about Horn DNF representations of a Horn function. Since these results may be of independent interest, we present them in this section, before turning to Horn minimization in Section 6.7.

Definition 6.6. *A set T of terms (elementary conjunctions) is said to be closed under consensus if, for any two terms $T, T' \in T$, their consensus, when it exists, also belongs to T.*

Let us note the difference between this definition and a similar one introduced in Section 3.2.2. In Definition 6.6, we consider a set of terms (without absorptions), and not their disjunction. This is an important detail, since our purpose is to understand the structure of different DNF representations of a given function.

Clearly, the intersection of closed sets of terms is closed again; hence, every set of terms has a unique smallest closed set containing it.

Definition 6.7. *The consensus closure T^c of a set of terms T is the smallest closed set containing T.*

Given a Boolean function f, let us denote by $\mathcal{I}_f$ the set of all implicants of f, and let $\mathcal{P}_f$ denote the set of its prime implicants. Clearly, $\mathcal{I}_f$ is a closed set, and $\mathcal{P}_f^c$ is a subset of $\mathcal{I}_f$ (typically, a proper subset).

Definition 6.8. *Let T be a closed set of terms. A partition $(\mathcal{R}, \mathcal{D})$ of T ($\mathcal{R} \cup \mathcal{D} = T$ and $\mathcal{R} \cap \mathcal{D} = \emptyset$) is called a* recessive-dominant *partition (or, in short, an RD-partition) of T if*

- *both $\mathcal{R}$ and $\mathcal{D}$ are closed under consensus, and*
- *if two terms $T_1 \in \mathcal{R}$ and $T_2 \in \mathcal{D}$ have a consensus T_3, then $T_3 \in \mathcal{D}$.*

This terminology is inspired by a biological analogy: We can view the set of implicants as a "population," and the consensus operation as "mating" between the members of this population. Then the above definition expresses that siblings inherit a "dominant" strain when at least one of the parents possesses it, and they inherit a "recessive" strain exactly when both parents have it.

Example 6.7. *Let h be a Horn function, and consider the partition of its implicants into positive and pure Horn terms, defined by*

$$\mathcal{R} = \{T \in \mathcal{I}_h \mid T \text{ is pure Horn}\} \quad \text{and} \quad \mathcal{D} = \{T \in \mathcal{I}_h \mid T \text{ is positive}\}.$$

It is easy to verify that this is an RD-partition of $\mathcal{I}_h$ (cf. Exercise 26).   □

For further examples of RD-partitions, we refer the reader to Čepek [173] and Boros, Čepek, and Kogan [108] (see also Exercise 27). Let us add that RD-partitions have nice algebraic properties (see, e.g., Exercise 25) which allow the generation of even larger families of RD-partitions.

The significance of RD-partitions for Horn minimization is that RD-partitions allow the decomposition of the minimization problem into a sequence of smaller minimization problems.

To simplify our notations for the rest of this section, and with a slight abuse of terminology, we shall view a DNF as a set of terms.

Definition 6.9. *Given a DNF η representing the Horn function h and a family $\mathcal{T}$ of terms, let us denote by $\eta^{\mathcal{T}} = \eta \cap \mathcal{T}$ the DNF formed by those terms of η that belong to $\mathcal{T}$, and let us call it the $\mathcal{T}$-component of the DNF η. Let us further denote by $h^{\mathcal{T}}$ the Horn function defined by the disjunction of the terms in $\mathcal{P}_h \cap \mathcal{T}$, and let us call it the $\mathcal{T}$-component of h.*

The following result of Čepek [173] implies that the $\mathcal{R}$-component of an arbitrary prime DNF representation η of a Horn function h defines the same Boolean function, namely, the $\mathcal{R}$-component of h, for any RD-partition $(\mathcal{R}, \mathcal{D})$ of $\mathcal{P}_h^c$. As a consequence, one can start the minimization of h by minimizing first $h^{\mathcal{R}}$, represented by $\eta^{\mathcal{R}}$, and then replacing in η the terms of $\eta^{\mathcal{R}}$ by the obtained minimal representation of $h^{\mathcal{R}}$, yielding a new, "shorter" DNF representation of h (in fact, this scheme works for several different measures of "size").

Theorem 6.15. *Let h be a Horn function represented by a Horn DNF $\eta \subseteq \mathcal{P}_h^c$, and let $(\mathcal{R}, \mathcal{D})$ be an RD-partition of $\mathcal{P}_h^c$. Then, the $\mathcal{R}$-component $\eta^{\mathcal{R}}$ of η is a Horn DNF representation of the $\mathcal{R}$-component $h^{\mathcal{R}}$ of h.*

Proof. Let us first note that $(\eta)^c = \mathcal{P}_h^c$ by Theorem 3.5 and by the properties of the consensus closure. Since we obviously have $(\eta \cap \mathcal{R})^c \subseteq \mathcal{R}^c = \mathcal{R}$, the above equality and the definition of an RD-partition by Definition 6.8 imply $(\eta \cap \mathcal{R})^c = \mathcal{R}$. Applying this for the particular DNF representation $\mathcal{P}_h$ of h, instead of η, we also get $(\mathcal{P}_h \cap \mathcal{R})^c = \mathcal{R}$. Consequently, both $\eta^{\mathcal{R}}$ and $h^{\mathcal{R}}$ have the same set of prime implicants, namely, $\mathcal{P}_h \cap \mathcal{R}$ (since no prime implicant of h is absorbed by a term of $\mathcal{P}_h^c$, and since $(\mathcal{P}_h \cap \mathcal{R})^c \subseteq \mathcal{P}_h^c$, obviously), which proves the statement. $\square$

For the special case in which $\mathcal{R}$ is the set of pure Horn terms of a Horn function (as in Example 6.7), this result was established by Hammer and Kogan [446].

Let us remark that the condition $\eta \subseteq \mathcal{P}_h^c$ in Theorem 6.15 can easily be fulfilled by requiring η to be prime. Note, however, that this condition cannot be relaxed completely; for instance, it cannot be simply replaced by the irredundancy of η. Indeed, irredundant Horn DNFs may contain terms that cannot be obtained from the prime implicants by consensus; moreover, there may exist an irredundant DNF representation of a Horn function, which is perfectly disjoint from the consensus closure of its prime implicants. To illustrate this, let us consider the following example.

Example 6.8. *Consider the Horn DNFs*

$$\eta = x_1 \bar{x}_2 \vee x_1 \bar{x}_3 \vee x_1 x_2 x_3 \vee \bar{x}_1 x_2 x_3,$$

$$\phi = x_1 \vee x_2 x_3.$$

It is easy to verify that both DNFs are irredundant Horn representations of the same function h. However, when $\mathcal{R}$ is the family of all pure Horn terms in $\mathcal{P}_h^c$ (as in Example 6.7), we have $\eta^{\mathcal{R}} = x_1\overline{x}_2 \vee x_1\overline{x}_3 \vee \overline{x}_1x_2x_3 \neq \mathbf{0} = \phi^{\mathcal{R}}$. The main reason for the equality $\eta^{\mathcal{R}} = h^{\mathcal{R}}$ to fail in this case is that none of the terms of η belongs to $\mathcal{P}_h^c$.   □

Let us further remark that a result analogous to Theorem 6.15 does not hold for $\mathcal{D}$-components: The $\mathcal{D}$-components of different Horn DNF representations of a same Horn function may represent different Boolean functions, as the following example shows:

Example 6.9. *Consider the following Horn DNFs*

$$\eta = x_1x_2 \vee x_1\overline{x}_3 \vee x_2\overline{x}_4 \vee x_3\overline{x}_1 \vee x_4\overline{x}_2,$$

$$\phi = x_3x_4 \vee x_1\overline{x}_3 \vee x_2\overline{x}_4 \vee x_3\overline{x}_1 \vee x_4\overline{x}_2.$$

It is easy to verify that η and ϕ are equivalent irredundant prime Horn DNFs of the Horn function h having the following prime implicants $\mathcal{P}_h = \{x_1x_2, x_1x_4, x_2x_3, x_3x_4, x_1\overline{x}_3, x_2\overline{x}_4, x_3\overline{x}_1, x_4\overline{x}_2\}$. If we partition the implicants of h into pure Horn and positive terms, we obtain an RD-partition (as in Example 6.7). However, the $\mathcal{D}$-components of the above DNFs, $\eta^{\mathcal{D}} = x_1x_2$ and $\phi^{\mathcal{D}} = x_3x_4$, are not equivalent, and none of them represents the disjunction $x_1x_2 \vee x_1x_4 \vee x_2x_3 \vee x_3x_4$ of all positive prime implicants of h.   □

Although the concept of $\mathcal{D}$-component of of a Horn *function* does not appear to be very useful, $\mathcal{D}$-components of Horn DNFs turn out to have a remarkable property, at least, for certain RD-partitions: Namely, for such RD-partitions, the $\mathcal{D}$-components of all irredundant and prime DNF representations of a Horn function h contain the same number of terms. Since a representation involving the minimum number of terms can be assumed to be irredundant and prime, this property implies that it is enough to minimize the $\mathcal{R}$-component of irredundant and prime representations in order to find a term-minimal representation of h. We now establish the property (see [173]).

Theorem 6.16. *Let h be a Horn function, let $\eta_1, \eta_2 \subseteq \mathcal{P}_h^c$ be two irredundant DNFs of h, and let $(\mathcal{R}, \mathcal{D})$ be an RD-partition of $\mathcal{P}_h^c$ such that no two terms in $\mathcal{D}$ have a consensus. Then, the number of terms in $\eta_1^{\mathcal{D}}$ and $\eta_2^{\mathcal{D}}$ is the same.*

Proof. Let us associate with h a directed graph $G = (\mathcal{D}, A)$, where

$$A = \{(T, T') \mid T, T' \in \mathcal{D}, \ T' \text{ is an implicant of } h^{\mathcal{R}} \vee T\}.$$

Clearly, G is a transitively closed directed graph, and its definition depends only on h and the considered RD-partition $(\mathcal{R}, \mathcal{D})$, but not on any particular representation of h. Let us denote by $C_1, \ldots, C_q$ the strong components of G and assume that $C_1, \ldots, C_t$ $(t \leq q)$ are its source components, that is, those components that have no incoming arcs.

By Theorem 6.15 we know that if $\eta \subseteq \mathcal{P}_h^c$, then $\eta^{\mathcal{R}} \equiv h^{\mathcal{R}}$. Using this fact, we can show that η must contain exactly one term from each of the components $C_1, \ldots, C_t$, and no other terms from $\mathcal{D}$. Applying this claim to η_1 and η_2 will then prove the statement.

To show the claim, let us consider an arbitrary irredundant DNF $\eta \subseteq \mathcal{P}_h^c$ of h. Since every implicant of h belonging to $\mathcal{D} \subseteq \mathcal{P}_h^c = (\eta)^c$ can be obtained by a series of consensus operations from η, and since no consensus operation can be performed between the terms of $\mathcal{D}$ by our assumption, only one term of $\eta^{\mathcal{D}}$ is used in such a consensus chain; all other terms must be from $\eta^{\mathcal{R}}$. Thus, for every $P \in \mathcal{D}$, there exists a term T in $\eta^{\mathcal{D}}$ such that P is an implicant of $\eta^{\mathcal{R}} \vee T \equiv h^{\mathcal{R}} \vee T$. In other words, for every $P \in \mathcal{D}$, there must exist a directed path in G from a term of $\eta^{\mathcal{D}}$, implying that $C_j \cap \eta^{\mathcal{D}} \neq \emptyset$ for $j = 1, \ldots, t$. On the other hand, if T is a term of $\eta^{\mathcal{D}}$, then for all other terms $P \in \mathcal{D}$ for which there exists a directed path from T to P in G, we have that $P \leq \eta^{\mathcal{R}} \vee T \equiv h^{\mathcal{R}} \vee T$; thus those terms cannot appear in the irredundant DNF η. This implies that $|\eta^{\mathcal{D}} \cap C_j| = 1$ for $j = 1, \ldots, t$ and $|\eta^{\mathcal{D}} \cap C_j| = 0$ for $j = t+1, \ldots, q$, proving the claim, and completing the proof of the theorem. $\square$

This theorem was proved for the positive terms of an irredundant prime Horn DNF (see Example 6.7) by Hammer and Kogan [446]. The statement implies in this case that the number of positive terms is the same constant in all irredundant and prime DNF representations of a Horn function; see Example 6.9 for an illustration of this.

Let us note again that the conditions $\eta_1, \eta_2 \subset \mathcal{P}_h^c$ cannot be simply disregarded, since the statement does not remain true, in general, even for irredundant Horn DNFs, as the following example shows:

Example 6.10. *Consider the Horn DNFs of Example 6.8. The DNF η contains only one positive term, while ϕ contains two such terms, and, in fact, ϕ is the (unique) shortest DNF of the corresponding Horn function. The conclusion of Theorem 6.16 fails here because η contains implicants that do not belong to $\mathcal{P}_h^c$. It is possible to perform consensus operations with these implicants that introduce extra arcs in the corresponding digraph G, and in effect reduce the number of source components from 2 to 1 (cf. Exercise 28).* $\square$

Theorems 6.15 and 6.16 provide the basis for a very useful decomposition technique of Horn minimization problems. For Horn functions, and especially for pure Horn functions, there are several different RD-partitions that could be utilized in such decomposition methods (see, e.g., [108] and Exercise 27). Similar structural properties of Horn CNFs also play an important role in decomposability of Horn functions, and in an AI context, in Horn belief revision (see [595]).

As we shall see in the rest of this chapter, the above results alone provide efficient minimization techniques for several special classes of Horn functions. We also refer the reader to [109] for a more thorough treatment of this topic.

6.7 Minimization of Horn DNFs

We now turn our attention to the problem of finding a "shortest" DNF representation of a given Horn function. We present here a number of related results from several different sources (see, e.g., [37, 108, 173, 446, 447, 646]. The word "shortest" may in fact refer to several different objectives here (cf. Chapter 3). Given a Horn function h, represented by the Horn DNF

$$\eta = \left(\bigvee_{P \in \mathcal{P}_0} P \right) \vee \left(\bigvee_{i=1}^{n} \bigvee_{P \in \mathcal{P}_i} P \overline{x}_i \right)$$

as in (6.3), we can consider the number of terms

$$\tau(\eta) = \|\eta\| = |\mathcal{P}_0| + \sum_{i=1}^{n} |\mathcal{P}_i|, \tag{6.10}$$

and the number of literals

$$\lambda(\eta) = |\eta| = \sum_{P \in \mathcal{P}_0} |P| + \sum_{i=1}^{n} \sum_{P \in \mathcal{P}_i} (1 + |P|) \tag{6.11}$$

as measures of the size of η. For a function h and $\mu \in \{\lambda, \tau\}$, we define

$$\mu(h) = \min\{\mu(\eta) \mid \eta \text{ is a DNF of } h\}.$$

Let us recall from Section 6.2 that a Horn function h can also be represented as a set of implications of the form

$$\begin{aligned} P &\Longrightarrow & \text{for } P \in \mathcal{P}_0, \text{ and} \\ P &\Longrightarrow x_i & \text{for } P \in \mathcal{P}_i, \ i = 1, \dots, n. \end{aligned}$$

The sets of positive literals $P \in \mathcal{P}_0 \cup \mathcal{P}_1 \cup \cdots \mathcal{P}_n$ are called the *source sides* of these implications. Let us also observe that, if $P \in \mathcal{P}_i \cap \mathcal{P}_j \cap \cdots \cap \mathcal{P}_k$, then the corresponding implications can be written as a single implication of the form $P \Longrightarrow (x_i \wedge x_j \wedge \cdots \wedge x_k)$. Thus, h can also be represented as the collection of such implications by a DNF of the form:

$$\Phi = \bigvee_{P \in \mathcal{P}} \left(P \Longrightarrow \left(\bigwedge_{j \in R(P)} x_j \right) \right). \tag{6.12}$$

The number $\sigma(\Phi) = |\mathcal{P}|$ is called the *number of source sides* in such an implication representation Φ, and can also be used as a measure of the size of the representation (see, e.g., [37, 646]). For a Horn function h we define $\sigma(h) = \min \sigma(\Phi)$, where the minimization is over all possible implication representations Φ of h, as in (6.12).

We also consider for each $\mu \in \{\lambda, \tau, \sigma\}$ the decision variant of the problem of finding a shortest representation of a given Horn function:

HORN μ-MINIMIZATION
Instance: A Horn DNF η of a Horn function h and an integer K.
Output: A (Horn) DNF or implication representation η^* of the Horn function h such that $\mu(\eta^*) \leq K$, if there is one.

Note that we do not have to require the output to be Horn in case of $\mu \in \{\lambda, \tau\}$. In fact, by substituting the non-Horn terms of η^* by prime implicants of η (which can easily be done in polynomial time in the size of η according to Lemma 6.3), we can always obtain a Horn DNF η^{**} such that $\eta \geq \eta^{**} \geq \eta^*$ and $\mu(\eta^{**}) \leq \mu(\eta^*)$ for both measures $\mu \in \{\tau, \lambda\}$. It is also easy to see that $\eta^* \geq \eta$ holds if and only if $\eta^{**} \geq \eta$ holds, and the latter can be checked in polynomial time by Lemma 6.3 (see [173]). Thus, we can assume in the sequel, without any loss of generality, that η^* is a Horn DNF when $\mu \in \{\lambda, \tau\}$.

6.7.1 Minimizing the number of terms

Since partitioning the implicants into pure Horn and positive terms provides an RD-partition, and since there is no consensus between positive terms, Theorems 6.15 and 6.16 immediately imply the following decomposition, as shown by Hammer and Kogan [446]:

Corollary 6.9. *Given a Horn function h and an irredundant prime DNF η of h, consider the RD-partition of its implicants into the sets of pure Horn and positive terms. Then we have $\tau(\eta^{\mathcal{D}}) = \tau(h) - \tau(h^{\mathcal{R}})$; that is, $\tau(\eta^{\mathcal{D}})$ is a constant, independent of η. Furthermore, $h = \eta^{\mathcal{D}} \vee \eta'$ holds for an arbitrary DNF η' of the pure Horn component $h^{\mathcal{R}}$ of h. Thus, the problem of finding a τ-minimal (shortest) DNF of h can be reduced in polynomial time to finding a τ-minimal DNF of its pure Horn component $h^{\mathcal{R}}$.*

Proof. Theorem 6.16 claims that for any RD-partition $(\mathcal{R}, \mathcal{D})$ of $\mathcal{P}_h^c$ such that there is no consensus between the terms of $\mathcal{D}$, the number of terms $|\eta \cap \mathcal{D}|$ in the $\mathcal{D}$-component of an irredundant DNF $\eta \subseteq \mathcal{P}_h^c$ of h is a constant. Applying this for the pure Horn versus positive RD-partition, we can conclude that even the "shortest" Horn DNF of h contains exactly the same constant number of positive terms.

Furthermore, Theorem 6.15 states that the $\mathcal{R}$-component of η represents the function $h^{\mathcal{R}}$, namely, the $\mathcal{R}$-component of h, for all representations $\eta \subseteq \mathcal{P}_h^c$ of h. Consequently, if η' is an arbitrary Horn DNF representation of $h^{\mathcal{R}}$, then $\eta'' = \eta^{\mathcal{D}} \vee \eta'$ is a Horn DNF representation of h. Thus, if η' is a "shortest" Horn DNF of $h^{\mathcal{R}}$, then η'' is a "shortest" Horn DNF of h. $\qquad\square$

Given a pure Horn function h, finding a τ-minimal pure Horn DNF of h, is however, a difficult problem. This problem was first considered in the slightly different context of directed hypergraphs, and its hardness was shown by Ausiello,

D'Atri, and Saccà [37] using a reduction from set covering. We sketch their proof in the context of pure Horn τ-minimization:

Theorem 6.17. *Horn τ-minimization is NP-complete, even if the input is restricted to pure Horn expressions.*

Proof. Let us consider a hypergraph $\mathcal{H} = (V, \mathcal{E})$ over the base set $V = \{1, 2, ..., n\}$ such that $\bigcup_{H \in \mathcal{E}} H = V$. It is well-known that, for a given integer $k < m = |\mathcal{E}|$, it is NP-complete to decide the existence of a subset of hyperedges $\mathcal{S} \subseteq \mathcal{E}$ that is a cover of $\mathcal{H}$ of cardinality at most k, that is, such that $|\mathcal{S}| \le k$ and $\bigcup_{H \in \mathcal{S}} H = V$ (see, e.g., [371]).

With the hypergraph $\mathcal{H}$ and with every subset of hyperedges $\mathcal{S} \subseteq \mathcal{E}$, we now associate pure Horn DNFs Φ and $\eta_\mathcal{S}$, depending on the Boolean variables z, x_j for $j \in V$, and y_H for $H \in \mathcal{E}$, where

$$\Phi = \left(\bigvee_{H \in \mathcal{E}} \bigvee_{j \in H} \overline{x}_j y_H \right) \vee \left(\bigvee_{H \in \mathcal{E}} \bigwedge_{j=1}^{n} x_j \overline{y}_H \right),$$

and

$$\eta_\mathcal{S} = \left(\bigvee_{H \in \mathcal{E}} z \overline{y}_H \right) \vee \Phi.$$

Let us further denote by h the Horn function represented by the pure Horn DNF $\eta_\mathcal{H}$. We claim that h has a DNF with no more than $k + \tau(\Phi)$ terms if and only if $\mathcal{H}$ has a cover of cardinality no more than k.

To see this, let us observe first that since $\eta_\mathcal{H}$ does not involve the literal $\overline{z}$, no term in $\mathcal{P}_h^c$ contains $\overline{z}$ (all those terms can be obtained from $\eta_\mathcal{H}$ by consensus). Let us then define $\mathcal{D}$ as the set of those terms in $\mathcal{P}_h^c$ involving the literal z, and let $\mathcal{R} = \mathcal{P}_h^c \setminus \mathcal{D}$. Any consensus involving a term in $\mathcal{D}$ will result in a term also containing z. Hence, $(\mathcal{R}, \mathcal{D})$ forms an RD-partition for h and thus Φ represents $h^\mathcal{R}$, the $\mathcal{R}$-component of h. Furthermore, Φ is a τ-minimal representation of $h^\mathcal{R}$. This is because all quadratic terms in Φ must appear in all representations of Φ, and all such representations must also contain at least one term including $\overline{y}_H$ for all $H \in \mathcal{H}$. Since the only prime implicants in $\mathcal{D}$ are $z \overline{y}_H$, $H \in \mathcal{H}$, and $z \overline{x}_j$, $j \in V$, and since a term $z \overline{x}_j$ can always be replaced by $z \overline{y}_H$ for $H \in \mathcal{H}$ such that $j \in H$ without changing the size of the representation, Theorem 6.15 implies that a τ-minimal prime DNF of h looks like $\eta_\mathcal{S}$ for some subhypergraph $\mathcal{S} \subseteq \mathcal{E}$. Since $\eta_\mathcal{S}$ represents h if and only if $\mathcal{S}$ is a cover, our main claim follows. $\square$

This result can further be improved, as observed by Boros, Čepek, and Kučera [110].

Theorem 6.18. *Horn τ-minimization remains NP-complete even if the input is restricted to cubic pure Horn expressions.*

Proof. Let us try to repeat the above proof with a small modification in the definition of Φ. Namely, let us introduce $n - 1$ additional variables and replace the high-degree terms by a chain of cubic and quadratic terms, as follows:

$$
\Psi = \left(\bigvee_{\substack{H \in \mathcal{H} \\ j \in H}} \overline{x}_j \, y_H \right) \vee x_1 x_2 \overline{u}_1 \vee u_1 x_3 \overline{u}_2 \vee \cdots \vee u_{n-2} x_n \overline{u}_{n-1} \vee \left(\bigvee_{H \in \mathcal{H}} u_{n-1} \overline{y}_H \right),
$$

and set

$$
\eta_S = \left(\bigvee_{H \in \mathcal{S}} z \overline{y}_H \right) \vee \Psi.
$$

As in the proof of Theorem 6.17, we denote by h the function represented by the cubic pure Horn DNF $\eta_{\mathcal{H}}$. We can then repeat the preceding proof, with Ψ playing the role of Φ. $\square$

Note that for quadratic pure Horn DNFs, τ-minimization is equivalent to finding the *transitive reduction* of a directed graph (that is, finding the smallest subset of arcs, the transitive closure of which is the same as that of the original graph), which is a polynomially solvable problem; see Sections 5.4.1 and 5.8.4.

On the positive side, for an arbitrary Horn function h, Hammer and Kogan [447] proved that $\tau(h)$ is approximated within a reasonable factor by the size of any irredundant prime DNF of h.

Theorem 6.19. *If h is a Horn function on $\mathcal{B}^n$ and $\eta \subseteq \mathcal{P}_h^c$ is an irredundant Horn DNF of h, then $\tau(\eta) \le (n-1)\tau(h)$.*

Proof. Let us consider the RD-partition $\mathcal{R} \cup \mathcal{D} = \mathcal{P}_h^c$ into pure Horn and positive terms, and let ζ denote a τ-optimal irredundant, prime DNF of h. Then, $\tau(\eta^{\mathcal{D}}) = \tau(\zeta^{\mathcal{D}})$ holds for the positive components according to Theorem 6.16, and $\eta_1 = \eta^{\mathcal{R}} \equiv \zeta_1 = \zeta^{\mathcal{R}} = h^{\mathcal{R}}$ must hold for the pure Horn components by Theorem 6.15. Let us further divide $\mathcal{R}$ into $\mathcal{R} = \mathcal{P}_{h^{\mathcal{R}}}^c = \mathcal{R}' \cup \mathcal{D}'$, where $\mathcal{D}'$ is the set of linear terms and $\mathcal{R}'$ is the set of nonlinear pure Horn terms in $\mathcal{R}$. This yields an RD-partition of the closure of the prime implicants of $h^{\mathcal{R}}$ (see Exercise 27), and by the same theorems, we get that $\tau(\eta_1^{\mathcal{D}'}) = \tau(\zeta_1^{\mathcal{D}'})$ and $\eta_2 = \eta_1^{\mathcal{R}'} \equiv \zeta_2 = \zeta_1^{\mathcal{R}'} = h^{\mathcal{R}'}$.

Let us consider next a term $A\overline{y}$ of ζ_2. Since $\eta_2 \equiv \zeta_2$, this term is an implicant of η_2, and thus, by Lemma 6.3 variable y must belong to the forward chaining closure A^{η_2} of A. Let $A^{\eta_2} \setminus A = \{x_{i_1}, x_{i_2}, \ldots, x_{i_k}\}$ be indexed according to the order in which forward chaining adds these variables to A, and let $A_{i_j} \overline{x}_{i_j}$ be the term of η_2 used in this process when adding x_{i_j} to A, for $j = 1, \ldots, k$. (We have $y = x_{i_t}$ for some $t \le k$.) It is easy to see that performing consensuses between these terms, we can derive the prime implicant $A\overline{y}$.

Thus we need at most $|A^{\eta_2} \setminus A|$ terms of η_2 to derive a term $A\overline{y} \in \zeta_2$. Due to the fact that η is irredundant, η_2 must also be irredundant (this follows by Theorem 6.15), and thus, all terms of η_2 must appear in such derivation for some terms of ζ_2. Therefore, we have $\tau(\eta_2) \le \sum_{A\overline{y} \in \zeta_2} |A^{\eta_2} \setminus A| \le (n-1)\tau(\zeta_2)$, since ζ_2 does not contain linear pure Horn terms by our construction.

Putting all the above together, we obtain

$$\tau(\eta) = \tau(\eta^{\mathcal{D}}) + \tau(\eta_1^{\mathcal{D}'}) + \tau(\eta_2) = \tau(\zeta^{\mathcal{D}}) + \tau(\zeta_1^{\mathcal{D}'}) + \tau(\eta_2)$$

$$\leq \tau(\zeta^{\mathcal{D}}) + \tau(\zeta_1^{\mathcal{D}'}) + (n-1)\tau(\zeta_2) \leq (n-1)\tau(\zeta),$$

which completes the proof.   □

Let us again observe that in this theorem, $\eta \subseteq \mathcal{P}_h^c$ is an important condition without which the claim does not remain true, as illustrated by the following example.

Example 6.11. *Consider the irredundant DNF representation* $\eta = x_1\overline{x}_2 \vee x_1\overline{x}_3 \vee x_1x_2x_3$ *of the Horn function* $h = x_1$. *In this case we have* $n = 3$, $\tau(\eta) = 3$, *and* $\tau(h) = 1$.   □

Let us finally remark that much better polynomial time approximation may not be achievable, as shown by a recent inapproximability result of Bhattacharya, DasGupta, Mubayi, and Turán [77]:

Theorem 6.20. *For any fixed* $0 < \epsilon < 1$, *one cannot guarantee a* $2^{\log^{1-\epsilon} n}$- *approximation for Horn* τ-*minimization in polynomial time, unless* $NP \subseteq DTIME(n^{polylog(n)})$.

6.7.2 Minimizing the number of literals

We turn next to the minimization of the number of literals in a Horn representation. The first related result, due to Maier [646], establishes the hardness of minimization for a somewhat different measure; its proof, however, carries easily over to the case of λ-minimization (see, e.g., [173]). A simpler and more elegant reduction from set covering to λ-minimization was presented by Hammer and Kogan [447]. This result can further be strengthened, as noted by Boros, Čepek, and Kučera [110]:

Theorem 6.21. *Horn* λ-*minimization is NP-complete, even if the input is restricted to cubic pure Horn DNFs.*

Proof. Given a hypergraph $(V, \mathcal{E})$, let us consider the cubic Horn DNF $\eta_\mathcal{S}$, defined as in the proof of Theorem 6.18, for any subfamily $\mathcal{S} \subseteq \mathcal{E}$. It can be verified that $\eta_\mathcal{S}$ is not only τ-minimal but also λ-minimal if and only if $\mathcal{S}$ is a minimal cover.   □

For quadratic pure Horn DNFs, λ-minimization is easily seen to be equivalent to τ-minimization, and hence, it is polynomially solvable, as we remarked earlier.

On the positive side, Hammer and Kogan [447] proved that $\lambda(h)$ is approximated within a reasonable factor by any irredundant and prime Horn DNF representation. More precisely, we can show the following:

Theorem 6.22. *If h is a Horn function on $\mathcal{B}^n$ and $\eta \subseteq \mathcal{P}^c_h$ is an irredundant Horn DNF of h, then $\lambda(\eta) \leq \binom{n}{2}\lambda(h)$.*

Proof. Let us consider the RD-partition $\mathcal{D} \cup \mathcal{R} = \mathcal{P}^c_h$ into linear and nonlinear terms of the set $\mathcal{P}^c_h$ (see Exercise 27), and let ζ denote a λ-minimal DNF of h. Then, by Theorems 6.15 and 6.16, we have $\lambda(\eta^{\mathcal{D}}) = \tau(\eta^{\mathcal{D}}) = \tau(\zeta^{\mathcal{D}}) = \lambda(\zeta^{\mathcal{D}})$ and $\eta^{\mathcal{R}} \equiv \zeta^{\mathcal{R}} \equiv h^{\mathcal{R}}$. Since $\zeta^{\mathcal{R}}$ does not contain any linear terms, we have $\lambda(\zeta^{\mathcal{R}}) \geq 2\tau(\zeta^{\mathcal{R}}) \geq 2\tau(h^{\mathcal{R}})$. Furthermore, by Theorem 6.19, we have $\tau(\eta^{\mathcal{R}}) \leq (n-1)\tau(h^{\mathcal{R}})$. Putting all these together with the trivial inequality $\lambda(\phi) \leq n\tau(\phi)$, we obtain

$$\lambda(\eta) = \lambda(\eta^{\mathcal{D}}) + \lambda(\eta^{\mathcal{R}}) = \lambda(\zeta^{\mathcal{D}}) + \lambda(\eta^{\mathcal{R}}) \leq \lambda(\zeta^{\mathcal{D}}) + n\tau(\eta^{\mathcal{R}})$$

$$\leq \lambda(\zeta^{\mathcal{D}}) + n(n-1)\tau(h^{\mathcal{R}}) \leq \lambda(\zeta^{\mathcal{D}}) + \tfrac{1}{2}n(n-1)\lambda(\zeta^{\mathcal{R}})$$

$$\leq \binom{n}{2}\lambda(\zeta) = \binom{n}{2}\lambda(h).$$

$\square$

Here again, condition $\eta \subseteq \mathcal{P}^c_h$ is important because Theorem 6.22 does not hold for arbitrary irredundant Horn DNFs.

Example 6.12. *Consider the DNF η of the Horn function $h = x_1$, as in Example 6.11. In this case, we have $n = 3$, $\lambda(\eta) = 7$, while $\lambda(h) = 1$.* $\square$

6.7.3 Minimization of the number of source sides

An arbitrary Horn DNF η can be rewritten straightforwardly as an implication expression Φ of the form (6.12), and $\sigma(\Phi)$ will be exactly the number of different sets of positive variables appearing in η. Conversely, any implication expression Φ can be rewritten as a Horn DNF η, such that the number of different sets of positive variables appearing in η is exactly $\sigma(\Phi)$. Thus, we can denote by $\sigma(\eta)$ the number of different sets of positive variables appearing in an arbitrary Horn DNF η, and restate Horn σ-minimization as the problem of finding a Horn DNF representation η of a given Horn function h minimizing $\sigma(\eta)$.

Horn σ-minimization was shown to be solvable in polynomial time by Maier [646] and by Ausiello, D'Atri, and Saccà [37]. In the rest of this section, we provide a proof of this lone, truly positive result in the area of Horn DNF minimization.

We first show that it is enough to consider the problem for pure Horn functions.

Lemma 6.8. *If π and π' are positive DNFs on $\{x_1, x_2, \dots, x_n\}$, and η and η' are pure Horn DNFs on $\{x_1, x_2, \dots, x_n\}$, then $\pi \vee \eta$ and $\pi' \vee \eta'$ represent the same Horn function h if and only if the DNFs $\eta \vee (\pi \wedge \overline{x}_{n+1})$ and $\eta' \vee (\pi' \wedge \overline{x}_{n+1})$ represent the same pure Horn function h' on $n+1$ variables.*

Proof. The claimed equivalence trivially holds if $x_{n+1} = 0$, and follows by the existence of a unique pure Horn component (see Theorem 6.15) when $x_{n+1} = 1$. $\square$

Lemma 6.8 implies that we can associate a unique pure Horn function h' in $n + 1$ variables with every Horn function h in n variables, so that $\sigma(h) = \sigma(h')$. Therefore, in the sequel, we shall consider source minimization only for pure Horn functions.

Recall from Section 6.4 that the forward chaining closure S^η of a subset S of the variables is uniquely defined for every (pure) Horn DNF η, and that this closure is the same for every (pure) Horn DNF representing a given function h, so that we can also denote S^η as S^h. It follows from Lemma 6.3 that a pure Horn term $A\bar{x} \in \mathcal{P}_h^c$ is an implicant of a Horn function h if and only if $x \in A^h$.

Note further that, since we view a DNF as a set of terms, we consider $\eta = x\bar{z}$ to be different from $\eta' = x\bar{z} \vee xy\bar{z}$, even if they represent the same Boolean function; but η is considered to be the same as $\eta'' = x\bar{z} \vee x\bar{z}$, even if they are written differently.

Definition 6.10. *Given an implicant $T\bar{x} \in \mathcal{P}_h^c$ of a pure Horn function h, the set of terms $\Sigma(T) = \{T\bar{y} \mid y \in T^h \setminus T\} \subseteq \mathcal{P}_h^c$ is called the h-star of T.*

Note that if $T\bar{x} \in \mathcal{P}_h^c$, then we have $\Sigma(T) \subseteq \mathcal{P}_h^c$, by Lemma 6.3.

Definition 6.11. *For a pure Horn DNF η, we denote by $\mathcal{S}(\eta)$ the family of all those subsets of variables which appear as sets of positive variables of a term of η. We call $\mathcal{S}(\eta)$ the family of source sets of η.*

With this definition, we have $\sigma(\eta) = |\mathcal{S}(\eta)|$ for every pure Horn DNF η.

Definition 6.12. *Given a DNF $\eta \subseteq \mathcal{P}_h^c$ of the pure Horn function h, we associate to it another DNF defined by $\eta^* = \bigcup_{T \in \mathcal{S}(\eta)} \Sigma(T)$. We say that η^* is the star closure of η, and we say that η is star closed if $\eta = \eta^*$.*

The star closure η^* represents h, and we have $\mathcal{S}(\eta) = \mathcal{S}(\eta^*)$ by the preceding definitions.

Definition 6.13. *A star closed pure Horn DNF η representing the pure Horn function h is called star irredundant if the DNF $\bigcup_{T \in \mathcal{S}'} \Sigma(T)$ does not represent h for any proper subset $\mathcal{S}' \subsetneq \mathcal{S}(\eta)$.*

Lemma 6.9. *Given a DNF $\eta \subseteq \mathcal{P}_h^c$ representing a pure Horn function h, a star closed and star irredundant DNF $\tilde{\eta}$ representing h can be constructed in $O(n|\eta|^2)$ time.*

Proof. Since T^η can be computed by forward chaining in $O(n + |\eta|)$ time for an arbitrary subset T of the variables (see Section 6.4), we can compute the star closure η^* of η, namely, the sets $\Sigma(T)$ for $T \in \mathcal{S}(\eta)$ in $O(|\mathcal{S}(\eta)|(n + |\eta|)) = O(n|\eta| + |\eta|^2)$ time.

Let us next initialize $\tilde{\eta} = \eta^*$ and label the sets $\mathcal{S}(\eta) = \{T_1, T_2, \ldots, T_k\}$ (where $k = |\mathcal{S}(\eta)|$). Then, repeat the following for $j = 1, \ldots, k$: define the DNF $\phi_j = \bigcup_{Q \in \mathcal{S}(\tilde{\eta}) \setminus \{T_j\}} \Sigma(Q)$, and compute the forward chaining closure $T_j^{\phi_j}$ in $O(n + |\phi_j|) = O(n|\eta|)$ time. Clearly, if $T_j^{\phi_j} = T_j^\eta$, then ϕ_j also represents h;

hence, the star $\Sigma(T_j)$ is redundant in $\widetilde{\eta}$. In this case, update $\widetilde{\eta} = \phi_j$. Otherwise, keep the star of T_j in the representation $\widetilde{\eta}$.

At the end of this loop, $\widetilde{\eta}$ is a star irredundant (and star closed) representation of h, as claimed. Since we have $|\mathcal{S}(\eta)| \leq |\eta|$ steps in the loop, we can complete this part in $O(n|\eta|^2)$ time. Thus, the total time required by the procedure is $O(n|\eta|^2)$, as stated. $\square$

The main result of this subsection, then, states that any star irredundant and star closed DNF representation of a pure Horn function is also σ-minimal.

Theorem 6.23. *If h is a pure Horn function, and $\eta \subseteq \mathcal{P}_h^c$ is a star closed, star irredundant DNF of h, then $\sigma(h) = \sigma(\eta)$.*

Before we prove this statement, we need a few more definitions and lemmas. Observe first that if h is a pure Horn function, and S is subset of its variables such that $S^h = S$, then the partition $\mathcal{R}_S = \{A\overline{x} \in \mathcal{P}_h^c \mid A \subseteq S\}$ and $\mathcal{D}_S = \{A\overline{x} \in \mathcal{P}_h^c \mid A \not\subseteq S\}$ is an RD-partition of $\mathcal{P}_h^c$ (see Exercise 27). To simplify our notations, we denote respectively by h^S and η^S the $\mathcal{R}_S$-components of h and η, when $\eta \subseteq \mathcal{P}_h^c$ is a DNF representation of h; we call h^S and η^S the S-*components* of h and η, respectively. Note that h^S could equivalently be defined by the disjunction of all terms $A\overline{x} \in \mathcal{P}_h^c$ for which $A^h \subseteq S$, and that η^S is a DNF representation of h^S for every DNF $\eta \subseteq \mathcal{P}_h^c$ of h, by Theorem 6.15.

Definition 6.14. *For a pure Horn function h and a subset S of its variables such that $S^h = S$, we denote by $h^{\underline{S}}$ the function defined by the disjunction of all those terms $T\overline{x} \in \mathcal{P}_h^c$ such that $T^h \subsetneqq S$. Analogously, for a DNF $\eta \subseteq \mathcal{P}_h^c$ of h, we denote by $\eta^{\underline{S}}$ the disjunction of all those terms $T\overline{x} \in \eta$ such that $T^h \subsetneqq S$.*

The next lemma is instrumental in our proof of Theorem 6.23, and it leads to the identification of another type of "subfunction" of pure Horn functions, not implied by RD-partitions.

Lemma 6.10. *Let h be a pure Horn function, let S be a subset of its variables such that $S^h = S$, and let $\eta \subseteq \mathcal{P}_h^c$ be a Horn DNF of h. Then, for every implicant $A\overline{x} \leq h^S$, either $A\overline{x} \leq \eta^{\underline{S}}$ or $A^h = S$.*

Proof. Let us consider an arbitrary implicant $A\overline{x} \leq h^S$ for which $A\overline{x} \not\leq \eta^{\underline{S}}$. We claim that $A^h \supseteq S$, which will imply the lemma, since $S \supseteq A$ and $S^h = S$ by our assumptions. To see this claim, we consider the partial assignment that sets all variables in A to 1 and assigns 0 to x. Since $A\overline{x} \not\leq \eta^{\underline{S}}$, the Horn function obtained from $\eta^{\underline{S}} \equiv h^{\underline{S}}$ by substituting this partial assignment has some false points, and thus it has a unique minimal false point by Corollary 6.3. Let X^* denote this unique binary assignment, extended with the values assigned to the variables of A and to x, and let us denote by Q the subset of variables which are assigned value 1 in X^*. It is easy to see by the definition of forward chaining that we have $Q \subseteq A^{\eta^{\underline{S}}}$ (since $x = 0$ limits the forward chaining procedure). Since $A\overline{x} \leq h^S \equiv \eta^S$, and the term $A\overline{x}$ evaluates to 1 at X^*, by our construction, there must exist a term $B\overline{y}$ of

η^S that also evaluates to 1 at X^*, that is, for which $B \subseteq Q$ and $y \notin Q$. Clearly, this term of η^S does not belong to $\eta^{\underline{S}}$, since all terms of $\eta^{\underline{S}}$ vanish at X^*; thus, $B^h = S$ is implied by the definition of $\eta^{\underline{S}}$. Since we have $A^h = A^\eta \supseteq A^{\eta^{\underline{S}}} \supseteq Q \supseteq B$, the relations $A^h = (A^h)^h \supseteq B^h = S$ follow, concluding the proof of the claim. $\qquad\square$

Corollary 6.10. *Let h be a pure Horn function, let S be a subset of its variables such that $S^h = S$, and let $\eta \subseteq \mathcal{P}_h^c$ be a Horn DNF of h. Then, $\eta^{\underline{S}}$ represents the function $h^{\underline{S}}$.*

Proof. For any term $T\overline{x} \in \mathcal{P}_h^c$ for which $T^h \subsetneq S$ it follows by Lemma 6.10 that $T\overline{x} \leq \eta^{\underline{S}}$, which then implies $h^{\underline{S}} \leq \eta^{\underline{S}}$ by Definition 6.14. For the converse direction, the terms of $\eta^{\underline{S}}$ are also implicants of $h^{\underline{S}}$ by Definition 6.14, since $\eta \subseteq \mathcal{P}_h^c$ is assumed. $\qquad\square$

We are now ready to prove the main theorem of this subsection.

Proof of Theorem 6.23. Consider two star closed, star irredundant DNFs $\eta \subseteq \mathcal{P}_h^c$ and $\zeta \subseteq \mathcal{P}_h^c$ of the pure Horn function h, and fix an arbitrary subset S of the variables for which $S^h = S$. Clearly, both η^S and ζ^S represent the S-component h^S of h; thus, they both must be star closed and star irredundant because both η and ζ are assumed to be star closed and star irredundant. Let us further denote by

$$S(\eta^S) \setminus S(\eta^{\underline{S}}) = \{A_1, \ldots, A_k\} \quad \text{and} \quad S(\zeta^S) \setminus S(\zeta^{\underline{S}}) = \{B_1, \ldots, B_\ell\}$$

the source sets of η and ζ, respectively, for which $A_i^h = B_j^h = S$ holds for $i = 1, \ldots, k$ and $j = 1, \ldots, \ell$.

We claim that $k = \ell$. Since every source set of $S(\eta)$ and $S(\zeta)$ corresponds to exactly one subset S of the variables, satisfying $S^h = S$, this claim implies the statement of the theorem, for example, by assuming that ζ is a σ-optimal representation.

To prove the claim, let us assume indirectly that, for instance, $k > \ell$. Note first that according to Corollary 6.10, both $\eta^{\underline{S}}$ and $\zeta^{\underline{S}}$ represent the same function $h^{\underline{S}}$. Furthermore, the star irreducibility of η^S and ζ^S implies that $A_i\overline{x} \not\leq \eta^{\underline{S}}$, and $B_j\overline{y} \not\leq \zeta^{\underline{S}}$ for some variables $x \in A_i^h \setminus A_i$, and $y \in B_j^h \setminus B_j$ for all $i = 1, \ldots, k$ and $j = 1, \ldots, \ell$.

Thus it follows, as in the proof of Lemma 6.10, that for every index i, there exists a corresponding index j, such that $A_i^{h^{\underline{S}}} \supseteq B_j$, and conversely, for every index j, there exists a corresponding index i such that $B_j^{h^{\underline{S}}} \supseteq A_i$. Since $k > \ell$, we must have indices i_1, i_2 and j for which $A_{i_1}^{h^{\underline{S}}} \supseteq B_j$ and $A_{i_2}^{h^{\underline{S}}} \supseteq B_j$. Let us denote by i_3 one of the indices for which $B_j^{h^{\underline{S}}} \supseteq A_{i_3}$ holds. Since $i_1 \neq i_2$, we can assume, without any loss of generality, that $i_3 \neq i_1$. Thus,

$$A_{i_1}^{\eta^{\underline{S}}} = \left(A_{i_1}^{\eta^{\underline{S}}}\right)^{\eta^{\underline{S}}} \supseteq B_j^{h^{\underline{S}}} \supseteq A_{i_3}$$

follows, from which we can derive

$$A_{i_1}^{\eta^{\underline{S}} \cup \Sigma(A_{i_3})} = \left(A_{i_1}^{\eta^{\underline{S}}}\right)^{\Sigma(A_{i_3})} \supseteq A_{i_3}^{\Sigma(A_{i_3})} = S.$$

This last relation implies by Lemma 6.3 that every term of $\Sigma(A_{i_1})$ is an implicant of $\eta^{\underline{S}} \cup \Sigma(A_{i_3})$, contradicting the fact that η was chosen as a star irredundant expression. This contradiction proves that $k = \ell$, finishing the proof of the claim and of the theorem. $\square$

We close this section by mentioning that a remarkable directed graph can be associated quite naturally with pure Horn DNFs (and with pure Horn functions, as well; see [448]), and this directed graph plays an important role (explicitly or implicitly) in many of the related results obtained in this area (see, e.g., [20, 37, 108, 173, 383, 448, 449, 711]). We refer the reader to Section 6.9.4 for further details. We also note that, besides the "minimality" of Horn expressions (in various senses), several other extremal properties of Horn representations lead to interesting combinatorial results (see, e.g., [594]).

6.8 Dualization of Horn functions

The dual of a Boolean function $f(X)$ has been defined as $f^d(X) = \overline{f}(\overline{X})$, where $\overline{X}$ denotes the componentwise negation of X; see Section 1.3 and Chapter 4. In this section, we consider the problems of characterizing and generating f^d when f is a Horn DNF.

Despite the fact that duals of Horn functions must be very special, since Horn functions are special, it is not immediate to obtain a simple characterization. Certainly, the dual of a Horn function is not necessarily Horn, as shown by the example $h(x_1, x_2) = \overline{x}_1 \vee \overline{x}_2$, for which $h^d(x_1, x_2) = \overline{x}_1 \overline{x}_2$ is not Horn. Generalizing slightly a result of Eiter, Ibaraki, and Makino [298], we can obtain the following characterization of DNF expressions of the dual of a Horn function:

Theorem 6.24. *Consider a DNF ϕ of a Boolean function f, where*

$$\phi(X) = \bigvee_{i=1}^{m} \left(\bigwedge_{j \in P_i} x_j \bigwedge_{k \in N_i} \overline{x}_k \right).$$

Then, f is the dual of a Horn function if and only if for any two distinct indices $i \neq i'$,

$$\phi|_{\{x_j = 1 | j \in P_i \cup P_{i'}\} \cup \{x_k = 0 | k \in N_i \cap N_{i'}\}} \equiv \mathbf{1}. \tag{6.13}$$

Proof. Recall (see Theorem 4.7) that a nontrivial term

$$T = \bigwedge_{j \in P} x_j \bigwedge_{k \in N} \overline{x}_k, \tag{6.14}$$

(where $P \cap N = \emptyset$) is a prime implicant of the dual function f^d if and only if $(P \cap P_i) \cup (N \cap N_i) \neq \emptyset$ for all $i = 1, \ldots, m$, and the set of literals in T is minimal with respect to these conditions. In other words, the prime implicants of the dual f^d, as subsets of literals, are in one-to-one correspondence with those minimal transversals of the hypergraph on the set of literals formed by the terms of ϕ, which do not contain complementary pairs of literals. Thus, f is the dual of a Horn function if and only if all such minimal transversals of the terms of ϕ contain at most one negative literal.

To prove the theorem, let us assume first that there exists a non-Horn prime implicant of f^d of the form (6.14) with $|N| \geq 2$, and let us prove that, in this case, condition (6.13) is violated.

Since T is prime, for every $\ell \in N$ there must exist a term $i(\ell)$ of ϕ such that $(P \cup N) \cap (P_{i(\ell)} \cup N_{i(\ell)}) = \{\ell\}$. Thus, for any two distinct indices $\ell \neq \ell'$, $\ell, \ell' \in N$, we have $\ell \in N_{i(\ell)} \setminus N_{i(\ell')}$, $\ell' \in N_{i(\ell')} \setminus N_{i(\ell)}$ and $P \cap (P_{i(\ell)} \cup P_{i(\ell')}) = \emptyset$. On the other hand, $P \cap P_i \neq \emptyset$ must hold for all terms of ϕ such that $N_i \subseteq (N_{i(\ell)} \cup N_{i(\ell')}) \setminus \{\ell, \ell'\}$, since $N \cap N_i = \emptyset$ for such terms.

It follows from these observations that the assignment $\{x_j = 0 \mid j \in P\} \cup \{x_k = 1 \mid k \in N\}$ is compatible with the assignment $\{x_j = 1 \mid j \in P_{i(\ell)} \cup P_{i(\ell')}\} \cup \{x_k = 0 \mid k \in N_{i(\ell)} \cap N_{i(\ell')}\}$. However, since T is an implicant of f^d, ϕ vanishes when $x_j = 0$ for $j \in P$ and $x_k = 1$ for $k \in N$, contradicting (6.13).

For the reverse direction, let us assume indirectly that there exist two distinct indices i and i' such that

$$\phi|_{\{x_j = 1 \mid j \in P_i \cup P_{i'}\} \cup \{x_k = 0 \mid k \in N_i \cap N_{i'}\}} \not\equiv \mathbf{1}.$$

Let X be an assignment of the variables x_j, $j \notin P_i \cup P_{i'} \cup (N_i \cap N_{i'})$, at which the left-hand side vanishes, and let us define $P = \{j \mid x_j = 0\}$ and $N = \{k \mid x_k = 1\}$. Then the term T corresponding to these sets P and N is a transversal of the terms in ϕ; thus, it contains a minimal transversal. All such minimal transversals, however, must have a literal from both terms i and i', which can only be from the sets $N_i \setminus N_{i'}$ and $N_{i'} \setminus N_i$, respectively, implying that all such minimal transversals must contain at least two negative literals.   □

For general DNFs ϕ the above characterization is not computationally efficient, since (6.13) is a tautology problem (and any tautology problem can arise in this way). Actually, we have:

Theorem 6.25. *It is co-NP-complete to decide whether a given DNF ϕ represents the dual of a Horn function.*

Proof. In view of Theorem 6.24, the recognition problem is in co-NP: Indeed, to show that ϕ does not represent the dual of a Horn function, it suffices to exhibit two indices i, i' and a point X such that the left-hand side of (6.13) evaluates to 0 at X.

Moreover, NP-hardness immediately follows from Theorem 1.30 in Section 1.11: If $\mathcal{C}$ denotes the class of duals of Horn functions, then $\mathcal{C}$ does not contain all Boolean functions, the constant function $\mathbf{1}_n$ is in $\mathcal{C}$, and all restrictions of a member of $\mathcal{C}$ are in $\mathcal{C}$.   □

However, for special classes of DNFs for which tautology is tractable and remains so after fixing some of the variables, Theorem 6.24 provides a computationally efficient way of recognizing whether the dual of the input DNF is indeed Horn. This applies, for instance, when ϕ itself is a Horn DNF; see also [298] and Section 6.9.2.

We turn now to the problem of generating a DNF of the dual f^d of a Horn function f. It is clear that this problem is at least as hard as the generation of the dual of a monotone function, since monotone functions are Horn. It is not so clear, however, whether "Horn dualization" is strictly harder than "monotone dualization." Recall that a prime DNF of the dual of a monotone function can be generated incrementally efficiently (see Fredman and Khachiyan [347] and Section 4.4.2 in Chapter 4). We explain next that a similar claim can be made for Horn dualization, as well.

While it is hard to recognize whether a given conjunction is an implicant of a function expressed in DNF (see Theorem 3.7), we can show that the same problem is tractable for the dual function (see also Theorem 4.7).

Theorem 6.26. *Given a DNF ϕ of a Boolean function f and an elementary conjunction T, we can test in $O(|T| + |\phi|)$ time whether T is an implicant or a prime implicant of the dual function f^d.*

Proof. By definition, T is an implicant of f^d if $T \leq f^d$ or, equivalently, if $T^d \geq f = \phi$. The latter inequality is easy to test, by simply fixing the literals in $T \neq \mathbf{0}$ at zero, and checking whether this partial assignment makes the DNF ϕ vanish, that is, whether every term of ϕ has a common literal with T. It is also clear that T is a prime implicant of f^d if for every literal u of T, the DNF ϕ contains a term that has only u as a common literal with T. These conditions can be checked by simply reading through ϕ and maintaining a counter for all literals in T.   □

In contrast, note that checking whether f^d has no prime implicant, that is, whether $\phi^d \equiv 0$, is co-NP-complete for general DNFs. However, even this case becomes easy when ϕ is a Horn DNF (see Theorems 6.10 and 6.11). This implies that the following special variant of the DUAL RECOGNITION problem may be easier than the general case:

HORN DUAL RECOGNITION
Instance: A Horn DNF η and a disjunction ϕ of some of the prime implicants of η^d.
Output: YES if $\eta^d = \phi$, and NO otherwise.

In fact, it was observed by Khardon [564] that the quasi-polynomial algorithm introduced by Fredman and Khachiyan [347] for monotone dualization (see Chapter 4) can be straightforwardly applied in this case, too.

Theorem 6.27. *The* HORN DUAL RECOGNITION *problem can be solved in* $N^{O(\log^2 N)}$ *time, where* $N = |\eta| + |\phi|$.

Proof. Clearly, $\eta^d \neq \phi$ only if there exists a binary assignment $X \in \mathcal{B}^n$ such that $\eta(X) \vee \phi(\overline{X}) = 0$, where n denotes the number of variables in η and ϕ. By the same reasoning as in Section 4.4.2, either such a vector is easy to find or there must exist a variable appearing in the DNF $\eta(X) \vee \phi(\overline{X})$ with high frequency, in which case recursion can be applied. Correctness and complexity of this procedure can be proved as in [347] (see also Section 4.4.2). $\square$

This result shows that HORN DUAL RECOGNITION is unlikely to be NP-hard, unless all NP-hard problems can be solved in quasi-polynomial time. It is also important to note that, though essentially the "same" algorithm works for Horn dualization as for monotone dualization, it remains an open question whether these two problems, that is DUAL RECOGNITION for Horn and monotone inputs, are indeed polynomially equivalent.

Finally, Theorem 6.27 implies that the dual of a Horn function (expressed in DNF) can be generated in quasi-polynomial total time; here again, the proof follows the same arguments as in Section 4.4.2.

6.9 Special classes

In this section, we discuss several interesting special classes of Horn functions which have been considered in the literature (see, e.g., [107, 108, 296, 298, 308, 449]).

6.9.1 Submodular functions

A Boolean function $f(X)$ on $\mathcal{B}^n$ is called *submodular* if

$$f(X \vee Y) \vee f(X \wedge Y) \leq f(X) \vee f(Y) \tag{6.15}$$

for all $X, Y \in \mathcal{B}^n$. A function $f(X)$ is called *co-Horn* if $g(X) = f(\overline{X})$ is Horn. Ekin, Hammer, and Peled [308] observed the following relation between Horn, co-Horn, and submodular functions:

Theorem 6.28 ([308]). *A Boolean function is submodular if and only if it is both Horn and co-Horn. All prime implicants of a submodular function are either linear or quadratic pure Horn.*

Proof. It is easy to verify that for a function f both conditions – namely, being submodular or being simultaneously Horn and co-Horn – are equivalent to the fact that $F(f)$ is closed with respect to both componentwise conjunction and componentwise disjunction; see Corollary 6.2. $\square$

Because submodular functions are quadratic, many of their properties immediately follow from the results established in Chapter 5. We simply recall them here briefly.

Consider a submodular function f, and (since linear prime implicants do not have common variables with other prime implicants) assume for simplicity that f is purely quadratic. If ϕ is a prime DNF of f, we can associate with it a directed graph $G_\phi = (V, A)$, where $V = \{1, 2, \ldots, n\}$, and $(i, j) \in A$ if $x_i \overline{x}_j$ is a term in ϕ. It is easy to see that $x_i \overline{x}_j$ is a prime implicant of f if and only if there is a directed path from i to j in G_ϕ. Thus, the transitive closure G_f of G_ϕ corresponds to f in the sense that the quadratic prime implicants of f are in one-to-one correspondence with the arcs of G_f; see Section 5.3 (and see Appendix A for the definition of the transitive closure).

As we observed in Sections 6.7.1 and 6.7.2 (see also Section 5.8.4), the number of terms (or the number of literals) in a DNF representation of a submodular function f can be minimized in polynomial time, since it is easy to find a minimum cardinality subset of the arcs of G_f that induces the same transitive closure.

We remark next that the dual f^d of a submodular function f can also be characterized with the help of the associated directed graph $G_f = (V, A)$ (see Ekin, Hammer, and Peled [308]). We write $i \prec j$ or, equivalently, $j \succ i$, if there is a directed path from i to j in G_f. We say that two vertices i and j are *comparable* in G_f if either $i \prec j$ or $i \succ j$. A set of pairwise incomparable vertices is called an *antichain*. Let $\mathcal{I}(G_f)$ denote the family of maximal antichains of G_f. The following characterization is established in [308]:

Theorem 6.29. *Let f be a submodular function without linear prime implicants, and let G_f be the associated directed graph.*

- *If G_f is strongly connected, then*

$$f^d = \bigwedge_{j=1}^{n} x_j \vee \bigwedge_{k=1}^{n} \overline{x}_k.$$

- *If G_f is acyclic, then*

$$f^d = \bigvee_{I \in \mathcal{I}(G_f)} \; \bigwedge_{\substack{j \notin I: j \prec a \\ \text{for some } i \in I}} x_j \; \bigwedge_{\substack{k \notin I: k \succ a \\ \text{for some } i \in I}} \overline{x}_k. \qquad \square$$

In the general case, when G_f has c strong components ($c > 1$), we can write $f = f_0 \vee f_1 \vee \cdots \vee f_c$, where f_0 is the disjunction of those prime implicants that involve variables from different strong components, and f_i is the disjunction of those prime implicants that involve variables only from the ith strong component of f, for $i = 1, \ldots, c$. Then, we have $f^d = f_0^d \wedge f_1^d \wedge \cdots \wedge f_c^d$, where each of these functions can be determined by Theorem 6.29, since G_{f_0} is acyclic, and G_{f_i} is strongly connected for $i = 1, \ldots, c$.

Let us finally mention that, if ϕ is an arbitrary DNF, then it is co-NP-complete to recognize whether ϕ represents a submodular function; this follows easily from Theorem 1.30 (see also [308]).

6.9.2 Bidual Horn functions

A Boolean function f is called *bidual Horn* if both f and f^d are Horn. We mention some interesting properties of bidual Horn functions established by Eiter, Ibaraki, and Makino [298], who were the first to consider this class of functions.

As we recall from Section 6.3, a function f is Horn if and only if its set of false points $F(f)$ is closed under componentwise conjunction (see Theorem 6.6). From this fact and from the definition of the dual function, it is easy to derive that f^d is Horn if and only if its set of true points $T(f)$ is closed under componentwise disjunction.

A special case of Theorem 6.24 can be used to recognize whether a Horn DNF represents a bidual Horn:

Theorem 6.30 ([298]). *A Horn DNF η represents a bidual Horn function if and only if for any two pure Horn terms $A\overline{x}$ and $B\overline{y}$ of η with $x \neq y$, the term AB is an implicant of η.*

Proof. Let us apply Theorem 6.24 for the DNF η. If for two terms $A\overline{x}$ and $B\overline{y}$ of η we have $x = y$, then (6.13) trivially holds, since all literals in these terms are assigned value 1. If $x \neq y$, then (6.13) means that AB is an implicant of η. $\quad\square$

Since testing $AB \leq \eta$ can be done in linear time when η is Horn (see, e.g., Corollary 6.6), the above characterization provides an $O(|\eta|^2 \|\eta\|)$ algorithm to test whether a Horn DNF represents a bidual Horn function.

Unfortunately, this positive result does not extend to general DNF representations. Namely, it was shown in [298] that it is co-NP-complete to recognize whether an arbitrary DNF represents a bidual Horn function (this is again a corollary of Theorem 1.30).

Recall from Definition 3.3 in Section 3.3.2 that a prime implicant of a Boolean function f is *essential* if it is present in all prime DNF representations of f. An interesting property of bidual Horn functions is stated next.

Theorem 6.31 ([298]). *If f is bidual Horn, then all pure Horn prime implicants of f are essential.* $\quad\square$

In light of Theorem 6.16, this implies that every irredundant prime DNF of a bidual Horn function has the same number of terms; thus, minimizing the number of terms in a DNF representation of a bidual Horn function given by a Horn DNF is polynomially solvable, by Theorem 6.12.

The foregoing does not imply that all irredundant prime DNfs of a bidual Horn function f should involve the same number of literals. Still, finding a representation with the minimum number of literals can also be solved efficiently in

$O(l(m_h^2 m_p + l))$ time, where l is the number of literals in a given Horn DNF η of f, and m_h and m_p denote respectively the number of Horn and positive terms in η. Furthermore, the number of positive prime implicants of f cannot be more than $2m_h^2 + m_p(m_h + 1)$, and thus the consensus algorithm generates from η all prime implicants of f in polynomial time (see [298]).

Let us further observe that generating the dual of a bidual Horn function f represented by a Horn DNF η is not easier than dualizing a monotone DNF, since bidual DNFs include all monotone DNFs as special cases.

Finally, we remark that the existence of a bidual extension for a given partially defined Boolean function (T, F) (see Chapter 12 for definitions) can be checked in $O(n|T||F|)$ time, where n is the number of variables. Interestingly, listing all bidual extensions of (T, F) is computationally equivalent (i.e., as easy or difficult) as generating all prime implicants of the dual of a monotone DNF (see [298]). In particular, deciding whether a given partially defined Boolean function (T, F) has a unique bidual extension is equivalent to DUAL RECOGNITION (see Chapter 4), and hence can be solved in quasi-polynomial time (see [347]).

6.9.3 Double Horn functions

A Boolean function f is called *double Horn* if both f and $\overline{f}$ (the negation of f) are Horn. This class of functions was studied by Eiter, Ibaraki, and Makino [296] who provided many interesting properties and nice characterizations, some of which we recall here without proofs.

First, as follows easily from Theorem 6.6, a function f is double Horn if and only if both its set of false points $F(f)$ and its set of true points $T(f)$ are closed under componentwise conjunction.

Theorem 6.32 ([296]). *A Boolean function f on $\mathcal{B}^n$ is double Horn if and only if it can be represented by a DNF of the form*

$$\phi = \bigvee_{i \in S} \left(\bigwedge_{k=1}^{i-1} x_{j_k} \right) \overline{x}_{j_i},$$

where $S \subseteq \{1, 2, \ldots, n\}$ and $(j_1, j_2, \ldots, j_n)$ is a permutation of $\{1, 2, \ldots, n\}$. $\square$

Note that the preceding DNF is an orthogonal expression (i.e., no two of its terms can take value 1 simultaneously; see Section 1.6 and Chapter 7) which is short, since it consists of at most $n + 1$ terms, where n is the number of variables. In fact, a much stronger statement can be established:

Theorem 6.33 ([296]). *If f is a double Horn function on $n > 1$ variables, then f, $\overline{f}$, and f^d all have unique prime DNF representations, each having at most n terms and n^2 literals. Given any of these DNFs, the other ones can be obtained in $O(n^2)$ time. Furthermore, the number of nonisomorphic (up to relabeling of the variables) double Horn functions on n variables is exactly 2^{n+1}.* $\square$

It can also be shown (e.g., by Theorem 6.32) that double Horn functions are *read-once*, that is they can be represented by a Boolean expression in which every variable appears at most once (see Chapter 10).

Despite the fact that this class of functions is very "small" and quite well characterized, recognizing whether a given DNF ϕ represents a double Horn function is still co-NP-complete in view of Theorem 1.30. However, the recognition problem is polynomially solvable under appropriate conditions on the input DNFs.

Theorem 6.34 ([296]). *Let $\mathcal{F}$ be a class of formulae that is closed under restrictions (i.e., variable fixing) and for which checking $\varphi \equiv 1$ and $\varphi \equiv 0$ can both be done in $t(n, |\varphi|)$ time, where n is the number of variables and $|\varphi|$ denotes the input length of formula $\varphi \in \mathcal{F}$. Then, deciding whether $\varphi \in \mathcal{F}$ represents a double Horn function can be performed in $O(n^2 t(n, |\varphi|))$ time.* □

Thus, in particular, if f is represented by a Horn DNF η, then we can recognize in $O(n^2 \|\eta\|)$ time if f is a double Horn function.

We finally mention that the existence of a double Horn extension of a partially defined Boolean function (T, F) can be decided in polynomial $O(n(|T| + |F|))$ time (see Chapter 12 for definitions). Furthermore, DNF expressions for all such extensions can be generated with $O(n^3(|T| + |F|))$ delay (namely, DNF expressions $\phi_1, \phi_2, \ldots$, can be produced so that the computing time between successive outputs ϕ_i and ϕ_{i+1} is never more than $O(n^3(|T| + |F|)))$. In particular, deciding if a given partially defined Boolean function has a unique double Horn extension can be done in polynomial time. Unfortunately, the number of double Horn extensions of a given partially defined Boolean function (T, F) can be exponential in terms of n, $|T|$, and $|F|$, and finding a "shortest" double Horn extension is NP-hard. We refer the reader to [296] for details.

6.9.4 Acyclic Horn functions

Graph-based special classes generalizing some subclasses of Horn formulae (see [20, 383, 711]) were introduced by Hammer and Kogan [448, 449]. We present here a few interesting properties of one of these classes.

Given a pure Horn DNF η, let us associate to it a directed graph $G_\eta = (V, A_\eta)$, where $V = \{1, 2, \ldots, n\}$ is the set of indices of the variables, and $(i, j) \in A_\eta$ if η has a term involving both x_i and $\overline{x}_j$. Analogously, if h is a pure Horn function, let us associate to it a directed graph $G_h = (V, A_h)$ by including an arc $(i, j) \in A_h$ if h has a prime implicant involving both x_i and $\overline{x}_j$. We call G_h the *implicant graph* of h.

Clearly, if η is a prime DNF of h, then G_η is a subgraph of G_h. A very useful property of these graphs is formulated in the following statement:

Theorem 6.35 ([448]). *If η is a prime DNF representing the pure Horn function h, and if $Ax_i\overline{x}_j$ is a prime implicant of h, then G_η has a directed path from vertex i to vertex j. In other words, G_h is a subgraph of the transitive closure of G_η.*

Proof. See Exercises 29 and 30.   □

A pure Horn function h is called *acyclic* if G_h is an acyclic directed graph. In view of Theorem 6.35, it follows that h is acyclic if and only if G_η is acyclic for an arbitrary prime DNF η of h.

Recall again from Definition 3.3 in Section 3.3.2 that a prime implicant of a Boolean function f is called *essential* if it is present in all prime DNF representations of f, and *redundant* if no irredundant prime DNF of f includes it.

Theorem 6.36 ([448]). *If h is an acyclic pure Horn function, then every prime implicant of h is either essential or redundant.*   □

This remarkable property of acyclic Horn functions implies that they have a unique irredundant prime DNF representation. Thus, in light of the preceding results and of Theorem 6.12, we can check whether a given pure Horn DNF η is acyclic, and if yes, we can find the unique irredundant prime DNF representing the same acyclic Horn function in $O(\|\eta\|^2)$ time (where, actually, the majority of the time will be spent on transforming η into an irredundant prime DNF). Clearly, the unique irredundant prime DNF of an acyclic Horn function minimizes all usual measures of complexity (see Chapter 3 and Section 6.7).

Further properties and generalizations of acyclic functions based on the structure of associated graphs can be found in [108, 173, 448, 449].

6.10 Generalizations

6.10.1 Renamable Horn expressions and functions

In most applications of Boolean functions, the meaning of a variable and its negation are interchangeable, since a particular variable x could equally well denote the truth value of a logical proposition or its negation. Thus, it is natural to consider logical expressions obtained from a given expression after replacing some of the variables by their negations. More formally, given a DNF

$$\phi = \bigvee_{i=1}^{m} \left(\bigwedge_{j \in P_i} x_j \bigwedge_{k \in N_i} \overline{x}_k \right) \tag{6.16}$$

and a subset $S \subseteq \{1, 2, \ldots, n\}$, we say that the DNF ϕ^S is obtained from ϕ by *switching* (or *renaming*; see also Chapter 5) the variables in the subset S if

$$\phi^S = \bigvee_{i=1}^{m} \left(\bigwedge_{j \in (P_i \setminus S) \cup (N_i \cap S)} x_j \bigwedge_{k \in (N_i \setminus S) \cup (P_i \cap S)} \overline{x}_k \right).$$

We say that the DNF ϕ is *renamable Horn* if ϕ^S is a Horn DNF for some subset S of the variables (as before, we do not distinguish between sets of variables and sets of indices whenever this does not cause any confusion).

The problem of recognizing whether a given DNF ϕ is renamable Horn was considered first by Lewis [612], who provided an elegant proof showing that this problem is polynomially solvable, namely, that it can be reduced to a quadratic Boolean equation.

Theorem 6.37 ([612]). *Let ϕ be a DNF given as in (6.16) and let $S \subseteq \{1, 2, \ldots, n\}$. Then ϕ^S is Horn if and only if the following implications hold for every $i = 1, \ldots, m$:*

$$P_i \cap S \neq \emptyset \implies N_i \subseteq S \quad and \quad |P_i \cap S| = 1,$$
$$N_i \setminus S \neq \emptyset \implies |N_i \setminus S| = 1 \quad and \quad P_i \cap S = \emptyset.$$

Proof. If some of the above implications were not valid for the ith term of ϕ, then after switching the variables in S, this term would have more than one negated variables. On the other hand, if all of the above implications hold for term i, then it will have at most one negated variable after switching. $\qquad\qquad\square$

Introducing the binary characteristic vector $Y^S = (y_1, \ldots, y_n)$, where $y_j = 1$ if and only if $j \in S$, we can rewrite the foregoing implications as a single quadratic Boolean condition

$$\bigvee_{i=1}^{m} \left(\left(\bigvee_{\substack{j_1 \in P_i \\ j_2 \in N_i}} y_{j_1} \overline{y}_{j_2} \right) \vee \left(\bigvee_{\substack{j_1 \in P_i \\ j_2 \in P_i \setminus \{j_1\}}} y_{j_1} y_{j_2} \right) \vee \left(\bigvee_{\substack{j_1 \in N_i \\ j_2 \in N_i \setminus \{j_1\}}} \overline{y}_{j_1} \overline{y}_{j_2} \right) \right) = 0 \quad (6.17)$$

If S is not given, then Y^S can be viewed as a vector of unknowns, and the condition for Horn renamability translates into the quadratic Boolean equation (6.17), involving n variables and $\sum_{i=1}^{m} \binom{|P_i \cup N_i|}{2}$ quadratic terms. This equation can be solved in $O(n^2)$ time (see Chapter 5), and the reduction provides a quadratic-time recognition algorithm for renamable Horn DNFs.

It was observed by Aspvall [33] that, by using some auxiliary variables, an equivalent quadratic system can be constructed that involves only $O(|\phi|)$ terms, thus providing the first linear-time recognition algorithm for renamable Horn DNFs. Further linear-time recognition algorithms were proposed by Chandru et al. [181], Mannila and Mehlhorn [662], and Sykora [853]. A linear-time recognition algorithm for a more general class of expressions was presented by Boros, Hammer, and Sun [135]; as a special case, this algorithm also detects in linear time if a given DNF is Horn renamable (see Section 6.10.2 for more details). We further add that recognizing whether a given DNF has a unique Horn renaming can also be detected in linear time (see Hebrard [480]), and that an iterative Horn renaming-based algorithm was presented by Boros, Čepek, and Kogan [108] to find a short DNF representation of a given Horn function.

In some applications, most notably when solving Boolean DNF equations, it may be advantageous to have many Horn terms in the input DNF (see, e.g.,

Section 2.5.2). The problem of switching a subset of variables so as to maximize the number of Horn terms of a given DNF was considered by several authors. We can observe, for instance, that for a cubic DNF, at least half of its terms can always be switched to Horn (see Exercise 33 at the end of this chapter). Chandru and Hooker [183] showed that finding the maximum number of terms of a given DNF that can be switched simultaneously to Horn is an NP-hard optimization problem, and Crama, Ekin, and Hammer [229] observed that it remains NP-hard even for quadratic DNFs. In Boros [106], a simple polynomial time approximation algorithm is presented for this hard optimization problem, guaranteeing that at least $\frac{40}{67}$ of the maximum possible number of terms can be renamed to Horn in polynomial time. It was shown by Zwick [941] that guaranteeing more than $\frac{2}{3}$ for cubic Horn DNFs is not possible, unless P=NP, and that $\frac{2}{3}$ is achievable by a semidefinite programming-based approximation algorithm.

In this section, so far, we have focused on the renamability of (DNF) *expressions*. We should note, however, that variable switching is not only an operation on expressions, but also defines a mapping (a bijection) on the set of Boolean *functions*. Namely, for a subset $S \subseteq \{1, 2, \ldots, n\}$, a binary point $X = (x_1, x_2, \ldots, x_n) \in \mathcal{B}^n$ and a Boolean function f, let us define the point $X[S]$ by

$$x_j[S] = \begin{cases} x_j & \text{if } j \notin S, \\ \overline{x}_j & \text{if } j \in S, \end{cases}$$

and $f^S(X) = f(X[S])$. Clearly, $X \longleftrightarrow X[S]$ is a bijection over $\mathcal{B}^n$, and thus $f \longleftrightarrow f^S$ is an induced bijection over the set of Boolean functions on $\mathcal{B}^n$. Accordingly, we say that a Boolean function f is *renamable Horn* if f^S is a Horn function for some subset S.

Note that even Horn functions (which clearly form a subfamily of renamable Horn functions) may have DNF representations that cannot be renamed to Horn.

Example 6.13. *The (monotone) Horn function h defined by the DNF*

$$\eta = x_1 \lor x_2 \lor x_3,$$

can also be represented by the irredundant DNF

$$\phi = x_1 x_2 \lor x_1 x_3 \lor x_2 x_3 \lor \overline{x}_1 \overline{x}_2 x_3 \lor \overline{x}_1 x_2 \overline{x}_3 \lor x_1 \overline{x}_2 \overline{x}_3,$$

which is not Horn renamable.   □

In fact, Theorem 1.30 implies that it is NP-hard to recognize whether an arbitrary DNF represents a Horn-renamable function.

However, if f^S is Horn, then the same switching set S turns all the prime implicants of f into Horn terms, and thus any DNF $\phi \subseteq \mathcal{P}_f^c$ representing f is also Horn renamable (where $\mathcal{P}_f^c$ denotes, as usual, the consensus closure of the prime implicants of f).

6.10.2 Q-Horn functions

A further generalization, the family of so-called *Q-Horn* functions, was introduced by Boros, Crama, and Hammer [112]. This class includes Horn and renamable Horn as well as quadratic functions.

With a DNF ϕ, given as in (6.16), let us associate a polyhedron $P_\phi \subseteq \mathbb{R}^n$, defined by

$$P_\phi = \left\{ \alpha \in \mathbb{R}^n \; \middle| \; \begin{array}{ll} \sum_{j \in P_i} \alpha_j + \sum_{k \in N_i} (1 - \alpha_k) \le 1 & \text{for } i = 1, \dots, m \\ 0 \le \alpha_j \le 1 & \text{for } j = 1, \dots, n \end{array} \right\}. \tag{6.18}$$

We say that ϕ is a *Q-Horn DNF* if $P_\phi \ne \emptyset$. It is easy to see that

- $\alpha = (0, 0, \dots, 0) \in P_\phi$ whenever ϕ is Horn;
- $\alpha = X^S \in P_\phi$ whenever ϕ can be turned into a Horn formula by switching the variables in S; and
- $\alpha = (\frac{1}{2}, \frac{1}{2}, \dots, \frac{1}{2}) \in P_\phi$ whenever ϕ is a quadratic DNF.

Example 6.14. *The following DNF*

$$\phi = x_1 x_2 x_3 \vee x_1 \overline{x}_2 x_4 \vee \overline{x}_1 x_2 \overline{x}_5 \vee \overline{x}_1 \overline{x}_2 x_6 \vee x_3 \overline{x}_4 \overline{x}_5 \vee x_3 x_6 \vee x_4 \overline{x}_5$$

is Q-Horn, since $(\frac{1}{2}, \frac{1}{2}, 0, 0, 1, 1) \in P_\phi$ in this case. In fact $P_\phi = \{(\frac{1}{2}, \frac{1}{2}, 0, 0, 1, 1)\}$, and thus, this DNF is neither quadratic, nor Horn, nor renamable Horn. □

Definition 6.15. *Given a real vector $\alpha \in \mathbb{R}^n$, let us define $[\alpha] \in \mathbb{R}^n$ by*

$$[\alpha]_j = \begin{cases} 1 & \text{if } \alpha_j > \frac{1}{2}, \\ \frac{1}{2} & \text{if } \alpha_j = \frac{1}{2}, \\ 0 & \text{if } \alpha_j < \frac{1}{2}. \end{cases}$$

Furthermore, let $H(\alpha) = \{ j \mid \alpha_j = \frac{1}{2} \}$.

Lemma 6.11 ([112]). *If $\alpha \in P_\phi$, then $[\alpha] \in P_\phi$, and hence, $P_\phi \ne \emptyset$ if and only if $P_\phi \cap \{0, \frac{1}{2}, 1\}^n \ne \emptyset$. Furthermore, if $P_\phi \ne \emptyset$, then there exists a unique minimal subset $H = H_\phi$ such that $H = H(\alpha)$ for some $\alpha \in P_\phi \cap \{0, \frac{1}{2}, 1\}^n$ and $H \subseteq H(\beta)$ for all $\beta \in P_\phi$.*

Proof. Let us first note that if $0 \le r \le 1$, then $[1 - r] = 1 - [r]$ by Definition 6.15, thus $\sum_{j \in P} [\alpha_j] + \sum_{j \in N} (1 - [\alpha_j]) = \sum_{j \in P} [\alpha_j] + \sum_{j \in N} [1 - \alpha_j]$. Observe next that, if the sum of some nonnegative reals is not larger than 1, then at most one of these numbers is larger than $\frac{1}{2}$ (and then all others are smaller than $\frac{1}{2}$), or at most two of them are equal to $\frac{1}{2}$ (and then all others are equal to 0). Thus $\sum_{j \in P} \alpha_j + \sum_{j \in N} (1 - \alpha_j) \le 1$ implies $\sum_{j \in P} [\alpha_j] + \sum_{j \in N} [1 - \alpha_j] \le 1$, proving that if $\alpha \in P_\phi$, then $[\alpha] \in P_\phi$, too.

For the second half of the lemma, let us observe that, if $\alpha, \beta \in P_\phi$, then for almost all reals $0 < \lambda < 1$ (except finitely many values), we have $H(\lambda \alpha + (1 - \lambda)\beta) =$

$H(\alpha) \cap H(\beta)$. Since there are only finitely many different subsets $H \subseteq \{1, 2, \ldots, n\}$, it follows that there exists a vector $\alpha \in P_\phi$ such that $H(\alpha) \subseteq H(\beta)$ for all $\beta \in P_\phi$. Thus the lemma follows by $H(\alpha) = H([\alpha])$. $\qquad\square$

Lemma 6.11 implies easily that if ϕ is a Q-Horn DNF, and $\alpha \in P_\phi$, then $\alpha \in P_{\phi^c}$, where ϕ^c denotes the DNF formed by the disjunction of all terms obtainable from ϕ by consensuses (i.e., ϕ^c is the consensus closure of ϕ; see Section 6.6).

Thus, we can define *Q-Horn functions* as those Boolean functions whose complete DNF (the disjunction of all their prime implicants) is Q-Horn. The family of Q-Horn functions properly includes all quadratic, Horn, and renamable Horn functions.

Using linear programming, we can recognize efficiently whether a given DNF ϕ is Q-Horn or not; moreover, a half-integral vector in $P_\phi \cap \{0, \frac{1}{2}, 1\}^n$ can be found in polynomial time. A linear time recognition algorithm was given by Boros, Hammer, and Sun [135].

It was shown in [112] that the Boolean equation $\phi = 0$ can be solved in linear time for a Q-Horn DNF ϕ whenever a vector $\alpha \in P_\phi \cap \{0, \frac{1}{2}, 1\}^n$ is known. More precisely, we can find $\alpha \in P_\phi \cap \{0, \frac{1}{2}, 1\}^n$, for which $H(\alpha) = H_\phi$, whenever $P_\phi \neq \emptyset$ or recognize that $P_\phi = \emptyset$, in $O(\|\phi\|)$ time. Since, in particular, ϕ is renamable Horn if and only if $H_\phi = \emptyset$, the same algorithm also recognizes in linear time whether or not a given DNF is renamable Horn.

Consequently, Q-Horn equations can be solved in linear time. More precisely, for every Boolean equation $\phi = 0$, we can either recognize that ϕ is not Q-Horn or solve the equation in linear time.

If we associate a $(0, \pm 1)$ matrix with a given DNF, as in Section 6.2.4, then the family of DNFs for which the corresponding $(0, \pm 1)$ matrix has a so-called *monotone decomposition*, as introduced by Truemper [871], includes Q-Horn DNFs. A linear time algorithm to find a monotone decomposition of a given $(0, \pm 1)$ matrix is also presented in [871].

Finally, by relaxing the definition of Q-Horn DNFs, we can introduce a useful index associated with a DNF ϕ, which is related to the difficulty of solving the Boolean equation $\phi = 0$. For a DNF ϕ defined by (6.16), we define the index $z(\phi)$ as the optimal value of the linear programming problem

$$z(\phi) \quad = \min z$$

$$\text{s.t.} \quad z \geq \sum_{j \in P_i} \alpha_j + \sum_{k \in N_i} (1 - \alpha_k) \quad \text{for } i = 1, \ldots, m,$$

$$0 \leq \alpha_j \leq 1 \quad\quad\quad\quad\quad \text{for } j = 1, \ldots, n.$$

Clearly, ϕ is Q-Horn if and only if $z(\phi) \leq 1$. Boros et al. [116] showed that if $z(\phi) \leq 1 + (c \log n)/n$, then the Boolean equation $\phi = 0$ can be solved in $O(n^c)$ time. On the other hand, the tautology problem remains NP-complete for any fixed $\epsilon < 1$ when restricted to instances for which $z(\phi) \leq 1 + n^{-\epsilon}$.

6.10.3 Extended Horn expressions

Another generalization of Horn formulae was introduced by Chandru and Hooker [182]. The motivation behind this generalization is the integer programming rounding result by Chandrasekaran [180] mentioned in Section 6.2.4, and the possibility of using linear programming to solve Boolean equations (see Section 2.8).

For a formal definition, let us consider an arborescence T rooted at vertex r (i.e., a directed tree with all arcs oriented away from the root) that has n arcs, labeled by $\{1, 2, \ldots, n\}$. We say that a term $\bigwedge_{j \in P} x_j \bigwedge_{k \in N} \overline{x}_k$ is *extended Horn with respect to* T if the set N is a directed path of T, and if the set P is a union of directed paths in T with the property that either (i) all paths in P start at the root or (ii) one of them starts where N starts, and all others start at the root. The same term is called *simple extended Horn with respect to* T if (ii) does not occur. Accordingly, a DNF ϕ is called *(simple) extended Horn* if all of its terms are (simple) extended Horn with respect to the same arborescence T.

Theorem 6.38 ([182]). *If ϕ is extended Horn, then the Boolean equation $\phi = 0$ has a solution if and only if the polyhedron*

$$Q_\phi = \left\{ X \in \mathbb{R}^n \;\middle|\; \begin{array}{ll} \sum_{j \in P_i} x_j + \sum_{k \in N_i} (1 - x_k) \leq |P_i \cup N_i| - 1 & \text{for } i = 1, \ldots, m \\ 0 \leq x_j \leq 1 & \text{for } j = 1, \ldots, n \end{array} \right\}$$

is not empty. Furthermore, repeated application of the unit literal rule allows us to detect whether $\phi \equiv \mathbf{1}$.   □

Note that by Theorem 2.10, an arbitrary DNF equation $\phi = 0$ has a solution if and only if the polyhedron Q_ϕ contains an integral point. The strength of the preceding statement is that for an extended Horn DNF ϕ, the integrality requirement can be disregarded, and hence, the consistency question can be decided in polynomial time by linear programming.

It was also shown by Schlipf et al. [809] that extended Horn equations (and many others, including renamable extended Horn equations) can be solved by the *single look-ahead unit literal rule*. In this algorithm, variables are assigned binary values one-by-one, and the unit literal rule is applied right after each assignment has been made. If a contradiction is found, then the last assignment is reversed; otherwise, the last assignment is accepted permanently.

The recognition of extended Horn DNFs is strongly related to the so-called *arborescence realization* problem (given a hypergraph $\mathcal{H}$ on a base set E, find an arborescence T with arc set E such that all hyperedges of $\mathcal{H}$ are directed paths in T), and, in fact, a polynomial time recognition algorithm for simple extended Horn DNFs was derived via arborescence realization by Swaminathan and Wagner [852]. This was later improved to a linear time algorithm by Benoist and Hebrard [59]. The problem of recognizing extended Horn DNFs is still open.

6.10.4 Polynomial hierarchies built on Horn expressions

A polynomial hierarchy of DNFs is a sequence of families of DNFs

$$\mathcal{D}_0 \subset \mathcal{D}_1 \subset \cdots \subset \mathcal{D}_k \subset \cdots$$

such that (i) the membership $\phi \in \mathcal{D}_k$ can be tested in time polynomial in $|\phi|^k$; (ii) if k is a fixed constant and $\phi \in \mathcal{D}_k$, then the Boolean equation $\phi = 0$ can be solved in polynomial time; and (iii) for every DNF ϕ, $\phi \in \mathcal{D}_k$ for some integer k.

Several such hierarchies were considered in the literature (see, e.g., [174, 253, 362, 756]), most of them built on Horn expressions or on some of their generalizations. To describe these, we need to introduce a few more notations.

With a DNF ϕ given by (6.16), let us associate the hypergraph $\mathcal{N}(\phi) = \{N_i \mid i = 1,\ldots,m\}$ consisting of the index sets of the negated variables of the terms of ϕ. Note that $\mathcal{N}(\phi)$ may not be a clutter; for example, it contains the empty set whenever ϕ includes a positive term. For an index j, consider two operations, defined by $\mathcal{N} \setminus \{j\} = \mathcal{N} \setminus \{N \in \mathcal{N} \mid N \ni j\}$ and $\mathcal{N} \div \{j\} = \{N \setminus \{j\} \mid N \in \mathcal{N}\}$, respectively, called the *deletion* and the *contraction* of element j (note the slight difference with the similar terminology introduced in Section 1.13.5).

One of the earliest polynomial hierarchies $\Gamma_0 \subset \Gamma_1 \subset \cdots \subset \Gamma_k \subset \cdots$, where Γ_0 is the family of Horn expressions, was proposed by Gallo and Scutellà [362]. To describe this hierarchy, first we need to define a hierarchy of hypergraphs $\Sigma_0 \subset \Sigma_1 \subset \cdots$ by

- $\mathcal{N} \in \Sigma_0$ if $|N| \leq 1$ for all $N \in \mathcal{N}$; and
- for $k > 0$, $\mathcal{N} \in \Sigma_k$ if there exists an index j such that $\mathcal{N} \setminus \{j\} \in \Sigma_{k-1}$ and $\mathcal{N} \div \{j\} \in \Sigma_k$.

Note that class Σ_k for $k > 0$ is initialized by the condition $\Sigma_{k-1} \subset \Sigma_k$. Then, classes of DNFs Γ_k, $k = 0,1,\ldots$ are defined by $\phi \in \Gamma_k$ if and only if $\mathcal{N}(\phi) \in \Sigma_k$.

Clearly, Γ_0 is the family of Horn DNFs. The class Γ_1 is the family of so-called *generalized Horn* DNFs, introduced earlier by Yamasaki and Doshita [931]. It was shown in [362] that the membership $\phi \in \Gamma_k$ can be tested in $O(|\phi|n^k)$ time. Furthermore, the membership algorithm in [362] provides the index j appearing in the recursive definition of Σ_k. When k is a fixed constant, a polynomial time algorithm to solve the Boolean equation $\phi = 0$, with $\phi \in \Sigma_k$, follows easily from these results. Indeed, branching on the j-th variable results in two subproblems, one from Γ_{k-1} and one from Γ_k, both having one variable less than the original problem. (The same results were obtained by [931] when $k = 1$.)

The previous hierarchy was somewhat improved by Dalal and Etherington [253], so that both Horn and quadratic formulae could be included at the lowest level of the hierarchy. Furthermore, it was shown by Kleine Büning [570] that, to prove $\phi \equiv \mathbf{1}$ for a DNF $\phi \in \Gamma_k$, it is enough to use a restricted version of the consensus algorithm in which the consensus of two terms is computed only if at least one of the terms is of degree at most k.

Pretolani [756] observed that many other classes of DNFs could be used in place of Γ_0, resulting in a similar polynomial hierarchy. Unfortunately, renamable extensions of otherwise simple classes may not always be included at low levels of such hierarchies. For instance, Eiter, Kilpelainen, and Mannila [301] showed that recognizing renamable generalized Horn DNFs is an NP-complete problem.

Recent work of Čepek and Kučera [174] provides a quite general framework for more general polynomial hierarchies. Let $\mathcal{D}_0$ be a class of DNFs, and

- for $k > 0$, let $\phi \in \mathcal{D}_k$ if and only if there exists a literal u of ϕ such that $\phi|_{u=0} \in \mathcal{D}_{k-1}$ and $\phi|_{u=1} \in \mathcal{D}_k$.

Theorem 6.39 ([174]). *If $\mathcal{D}_0$ is a nontrivial class that is closed under (i) switching a subset of the variables and (ii) fixing a subset of the variables at binary values, and if the Boolean equation $\phi = 0$ for $\phi \in \mathcal{D}_0$ can be solved in polynomial $p(|\phi|)$ time, then the classes $\mathcal{D}_0 \subset \mathcal{D}_1 \subset \cdots$ define a polynomial hierarchy. In particular, for a DNF ϕ, membership in $\mathcal{D}_k$ can be tested in $O(p(|\phi|n^{k+1})$ time, and if $\phi \in \mathcal{D}_k$, then the Boolean equation $\phi = 0$ can also be solved in $O(|\phi|n^{k+1})$ time.* □

For example, the class $\mathcal{D}_0$ can be chosen to be the family of renamable Horn DNFs or the family of Q-Horn DNFs, and so on, with each choice resulting in a different polynomial hierarchy.

6.11 Exercises

1. Let f and g denote arbitrary Horn functions. Decide whether the following claims are true or false:
   - $f \vee g$ is Horn.
   - $f \wedge g$ is Horn.
   - $\overline{f}$ is Horn.

2. Find a Boolean function, in n variables for which the number of minimal Horn majorants is exponential in n.

3. Find a Horn function in n variables for which the number of prime implicants is polynomial in n, but the number of different Horn DNF representations is exponential in n.

4. Let f be a Boolean function, and let $P_i, i = 1, \ldots, m$, be its Horn prime implicants. Prove that $\eta(X) = \bigvee_{i=1}^{m} P_i(X)$ is the unique maximal Horn minorant of f. Does this claim remain true if $P_i, i = 1, \ldots, m$ are the Horn terms of an arbitrary DNF of f?

5. Let

$$f = \bigvee_{i=1}^{n} \bigvee_{P \in \mathcal{P}_i} P\overline{x}_i$$

and

$$g = \bigvee_{i=1}^{n} \bigvee_{Q \in \mathcal{Q}_i} Q\overline{x}_i$$

be the complete DNFs of two pure Horn functions. We then define

$$f \otimes g = \bigvee_{i=1}^{n} \bigvee_{\substack{P \in \mathcal{P}_i \\ Q \in \mathcal{Q}_i}} P Q\overline{x}_i.$$

- Prove that the family of pure Horn functions with the operations $\otimes$ and $\vee$ form a lattice.
- Prove that $f \otimes g$ is the unique largest Horn minorant of $f \wedge g$.
- Can you generalize this for the family of Horn functions?

6. Prove that for a nonempty subset $S \subseteq \mathcal{B}^n$ and for the characteristic models $Q(S)$ of this set (see Corollary 6.4), we have

$$Q(S) = \{X \in S \mid X \notin (S \setminus \{X\})^\wedge\}. \tag{6.19}$$

7. Let h be a Horn function in n variables, and let m^* and l^* denote, respectively, the numbers of terms and literals in a DNF representation of h^d. Prove that the following inequality holds:

$$|Q(F(h))| \leq m^*(n+1) - l^*. \tag{6.20}$$

8. Construct examples of Horn functions h for which there is an exponential gap in inequality (6.20).

9. Find examples of Horn DNFs η such that $Q(F(\eta))$ and η^d are simultaneously exponentially larger than η.

10. Find examples of Horn functions h for which any DNF representation of both functions h and h^d are exponentially larger than the cardinality $|Q(F(h))|$.

11. Given $X, Y, A \in \mathcal{B}^n$, let us write $X \geq_A Y$ if $x_i \oplus a_i \geq y_i \oplus a_i$ for all $i = 1, \ldots, n$, where $\oplus$ denotes the modulo 2 addition. For a subset $S \subseteq \mathcal{B}^n$ let

$$S^A = \{X \mid X \geq_A Y \text{ for some } Y \in S\}$$

denote the *A-monotone closure* of S.

- Prove that, for every subset $S \subseteq \mathcal{B}^n$,

$$S = \bigcap_{A \in \mathcal{B}^n} S^A.$$

- Let $\mathcal{A}$ denote the set of those $n+1$ binary vectors from $\mathcal{B}^n$ that contain at least $n - 1$ ones. Prove that, for every Horn function h, we have

$$F(h) = \bigcap_{A \in \mathcal{A}} F(h)^A.$$

(See more in [161].)

12. Given a DNF

$$\phi = \bigvee_{i=1}^{m} \left(\bigwedge_{j \in P_i} x_j \right) \wedge \left(\bigwedge_{j \in N_i} \overline{x}_j \right)$$

in n variables, let us call a mapping $\sigma : [m] \longrightarrow [n]$ a *selector* if $\sigma(i) \in N_i$ whenever $N_i \neq \emptyset$. With ϕ and the selector σ, let us associate a DNF ϕ_σ defined by

$$\phi_\sigma = \left(\bigvee_{i:N_i=\emptyset} \bigwedge_{j \in P_i} x_j \right) \vee \left(\bigvee_{i:N_i\neq\emptyset} \left(\bigwedge_{j \in P_i} x_j \right) \wedge \overline{x}_{\sigma(i)} \right).$$

- Prove that, ϕ_σ is a Horn majorant of ϕ, for every selector σ.
- Prove that, for every Horn majorant η of ϕ, there exists a selector σ such that $\phi \leq \phi_\sigma \leq \eta$.

13. Let h_i, $i = 1,\ldots,N$, be the set of minimal Horn majorants of the Boolean function f. Prove that $f = \bigwedge_{i=1}^{N} h_i$.

14. Which Boolean functions have a unique minimal Horn majorant?

15. Can you characterize those Boolean functions that have exactly two minimal Horn majorants?

16. Let ϕ be a DNF, and let η be a Horn DNF. How difficult is it to decide whether or not $\eta \leq \phi$ holds? What is the complexity of this problem if we assume that ϕ contains all prime implicants of the Boolean function it represents?

17. Let η be a Horn DNF representing the unique maximal Horn minorant of the DNF ϕ. Prove that deciding the consistency of the Boolean equations $\eta = 0$ and $\phi = 0$ are computationally equivalent problems. What is the complexity of finding the maximal Horn minorant of a DNF?

18. Given Horn DNFs η_j, $j = 1,\ldots,k$, what is the complexity of finding the maximal Horn minorant of $\eta_1 \wedge \eta_2 \wedge \cdots \wedge \eta_k$?

19. Let η be a Horn DNF representing a minimal Horn majorant of the DNF ϕ. Prove that deciding the consistency of the Boolean equations $\eta = 0$ and $\phi = 0$ are computationally equivalent problems. What is the complexity of finding a minimal Horn majorant of a DNF?

20. Given a pure Horn function h in variables $V = \{x_1, x_2,\ldots,x_n\}$, find a minimal subset $S \subseteq V$ for which $S^h = V$, that is, for which the forward chaining closure of S includes all variables. How difficult is this problem? Is such a minimal subset unique?

21. Prove that for two Horn functions h and h', we have $S^h = S^{h'}$ for every subset S of variables if and only if $h = h'$.

22. Given a Horn function h of n variables, let us denote by $h^{(k)}$ the disjunction of those prime implicants of h having degree at most k. Note that $h^{(1)} \leq h^{(2)} \leq \cdots \leq h^{(n)} = h$.

- Is it true that $h^{(1)}$ has a DNF representation not longer than the shortest DNF of h?

- Construct a Horn DNF η representing the Horn function h such that, for every DNF representation $\eta^{(2)}$ of $h^{(2)}$, we have $|\eta^{(2)}| > |\eta|$ (cf. [172]).

23. Let us call a consensus k-*restricted* if at least one of the terms involved in the consensus has degree at most k.

   - Prove that all linear prime implicants of a pure Horn DNF η can be obtained by a sequence of 1-restricted consensuses.
   - Generalize this statement for any $k \geq 2$ (see [173]).

24. Consider a Horn function h given by a prime DNF η, and let T be an implicant of h. How difficult is it to decide whether T can be derived from the prime implicants of h by a sequence of consensuses? How many prime implicants of h are needed for such a consensus derivation of T when it exists?

25. Let T and $Q \subseteq T$ be two sets of terms, both closed under consensus, and let $(\mathcal{R}_1, \mathcal{D}_1)$ and $(\mathcal{R}_2, \mathcal{D}_2)$ be two RD-partitions of T.

   - Prove that $(\mathcal{R}_3, \mathcal{D}_3)$ is also an RD-partition of T if $\mathcal{R}_3 = \mathcal{R}_1 \cap \mathcal{R}_2$ and $\mathcal{D}_3 = \mathcal{D}_1 \cup \mathcal{D}_2$.
   - Prove that $(\mathcal{R}_4, \mathcal{D}_4)$ is an RD-partition of Q if $\mathcal{R}_4 = \mathcal{R}_1 \cap Q$ and $\mathcal{D}_4 = \mathcal{D}_1 \cap Q$.

26. Prove that the partition in Example 6.7 is an RD-partition.

27. Prove that, if h is a Horn function, then each of the following defines an RD-partition of $\mathcal{P}_h^c$:

   (a) $\mathcal{R} = \{T \in \mathcal{P}_h^c \mid |T| \geq 2\}$ and $\mathcal{D} = \mathcal{P}_h^c \setminus \mathcal{R}$.
   (b) $\mathcal{R} = \{T \in \mathcal{P}_h^c \mid |T| \leq 2\}$ and $\mathcal{D} = \mathcal{P}_h^c \setminus \mathcal{R}$.
   (c) $\mathcal{R} = \{T \in \mathcal{P}_h^c \mid T(X) = 0\}$ and $\mathcal{D} = \mathcal{P}_h^c \setminus \mathcal{R}$, if $X \in \mathcal{B}^n$ is a point at which every prime implicant of h contains at most one literal that evaluates to zero. (How easy is it to check for the existence of such a binary vector $X \in \mathcal{B}^n$?)
   (d) $\mathcal{R} = \{T \in \mathcal{P}_h^c \mid$ all variables of T belong to $S\}$ and $\mathcal{D} = \mathcal{P}_h^c \setminus \mathcal{R}$, where S is a subset of the variables that is closed under forward chaining, namely, $S^h = S$ (see Section 6.4).

28. Prove that the minimum number of positive terms in a Horn DNF of a Horn function h is always at most 1. For which Horn functions is it 0? How difficult is to find such an "optimal" Horn DNF, having the minimum number of positive terms?

29. Consider a pure Horn DNF η of a pure Horn function h, and the associated directed graph $G_\eta = (V, A_\eta)$ defined in Section 6.9.4. Prove that if $A\overline{y}$ is an implicant of h (not necessarily present in η) and $x \in A$, then there is a directed path from x to y in G_η.

30. Consider two prime DNFs η_1 and η_2 of the pure Horn function h. Prove that the transitive closures of the directed graphs G_{η_1} and G_{η_2} are the same, and that they coincide with the transitive closure of G_h (see Appendix A for definitions).

31. Let us consider a pure Horn function h, the associated transitively closed directed graph $G_h = (V, A_h)$, as defined in the previous exercise, and let us assume that $S \subseteq V$ is an initial set of the vertices (namely, there is no arc (x, y) with $x \in V \setminus S$ and $y \in S$). Define

$$\mathcal{R} = \{T \mid T \in \mathcal{P}_h^c, \text{ the head of } T \text{ belongs to } S\}.$$

Prove that $\mathcal{R}$ and $\mathcal{D} = \mathcal{P}_h^c \setminus \mathcal{R}$ form an RD-partition of $\mathcal{P}_h^c$.

32. Consider a transitively closed directed graph $D = (V, A)$, and let $\mathcal{H}_D$ denote the set of those pure Horn functions h for which $D = G_h$. Prove that if $h, h' \in \mathcal{H}_D$, then both $h \vee h'$ and $h \otimes h'$ (as defined in Exercise 5) belong to $\mathcal{H}_D$. Prove also that $\mathcal{H}_D$ contains a unique minimal function and a unique maximal function. Can you write a DNF of these unique minimal and maximal members of $\mathcal{H}_D$?

33. Let ϕ be a DNF of m terms, as given in (6.16). Prove that at least $\lceil \sum_{i=1}^{m} \frac{|P_i \cup N_i| + 1}{2^{|P_i \cup N_i|}} \rceil$ of its terms can be switched to Horn by renaming some of its variables. Can you give a polynomial time algorithm to accomplish this?

34. Prove that the lower bound in the previous exercise is tight (cf. [585]).

7

Orthogonal forms and shellability

The concept of *orthogonal disjunctive normal form* (or ODNF, sometimes called sum of disjoint products) was introduced in Chapter 1. Orthogonal forms are a classic object of investigation in the theory of Boolean functions, where they were originally introduced in connection with the solution of Boolean equations (see Kuntzmann [589], Rudeanu [795]). More recently, they have also been extensively studied in the reliability literature (see, e.g., Colbourn [205, 206]; Provan [759]; Schneeweiss [811]).

In general, however, orthogonal forms are difficult to compute, and few classes of disjunctive normal forms are known for which orthogonalization can be efficiently performed. An interesting class with this property, called the class of *shellable* DNFs, has been introduced and investigated by Ball and Provan [49, 760]. As these authors established, the DNFs describing several important classes of reliability problems (all-terminal reliability, all-point reachability, k-out-of-n systems, etc.) are shellable. Moreover, besides its unifying role in reliability theory, shellability also provides a powerful theoretical and algorithmic tool of combinatorial geometry, where it originally arose in the study of abstract simplicial complexes (see [96, 97, 205, 206, 254, 569, etc.]; let us simply mention here, without further details, that an abstract simplicial complex can be viewed as the set of true points of a positive Boolean function).

In this chapter, we first review some basic facts concerning orthogonal forms and describe a simple orthogonalization procedure for DNFs. Then, we introduce shellable DNFs and establish some of their most remarkable properties: In particular, we prove that shellable DNFs can be orthogonalized and dualized in polynomial time. Finally, we define and investigate a fruitful strengthening of shellability, namely, the lexico-exchange property.

7.1 Computation of orthogonal DNFs

Recall from Chapter 1, Section 1.6, that the DNF

$$\phi = \bigvee_{k=1}^{m} \left(\bigwedge_{i \in A_k} x_i \bigwedge_{j \in B_k} \overline{x}_j \right), \tag{7.1}$$

is orthogonal if no two terms of ϕ can be simultaneously equal to 1, that is, if

$$\left(A_k \cap B_\ell \right) \cup \left(A_\ell \cap B_k \right) \neq \emptyset \quad \text{for all } 1 \leq k < \ell \leq m,$$

or, equivalently,

$$\left(\bigwedge_{i \in A_k} x_i \bigwedge_{j \in B_k} \overline{x}_j \right) \left(\bigwedge_{i \in A_\ell} x_i \bigwedge_{j \in B_\ell} \overline{x}_j \right) \equiv 0 \quad \text{for all } 1 \leq k < \ell \leq m.$$

As described in Section 1.6, one of the main applications of ODNFs is in enumerating the true points of a Boolean function or, more generally, in computing the probability that a Boolean function takes the value 1 when each of its variables takes the value 0 or 1 randomly and independently of the values of the other variables. Indeed, for functions in orthogonal form, this probability is very easily computed by summing the probabilities associated with all individual terms, since any two terms correspond to a pair of disjoint events. This explains, in particular, why ODNFs have become an object of study in reliability theory (see Section 1.13.4).

As noted earlier, however, computing an ODNF of a Boolean function often turns out to be a difficult computational task. In previous chapters, we described different ways of obtaining an ODNF of a given function, for instance, by computing its minterm expression (see Section 2.11.2 and the "complete state enumeration scheme" in Provan's classification [759]), by iterative applications of the Shannon expansion (see Section 1.8 and the "pivotal decomposition scheme" in [759]), or as a byproduct of binary decision diagrams (see Section 1.12.3; Ball and Nemhauser [48]; Birnbaum and Lozinskii [90]; Wegener [903], etc.). We now present another classical approach, which relies on the following simple observations.

Theorem 7.1. *Let $\phi = \bigvee_{k=1}^{m} C_k$ be a DNF. Then,*

(i) *the expression*

$$\psi = C_1 \vee \overline{C_1} C_2 \vee \overline{C_1}\,\overline{C_2} C_3 \vee \ldots \vee \overline{C_1}\,\overline{C_2} \ldots \overline{C_{m-1}} C_m$$

is equivalent to ϕ;

(ii) *if ψ_k is an ODNF of $\overline{C_1}\,\overline{C_2} \ldots \overline{C_{k-1}} C_k$ for $k = 1, 2, \ldots, m$, then $\bigvee_{k=1}^{m} \psi_k$ is an ODNF of ϕ.*

Proof. The expression ψ is clearly equivalent to ϕ. Let T_1 be a term of ψ_k and T_2 be a term of ψ_j, where $T_1 \neq T_2$ and $k \leq j$. If $k < j$, then $T_1 T_2 \equiv 0$, since $\psi_k \psi_j \equiv 0$. On the other hand, if $k = j$, then $T_1 T_2 \equiv 0$ by orthogonality of ψ_k. $\square$

Theorem 7.1 suggests the recursive procedure described in Figure 7.1 for computing an ODNF of an arbitrary DNF (see, e.g., Kuntzmann [589]).

```
Procedure ORTHOGONALIZE(φ)
Input: A DNF φ = ⋁ᵐₖ₌₁ Cₖ.
Output: An orthogonal DNF ψ equivalent to φ.

begin
for k := 1 to m do
    begin
        compute a DNF φₖ of C̄₁ C̄₂ ... C̄ₖ₋₁Cₖ;
        ψₖ := ORTHOGONALIZE(φₖ);
    end;
    ψ := ⋁ᵐₖ₌₁ ψₖ;
end
```

Figure 7.1. Procedure ORTHOGONALIZE.

There are many ways of implementing this algorithm, thus giving rise to different variants of ORTHOGONALIZE, such as those proposed by Fratta and Montanari [346]; Abraham [2]; Aggarwal, Misra, and Gupta [7]; Locks [619]; Bruni [158]; and so on; see also the surveys [206, 776]. (Note that most authors restrict their attention to positive Boolean functions, although there is no need to be so restrictive.)

A specific difficulty with ORTHOGONALIZE is to work around the recursive call to the procedure, since orthogonalizing ϕ_k may, in general, be as difficult as orthogonalizing ϕ itself. One way to resolve this difficulty is to produce ϕ_k directly in orthogonal form, as this suppresses the need for the recursive call. To achieve this goal, we write $C_j = \bigwedge_{i=1}^{n_j} \ell_{ij}$, where $\ell_{1j}, \ell_{2j}, \ldots, \ell_{n_j j}$ are literals, for $j = 1, 2, \ldots, m$. Then,

$$\overline{C_1}\,\overline{C_2}\ldots\overline{C_{k-1}}C_k = \bigwedge_{j=1}^{k-1}\left(\bigvee_{i=1}^{n_j}\overline{\ell_{ij}}\right)C_k$$

$$= \bigwedge_{j=1}^{k-1}\left(\overline{\ell_{1j}} \vee \ell_{1j}\overline{\ell_{2j}} \vee \ell_{1j}\ell_{2j}\overline{\ell_{3j}} \vee \ldots \vee \ell_{1j}\ell_{2j}\ldots\ell_{n_j-1,j}\overline{\ell_{n_j,j}}\right)C_k.$$

Using distributivity to "multiply out" its $k-1$ factors, the latter expression can easily be transformed into an orthogonal DNF ψ_k.

Abraham [2] suggested to implement this approach in an iterative fashion, by successively computing an ODNF expression φ_j of $\overline{C_1}\,\overline{C_2}\ldots\overline{C_j}C_k$ for $j = 1, 2, \ldots, k-1$, until $\varphi_{k-1} = \phi_k = \psi_k$ is obtained. Suppose that the ODNF φ_{j-1} is in the form $\varphi_{j-1} = \bigvee_{t \in T} P_t$, where P_t ($t \in T$) are elementary conjunctions. Then,

$$\overline{C_1}\,\overline{C_2}\ldots\overline{C_j}C_k = \overline{C_j}\,\varphi_{j-1} = \left(\bigvee_{i=1}^{n_j}\overline{\ell_{ij}}\right)\left(\bigvee_{t \in T} P_t\right), \qquad (7.2)$$

and the right-hand side of (7.2) can be transformed to produce the ODNF

$$\varphi_j = \left(\bigvee_{i=1}^{n_j} \overline{\ell_{ij}} \right) \left(\bigvee_{t \in T} P_t \right)$$

$$= \bigvee_{t \in T} \left(\overline{\ell_{1j}} P_t \vee \ell_{1j} \overline{\ell_{2j}} P_t \vee \ell_{1j} \ell_{2j} \overline{\ell_{3j}} P_t \vee \ldots \vee \ell_{1j} \ell_{2j} \ldots \ell_{n_j-1,j} \overline{\ell_{n_j,j}} P_t \right).$$

$$(7.3)$$

Abraham [2] proposed to accelerate this procedure by various types of computational shortcuts (similar to those described in the context of dualization algorithms – see Theorem 4.31). For instance, if some term P_t contains the complement of one of the literals ℓ_{ij}, then $\overline{C_j} P_t = P_t$, and the t-th subexpression in (7.3) can be replaced by P_t. If P_t contains a subset of $\{\ell_{1j}, \ell_{2j}, \ldots, \ell_{n_j j}\}$, say, without loss of generality $\{\ell_{r+1,j}, \ell_{r+2,j}, \ldots, \ell_{n_j j}\}$, then $\overline{C_j} P_t = \left(\bigvee_{i=1}^{r} \overline{\ell_{ij}} \right) P_t$ and the right-hand side of (7.3) simplifies accordingly. Also, absorption can be applied at any stage of the procedure (the previous two simplifications can actually be viewed as resulting from absorption). Finally, as noted in [2, 7, 346, etc.], the efficiency of the procedure is usually improved if the terms of ϕ are reordered by nondecreasing degree.

Example 7.1. Let $\phi = x_1 x_2 \vee x_2 \overline{x_3} \vee x_3 x_4$, and let us apply Abraham's method. First, we let $\phi_1 = \psi_1 = x_1 x_2$. Next, we find

$$\phi_2 = \psi_2 = (\overline{x_1} \vee \overline{x_2}) x_2 \overline{x_3} = \overline{x_1} x_2 \overline{x_3}.$$

Finally, we need an orthogonal DNF of $(\overline{x_1} \vee \overline{x_2})(\overline{x_2} \vee x_3) x_3 x_4$. We first produce

$$\varphi_1 = (\overline{x_1} \vee x_1 \overline{x_2}) x_3 x_4 = \overline{x_1} x_3 x_4 \vee x_1 \overline{x_2} x_3 x_4.$$

Then, we produce (note that both terms of φ_1 conflict with a literal of $x_2 \overline{x_3}$)

$$\varphi_2 = \phi_3 = \psi_3 = (\overline{x_2} \vee x_3) \varphi_1 = \overline{x_1} x_3 x_4 \vee x_1 \overline{x_2} x_3 x_4,$$

and we eventually obtain the following ODNF of ϕ:

$$\psi = \psi_1 \vee \psi_2 \vee \psi_3 = x_1 x_2 \vee \overline{x_1} x_2 \overline{x_3} \vee \overline{x_1} x_3 x_4 \vee x_1 \overline{x_2} x_3 x_4.$$

$$\square$$

Another way to look at the **for** loop of the procedure in Figure 7.1 relies on the observation that computing a DNF of $\overline{C_1 C_2 \ldots C_{k-1}} C_k$ is essentially equivalent to dualizing the function $f_{k-1} = \bigvee_{i=1}^{k-1} C_i$. Indeed, if $\theta_{k-1}(X)$ is a DNF of $f_{k-1}^d(X)$, then the required DNF ϕ_k is easily derived from the expression $\theta_{k-1}(\overline{X}) C_k(X)$. Note, however, that the resulting DNF is usually not orthogonal, so that the recursive call to ORTHOGONALIZE is needed here.

Incidentally, for positive functions, this relation between dualization and orthogonalization procedures prompts an intriguing conjecture.

Conjecture 7.1. *Every positive Boolean function f has an ODNF ψ whose length is polynomially related to the length of the complete (i.e., prime irredundant) DNFs of f and f^d: More precisely, there exist positive constants α and β such that, if p, q, and r respectively denote the number of terms of f, f^d, and a shortest ODNF of f, then asymptotically*

$$\alpha(p+q) \leq r \leq (p+q)^{\beta}.$$

Weaker forms of the lower bound conjecture have been informally stated by Ball and Nemhauser [48] and Boros et al. [111] (see also Jukna et al. [541] for related considerations and negative results in the context of decision trees and branching programs). Note that if m denotes the number of terms of an *arbitrary* DNF of f, then the bound $p \leq m$ holds in view of the unicity of the prime irredundant representation of positive functions.

An interesting result concerning the length of ODNFs was established by Ball and Nemhauser [48]. (The proof of this result involves arguments based on linear programming duality. It is rather lengthy and we omit it here.)

Theorem 7.2. *For all $n \geq 1$, the shortest ODNF of $f(x_1, x_2, \ldots, x_n, y_1, y_2, \ldots, y_n) = \bigvee_{i=1}^{n} x_i y_i$ contains $2^n - 1$ terms.*

Observe that the dual of the function mentioned in Theorem 7.2 has 2^n prime implicants, in agreement with Conjecture 7.1.

7.2 Shellings and shellability

7.2.1 Definition

An extreme simplification of the procedure ORTHOGONALIZE is achieved when each of the expressions $\overline{C_1} \, \overline{C_2} \ldots \overline{C_{k-1}} C_k$ ($k = 1, 2, \ldots, m$) reduces to an elementary conjunction. This observation motivated Ball and Provan [49] to introduce and to investigate the properties of shellable disjunctive normal forms.

Definition 7.1. *A shelling of the DNF $\bigvee_{k=1}^{m} C_k$ is a permutation $(C_{\pi(1)}, C_{\pi(2)}, \ldots, C_{\pi(m)})$ of its terms such that, for each $k = 1, 2, \ldots, m$, the expression*

$$\overline{C_{\pi(1)}} \, \overline{C_{\pi(2)}} \ldots \overline{C_{\pi(k-1)}} C_k$$

is equivalent to an elementary conjunction. A DNF is called shellable *if it admits a shelling.*

Note that the definition in [49] is given for positive DNFs only, but it extends in a straightforward way to arbitrary DNFs. It should also be stressed that, as usual, we identify the constant 1 with the empty elementary conjunction, but the constant 0 is not an elementary conjunction. However, we could slightly generalize Definition 7.1 to include the case where $\overline{C_{\pi(1)}} \, \overline{C_{\pi(2)}} \ldots \overline{C_{\pi(k-1)}} C_k = 0$, and all results in forthcoming sections could be adapted accordingly without much difficulty.

It can be shown that several natural classes of DNFs are shellable, but we delay our presentation of such generic examples until the end of the chapter

(Section 7.6), when we shall have more tools at hand with which to establish shellability.

For now, we just provide a couple of small examples showing that shellable DNFs exist, that some DNFs are not shellable, and that an arbitrary permutation of the terms of a shellable DNF is not necessarily a shelling.

Example 7.2. *Consider again the DNF* $\phi = x_1x_2 \vee x_2\overline{x}_3 \vee x_3x_4$, *as in Example 7.1. The permutation* $(x_1x_2, x_2\overline{x}_3, x_3x_4)$ *is not a shelling of its terms, since*

$$(\overline{x_1} \vee \overline{x_2})\,(\overline{x_2} \vee x_3)\,x_3\,x_4 = \overline{x_1}\,x_3\,x_4 \vee \overline{x_2}\,x_3\,x_4$$

is not equivalent to an elementary conjunction. However, ϕ *is shellable. Indeed, when we consider its terms in the order* $(x_2\overline{x}_3,\ x_3x_4,\ x_1x_2)$, *we successively obtain*

$$(\overline{x}_2 \vee x_3)\,x_3x_4 = x_3x_4,$$

and

$$(\overline{x}_2 \vee x_3)\,(\overline{x}_3 \vee \overline{x_4})\,x_1x_2 = x_1\,x_2\,x_3\,\overline{x_4}.$$

Thus, in particular, ϕ *is equivalent to the orthogonal DNF* $x_2\overline{x}_3 \vee x_3x_4 \vee x_1x_2x_3\overline{x_4}$.

Finally, the positive DNF $x_1x_2 \vee x_3x_4$ *is not shellable, since neither* $(\overline{x_1} \vee \overline{x_2})\,x_3x_4$ *nor* $(\overline{x_3} \vee \overline{x_4})\,x_1x_2$ *is equivalent to an elementary conjunction.* □

As should be clear from the introductory discussion, and as illustrated by Example 7.2, the following statement holds:

Theorem 7.3. *If* ϕ *is a shellable DNF on m terms, then* ϕ *is equivalent to an orthogonal DNF on m terms.*

Proof. This follows from Definition 7.1 and Theorem 7.1. □

However, it is absolutely not obvious that the "short" ODNF whose existence is guaranteed by Theorem 7.3 can always be computed efficiently (say, in polynomial time) for every shellable DNF. This question actually raises multiple side issues: How difficult is it to recognize whether a DNF is shellable? How difficult is it to find a shelling of a shellable DNF? How difficult is it to recognize whether a given permutation of the terms of a DNF is a shelling? Given a shelling of a DNF, how difficult is it to compute an equivalent ODNF? and so on. We tackle most of these questions in forthcoming sections. From here on, however, we restrict our attention to positive DNFs, since all published results concerning shellability have been obtained for such DNFs.

7.2.2 Orthogonalization of shellable DNFs

For positive DNFs, Ball and Provan [49] proposed an alternative approach to the concept of shellability, based again on the consideration of the procedure ORTHOGONALIZE. To motivate this approach, let us consider a positive DNF $\phi = \bigvee_{k=1}^{m} C_k$, where

$$C_k = \bigwedge_{i \in A_k} x_i, \quad k = 1, 2, \ldots, m, \tag{7.4}$$

and $A_k \neq \emptyset$ for $k = 1, 2, \ldots, m$. Since computing an ODNF of $\phi_k = \overline{C_1}\,\overline{C_2} \ldots \overline{C_{k-1}} C_k$ is usually a rather costly process, Ball and Provan suggest computing instead an elementary conjunction U_k such that $\phi_k \leq U_k$. The disjunction of these elementary conjunctions yields a DNF $\phi_U = \bigvee_{k=1}^{m} U_k$ such that $\phi \leq \phi_U$. If the ultimate goal is to compute the probability that $\phi = 1$, then the conjunctions U_k can be used to produce an upper-bound on the target value, since

$$\mathrm{Prob}[\phi = 1] \leq \mathrm{Prob}[\phi_U = 1] \leq \sum_{k=1}^{m} \mathrm{Prob}[U_k = 1].$$

We now describe conditions that must be fulfilled by any upper-bounding elementary conjunctions U_k. We start with an easy lemma, for further reference.

Lemma 7.1. *For $k = 1, 2, \ldots, m$, the expression $\phi_k = \overline{C_1}\,\overline{C_2} \ldots \overline{C_{k-1}} C_k$ is identically zero if and only if there exists $\ell < k$ such that $A_\ell \subseteq A_k$.*

Proof. The expression ϕ_k is identically zero if and only if

$$C_k = 1 \Rightarrow \overline{C_1}\,\overline{C_2} \ldots \overline{C_{k-1}} = 0$$

or, equivalently, if and only if

$$C_k = 1 \Rightarrow C_1 \vee C_2 \vee \ldots \vee C_{k-1} = 1,$$

which means that C_k is an implicant of $C_1 \vee C_2 \vee \ldots \vee C_{k-1}$. This completes the proof, since all conjunctions $C_1, C_2, \ldots, C_m$ are positive. $\square$

The lemma shows, in particular, that if $A_\ell \subseteq A_k$, then C_k must precede C_ℓ in every shelling (since 0 is not an elementary conjunction).

We need yet another definition ([49, 105, 111]).

Definition 7.2. *Let $A_1, A_2, \ldots, A_m$ be an ordered list of subsets of $\{1, 2, \ldots, n\}$. For $k = 1, 2, \ldots, m$, the shadow of A_k is the set*

$$S(A_k) = \{ j \in \{1, 2, \ldots, n\} : \text{there exists } \ell < k \leq m \text{ such that } A_\ell \setminus A_k = \{j\} \}. \quad (7.5)$$

Note that the shadow of A_k depends on the order in which the sets $A_1, A_2, \ldots, A_m$ are listed, so that a notation like $S(A_1, A_2, \ldots, A_k)$ may be more appropriate than $S(A_k)$. However, we adhere to the shorter notation for the sake of brevity.

Example 7.3. *Consider the sets $A_1 = \{1, 2\}$, $A_2 = \{1, 3, 5\}$, $A_3 = \{2, 3, 5\}$, $A_4 = \{3, 4, 5\}$, in this order. Their shadows are, respectively, $S(A_1) = \emptyset$, $S(A_2) = \{2\}$, $S(A_3) = \{1\}$, and $S(A_4) = \{1, 2\}$.* $\square$

Lemma 7.2. *Let $C_\ell = \bigwedge_{i \in A_\ell} x_i$ for $\ell = 1, 2, \ldots, m$, let $k \in \{1, 2, \ldots, m\}$, and assume that $\phi_k = \overline{C_1}\,\overline{C_2} \ldots \overline{C_{k-1}} C_k$ is not identically zero. For an arbitrary elementary conjunction*

$$U_k = \bigwedge_{i \in O_k} x_i \bigwedge_{j \in F_k} \overline{x}_j, \quad (7.6)$$

the implication $\phi_k \leq U_k$ holds if and only if

(a) $O_k \subseteq A_k$; *and*
(b) $F_k \subseteq S(A_k)$.

Proof. Sufficiency. Assume that conditions (a)–(b) hold and assume that $U_k(X^*) = 0$ for some $X^* \in \mathcal{B}^n$. We want to show that $\phi_k(X^*) = 0$. If there is $i \in A_k$ such that $x_i^* = 0$, then $C_k(X^*) = 0$; hence, $\phi_k(X^*) = 0$. On the other hand, if $x_i^* = 1$ for all $i \in A_k$, then condition (a) implies that $x_i^* = 1$ for all $i \in O_k$. Since $U_k(X^*) = 0$, there must be an index $j \in F_k$ such that $x_j^* = 1$ and, by condition (b), there exists $\ell < k$ such that $A_\ell \setminus A_k = \{j\}$. This implies that $x_i^* = 1$ for all $i \in A_\ell$, hence $C_\ell(X^*) = 1$ and $\phi_k(X^*) = 0$, as required.

Necessity. Conversely, if $\phi_k \leq U_k$, let $X^* \in \mathcal{B}^n$ denote the characteristic vector of A_k. Then $\phi_k(X^*) = 1$ (because ϕ_k is not identically 0); hence, $U_k(X^*) = 1$, which implies condition (a).

Suppose now that condition (b) does not hold, that is, suppose that there is an index $j \in F_k$ such that $A_\ell \setminus A_k \neq \{j\}$ for all $\ell < k$. Note that $A_\ell \setminus A_k \neq \emptyset$ (by Lemma 7.1). Hence, for all $\ell < k$, there exists $i_\ell \neq j$ such that $i_\ell \in A_\ell \setminus A_k$. Define a point $Y^* \in \mathcal{B}^n$ by setting $y_i^* = 0$ if $i \in \{i_1, i_2, \ldots, i_{k-1}\}$ and $y_i^* = 1$ otherwise. In particular, $y_j^* = 1$, and therefore $U_k(Y^*) = 0$. On the other hand, $C_k(Y^*) = 1$ and $C_\ell(Y^*) = 0$ for all $\ell < k$, so that $\phi_k(Y^*) = 1$. This contradicts the assumption that $\phi_k \leq U_k$, and the proof is complete. $\qquad\square$

As an easy corollary, we obtain [49, 111]:

Lemma 7.3. *If $\phi = \bigvee_{k=1}^m \left(\bigwedge_{i \in A_k} x_i \right)$ and $\phi^{sh} = \bigvee_{k=1}^m \left(\bigwedge_{i \in A_k} x_i \bigwedge_{j \in S(A_k)} \overline{x}_j \right)$, then ϕ and ϕ^{sh} are equivalent DNFs.*

Proof. Comparing the DNFs termwise, it is obvious that $\phi^{sh} \leq \phi$. The inequality $\phi \leq \phi^{sh}$ follows from Theorem 7.1 and Lemma 7.2. $\qquad\square$

Example 7.4. *Observe that the DNF ϕ^{sh} is not necessarily orthogonal. For instance, when $\phi = x_1 x_2 \vee x_3 x_4$, we find $S(A_1) = S(A_2) = \emptyset$, and $\phi = \phi^{sh}$.* $\qquad\square$

We are now ready to establish several characterizations of shellable positive DNFs due to Ball and Provan [49] (see also [111]).

Theorem 7.4. *Let $\phi = \bigvee_{k=1}^m C_k$, where $C_k = \bigwedge_{i \in A_k} x_i$ for $k = 1, 2, \ldots, m$. The following statements are equivalent:*

(a) $(C_1, C_2, \ldots, C_m)$ *is a shelling of ϕ.*
(b) *For $k = 1, 2, \ldots, m$,*

$$\overline{C_1}\,\overline{C_2}\ldots\overline{C_{k-1}}C_k = \bigwedge_{i \in A_k} x_i \bigwedge_{j \in S(A_k)} \overline{x}_j. \tag{7.7}$$

(c) *The DNF*

$$\phi^{sh} = \bigvee_{k=1}^{m} (\bigwedge_{i \in A_k} x_i \bigwedge_{j \in S(A_k)} \overline{x}_j) \tag{7.8}$$

     is orthogonal.

(d) $A_\ell \cap S(A_k) \neq \emptyset$ *for all* $1 \leq \ell < k \leq m$.

(e) *For all* $1 \leq \ell < k \leq m$, *there exists* $j \in A_\ell$ *and* $h < k$ *such that* $A_h \setminus A_k = \{j\}$.

Proof. (a) $\Longleftrightarrow$ (b). Statement (b) implies (a), by definition of shellings. Conversely, assume that $(C_1, C_2, \ldots, C_m)$ is a shelling of ϕ. Then, $\overline{C_1} \overline{C_2} \ldots \overline{C_{k-1}} C_k$ must be an elementary conjunction. But Lemma 7.2 implies that the right-hand side of (7.7) is the smallest elementary conjunction implied by $\overline{C_1} \overline{C_2} \ldots \overline{C_{k-1}} C_k$, and hence, equality must hold in (7.7).

     (b) $\Longleftrightarrow$ (c). If (b) holds, then ϕ^{sh} is orthogonal, since the expressions $\overline{C_1} \overline{C_2} \ldots \overline{C_{k-1}} C_k$ are pairwise orthogonal. Conversely, suppose that ϕ^{sh} is orthogonal. By Lemma 7.2, we know that

$$\overline{C_1} \overline{C_2} \ldots \overline{C_{k-1}} C_k \leq \bigwedge_{i \in A_k} x_i \bigwedge_{j \in S(A_k)} \overline{x}_j \tag{7.9}$$

for every $k = 1, 2, \ldots, m$. If the ℓ-th inequality is strict, then there exists $X^* \in \mathcal{B}^n$ such that the left-hand side of (7.9) is 0 and the right-hand side of (7.9) is 1 at the point X^*, for $k = \ell$. Moreover, since ϕ^{sh} is orthogonal, the right-hand side (and therefore, the left-hand side) of (7.9) is 0 at the point X^* for all $k \neq \ell$. Thus, we conclude that $\phi(X^*) = \bigvee_{k=1}^{m} \overline{C_1} \overline{C_2} \ldots \overline{C_{k-1}} C_k = 0$, while $\phi^{sh}(X^*) = 1$, contradicting Lemma 7.3.

     (c) $\Longleftrightarrow$ (d). Condition (d) trivially implies (c). Conversely, suppose that ϕ^{sh} is orthogonal, and that condition (d) does not hold, that is, $A_\ell \cap S(A_k) = \emptyset$ for some pair (l, k) with $l < k$. Choose ℓ as small as possible with this property. Since ϕ^{sh} is orthogonal, it must be the case that $S(A_\ell) \cap A_k \neq \emptyset$, say, $j \in S(A_\ell) \cap A_k$. So, by definition of $S(A_\ell)$, there exists $h < \ell$ such that $A_h \setminus A_\ell = \{j\}$. Moreover, $j \notin S(A_k)$, since A_k and $S(A_k)$ are disjoint. Therefore, $A_h \cap S(A_k) \subseteq (A_\ell \cup \{j\}) \cap S(A_k) = \emptyset$. Since $h < \ell$, this contradicts our choice of ℓ.

     (d) $\Longleftrightarrow$ (e). The equivalence of these conditions is obvious in view of the definition of shadows. $\square$

Example 7.5. *Consider the DNF* $\phi = x_1 x_2 \vee x_1 x_3 x_5 \vee x_2 x_3 x_5 \vee x_3 x_4 x_5$ *and the corresponding sets* $A_1 = \{1, 2\}$, $A_2 = \{1, 3, 5\}$, $A_3 = \{2, 3, 5\}$, $A_4 = \{3, 4, 5\}$. *We computed the shadows of these sets in Example 7.3. The reader will check that Equation (7.7) holds for* $k = 1, 2, 3, 4$, *so that* ϕ *is shellable and is represented by the orthogonal DNF* $\phi = x_1 x_2 \vee x_1 \overline{x}_2 x_3 x_5 \vee \overline{x}_1 x_2 x_3 x_5 \vee \overline{x}_1 \overline{x}_2 x_3 x_4 x_5$. $\square$

As a corollary of Theorem 7.4, we can now answer some of the questions posed at the end of Section 7.2.1 (compare with Theorem 7.3).

Theorem 7.5. *If* $\phi = \bigvee_{k=1}^{m} C_k$ *is a positive DNF on n variables, there is an* $O(nm^2)$*-time algorithm to test whether* $(C_1, C_2, \ldots, C_m)$ *is a shelling of* ϕ *and, when this is the case, to compute an orthogonal DNF of* ϕ.

Proof. Given a permutation of the terms, it suffices to compute the expression (7.8) and to test whether it is orthogonal. $\qquad\square$

In contrast with Theorem 7.5, the complexity of recognizing shellable DNFs is an important and intriguing open problem, already mentioned, for instance, in [49, 254].

7.2.3 Shellable DNFs versus shellable functions

So far, we have defined and investigated shellable DNFs, rather than the functions they represent. We now consider the following definitions.

Definition 7.3. *A positive Boolean function is* shellable *if its complete DNF is shellable. It it* weakly shellable *if it can be represented by a shellable DNF.*

We have already seen (in Example 7.2) that certain positive functions are not shellable: A minimal example is provided by the function

$$f(x_1, \ldots, x_4) = x_1 x_2 \vee x_3 x_4.$$

On the other hand, the concept of weak shellability is rather vacuous, since it can be shown that every positive Boolean function is weakly shellable (Boros et al. [111]).

Theorem 7.6. *Every positive Boolean function can be represented by a shellable DNF.*

Proof. Let f be a positive function, let $\{C_I = \bigwedge_{j \in I} x_j \mid I \in \mathcal{I}\}$ denote the set of all implicants (not necessarily prime) of f, and let π be a permutation that orders the implicants by nonincreasing degree. Then, the DNF

$$\phi = \bigvee_{I \in \mathcal{I}} \left(\bigwedge_{j \in I} x_j \right)$$

represents f, and condition (d) in Theorem 7.4 can be used to verify that π is a shelling of ϕ. Indeed, if $I_\ell, I_k \in \mathcal{I}$ and C_{I_ℓ} precedes C_{I_k} in π, then there is an index $j \in I_\ell \setminus I_k$. The set $I = I_k \cup \{j\}$ is in $\mathcal{I}$ and C_I precedes C_{I_k} in π. Therefore, $j \in S(I_k)$, and we conclude that $j \in I_\ell \cap S(I_k)$, as required by condition (d). $\qquad\square$

Since the size of the DNF produced in the proof of Theorem 7.6 can generally be very large relative to the number of prime implicants of f, let us provide another construction that uses a smaller subset of the implicants.

We first recall a well-known definition.

Definition 7.4. *If I, J are two subsets of $N = \{1,2,\ldots,n\}$, we say that I precedes J in the lexicographic order, and we write $I <_L J$ if*

$$\min\{j \in N \mid j \in I \setminus J\} < \min\{j \in N \mid j \in J \setminus I\}.$$

Now, for $I \in \mathcal{I}$, let $h(I)$ denote the largest element of the subset $I \subseteq \{1,2,\ldots,n\}$, and let $H(I) = I \setminus \{h(I)\}$. We call *leftmost implicant* of f any implicant C_I of f for which $C_{H(I)}$ is not an implicant of f, and we denote by $\mathcal{L}$ the family of leftmost implicants of f. Clearly, all prime implicants of f are in $\mathcal{L}$, therefore f is represented by the DNF $\psi_{\mathcal{L}} = \bigvee_{I \in \mathcal{L}} \left(\bigwedge_{j \in I} x_j \right)$. Boros et al. [111] showed that the lexicographic order $<_L$ defines a shelling of $\psi_{\mathcal{L}}$. We leave the proof of this claim as an end-of-chapter exercise and simply illustrate it on an example.

Example 7.6. *We know that the function $f(x_1,\ldots,x_4) = x_1 x_2 \vee x_3 x_4$ is not shellable. Its leftmost implicants are $x_1 x_2$, $x_1 x_3 x_4$, $x_2 x_3 x_4$ and $x_3 x_4$, listed here in lexicographic order. The corresponding DNF*

$$\psi_{\mathcal{L}} = x_1 x_2 \vee x_1 x_3 x_4 \vee x_2 x_3 x_4 \vee x_3 x_4$$

represents f and is shellable, since the DNF

$$\psi_{\mathcal{L}}^{sh} = x_1 x_2 \vee x_1 \overline{x}_2 x_3 x_4 \vee \overline{x}_1 x_2 x_3 x_4 \vee \overline{x}_1 \overline{x}_2 x_3 x_4$$

is orthogonal.   □

Let us finally observe that there exist families of positive functions for which the smallest shellable DNF representation involves a number of terms that grows exponentially with the number of its prime implicants:

Theorem 7.7. *For all $n \geq 1$, every shellable DNF of $f(x_1, x_2,\ldots,x_n,y_1,y_2,\ldots,y_n) = \bigvee_{i=1}^{n} x_i y_i$ contains at least $2^n - 1$ terms.*

Proof. This is an immediate corollary of Theorems 7.2 and 7.3.   □

7.3 Dualization of shellable DNFs

The formal similarity between certain dualization and orthogonalization procedures was noted in Section 7.1. Since shellable DNFs have short orthogonal forms, it is quite natural to wonder whether they also have short dual expressions (remember Conjecture 7.1). In this section, we provide an affirmative answer to this question and prove a result due to Boros et al. [111] stating that shellable positive DNFs can be dualized in time polynomial in their input size. This result implies, in particular, that for shellable positive functions, the number of prime implicants of the dual is polynomially bounded in the number of prime implicants of the function.

Theorem 7.8. *If a Boolean function f in n variables can be represented by a shellable positive DNF ϕ involving m terms, then f^d has at most nm prime implicants. If a shelling of ϕ is available, then the prime implicants of f^d can be generated in $O(nm^2)$ time.*

Proof. We prove the first statement by induction on m.

If $m = 1$, then f is an elementary conjunction and its dual is an elementary disjunction that has at most n prime implicants.

Let us now assume that the statement has been established for shellable DNFs of at most $m - 1$ terms, let f be represented by the DNF $\phi = \bigvee_{k=1}^{m} C_k = \bigvee_{k=1}^{m} \bigwedge_{i \in A_k} x_i$, where $(C_1, C_2, \dots, C_m)$ is a shelling of ϕ, and let $g = \bigvee_{k=1}^{m-1} C_k$. Observe that $(C_1, C_2, \dots, C_{m-1})$ is a shelling of g. Therefore, by the induction hypothesis, g^d has at most $n(m - 1)$ prime implicants. Let us denote the complete DNF of g^d by

$$\psi = \bigvee_{k=1}^{p} P_k = (\bigvee_{k=1}^{p} \bigwedge_{j \in J_k} x_j), \tag{7.10}$$

where $P_1, P_2, \dots, P_p$ are all prime implicants of g^d, and $p \leq n(m - 1)$. Then,

$$f^d = g^d \wedge \left(\bigvee_{i \in A_m} x_i \right) = \left(\bigvee_{k=1}^{p} \bigwedge_{j \in J_k} x_j \right) \wedge \left(\bigvee_{i \in A_m} x_i \right). \tag{7.11}$$

On the other hand, $\overline{g} = \overline{C_1}\,\overline{C_2} \dots \overline{C_{m-1}} = \bigvee_{k=1}^{p} (\bigwedge_{j \in J_k} \overline{x_j})$, so that

$$\overline{C_1}\,\overline{C_2} \dots \overline{C_{m-1}} C_m = \left(\bigvee_{k=1}^{p} \bigwedge_{j \in J_k} \overline{x_j} \right) \wedge \left(\bigwedge_{i \in A_m} x_i \right)$$

$$= \bigvee_{k=1}^{p} \left(\bigwedge_{j \in J_k} \overline{x_j} \bigwedge_{i \in A_m} x_i \right). \tag{7.12}$$

By definition of shellings, the DNF (7.12) is equivalent to a single conjunction. Since no absorption can take place in (7.12), and no two terms of (7.12) form a consensus, it must be the case that all its terms are identically zero, except one. In other words, there is an index $\ell \in \{1, 2, \dots, p\}$ such that $J_\ell \cap A_m = \emptyset$ and $J_k \cap A_m \neq \emptyset$ for all $k \neq \ell$. (The same conclusion can be reached by noting that J_ℓ is exactly the shadow of A_m.)

Thus, from (7.11),

$$f^d = \left(\bigvee_{\substack{k=1 \\ k \neq \ell}}^{p} \bigwedge_{j \in J_k} x_j \right) \vee \left(\bigvee_{i \in A_m} \bigwedge_{j \in J_\ell \cup \{i\}} x_j \right), \tag{7.13}$$

and we conclude that f^d has at most nm prime implicants.

Using relation (7.13), all prime implicants of f^d can easily be generated in $O(nm)$ time once the prime implicants of g^d are known. The overall $O(nm^2)$ time bound follows.   □

Note that the dualization procedure sketched in the proof of Theorem 7.8 is exactly the classical algorithm SD-DUALIZATION presented in Chapter 4, Section 4.3.2.

In Theorem 7.6, we established that every positive function can be represented by a shellable DNF. This result, combined with Theorem 7.8, might raise the impression that every positive function can be dualized in polynomial time. This is, of course, a fallacy because, as shown in Theorem 7.7, the shortest shellable representation of a positive Boolean function may be extremely large.

Finally, we mention that Theorem 7.8 generalizes a sequence of earlier results on regular functions [74, 225, 735, 736] and on aligned functions [105], since these are special classes of shellable positive DNFs (see Chapter 8 and the end-of-chapter exercises). For aligned and regular functions, efficient dualization algorithms with running time $O(n^2m)$ have been proposed in [74, 105, 225, 736]. None of those procedures, however, seems to be generalizable to shellable functions.

7.4 The lexico-exchange property

7.4.1 Definition

We now introduce a subclass of shellable DNFs, whose definition can best be viewed as a specialization of condition (e) in Theorem 7.4. As in Definition 7.4, $<_L$ denotes the lexicographic order on the subsets of $N = \{1, 2, \ldots, n\}$.

Definition 7.5. *A positive DNF*

$$\phi(x_1, x_2, \ldots, x_n) = \bigvee_{k=1}^{m} \bigwedge_{j \in A_k} x_j \qquad (7.14)$$

has the lexico-exchange (LE) *property with respect to* $(x_1, x_2, \ldots, x_n)$ *if, for every pair* $\ell, k \in \{1, 2, \ldots, m\}$ *such that* $A_\ell <_L A_k$, *there exists* $h \in \{1, 2, \ldots, m\}$ *such that* $A_h <_L A_k$ *and* $A_h \setminus A_k = \{j\}$, *where* $j = \min\{i \mid i \in A_\ell \setminus A_k\}$.

We say that ϕ *has the* LE *property with respect to a permutation* $(\sigma(x_1), \sigma(x_2), \ldots, \sigma(x_n))$ *of its variables, or that* σ *is an LE order for* ϕ, *if the DNF* ϕ^σ *defined by*

$$\phi^\sigma(\sigma(x_1), \sigma(x_2), \ldots, \sigma(x_n)) = \phi(x_1, x_2, \ldots, x_n)$$

has the LE property with respect to $(\sigma(x_1), \sigma(x_2), \ldots, \sigma(x_n))$.

Finally, we simply say that ϕ *has the LE property if* ϕ *has the LE property with respect to some permutation of its variables.*

Note that these definitions can be extended to positive functions by applying them to the complete DNF of such functions (as in Section 7.2.3).

The LE property was introduced by Ball and Provan in [49] and further investigated in [111, 760]. Interest in this concept is motivated by the observation that
every DNF with the LE property is also shellable.

Theorem 7.9. *If the DNF* $\phi(x_1,x_2,\ldots,x_n)$ *given by equation (7.14) has the
LE property with respect to* $(x_1,x_2,\ldots,x_n)$, *then the lexicographic order on*
$\{A_1,A_2,\ldots,A_m\}$ *induces a shelling of the terms of* ϕ.

Proof. This follows by comparing Definition 7.5 and condition (e) in
Theorem 7.4.   □

In fact, most classes of shellable DNFs investigated in the literature have the
LE property (see [49, 105] and the examples in Section 7.6).

It is interesting to observe that the converse of Theorem 7.9 does not hold:
Namely, the lexicographic order may induce a shelling of the terms of a DNF,
even when this DNF *does not* have the LE property with respect to $(x_1,x_2,\ldots,x_n)$.
This is because Definition 7.5 not only determines the order of the terms of ϕ but
also imposes the choice of the element j in $A_\ell \setminus A_k$.

Example 7.7. *The DNF* $\phi = x_1x_2 \vee x_2x_3 \vee x_3x_4$ *is shellable with respect to
the lexicographic order of its terms. However,* ϕ *does not have the LE property with respect to* (x_1,x_2,x_3,x_4): *With* $A_\ell = \{1,2\}$ *and* $A_k = \{3,4\}$, *we obtain*
$j = \min\{i \mid i \in A_\ell \setminus A_k\} = 1$, *and there is no* h *such* $A_h \setminus A_k = \{1\}$. *(But the reader
may check that* ϕ *has the LE property with respect to the permutation* (x_2,x_3,x_1,x_4)
of its variables.)   □

7.4.2 LE property and leaders

In the remainder of this section, when $\phi(x_1,x_2,\ldots,x_n)$ is a positive DNF, we denote
by ϕ_1 (respectively, ϕ_0) the disjunction of the terms of ϕ involving x_1 (respectively,
not involving x_1), so that

$$\phi(x_1,x_2,\ldots,x_n) = x_1\phi_1 \vee \phi_0. \tag{7.15}$$

Definition 7.6. *We say that* x_1 *is a leader for a positive DNF* $\phi(x_1,x_2,\ldots,x_n)$ *if*
$\phi_1 \geq \phi_0$. *Equivalently,* x_1 *is a leader if every term of* ϕ_0 *is absorbed by a term of*
ϕ_1, *or if every term of* ϕ_0 *is an implicant of* ϕ_1.

The next theorem clarifies the relationship between the LE property and the
existence of leaders.

Theorem 7.10. *A positive DNF* $\phi(x_1,x_2,\ldots,x_n) = x_1\phi_1 \vee \phi_0$ *has the LE property
with respect to* $(x_1,x_2,\ldots,x_n)$ *if and only if*

    (a) *both* ϕ_1 *and* ϕ_0 *have the LE property with respect to* $(x_2,x_3,\ldots,x_n)$; *and*
    (b) *either* x_1 *is a leader for* ϕ *or* ϕ *does not involve* x_1.

Proof. Let $\phi = \bigvee_{k=1}^m C_k = \bigvee_{k=1}^m \bigwedge_{j \in A_k} x_j$.

Necessity. Property (a) is an immediate consequence of Definition 7.5. To establish property (b), we must show that, if ϕ involves x_1 and C_k is any term of ϕ_0, then C_k is absorbed by some term of ϕ_1. Let C_ℓ be any term of ϕ_1, and observe that $A_\ell <_L A_k$ and $\min\{i \mid i \in A_\ell \setminus A_k\} = 1$. Since ϕ has the LE property, there exists $h \in \{1, 2, \ldots, m\}$ such that $A_h <_L A_k$ and $A_h \setminus A_k = \{1\}$. Then, $\bigwedge_{j \in A_h \setminus \{1\}} x_j$ is a term of ϕ_1 which absorbs C_k, as required.

Sufficiency. Suppose that (a) and (b) hold. If ϕ does not involve x_1, then (a) implies that $\phi = \phi_0$ has the LE property. So, assume that x_1 is a leader for ϕ, let $A_\ell <_L A_k$, and let $j = \min\{i \mid i \in A_\ell \setminus A_k\}$. If C_ℓ and C_k are both in ϕ_1 or both in ϕ_0, then condition (a) implies that ϕ has the LE property. Otherwise, it must be the case that C_ℓ is a term of ϕ_1, C_k is a term of ϕ_0, and $j = 1$. By definition of leaders, there is a term in ϕ_1, say, $\bigwedge_{j \in A_h \setminus \{1\}} x_j$, which absorbs C_k. Then, however, $A_h <_L A_k$ and $A_h \setminus A_k = \{1\}$, showing that ϕ has the LE property. □

7.4.3 Recognizing the LE property

In view of Definition 7.5, verifying whether a positive DNF $\phi(x_1, x_2, \ldots, x_n)$ has the LE property with respect to the identity permutation $(x_1, x_2, \ldots, x_n)$ can easily be done in polynomial time, say, in $O(nm^3)$ time, where m is the number of terms of ϕ. Provan and Ball [760] presented another procedure with $O(n^2 m)$ time complexity for this problem. Since m is typically much larger than n, we expect their procedure to be more efficient than the trivial one. We now describe this procedure, which also turns out to be useful for recognizing regular Boolean functions (in Chapter 8).

The procedure can be seen as relying on Theorem 7.10, which characterizes the LE property in terms of leaders (although the description in [760] does not explicitly use this characterization). Let us, therefore, momentarily concentrate on the algorithmic complexity of the following type of queries: For a positive DNF $\phi(x_1, x_2, \ldots, x_n)$, and for a subset $A \subseteq \{1, 2, \ldots, n\}$, is $\bigwedge_{j \in A} x_j$ absorbed by a term of ϕ? (Remember the definition of leaders.)

Such a query can easily be answered in $O(nm)$ time for a DNF on m terms. But in fact, this time complexity is far from optimal when ϕ is a fixed DNF possessing the LE property: Then, for each input subset A, it becomes possible to answer the query in time $O(n)$. This complexity can be achieved by using an appropriate data structure to represent ϕ. The fixed overhead incurred in setting up the data structure amounts to $O(nm)$ operations but can be amortized if the number of queries to be answered for ϕ is large enough.

The data structure to be used is a rooted, labeled binary tree $T(\phi)$. The tree $T(\phi)$ is defined for an arbitrary positive DNF $\phi(x_1, x_2, \ldots, x_n)$. (As we will see in Section 7.4.4, $T(\phi)$ is essentially equivalent to a decision tree for the function

represented by ϕ when ϕ has the LE property with respect to $(x_1, x_2, \ldots, x_n)$).
For $n \geq 1$, the tree $T(\phi)$ is recursively defined as follows (we denote its root
by $r(\phi)$):

(a) If ϕ is identically 0, then $T(\phi)$ is empty, that is, $T(\phi)$ has no vertices.
(b) If ϕ is identically 1, then $T(\phi)$ has exactly one unlabeled vertex, namely,
   its root $r(\phi)$.
(c) If $\phi(x_1, x_2, \ldots, x_n)$ is not identically 1, then let $\phi = x_1 \phi_1 \vee \phi_0$ (where ϕ_0
   and ϕ_1 do not involve x_1, as usual); build $T(\phi)$ by introducing a root $r(\phi)$
   labeled by x_1, creating disjoint copies of $T(\phi_0)$ and $T(\phi_1)$, and making
   $r(\phi_1)$ (respectively, $r(\phi_0)$) the left son (respectively, the right son) of $r(\phi)$.
   (If either $r(\phi_1)$ or $r(\phi_0)$ is not defined, i.e., if either ϕ_1 or ϕ_0 is identically
   zero, then the corresponding son of $r(\phi)$ does not exist.)

Example 7.8. *Consider the DNF* $\phi(x_1, x_2, x_3, x_4, x_5) = x_1 x_2 \vee x_1 x_3 \vee x_1 x_4 x_5 \vee$
$x_2 x_3 x_4$. *The corresponding tree* $T(\phi)$ *is represented in Figure 7.2. The leaves are
indexed by terms as explained in Theorem 7.11 hereunder.* $\square$

It is obvious that $T(\phi)$ has height at most n. Except for the leaves, all
vertices of $T(\phi)$ are labeled by a variable. Moreover, the leaves themselves correspond in a natural way to the terms of ϕ. Indeed, for an arbitrary leaf v, let

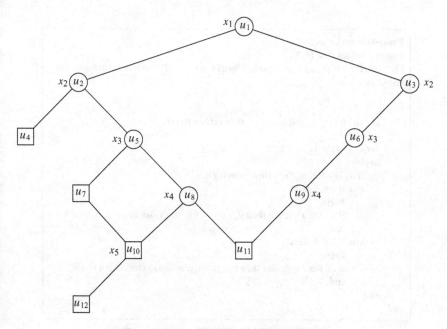

Figure 7.2. The binary tree for Example 7.8.

$r(\phi) = u_1, u_2, \ldots, u_q = v$ be the vertices of $T(\phi)$ lying on the unique path from the root $r(\phi)$ to v. Define

$$P(v) = \bigcup_{k=1}^{q-1} \{ j \mid u_k \text{ is labeled by } x_j \text{ and } u_{k+1} \text{ is the left son of } u_k \}. \qquad (7.16)$$

Theorem 7.11. *For every positive DNF ϕ, the mapping $v \mapsto \bigwedge_{j \in P(v)} x_j$ defines a one-to-one correspondence between the leaves of $T(\phi)$ and the terms of ϕ.*

Proof. The proof is left to the reader. □

Thus, $T(\phi)$ has exactly m leaves and at most nm vertices. It is actually easy to see that $T(\phi)$ sorts the terms of ϕ in lexicographic order, from left to right. Moreover, $T(\phi)$ can be set up in time $O(nm)$.

With the data structure $T(\phi)$ at hand, let us now revert to the query: "Is $\bigwedge_{j \in A} x_j$ absorbed by a term of ϕ?" Our next goal is to show that, when ϕ has the LE property with respect to $(x_1, x_2, \ldots, x_n)$, the query is correctly answered by the procedure IMPLICANT(A) in Figure 7.3, consisting of one traversal of $T(\phi)$ along a path from root to leaf.

Example 7.9. *The reader may want to apply the procedure* IMPLICANT(A) *to the tree $T(\phi)$ displayed in Figure 7.2, with $A = \{2, 4, 5\}$, and check that it returns the answer* FALSE. □

Procedure IMPLICANT(A)
Input: A subset A of $\{1, 2, \ldots, n\}$.
Output: TRUE if $\bigwedge_{j \in A} x_j$ is absorbed by a term of the DNF ϕ represented by $T(\phi)$, FALSE otherwise.

begin
    if $T(\phi)$ is empty (that is, $\phi = 0$) **then return** FALSE;
    $u_1 := r(\phi)$;
    for $k = 1$ **to** n **do**
    begin
        if u_k is a leaf of $T(\phi)$ **then return** TRUE
        else if $k \in A$ **then**
            begin
                if u_k has a left son **then** $u_{k+1} :=$ leftson(u_k) **else** $u_{k+1} :=$ rightson(u_k)
            end
        else if $k \notin A$ **then**
            begin
                if u_k has a right son **then** $u_{k+1} :=$ rightson(u_k) **else return** FALSE
            end
    end
end

Figure 7.3. Procedure IMPLICANT.

The procedure IMPLICANT can be implemented to run in time $O(n)$. It is certainly worth stressing that it does *not* necessarily return the correct answer when ϕ does not have the LE property with respect to $(x_1, x_2, \ldots, x_n)$, as the next example illustrates.

Example 7.10. *Consider the DNF* $\phi(x_1, x_2, x_3, x_4, x_5) = x_1 x_2 x_3 \vee x_1 x_4 \vee x_1 x_5 \vee x_2 x_4 x_5$. *We leave it to the reader to verify that* IMPLICANT$(\{1, 2, 5\})$ *returns the answer* FALSE, *despite the fact that* $x_1 x_2 x_5$ *is an implicant of* ϕ. $\square$

However, even for an arbitrary DNF ϕ, IMPLICANT works correctly "in half of the cases": Namely, it never errs on the answer TRUE.

Theorem 7.12. *Let* $\phi(x_1, x_2, \ldots, x_n)$ *be a positive DNF, let* $T(\phi)$ *be the associated binary tree, and let* $A \subseteq \{1, 2, \ldots, n\}$. *If the procedure* IMPLICANT(A) *returns the answer* TRUE *and terminates at the leaf* v *of* $T(\phi)$, *then* $\bigwedge_{j \in A} x_j$ *is absorbed by the term of* ϕ *associated with* v.

Proof. Suppose that the procedure eventually reaches the leaf $v = u_q$ and returns the answer TRUE. Let $\bigwedge_{j \in P(v)} x_j$ be the term of ϕ associated with v by (7.16). From the description of IMPLICANT, we see that, if $k \notin A$, then u_{k+1} is the right son of u_k, $k = 1, 2, \ldots, q$. Hence, by construction of $P(v)$, $k \notin P(v)$. Thus, $P(v) \subseteq A$, and $\bigwedge_{j \in A} x_j$ is absorbed by the term $\bigwedge_{j \in P(v)} x_j$ of ϕ. $\square$

More interestingly for our purpose, Provan and Ball [760] proved that IMPLI-CANT works correctly when ϕ has the LE property with respect to $(x_1, x_2, \ldots, x_n)$. Note that the DNF ϕ considered in Example 7.10 does not have the LE property with respect to $(x_1, x_2, x_3, x_4, x_5)$ (although it has it with respect to the permutation $(x_1, x_4, x_5, x_2, x_3)$ of its variables).

Theorem 7.13. *Let* $\phi(x_1, x_2, \ldots, x_n)$ *be a positive DNF having the LE property with respect to* $(x_1, x_2, \ldots, x_n)$, *let* $T(\phi)$ *be the associated binary tree, and let* $A \subseteq \{1, 2, \ldots, n\}$. *The procedure* IMPLICANT(A) *returns the answer* TRUE *if and only if* $\bigwedge_{j \in A} x_j$ *is absorbed by a term of* ϕ.

Proof. The "only if" statement follows from Theorem 7.12.

We prove the converse statement by induction on n. If $n = 1$, then the statement is easily verified. Assume next that $n \geq 2$ and that $\bigwedge_{j \in A} x_j$ is absorbed by a term of ϕ. If ϕ is identically 1, then $T(\phi)$ has exactly one vertex, namely, $r(\phi)$, and we are done. Otherwise, write $\phi = x_1 \phi_1 \vee \phi_0$. By Theorem 7.10, x_1 is either a leader for ϕ or does not appear in ϕ.

If $1 \notin A$ or if x_1 does not appear in ϕ, then $\bigwedge_{j \in A} x_A$ is absorbed by a term of ϕ_0. Hence, ϕ_0 is not identically 0, and $r(\phi)$ has a right son, which can be identified with the root of $T(\phi_0)$, say, $r(\phi_0)$. In the execution of IMPLICANT(A), u_2 is set equal to $r(\phi_0)$ (note that if x_1 does not appear in ϕ, then $r(\phi)$ has no left son). The next steps of the procedure are identical to those performed by IMPLICANT(A) on the subtree $T(\phi_0)$. Note that, by Theorem 7.10, ϕ_0 has the LE property with

Procedure LE-PROPERTY(ϕ)
Input: A DNF $\phi(x_1,x_2,\ldots,x_n) = \bigvee_{k=1}^{m} \bigwedge_{j \in A_k} x_j$, where $A_1 <_L A_2 <_L \cdots <_L A_m$.
Output: TRUE if ϕ has the LE property with respect to $(x_1,x_2,\ldots,x_n)$, FALSE
otherwise.

begin
    set up the binary tree $T(\phi)$;
    for $k = 2$ **to** m **do**
    begin
        find the leaf of $T(\phi)$, say v_k, associated with the term $\bigwedge_{j \in A_k} x_j$;
        for each vertex u on the path from $r(\phi)$ to v_k
            if v_k is a successor of the right son of u **and if** u has a left son **then**
            begin
                let x_i be the label of u;
                if IMPLICANT($A_k \cup \{i\}$) = FALSE **then return** FALSE;
            end
    end
    return TRUE;
end

Figure 7.4. Procedure LE-PROPERTY.

respect to $(x_2,x_3,\ldots,x_n)$. Hence, by induction, the procedure returns the output
TRUE.

Assume now, on the other hand, that $1 \in A$ and that x_1 is a leader. Then, $\bigwedge_{j \in A} x_A$
is absorbed by a term of ϕ_1 or by a term of ϕ_0. In both cases, however, the definition
of leaders implies that $\bigwedge_{j \in A} x_A$ is absorbed by a term of ϕ_1. Hence, u_2 is set equal
to left son of $r(\phi)$, namely, $r(\phi_1)$, and the proof is complete by induction as in the
previous case. $\square$

We can now state (in Figure 7.4) the efficient procedure proposed by Provan
and Ball [760] to test whether a DNF $\phi(x_1,x_2,\ldots,x_n)$ has the LE property with
respect to the identity permutation $(x_1,x_2,\ldots,x_n)$.

Theorem 7.14. *The procedure* LE-PROPERTY *is correct and can be implemented
to run in* $O(n^2 m)$ *time.*

Proof. Assume first that ϕ has the LE property with respect to $(x_1,x_2,\ldots,x_n)$.
Trivially, for all $k \in \{1,2,\ldots,m\}$ and all $i \in \{1,2,\ldots,n\}$, $\bigwedge_{j \in A_k \cup \{i\}} x_j$ is absorbed by
the term $\bigwedge_{j \in A_k} x_j$. Hence, by Theorem 7.13, IMPLICANT($A_k \cup \{i\}$) always returns
the answer TRUE, and LE-PROPERTY eventually returns TRUE.

Conversely, assume that LE-PROPERTY returns TRUE, and consider two sets
A_ℓ, A_k with $A_\ell <_L A_k$ and $i = \min\{j \mid j \in A_\ell \setminus A_k\}$. Let v_k and v_ℓ be the leaves
of $T(\phi)$ associated with A_k and A_ℓ, respectively. On the path from $r(\phi)$ to v_k,
consider the last vertex u that is an ancestor of v_ℓ. Then, u is labeled by x_i, and
IMPLICANT($A_k \cup \{i\}$) is called in the innermost **for** loop of the procedure. When
running on the input $A_k \cup \{i\}$, IMPLICANT traverses $T(\phi)$ until vertex u, then visits

the left son of u, and eventually returns the value TRUE (by assumption). By Theorem 7.12, this means that $A_k \cup \{i\}$ is absorbed by the term $C_h = \bigwedge_{j \in A_h} x_j$ associated with the leaf reached by IMPLICANT. It follows that $A_h <_L A_k$ and $A_h \setminus A_k = \{i\}$; hence, ϕ has the LE property.

We have mentioned that $T(\phi)$ can be set up in time $O(nm)$. LE-PROPERTY makes at most nm calls on IMPLICANT, and each of these calls can be executed in time $O(n)$. Hence, the overall running time of LE-PROPERTY is $O(n^2 m)$.   □

In contrast with the previous results, Provan and Ball [760] pointed out that the existence of an efficient procedure to determine whether a DNF has the LE property with respect to *some unknown permutation* of its variables, is far from obvious. Boros et al. [111] settled this question in the negative by proving the following result:

Theorem 7.15. *It is NP-complete to decide whether a positive DNF ϕ has the LE property, even when ϕ has degree at most 5.*

We omit the (rather technical) proof of this result. It should be noted, however, that the complexity of this recognition problem remains open for DNFs of degree 3 or 4. The case of quadratic DNFs is the topic of Section 7.5.

7.4.4 Dualization of functions having the LE property

In Section 1.12.3 of Chapter 1, we saw that, given a Boolean function f and an arbitrary order of the variables, say, $(x_1, x_2, \ldots, x_n)$, a decision tree $D(f)$ for f can be recursively constructed as follows (see Figure 1.5):

(a) If f is constant, then $D(f)$ has a unique vertex (which is both its root and its leaf) labeled with the constant value of f (either 0 or 1).
(b) Otherwise, let $f_0 = f_{|x_1=0}$, $f_1 = f_{|x_1=1}$, and build $D(f)$ by introducing a root $r(f)$ labeled by x_1, creating disjoint copies of $D(f_0)$ and $D(f_1)$ and making $r(f_1)$ (respectively, $r(f_0)$) the left son (respectively, the right son) of $r(f)$.

When f is represented by a positive DNF $\phi(x_1, x_2, \ldots, x_n) = x_1 \phi_1 \vee \phi_0$, and when ϕ has the LE property with respect to $(x_1, x_2, \ldots, x_n)$, Theorem 7.10 implies that $f_1 = \phi_1$ and $f_0 = \phi_0$ (unless $\phi_1 = 0$, in which case $f_1 = f_0 = \phi_0$). It is then easy to see that the decision tree $D(f)$ produced by the above procedure is essentially identical to the binary tree $T(\phi)$ defined in Section 7.4.3, up to some minor differences. In particular, $D(f)$ has at most $2nm$ leaves and can be set up in time $O(nm)$.

Therefore, as a corollary of Theorem 1.35 and Theorem 1.36, we obtain (E. Boros, personal communication):

Theorem 7.16. *If a Boolean function $f(x_1, x_2, \ldots, x_n)$ is expressed by a positive DNF ϕ such that ϕ has the LE property with respect to $(x_1, x_2, \ldots, x_n)$, then a* *decision tree $D(f)$ representing f can be built in time $O(nm)$. Moreover, an ODNF of f, an ODNF of f^d, and the prime implicants of f^d can be generated from $D(f)$ in time $O(n^2 m)$.*

Proof. We leave the details of the proof to the reader.   □

Although this theorem follows in a rather straightforward way from well-known properties of decision trees and from the results in Provan and Ball [760], it does not seem to have been formulated explicitly in the literature; see Boros [105] for related considerations.

7.5 Shellable quadratic DNFs and graphs

In this section, we concentrate on the case in which $\phi(x_1, x_2, \ldots, x_n)$ is a pure quadratic positive DNF, that is, a DNF of the form $\phi = \bigvee_{\{i,j\} \in E} x_i x_j$, where E is a set of pairs of elements of $N = \{1, 2, \ldots, n\}$. We assume that all members of E are distinct, so that ϕ can be viewed as the complete DNF of a quadratic positive function f and $G = (N, E)$ is a simple, undirected graph. For simplicity, we transpose from DNFs to graphs the terminology introduced in this chapter. Thus, a graph $G = (N, E)$ is shellable if and only if the corresponding quadratic positive DNF $\phi = \bigvee_{\{i,j\} \in E} x_i x_j$ is shellable. Similarly, we speak of shelling of the edges of G, of the LE property for G, and so on.

The purpose of this section is to present some results characterizing shellable graphs from Benzaken et al. [66]. Let us first recall a few graph-theoretic definitions (we follow the terminology in Appendix A and in Golumbic [398]). We denote by C_k a chordless cycle on k vertices and k edges ($k \geq 3$), and by $2K_2$ the graph on four vertices consisting of two disjoint edges. So, $2K_2$ is the complement of C_4. A graph is called *triangulated* (or *chordal*) if it contains no induced cycle of length 4 or more. Triangulated graphs constitute one of the fundamental, and most extensively studied, classes of perfect graphs. They have been characterized in numerous ways; see, for example, Berge [71], Brandstädt, Le and Spinrad [152], Duchet [284], Golumbic [398], and so on. We shall use the fact that a graph $G = (N, E)$ is the complement of a triangulated graph if and only if every induced subgraph of G contains a *cosimplicial vertex*, that is, a vertex v such that $\{u \in N \mid \{u, v\} \notin E\}$ is a stable set.

Benzaken et al. [66] observed that shellable graphs can be built up, one edge at a time, without ever producing $2K_2$.

Theorem 7.17. *Let $G = (V, E)$ and $E = \{e_1, e_2, \ldots, e_m\}$. The permutation $(e_1, e_2, \ldots, e_m)$ is a shelling of G if and only if, for every $k = 1, 2, \ldots, m$, the graph $G_k = (N, \{e_1, e_2, \ldots, e_k\})$ has no induced subgraph isomorphic to $2K_2$.*

Proof. By Theorem 7.4(e), $(e_1, e_2, \ldots, e_m)$ is a shelling of G if and only if, for all $e_\ell, e_k \in E$ with $\ell < k$, there exists $j \in e_\ell$ and there exists $h < k$ such that $e_h \setminus e_k = \{j\}$. The latter condition means that the edge e_h shares at least one vertex (namely, vertex j) with e_ℓ, and shares exactly one vertex with e_k. This is easily seen to be equivalent to the condition that $e_\ell \cup e_k$ does not induce $2K_2$ in G_k. □

We are now ready for our main characterization of shellable graphs [66].

Theorem 7.18. *For a graph G, the following statements are equivalent:*

(a) *G has the LE property.*
(b) *G is shellable.*
(c) *The complement of G is triangulated.*

Proof. (a) $\Rightarrow$ (b). This implication holds by Theorem 7.9.

(b) $\Rightarrow$ (c). Assume first that $G = (N, E)$ is the complement of a chordless cycle on n vertices, that is, $G = \overline{C_n}$. Then, we show by induction on n that G is not shellable. Indeed, if $n = 4$, then $G = \overline{C_4}$ is isomorphic to $2K_2$, hence G is not shellable by Theorem 7.17. For $n > 4$, assume by contradiction that $(e_1, e_2, \ldots, e_m)$ is a shelling of G, and let $H = (N, E \setminus \{e_m\})$. Then, $(e_1, e_2, \ldots, e_{m-1})$ is a shelling of H, and hence, H is shellable. Note that H contains $\overline{C_k}$, the complement of a chordless cycle on k vertices, with $4 \leq k < n$ (indeed, the complement of H is a cycle on n vertices with exactly one chord). Now, by induction, $\overline{C_k}$ is not shellable. On the other hand, in view of Theorem 7.17, all the induced subgraphs of a shellable graph are shellable; hence, $\overline{C_k}$ (as a subgraph of H) should be shellable. The contradiction shows that the complement of a chordless cycle is not shellable. Now, if G is not the complement of a triangulated graph, then G contains the complement of a chordless cycle as an induced subgraph, implying that G is not shellable.

(c) $\Rightarrow$ (a). Let G be the complement of a triangulated graph, and let $(v_1, v_2, \ldots, v_n)$ be a permutation of N such that, for $j = 1, 2, \ldots, n$, v_j is a cosimplicial vertex in the subgraph G_j of G induced by $\{v_j, v_{j+1}, \ldots, v_n\}$. We want to prove that G has the LE property with respect to $(v_1, v_2, \ldots, v_n)$.

Consider two edges $e_\ell = \{v_j, v_i\}$ and $e_k = \{v_r, v_s\}$ with $e_\ell <_L e_k$, $j < i$ and $j \leq r < s$. We must show that there exists $e_h <_L e_k$ such that $e_h \setminus e_k = \min\{v_t \mid v_t \in e_\ell \setminus e_k\}$. If $v_j = v_r$, or $v_i = v_r$, or $v_i = v_s$, then it is easy to check that $e_h = e_\ell$ satisfies the condition. Hence, we can assume that all four vertices v_j, v_i, v_r, and v_s are distinct. Consider now the subgraph G_j: Since v_j is cosimplicial in this graph, and $\{v_r, v_s\}$ is not stable, either v_r or v_s must be a neighbor of v_j. Suppose, for instance, that $e_h = \{v_j, v_r\} \in E$ (the other case is similar). Then e_h is as required. □

As a consequence of Theorem 7.18, we note that quadratic shellable DNFs can be recognized in $O(n^2)$ time, since the same result holds for triangulated graphs.

7.6 Applications

We conclude this chapter with a brief presentation of three generic classes of shellable DNFs arising in reliability theory and in game theory. We refer to Ball and Nemhauser [48], Ball and Provan [49, 760], or Colbourn [205, 206] for a more detailed discussion.

Application 7.1. (Undirected all-terminal reliability.) *Let $G = (N, E)$ be a connected undirected graph with $E = \{e_1, e_2, \ldots, e_m\}$, and let $\mathcal{T}$ be the collection of all spanning trees of G, viewed as subsets of E. Let us associate with every edge e_i, $i = 1, 2, \ldots, m$, a Boolean variable x_i indicating whether the edge is operational or failed. Then, the DNF $\phi = \bigvee_{T \in \mathcal{T}} \bigwedge_{e_i \in T} x_i$ takes value 1 exactly when the graph formed by the operational edges is connected. In the terminology of Section 1.13.4, ϕ represents the structure function of the reliability system whose minimal pathsets are the spanning trees of G.*

We claim that ϕ satisfies the LE property with respect to $(x_1, x_2, \ldots, x_m)$ (i.e., with respect to an arbitrary permutation of its variables). Indeed, let T_ℓ, T_k be two spanning trees with $T_\ell <_L T_k$, and let $j = \min\{i \mid e_i \in T_\ell \setminus T_k\}$. From elementary properties of trees, there exists an edge $e_i \in T_k \setminus T_\ell$ such that $T_k \cup \{e_j\} \setminus \{e_i\}$ is a spanning tree. Call this spanning tree T_h. Then, $T_h \setminus T_k = \{e_j\}$, and $T_h <_L T_k$, as required by the LE property.

This result implies, in particular, that the all-terminal reliability of a graph can be computed in time polynomial in the number of spanning trees of the graph (see Ball and Nemhauser [48] for details). □

Application 7.2. (Matroids.) *This is a generalization of the previous example. A collection $\mathcal{M}$ of subsets of $N = \{1, 2, \ldots, n\}$ is the set of bases of a matroid if it satisfies the following condition: For all $B_\ell, B_k \in \mathcal{M}$ and for all $j \in B_\ell \setminus B_k$, there exists $i \in B_k \setminus B_\ell$ such that $B_k \cup \{j\} \setminus \{i\}$ is in $\mathcal{M}$ (see, e.g., Welsh [905, 906]). It is well-known that the spanning trees of a connected graph are the bases of a matroid.*

Now, when $\mathcal{M}$ is the set of bases of a matroid on N, let $\phi_{\mathcal{M}}(x_1, x_2, \ldots, x_n) = \bigvee_{B \in \mathcal{M}} \bigwedge_{i \in B} x_i$. From the foregoing definition, it is easy to check that $\phi_{\mathcal{M}}$ satisfies the LE property with respect to every ordering of $(x_1, x_2, \ldots, x_n)$. □

Application 7.3. (Threshold functions and weighted majority games.) *Suppose that $f(x_1, x_2, \ldots, x_n)$ is a threshold function representing a weighted majority game on $N = \{1, 2, \ldots, n\}$, as defined in Chapter 1, Section 1.13.3. Thus, each player i carries a positive weight $w_i \in \mathbb{R}$, and the point $X^* \in \mathcal{B}^n$ is a true point of f (i.e., X^* is the characteristic vector of a winning coalition of players) if and only if $\sum_{i=1}^{n} w_i x_i^* > t$, where $t \in \mathbb{R}$ is a predetermined quota. If the weights are sorted so that $w_1 \geq w_2 \geq \ldots \geq w_n$, then the function f (or, equivalently, the complete DNF of f) has the LE property with respect to $(x_1, x_2, \ldots, x_n)$.*

Indeed, observe that the prime implicants of f correspond to the minimal winning coalitions of the game, and let A_ℓ and A_k be two minimal winning coalitions such that $A_\ell <_L A_k$. If $j = \min\{i \mid i \in A_\ell \setminus A_k\}$ and i is any index in $A_k \setminus A_\ell$, then $A_k \cup \{j\} \setminus \{i\}$ is a winning coalition (since $j < i$ implies $w_j \geq w_i$). There-fore, there exists $A_h \subseteq A_k \cup \{j\} \setminus \{i\}$ such that A_h is a minimal winning coalition, $A_h \setminus A_k = \{j\}$, and $A_h <_L A_k$. This shows that f has the LE property.

As a consequence, the number of true points (namely, winning coalitions) of f can be efficiently computed when the list of all its minimal true points (namely, minimal winning coalitions) is available. In view of the relation between Chow parameters and Banzhaf indices (as discussed in Chapter 1, Section 1.13.3 and Section 1.13.4), this also implies that the Banzhaf indices of a weighted major-ity game can be computed in time polynomial in the number of its minimal true points. We return to these topics in subsequent chapters (Chapter 8 and Chapter 9). In particular, Chapter 8 is devoted to the investigation of an important class of shellable functions generalizing threshold functions, namely, the class of regular functions.   □

7.7 Exercises

1. Let $C_1, C_2, \ldots, C_k$ and U be elementary conjunctions. Prove that it is co-NP-complete to decide whether $\overline{C_1} \, \overline{C_2} \ldots \overline{C_{k-1}} C_k \leq U$. (Compare with Lemma 7.2.) *Hint:* Let $C_k = y$ and $U = z$, where y, z are variables not occuring in $C_1, C_2, \ldots, C_{k-1}$.

2. Complete the argument following the proof of Theorem 7.6: Show that the DNF $\psi_\mathcal{L}$ is shellable, where $\psi_\mathcal{L}$ is the disjunction of all leftmost implicants of a positive function.

3. A positive DNF $\phi = \bigvee_{k=1}^m \bigwedge_{i \in A_k} x_i$ is *aligned* if, for every $k = 1, 2, \ldots, m$ and for every $j \notin A_k$ such that $j < h_k = \max\{i : i \in A_k\}$, there exists $A_\ell \subseteq (A_k \cup \{j\}) \setminus \{h_k\}$. Prove that every aligned DNF has the LE property (see Boros [105] and Section 8.9.2).

4. Complete the proof of Theorem 7.16 (see Boros [105]).

5. Let $\phi(x_1, x_2, \ldots, x_n) = \bigvee_{k=1}^m \bigwedge_{i \in A_k} x_i$ be a DNF such that $|A_k| = n - 2$ for $k = 1, 2, \ldots, m$. Show that ϕ is shellable if and only if the graph $G = (N, E)$ is connected, where $N = \{1, 2, \ldots, n\}$ and $E = \{N \setminus A_k \mid k = 1, 2, \ldots, m\}$.

Questions for thought

6. Find a small, shellable DNF that does not have the LE property with respect to *any* order of its variables.

7. Prove or disprove: If a DNF $\phi(x_1, x_2, \ldots, x_n)$ is shellable with respect to the lexicographic order of its terms, then it has the LE property with respect to *some* order of its variables. (Compare with Example 7.7.)

8. Determine the complexity of SHELLABILITY:
   Instance: A positive DNF ϕ.
   Output: Yes if ϕ is shellable, No otherwise.

9. The article [111] states a stronger form of Theorem 7.8, namely, it claims that:

   Claim. *If a Boolean function in n variables can be represented by a shellable positive DNF of m terms, then its dual can be represented by a shellable DNF of at most nm terms.*

   Unfortunately, the proof given in [111] is flawed, so that the validity of the claim (namely, the existence of a short, *shellable* DNF of the dual) remains open. Can you prove or disprove it?

8

Regular functions

In this chapter we investigate the main properties of regular Boolean functions. This class of functions constitutes a natural extension of the class of threshold functions, and, as such, has repeatedly and independently been "rediscovered" by several researchers over the last 40 years. It turns out that regular functions display many of the most interesting properties of threshold functions, and that these properties are, accordingly, best understood by studying them in the appropriate context of regularity. From an algorithmic viewpoint, regular functions constitute one of the most tractable classes of Boolean functions: Indeed, fundamental problems such as dualization, computation of reliability, or set covering are efficiently solvable when associated with regular functions. Besides its more obvious implications, this nice algorithmic behavior will eventually pave the way for the efficient recognition of threshold functions, which are discussed in the next chapter.

8.1 Relative strength of variables and regularity

In Chapter 1 (Definition 1.31), we defined the class of threshold Boolean functions as follows:

Definition 8.1. *A Boolean function f on B^n is a* threshold *(or* linearly separable*) function if there exist n weights $w_1, w_2, \ldots, w_n \in \mathbb{R}$ and a threshold $t \in \mathbb{R}$ such that, for all $(x_1, x_2, \ldots, x_n) \in B^n$,*

$$f(x_1, x_2, \ldots, x_n) = 0 \text{ if and only if } \sum_{i=1}^{n} w_i x_i \leq t.$$

The $(n+1)$-tuple $(w_1, w_2, \ldots, w_n, t)$ is called a (separating) structure *of f.*

One of the most remarkable properties of a threshold function is that the weights $w_1, w_2, \ldots, w_n$ naturally determine an ordinal ranking of the variables, translating the relative "influence" of the variables on the value of the function: Namely, if $w_i \geq w_j$, then the function is "more likely" to take the value 1 when $x_i = 1$ and $x_j = 0$ than when $x_i = 0$ and $x_j = 1$.

This notion of relative influence, or relative strength, of variables can be extended to more general Boolean functions, as expressed by the following definition, which was independently introduced by Isbell [520]; Muroga, Toda, and Takasu [700]; Paull and McCluskey [732]; Winder [916]; Maschler and Peleg [673]; Neumaier [709], and so on. In this definition, as usual, we denote by e_k the n–dimensional unit vector with k-th component equal to 1 ($k = 1, 2, \ldots, n$).

Definition 8.2. *Let* $f(x_1, x_2, \ldots, x_n)$ *be a Boolean function, and let* $i, j \in \{1, 2, \ldots, n\}$. *We say that* variable x_i is stronger than variable x_j with respect to f, *and we write* $x_i \succeq_f x_j$ *if and only if, for all* $X^* \in \mathcal{B}^n$,

$$x_i^* = x_j^* = 0 \;\Rightarrow\; f(X^* \vee e_i) \geq f(X^* \vee e_j).$$

Equivalently, $x_i \succeq_f x_j$ *if either* $i = j$ *or* $f_{|x_i=1, x_j=0} \geq f_{|x_i=0, x_j=1}$.

The subscript f appearing in the symbol $\succeq_f$ is a reminder that the strength relation depends on f. To simplify the notations, we sometimes write "$x_i \succeq x_j$ with respect to f", instead of $x_i \succeq_f x_j$.

Let us illustrate Definition 8.2 with a couple of examples.

Example 8.1. *Let* $f(x_1, x_2) = x_1 \overline{x_2} \vee \overline{x_1} x_2$. *There holds* $x_1 \succeq_f x_2$ *and* $x_2 \succeq_f x_1$, *since* $f(0, 1) = f(1, 0) = 1$. □

Example 8.2. *If* $f(x_1, x_2, x_3, x_4) = x_1 x_2 \vee x_2 x_3 \vee x_3 x_4$, *then* $x_2 \succeq_f x_1$. *Indeed, for all values of* x_3 *and* x_4, $f(0, 1, x_3, x_4) = x_3 \geq f(1, 0, x_3, x_4) = x_3 x_4$.

One similarly verifies that $x_2 \succeq_f x_4$, $x_3 \succeq_f x_1$, *and* $x_3 \succeq_f x_4$. *No other pairs of variables are comparable with respect to* $\succeq_f$. *For instance,* x_1 *and* x_4 *are not comparable, since* $f(1, x_2, x_3, 0) = x_2$ *and* $f(0, x_2, x_3, 1) = x_3$. □

A bit of additional terminology and notations comes in handy when dealing with the strength relation. We say that

- x_i is *strictly stronger* than x_j ($x_i \succ_f x_j$) if $x_i \succeq_f x_j$ but not $x_j \succeq_f x_i$;
- x_j is *weaker* than x_i ($x_j \preceq_f x_i$) if $x_i \succeq_f x_j$, and x_j is *strictly weaker* than x_i ($x_j \prec_f x_i$) if $x_i \succ_f x_j$;
- x_i and x_j are *comparable* if $x_i \succeq_f x_j$ or $x_j \succeq_f x_i$;
- x_i and x_j are *equivalent* or *symmetric* ($x_i \approx_f x_j$) if $x_i \succeq_f x_j$ and $x_j \succeq_f x_i$.

The qualifier "symmetric" is justified by the following easy observation.

Theorem 8.1. *For a Boolean function* $f(x_1, x_2, \ldots, x_n)$, *and for* $i, j \in \{1, 2, \ldots, n\}$, $x_i \approx_f x_j$ *if and only if*

$$f(Y) = f(Y^{ij}) \text{ for all } Y \in \mathcal{B}^n, \tag{8.1}$$

where Y^{ij} *is the point obtained by exchanging the values of components i and j in Y.*

Proof. Assume that $x_i \approx_f x_j$, and let $Y \in \mathcal{B}^n$. If $y_i = y_j$ then (8.1) trivially holds. Else, suppose, for instance, that $y_i = 1$ and $y_j = 0$. Then, since $x_i \succcurlyeq_f x_j$, $f(Y) \geq f(Y^{ij})$, and since $x_j \succcurlyeq_f x_i$, $f(Y) \leq f(Y^{ij})$. Thus (8.1) holds again.

One similarly shows that (8.1) implies $x_i \approx_f x_j$.   □

As Example 8.2 illustrates, certain pairs of variables may turn out to be incomparable with respect to the strength relation $\succcurlyeq_f$; in other words, $\succcurlyeq_f$ is generally not a *complete* relation. On the other hand, as we prove now, the strength relation always defines a *preorder*, that is, a reflexive and transitive relation.

Theorem 8.2. *The strength relation is a preorder on the set of variables of every Boolean function.*

Proof. The strength relation is obviously reflexive. To see that it is also transitive, consider a function $f(x_1, x_2, \ldots, x_n)$, three indices i, j, k such that $x_i \succcurlyeq_f x_j$ and $x_j \succcurlyeq_f x_k$, and a point $X^* \in \mathcal{B}^n$ with $x_i^* = x_k^* = 0$. We must show that $f(X^* \vee e_k) \leq f(X^* \vee e_i)$.

If $x_j^* = 0$, then $f(X^* \vee e_k) \leq f(X^* \vee e_j) \leq f(X^* \vee e_i)$ and we are done. If $x_j^* = 1$, then let $Y^* \in \mathcal{B}^n$ be the point obtained by switching the j-th component of X^* from 1 to 0; thus, $y_j^* = 0$ and $X^* = Y^* \vee e_j$. Then,

$$
\begin{aligned}
f(X^* \vee e_k) &= f(Y^* \vee e_j \vee e_k) \\
&\leq f(Y^* \vee e_i \vee e_k) &&\text{(since } x_i \succcurlyeq_f x_j) \\
&\leq f(Y^* \vee e_i \vee e_j) &&\text{(since } x_j \succcurlyeq_f x_k) \\
&= f(X^* \vee e_i),
\end{aligned}
$$

and the proof is complete.   □

We pointed out in the introductory paragraphs of this section that the strength relation associated with a threshold function is always complete. More precisely, we can state:

Theorem 8.3. *If $f(x_1, x_2, \ldots, x_n)$ is a threshold function with separating structure $(w_1, w_2, \ldots, w_n, t)$ and $w_1 \geq w_2 \geq \ldots \geq w_n$, then $x_1 \succcurlyeq_f x_2 \succcurlyeq_f \cdots \succcurlyeq_f x_n$.*

Proof. Let $1 \leq i < j \leq n$. If $X^* \in \mathcal{B}^n$ and $x_i^* = x_j^* = 0$, then $\sum_{k=1}^n w_k x_k^* + w_i \geq \sum_{k=1}^n w_k x_k^* + w_j$, and hence, $f(X^* \vee e_i) \geq f(X^* \vee e_j)$.   □

Threshold functions are not the only Boolean functions featuring a complete strength preorder. For instance, the function displayed in Example 8.1 has a complete strength preorder but is not a threshold function, since it is not monotone (the reader will easily verify that every threshold function is monotone). If we restrict our attention to monotone functions, then it can be shown that all functions of five variables for which the strength preorder is complete are threshold functions, but this implication fails for functions of six variables or more (see Winder [917] and Exercise 11 at the end of this chapter).

The foregoing observations motivate the main definition of this chapter.

Definition 8.3. *A positive Boolean function* f *is* regular *if its strength preorder is complete. In particular, we say that* $f(x_1,x_2,\ldots,x_n)$ *is* regular with respect to $(x_1,x_2,\ldots,x_n)$ *if* $x_1 \succeq_f x_2 \succeq_f \cdots \succeq_f x_n$.

Example 8.3. *The function* f *in Example 8.2 is not regular since* x_1 *and* x_4 *are not comparable in the preorder* $\succeq_f$.

On the other hand, the function $g(x_1,x_2,x_3) = x_1 x_2 \vee x_1 x_3$ *is regular with respect to* (x_1,x_2,x_3), *and the function* $h(x_1,x_2,\ldots,x_5) = x_1 x_2 \vee x_1 x_3 \vee x_1 x_4 x_5 \vee x_2 x_3 x_4$ *is* regular with respect to $(x_1,x_2,\ldots,x_5)$. $\square$

Because it is so natural and (as we will see) fruitful, the regularity concept has been "rediscovered" several times in various fields of applications (see Muroga, Toda, and Takasu [700]; Paull and McCluskey [732]; Winder [916]; Neumaier [709]; Golumbic [398]; Ball and Provan [49], etc.). It constitutes our main object of study in this chapter.

Before diving more deeply into this topic, however, let us first offer the impatient reader an illustration of how the notion of strength preorder can be used in a game-theoretical framework. More applications are presented at the end of Section 8.2, after we have become better acquainted with the elementary properties of the strength relation.

Application 8.1. (Political science, Game theory.) *The legislative body in Booleland consists of 45 representatives, 11 senators and a president. In order to be passed by this legislature, a bill must receive*

(1) *at least half of the votes in the House of Representatives and in the Senate, as well as the president's vote, or*
(2) *at least two-thirds of the votes in the House of Representatives and in the Senate.*

(The knowledgeable reader will recognize that this lawmaking process is a slightly simplified version of the system actually in use in the United States)

As usual, we can model this voting mechanism by a monotone Boolean function $f(r_1,\ldots,r_{45}, s_1,\ldots,s_{11}, p)$, *where variable* r_i *(respectively,* s_j, p*) takes value 1 if representative* i *(respectively, senator* j, *the president) casts a "Yes" vote, and takes value 0 otherwise* $(1 \le i \le 45$ *and* $1 \le j \le 11)$. *The true points of* f *correspond to the voting patterns described by rules (1) and (2) above.*

A more detailed description of f *can be obtained as follows: For* $k,n \ge 1$, *denote by* $g_k(x_1,x_2,\ldots,x_n)$ *the "k–majority" function on* n *variables, that is, the threshold function defined by*

$$g_k(x_1,x_2,\ldots,x_n) = 1 \iff \sum_{i=1}^{n} x_i \ge k.$$

Then f can be expressed as

$$f(r_1,\ldots,r_{45},s_1,\ldots,s_{11},p)$$
$$= \big(g_{23}(r_1,\ldots,r_{45}) \wedge g_6(s_1,\ldots,s_{11}) \wedge p \big) \vee \big(g_{30}(r_1,\ldots,r_{45}) \wedge g_7(s_1,\ldots,s_{11}) \big).$$

One can easily verify that, with respect to the strength preorder associated to f,

- *any two representatives are equivalent;*
- *any two senators are equivalent;*
- *a representative and a senator cannot be compared in terms of strength;*
- *the president is strictly stronger than any representative or any senator.*

In this political setting, the strength preorder can be straightforwardly interpreted as defining an ordinal measure of power *on the set of legislators. Indeed, what does it mean here for legislator i to be* (strictly) *stronger than legislator j? Simply that, if S is any coalition (that is, subset) of legislators who all decided to vote in the same way (either all "Yes" or all "No"), and if neither i nor j has committed her vote yet, then the members of S prefer i to j as an additional convert. Indeed, i's vote is more likely to influence the final outcome of the vote than j's vote. Thus, i is "more powerful" than j. It seems that the strength relation was first explicitly introduced in this context by Maschler and Peleg [673], although similar concepts can be found in Isbell [520].* □

8.2 Basic properties

We present in this section some of the fundamental properties of the strength preorder and of regular functions. To begin with, we address an issue which may already have come to the reader's mind during our discussion of Application 8.1, namely, the question of the relationship between the strength preorder and the Chow parameters of a function. As a matter of fact, we argued in Chapter 1 that the Chow parameters provide a numerical measure of the influence of each variable on the value of the function (see Sections 1.6, 1.13.3, 1.13.4). Since the strength relation also captures this influence, albeit in an ordinal setting, one would legitimately expect some connection between the two concepts. Such a connection indeed exists, as expressed by the next statement.

Theorem 8.4. *Let $f(x_1, x_2, \ldots, x_n)$ be a Boolean function, and let $(\omega_1, \omega_2, \ldots, \omega_n, \omega)$ denote its Chow parameters. If $x_i \succ_f x_j$, then $\omega_i > \omega_j$. If $x_i \approx_f x_j$, then $\omega_i = \omega_j$.*

Proof. Let T be the set of true points of f. If $x_i \succcurlyeq_f x_j$, then it follows immediately from the definitions of the Chow parameters and of the strength preorder that

$$\omega_i = |\{X \in T : x_i = 1\}|$$
$$= |\{X \in T : x_i = x_j = 1\}| + |\{X \in T : x_i = 1, x_j = 0\}|$$
$$\geq |\{X \in T : x_i = x_j = 1\}| + |\{X \in T : x_i = 0, x_j = 1\}|$$
$$= |\{X \in T : x_j = 1\}|$$
$$= \omega_j.$$

If $x_i \succ_f x_j$, then there exists at least one point $X^* \in \mathcal{B}^n$ such that $x_i^* = x_j^* = 0$, $f(X^* \vee e_i) = 1$ and $f(X^* \vee e_j) = 0$. Thus, the above inequality is strict. If $x_i \approx_f x_j$, then f is symmetric on x_i, x_j, and hence, $\omega_i = \omega_j$.   □

Having clarified this point, let us now turn to the issue of deciding whether two variables are comparable with respect to the strength relation. We only deal with positive functions expressed by their complete (i.e., prime irredundant) DNF, as the same question turns out to be NP-hard for arbitrary DNFs (see Exercise 2 at the end of this chapter).

Theorem 8.5. *Let $f(x_1, x_2, \ldots, x_n)$ be a positive Boolean function and let i, j be distinct indices in $\{1, 2, \ldots, n\}$. Write the complete DNF of f in the form $\alpha\, x_i\, x_j \vee \beta\, x_i \vee \gamma\, x_j \vee \delta$, where α, β, γ and δ are positive DNFs which do not involve x_i nor x_j. Then*

$$x_i \succcurlyeq_f x_j \text{ if and only if } \beta \geq \gamma.$$

Proof. Without loss of generality, suppose that $i = 1$ and $j = 2$. For $X = (0, 0, Y) \in \mathcal{B}^n$, we get $f(X \vee e_1) = \beta(Y) \vee \delta(Y)$, and $f(X \vee e_2) = \gamma(Y) \vee \delta(Y)$. Hence, by definition of the strength relation, $x_1 \succcurlyeq_f x_2$ if and only if $\beta(Y) \vee \delta(Y) \geq \gamma(Y) \vee \delta(Y)$ for all $Y \in \mathcal{B}^{n-2}$. To establish the theorem, note that $\beta \geq \gamma$ trivially implies $\beta \vee \delta \geq \gamma \vee \delta$. For the converse implication, assume that $\beta \vee \delta \geq \gamma \vee \delta$, and let C be a prime implicant of γ. Since $C \leq \gamma \leq \beta \vee \delta$, the DNF $\beta \vee \delta$ contains a term B which absorbs C. Note that B cannot be a term of δ (hence, of f), since B absorbs Cx_2, which is, by assumption, a prime implicant of f. Hence, B must be a term of β. We conclude that $\beta \geq \gamma$, and the proof is complete.   □

Theorem 8.5 can be rephrased as follows:

Theorem 8.6. *Let $f(x_1, x_2, \ldots, x_n)$ be a positive Boolean function. For all $i, j \in \{1, 2, \ldots, n\}$, the following statements are equivalent:*

(a) *$x_i \succcurlyeq_f x_j$.*
(b) *For each prime implicant of f, say, $\bigwedge_{k \in A} x_k$, such that $j \in A$ and $i \notin A$, $\bigwedge_{k \in (A \cup \{i\}) \setminus \{j\}} x_k$ is an implicant of f.*
(c) *For each prime implicant of f, say, $\bigwedge_{k \in A} x_k$, such that $j \in A$ and $i \notin A$, there is a prime implicant of f, say, $\bigwedge_{k \in P} x_k$, such that $P \subseteq (A \cup \{i\}) \setminus \{j\}$.*

Proof. This is an immediate consequence of Theorem 8.5 and Theorem 1.22.   □

Example 8.4. *Let* $f(x_1,x_2,x_3,x_4,x_5,x_6,x_7) = x_1x_2x_3 \lor x_1x_3x_4 \lor x_1x_3x_5 \lor x_2x_3x_4x_6 \lor x_2x_3x_4x_7 \lor x_2x_3x_5x_7 \lor x_4x_5x_6$. *Letting* $i = 1$ *and* $j = 2$ *in the statement of Theorem 8.5, we get*

$$\beta = x_3x_4 \lor x_3x_5 > \gamma = x_3x_4x_6 \lor x_3x_4x_7 \lor x_3x_5x_7,$$

and hence, $x_1 \succ_f x_2$. *On the other hand, for* $i = 1$ *and* $j = 4$, *we have*

$$\beta = x_2x_3 \lor x_3x_5 \text{ and } \gamma = x_2x_3x_6 \lor x_2x_3x_7 \lor x_5x_6.$$

Since neither $\beta \geq \gamma$ *nor* $\beta \leq \gamma$ *holds, we conclude that* x_1 *and* x_4 *are not comparable with respect to* $\succcurlyeq_f$. □

Let us now see how the strength preorder behaves under some fundamental transformations of Boolean functions, namely, restriction, composition, and dualization.

Theorem 8.7. *Let* $f(x_1,x_2,\ldots,x_n)$ *be a Boolean function, let* i,j,k *be distinct indices in* $\{1,2,\ldots,n\}$, *let* $g = f_{|x_k=1}$, *and let* $h = f_{|x_k=0}$. *Then,* $x_i \succcurlyeq_f x_j$ *if and only if both* $x_i \succcurlyeq_g x_j$ *and* $x_i \succcurlyeq_h x_j$. *Moreover, the following statements are equivalent:*

(a) f *is regular with respect to* $(x_1,x_2,\ldots,x_n)$.
(b) $x_1 \succcurlyeq_f x_2$, *and both* $f_{|x_1=1}$ *and* $f_{|x_1=0}$ *are regular with respect to* $(x_2,x_3,\ldots,x_n)$.

Proof. The first equivalence is an immediate consequence of Definition 8.2, and the second equivalence follows from it. □

Example 8.5. *As in Example 8.2, consider the function* $f(x_1,x_2,x_3,x_4) = x_1x_2 \lor x_2x_3 \lor x_3x_4$, *for which* $x_3 \succcurlyeq_f x_1$ *and* $x_3 \succcurlyeq_f x_4$. *The restriction of* f *to* $x_2 = 1$ *is the function* $g(x_1,x_3,x_4) = x_1 \lor x_3$, *and its restriction to* $x_2 = 0$ *is the function* $h(x_1,x_3,x_4) = x_3x_4$. *Theorem 8.7 implies that* $x_3 \succcurlyeq_g x_1$, $x_3 \succcurlyeq_g x_4$, $x_3 \succcurlyeq_h x_1$, $x_3 \succcurlyeq_h x_4$. *On the other hand,* x_1 *and* x_4 *are not comparable with respect to* f, *since* $x_1 \succ_g x_4$ *and* $x_4 \succ_h x_1$. *Thus,* f *is not regular (even though both* g *and* h *are regular; see also Exercise 1 at the end of the chapter).* □

We next establish an easy result concerning the composition of functions.

Theorem 8.8. *If* x_i *is stronger than* x_j *with respect to each of the Boolean functions* $f_k(x_1,x_2,\ldots,x_n)$ $(k = 1,2,\ldots,m)$, *and if* $g(y_1,y_2,\ldots,y_m)$ *is a positive function, then* x_i *is stronger than* x_j *with respect to the composite function* $h = g(f_1,f_2,\ldots,f_m)$, *for all* i,j *in* $\{1,2,\ldots,n\}$.

Proof. Let $h = g(f_1,f_2,\ldots,f_m)$, and let X^* be a point of $\mathcal{B}^n$ with $x_i^* = x_j^* = 0$. For $k = 1,2,\ldots,m$, $f_k(X^* \lor e_i) \geq f_k(X^* \lor e_j)$. Hence, by positivity of g, $h(X^* \lor e_i) \geq h(X^* \lor e_j)$. □

In particular, we observe that:

Theorem 8.9. *If x_i is stronger than x_j with respect to each of the Boolean functions $f_k(x_1, x_2, \ldots, x_n)$ $(k = 1, 2, \ldots, m)$, then x_i is stronger than x_j with respect to $f_1 f_2 \ldots f_m$ and with respect to $f_1 \vee f_2 \vee \ldots \vee f_m$.*

Proof. This is an immediate corollary of Theorem 8.8.   □

The strength preorder is invariant under dualization:

Theorem 8.10. *The strength preorders of a function and of its dual are identical. In particular, a function is regular if and only if its dual is regular.*

Proof. Let $f(x_1, x_2, \ldots, x_n)$ be a Boolean function and let i, j be distinct indices in $\{1, 2, \ldots, n\}$. We only have to show that, if x_i is stronger than x_j with respect to f, then x_i is stronger than x_j with respect to f^d (the converse implication follows by duality). For simplicity of presentation, assume that $i = 1$ and $j = 2$, and that $x_1 \succcurlyeq_f x_2$. Then, for all $X = (0, 0, Y) \in \mathcal{B}^n$,

$$f^d(1, 0, Y) = \overline{f}(0, 1, \overline{Y}) \geq \overline{f}(1, 0, \overline{Y}) = f^d(0, 1, Y).$$

Hence, x_1 is stronger than x_2 with respect to f^d.   □

Example 8.6. *Consider again $f(x_1, x_2, x_3, x_4) = x_1 x_2 \vee x_2 x_3 \vee x_3 x_4$, as in the previous example. Then, $f^d = x_1 x_3 \vee x_2 x_3 \vee x_2 x_4$, and the strength preorder of f^d is the same as that of f.*   □

In some of the subsequent developments, it will be of interest to know conditions which must hold when a variable is stronger than all the other ones. The next result states a simple necessary condition found in Winder [916].

Theorem 8.11. *Let $f(x_1, x_2, \ldots, x_n)$ be a positive Boolean function, not identically equal to 1, and let $i \in \{1, 2, \ldots, n\}$. Write the complete DNF of f in the form $\phi_1 x_i \vee \phi_0$, where ϕ_1 and ϕ_0 are positive DNFs that do not involve x_i. If $x_i \succcurlyeq_f x_j$ for $j = 1, 2, \ldots, n$, then $\phi_1 \geq \phi_0$.*

Proof. Suppose, for instance, that $i = 1$ and $x_1 \succcurlyeq_f x_j$ for $j = 1, 2, \ldots, n$. We only have to show that if $Y^* = (y_2^*, y_3^*, \ldots, y_n^*)$ is a minimal true point of ϕ_0, then Y^* is a true point of ϕ_1. If $Y^* = 0$, then ϕ_0 and f are identically 1, contradicting the hypothesis. Thus $Y^* \neq 0$, and we can assume that $Y^* = (1, Z^*)$, where $Z^* \in \mathcal{B}^{n-2}$. By assumption, $(0, Z^*)$ is a false point of ϕ_0. Therefore,

$$f(0, 1, Z^*) = \phi_0(1, Z^*) = \phi_0(Y^*) = 1,$$

$$f(1, 0, Z^*) = \phi_1(0, Z^*) \vee \phi_0(0, Z^*) = \phi_1(0, Z^*).$$

However, $x_1 \succcurlyeq_f x_2$ implies that $f(0, 1, Z^*) \leq f(1, 0, Z^*)$, and hence $\phi_1(0, Z^*) = 1$. By positivity of ϕ_1, we conclude that $\phi_1(Y^*) = \phi_1(1, Z^*) = 1$, as required.   □

Example 8.7. *Consider the function* $f(x_1, x_2, x_3) = x_1 x_2 \vee x_1 x_3$, *for which* $x_1 \succ_f x_2$ *and* $x_1 \succ_f x_3$. *Letting* $i = 1$ *in Theorem 8.11, we obtain* $\phi_1 = x_2 \vee x_3$, $\phi_0 = 0$, *and hence,* $\phi_1 \geq \phi_0$, *as expected.*

We can use the same example to show that the converse of Theorem 8.11 does not hold in general. Indeed, if we let $i = 2$ *in the statement of Theorem 8.11, then we get* $\phi_1 = x_1$, $\phi_0 = x_1 x_3$, *and hence* $\phi_1 \geq \phi_0$. *But* x_2 *is strictly weaker than* x_1. $\square$

In Chapter 7, when discussing shellability and the lexico-exchange (LE) property, we called "leader" a variable satisfying the necessary condition in Theorem 8.11 (see Definition 7.6), and we established the relation between the LE property and the existence of leaders in Theorem 7.10.

Combining these results, it is now rather straightforward to prove the following theorem due to Ball and Provan [49, 760] (see also Application 7.3 in Chapter 7).

Theorem 8.12. *If* $f(x_1, x_2, \ldots, x_n)$ *is regular with respect to* $(x_1, x_2, \ldots, x_n)$, *then* f *has the LE property with respect to* $(x_1, x_2, \ldots, x_n)$.

Proof. We use induction on n. When $n = 1$, the claim is trivial, so let us assume that $n > 1$. If f is regular with respect to $(x_1, x_2, \ldots, x_n)$, then, by Theorem 8.11, x_1 is a leader of f. Moreover, by Theorem 8.7, both $f_{|x_1=1}$ and $f_{|x_1=0}$ are regular, and hence, they have the LE property with respect to $(x_2, x_3, \ldots, x_n)$. Then, Theorem 7.10 implies that f also has the LE property with respect to $(x_1, x_2, \ldots, x_n)$. $\square$

We will return in subsequent sections to this connection between the LE property and regularity. For now, we describe some additional applications of the concepts of strength preorder and of regularity.

Application 8.2. (Integer programming.) *Consider an optimization problem in 0–1 variables of the form:*

$$\text{maximize } z(x_1, x_2, \ldots, x_n) = \sum_{i=1}^{n} c_i x_i \tag{8.2}$$

$$\text{subject to } f(x_1, x_2, \ldots, x_n) = 0 \tag{8.3}$$

$$(x_1, x_2, \ldots, x_n) \in \mathcal{B}^n, \tag{8.4}$$

where f *is a positive Boolean function (cf. Section 1.13.6 in Chapter 1). If* x_i *and* x_j *are two variables such that* $x_i \succeq_f x_j$ *and* $c_i \leq c_j$, *then one easily verifies that there exists an optimal solution* X^* *of (8.2)–(8.4) such that* $x_i^* \leq x_j^*$. *This fact can be used in an enumerative approach to the solution of (8.2)–(8.4). Indeed, as soon as variable* x_i *has been fixed to 1 in a branch of the enumeration tree, then* x_j *can automatically be fixed to 1. More generally, the conclusion that* $x_i \leq x_j$ *can also be handled as a logical condition to be satisfied by the optimal solution of the problem (see Application 2.4 in Section 2.1).*

In particular, if $c_1 \leq c_2 \leq \cdots \leq c_n$ *and if* f *is regular with* $x_1 \succeq_f x_2 \succeq_f \cdots \succeq_f x_n$, *then (8.2)–(8.4) has an optimal solution* X^* *satisfying* $x_1^* \leq x_2^* \leq \cdots \leq x_n^*$. *Under*

these assumptions, an optimal solution of (8.2)–(8.4) is given by the largest vector
X^* *of the form* $X^* = e_i \vee e_{i+1} \vee \cdots \vee e_n$ *which satisfies the constraint* $f(X^*) = 0$.
Such a solution is delivered by the greedy *procedure, which successively sets the*
variables $x_n, x_{n-1}, \ldots, x_1$ *to 1, while maintaining the feasibility of the solution thus*
produced.

In Section 8.6, we shall see that, when f *is a regular function given by the list*
of its prime implicants, problem (8.2)–(8.4) is always solvable in polynomial time,
without any further conditions on the coefficients $c_1, c_2, \ldots, c_n$. $\qquad\square$

Application 8.3. (Game theory). *Since a simple game is nothing but a positive*
Boolean function, we can speak of the strength preorder of a simple game (see
Section 1.13.3). What can be said about this preorder in a game-theoretic setting?

As discussed in Application 8.1, the strength preorder can be naturally inter-
preted as providing an ordinal *ranking of the players according to their relative*
power in the game. On the other hand, we have defined in Section 1.13.3 different
cardinal measures of power, or power indices, *associated with a simple game. In*
particular, we have observed that the Banzhaf indices are a monotone transfor-
mations of the Chow parameters of the associated Boolean function. Hence, it
follows from Theorem 8.4 that these power indices are consistent with the strength
preorder, in the following sense: If variable x_i *is (strictly) stronger than variable*
x_j *with respect to the strength preorder of the game, then the Banzhaf index of*
player i is (strictly) larger than the Banzhaf index of player j.

The notion of strength preorder has been extended by Maschler and Peleg [673]
to cooperative games in characteristic function form (i.e., pseudo-Boolean func-
tions, or real-valued functions of 0-1 variables; see Chapter 13). $\qquad\square$

Application 8.4. (Combinatorics). *A* tactical configuration *over the finite set* $N =$
$\{1, 2, \ldots, n\}$ *is a hypergraph* $\mathcal{H} = (N, \mathcal{E})$ *with the following two properties:*

1. *Each member of* $\mathcal{E}$ *has the same cardinality, say,* $k > 0$.
2. *Each element of* N *appears in the same number, say,* $r > 0$, *of members of* $\mathcal{E}$.

Neumaier [709] proved a result about tactical configurations, which is easily
stated and established in our Boolean-theoretic framework. Given the tactical
configuration $\mathcal{H} = (N, \mathcal{E})$, *let* $f_{\mathcal{H}}(x_1, x_2, \ldots, x_n)$ *be the positive Boolean function*
defined, as in Section 1.13.5, by

$$f_{\mathcal{H}}(x_1, x_2, \ldots, x_n) = \bigvee_{A \in \mathcal{E}} \bigwedge_{j \in A} x_j.$$

Note that $\mathcal{H}$ *is a tactical configuration if and only if all terms of* $f_{\mathcal{H}}$ *have the same*
degree k and every variable appears in r terms of $f_{\mathcal{H}}$. *Then, Neumaier's result*
states: If $\mathcal{H}$ *is a tactical configuration such that* $f_{\mathcal{H}}$ *is regular, then* $\mathcal{E} = \{A \subseteq N :$
$|A| = k\}$. *To see that this is indeed the case, consider any two variables* x_i *and* x_j
with $x_i \succcurlyeq_f x_j$ *and rewrite* $f_{\mathcal{H}}$ *in the form:* $f_{\mathcal{H}} = \alpha x_i x_j \vee \beta x_i \vee \gamma x_j \vee \delta$. *Theorem*
8.7 implies that $\beta \geq \gamma$. *But then, using the definition of a tactical configuration,*

it is easy to verify that $\beta = \gamma$. Since x_i and x_j are two arbitrary variables, we conclude that $f_{\mathcal{H}}$ is symmetric on all its variables, and Neumaier's result follows.

Euler [317] and Reiterman et al. [784] have investigated other classes of regular hypergraphs. □

Application 8.5. (Reliability.) *We have already mentioned that, in the terminology of reliability theory, every positive Boolean function $f(x_1,x_2,\ldots,x_n)$ can be interpreted as the structure function of a coherent binary system (see Section 1.13.4). The strength relation often has an obvious interpretation for complex engineering systems. For instance, if two resistors R_1 and R_2 are placed in series in an electrical circuit, and if R_1 has higher resistance than R_2, then R_1 is stronger than R_2 with respect to the structure function of the circuit.*

We have also mentioned that two of the fundamental algorithmic problems in reliability theory are the dualization of the structure function f and the computation of the reliability polynomial of f, that is, $\mathrm{Rel}_f = \mathrm{Prob}[f(x_1,x_2,\ldots,x_n)] = 1$, when the x_i's are viewed as independent Bernoulli random variables taking value 1 with probability p_i and value 0 with probability $1 - p_i$ $(i = 1,2,\ldots,n)$.

We already know that these two problems are computationally difficult for general functions but turn out to be polynomially solvable when f has the LE property (and is given as a complete DNF). Hence, by virtue of Theorem 8.12, they are also polynomially solvable when f is regular. Section 8.5 is devoted to the description of a streamlined, very efficient algorithm for the dualization of regular functions. As for the computation of Rel_f, the results described in Chapter 7 can be specialized as follows. □

Theorem 8.13. *Assume that $f(x_1,x_2,\ldots,x_n)$ is regular with respect to $(x_1,x_2,\ldots,x_n)$, and let $\bigvee_{k=1}^{m} \bigwedge_{j\in A_k} x_j$ denote the complete DNF of f. For $k = 1,2,\ldots,m$, let $\mu_k = \max\{j : j \in A_k\}$ and $S_k = \{1,2,\ldots,\mu_k\} \setminus A_k$. Then, f is represented by the orthogonal (sum of disjoint products) DNF*

$$\phi^{sh} = \bigvee_{k=1}^{m} \left(\bigwedge_{j\in A_k} x_j \right) \left(\bigwedge_{j\in S_k} \overline{x_j} \right) \tag{8.5}$$

and

$$\mathrm{Rel}_f(p_1,p_2,\ldots,p_n) = \mathrm{Prob}[f(X) = 1] = \sum_{k=1}^{m} \left(\prod_{j\in A_k} p_j \right) \left(\prod_{j\in S_k} (1 - p_j) \right). \tag{8.6}$$

Before proving Theorem 8.13, we illustrate it by means of a small example.

Example 8.8. Let $f = x_1x_2 \vee x_1x_3 \vee x_1x_4x_5 \vee x_2x_3x_4$. Then, $x_1 \succcurlyeq_f x_2 \succcurlyeq_f x_3 \succcurlyeq_f x_4 \succcurlyeq_f x_5$. We obtain

$$\mu_1 = 2,\ S_1 = \emptyset,\ \mu_2 = 3,\ S_2 = \{2\},\ \mu_3 = 5,\ S_3 = \{2,3\},\ \mu_4 = 4,\ S_4 = \{1\},$$

so that f can be written as the sum of disjoint products

$$f = x_1 x_2 \vee x_1 \overline{x_2} x_3 \vee x_1 \overline{x_2} \overline{x_3} x_4 x_5 \vee \overline{x_1} x_2 x_3 x_4,$$

and for all choices of $(p_1, p_2, p_3, p_4, p_5)$,

$\text{Prob}[f(X) = 1]$

$\quad = p_1 p_2 + p_1(1 - p_2)p_3 + p_1(1 - p_2)(1 - p_3)p_4 p_5 + (1 - p_1)p_2 p_3 p_4.$

Proof. Assume, without loss of generality, that the prime implicants of f are listed in lexicographic order, that is, $A_1 <_L A_2 <_L \ldots <_L A_m$ (remember Definition 7.4). Then, the statement is an immediate corollary of Theorem 7.4 if we can prove that, for $k = 1, 2, \ldots, m$, the set S_k is the shadow of A_k, that is,

$$S_k = \{j \in \{1, 2, \ldots, n\} : \text{there exists } \ell < k \leq m \text{ such that } A_\ell \setminus A_k = \{j\}\}. \quad (8.7)$$

Consider first an index $r \in S_k = \{1, 2, \ldots, \mu_k\} \setminus A_k$. By Theorem 8.6(c), since $r < \mu_k$, there exists a prime implicant $\bigwedge_{j \in A_\ell} x_j$ of f such that $A_\ell \subseteq (A_k \cup \{r\}) \setminus \{\mu_k\}$. Clearly, $A_\ell \setminus A_k = \{r\}$ and $A_\ell <_L A_k$. This shows that S_k is contained in the right-hand side of (8.7).

Conversely, suppose now that $A_\ell \setminus A_k = \{r\}$ for some $\ell < k \leq m$. From the definition of the lexicographic order, it follows that $r = \min\{j : j \in A_\ell \setminus A_k\} < \min\{j : j \in A_k \setminus A_\ell\} \leq \mu_k$. Hence, $r \in S_k$, and equality holds in (8.7). $\square$

Note that the computation of the expressions (8.5) and (8.6) does not require explicitly computing the lexicographic order of $A_1, A_2, \ldots, A_m$, that is, the shelling of f. All that is actually needed is the knowledge of the strength (complete) preorder on the variables of f.

As a corollary of Theorem 8.13, we observe that the number of true points and the Chow parameters of a regular Boolean function can be efficiently computed. Indeed, as pointed out in Section 1.13.4, the number of true points of a function f is equal to 2^n times the probability that f takes the value 1 when each variable takes value 0 or 1 with probability $\frac{1}{2}$. In view of equation (8.6), this probability is given by the expression

$$\text{Rel}_f(\frac{1}{2}, \ldots, \frac{1}{2}) = \sum_{k=1}^{m} \left(\prod_{j \in A_k} \frac{1}{2}\right) \left(\prod_{j \in S_k} \frac{1}{2}\right) = \sum_{k=1}^{m} \left(\frac{1}{2}\right)^{\mu_k}$$

(see Winder [920] for related observations).

8.3 Regularity and left-shifts

In this section, we briefly discuss a useful characterization of regular functions relying on the notion of left-shift of a Boolean point. Recall that the *support* of a point $Y \in \mathcal{B}^n$ is the set $supp(Y) = \{i \in \{1, 2, \ldots, n\} : y_i = 1\}$.

Definition 8.4. *For any two points $X^*, Y^* \in \mathcal{B}^n$, we say that Y^* is a left–shift of X^*, and we write $Y^* \curvearrowleft X^*$ if there exists a mapping $\sigma : supp(X^*) \to supp(Y^*)$ such that*

(a) *σ is injective, that is, $\sigma(i) \neq \sigma(j)$ when $i \neq j$; and*
(b) *$\sigma(i) \leq i$ for all $i = 1, 2, \ldots, n$.*

Intuitively speaking, $Y^* \curvearrowleft X^*$ if the 1's of X^* can be "shifted to the left" (from position i to position $\sigma(i)$) until they coincide with a subset of the 1's of Y^*. Notice that $\curvearrowleft$ is a preorder and that $\curvearrowleft$ is an extension of the preorder $\geq$, in the sense that $Y^* \geq X^*$ implies $Y^* \curvearrowleft X^*$.

Example 8.9. *In $\mathcal{B}^3$,*

$$(1,1,1) \curvearrowleft (1,1,0) \curvearrowleft (1,0,1) \curvearrowleft (1,0,0) \curvearrowleft (0,1,0) \curvearrowleft (0,0,1) \curvearrowleft (0,0,0)$$

and

$$(1,1,1) \curvearrowleft (1,1,0) \curvearrowleft (1,0,1) \curvearrowleft (0,1,1) \curvearrowleft (0,1,0) \curvearrowleft (0,0,1) \curvearrowleft (0,0,0),$$

but the points $(1,0,0)$ and $(0,1,1)$ are not comparable with respect to $\curvearrowleft$.   □

Theorem 8.14. *For a positive Boolean function $f(x_1, x_2, \ldots, x_n)$, the following statements are equivalent:*

(a) *f is regular, with $x_1 \succcurlyeq_f x_2 \succcurlyeq_f \cdots \succcurlyeq_f x_n$.*
(b) *Every left-shift of a true point is a true point: For all $Y, Z \in \mathcal{B}^n$, if $Z \curvearrowleft Y$, then $f(Y) \leq f(Z)$.*

Proof. Assume that f is regular with respect to $(x_1, x_2, \ldots, x_n)$, and consider two points $Y, Z \in \mathcal{B}^n$ with $Z \curvearrowleft Y$. Let σ be the mapping associated with Y and Z, as in Definition 8.4, and let Y^σ be the point with support $\{\sigma(i) : y_i = 1\}$. Then, $Y^\sigma \curvearrowleft Y$ and $Z \geq Y^\sigma$. Since f is positive, $f(Y^\sigma) \leq f(Z)$. On the other hand, the definition of the strength preorder easily implies that $f(Y) \leq f(Y^\sigma)$, since Y^σ is obtained by "shifting to the left" the nonzero entries of Y. Condition (b) follows.

Assume now that condition (b) is satisfied, and consider two indices $1 \leq i < j \leq n$. Let $X^* \in \mathcal{B}^n$ and $x_i^* = x_j^* = 0$. Then, $(X^* \vee e_i) \curvearrowleft (X^* \vee e_j)$ implies $f(X^* \vee e_j) \leq f(X^* \vee e_i)$. Hence, $x_i \succcurlyeq_f x_j$, and condition (a) follows.   □

Some authors prefer to take condition (b) in Theorem 8.14 as the defining property of regular functions (up to a permutation of the variables). In particular, consideration of the "left-shift" relation allows us to introduce in a natural way some special types of false points and true points that play an interesting role in computational manipulations of regular and threshold functions (see e.g., Bradley, Hammer, and Wolsey [148], Muroga [698], and Section 9.4.2).

Definition 8.5. *A point $X^* \in \mathcal{B}^n$ is a ceiling of the Boolean function $f(x_1, x_2, \ldots, x_n)$ if X^* is a false point of f and if no other false point of f is* *a left–shift of X^*. Similarly, X^* is a* floor *of f if X^* is a true point of f and if X^* is a left-shift of no other true point of f.*

Thus, a ceiling is a "leftmost" false point, and a floor is a "rightmost" true point. Observe that a ceiling X^* of f is necessarily a maximal false point of f, since, for all $Y^* \in \mathcal{B}^n$, $X^* \leq Y^*$ implies $Y^* \curvearrowright X^*$, and hence, either $X^* = Y^*$ or Y^* is a true point of f. Similarly, every floor of f must be a minimal true point of f.

Clearly, the notions of ceiling and floor depend on the labeling of the variables. In the sequel, when we refer to ceilings and floors of a regular function f, we always assume that f is regular with respect to $(x_1, x_2, \ldots, x_n)$, meaning that the variables have been preliminarily sorted by nonincreasing strength.

Example 8.10. *Consider the function $f = x_1 \vee x_2 x_3$. Its maximal false point $X^* = (0,0,1)$ is not a ceiling, since $Y^* = (0,1,0)$ is another false point of f and $Y^* \curvearrowright X^*$. One can check that Y^* is the unique ceiling of f. The floors of f are the minimal true points $(1,0,0)$ and $(0,1,1)$.* $\qquad\square$

An easy corollary of Theorem 8.14 is that a regular Boolean function is uniquely defined by the collection of its ceilings or its floors. This can be seen as the main motivation for introducing Definition 8.5. More precisely, we can state:

Theorem 8.15. *Let A be a subset of $\mathcal{B}^n$ such that no two points in A are comparable with respect to $\curvearrowright$. Then, there exists a unique function $r_A(x_1, x_2, \ldots, x_n)$ that is regular with respect to $(x_1, x_2, \ldots, x_n)$, and for which A is the set of ceilings. Similarly, there exists a unique function $r^A(x_1, x_2, \ldots, x_n)$ that is regular with respect to $(x_1, x_2, \ldots, x_n)$ and for which A is the set of floors.*

Proof. We only establish the statement concerning ceilings, since the argument is easily adapted to prove the statement about floors. Let $r_A(x_1, x_2, \ldots, x_n)$ be the Boolean function defined as follows:

For all $X^* \in \mathcal{B}^n$, $r_A(X^*) = 0$ if and only if there exists $Y^* \in A$ such that $Y^* \curvearrowright X^*$.
$$(8.8)$$

Now, let $Y, Z \in \mathcal{B}^n$ with $Z \curvearrowright Y$ and $r_A(Z) = 0$. Then, it follows from (8.8) and from the transitivity of $\curvearrowright$ that $r_A(Y) = 0$. Hence, by Theorem 8.14, r_A is regular with respect to $(x_1, x_2, \ldots, x_n)$. Moreover, it is easy to verify that A is exactly the set of ceilings of r_A. To establish the unicity of r_A, consider now a regular function $f(x_1, x_2, \ldots, x_n)$ with $x_1 \succcurlyeq_f x_2 \succcurlyeq_f \cdots \succcurlyeq_f x_n$, which admits A for set of ceilings. We want to show that f necessarily is the unique function satisfying (8.8). Suppose first that $Y^* \in A$ and that $Y^* \curvearrowright X^*$. Since Y^* is a ceiling of f, $f(Y^*) = 0$, and hence, by Theorem 8.14, $f(X^*) = 0$. Conversely, if $f(X^*) = 0$, then there exists a "leftmost" point Y^* such that $Y^* \curvearrowright X^*$ and $f(Y^*) = 0$. By definition, Y^* is a ceiling of f, and hence $Y^* \in A$. $\qquad\square$

Example 8.11. *Let $A = \{(0,1,0)\}$. If $r_A(x_1, x_2, x_3)$ is a function with $x_1 \succcurlyeq_f x_2 \succcurlyeq_f x_3$ and such that $(0,1,0)$ is its unique ceiling, then, by Theorem 8.14, all points X^**

such that $(0,1,0) \curlywedge X^*$ *must be false points of* r_A. *Moreover, by definition of a ceiling, all left-shifts of* $(0,1,0)$ *are true points of* r_A. *A look at Example 8.9 indicates that this classification exhausts all points of* $\mathcal{B}^3$. *Hence,* r_A *is uniquely determined. One easily verifies that* $r_A(x_1,x_2,x_3) = x_1 \vee x_2 x_3$. $\square$

Peled and Simeone [735] used Theorem 8.15 to show that, if $r(n)$ is the number of regular functions on n variables, then $\log_2 r(n) \geq cn^{-\frac{3}{2}} 2^n$ for some constant c.

8.4 Recognition of regular functions

We tackle in this section the algorithmic problem of recognizing regular Boolean functions, mostly concentrating on the case in which the input function f is positive and is represented by its complete DNF, that is, on the problem:

REGULARITY RECOGNITION
Instance: The complete DNF of a positive Boolean function f.
Output: TRUE if f is regular, FALSE otherwise.

It is not too hard to see that REGULARITY RECOGNITION can be solved in polynomial time. Indeed, each question of the form:

"Is $x_i \succcurlyeq_f x_j$, or is $x_j \succcurlyeq_f x_i$, or are x_i and x_j incomparable with respect to $\succcurlyeq_f$?"

$$(8.9)$$

can be answered in time $O(nm^2)$ by virtue of Theorem 8.6, where n is the number of variables and m is the number of prime implicants of f. By asking enough questions of this type, we can either find a pair of incomparable variables, or determine a permutation $(x_{i_1}, x_{i_2}, \ldots x_{i_n})$ of the variables such that $x_{i_1} \succcurlyeq_f x_{i_2} \succcurlyeq_f \cdots x_{i_n}$ in case f is regular. Therefore, we can state the following result:

Theorem 8.16. *There is an* $O(n^2 m^2 \log n)$ *algorithm to decide whether a positive Boolean function given by its complete DNF is regular, where* n *is the number of variables and* m *is the number of prime implicants of the function.*

Proof. Using an optimal sorting strategy (like Mergesort [11]), one can determine whether the input function is regular by asking $O(n \log n)$ questions of the form (8.9), and each question can be answered in time $O(nm^2)$. $\square$

Although polynomially bounded, the complexity of this simple procedure is quite high. In particular, the factor m^2 in the time bound is unsatisfactory since we generally expect m to be large with respect to n. In the remainder of this section, we present several results due to Winder [916, 917] and Provan and Ball [760] that will allow us to derive an improved recognition procedure for regular functions with time complexity $O(n^2 m)$.

The improvements will be achieved on two separate fronts. First, we will show how to quickly obtain a complete ordering σ of the variables of f, with the property that f is regular if and only if σ coincides with the strength preorder of f. "Quickly" means here in $O(n^2 + nm)$ operations. Next, making use of an appropriate data structure, we explain how to check in $O(n^2m)$ steps whether σ actually is the strength preorder of f and, hence, whether f is regular.

Strength preorder and Winder matrix

We start with an elegant result due to Winder [916, 917], which makes use of the concept of lexicographic order of points in $\mathbb{R}^n$.

Definition 8.6. *For* $X, Y \in \mathbb{R}^n$, *we say that* X *precedes* Y *in the lexicographic order; and we write* $X <_L Y$ *if* $x_k < y_k$, *where* $k = \min\{j : x_j \neq y_j, 1 \leq j \leq n\}$. *We write* $X \leq_L Y$ *if either* $X = Y$ *or* $X <_L Y$.

Definition 8.7. *The* Winder matrix *of a positive Boolean function* $f(x_1, x_2, \ldots, x_n)$ *is the* $n \times n$ *matrix* $R = (r_{id})$, *where* r_{id} *denotes the number of prime implicants of* f *that involve* x_i *and whose degree is exactly* d $(i, d = 1, 2, \ldots, n)$.

Theorem 8.17. *Let* $f(x_1, x_2, \ldots, x_n)$ *be a positive Boolean function, and denote by* R^i *the* i-*th row of its Winder matrix* $(i = 1, 2, \ldots, n)$. *For* $i, j = 1, 2, \ldots, n$,

(a) *if* $x_i \approx_f x_j$ *then* $R^i = R^j$;
(b) *if* $x_i \succ_f x_j$ *then* $R^i >_L R^j$.

Proof. Consider two variables x_i, x_j, and write the complete DNF of f in the form $\alpha x_i x_j \vee \beta x_i \vee \gamma x_j \vee \delta$ as in Theorem 8.5. If $x_i \approx_f x_j$, then $\beta = \gamma$, and hence, $R^i = R^j$. So, assume now that $x_i \succ_f x_j$. Then $\beta > \gamma$. For $d = 0, 1, \ldots, n-1$, define

$$B(d) = \{P : |P| = d \text{ and } \bigwedge_{k \in P} x_k \text{ is a term of } \beta\}$$

$$C(d) = \{P : |P| = d \text{ and } \bigwedge_{k \in P} x_k \text{ is a term of } \gamma\}.$$

If $B(d) = C(d)$ for all d, then $\beta = \gamma$, which contradicts our assumption. Thus, there exists a smallest d^* such that $B(d^*) \neq C(d^*)$. We claim that $C(d^*) \subset B(d^*)$.

Indeed, let $P \in C(d^*)$. Since $\beta > \gamma$, there exists a term of β, say $\bigwedge_{k \in Q} x_k$, such that $Q \subseteq P$. If Q is not equal to P, then $|Q| < |P| = d^*$, and hence, $Q \in B(d)$ for some $d < d^*$. By our choice of d^*, this implies that $Q \in C(d)$. But, then both $\bigwedge_{k \in P} x_k$ and $\bigwedge_{k \in Q} x_k$ are terms of γ, a contradiction. So, we conclude that $Q = P$, and hence, $P \in B(d^*)$ as required.

From the assertions $B(d) = C(d)$ for $d < d^*$ and $C(d^*) \subset B(d^*)$, one easily derives $r_{id} = r_{jd}$ for $d < d^*$ and $r_{jd^*} < r_{id^*}$, which completes the proof. □

Example 8.12. *Let* $f = x_1 x_2 \vee x_1 x_3 \vee x_1 x_4 x_5 \vee x_2 x_3 x_4$. *One checks for instance that* $r_{2,3} = 1$, *since* x_2 *occurs in exactly one prime implicant of degree 3. The complete matrix* R *associated with* f *is*

$$R = \begin{bmatrix} 0 & 2 & 1 & 0 & 0 \\ 0 & 1 & 1 & 0 & 0 \\ 0 & 1 & 1 & 0 & 0 \\ 0 & 0 & 2 & 0 & 0 \\ 0 & 0 & 1 & 0 & 0 \end{bmatrix}.$$

Since $x_1 \succcurlyeq_f x_2$, the first row of R is lexicographically larger than its second row. Also, the second and third rows of R are identical, since $x_2 \approx_f x_3$. $\square$

Note that the strength preorder $\succcurlyeq_f$ does not coincide perfectly, in general, with the lexicographic order $\geq_L$ on the rows of R (in particular, $\geq_L$ completely orders the rows of R, whereas $\succcurlyeq_f$ is generally incomplete). When f is regular, however, we obtain as an immediate corollary of Theorem 8.17:

Theorem 8.18. *Let $f(x_1, x_2, \ldots, x_n)$ be a regular function and denote by R^i the i-th row of its Winder matrix $(i = 1, 2, \ldots, n)$. Then,*

$$R^1 \geq_L R^2 \geq_L \cdots \geq_L R^n \text{ if and only if } x_1 \succcurlyeq_f x_2 \succcurlyeq_f \cdots \succcurlyeq_f x_n.$$

Proof. This immediately follows from Theorem 8.17. $\square$

For a positive function f expressed in complete DNF, with n variables and m prime implicants, the Winder matrix R can be computed in time $O(n^2 + nm)$ and its rows can be lexicographically ordered in time $O(n^2)$ (see [11]). Assuming for simplicity that $R^1 \geq_L R^2 \geq_L \cdots \geq_L R^n$, one can then decide whether f is regular by checking whether $x_1 \succcurlyeq_f x_2 \succcurlyeq_f \cdots \succcurlyeq_f x_n$. This requires $(n-1)$ pairwise comparisons of variables, and each of these can be performed in time $O(nm^2)$ (using Theorem 8.5). Thus, Theorem 8.18 directly leads to an $O(n^2 m^2)$ recognition algorithm for regular functions.

To get rid of a factor of m in this time complexity, more work is needed.

Efficient comparison of variables

Given two variables x_i, x_j of a positive function $f(x_1, x_2, \ldots, x_n)$, deciding whether $x_i \succcurlyeq_f x_j$ amounts (by Theorem 8.6) to testing whether $\bigwedge_{k \in (A \cup \{i\}) \setminus \{j\}} x_k$ is an implicant of f, for each prime implicant of f of the form $\bigwedge_{k \in A} x_k$ such that $j \in A$ and $i \notin A$. This observation motivates us to momentarily concentrate on the algorithmic complexity of the following type of queries: For a positive function $f(x_1, x_2, \ldots, x_n)$ expressed in complete DNF and for a subset $A \subseteq \{1, 2, \ldots, n\}$, is $\bigwedge_{k \in A} x_k$ an implicant of f?

Now, we have already seen in Chapter 7, Section 7.4, that queries of this type can be answered efficiently when f has the LE property. Since Theorem 8.12 asserts that regular functions have the LE property, all results in Section 7.4 apply to regular functions as well. (Note that it is not necessary to master all of Chapter 7 in order to appreciate the contents of Section 7.4: The reader can still study Section 7.4, now simply substituting the words "regularity property" for "LE property" everywhere in the section.)

Procedure REGULAR(f)
Input: A positive Boolean function $f(x_1, x_2, \ldots, x_n)$ in complete DNF.
Output: TRUE if f is regular, FALSE otherwise.

begin
        compute R, the Winder matrix of f;
        order the rows of R lexicographically;
        {comment: assume without loss of generality that $R^1 \geq_L R^2 \geq_L \cdots \geq_L R^n$ }
        set up the binary tree $T(f)$;
        for $i = 1$ to $n - 1$ **and**
            for every prime implicant $\bigwedge_{k \in A} x_k$ of f such that $i \notin A$ and $i + 1 \in A$ **do**
                if IMPLICANT($A \cup \{i\} \setminus \{i + 1\}$) = FALSE **then return** FALSE;
        return TRUE;
end

Figure 8.1. Procedure REGULAR.

More precisely, denote by $T(f)$ the binary tree associated with (the complete DNF of) a positive function f as on page 341, and consider the procedure IMPLICANT(A) defined on page 342. Then, we can state:

Theorem 8.19. *Let* $f(x_1, x_2, \ldots, x_n)$ *be a positive function, and let* $A \subseteq \{1, 2, \ldots, n\}$.

(a) *If the procedure* IMPLICANT(A) *returns the answer* TRUE, *then* $\bigwedge_{j \in A} x_j$ *is an implicant of* f.

(b) *When* f *is regular with respect to* $(x_1, x_2, \ldots, x_n)$, *the procedure* IMPLICANT(A) *returns the answer* TRUE *if and only if* $\bigwedge_{j \in A} x_j$ *is an implicant of* f.

Proof. This is a corollary of Theorem 7.12 and Theorem 7.13. $\square$

We are now ready to state an efficient algorithm due to Provan and Ball [760] for the recognition of regular functions; see Figure 8.1 for a formal statement of the algorithm.

Theorem 8.20. *Algorithm* REGULAR *correctly recognizes regular functions given by their complete DNF. It can be implemented to run in time* $O(n^2 m)$, *where n is the number of variables and m is the number of prime implicants of the function to be tested.*

Proof. If f is regular and $R^1 \geq_L R^2 \geq_L \cdots \geq_L R^n$, then $x_1 \succcurlyeq_f x_2 \succcurlyeq_f \cdots \succcurlyeq_f x_n$ by Theorem 8.18. So, for every $i \in \{1, 2, \ldots, n - 1\}$ and for every prime implicant $\bigwedge_{k \in A} x_k$ of f such that $i \notin A$ and $i + 1 \in A$, Theorem 8.6 implies that $\bigwedge_{k \in (A \cup \{i\}) \setminus \{i+1\}} x_k$ is an implicant of f. Hence, by Theorem 8.19, REGULAR(f) returns the answer TRUE.

Conversely, if f is not regular, then there is a smallest index i such that $x_i \not\succcurlyeq_f x_{i+1}$. For this i, there is a prime implicant $\bigwedge_{k \in A} x_k$ of f such that $i \notin A, i + 1 \in A$, and $\bigwedge_{k \in (A \cup \{i\}) \setminus \{i+1\}} x_k$ is not an implicant of f. But then, IMPLICANT$(A \cup \{i\} \setminus \{i + 1\})$ returns FALSE, by Theorem 8.19. This establishes that the procedure is correct.

As for the complexity of the procedure, we have already observed that its first and second steps can be performed in time $O(n^2 + nm)$. Setting up the tree $T(f)$ takes time $O(nm)$ (see Section 7.4). The nested loops require at most nm calls on the procedure IMPLICANT, and each of these calls can be executed in time $O(n)$. Hence, the overall running time of REGULAR is $O(n^2m)$.   □

Example 8.13. *Consider the function* $f(x_1, x_2, x_3, x_4, x_5) = x_1 x_2 \vee x_1 x_3 \vee x_1 x_4 x_5 \vee x_2 x_3 x_4$. *We computed the Winder matrix of* f *in Example 8.12. The tree* $T(f)$ *is represented in Figure 7.2. The reader can check that* REGULAR *returns the answer* TRUE *when running on* f.   □

The $O(n^2m)$ time complexity stated in Theorem 8.20 has been further improved to $O(nm)$ by Makino [649]. Makino's algorithm makes use of an improved binary tree data structure in order to achieve this time complexity; we refer to the paper [649] for details.

Before closing this section on the recognition of regular functions, let us address the complexity of a more general version of the problem: Namely, given an arbitrary DNF (as opposed to a positive one), how difficult is it to determine whether this DNF represents a regular Boolean function? Peled and Simeone [735] showed:

Theorem 8.21. *Deciding whether a DNF represents a regular Boolean function is co-NP-complete, even if the DNF has degree at most three.*

Proof. NP-hardness follows immediately from Theorem 1.30 and from the observation that not all functions are regular. The decision problem is in co-NP since we can show that a function is not regular by exhibiting a pair of incomparable variables.   □

The problem of recognizing regular functions given by an oracle, rather than by a Boolean expression, has also been considered in a number of publications; see, for instance, Boros, Hammer, Ibaraki, and Kawakami [129] or Makino and Ibaraki [653].

8.5 Dualization of regular functions

In this section, we consider the problem of dualizing regular Boolean functions expressed in complete (prime irredundant) disjunctive normal form, namely, the problem:

REGULAR DUALIZATION
Instance: The complete DNF of a regular function f or, equivalently, the list of all minimal true points of f.

Output: The complete DNF of f^d or, equivalently, the list of all maximal false points of f.

Motivation for this problem can be found in Chapter 4, as well as in Application 8.5. Also, and perhaps most importantly, the efficient dualization of regular functions will turn out to be an essential step for the efficient recognition of threshold functions in Chapter 9. As a consequence, this problem has a rather complex and interesting history.

The first specialized dualization algorithm for regular functions was proposed by Hammer, Peled, and Pollatschek [455]. This algorithm runs in "polynomial total time" in the sense of Appendix B, meaning that its running time is bounded by a polynomial in the size of its input *and of its output*. Denote by n, m, and p, respectively, the number of variables, minimal true points, and maximal false points of the function f. So, the algorithm of Hammer, Peled, and Pollatschek [455] is polynomial in n, m and p. However, the authors did not carry out a more detailed complexity analysis of their algorithm and, in particular, they did not provide any precise bound on the magnitude of p.

Similar comments hold for the general dualization scheme of Lawler, Lenstra, and Rinnooy Kan [605] sketched in Theorem 4.39. Indeed, as noticed by Peled and Simeone [735], the approach proposed by Lawler, Lenstra, and Rinnooy Kan for the enumeration of all maximal feasible solutions of knapsack problems can be generalized for the dualization of regular functions. It leads to an $O(n^2 p)$ dualization algorithm for regular functions, but again, the approach does not seem to imply any reasonable bound on p.

Peled and Simeone [735] presented the first dualization algorithm for regular functions whose running time could be proved to be polynomially bounded in n and m only. More precisely, their algorithm outputs the maximal false points of f in time $O(n^3 m)$. Clearly, such a result is only possible if the number of maximal false points of f, namely, p, is itself polynomially bounded in n and m. And indeed, as a by-product of the complexity analysis of their algorithm, Peled and Simeone established that the bound $p \leq nm + m + n$ always holds for regular functions. Therefore, in particular, the algorithms of Hammer, Peled, and Pollatschek [455] and Lawler, Lenstra, and Rinnooy Kan [605] mentioned above also have their running time bounded by a polynomial in n and m.

In spite of its low computational complexity, Peled and Simeone's algorithm is quite intricate. By contrast, Crama [225] proposed a straightforward $O(n^2 m)$ dualization algorithm for regular functions, based on a simple recursive characterization of the maximal false points of these functions in terms of their minimal true points (see Theorem 8.22 hereunder). His characterization also implies a stronger bound on the number of maximal false points: Namely, $p \leq (n-1)m$ when $m > 1$.

Bertolazzi and Sassano [74, 75] independently rediscovered these same results and extended them to a more compact characterization of the maximal false points of regular functions (see Theorem 8.27). Their characterization also leads to an

$O(n^2m)$ dualization algorithm and lends itself to an $O(nm)$ algorithm for the solution of "regular set covering problems" to be discussed in Section 8.6. Later on, Peled and Simeone [736] proposed yet another $O(n^2m)$ regular dualization algorithm.

Finally, we note that an $O(n^2m)$ dualization algorithm for regular functions can be obtained as a corollary of Theorem 7.16, since regular functions have the LE property by Theorem 8.12. This algorithm was first described by Boros [105], within the framework of his analysis of so-called *aligned functions* (see Section 8.9.2).

The presentation hereunder combines ideas from Crama [225] and Bertolazzi and Sassano [74]. It mostly rests on a key result from Crama [225]:

Theorem 8.22. *Assume that $f(x_1,x_2,\ldots,x_n)$ is regular with respect to $(x_1,x_2,\ldots,x_n)$ and let $X^* \in \mathcal{B}^{n-1}$. Then, $(X^*,0)$ is a maximal false point of f if and only if $(X^*,1)$ is a minimal true point of f.*

Proof. Assume that $(X^*,0)$ is a maximal false point of f. Then, $(X^*,1)$ is a true point of f. To see that $(X^*,1)$ actually is a *minimal* true point of f, consider any index $i < n$ such that $x_i^* = 1$. Since $x_i \succcurlyeq_f x_n$, $(x_1^*,x_2^*,\ldots,x_{i-1}^*,0,x_{i+1}^*,\ldots,x_{n-1}^*,1)$ is a false point of f, as required.

Conversely, if $(X^*,1)$ is a minimal true point of f, then $(X^*,0)$ is a false point of f. To see that $(X^*,0)$ is a *maximal* false point, consider $i < n$ such that $x_i^* = 0$. Since $x_i \succcurlyeq_f x_n$, $(x_1^*,x_2^*,\ldots,x_{i-1}^*,1,x_{i+1}^*,\ldots,x_{n-1}^*,0)$ is a true point of f, as required.   □

Theorem 8.22 provides a simple and tractable characterization of those maximal false points of a regular function that have their last component equal to 0. On the other hand, the maximal false points with last component equal to 1 can easily be treated recursively. To see this, let us introduce a new notation: For a function $f(x_1,x_2,\ldots,x_n)$ and an index $i \in \{1,2,\ldots,n\}$, let us denote by f_i the restriction of f to $x_i = x_{i+1} = \cdots = x_n = 1$. We look at f_i as a function of $(x_1,x_2,\ldots,x_{i-1})$. By convention, we also set $f_{n+1} = f$.

Theorem 8.23. *Let $f(x_1,x_2,\ldots,x_n)$ be a positive function and let $X^* \in \mathcal{B}^{n-1}$. Then, $(X^*,1)$ is a maximal false point of f if and only if X^* is a maximal false point of f_n.*

Proof. This is trivial.   □

Note that, in contrast with Theorem 8.22, Theorem 8.23 is valid for all positive functions, whether regular or not. Taken together, these theorems immediately suggest a recursive dualization procedure for regular functions. This procedure, which we call DUALREGO, is described in Figure 8.2.

The procedure is obviously correct in view of Theorem 8.22 and Theorem 8.23. Moreover, it can actually be implemented recursively, since f_n is regular when f is regular (by Theorem 8.7).

Procedure DUALREG0(f)
Input: The list of minimal true points of a regular function $f(x_1, x_2, \ldots, x_n)$
     such that $x_1 \succcurlyeq_f x_2 \succcurlyeq_f \cdots \succcurlyeq_f x_n$.
Output: All maximal false points of f.

begin
       identify all minimal true points of f with last component equal to 1,
          say $(X_1^*, 1)$, $(X_2^*, 1)$, ..., $(X_k^*, 1)$;
       fix x_n to 1 in f and determine the minimal true points of f_n;
       generate (recursively) all maximal false points of f_n,
          say $X_{k+1}^*, X_{k+2}^*, \ldots, X_p^*$;
       return $(X_1^*, 0), (X_2^*, 0), \ldots, (X_k^*, 0)$ and $(X_{k+1}^*, 1), (X_{k+2}^*, 1), \ldots, (X_p^*, 1)$;
end

Figure 8.2. Procedure DUALREG0.

Example 8.14. *Consider the function* $f(x_1, x_2, x_3, x_4, x_5) = x_1 x_2 \vee x_1 x_3 \vee x_1 x_4 \vee x_2 x_3 \vee x_2 x_4 x_5$, *which is regular with* $x_1 \succ_f x_2 \succ_f x_3 \succ_f x_4 \succ_f x_5$. *The minimal true points of f are (in lexicographic order):* $Y^1 = (0, 1, 0, 1, 1)$, $Y^2 = (0, 1, 1, 0, 0)$, $Y^3 = (1, 0, 0, 1, 0)$, $Y^4 = (1, 0, 1, 0, 0)$ *and* $Y^5 = (1, 1, 0, 0, 0)$. *Let us execute the procedure* DUALREG0 *on* f.

Step 1. *The only maximal false point of f with 0 as last component is* $X^1 = (0, 1, 0, 1, 0)$ *(derived from* Y^1 *via Theorem 8.22).*
Step 2. *The restriction of f to* $x_5 = 1$ *is* $f_5(x_1, x_2, x_3, x_4) = x_1 x_2 \vee x_1 x_3 \vee x_1 x_4 \vee x_2 x_3 \vee x_2 x_4$, *which has the minimal true points:* $Z^1 = (0, 1, 0, 1)$, $Z^2 = (0, 1, 1, 0)$, $Z^3 = (1, 0, 0, 1)$, $Z^4 = (1, 0, 1, 0)$, $Z^5 = (1, 1, 0, 0)$.
Step 3. *We now recursively apply* DUALREG0 *to* f_5.

Step 1. *The maximal false points of* f_5 *with last component equal to 0 are* $(0, 1, 0, 0)$ *and* $(1, 0, 0, 0)$ *(derived from* Z^1 *and* Z^3 *by Theorem 8.22). Thus, f has the maximal false points* $X^2 = (0, 1, 0, 0, 1)$ *and* $X^3 = (1, 0, 0, 0, 1)$ *(by Theorem 8.23).*
Step 2. *The restriction of* f_5 *to* $x_4 = 1$ *is* $f_4(x_1, x_2, x_3) = x_1 \vee x_2$, *with minimal true points* $V^1 = (0, 1, 0)$ *and* $V^2 = (1, 0, 0)$.
Step 3. *We recursively apply* DUALREG0 *to* f_4.

Step 1. f_4 *has no maximal false points with* $x_3 = 0$.
Step 2. *Setting* $x_3 = 1$ *in* f_4, *we get* $f_3(x_1, x_2) = x_1 \vee x_2$, *with minimal true points* $W^1 = (0, 1)$ *and* $W^2 = (1, 0)$.
Step 3. *We recursively apply* DUALREG0 *to* f_3.

Step 1. *Using Theorem 8.22, we see that* f_3 *has the maximal false point* $(0, 0)$ *with last component equal to 0. Thus, f has the maximal false point* $X^4 = (0, 0, 1, 1, 1)$ *(by repeated applications of Theorem 8.23).*
Step 2. *Fixing* $x_2 = 1$ *in* f_3, *we obtain* $f_2(x_1) \equiv 1$.

Step 3. *Since f_2 has no maximal false points, the procedure terminates here: all maximal false points of f have been listed.*   □

DUALREG0 requires generating the minimal true points of f_n from the minimal true points of f. To carry out this step efficiently, one may rely on the next observation.

Theorem 8.24. *Let $f(x_1, x_2, \ldots, x_n)$ be a positive function and let $Y \in \mathcal{B}^{n-1}$. Then, Y is a minimal true point of f_n if and only if*

(a) *either $(Y, 1)$ is a minimal true point of f, or*
(b) *$(Y, 0)$ is a minimal true point of f, and f has no minimal true point of the form $(Z, 1)$ with $Z < Y$.*

Proof. We leave this easy proof to the reader.   □

A straightforward implementation of DUALREG0 based on Theorem 8.24 yields an $O(n^2 m^2)$ dualization algorithm for regular functions with n variables and m minimal true points. Our next goal in this section will be to reduce this complexity by a factor of m. We now briefly sketch the line of attack that we will follow in order to achieve this goal.

We first derive an accurate characterization of certain minimal true points of the restricted functions $f_1, f_2, \ldots, f_n$ in terms of the minimal true points of f, under the assumption that f is regular (see Theorem 8.26; notice that Theorem 8.24 does not rest on any regularity assumption). This result will then lead to a compact description of the maximal false points of a regular function (Theorem 8.27) and, finally, to the announced $O(n^2 m)$ dualization algorithm (Theorem 8.28).

We now launch this programme with a first refinement of Theorem 8.24. We use the following notations: If Y is a nonzero point in $\mathcal{B}^n$, we denote by $\mu(Y)$ the largest index k such that $y_k = 1$, and we denote by $Y - e_i$ the point $(y_1, \ldots, y_{i-1}, 0, y_{i+1}, \ldots, y_n)$, for $i = 1, 2, \ldots, n$. Then (Crama [225]):

Theorem 8.25. *Assume that $f(x_1, x_2, \ldots, x_n)$ is regular with respect to $(x_1, x_2, \ldots, x_n)$, and let Y be a nonzero point in $\mathcal{B}^{n-1}$. Then, Y is a minimal true point of f_n if and only if*

(a) *either $(Y, 1)$ is a minimal true point of f, or*
(b) *$(Y, 0)$ is a minimal true point of f, but $(Y - e_{\mu(Y)}, 1)$ is not.*

Proof. Necessity. This is a corollary of Theorem 8.24.

Sufficiency. If $(Y, 1)$ is a minimal true point of f, then Y is a minimal true point of f_n by Theorem 8.24. So, assume now that $(Y, 0)$ is a minimal true point of f, but that Y is not a minimal true point of f_n. We will deduce from these assumptions that $(Y - e_{\mu(Y)}, 1)$ is a minimal true point of f, thus completing the proof.

By Theorem 8.24, f must have a minimal true point of the form $(Z, 1)$, with $Z < Y$. Let j be any index in $\{1, 2, \ldots, n-1\}$ such that $z_j = 0$ and $y_j = 1$. By regularity, $(Z \vee e_j, 0)$ is a true point of f, and, by minimality of $(Y, 0)$, it follows that $Y = Z \vee e_j$. So, $(Z, 1) = (Y - e_j, 1)$ is a minimal true point of f.

If $j = \mu(Y)$, then we are done. Otherwise, $j < \mu(Y)$, and, by regularity, $(Y - e_{\mu(Y)}, 1)$ is a true point of f, as required. To see that $(Y - e_{\mu(Y)}, 1)$ actually is a *minimal* true point of f, observe first that $(Y - e_{\mu(Y)}, 0)$ is a false point of f, since $(Y, 0)$ is a minimal true point. Next, consider any index $k < \mu(Y)$ such that $y_k = 1$. Since $(Y - e_{\mu(Y)}, 0)$ is a false point of f, $(Y - e_k - e_{\mu(Y)}, 1)$ also is a false point, by regularity. Thus, $(Y - e_{\mu(Y)}, 1)$ is a minimal true point. $\square$

As shown in Crama [225], Theorem 8.25 can already be used to produce an $O(n^2 m)$ implementation of DualReg0. But we now go one step further and establish a more precise characterization of those minimal true points of f_j $(j = 1, 2, \ldots, n)$ that have a 1 as last component (observe that these are the only minimal true points of f_j that we need to know to carry out DualReg0).

For two minimal true points Y and Z of f, let us say that Z *immediately precedes* Y if $Z <_L Y$ and if there is no minimal true point of f between Z and Y in the lexicographic order $<_L$ (see Definition 8.6). The following result is essentially due to Bertolazzi and Sassano (see Theorem 4.2 in [74]):

Theorem 8.26. *Let* $f(x_1, x_2, \ldots, x_n)$ *be regular with respect to* $(x_1, x_2, \ldots, x_n)$, *let* $j \in \{1, 2, \ldots, n - 1\}$, *and let* $(y_1, y_2, \ldots, y_j)$ *be a point in* $\mathcal{B}^j$ *such that* $y_j = 1$. *The point* $(y_1, y_2, \ldots, y_j)$ *is a minimal true point of* f_{j+1} *if and only if there exists* $(y_{j+1}, y_{j+2}, \ldots, y_n) \in \mathcal{B}^{n-j}$ *such that*

(a) $Y = (y_1, y_2, \ldots, y_n)$ *is a minimal true point of* f, *and*
(b) *if* Z *is the minimal true point of* f *immediately preceding* Y, *then* $(y_1, y_2, \ldots, y_{j-1}) \neq (z_1, z_2, \ldots, z_{j-1})$.

Proof. Necessity. If $(y_1, y_2, \ldots, y_j)$ is a minimal true point of f_{j+1}, then it follows from Theorem 8.24 that f must have a minimal true point of the form $Y = (y_1, y_2, \ldots, y_n)$ for some appropriate values of $y_{j+1}, y_{j+2}, \ldots, y_n$. Choose $y_{j+1}, y_{j+2}, \ldots, y_n$ in such a way that Y is lexicographically smallest among all minimal true points of f of this form.

Let now Z be the minimal true point of f immediately preceding Y and assume by contradiction that $(y_1, y_2, \ldots, y_{j-1}) = (z_1, z_2, \ldots, z_{j-1})$. If $z_j = 1$, then $(y_1, y_2, \ldots, y_j) = (z_1, z_2, \ldots, z_j)$, contradicting the choice of Y. So, $z_j = 0$. On the other hand, $(z_1, z_2, \ldots, z_j)$ is a true point of f_{j+1}, since

$$f_{j+1}(z_1, z_2, \ldots, z_j) = f(z_1, z_2, \ldots, z_j, 1, \ldots, 1) \geq f(Z) = 1.$$

Hence, $(y_1, y_2, \ldots, y_j)$ is not a *minimal* true point of f_{j+1}, a contradiction.

Sufficiency. Assume now that f has a minimal true point of the form $Y = (y_1, y_2, \ldots, y_n)$ but that $(y_1, y_2, \ldots, y_j)$ is not a minimal true point of f_{j+1}. Then, there exists an index k, $j < k \leq n$, such that $(y_1, y_2, \ldots, y_k)$ is a minimal true point of f_{k+1}, but $(y_1, y_2, \ldots, y_{k-1})$ is not a minimal true point of f_k. In view of Theorem 8.25, this means that $y_k = 0$, and that $V = (y_1, y_2, \ldots, y_{k-1}, 1) - e_\mu$ is a minimal true point of f_{k+1}, where $\mu = \mu(y_1, y_2, \ldots, y_{k-1})$. Observe that $j \leq \mu$, since $y_j = 1$ by assumption.

Since V is a minimal true point of f_{k+1}, there exists (by Theorem 8.24) a minimal true point of f of the form (V, W). Moreover, $(V, W) <_L Y$, since $v_1 = y_1, v_2 = y_2, \ldots, v_{\mu-1} = y_{\mu-1}, v_\mu = 0 < 1 = y_\mu$. Hence, if Z denotes the minimal true point of f immediately preceding Y, then $(V, W) \leq_L Z <_L Y$. From $j \leq \mu$, it now follows easily that $(v_1, v_2, \ldots, v_{j-1}) = (z_1, z_2, \ldots, z_{j-1}) = (y_1, y_2, \ldots, y_{j-1})$, as required.   □

As announced earlier in the section, we are now ready to present a complete characterization of the maximal false points of a regular function in terms of its minimal true points.

Theorem 8.27. *Let $f(x_1, x_2, \ldots, x_n)$ be regular with respect to $(x_1, x_2, \ldots, x_n)$. Assume that f is not identically equal to 0, and let $Y^1, Y^2, \ldots, Y^m$ be its minimal true points, labeled in such a way that $Y^1 <_L Y^2 <_L \ldots <_L Y^m$. A point $X^* \in \mathcal{B}^n$ is a maximal false point of f if and only if there exists a minimal true point Y^i $(1 \leq i \leq m)$ and an index $j \in \{1, 2, \ldots, n\}$ such that*

(a) *either $i = 1$ or $(y_1^{i-1}, y_2^{i-1}, \ldots, y_{j-1}^{i-1}) \neq (y_1^i, y_2^i, \ldots, y_{j-1}^i)$;*
(b) *$x_k^* = y_k^i$ for $k = 1, 2, \ldots, j-1$;*
(c) *$x_j^* = 0$ and $y_j^i = 1$;*
(d) *$x_k^* = 1$ for $k = j+1, \ldots, n$.*

Proof. Let $X^* \in \mathcal{B}^n$, $X^* \neq (1, \ldots, 1)$, and let j be the largest index such that $x_j^* = 0$. By Theorem 8.23, X^* is a maximal false point of f if and only if $(x_1^*, \ldots, x_j^*)$ is a maximal false point of f_{j+1}. Hence, by Theorem 8.22, X^* is a maximal false point of f if and only if $(x_1^*, \ldots, x_{j-1}^*, 1)$ is a minimal true point of f_{j+1}. The proof is now easily completed by referring to Theorem 8.26.   □

An efficient dualization algorithm for regular functions can be immediately deduced from Theorem 8.27. To efficiently test condition (a) in the statement of this theorem, it is convenient to compute, in a preprocessing phase of the algorithm, the smallest index ν_i on which Y^i differs from Y^{i-1}, for $i = 2, 3, \ldots, m$. By convention, we let $\nu_1 = 0$. Then, condition (a) can be simply replaced by

(a') $\nu_i < j$,

and the algorithm can be stated as in Figure 8.3.

We are now finally ready for the main result of this section.

Theorem 8.28. *The procedure DUALREG(f) is correct and can be implemented to run in time $O(n^2 m)$, where n is the number of variables and m is the number of minimal true points of f.*

Proof. The correctness of the procedure follows from Theorem 8.27. As for its complexity, notice that the minimal true points of f can be lexicographically ordered

---

Procedure DUALREG(f)
Input: The list of minimal true points $Y^1, Y^2, \ldots, Y^m$ of a regular function
$f(x_1, x_2, \ldots, x_n)$ such that $x_1 \succcurlyeq_f x_2 \succcurlyeq_f \cdots \succcurlyeq_f x_n$.
Output: The list L of all maximal false points of f.

begin
    sort the points Y^i ($i = 1, 2, \ldots, m$) in lexicographic order;
    {comment: assume without loss of generality that $Y^1 <_L Y^2 <_L \ldots <_L Y^m$ };
    $v_1 := 0$;
    for $i = 2$ **to** m **do** $v_i := \min\{k : y_k^{i-1} < y_k^i\}$;
    initialize $L :=$ empty list;
    for $i = 1$ **to** m **and for** $j = 1$ **to** n **do**
        if $y_j^i = 1$ **and** $v_i < j$ **then**
            begin
                for $k = 1$ **to** $j - 1$ **do** $x_k^* := y_k^i$;
                $x_j^* := 0$;
                for $k = j + 1$ **to** n **do** $x_k^* := 1$;
                add X^* to L;
            end
    return L;
end

Figure 8.3. Procedure DUALREG.

in time $O(nm)$ (see for instance Aho, Hopcroft, and Ullman [11]). The parameters $v_1, v_2, \ldots, v_m$ can be simultaneously computed on the run. Each execution of the **(for i, for j)**–loop requires $O(n)$ operations, thus leading to the overall $O(n^2 m)$ time bound. □

An interesting feature of the procedure DUALREG is worth stressing here. Namely, in each execution of the **(for i, for j)**–loop, at most one maximal false point is identified and added to the list L. Explicitly producing, that is, writing up this false point, requires $O(n)$ operations. But in fact, the point is implicitly identified in constant time by simply testing whether $y_j^i = 1$ and $v_i < j$ (this is, of course, a direct consequence of Theorem 8.27). It is this feature of DUALREG that allows Bertolazzi and Sassano [74] to solve regular set covering problems in time $O(nm)$, as we explain in Section 8.6.

We close this section with a bound on the size of the dual of a regular function (see Bertolazzi and Sassano [74] and Crama [225]):

Theorem 8.29. *If $f(x_1, x_2, \ldots, x_n)$ is a regular function with minimal true points $Y^1, Y^2, \ldots, Y^m$, then the number of maximal false points of f is exactly*

$$p = \sum_{i=1}^{m} \sum_{j=1}^{n} \{y_j^i : v_i < j\},$$

where v_i is defined as in DUALREG. In particular, $p \leq |f|$, where $|f|$ is the number of literals in the complete DNF of f, and $p \leq (n-1)m$ when $m > 1$.

Proof. This is a straightforward corollary of Theorem 8.27.   □

Theorem 8.29 strengthens the result of Peled and Simeone [735] mentioned in the introduction of this section. Further refinements of the bound can be found in [105, 225, 735].

8.6 Regular set covering problems

We deal in this section with the set covering problem $(\mathcal{SCP})$:

$$\text{maximize}\quad z(x_1, x_2, \ldots, x_n) = \sum_{i=1}^{n} c_i x_i \tag{8.10}$$

$$\text{subject to}\quad f(x_1, x_2, \ldots, x_n) = 0 \tag{8.11}$$

$$(x_1, x_2, \ldots, x_n) \in \mathcal{B}^n, \tag{8.12}$$

where f is a positive Boolean function expressed in complete (prime irredundant) disjunctive normal form and $c_i \geq 0$ for $i = 1, 2, \ldots, n$ (see, e.g., Section 1.13.6, Application 4.4 in Section 4.2, and Application 8.2 in Section 8.2). We are more particularly interested in the special case in which f is regular. When this is the case, we say that $\mathcal{SCP}$ is a *regular set covering problem* $(\mathcal{RSCP})$.

Since we know that at least one optimal solution of $\mathcal{RSCP}$ is to be found among the maximal false points of f, we immediately conclude that $\mathcal{RSCP}$ is solved in polynomial time by the procedure REGCOVER0 in Figure 8.4.

Theorem 8.30. *The procedure* REGCOVER0(c, f) *is correct and can be implemented to run in time* $O(n^2 m)$, *where n is the number of variables and m is the number of prime implicants of f.*

Proof. The procedure is obviously correct. It can easily be implemented to run in time $O(n^2 m)$ if the dualization algorithm DUALREG is used to generate the maximal false points of f (see Theorem 8.28).   □

Procedure REGCOVER0(c, f)
Input: A vector $(c_1, c_2, \ldots, c_n)$ of integer coefficients and a regular function
        $f(x_1, x_2, \ldots, x_n)$ in complete disjunctive normal form.
Output: An optimal solution of the instance of $\mathcal{RSCP}$ defined by $(c_1, c_2, \ldots, c_n)$ and f.

begin
        generate all maximal false points of f;
        evaluate the value of each maximal false point and return the best one;
end

Figure 8.4.   Procedure REGCOVER0.

Example 8.15. *Consider the regular set covering problem:*

$$\text{maximize} \quad z(x_1, x_2, x_3, x_4, x_5) = 3x_1 + 2x_2 + x_3 + x_4 + 2x_5$$

$$\text{subject to} \quad f = x_1 x_2 \vee x_1 x_3 \vee x_1 x_4 \vee x_2 x_3 \vee x_2 x_4 x_5 = 0$$

$$(x_1, x_2, x_3, x_4, x_5) \in \mathcal{B}^5.$$

The maximal false points of f have been computed in Example 8.14; they are $X^1 = (0,1,0,1,0)$, $X^2 = (0,1,0,0,1)$, $X^3 = (1,0,0,0,1)$ and $X^4 = (0,0,1,1,1)$. Their respective values are $z(X^1) = 3$, $z(X^2) = 4$, $z(X^3) = 5$ and $z(X^4) = 4$. So, X^3 is an optimal solution for this instance of $\mathcal{RSCP}$. $\square$

The first polynomial-time algorithm for $\mathcal{RSCP}$ was obtained by Peled and Simeone [735], based on the general approach outlined in procedure DUALREGO. The complexity of their algorithm is $O(n^3 m)$, since this is also the complexity of the dualization algorithm proposed in [735]. The better time bound mentioned in Theorem 8.30 immediately results from the improvements brought by Crama [225] or Bertolazzi and Sassano [74] to the efficiency of dualization procedures for regular functions.

However, as shown by Bertolazzi and Sassano [74], Hammer and Simeone [462], or Peled and Simeone [736], even faster algorithms (with complexity $O(nm)$) can be obtained for $\mathcal{RSCP}$ by exploiting a slightly different idea: Namely, these authors manage to replace the *explicit* generation of the maximal false points of f by their *implicit* generation, and to compute in *constant time* the value $z(X)$ of each such point. In Bertolazzi and Sassano [74], this idea is implemented via a simple adaptation of the dualization algorithm DUALREG. This leads to the procedure REGCOVER shown in Figure 8.5. In this procedure, the variable *best* keeps track of the value of the best point found so far, and i^*, j^* are the values of i and j describing this point, as in Theorem 8.27.

The meaning of the computations carried out in REGCOVER is revealed in the following proof.

Theorem 8.31. *The procedure* REGCOVER(c, f) *is correct and can be implemented to run in time $O(nm)$, where n is the number of variables and m is the number of prime implicants of f.*

Proof. It is trivial to verify that, at the beginning of an arbitrary **(for** j)–loop (that is, just after the counter j has been increased), the value of C is given as

$$C = \sum_{k<j} c_k y_k^i + \sum_{k \geq j} c_k.$$

On the other hand, in view of Theorem 8.27 and the comments that follow it, we know that the maximal false points of f are all points of the form

$$X^* = (y_1^i, y_2^i, \ldots, y_{j-1}^i, 0, 1, \ldots, 1)$$

Procedure REGCOVER(c, f)
Input: A vector $(c_1, c_2, \ldots, c_n)$ of nonnegative integer coefficients and the list of
    minimal true points $Y^1, Y^2, \ldots, Y^m$ of a regular function $f(x_1, x_2, \ldots, x_n)$
    such that $x_1 \succeq_f x_2 \succeq_f \cdots \succeq_f x_n$.
Output: An optimal solution of the instance of $\mathcal{RSCP}$ defined by $(c_1, c_2, \ldots, c_n)$ and f.

begin
    $best := -1$;
    $S := \sum_{j=1}^n c_j$;
    sort the points Y^i $(i = 1, 2, \ldots, m)$ in lexicographic order;
    {comment: assume without loss of generality that $Y^1 <_L Y^2 <_L \cdots <_L Y^m$ };
    $v_1 := 0$;
    for $i = 2$ **to** m **do** $v_i := \min\{k : y_k^{i-1} < y_k^i\}$;
    {comment: compute the value of each maximal false point};
    for $i = 1$ **to** m **do**
    begin
        $C := S$;
        for $j = 1$ **to** n **do**
        begin
            if $y_j^i = 0$ **then** $C := C - c_j$;
            if $y_j^i = 1$ **and** $v_i < j$ **and** $C - c_j > best$ **then**
            begin
                $best := C - c_j$;
                $i^* := i$;
                $j^* := j$;
            end
        end
    end
    return $(y_1^{i^*}, y_2^{i^*}, \ldots, y_{j^*-1}^{i^*}, 0, 1, 1, \ldots, 1)$;
end

Figure 8.5. Procedure REGCOVER.

such that $y_j^i = 1$ and $v_i < j$. Thus, if $X^* = (y_1^i, y_2^i, \ldots, y_{j-1}^i, 0, 1, \ldots, 1)$ is such a point, then $C - c_j$ is precisely the value of $z(X^*)$. It follows easily that REGCOVER returns a maximal false point with maximum value.

The complexity analysis is straightforward. $\square$

Example 8.16. *Let us consider again the set covering instance given in Example 8.15, and let us run* REGCOVER *on this instance. The minimal true points of f are (in lexicographic order):* $Y^1 = (0, 1, 0, 1, 1)$, $Y^2 = (0, 1, 1, 0, 0)$, $Y^3 = (1, 0, 0, 1, 0)$, $Y^4 = (1, 0, 1, 0, 0)$, *and* $Y^5 = (1, 1, 0, 0, 0)$. *So,* $v_1 = 0$, $v_2 = 3$, $v_3 = 1$, $v_4 = 3$, $v_5 = 2$. *The sum of the objective function coefficients is* $S = 9$, *and we initially set* $best := -1$.

For $i = 1$ *and for* $j = 1$ *to* 5, *we successively obtain*
$j = 1$: $y_1^1 = 0 \implies C := 9 - c_1 = 6$;
$j = 2$: $y_2^1 = 1$ *and* $v_1 < 2$ *and* $C - c_2 = 4 > best \implies best := 4$, $i^* := 1$, $j^* := 2$;

$j = 3:\ y_3^1 = 0 \implies C := 6 - c_3 = 5;$
$j = 4:\ y_4^1 = 1$ and $C - c_4 = 4 \leq best \implies$ no update;
$j = 5:\ y_5^1 = 1$ and $C - c_5 = 3 \leq best \implies$ no update.
   No better solution is found for $i = 2$, since $v_2 \geq j$ whenever $y_j^2 = 1$.
   For $i = 3$, we get:
$j = 1:\ y_1^3 = 1$ and $v_3 \geq j \implies$ no update;
$j = 2: \implies C := 9 - c_2 = 7;$
$j = 3: \implies C := 7 - c_3 = 6;$
$j = 4:\ y_4^3 = 1$ and $v_3 < 4$ and $C - c_4 = 5 > best \implies best := 5,\ i^* := 3,\ j^* := 4.$

   We leave it to the reader to continue the execution of REGCOVER on this example
and to verify that no further updates of best, i^* and j^* take place. So, the solution
returned by the algorithm is

$$(y_1^{i^*}, y_2^{i^*}, \ldots, y_{j^*-1}^{i^*}, 0, 1, \ldots, 1) = (y_1^3, y_2^3, y_3^3, 0, 1) = (1, 0, 0, 0, 1),$$

with an objective function value of 5.   □

   Further connections between regular functions and set covering problems can be
found, for instance, in Balas [42], Hammer, Johnson and Peled [443, 444], Laurent
and Sassano [602], Wolsey [922], etc. (see also Section 8.7.3 and Chapter 9).

8.7 Regular minorants and majorants

In view of the computational tractability of regular functions, it may be of interest
to approximate a given nonregular function by a regular one. We deal in this section
with a restricted form of these problems in which the approximating function is
required to be either a *majorant* or a *minorant* of the original one, and in which
the strength ordering of the approximant is imposed. Thus, we state as follows
the problem to be tackled: Given a positive function $f(x_1, x_2, \ldots, x_n)$, find two
positive functions $f_-(x_1, x_2, \ldots, x_n)$ and $f^+(x_1, x_2, \ldots, x_n)$ such that

- f_- and f^+ are both regular with respect to $(x_1, x_2, \ldots, x_n)$;   (8.13)

- $f_- \leq f \leq f^+$;   (8.14)

- f_- and f^+ are "closest" to f among all functions satisfying (8.13) and (8.14).

                                                                                     (8.15)

   The word "closest" in condition (8.15) needs to be further clarified: Before
we can speak of closeness, it may seem necessary to introduce first a notion of
distance between Boolean functions. However, this difficulty is easily avoided
in the present context. Indeed, as we prove next, there always exists a *smallest
majorant* and a *largest minorant* satisfying conditions (8.13) and (8.14). They will
play for us the roles of "closest majorant" and "closest minorant." (Compare with
Exercise 13 in Chapter 1.)

Theorem 8.32. *For every Boolean function $f(x_1,x_2,\ldots,x_n)$, there exist two positive functions $f^R(x_1,x_2,\ldots,x_n)$ and $f_R(x_1,x_2,\ldots,x_n)$ such that*

(a) *f_R and f^R are both regular with respect to $(x_1,x_2,\ldots,x_n)$;*

(b) *$f_R \leq f \leq f^R$;*

(c) *if f_- and f_+ are any two functions satisfying conditions (8.13) and (8.14), then $f_- \leq f_R$ and $f^R \leq f^+$.*

Proof. Let us denote by L and U the sets of all positive functions such that (8.13) and (8.14) are satisfied for all $f_- \in L$ and $f^+ \in U$. Observe that L and U are both nonempty, since $\mathbf{0}_n \in L$ and $\mathbf{1}_n \in U$. Define

$$f_R = \bigvee \{f_- : f_- \in L\}, \ f^R = \bigwedge \{f^+ : f^+ \in U\}.$$

Then, f_R and f^R trivially satisfy conditions (b) and (c), and Theorem 8.9 implies (a).   □

The functions f_R and f^R introduced in Theorem 8.32 will be called the *largest regular minorant* and the *smallest regular majorant* of f *with respect to* $(x_1,x_2,\ldots,x_n)$, respectively.

Note that condition (a) in Theorem 8.32 cannot be replaced by the weaker condition "f_R and f^R are regular" without further specification of the strength ordering. This is illustrated by the next example.

Example 8.17. *The function $f = x_1x_2 \vee x_3x_4$ is not regular. The largest regular minorant of f with respect to (x_1,x_2,x_3,x_4) is $f_R = x_1x_2 \vee x_1x_3x_4 \vee x_2x_3x_4$ (see Example 8.20 hereunder). Another regular minorant of f is $g = x_1x_2x_3 \vee x_1x_2x_4 \vee x_3x_4$, which is such that $x_3 \approx_g x_4 \succ_g x_1 \approx_g x_2$. But there is no regular minorant of f which is larger than both f_R and g. Indeed, assume that h is a minorant of f such that $f_R \leq h$ and $g \leq h$. Then, $f_R \vee g \leq h \leq f$. However, $f_R \vee g = f$. Hence, $h = f$, and h is not regular.*   □

A useful characterization of the functions f_R and f^R can be derived from results in Section 8.3 (recall in particular Theorem 8.15, and compare with Exercise 14 in Chapter 1).

Theorem 8.33. *For every Boolean function $f(x_1,x_2,\ldots,x_n)$,*

(i) *f_R is the unique function that is regular with respect to $(x_1,x_2,\ldots,x_n)$ and that has the same set of ceilings as f; and*

(ii) *f^R is the unique function that is regular with respect to $(x_1,x_2,\ldots,x_n)$ and that has the same set of floors as f.*

Proof. Let A be the set of ceilings of f, and let τ_A be defined as in Theorem 8.15. We want to prove that $f_R = r_A$, that is, we want to prove that r_A satisfies conditions (a)-(c) in Theorem 8.32. Condition (a) follows from the definition of τ_A.

To obtain condition (b), let X^* be a false point of f. Then, by Definition 8.5, there exists a ceiling $Y^* \in A$ such that $Y^* \curvearrowright X^*$. It follows from (8.8) in the proof of Theorem 8.15 that $r_A(X^*) = 0$, and hence $r_A \leq f$.

Since r_A is a regular minorant of f, we have $r_A \leq f_R$. To see that $r_A = f_R$, consider now any point X^* such that $r_A(X^*) = 0$. By (8.8) in Theorem 8.15, there exists $Y^* \in A$ such that $Y^* \curvearrowright X^*$. Since Y^* is a ceiling of f, $f(Y^*) = 0$, and hence, $f_R(Y^*) = 0$. By Theorem 8.14, we conclude that $f_R(X^*) = 0$.

This completes the proof of statement (i). The proof of the second statement is similar. $\square$

In the next subsections, we propose some algorithms for the computation of f_R and f^R when f is positive (observe that Theorem 8.32 and Theorem 8.33 do not depend on the positivity assumption). Before we turn to these problems, however, we note that the size of the complete DNF of f_R and of the complete DNF of f^R can be exponentially large in the size of the complete DNF of f, so that there is no hope of computing f_R and f^R in polynomial time.

Example 8.18. *Consider the function* $f = x_{n+1} \vee \ldots \vee x_{2n}$ *on* $\mathcal{B}^{2n}$. *Its unique maximal false point (and unique ceiling) is the characteristic vector of* $\{1, 2, \ldots, n\}$, *that is,* $Y^* = (1, \ldots, 1, 0, \ldots, 0)$. *Let* $A = \{Y^*\}$ *and let* F *be the set of all points of* $\mathcal{B}^{2n}$ *with exactly n components equal to 1. Then,* Y^* *is a left-shift of every point in* F, *and it follows from condition* (8.8) *in Theorem 8.15 that* F *is exactly the set of maximal false points of* $r_A = f_R$. *Hence, the minimal true points of* f_R *are the points of* $\mathcal{B}^{2n}$ *with exactly* $n + 1$ *components equal to 1, and their number is exponential in the size of* f. $\square$

8.7.1 Largest regular minorant with respect to a given order

Consider a positive function $f(x_1, x_2, \ldots, x_n) = \alpha x_i x_j \vee \beta x_i \vee \gamma x_j \vee \delta$, where $i, j \in \{1, 2, \ldots, n\}$ and $\alpha, \beta, \gamma, \delta$ are positive DNFs that not involve either x_i or x_j. Hammer, Johnson, and Peled [443] introduced an operation (to be called (i, j)–*minorization*) that transforms the function f into another positive function f_{ij} defined by any of the following equivalent expressions:

$$f_{ij} = f \wedge (x_i \vee f_{|x_i=1, \, x_j=0}) \tag{8.16}$$

$$= f \wedge (x_i \vee \beta \vee \delta) \tag{8.17}$$

$$= (\alpha \vee \gamma) x_i x_j \vee \beta x_i \vee \beta \gamma x_j \vee \delta, \tag{8.18}$$

where we look at $f_{|x_i=1, \, x_j=0}$ as a function of $(x_1, x_2, \ldots, x_n)$. We leave it to the reader to verify that these expressions actually are equivalent. We say that f_{ij} is the (i, j)–*minor* of f.

Example 8.19. *The* $(1, 3)$–*minor of* $f = x_1 x_2 \vee x_3 x_4$ *is*

$$f_{13} = (x_1 x_2 \vee x_3 x_4)(x_1 \vee x_2) = x_1 x_2 \vee x_1 x_3 x_4 \vee x_2 x_3 x_4.   \square$$

The next result shows that f_{ij} is the largest positive minorant of f for which x_i is stronger than x_j.

Theorem 8.34. *Let* $f(x_1,x_2,\ldots,x_n)$ *be a positive Boolean function and let* i,j *be any two indices in* $\{1,2,\ldots,n\}$. *Then,*

(a) $f_{ij} \leq f$;
(b) x_i *is stronger than* x_j *with respect to* f_{ij};
(c) *if* $g(x_1,x_2,\ldots,x_n)$ *is a positive function such that* $g \leq f$ *and* $x_i \succeq_g x_j$, *then*
    $g \leq f_{ij}$.

Proof. Assertions (a) and (b) are easily verified. Suppose now that $g \leq f$ and $x_i \succeq_g x_j$. Let $Y \in \mathcal{B}^n$. We must show that $g(Y) \leq f_{ij}(Y)$.
If $y_i = 1$, then $f_{ij}(Y) = f(Y)$ by (8.16), and hence, $g(Y) \leq f_{ij}(Y)$.
If $y_i = y_j = 0$, then

$$\begin{aligned}
f_{ij}(Y) &= f(Y) \wedge f(Y \vee e_i) && \text{(by (8.16))} \\
&= f(Y) && \text{(by positivity of } f) \\
&\geq g(Y).
\end{aligned}$$

If $y_i = 0$ and $y_j = 1$, let $Z \in \mathcal{B}^n$ be such that $Y = Z \vee e_j$ and $z_j = 0$. Then,

$$\begin{aligned}
f_{ij}(Y) &= f(Y) \wedge f(Z \vee e_i) && \text{(by (8.16))} \\
&\geq g(Y) \wedge g(Z \vee e_i) && \\
&\geq g(Y) \wedge g(Z \vee e_j) && \text{(because } x_i \succeq_g x_j) \\
&= g(Y).
\end{aligned}$$

$\square$

Theorem 8.34 suggests the procedure REGMINOR0 displayed in Figure 8.6. Hammer, Johnson, and Peled [443] proved:

Procedure REGMINOR0(f)
Input: A positive Boolean function $f(x_1,x_2,\ldots,x_n)$.
Output: f_R, the largest regular minorant of f with respect to $(x_1,x_2,\ldots,x_n)$.

begin
    $f_R := f$;
    while there is a pair of variables x_i,x_j such that $i < j$ and
        x_i is not stronger than x_j with respect to f_R
        do $f_R := f_{ij}$;
    return f_R;
end

Figure 8.6. Procedure REGMINOR0.

Theorem 8.35. *The procedure* REGMINOR0(f) *is correct, that is, it stops for every input, and it returns the largest regular minorant of f with respect to* $(x_1, x_2, \ldots, x_n)$.

Proof. It follows from Theorem 8.34 that, if x_i is not stronger than x_j with respect to f, then $f_{ij} < f$. Thus, the sequence of functions produced in the **while** loop is strictly decreasing, and it must terminate. Denote by g the output of the procedure, and denote by f_R the largest regular minorant of f with respect to $(x_1, x_2, \ldots, x_n)$ (its existence is guaranteed by Theorem 8.32). We must show that $g = f_R$.

By construction, g is regular with respect to $(x_1, x_2, \ldots, x_n)$. Thus, $g \leq f_R$ by definition of f_R.

On the other hand, Theorem 8.34(c) implies (by induction) that $f_R \leq f^*$ for each of the functions f^* produced in the course of the procedure. In particular, $f_R \leq g$, and this completes the proof.   □

Example 8.20. *Consider the function $f = x_1 x_2 \vee x_3 x_4$, and note that x_1 and x_3 are not comparable with respect to $\succcurlyeq_f$. The $(1,3)$–minor of f has already been computed in Example 8.19. Since f_{13} is regular with respect to (x_1, x_2, x_3, x_4), we conclude that f_{13} is the largest regular minorant of f with respect to this order of the variables.*   □

Regarding the computational complexity of REGMINOR0, note that the number of iterations of the **while** statement, that is, the number of (i, j)–minorization steps to be executed, does not appear to be polynomially bounded. Indeed, it may very well happen that a pairwise strength relation imposed in some minorization step is destroyed in a further step, and hence, needs to be reestablished later on. This possibility is illustrated by the following example.

Example 8.21. *The strength preorder of the function $f(x_1, x_2, x_3, x_4) = x_1 x_2 \vee x_3 \vee x_4$ is given as $x_3 \approx_f x_4 \succ_f x_1 \approx_f x_2$. Assume that we want to produce f_R, namely, the largest minorant of f with respect to (x_1, x_2, x_3, x_4). We start the execution of REGMINOR0 by performing a $(1,3)$–minorization step on f, thus producing $f_{13} = x_1 x_2 \vee x_1 x_3 \vee x_2 x_3 \vee x_4$. Observe now that x_4 is strictly stronger than x_3 with respect to f_{13}. Thus, the relation "x_3 is stronger than x_4," which holds for f and is to hold for f_R, has been temporarily lost for f_{13}.*   □

It is possible, however, to carry out the (i, j)–minorization steps in such a way that, once established, the strength relation among a pair of variables will not be spoiled in a later stage. More precisely, consider the following minorization strategy: Impose first $x_1 \succcurlyeq_f x_j$ for all $j \geq 2$, then $x_2 \succcurlyeq_f x_j$ for all $j \geq 3$, then $x_3 \succcurlyeq_f x_j$ for all $j \geq 4$, and so on. The next result shows that, if we adopt this strategy, then no relation $x_i \succcurlyeq_f x_j$ ever needs to be imposed twice; indeed, all relations valid at some stage of the procedure (and expressed by conditions (a), (b), (c) in the next statement) remain valid after a subsequent minorization step.

Theorem 8.36. *Let $f(x_1, x_2, \ldots, x_n)$ be a positive function, and let i, j be two indices in $\{1, 2, \ldots, n\}$, $i < j$, such that the following conditions hold with respect to f:*

(a) $x_1 \succcurlyeq x_2 \succcurlyeq \cdots \succcurlyeq x_{i-j}$;

(b) *For all $k \in \{i, i+1, \ldots, n\}$, $x_{i-1} \succcurlyeq x_k$;*

(c) *For all $k \in \{i+1, i+2, \ldots, j-1\}$, $x_i \succcurlyeq x_k$.*

Then, conditions (a), (b), and (c) also hold with respect to f_{ij}.

Proof. Let $h = f_{ij}$ and $g = x_i \vee f_{|x_i=1, x_j=0}$, so that $h = f \wedge g$ (see (8.16)).

(a) To see that condition (a) holds with respect to h, suppose that $1 \leq k < r \leq i-1$. Theorem 8.7 implies that $x_k \succcurlyeq x_r$ with respect to $f_{|x_i=1, x_j=0}$. So, $x_k \succcurlyeq_f x_r$, $x_k \succcurlyeq_g x_r$, and $x_k \succcurlyeq_h x_r$ follows by virtue of Theorem 8.9.

(b) To establish condition (b), consider first the case in which $k \neq i$ and $k \neq j$. Then, the same argument used in (a) shows that $x_{i-1} \succcurlyeq_h x_k$.

Consider next the case $k = i$, and let $Y \in \mathcal{B}^n$ be such that $y_{i-1} = y_i = 0$. We must show that $h(Y \vee e_i) \leq h(Y \vee e_{i-1})$. Now,

$$h(Y \vee e_i) = f(Y \vee e_i)$$

and

$$h(Y \vee e_{i-1}) = f(Y \vee e_{i-1}) \wedge g(Y \vee e_{i-1})$$
$$= f(Y \vee e_{i-1}) \wedge f_{|x_i=1, x_j=0}(Y \vee e_{i-1})$$
$$= f(Y \vee e_{i-1}) \wedge f_{|x_j=0}(Y \vee e_{i-1} \vee e_i).$$

There are two distinct subcases. If $y_j = 0$, then by positivity of f

$$f_{|x_j=0}(Y \vee e_{i-1} \vee e_i) = f(Y \vee e_{i-1} \vee e_i) \geq f(Y \vee e_i).$$

On the other hand, if $y_j = 1$, then

$$f_{|x_j=0}(Y \vee e_{i-1} \vee e_i) \geq f(Y \vee e_i)$$

since $x_{i-1} \succcurlyeq_f x_j$. In either case, we get

$$h(Y \vee e_{i-1}) \geq f(Y \vee e_{i-1}) \wedge f(Y \vee e_i)$$
$$= f(Y \vee e_i) \qquad \text{(since } x_{i-1} \succcurlyeq_f x_i)$$
$$= h(Y \vee e_i),$$

so that $x_{i-1} \succcurlyeq_h x_i$ as required.

Consider finally the case $k = j$. Here, the relation $x_{i-1} \succcurlyeq_h x_j$ directly follows from $x_{i-1} \succcurlyeq_h x_i$ (which we just established) and from $x_i \succcurlyeq_h x_j$ (which follows from the definition of $h = f_{ij}$).

Procedure REGMINOR(f)
Input: A positive Boolean function $f(x_1, x_2, \ldots, x_n)$.
Output: f_R, the largest regular minorant of f with respect to $(x_1, x_2, \ldots, x_n)$.

begin
    $f_R := f$;
    for $i = 1$ **to** $n - 1$ **do**
        for $j = i + 1$ **to** n **do**
            if x_i is not stronger than x_j with respect to f_R **then** $f_R := f_{ij}$;
    return f_R;
end

Figure 8.7. Procedure REGMINOR.

(c) We want to show that, for $k \in \{i+1, i+2, \ldots, j-1\}$, $x_i \succeq_h x_k$. Let $Y \in \mathcal{B}^n$ be such that $y_i = y_k = 0$. Then,

$$
\begin{aligned}
h(Y \vee e_k) &= f(Y \vee e_k) \wedge f_{|x_i=1, x_j=0}(Y \vee e_k) \\
&\le f(Y \vee e_k) \\
&\le f(Y \vee e_i) \qquad \text{(since } x_i \succeq_f x_k) \\
&= h(Y \vee e_i),
\end{aligned}
$$

and hence, $x_i \succeq_h x_k$ as required. $\qquad\qquad \square$

Theorem 8.36 suggests the specialization of REGMINOR0 described in Figure 8.7.

Theorem 8.37. *The procedure* REGMINOR(f) *is correct and performs* $O(n^2)$ *minorization steps, where n is the number of variables of f.*

Proof. The correctness of the procedure is implied by Theorem 8.36, and the bound on the number of minorization steps is trivial. $\qquad\qquad \square$

Note that, despite the fact that the number of minorization steps performed by REGMINOR is small, this procedure necessarily runs in exponential (input) time, in view of Example 8.18. It is not clear, however, whether the procedure runs in polynomial total time, that is, in time polynomial in $|f| + |f_R|$ (see Appendix B).

8.7.2 Smallest regular majorant with respect to a given order

We start with an easy observation:

Theorem 8.38. *The smallest regular majorant of a function $f(x_1,x_2,\ldots,x_n)$ with respect to a given order of the variables is the dual of the largest regular minorant of f^d with respect to the same order of the variables.*

Proof. Let $(x_1,x_2,\ldots,x_n)$ be the given order, and let g^d be the largest regular minorant of f^d with respect to $(x_1,x_2,\ldots,x_n)$. We must prove that g is the smallest regular majorant of f with respect to $(1,2,\ldots,n)$.

First, $g^d \leq f^d$ implies $f \leq g$. Next, since g^d is regular with respect to $(x_1,x_2,\ldots,x_n)$, g is regular with the same strength preorder (by Theorem 8.10). Finally, if g is not the *smallest* regular majorant of f with respect to $(x_1,x_2,\ldots,x_n)$, then there exists another regular function h, with the same strength preorder, such that $f \leq h < g$. But then, $g^d < h^d \leq f^d$, contradicting the definition of g^d. $\square$

According to Theorem 8.38, everything there is to know about smallest regular majorants can easily be derived from the corresponding results concerning largest regular minorants. In particular, a procedure similar to REGMINOR can be developed for the computation of the smallest regular majorant of a positive function with respect to a given order, based on the following ideas.

Consider a positive function $f(x_1,x_2,\ldots,x_n) = \alpha x_i x_j \vee \beta x_i \vee \gamma x_j \vee \delta$, where $i,j \in \{1,2,\ldots,n\}$ and $\alpha,\beta,\gamma,\delta$ are positive DNFs that do not involve either x_i or x_j. We define the (i,j)–*major* of f as the function $f^{ij}(x_1,x_2,\ldots,x_n)$ represented by any of the following equivalent expressions:

$$f^{ij} = f \vee (x_i \wedge f_{|x_i=0,\,x_j=1}) \tag{8.19}$$

$$= f \vee \gamma x_i \tag{8.20}$$

$$= \alpha x_i x_j \vee (\beta \vee \gamma) x_i \vee \gamma x_j \vee \delta. \tag{8.21}$$

Paraphrasing the statement of Theorem 8.34, we obtain the following result due to Hammer and Mahadev [452].

Theorem 8.39. *Let $f(x_1,x_2,\ldots,x_n)$ be a positive Boolean function, and let i,j be two indices in $\{1,2,\ldots,n\}$. Then,*

(a) $f \leq f^{ij}$;
(b) x_i *is stronger than* x_j *with respect to* f^{ij};
(c) *if* $g(x_1,x_2,\ldots,x_n)$ *is a positive function such that* $f \leq g$ *and* $x_i \succeq_g x_j$, *then* $f^{ij} \leq g$.

Proof. This can be proved either by a duality argument or by adapting the proof of Theorem 8.34. Details are left to the reader. $\square$

Similarly to the minorization case, Theorem 8.39 leads to an algorithm that produces the smallest regular majorant of an arbitrary positive function, and this algorithm can be implemented to perform $O(n^2)$ majorization steps.

8.7.3 Regular minorization and set covering problems

We conclude this section by indicating how the concept of regular minorization can be used to transform an arbitrary set covering problem into an equivalent (in some sense to be made precise) *regular* set covering problem. Most results in this section are due to Hammer, Johnson, and Peled [443].

Let us consider an instance of the problem SCP:

$$\text{maximize} \quad z(x_1, x_2, \ldots, x_n) = \sum_{i=1}^{n} c_i x_i \tag{8.22}$$

$$\text{subject to} \quad f(x_1, x_2, \ldots, x_n) = 0 \tag{8.23}$$

$$(x_1, x_2, \ldots, x_n) \in \mathcal{B}^n, \tag{8.24}$$

where f is a positive Boolean function, and let us define the set covering problem SCP_{12} as follows:

$$\text{maximize} \quad z(x_1, x_2, \ldots, x_n) = \sum_{i=1}^{n} c_i x_i$$

$$\text{subject to} \quad f_{12}(x_1, x_2, \ldots, x_n) = 0$$

$$(x_1, x_2, \ldots, x_n) \in \mathcal{B}^n,$$

where f_{12} is the (1,2)-minor of f defined in Section 8.7.1.

Since $f_{12} \leq f$ (Theorem 8.34), every feasible solution of SCP also is a feasible solution of SCP_{12}. Hence, the optimal value of SCP_{12} is at least as large as the optimal value of SCP. However, more is actually true, namely:

Theorem 8.40. *If $c_1 > c_2$, then SCP and SCP_{12} have the same set of optimal solutions.*

Proof. We only need to show that every optimal solution of SCP_{12} is feasible for SCP. Let $X^* = (x_1^*, x_2^*, \ldots, x_n^*)$ be an optimal solution of SCP_{12}. Since $f_{12}(X^*) = 0$, Equation (8.16) implies that $f(X^*) = 0$ or

$$x_1^* \vee f_{|x_1=1, x_2=0}(X^*) = 0. \tag{8.25}$$

If $f(X^*) = 0$, then we are done. Otherwise, (8.25) holds, or equivalently,

$$x_1^* = 0 \text{ and } f(1, 0, x_3^*, x_4^*, \ldots, x_n^*) = 0. \tag{8.26}$$

Now, there are two cases. If $x_2^* = 0$, then

$$f(X^*) = f(0, 0, x_3^*, x_4^*, \ldots, x_n^*) \leq f(1, 0, x_3^*, x_4^*, \ldots, x_n^*) = 0,$$

and X^* is feasible for SCP.

If $x_2^* = 1$, then define $Y^* = (1, 0, x_3^*, x_4^*, \ldots, x_n^*)$. In view of (8.26), Y^* is feasible for SCP, and hence, for SCP_{12}. However, since $c_2 < c_1$,

$z(X^*) = z(0,1,x_3^*,x_4^*,\ldots,x_n^*) < z(1,0,x_3^*,x_4^*,\ldots,x_n^*) = z(Y^*)$. But then X^* is not an optimal solution of $\mathcal{SCP}_{12}$, and we reach a contradiction.   □

Example 8.22. *Consider the set covering problem:*

$$maximize \quad z = 5x_1 + 4x_2 + 3x_3 + 2x_4 + x_5 \tag{8.27}$$

$$subject\ to \quad f = x_1x_2 \vee x_1x_3 \vee x_1x_4 \vee x_2x_3 \vee x_2x_5 = 0 \tag{8.28}$$

$$(x_1,x_2,x_3,x_4,x_5) \in \mathcal{B}^5. \tag{8.29}$$

The $(1,2)$–minor of f is the regular function:

$$f_{12} = x_1x_2 \vee x_1x_3 \vee x_1x_4 \vee x_2x_3 \vee x_2x_4x_5.$$

As shown in Example 8.14, the maximal false points of f_{12} are $X^1 = (0,1,0,1,0)$, $X^2 = (0,1,0,0,1)$, $X^3 = (1,0,0,0,1)$, and $X^4 = (0,0,1,1,1)$. The objective function value of X^1, X^3 and X^4 is 6, and the value of X^2 is 5. So, Theorem 8.40 implies that $\{X^1,X^3,X^4\}$ is the set of optimal solutions of the original problem (8.27)–(8.29). One easily verifies that this is indeed the fact, since X^1, X^3, and X^4 are all the maximal false points of f.   □

In view of Theorem 8.40 and of the results obtained in Section 8.7, it is now natural to associate with $\mathcal{SCP}$ the set covering problem $\mathcal{SCP}_R$, defined as follows:

$$maximize \quad z(x_1,x_2,\ldots,x_n) = \sum_{i=1}^{n} c_i x_i$$

$$subject\ to \quad f_R(x_1,x_2,\ldots,x_n) = 0$$

$$(x_1,x_2,\ldots,x_n) \in \mathcal{B}^n,$$

where f_R is the largest regular minorant of f with respect to $(x_1,x_2,\ldots,x_n)$.

Theorem 8.41. *If $c_1 > c_2 > \cdots > c_n$, then $\mathcal{SCP}$ and $\mathcal{SCP}_R$ have the same set of optimal solutions.*

Proof. The statement easily follows from Theorem 8.35 and 8.40, by induction on the number of (i,j)–minorization steps that are necessary to derive f_R from f. □

Since the coefficients $c_1,c_2,\ldots,c_n$ can always be sorted in nonincreasing order, Theorem 8.41 provides a constructive transformation of an arbitrary set covering problem into an equivalent regular one, under the assumption that all coefficients of the objective function are distinct. (Compare with Exercise 15 in Chapter 1.) As illustrated by the next example, the conclusion of Theorem 8.41 may fail to hold when the coefficients are not distinct.

Example 8.23. *Consider the set covering problem:*

$$maximize z = 4x_1 + 4x_2 + 2x_3 + x_4 + x_5$$

$$subject\ to\ f = x_1 x_2 \lor x_1 x_3 \lor x_1 x_4 \lor x_2 x_3 \lor x_2 x_5 = 0$$

$$(x_1, x_2, x_3, x_4, x_5) \in \mathcal{B}^5$$

(compare with Example 8.22). It is easy to check that $\mathcal{SCP}$ has exactly two optimal solutions, namely, $X^1 = (0,1,0,1,0)$ and $X^3 = (1,0,0,0,1)$, whereas the associated problem $\mathcal{SCP}_R$ has three optimal solutions, namely, X^1, X^3, and $X^2 = (0,1,0,0,1)$. □

Nevertheless, the following result can be proved:

Theorem 8.42. *If $c_1 \geq c_2 \geq \cdots \geq c_n$, then the lexicographically largest optimal solution of $\mathcal{SCP}_R$ is an optimal solution of $\mathcal{SCP}$.*

Proof. As was the case for Theorem 8.41, we only need to show that the statement is correct with $\mathcal{SCP}_R$ replaced by $\mathcal{SCP}_{12}$. So, let $X^* = (x_1^*, x_2^*, \ldots, x_n^*)$ be the lexicographically largest optimal solution of $\mathcal{SCP}_{12}$. If $f(X^*) = 0$, then we are done. Otherwise, as in the proof of Theorem 8.40, one shows that $x_1^* = 0$, and that X^* is feasible (hence, optimal) for $\mathcal{SCP}$ if $x_2^* = 0$. So, assume that $x_2^* = 1$. Note that the point $Y^* = (1, 0, x_3^*, x_4^*, \ldots, x_n^*)$ is feasible for $\mathcal{SCP}_{12}$. Moreover, since $c_2 \leq c_1$, $z(X^*) \leq z(Y^*)$. Hence, Y^* is an optimal solution of $\mathcal{SCP}_{12}$ and $X^* <_L Y^*$, contradicting the choice of X^*. □

Theorem 8.42 suggests an approach to the solution of an arbitrary set covering problem $\mathcal{SCP}$: First, transform $\mathcal{SCP}$ into the regular set covering problem $\mathcal{SCP}_R$, then compute the lexicographically largest optimal solution of $\mathcal{SCP}_R$. This approach may look attractive, since the procedures REGCOVER0 or REGCOVER (presented in Section 8.6) are easily adapted to carry out its second phase in polynomial time. However, the first phase involves the computation of f_R, and hence, as observed earlier, it cannot be performed in polynomial time (which does not come as a surprise in view of the fact that the set covering problem is NP-hard).

As shown by Hammer, Johnson, and Peled [443], most results in this section can easily be extended to optimization problems of the form

$$\text{maximize} \quad g(X) \qquad\qquad (8.30)$$

$$\text{subject to} \quad f(X) = 0 \qquad\qquad (8.31)$$

$$X \in \mathcal{B}^n, \qquad\qquad (8.32)$$

where f is a Boolean function and g is a pseudo-Boolean function, that is, a real-valued function on $\mathcal{B}^n$ (see Chapter 13), if the assumption $c_1 > c_2 > \ldots > c_n$ is replaced by an appropriate "generalized regularity" condition, namely,

for all $1 \leq i < j \leq n$, and for $X^* \in \mathcal{B}^n$,

$$\text{if } x_i^* = x_j^* = 0 \text{ then } g(X^* \lor e_i) > g(X^* \lor e_j). \qquad\qquad (8.33)$$

8.8 Higher-order monotonicity

The notion of strength preorder among variables can be generalized to a notion of strength relation among subsets of variables. Various such generalizations have been introduced for instance by Muroga, Toda, and Takasu [700]; Paull and McCluskey [732]; Winder [916] in the context of switching theory; and by Lapidot [597] in his investigations of simple games (as cited by Einy [291]). Most of these proposals originally stemmed from attempts to provide purely Boolean or combinatorial characterizations of threshold functions (or weighted majority games) in contrast with the numerical flavor of Definition 8.1. These efforts, where the study of regularity also found its origins (see Section 8.1), eventually resulted in the unearthing of several important properties of threshold functions, that is, necessary conditions for a function to be threshold.

We adopt here Lapidot's approach [597, 291] which rests on a natural extension of Definition 8.2. Recall the following notation: if T is a subset of $\{1, 2, \ldots, n\}$, then the characteristic vector of T is denoted

$$e_T = \sum_{i \in T} e_i \quad (\text{and } e_\emptyset = 0).$$

Definition 8.8. Let $f(x_1, x_2, \ldots, x_n)$ be a Boolean function and let S, T be two subsets of $\{1, 2, \ldots, n\}$. We say that S is stronger than T with respect to f, and we write $S \succcurlyeq_f T$ if and only if, for all $X^* \in \mathcal{B}^n$,

$$x_i^* = 0 \text{ for all } i \in S \cup T \implies f(X^* \vee e_S) \geq f(X^* \vee e_T).$$

We say that S and T are comparable with respect to $\succcurlyeq_f$ if either $S \succcurlyeq_f T$ holds or $T \succcurlyeq_f S$ holds.

As usual, we drop the subscript f from the symbol $\succcurlyeq_f$ when no confusion can result.

Application 8.6. (Game theory.) *The strength relation among subsets of variables has a clear interpretation in the context of game theory. If f represents a simple game, and S is stronger than T with respect to f, then a coalition C (disjoint from S and T) can more easily form a winning coalition by joining S than by joining T (remember Application 8.1). Therefore, Lapidot [597] says that S is "more desirable" than T when $S \succcurlyeq_f T$. This relation among coalitions was used by Peleg [737, 738] to develop a theory of coalition formation in simple games, and its game-theoretic properties have been further investigated by Einy [291].* □

It is easily checked that, for a function $f(x_1, x_2, \ldots, x_n)$ and for $i, j \in \{1, 2, \ldots, n\}$,

- $\{i\} \succcurlyeq_f \emptyset$ if and only if f is positive in x_i;
- $\emptyset \succcurlyeq_f \{i\}$ if and only if f is negative in x_i;
- $\{i\} \succcurlyeq_f \{j\}$ if and only if x_i is stronger than x_j in the sense of Definition 8.2.

Thus, in particular, monotone functions are precisely those functions such that S and T are comparable whenever $|S \cup T| \leq 1$. Similarly, a positive function is regular if and only if S and T are comparable whenever $|S \cup T| = 2$.

Definition 8.9. *A Boolean function f on $\mathcal{B}^n$ is k-monotone ($1 \leq k \leq n$) if, for all pairs of subsets $S, T \subseteq \{1, 2, \ldots, n\}$ such that $|S \cup T| \leq k$, S and T are comparable with respect to $\succcurlyeq_f$. A function on $\mathcal{B}^n$ is* completely monotone *if it is n-monotone, that is, if the strength relation is complete on the power set of $\{1, 2, \ldots, n\}$.*

So, 1-monotonicity is equivalent to monotonicity, and, up to switching the negative variables, 2-monotonicity is equivalent to regularity. The motivation for introducing k-monotonicity in connection with the study of threshold functions is provided by the following result, which extends Theorem 8.3.

Theorem 8.43. *Every threshold function is completely monotone. More precisely, if $f(x_1, x_2, \ldots, x_n)$ is a threshold function with structure $(w_1, w_2, \ldots, w_n, t)$, and if S, T are two subsets of $\{1, 2, \ldots, n\}$ such that $\sum_{i \in S} w_i \geq \sum_{i \in T} w_i$, then $S \succcurlyeq_f T$.*

Proof. This is straightforward.   □

Properties of k-monotone and completely monotone functions have been extensively studied in the threshold logic literature (see, e.g., Winder [917] and Muroga [698] for an account). Some of them have been independently rediscovered in the framework of game theory (Einy [291]). We present now a sample of such properties.

In view of Definition 8.9, k-monotonicity implies h-monotonicity for all $h \leq k$. Winder [916, 917] showed that this implication cannot be reversed in general: Namely, for each k, there exists a $(k-1)$-monotone function of n variables that is not k-monotone (we omit the proof of this result, but see the end-of-chapter exercises for the case $k = 3$). Also, all completely monotone functions of eight or fewer variables are threshold functions, but in Chapter 9 we provide an example of a nonthreshold completely monotone function of nine variables (Theorem 9.15). In other words, complete monotonicity fails to be a sufficient condition for thresholdness.

Thus, if we denote by $\mathcal{T}h$ the set of all threshold functions, by $\mathcal{M}_k$ the set of k-monotone functions, and by $\mathcal{CM}$ the set of completely monotone functions, we obtain the picture in Figure 8.8 for all $k \geq 1$, where all inclusions are strict.

However, Winder [916, 917] proved that, *for fixed n*, the hierarchy of k-monotone functions collapses at level $\lfloor n/2 \rfloor$.

$$\mathcal{T}h \subset \mathcal{CM} \subset \ldots \subset \mathcal{M}_{k+1} \subset \mathcal{M}_k \subset \ldots \subset \mathcal{M}_1$$

Figure 8.8. The hierarchy of k-monotone Boolean functions.

Theorem 8.44. *A Boolean function of n variables is completely monotone if and only if it is $\lfloor n/2 \rfloor$-monotone.*

Proof. Assume that $f(x_1, x_2, \ldots, x_n)$ is not completely monotone. Then, there exist $S, T \subseteq \{1, 2, \ldots, n\}$ such that S and T are not comparable with respect to $\succcurlyeq_f$, meaning that there exist $X^*, Y^* \in \mathcal{B}^n$ such that

$$x_i^* = 0 \text{ for all } i \in S \cup T, \; f(X^* \vee e_S) = 0, \; f(X^* \vee e_T) = 1,$$

and

$$y_i^* = 0 \text{ for all } i \in S \cup T, \; f(Y^* \vee e_S) = 1, \; f(Y^* \vee e_T) = 0.$$

We can assume without loss of generality that S and T are disjoint (see the end-of-chapter exercises).

Let now $I = \{i : x_i^* = 0, y_i^* = 1\}$, $J = \{i : x_i^* = 1, y_i^* = 0\}$, $K = \{i : x_i^* = y_i^*, i \notin S \cup T\}$, and define two points W^*, Z^* as follows:

$$w_i^* = \begin{cases} 1 & \text{if } i \in T, \\ x_i^* & \text{if } i \in K, \\ 0 & \text{otherwise,} \end{cases}$$

and

$$z_i^* = \begin{cases} 1 & \text{if } i \in S, \\ x_i^* & \text{if } i \in K, \\ 0 & \text{otherwise.} \end{cases}$$

It is trivial to check that $X^* \vee e_S = Z^* \vee e_J$, $X^* \vee e_T = W^* \vee e_J$, $Y^* \vee e_S = Z^* \vee e_I$, and $Y^* \vee e_T = W^* \vee e_I$. As a consequence, we see that

$$f(W^* \vee e_I) = 0, \; f(W^* \vee e_J) = 1,$$

and

$$f(Z^* \vee e_I) = 1, \; f(Z^* \vee e_J) = 0.$$

Hence, I and J are not comparable with respect to $\succcurlyeq_f$. Since $|I \cup J| + |T \cup S| \le n$, we conclude that f is not k-monotone for some $k \le \lfloor n/2 \rfloor$. $\square$

So, if we restrict our attention to functions of n variables (for fixed n), the hierarchy of k-monotone functions boils down to

$$\mathcal{T}h \subseteq \mathcal{CM} = \mathcal{M}_{\lfloor n/2 \rfloor} \subset \mathcal{M}_{\lfloor n/2 \rfloor - 1} \subset \ldots \subset \mathcal{M}_1,$$

and all inclusions are strict when $n \ge 9$.

We now investigate the behavior of the strength relation with respect to the fixation of variables (compare with Theorem 8.7).

Theorem 8.45. *Let $f(x_1, x_2, \ldots, x_n)$ be a Boolean function, let $i \in \{1, 2, \ldots, n\}$, and let $S, T \subseteq \{1, 2, \ldots, n\} \setminus \{i\}$. If S is stronger than T with respect to f, then S is stronger than T with respect to $f_{|x_i=1}$ and with respect to $f_{|x_i=0}$.*

Proof. This immediately follows from Definition 8.8.   □

Recall that, for $0 \le d \le n$, a face of $\mathcal{B}^n$ of dimension d is a subset of $\mathcal{B}^n$ of the form

$$F(I,J) = \{X \in \mathcal{B}^n \mid x_i = 1 \text{ for all } i \in I \text{ and } x_j = 0 \text{ for all } j \in J\}$$

where I,J are disjoint subsets of $\{1,2,\ldots,n\}$ such that $|I \cup J| = n - d$. Two faces F_1, F_2 are *complementary* if $F_1 = F(I,J)$ and $F_2 = F(J,I)$ for some $I,J \subseteq \{1,2,\ldots,n\}$. We denote by $f_{|I,J}$ or by $f_{|F}$ the restriction of a function $f(x_1,x_2,\ldots,x_n)$ to a face $F = F(I,J)$ of $\mathcal{B}^n$. As usual, we sometimes consider $f_{|F}$ as a Boolean function of d variables, where d is the dimension of F.

Theorem 8.46. *For $k \le n$, a Boolean function f on $\mathcal{B}^n$ is k-monotone if and only if one of the implications $f_{|I,J} \le f_{|J,I}$ or $f_{|J,I} \le f_{|I,J}$ holds for all pairs of complementary faces $F(I,J)$ and $F(J,I)$ of dimension at least $n - k$.*

Proof. Necessity. Assume first that f is k-monotone, and consider two complementary faces $F(I,J)$ and $F(J,I)$, with $|I \cup J| \le k$. By definition of k-monotonicity, we can assume without loss of generality that $I \succcurlyeq_f J$. But this easily implies that $f_{|I,J} \ge f_{|J,I}$.

Sufficiency. To prove the reverse implication, consider two (disjoint) subsets S,T of $\{1,2,\ldots,n\}$ such that $|S \cup T| \le k$. Then, if we assume for instance that $f_{|S,T} \ge f_{|T,S}$, it is straightforward to check that $S \succcurlyeq_f T$.   □

As a corollary, we obtain:

Theorem 8.47. *A Boolean function is k-monotone if and only if its dual is k-monotone.*

Proof. This follows from Theorem 8.46 and from Theorem 4.2 in Section 4.1.   □

Muroga, Toda, and Takasu [700] observed that completely monotone functions are dual-comparable (see Section 4.1.3 for definitions).

Theorem 8.48. *Every completely monotone Boolean function is either dual-minor or dual-major.*

Proof. Assume that f is neither dual-minor nor dual-major. Then, there exist $X^*, Y^* \in \mathcal{B}^n$ such that $f(X^*) = 1$, $f^d(X^*) = 0$, $f(Y^*) = 0$, and $f^d(Y^*) = 1$.

Let $S = \{i : x_i^* = y_i^* = 1\}$ and $T = \{i : x_i^* = y_i^* = 0\}$. Define two points $W^*, Z^* \in \mathcal{B}^n$ as follows:

$$w_i^* = 0 \quad \text{if } i \in S \cup T,$$
$$= x_i^* \quad \text{otherwise,}$$

and

$$z_i^* = 0 \text{ if } i \in S \cup T,$$
$$= y_i^* \text{ otherwise.}$$

One easily verifies that

$$f(W^* \vee e_S) = f(X^*) = 1, \quad f(W^* \vee e_T) = f(\overline{Y^*}) = 0,$$

and

$$f(Z^* \vee e_S) = f(Y^*) = 0, \quad f(Z^* \vee e_T) = f(\overline{X^*}) = 1.$$

Hence, S and T are not comparable with respect to $\succcurlyeq_f$, and f is not completely monotone. $\square$

Taken together, Theorem 8.45 and 8.48 imply that the restriction of a completely monotone function to any face of $\mathcal{B}^n$ is either dual-minor or dual-major. Ding [272] established that this property actually characterizes completely monotone functions.

Theorem 8.49. *A Boolean function f on $\mathcal{B}^n$ is completely monotone if and only if, for every face F of $\mathcal{B}^n$, $f_{|F}$ is either dual-minor or dual-major.*

Proof. Assume that f is not completely monotone. Then, there exist $S, T \subseteq \{1, 2, \ldots, n\}$ and $X^*, Y^* \in \mathcal{B}^n$ such that

$$x_i^* = 0 \text{ for all } i \in S \cup T, \quad f(X^* \vee e_S) = 0, \quad f(X^* \vee e_T) = 1,$$

and

$$y_i^* = 0 \text{ for all } i \in S \cup T, \quad f(Y^* \vee e_S) = 1, \quad f(Y^* \vee e_T) = 0.$$

Moreover, we can again assume, without loss of generality, that S and T are disjoint.

Let now $I = \{i : x_i^* = y_i^* = 1\}$, $J = \{i \notin (S \cup T) : x_i^* = y_i^* = 0\}$, $F = F(I, J)$ and $g = f_{|F}$. We claim that g is neither dual-minor nor dual-major, that is, there exist $W^*, Z^* \in F$ such that $g(W^*) = 1$, $g^d(W^*) = 0$ and $g(Z^*) = 0$, $g^d(Z^*) = 1$. We leave it to the reader to verify that $W^* = X^* \vee e_T$ and $Z^* = X^* \vee e_S$ are as required. $\square$

We conclude this section with a last characterization of complete monotonicity that rests on the concept of 2-summability.

Definition 8.10. *A Boolean function f on $\mathcal{B}^n$ is 2-summable if there exist two (not necessarily distinct) false points of f, say, $X^*, W^* \in \mathcal{B}^n$, and two (not necessarily distinct) true points of f, say, $Y^*, Z^* \in \mathcal{B}^n$, such that $X^* + W^* = Y^* + Z^*$ (where the summation is over $\mathbb{R}^n$). Otherwise, f is 2-asummable.*

Example 8.24. *The function $f(x_1, x_2) = x_1 x_2 \vee \overline{x_1}\,\overline{x_2}$ is 2-summable. Indeed, if we let $X^* = (0, 1)$, $W^* = (1, 0)$, $Y^* = (0, 0)$ and $Z^* = (1, 1)$, then $X^* + W^* = Y^* + Z^*$.*

The function $f(x_1, x_2, x_3, x_4) = x_1 x_2 \vee x_3 x_4$ *is also 2-summable. On the other hand, it is easy to see that every threshold function is 2-asummable (see Theorem 9.14 in Chapter 9).* □

Elgot [310] proved:

Theorem 8.50. *A Boolean function is completely monotone if and only if it is 2-asummable.*

Proof. Sufficiency. Assume that $f(x_1, x_2, \ldots, x_n)$ is not completely monotone, that is, there exist $S, T \subseteq \{1, 2, \ldots, n\}$ and $X^*, Y^* \in \mathcal{B}^n$ such that $x_i^* = y_i^* = 0$ for $i \in S \cup T$ and $f(X^* \vee e_S) = f(Y^* \vee e_T) = 0$, $f(X^* \vee e_T) = f(Y^* \vee e_S) = 1$. Then f is 2-summable, since

$$(X^* \vee e_S) + (Y^* \vee e_T) = (X^* \vee e_T) + (Y^* \vee e_S).$$

Necessity. Assume that $f(x_1, x_2, \ldots, x_n)$ is 2-summable, and let X^*, W^*, Y^*, Z^* be as in Definition 8.10. Let $S = \{i : x_i^* = 1, y_i^* = 0\}$, $T = \{i : x_i^* = 0, y_i^* = 1\}$, and define two points $U^*, V^* \in \mathcal{B}^n$ as follows:

$$u_i^* = \begin{cases} 0 & \text{if } i \in S \cup T, \\ x_i^* & \text{otherwise,} \end{cases}$$

$$v_i^* = \begin{cases} 0 & \text{if } i \in S \cup T, \\ w_i^* & \text{otherwise.} \end{cases}$$

From the equality $X^* + W^* = Y^* + Z^*$, one easily derives that
- for $i \notin S \cup T$, $x_i^* = y_i^* = u_i^*$ and $v_i^* = w_i^* = z_i^*$;
- for $i \in S$, $x_i^* = z_i^* = 1$ and $y_i^* = w_i^* = u_i^* = v_i^* = 0$;
- for $i \in T$, $x_i^* = z_i^* = u_i^* = v_i^* = 0$ and $y_i^* = w_i^* = 1$.

This, in turn, implies that

$$f(U^* \vee e_S) = f(X^*) = 0, \ f(U^* \vee e_T) = f(Y^*) = 1,$$

$$f(V^* \vee e_S) = f(Z^*) = 1, \ f(V^* \vee e_T) = f(W^*) = 0.$$

Hence, S and T are not comparable with respect to $\succcurlyeq_f$, and f is not completely monotone. □

As a corollary of Theorem 8.50, we conclude that completely monotone functions can be recognized in polynomial time (Ding [272]).

Theorem 8.51. *There exists an $O(n^3 m^4)$ algorithm to determine whether a positive function expressed in complete DNF is completely monotone, where n is the number of variables and m is the number of prime implicants of the function.*

Proof. Given a positive function f, we first test in time $O(n^2 m)$ whether f is regular (see Theorem 8.20). If f is not regular, then it is not completely monotone. Otherwise, we generate in time $O(n^2 m)$ the maximal false points of f (see

Theorem 8.28). Because f is positive, it follows from Definition 8.10 that f is 2-summable if and only if there exists a pair of maximal false points X^*, W^* and a pair of minimal true points Y^*, Z^* such that $Y^* + Z^* \leq X^* + W^*$ (see Exercise 13). In view of Theorem 8.29, f has at most nm maximal false points, and the claim follows.   □

More properties of k-monotone and completely monotone functions can be found for instance in Ding [272], Einy [291], Giles and Kannan [379], Muroga [698], Winder [917], and so on. We return to the topic of asummability in Chapter 9.

8.9 Generalizations of regularity

In this section, we briefly introduce several extensions of the class of regular functions and describe their main properties.

8.9.1 Weakly regular functions

The next result generalizes Theorem 8.22.

Theorem 8.52. *For a positive Boolean function $f(x_1, x_2, \ldots, x_n)$, the following properties are equivalent:*

(a) $x_j \succeq_f x_n$ *for all $j \in \{1, 2, \ldots, n\}$.*
(b) *For all $X^* \in B^{n-1}$, $(X^*, 0)$ is a maximal false point of f if and only if $(X^*, 1)$ is a minimal true point of f.*

Proof. The proof of Theorem 8.22 establishes that (a) implies (b). We leave the proof of the reverse implication as an easy end-of-chapter exercise.   □

Based on this observation, and on the fact that many of the remarkable features of regular functions actually rest on Theorem 8.22 and Theorem 8.23, Crama [224] introduced the following class of functions:

Definition 8.11. *A positive Boolean function $f(x_1, x_2, \ldots, x_n)$ is* weakly regular *with respect to $(x_1, x_2, \ldots, x_n)$ if f is constant, or if*

(a) $x_j \succeq_f x_n$ *for all $j \in \{1, 2, \ldots, n\}$, and*
(b) $f_{|x_n=1}$ *is weakly regular with respect to $(x_1, x_2, \ldots, x_{n-1})$.*

We simply say that f is weakly regular *if f is weakly regular with respect to some permutation of its variables.*

So, when f is weakly regular with respect to $(x_1, x_2, \ldots, x_n)$, x_i is a "weakest" variable in the preorder associated with $f_{|x_{i+1}=\cdots=x_n=1}$, for all $i \in \{1, 2, \ldots, n\}$.

Clearly, regular functions are weakly regular, but the converse is not necessarily true.

Example 8.25. *The function $x_1 x_3 \vee x_1 x_4 \vee x_2 x_3 \vee x_1 x_2 x_5$ is weakly regular with respect to $(x_1, x_2, \ldots, x_6)$, but it is not regular (x_1 and x_3 are not comparable).* □

Many results from previous sections extend in a straightforward way to weakly regular functions. For instance, algorithm DUALREGO allows us to dualize these functions in $O(n^2 m^2)$ time, and REGCOVER0 solves weakly regular set covering problems in the same time complexity. Regularization can be extended in an obvious way to weak regularization and can be used to solve set covering problems as explained in Section 8.7.

Finally, Crama [224] noted that a function $f(x_1, x_2, \ldots, x_n)$ can be tested for weak regularity in polynomial time by a simple greedy procedure: if there is no variable x_i such that $x_j \succcurlyeq_f x_i$ for all $j \in \{1, 2, \ldots, n\}$, then f is not weakly regular; otherwise, we can fix x_i to 1 in f and repeat the test with $f_{|x_i=1}$. The procedure is correct because, when several variables qualify as "weakest" variables, f is symmetric on these variables and hence, the choice among them is immaterial (see Theorem 8.1).

8.9.2 Aligned functions

Boros [105] has introduced and investigated the class of *aligned* functions, which provide another generalization of regular functions:

Definition 8.12. *A positive Boolean function $f(x_1, x_2, \ldots, x_n)$ is* aligned *(with respect to $(x_1, x_2, \ldots, x_n)$) if its dual f^d is weakly regular (with respect to $(x_1, x_2, \ldots, x_n)$).*

Equivalently, the function f is aligned with respect to $(x_1, x_2, \ldots, x_n)$ if, for all $i \in \{1, 2, \ldots, n\}$, x_i is weakest in the preorder associated with $(f^d)_{|x_{i+1}=\cdots=x_n=1}$, which is identical to the preorder associated with $f_{|x_{i+1}=\cdots=x_n=0}$. This implies, in particular, that aligned functions can be recognized in polynomial time by the same type of procedure described for the recognition of weakly regular functions.

Boros [105] established yet another characterization of aligned functions:

Theorem 8.53. *A positive Boolean function $f(x_1, x_2, \ldots, x_n)$ is aligned with respect to $(x_1, x_2, \ldots, x_n)$ if and only if, for every prime implicant of f, say $\bigwedge_{k \in A} x_k$, and for every $j \notin A$ such that $j < \mu = \max\{k \mid k \in A\}$, $\bigwedge_{k \in (A \cup \{j\}) \setminus \{\mu\}} x_k$ is an implicant of f.*

Proof. We leave this proof as an exercise at the end of the chapter.   □

Comparing this statement with Definition 7.5 in Section 7.4, it is easy to conclude that aligned functions have the LE property. As a consequence, aligned functions can be dualized in $O(n^2 m)$ time (see Theorem 7.16 in Section 7.4.4).

Example 8.26. *To see that the class of aligned functions is distinct from previously introduced classes, consider $f_B = x_1 x_2 \vee x_1 x_3 \vee x_1 x_4 \vee x_2 x_3 \vee x_2 x_4 x_5 \vee x_3 x_4 x_5 x_6 \vee x_4 x_5 x_6 x_7$. This function is aligned with respect to $(x_1, x_2, \ldots, x_7)$, but it* *is not weakly regular, and hence, it is not regular either (Boros [105]). By duality, f^d is weakly regular, but it is not aligned.*

On the other hand, the function $f_{C_4} = x_1x_3 \vee x_1x_4 \vee x_2x_3 \vee x_2x_4$ has the LE property by virtue of Theorem 7.18, but it is not aligned since it does not have a weakest variable.   □

8.9.3 Ideal functions

Bertolazzi and Sassano [75] defined the class of *ideal* functions.

Definition 8.13. *Let $f = \bigvee_{k=1}^{m} \bigwedge_{i \in A_k} x_i$ be the complete DNF of a positive function. We say that x_n is a* last variable *of f if, for all $k,\ell \in \{1,2,\dots,m\}$ such that $n \in A_k \setminus A_\ell$, there exists $j \in A_\ell$ such that $\bigwedge_{i \in (A_k \cup \{j\}) \setminus \{n\}} x_i$ is an implicant of f. We say that f is* ideal *with respect to $(x_1,x_2,\dots,x_n)$ if x_i is a last variable of $f_{|x_{i+1}=\cdots=x_n=1}$ for all $i \in \{1,2,\dots,n\}$, and we say that f is* ideal *if f is ideal with respect to some permutation of its variables.*

Bertolazzi and Sassano proved that ideal functions can be recognized efficiently (in time $O(n^3 m^2)$; see [75] and Exercise 18 at the end of this chapter). They also observed that regular function are ideal. More precisely:

Theorem 8.54. *If $x_i \succcurlyeq_f x_n$ for all $i = 1,2,\dots,n$, then x_n is a last variable of f. In particular, every weakly regular function is ideal.*

Proof. To show that x_n is last, suppose that $k,\ell \in \{1,2,\dots,m\}$ and that $n \in A_k \setminus A_\ell$. Choose j arbitrarily in $A_\ell \setminus A_k$. Since $x_j \succcurlyeq_f x_n$, $\bigwedge_{i \in (A_k \cup \{j\}) \setminus \{n\}} x_i$ is an implicant of f, and this proves the first part of the statement. The second part follows immediately.   □

The converse of this theorem is false: An ideal function is not necessarily weakly regular.

Example 8.27. *Each of x_1,x_2,x_3,x_4 is a last variable of $f_{C_4} = x_1x_3 \vee x_1x_4 \vee x_2x_3 \vee x_2x_4$, so that the function is ideal. But f_{C_4} is neither weakly regular nor aligned because it has no weakest variable.*

Similarly, the function $f_{P_4} = x_1x_2 \vee x_1x_3 \vee x_2x_4$ is ideal with respect to the order (x_1,x_2,x_3,x_4), but f_{P_4} is neither weakly regular nor aligned.   □

The main motivation for considering ideal functions is that they can be dualized in polynomial time. To describe this result, let us introduce the following notation: If $\bigwedge_{i \in A} x_i$ is a prime implicant of f and if $j \in A$, we let

$$P(A,j) = \{h \in \{1,2,\dots,n\} \mid \bigwedge_{i \in (A \cup \{h\}) \setminus \{j\}} x_i \text{ is an implicant of } f\}$$

and

$$Q(A,j) = P(A,j) \cup \{j\}.$$

Note for futher reference that $P(A,j) \cap A = \emptyset$.

Bertolazzi and Sassano [75] proved:

Theorem 8.55. *Let x_n be a last variable of $f = \bigvee_{k=1}^{m} \bigwedge_{i \in A_k} x_i$. The prime implicants of f^d containing x_n are exactly the elementary conjunctions of the form $\bigwedge_{i \in Q(A_k,n)} x_i$ for all $k \in \{1,2,\ldots,m\}$ such that $n \in A_k$.*

Proof. Let $\mathcal{P} = \{A_k \mid k = 1,2,\ldots,m\}$, let $\{A_1, A_2,\ldots,A_q\} = \{A \in \mathcal{P} \mid n \in A\}$ and let $\mathcal{T} = \{Q(A_k,n) \mid k = 1,2,\ldots,q\}$. By Theorem 4.19, we must show that the sets in $\mathcal{T}$ are exactly the minimal transversals of $\mathcal{P}$ that contain n.

Fix $k \in \{1,2,\ldots,q\}$. We first want to show that $Q(A_k,n)$ is a transversal of $\mathcal{P}$, meaning that $Q(A_k,n) \cap A_\ell \neq \emptyset$ for all $\ell \in \{1,2,\ldots,m\}$.

  (i) If $n \in A_\ell$, then $n \in Q(A_k,n) \cap A_\ell$.
  (ii) If $n \notin A_\ell$, then, since x_n is a last variable, there exists $j \in A_\ell$ such that $\bigwedge_{i \in (A_k \cup \{j\}) \setminus \{n\}} x_i$ is an implicant of f. This, however, means that $j \in P(A_k,n)$, and hence, $j \in Q(A_k,n) \cap A_\ell$.

So, for all $k \in \{1,2,\ldots,q\}$, the set $Q(A_k,n)$ contains a minimal transversal of $\mathcal{P}$. Assume now that A is a minimal transversal of $\mathcal{P}$ such that $n \in A$, and assume that $Q(A_k,n) \neq A$ for all $k = 1,2,\ldots,q$. Fix $k \in \{1,2,\ldots,q\}$. Note that there exists $h \in Q(A_k,n) \setminus A$: Otherwise, $Q(A_k,n) \subset A$, contradicting the minimality of A. Clearly, $h \neq n$ and hence $h \in P(A_k,n)$. So, by definition of $P(A_k,n)$, there exists a prime implicant A_ℓ, $\ell \in \{1,2,\ldots,m\}$ such that $A_\ell \subseteq (A_k \cup \{h\}) \setminus \{n\}$. But $A \cap A_\ell \neq \emptyset$ (since A is a transversal) and $h \notin A$ (by choice of h). Hence, $A \cap (A_k \setminus \{n\}) \neq \emptyset$ or, equivalently, $(A \setminus \{n\}) \cap A_k \neq \emptyset$. This conclusion holds for all $k = 1,2,\ldots,q$, but it also holds trivially for $k = q+1,\ldots,m$ because A is a transversal of $\mathcal{P}$. Therefore, we obtain that $(A \setminus \{n\})$ is a transversal of $\mathcal{P}$, which contradicts the minimality of A. This concludes the proof. $\square$

Theorem 8.55, combined with Theorem 8.23, allows us to generate the dual of an ideal function in polynomial time. Details are left to the reader.

Example 8.28. *Note that the dual of an ideal function is generally not ideal: For instance, the dual of the function f_{C_4} defined in Example 8.27 is $f_{2K_2} = x_1 x_2 \vee x_3 x_4$, which is not ideal.* $\square$

8.9.4 Relations among classes

The mutual relations among the classes of Boolean functions introduced in this and previous sections have not been completely clarified in the literature. Figure 8.9 summarizes the relations we have explicitly identified in this chapter.

We have provided examples showing that none of the implications in Figure 8.9 can be reversed (see Examples 8.25, 8.26, and 8.27). In Example 8.29, we show that most of the missing implications cannot be added either. The exercises in Section 8.10 contain a number of additional open questions that may be worth

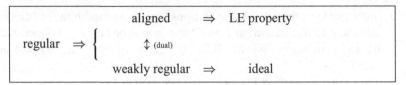

Figure 8.9. Generalizations of regular functions.

investigating (we do not claim that these are very difficult questions, but simply that their answers do not seem to appear readily in the literature).

Example 8.29. *We have already noted that the function* $f_{C_4} = x_1 x_3 \vee x_1 x_4 \vee x_2 x_3 \vee x_2 x_4$ *is ideal with respect to* (x_1, x_2, x_3, x_4)*, but it is not aligned. Conversely, aligned functions are not necessarily ideal, as illustrated by the function* f_B *in Example 8.26.*

The function $f_{P_4} = x_1 x_2 \vee x_1 x_3 \vee x_2 x_4$ *has the LE property, and it is ideal with respect to* (x_1, x_2, x_3, x_4)*, but it is neither aligned nor weakly regular.*

Ideal functions are not necessarily shellable: Indeed, $f = x_1 x_2 x_5 \vee x_1 x_3 \vee x_2 x_4$ *is ideal with respect to* $(x_1, x_2, x_3, x_4, x_5)$ *but it is not shellable with respect to any permutation of its terms (and it is not weakly regular either).*

Finally, the dual of ideal functions is not necessarily shellable: the function $f_{C_4}^d = f_{2K_2}$ *is a counter-example.* □

8.10 Exercises

1. Let $f(x_1, x_2, \ldots, x_n)$ be a positive Boolean function, let $g = f_{|x_1=1}$, let $h = f_{|x_1=0}$, and assume that both g and h are regular with respect to $(x_2, x_3, \ldots, x_n)$. Show that f is not necessarily regular (compare with Theorem 8.7).

2. Show that, for a function $f(x_1, x_2, \ldots, x_n)$ given in DNF, it is co-NP-complete to decide whether $x_1 \succcurlyeq_f x_2$.

3. Prove Theorem 8.12 by resorting only to the definitions of regularity and of the LE property.

4. Prove the validity of the claims in Application 8.2.

5. Prove the validity of the claims in Application 8.4.

6. Prove Theorem 8.24.

7. Show that Theorem 8.25 can be used to produce an $O(n^2 m)$ implementation of DualReg0 (Crama [225]).

8. Prove Theorem 8.39.

9. Prove that Theorem 8.41 extends to problem (8.30)–(8.32) if the objective function g satisfies the "generalized regularity" condition (8.33). (Hammer, Johnson, and Peled [443]).

10. Show that in Definition 8.9, comparisons can be restricted to pairs of disjoint subsets: A Boolean function f on $\mathcal{B}^n$ is k-monotone $(1 \le k \le n)$ if and only if S and T are comparable for all $S, T \subseteq \{1, 2, \ldots, n\}$ such that $S \cap T = \emptyset$ and $|S \cup T| \le k$.

11. Show that the function $f(x_1, x_2, \ldots, x_6)$ defined by

$$f = x_1 \, (x_2 \vee x_3 \vee x_4 x_5 x_6) \vee x_2 x_3 \, (x_4 x_5 \vee x_4 x_6 \vee x_5 x_6) \vee (x_2 \vee x_3) x_4 x_5 x_6$$

is regular, but is not 3-monotone (and, hence, is not threshold; Winder [917]).

12. Prove that
    (a) a function f of n variables is completely monotone if and only if its self-dual extension $f^{SD}(x_1, x_2, \ldots, x_n, x_{n+1}) = f \overline{x}_{n+1} \vee f^d x_{n+1}$ is completely monotone;
    (b) a self-dual function of n variables is completely monotone if and only if it is $\lfloor n/3 \rfloor$-monotone.
    (See [698, 918].)

13. Show that, for a positive function f,
    (a) f is 2-summable if and only if there exists a pair of maximal false points X^*, W^* and a pair of minimal true points Y^*, Z^* such that $Y^* + Z^* \le X^* + W^*$;
    (b) in the previous statement, the inequality $Y^* + Z^* \le X^* + W^*$ cannot be replaced by $Y^* + Z^* = X^* + W^*$. (Compare with Definition 8.10.)

14. Prove that a function f is completely monotone if and only if it is k-monotone, where k is the largest degree of a prime implicant of f. (See [272].)

15. When $S \succcurlyeq_f T$ holds, but $T \not\succcurlyeq_f S$ does not hold for two subsets $S, T \subseteq \{1, 2, \ldots, n\}$, we write $S \succ_f T$. Show that the relation $\succ_f$ may be cyclic in general, but is acylic for threshold functions. (See, e.g., Einy [291], but also Muroga [698, p. 200] and Winder [917] for additional results along this line.)

16. Complete the proof of Theorem 8.52.

17. Prove Theorem 8.53 and conclude that aligned functions have the LE property.

18. Prove that, if $f(x_1, x_2, \ldots, x_n)$ is ideal, then $f_{|x_j=1}$ is ideal for all $j \in \{1, 2, \ldots, n\}$. Use this result to derive a polynomial-time algorithm for the recognition of ideal functions. (See Bertolazzi and Sassano [75].)

19. Let $G = (V, E)$ be a graph and let $f_G(x_1, x_2, \ldots, x_n) = \bigvee_{(i,j) \in E} x_i x_j$ be the corresponding stability function (see Section 1.13.5 in Chapter 1). For $i \in V$, denote by $N(i)$ the neighborhood of vertex i, that is, $N(i) = \{j \in V : (i, j) \in E\}$.
    (a) Prove that $x_i \succcurlyeq x_j$ with respect to f_G if and only if $N(j) \setminus \{i\} \subseteq N(i) \setminus \{j\}$, for all $i, j \in V$.
    (b) Prove that f_G is regular if and only if G does not contain $2K_2$, P_4, or C_4 as induced subgraphs.
    (c) Prove that f_G is regular if and only if f_G is weakly regular.
    (See Chvátal and Hammer [201]; Crama [224].)

Questions for thought

20. Analyze the complexity of the procedure REGMINOR in Figure 8.7. Does it run in polynomial total time (that is, in time polynomial in $|f| + |f_R|$)?

21. Relations among function classes:
    (a) Is it true that weakly regular functions have the LE property? Are they shellable?
    (b) If a function is both aligned and weakly regular with respect to $(x_1, x_2, \ldots, x_n)$, is it also regular with respect to $(x_1, x_2, \ldots, x_n)$?
    (c) Characterize ideal quadratic functions.
    (d) Is there some unifying concept behind the definitions of "leaders" and of the LE property on one hand, and those of "last variables" and of ideal functions on the other hand? (Compare, e.g., Theorem 8.11 and Theorem 8.54.) Exploring these concepts in parallel may lead to fruitful insights.

22. (Due to Endre Boros.) What can be said about self-dual regular functions? Are they always threshold?

23. (Due to Endre Boros.) For a regular function f and an arbitrary maximum false point X^* of f, does there always exist a permutation σ of the variables such that f is regular and such that X^* is a ceiling with respect to the new order of the variable $\sigma(x_1), \ldots, \sigma(x_n)$? (Compare with Example 8.10.)

9

Threshold functions

In this chapter, we investigate the properties of threshold Boolean functions, an important class of functions which has already been mentioned several times in previous chapters. Threshold functions provide a simple but fundamental model for many questions investigated in electrical engineering, artificial intelligence, game theory, and many other areas. As such, their main properties have been investigated by countless researchers and frequently rediscovered in various guises. In particular, we present here a number of necessary conditions for a function to be threshold, and we establish a classical characterization of threshold functions based on their "asummability" properties. We also describe a polynomial-time recognition algorithm for threshold functions represented by positive disjunctive normal forms, we analyze the complexity of enumerating the prime implicants and of computing the Chow parameters of threshold functions, and we briefly examine the class of threshold graphs.

9.1 Definitions and applications

Let us first recall the definition of threshold functions.

Definition 9.1. *A Boolean function f on B^n is called a* threshold *(or* linearly separable*) function if there exist n weights $w_1, w_2, \ldots, w_n \in \mathbb{R}$ and a threshold $t \in \mathbb{R}$ such that, for all $(x_1, x_2, \ldots, x_n) \in B^n$,*

$$f(x_1, x_2, \ldots, x_n) = 0 \text{ if and only if } \sum_{i=1}^{n} w_i x_i \leq t.$$

The hyperplane $\{X \in \mathbb{R}^n : \sum_{i=1}^{n} w_i x_i = t\}$ is called a separator *of f, and the $(n+1)$–tuple $(w_1, w_2, \ldots, w_n, t)$ is called a* (separating) structure *of f. We say that the separator and the separating structure* represent *f.*

Note that in this definition, variables $x_1, x_2, \ldots, x_n$ have to be interpreted as *natural numbers* in $\{0, 1\} \subset \mathbb{N}$, rather than purely Boolean, meaningless symbols (remember the discussion in Section 1.1 of Chapter 1). In geometric terms, threshold functions are precisely those functions for which the set of true points can be separated from the set of false points by a hyperplane (the separator).

Example 9.1. *The function* $f(x,y,z) = x\overline{y} \vee z$ *is a threshold function, with separator* $\{(x,y,z) \in \mathbb{R}^3 : x - y + 2z = 0\}$ *and with structure* $(1,-1,2,0)$. *Observe that* f *admits many other separators (actually, an infinite number of them): For instance* $\{(x,y,z) \in \mathbb{R}^3 : \alpha x - \alpha y + 2\alpha z = 0\}$ *is a separator for all* $\alpha > 0$, *but so are* $\{(x,y,z) \in \mathbb{R}^3 : x - 2y + 3z = 0\}$, $\{(x,y,z) \in \mathbb{R}^3 : 5x - 5y + 10z = 3\}$, *and so on.* $\square$

Example 9.2. *The function* $f(x,y) = xy \vee \overline{x}\overline{y}$ *is not a threshold function. Indeed, its set of true points is* $\{(0,0),(1,1)\}$, *its set of false points is* $\{(0,1),(1,0)\}$, *and these two sets cannot be separated by a line in* $\mathbb{R}^2$. $\square$

Threshold functions constitute one of the most extensively investigated classes of Boolean functions. This interest in threshold functions has been stimulated by their central role in many fields of application, a role which is itself justified by the simplicity of their description (since a threshold function is completely characterized by a vector of $(n+1)$ numbers) and by their numerous nice properties.

Application 9.1. (Electrical engineering.) *A* switching gate *is an electrical device (i.e., a circuit consisting of resistors, transistors, etc.) which admits a number of input voltages* $V_1, V_2, \ldots, V_n$, *and which releases an output voltage* V_0. *In a simplified model, each of the input and output voltages can only assume two distinct values, say,* $V_i \in \{a_i, b_i\}$ *for* $i = 0, 1, 2, \ldots, n$.

A threshold gate *is a special type of switching gate characterized by a threshold value t and numerical weights* $w_1, w_2, \ldots, w_n$ *attached to the inputs; the value of the output voltage is equal to* a_0 *if* $\sum_i w_i V_i \le t$ *and is equal to* b_0 *otherwise. So, up to a simple transformation of variables, the functioning of a threshold gate is described by a threshold Boolean function.*

Threshold gates can be combined in various ways to produce switching networks, *that is, physical realizations of more general (not necessarily threshold) Boolean functions, as explained in Section 1.13.2. As a matter of fact, every Boolean function can be realized by a switching network of threshold gates (see Theorem 9.2 hereunder). For this reason, threshold gates were widely used as basic components in the design of early computers. This important application stimulated, in the late 1950s, a dense flow of research aimed at understanding the theoretical properties of threshold functions. This research eventually evolved into a coherent field known as threshold logic, an account of which can be found, for instance, in books by Dertouzos [269], Hu [511, 512], Mendelson [680], or Muroga [698, 699].*

More recently, the complexity of Boolean circuits made up of threshold gates has been investigated in the theoretical computer science literature; we refer, for instance, to the monograph Wegener [902] and to papers by Anthony [27, 28], Bruck [157], Krause and Wegener [583] for various aspects of this line of research.　□

Application 9.2. (Artificial neural networks.) *An artificial neural network consists of a directed graph D together with a collection of functions (or neurons) associated with the vertices of G. In one of the best known models, D is acyclic, each vertex of indegree 0 corresponds to a Boolean variable, and each neuron is a threshold Boolean function, sometimes called* perceptron *in this context. Then, each vertex of outdegree 0 can be viewed as computing a Boolean function obtained as a superposition of several threshold functions, via the same* feedforward *process described for combinational networks in Section 1.1.*

The reader will quickly notice that this model is extremely similar to the switching circuit model sketched in Application 9.1. Its interpretation as an abstract computational model, however, has given rise to an independent stream of research originating with the book by Minsky and Papert [684]. We refer to Anthony [25, 27] for a discussion of the links between Boolean threshold functions and neural network theory.　□

Application 9.3. (Reliability theory.) *In reliability theory, a complex system consisting of n components is called a* k-out-of-n *system (k ≤ n) if the system works whenever at least k of its components work and if it fails otherwise. Thus, the structure function (see Section 1.13.4) of a k-out-of-n system is the threshold function with separator* $\{(x_1, x_2, \ldots, x_n) \in \mathcal{B}^n : \sum_{i=1}^{n} x_i \leq k - 1\}$.

More general threshold systems, namely, systems whose structure function is an arbitrary threshold function, have been considered, for instance, by Ball and Provan [49].　□

Application 9.4. (Game theory.) *In the framework of game theory (recall Section 1.13.3), a positive threshold Boolean function is called a* weighted majority game. *Such games model the familiar situation in which each of n players (or voters) is assigned a number of votes, say, w_i (i = 1, 2, …, n), which she can decide to cast – or not – in favor of the issue at stake. The issue is adopted if the total number of votes cast in its favor exceeds a predetermined threshold t. In the simplest case (simple majority rule), every voter carries exactly one vote, and the threshold is equal to half the number of players. More elaborate voting rules arise, for instance, in legislatures, where the number of votes of each member is correlated to the size of the constituency that she represents, or in shareholder meetings, where the number of votes corresponds to the number of shares held by each member.*

Weighted majority procedures constitute the main paradigm in the theory of simple games and social choice. Many properties of these procedures and of their generalizations appear in the literature, for instance, in [79, 720, 777, 850, 861, 893], and so on.　□

Application 9.5. (Integer programming.) *A* knapsack problem *is an optimization problem of the form*

$$maximize \quad \sum_{i=1}^{n} c_i x_i$$

$$subject\ to \quad \sum_{i=1}^{n} w_i x_i \leq t$$

$$(x_1, x_2, \ldots, x_n) \in \mathcal{B}^n,$$

where c_i, w_i and t are nonnegative integers for $i = 1, 2, \ldots, n$. Knapsack problems have been extensively studied in integer programming (Kellerer, Pferschy, and Pisinger [561]; Martello and Toth [671]).

Remember from Section 1.13.6 that the resolvent of a system of constraints in 0–1 variables is the Boolean function whose false points are the feasible solutions of the system. So, by definition, the resolvent of the knapsack inequality $\sum_i w_i x_i \leq t$ is a threshold function. We shall see in Section 9.5 that fundamental Boolean concepts, such as that of prime implicant, prove useful in describing the solution set of a knapsack inequality.

Conversely, any system of inequalities in 0–1 variables whose resolvent is a threshold function, say $f(x_1, x_2, \ldots, x_n)$, is equivalent to a single linear inequality $\sum_i w_i x_i \leq t$, where $(w_1, w_2, \ldots, w_n, t)$ is a structure of f. In particular, a polynomial-time algorithm will be given in Section 9.4 to decide whether an instance of the set–covering problem can be transformed into an equivalent instance of the knapsack problem in the same variables (where the term "equivalent" means that both instances have the same set of feasible solutions). □

Application 9.6. (Distributed computing systems.) *Boolean functions can be used to prevent conflicts in distributed computing systems, as briefly sketched in Application 4.7 of Section 4.2. A popular way to implement mutual exclusion mechanisms in a distributed system relies on threshold functions. In this approach, a vote w_i is assigned to each site of the system, and a group of sites is allowed to perform an operation (such as updating a database) only if its members have a majority of the total number of votes (see, e.g., [258, 370]). Similar ideas have been proposed in other computational contexts, such as the synchronization of parallel processes [487, 888].* □

Before concluding this section, we mention the existence of a large body of literature dealing with higher-degree generalizations of threshold functions. Namely, a Boolean function f on $\mathcal{B}^n$ is called a *polynomial threshold function of degree k* if there exists a multilinear (pseudo-Boolean) polynomial $p(X) = \sum_{A \in \mathcal{P}(N)} c(A) \prod_{i \in A} x_i$ of degree k such that

$$f(X) = 0 \text{ if and only if } p(X) \leq 0$$

(recall the definitions in Section 1.12.2). So, a linearly separable function is a threshold function of degree 1, and it is easy to see that every Boolean function on $\mathcal{B}^n$ is a polynomial threshold function of degree n.

Polynomial threshold functions have been investigated in connection with circuit complexity and neural networks. The reader is referred to the monograph by Anthony [25] and to survey papers by Anthony [27], Bruck [157], or Saks [799] for more information.

9.2 Basic properties of threshold functions

In this section, we get acquainted with some of the elementary properties of threshold functions (see, e.g., [700, 732, 916, 917], as well as additional references cited in [698]). Many of these properties can best be seen as necessary conditions for a function to be threshold, but all turn out to be strictly weaker than thresholdness. Complete characterizations of threshold functions will be presented in Section 9.3.

We start with a few easy observations.

Theorem 9.1. *Elementary conjunctions and elementary disjunctions represent threshold functions.*

Proof. The equation $\sum_{i \in A} x_i + \sum_{j \in B}(1 - x_j) = |A| + |B| - 1$ defines a separator of the function $C_{AB} = \bigwedge_{i \in A} x_i \bigwedge_{j \in B} \overline{x}_j$, and the equation $\sum_{i \in A} x_i + \sum_{j \in B}(1 - x_j) = 0$ defines a separator of the function $C_{AB} = \bigvee_{i \in A} x_i \vee \bigvee_{j \in B} \overline{x}_j$. $\square$

An interesting corollary of Theorem 9.1 is that every Boolean function can be expressed as a composition of threshold functions.

Theorem 9.2. *Every Boolean function $f(X)$ on $\mathcal{B}^n$ can be expressed in the form*

$$f(X) = g(h_1(X), h_2(X), \ldots, h_m(X)),$$

where g and $h_1, h_2, \ldots, h_m$ are threshold functions.

Proof. This follows immediately from Theorem 9.1 and from the fact that every Boolean function has a disjunctive normal form. $\square$

As mentioned in Application 9.1, this observation motivates the realization of switching networks by threshold gates.

Another easy, but important, property of threshold functions is that their class is closed under restrictions.

Theorem 9.3. *If $f(x_1, x_2, \ldots, x_n)$ is a threshold function on $\mathcal{B}^n$ with separating structure $(w_1, w_2, \ldots, w_n, t)$, then $f_{|x_1=1}$ is a threshold function on $\mathcal{B}^{n-1}$ with separating structure $(w_2, w_3, \ldots, w_n, t - w_1)$ and $f_{|x_1=0}$ is a threshold function on $\mathcal{B}^{n-1}$ with structure $(w_2, w_3, \ldots, w_n, t)$.*

Proof. This is trivial. $\square$

We have already observed that a threshold function may have infinitely many separators (see Example 9.1). In fact, the set of separators can be characterized more precisely.

Theorem 9.4. *The separating structures of a threshold function of n variables constitute a full-dimensional convex cone in $\mathbb{R}^{n+1}$.*

Proof. If S and S' are two arbitrary separating structures of the threshold function f, and if α is a positive scalar, then αS and $S + S'$ are also separating structures of f: Thus, the set of separating structures is a convex cone.

To establish full-dimensionality, let $S = (w_1, w_2, \ldots, w_n, t)$. If f is identically 0, then the claim is easily checked. Otherwise, define

$$\mu = \min \left\{ \sum_{i=1}^{n} w_i x_i : \sum_{i=1}^{n} w_i x_i > t, X \in \mathcal{B}^n \right\},$$

and choose α arbitrarily in the interval $(0, \mu - t)$ so that $t + \alpha$ is nonzero (note that μ is well-defined, and that $\mu - t > 0$). Consider now the $(n+1)$ vectors $S^1, S^2, \ldots, S^{n+1}$, where $S^i = S + \alpha e_i + \alpha e_{n+1}$ $(i = 1, 2, \ldots, n)$, $S^{n+1} = S + \alpha e_{n+1}$, and e_j denotes the jth unit vector in $\mathbb{R}^{n+1}$. It is straightforward to check that the vectors $S^1, S^2, \ldots, S^{n+1}$ are linearly independent, and that each of them is a structure of f. $\square$

As a corollary of Theorem 9.4, we obtain the following useful property (which is also easily established from first principles).

Theorem 9.5. *Every threshold function has an integral separating structure.*

Proof. Details are left to the reader. $\square$

In the remainder of this section, we try to understand where threshold functions fit in the world of Boolean functions, in particular with respect to monotone, regular, or dual-comparable functions (this topic will be taken up further in Section 9.3).

First, we note that every threshold function is monotone, and hence, can be turned into a positive function by "switching" some of its variables. Moreover, the positivity or negativity of each variable is reflected in the sign of the corresponding weight.

Theorem 9.6. *Every threshold function is monotone. More precisely, if $f(x_1, x_2, \ldots, x_n)$ is a threshold function with structure $(w_1, w_2, \ldots, w_n, t)$, then, for $i = 1, 2, \ldots, n$:*

(1) *If $w_i = 0$, then f does not depend on x_i.*
(2) *If f does not depend on x_i, then $(w_1, \ldots, w_{i-1}, 0, w_{i+1}, \ldots, w_n, t)$ is a structure of f.*
(3) *If $w_i > 0$, then f is positive in x_i.*

(4) *If f is positive in x_i and f depends on x_i, then $w_i > 0$.*
(5) *If $w_i < 0$, then f is negative in x_i.*
(6) *If f is negative in x_i and f depends on x_i, then $w_i < 0$.*
(7) *Assume that $w_j \geq 0$ for $j = 1, 2, \ldots, k$, $w_j < 0$ for $j = k+1, k+2, \ldots, n$, and define the function $g(x_1, x_2, \ldots, x_n) = f(x_1, x_2, \ldots, x_k, \overline{x}_{k+1}, \ldots, \overline{x}_n)$. Then, g is a positive threshold function with structure $(w_1, w_2, \ldots, w_k, -w_{k+1}, \ldots, -w_n, t - \sum_{j=k+1}^{n} w_j)$.*

Proof. The proof is left as an exercise.   □

Example 9.3. *The function $f(x, y, z) = x\overline{y} \vee z$ considered in Example 9.1 is a threshold function with structure $(1, -2, 3, 0)$. The associated function $g(x, y, z) = xy \vee z$ is also a threshold function with structure $(1, 2, 3, 2)$.*   □

Let us stress the fact that, as the next example illustrates a variable can have nonzero weight in the separating structure of a threshold function even if the function does not depend on this variable (as a matter of fact, we will show later that it is NP-hard to determine whether or not a threshold function given by a separating structure depends on a particular variable; see Theorem 9.26).

Example 9.4. *The function $f(x, y, z, u) = xy \vee z$ is a threshold function with structure $(2, 4, 6, 1, 5)$. The variable u, which is inessential, has positive weight in this separating structure.*   □

One can check by complete enumeration that monotonicity is equivalent to thresholdness for functions of three variables or less. However, as one may expect, this statement does not hold for functions of more variables.

Example 9.5. *The functions $f(x, y, z, u) = xy \vee zu$, $g(x, y, z, u) = xy \vee yz \vee zu$, and $h(x, y, z, u) = xy \vee yz \vee zu \vee xu$ are not threshold. Up to permutations of their variables, f, g, and h are, in fact, the only positive nonthreshold functions of four variables [698].*   □

An easy way of proving that the functions f, g, and h in Example 9.5 are not threshold is to observe that they are not regular. Indeed:

Theorem 9.7. *Every threshold function has a complete strength preorder. More precisely, if $f(x_1, x_2, \ldots, x_n)$ is a threshold function, then,*

(1) *for every structure $(w_1, w_2, \ldots, w_n, t)$ of f, and for all i, j $= 1, 2, \ldots, n$, if $w_i \geq w_j$, then $x_i \succeq_f x_j$;*
(2) *there exists a structure $(w_1, w_2, \ldots, w_n, t)$ of f such that, for all $i, j = 1, 2, \ldots, n$, $w_i \geq w_j$ if and only if $x_i \succeq_f x_j$.*

Proof. The proof of (1) is straightforward (this statement was established as Theorem 8.3 in Section 8.1). As for statement (2), consider an arbitrary structure $(v_1, v_2, \ldots, v_n, t)$ of f. Denote by C an equivalence class of the relation $\approx_f$, say, without loss of generality, $C = \{x_1, x_2, \ldots, x_k\}$. By symmetry of the variables $x_1, x_2, \ldots, x_k$, it is clear that the vector

$$(V_i, t) = (v_i, v_{i+1}, \ldots, v_k, v_1, v_2, \ldots, v_{i-1}, v_{k+1}, v_{k+2}, \ldots, v_n, t)$$

is a structure of f, for $i = 1, 2, \ldots, k$. Therefore, by Theorem 9.4, $(V', t) = \frac{1}{k} \sum_{i=1}^{k} (V_i, t)$ is also a structure of f, for which all variables in C have the same weight. The same procedure can be repeated for all other equivalence classes of $\approx_f$ until we eventually obtain a structure (W, t) of f with the property that $x_i \approx_f x_j$ implies $w_i = w_j$, for $i, j = 1, 2, \ldots, n$. But then, (W, t) is a structure of f as described by (2), since, by statement (1), $x_i \succ_f x_j$ implies $w_i > w_j$, for $i, j = 1, 2, \ldots, n$. $\qquad\square$

Example 9.6. *Consider again the function $g(x, y, z) = xy \vee z$ defined in Example 9.3. This is a threshold function with structure $(1, 2, 3, 2)$. From statement (1) in Theorem 9.7, we conclude that $x \preceq_g y \preceq_g z$. On the other hand, applying the procedure described in the proof of statement (2) with $C = \{x, y\}$, we obtain the alternative structure $(3/2, 3/2, 3, 2)$ that gives equal weight to symmetric variables.* $\qquad\square$

It can be checked directly that every regular function of five variables or less is a threshold function, but there exist nonthreshold regular functions of six variables (see Winder [917] and Exercise 11 in Chapter 8).

We also recall Theorem 8.43 (Section 8.8).

Theorem 9.8. *Every threshold function is completely monotone. More precisely, if $f(x_1, x_2, \ldots, x_n)$ is a threshold function with structure $(w_1, w_2, \ldots, w_n, t)$, and if S, T are two subsets of $\{1, 2, \ldots, n\}$ such that $\sum_{i \in S} w_i \geq \sum_{i \in T} w_i$, then $S \succeq_f T$.*

Proof. Straightforward. $\qquad\square$

As mentioned in Section 8.8, all completely monotone functions of eight variables or less are threshold functions, but we shall present in Section 9.3 an example of a nonthreshold completely monotone function of nine variables (see Theorem 9.15). Let us also mention that Winder [917] has constructed a nonthreshold completely monotone function, for which the strength relation $\succeq_f$ is acyclic (see Einy [291] and Muroga [698] for additional information on this line of research).

We now investigate the behavior of threshold functions with respect to dualization (see Section 4.1.3 for definitions).

Theorem 9.9. *If f is a threshold function on $\mathcal{B}^n$ and $(w_1, w_2, \ldots, w_n, t)$ is an integral structure of f, then f^d is a threshold function with structure $(w_1, w_2, \ldots, w_n, \sum_{i=1}^{n} w_i - t - 1)$. If $t \leq \frac{1}{2}(\sum_{i=1}^{n} w_i - 1)$, then f is dual-major. If $t \geq \frac{1}{2}(\sum_i w_i - 1)$, then f is dual-minor.*

Proof. Let $t' = \sum_{i=1}^{n} w_i - t - 1$. Since t and $w_1, w_2, \ldots, w_n$ are integral, the following equivalences hold for all $X \in \mathcal{B}^n$:

$$f^d(X) = 0 \text{ if and only if } f(\overline{X}) = 1$$

$$\text{if and only if } \sum_{i=1}^{n} w_i (1 - x_i) > t$$

$$\text{if and only if } \sum_{i=1}^{n} w_i x_i \leq t'.$$

This proves the first part of the statement. For the second and third parts, simply notice that $f^d \leq f$ if $t \leq t'$ and $f \leq f^d$ if $t' \leq t$.   □

In view of Theorem 9.5, the requirement that the structure be integral is obviously not essential in the statement of Theorem 9.9; it merely simplifies its expression. Note also that the conditions for f to be dual-major or dual-minor are sufficient, but not necessary, in this statement, as illustrated by the next example. (Exercise 8 at the end of the chapter actually suggests that it may be hard to characterize self-dual threshold functions.)

Example 9.7. *The threshold function* $f(x, y, z, u) = xy \vee xz \vee xu \vee yzu$ *admits the structure* $(4, 2, 2, 2, 5)$. *Thus,* f^d *is a threshold function with structure* $(4, 2, 2, 2, 4)$, *and* f *is dual-minor. But another structure of* f *is* $(2, 1, 1, 1, 2)$, *which implies that the same vector* $(2, 1, 1, 1, 2)$ *is also a structure of* f^d, *and hence, that* f *is self-dual:* $f = f^d$.   □

The next property was independently observed in the context of threshold logic (see for instance [698]) and of game theory (see [291]). It involves the concept of self-dual extension, which we introduced in Section 4.1.3.

Theorem 9.10. *The function* $f(x_1, x_2, \ldots, x_n)$ *is a threshold function if and only if its self-dual extension* $f^{SD}(x_1, x_2, \ldots, x_n, x_{n+1}) = f \overline{x}_{n+1} \vee f^d x_{n+1}$ *is a threshold function.*

Proof. Assume that f is a threshold function and that $(w_1, w_2, \ldots, w_n, t)$ is an integral structure of f. Then, it follows from Theorem 9.9 that $(w_1, w_2, \ldots, w_n, 2t + 1 - \sum_{i=1}^{n} w_i, t)$ is a structure of f^{SD}. Conversely, if f^{SD} is a threshold function with structure $(w_1, w_2, \ldots, w_n, w_{n+1}, t)$, then $(w_1, w_2, \ldots, w_n, t)$ is a structure of f.   □

We conclude this section by stating some results regarding the number of threshold functions and the size of the weights required in a separating structure. The number of threshold functions of n variables is known quite precisely.

Theorem 9.11. *The number* τ_n *of threshold functions of n variables satisfies*

$$\frac{n^2}{2} - \frac{n}{2} \leq \log_2 \tau_n \leq n^2. \tag{9.1}$$

Moreover, for n sufficiently large,

$$n^2 \left(1 - \frac{10}{\ln n}\right) < \log_2 \tau_n. \tag{9.2}$$

The upper bound in (9.1) was independently proved by several authors and published by Winder in [916] (see [698, 917] for an account). The lower bound in (9.1) is due to Yajima and Ibaraki [928] and Smith [842]. The sharper asymptotic lower bound (9.2) was eventually established by Zuev [940], thus settling Winder's conjecture [917] that $(\log_2 \tau_n)/n^2$ approaches 1 as n goes to infinity. We do not prove these results here; the reader is referred to the original publications or to Anthony [25, 27] for extensions.

The size of weights in a separator can be bounded as follows:

Theorem 9.12. *For every threshold function of n variables, there exists an integral separating structure $(w_1, w_2, \ldots, w_n, t)$ such that*

$$\max\{|w_1|, |w_2|, \ldots, |w_n|, |t|\} \le (n+1)n^{n/2}. \tag{9.3}$$

Moreover, there are constants $k > 0$ and $c > 1$ such that, for n a power of 2, there is a threshold function of n variables, such that any integral separating structure representing f involves a weight of magnitude at least $kc^{-n}n^{n/2}$.

In Theorem 9.12 (that we quote directly from Anthony [25]), the upper bound is due to Muroga [698] and the lower bound is due to Håstad [477]. Observe that the dominating factor $n^{n/2}$ is identical in both bounds. Here, we again omit the proofs and refer the reader to [25, 27, 477, 698] for additional details; see also Diakonikolas and Servedio [271] for significant extensions.

9.3 Characterizations of threshold functions

In this section, we present two alternative characterizations of threshold functions and discuss related results.

The first characterization is a simple linear programming formulation which provides a useful computational tool for the recognition of threshold functions (see Section 9.4). For the sake of simplicity, we only state it for positive functions: Since every threshold function is monotone, this restriction does not entail any essential loss of generality.

Theorem 9.13. *A positive Boolean function with maximal false points $X^1, X^2, \ldots, X^p$ and minimal true points $Y^1, Y^2, \ldots, Y^m$ is a threshold function if and only if the system of inequalities*

$$(TS) \quad \begin{cases} \sum_{i=1}^n w_i x_i^j & \le \quad t \qquad (j = 1, 2, \ldots, p) \\ \sum_{i=1}^n w_i y_i^j & \ge \quad t+1 \quad (j = 1, 2, \ldots, m) \\ w_i & \ge \quad 0 \qquad (i = 1, 2, \ldots, n) \end{cases}$$

has a solution $(w_1, w_2, \ldots, w_n, t)$. *When this is the case, every solution of* (TS) *is a separating structure of the function.*

Proof. The statement follows directly from Definition 9.1 and Theorems 9.4 and 9.6.   □

Example 9.8. *Let* $f = x_1 x_2 \vee x_1 x_3 x_4 \vee x_2 x_3 x_4$. *The maximal false points of* f *are* $(1,0,1,0)$, $(1,0,0,1)$, $(0,1,1,0)$, $(0,1,0,1)$, $(0,0,1,1)$, *and its minimal true points are* $(1,1,0,0)$, $(1,0,1,1)$, $(0,1,1,1)$. *Thus, the system* (TS) *associated with* f *is*

$$
\begin{array}{ccccccccc}
w_1 & & & + & w_3 & & & \leq & t \\
w_1 & & & & & + & w_4 & \leq & t \\
& & w_2 & + & w_3 & & & \leq & t \\
& & w_2 & & & + & w_4 & \leq & t \\
& & & & w_3 & + & w_4 & \leq & t \\
w_1 & + & w_2 & & & & & \geq & t+1 \\
w_1 & & & + & w_3 & + & w_4 & \geq & t+1 \\
& & w_2 & + & w_3 & + & w_4 & \geq & t+1 \\
& & w_1, & w_2, & w_3, & w_4 & & \geq & 0.
\end{array}
$$

This system admits the solution $(w_1, w_2, w_3, w_4, t) = (5, 4, 3, 2, 8)$. *Hence,* f *is a threshold function with structure* $(5, 4, 3, 2, 8)$.   □

Theorem 9.13, like Definition 9.1, has a strong numerical flavor. The next result originated in the efforts devoted by researchers in switching logic to establish purely combinatorial, rather than numerical, characterizations of threshold functions (remember that the study of regularity and of k-monotonicity also originated in such attempts; see Chapter 8).

We start with a definition (due to Winder [917]) that extends the notions of 2-summability and 2-asummability already introduced in Definition 8.10 of Section 8.8.

Definition 9.2. *Let* $k \in \mathbb{N}$, $k \geq 2$. *A Boolean function* f *on* $\mathcal{B}^n$ *is* k-summable *if, for some* $r \in \{2, 3, \ldots, k\}$, *there exist* r *(not necessarily distinct) false points of* f, *say,* $X^1, X^2, \ldots, X^r$, *and* r *(not necessarily distinct) true points of* f, *say,* $Y^1, Y^2, \ldots, Y^r$, *such that* $\sum_{i=1}^{r} X^i = \sum_{i=1}^{r} Y^i$. *A function is* k-asummable *if it is not* k-summable, *and it is* asummable *if it is* k-asummable *for all* $k \geq 2$.

Example 9.9. *We have shown in Example 8.24 that the function* $f(x_1, x_2) = x_1 x_2 \vee \overline{x_1}\, \overline{x_2}$ *is 2-summable. We shall provide an example of a 2-asummable, 3-summable function in the proof of Theorem 9.15.*   □

The following characterization of threshold Boolean functions is due to Chow [193] and Elgot [310].

Theorem 9.14. *A Boolean function is a threshold function if and only if it is asummable.*

Proof. Let f be a threshold function on $\mathcal{B}^n$ with structure $(W, t) \in \mathbb{R}^{n+1}$, let $X^1, X^2, \ldots, X^r$ be r false points of f, and let $Y^1, Y^2, \ldots, Y^r$ be r true points of f. Then, for $i = 1, 2, \ldots, r$,

$$W X^i \le t < W Y^i,$$

and hence, $\sum_{i=1}^{r} X^i \ne \sum_{i=1}^{r} Y^i$. Therefore, f is asummable.

Conversely, assume that f is not threshold, meaning that the set $\{X^1, X^2, \ldots, X^p\}$ of false points of f cannot be separated from the set $\{Y^1, Y^2, \ldots, Y^m\}$ of its true points by a hyperplane of $\mathbb{R}^n$. Then, standard separation theorems (see, e.g., [788]) imply that the convex hulls of $\{X^1, X^2, \ldots, X^p\}$ and of $\{Y^1, Y^2, \ldots, Y^m\}$ have nonempty intersection. In other words, the following system has a feasible solution in the variables u_i, $i = 1, 2, \ldots, p$, and v_j, $j = 1, 2, \ldots, m$:

$$\sum_{i=1}^{p} u_i X^i = \sum_{j=1}^{m} v_j Y^j \tag{9.4}$$

$$\sum_{i=1}^{p} u_i = 1 \tag{9.5}$$

$$\sum_{j=1}^{m} v_j = 1 \tag{9.6}$$

$$u_i \ge 0 \quad (i = 1, 2, \ldots, p) \tag{9.7}$$

$$v_j \ge 0 \quad (j = 1, 2, \ldots, m). \tag{9.8}$$

Let $(U, V) \in \mathbb{Q}^{p+m}$ be a rational solution of (9.4)–(9.8) (such a solution exists, since the system has rational coefficients). For some positive integer k, all components of the vector (kU, kV) are nonnegative integers, and

$$\sum_{i=1}^{p} (k u_i) X^i = \sum_{j=1}^{m} (k v_j) Y^j \tag{9.9}$$

$$\sum_{i=1}^{p} k u_i = k \tag{9.10}$$

$$\sum_{j=1}^{m} k v_j = k. \tag{9.11}$$

Now the equalities (9.9)–(9.11) express that f is a k-summable function: Simply take $k u_i$ copies of the false point X^i for $i = 1, 2, \ldots, p$, and $k v_j$ copies of the true point Y^j for $j = 1, 2, \ldots, m$. $\square$

In this proof, we have stressed the connection of Theorem 9.14 with geometric separability theorems. Alternatively, this result could also be deduced directly by

$$Th \subseteq \mathcal{A}_{k+1} \subseteq \mathcal{A}_k \subseteq \ldots \subseteq \mathcal{A}_2 = \mathcal{CM}$$

Figure 9.1. The hierarchy of k-asummable Boolean functions.

applying the strong duality theorem of linear programming to the formulation (TS) (as in [310, 917]).

Thus, if we denote by Th the set of threshold functions, by $\mathcal{A}_k$ the set of k-asummable functions ($k \geq 2$), and by $\mathcal{CM}$ the set of completely monotone functions, we obtain the hierarchy displayed in Figure 9.1 for all $k \geq 2$. (Compare with the hierarchy of k-monotone functions pictured in Figure 8.8 of Section 8.8, and recall that $\mathcal{A}_2 = \mathcal{CM}$ by Theorem 8.50.)

It was once conjectured that this hierarchy may be finite, meaning that there would exist some possibly large, but fixed value k^* such that the equality $Th = \mathcal{A}_k = \mathcal{A}_{k^*}$ holds for all $k \geq k^*$. This conjecture was demolished by Winder [915, 917] who proved that, for every k, there exist k-asummable functions that are not linearly separable. We do not establish this result here, but simply prove the weaker statement that the inclusion $Th \subseteq \mathcal{A}_2$ is strict.

Theorem 9.15. *Some 2-asummable functions are not threshold functions.*

Proof. Moore [692] (cited in [698, 917]) first exhibited a 12-variable function establishing this statement. Gabelman [356] later produced a 9-variable example. We propose here a variant of Gabelman's example.

Consider first the vector $A = (14, 18, 24, 26, 27, 30, 31, 36, 37)$. We shall use the observation that the only points of $\mathcal{B}^9$ lying on the hyperplane $\mathcal{H} = \{X \in \mathcal{B}^9 : \sum_{i=1}^{9} a_i x_i = 81\}$ are the six points:

$$X^1 = (1,0,0,0,0,0,1,1,0), \quad X^2 = (0,1,0,1,0,0,0,0,1), \quad X^3 = (0,0,1,0,1,1,0,0,0),$$

$$Y^1 = (1,0,0,0,0,1,0,0,1), \quad Y^2 = (0,1,0,0,1,0,0,1,0), \quad Y^3 = (0,0,1,1,0,0,1,0,0).$$

Define now a Boolean function $f(x_1, x_2, \ldots, x_9)$ as follows: The false points of f are all points X such that $\sum_{i=1}^{9} a_i x_i \leq 80$, plus the three points X^1, X^2, and X^3. Notice that, in particular, Y^1, Y^2, and Y^3 are true points of f. We claim that f is 2-asummable but not a threshold function.

To see that f is not a threshold function, it suffices to observe that $X^1 + X^2 + X^3 = Y^1 + Y^2 + Y^3$, and hence, that f is 3-summable.

On the other hand, assume that f is 2-summable, and that

$$U^* + V^* = W^* + Z^*, \tag{9.12}$$

where U^*, V^* are false points of f, and W^*, Z^* are true points of f. Then, all four points U^*, V^*, W^*, and Z^* must lie on the hyperplane $\mathcal{H}$; otherwise,

$$\sum_{i=1}^{9} a_i u_i^* + \sum_{i=1}^{9} a_i v_i^* < \sum_{i=1}^{9} a_i w_i^* + \sum_{i=1}^{9} a_i z_i^*,$$

threshold		2-asummable	
$\Updownarrow$ $\quad\Rightarrow\quad$ k-asummable $\quad\Rightarrow\quad$		$\Updownarrow$ $\quad\Rightarrow\quad$ k-monotone	
asummable $\qquad\qquad (k > 2)$		completely $\qquad\qquad (k \geq 1)$	
		monotone	

Figure 9.2. A hierarchy of Boolean functions.

contradicting equation (9.12). So, $\{U^*, V^*\} \subset \{X^1, X^2, X^3\}$ and $\{W^*, Z^*\} \subset \{Y^1, Y^2, Y^3\}$. But this is easily seen to be incompatible with equation (9.12). $\quad\square$

The proof of Theorem 9.15 actually shows that the inclusion $\mathcal{A}_3 \subset \mathcal{A}_2$ is strict. This result was generalized by Taylor and Zwicker [860], who proved that $\mathcal{A}_{k+1} \neq \mathcal{A}_k$ for all $k \geq 2$. Another interesting generalization of Winder's result is provided by Theorem 11.14 in Chapter 11.

Figure 9.2 summarizes the relations between some of the classes of Boolean functions studied in this chapter and in the previous one. The one-way implications displayed in Figure 9.2 cannot be reversed. It may be useful to recall here that 1-monotone functions are exactly monotone functions, and that 2-monotone positive functions coincide with regular functions. Figure 9.2 will be enriched with one more class of functions in Section 9.6 (see Figure 9.5).

9.4 Recognition of threshold functions

9.4.1 A polynomial-time algorithm for positive DNFs

A fundamental algorithmic problem is to recognize whether a given Boolean function f is a threshold function, and, when the answer is affirmative, to produce a separating structure of f. As always, the complexity of this problem depends very much on the assumptions regarding the format of its input: For instance, it is easy to see that the problem can be solved by linear programming when f is given by its truth table, but that it may require an exponential number of steps when f is given by an oracle (see Exercise 10; note however that Matulef, O'Donnell, Rubinfeld, and Servedio [677] provide efficient algorithms for "approximately" recognizing threshold function in the oracle framework).

We focus here on the following formulation of the recognition problem:

THRESHOLD RECOGNITION
Instance: A Boolean function f represented by a Boolean expression.
Output: FALSE if f is not a threshold function; a separating structure of f otherwise.

This question has been extensively studied in the threshold logic literature under the name of *threshold synthesis* problem (see, e.g., Hu [511] or Muroga [698]). It has stimulated the discovery of properties of threshold functions that we discussed in Section 9.2 and Section 9.3. As we have seen, all early attempts to derive a "tractable" characterization of thresholdness were unsuccessful. In particular, none of the increasingly intricate conjectures linking threshold functions to k-monotonicity or to k-asummability has resisted a deeper examination. Note also that the asummability characterization in Theorem 9.14 does not seem to yield a straightforward, efficient thresholdness test.

In spite of this negative news, we are going to prove in this section that the threshold recognition problem is polynomially solvable when the input function is positive and is expressed by its complete (prime irredundant) disjunctive normal form. In Section 9.4.3, we briefly discuss the extent to which these assumptions are restrictive.

Like most classical approaches to the threshold recognition problem, the algorithm presented relies on the characterization of threshold functions and on the system of inequalities (TS) formulated in Theorem 9.13. We know that if a positive Boolean function is given by its complete DNF, then the list of its minimal true points is readily available. Thus, in order to generate the system (TS) for such a function, we only need to enumerate the maximal false points of the function, or, equivalently, to dualize it. But, as we know from Chapter 4, dualizing an arbitrary positive Boolean function is in general a difficult task, and the number of maximal false points may very well be exponential in the size of the input DNF. These difficulties originally motivated the quest for efficient dualization algorithms for regular functions, which eventually led to the results presented in Section 8.5. Indeed, these results are easily exploited to obtain a polynomial-time implementation of the recognition procedure displayed in Figure 9.3.

We thus obtain a remarkable result due to Peled and Simeone [735].

Theorem 9.16. *The procedure* THRESHOLD *is correct and can be implemented to run in time* $O(n^7 m^5)$, *where n is the number of variables and m is the number of prime implicants of the function to be tested.*

Procedure THRESHOLD(f)
Input: The complete DNF of a positive Boolean function $f(x_1, x_2, \ldots, x_n)$.
Output: FALSE if f is not a threshold function; a separating structure of f otherwise.

begin
    if f is not regular **then return** FALSE
    else begin
        dualize f;
        set up the system (TS);
        solve (TS);
        if (TS) has no solution **then return** FALSE
            else return a solution $(w_1, w_2, \ldots, w_n, t)$ of (TS);
    end
end

Figure 9.3. Procedure THRESHOLD.

Proof. Testing whether the input function f is regular can be accomplished in time $O(n^2m)$ by the procedure REGULAR presented in Section 8.4 (Theorem 8.20). If f is not regular, then f is not a threshold function (by Theorem 9.7). If f is regular, then it can be dualized in $O(n^2m)$ time by the procedure DUALREG (Theorem 8.28), and the system (TS) can be set up within the same time bound. Now, by Theorem 9.13, f is a threshold function if and only if the system (TS) is consistent, and every solution of (TS) is a structure of f. Using a polynomial-time algorithm for linear programming (see [76, 812]), (TS) can be solved in time $O(n^7m^5)$, since (TS) has $n + 1$ variables and $O(nm)$ constraints (by Theorem 8.29).   □

Example 9.10. *Let* $f = x_1x_2 \vee x_1x_3x_4 \vee x_2x_3x_4$. *This function is regular, with* $x_1 \approx_f x_2 \succ_f x_3 \approx_f x_4$, *and* $f^d = x_1x_2 \vee x_1x_3 \vee x_1x_4 \vee x_2x_3 \vee x_2x_4$. *The system* (TS) *associated with* f *was set up in Example 9.8, where we learned that* f *is a threshold function with structure* $(5,4,3,2,8)$.   □

Note that the worst-case time complexity of the procedure THRESHOLD is quite high due to the solution of the system of linear inequalities (TS) by a generic linear programming algorithm. This observation is somewhat disturbing in view of the fact that the other steps of the procedure require only $O(n^2m)$ operations. It may be interesting to know whether threshold functions can be recognized through an entirely combinatorial procedure without resorting to the solution of the system (TS) by a generic linear programming algorithm. An attempt in this direction can be found in Smaus [841], but some of the details missing in the proofs of this paper may not be easy to fill in.

9.4.2 A compact formulation

For practical computations, the system (TS) can be simplified considerably (even though these simplifications do not affect the worst-case complexity of THRESHOLD). To understand this, assume, for instance, that the input function is regular with $x_1 \approx_f x_2 \succ_f x_3$. Then, Theorem 9.7(2) can be used to introduce the additional constraints $w_1 = w_2 \geq w_3$ in (TS). (Actually, we could even add the constraints $w_2 \geq w_3 + 1$; check this!) As a consequence, some of the original constraints of (TS) become redundant and can be eliminated.

Example 9.11. *Consider again the function* $f = x_1x_2 \vee x_1x_3x_4 \vee x_2x_3x_4$ *as in Example 9.10. Since* $x_1 \approx_f x_2 \succ_f x_3 \approx_f x_4$, *we can add to* (TS) *the constraints* $w_1 = w_2 \geq w_3 = w_4$. *As a consequence,* w_2 *and* w_4 *can be eliminated from the system* (TS), *which reduces to*

$$(\text{TS}^*)\begin{cases} w_1 & + & w_3 & \leq & t \\ & & 2w_3 & \leq & t \\ 2w_1 & & & \geq & t+1 \\ w_1 & + & 2w_3 & \geq & t+1 \\ w_1 & \geq & w_3 & \geq & 0. \end{cases}$$

Moreover, since $2w_3 \leq w_1 + w_3$ *in every solution of* (TS*), *the second inequality of* (TS*) *is redundant and can be removed. A solution of* (TS*) *is, for instance,* $(2,2,1,1,3)$, *which is easily seen to be a separating structure of* f. $\square$

To describe more precisely what happens in Example 9.11, we first recall two definitions from Section 8.3.

Definition 9.3. *For any two points* $X^*, Y^* \in \mathcal{B}^n$, *we say that* Y^* *is a* left-shift *of* X^* *and we write* $Y^* \frown X^*$ *if there exists a mapping* $\sigma : supp(X^*) \to supp(Y^*)$ *such that*

(a) σ *is injective, that is,* $\sigma(i) \neq \sigma(j)$ *when* $i \neq j$, *and*
(b) $\sigma(i) \leq i$ *for all* $i = 1, 2, \ldots, n$.

Definition 9.4. *A point* $X^* \in \mathcal{B}^n$ *is a* ceiling *of the Boolean function* $f(x_1, x_2, \ldots, x_n)$ *if* X^* *is a false point of* f *and if no other false point of* f *is a left-shift of* X^*. *Similarly,* X^* *is a* floor *of* f *if* X^* *is a true point of* f *and if* X^* *is a left-shift of no other true point of* f.

Thus, a ceiling is a "leftmost" (maximal) false point, and a floor is a "rightmost" (minimal) true point. Theorem 8.14 implies that a regular function is completely characterized by the list of its ceilings or of its floors. As for threshold functions, we can refine the statement of Theorem 9.13 as follows:

Theorem 9.17. *Let* $f(x_1, x_2, \ldots, x_n)$ *be a regular Boolean function such that* $x_1 \succcurlyeq_f x_2 \succcurlyeq_f \cdots \succcurlyeq_f x_n$, *let* $X^1, X^2, \ldots, X^r$ *denote the ceilings of* f, *and let* $Y^1, Y^2, \ldots, Y^s$ *denote its floors. Then,* f *is a threshold function if and only if the system of inequalities*

$$
(TS^*) \quad
\begin{cases}
\sum_{i=1}^{n} w_i x_i^j \leq t & (j = 1, 2, \ldots, r) \\
\sum_{i=1}^{n} w_i y_i^j \geq t + 1 & (j = 1, 2, \ldots, s) \\
w_i \geq 0 & (i = 1, 2, \ldots, n) \\
w_i = w_j \quad if \quad x_i \approx_f x_j & (i, j = 1, 2, \ldots, n) \\
w_i \geq w_j \quad if \quad x_i \succ_f x_j & (i, j = 1, 2, \ldots, n)
\end{cases}
$$

has a solution $(w_1, w_2, \ldots, w_n, t)$. *When this is the case, every solution of* (TS*) *is a separating structure of the function.*

Proof. If f is a threshold function, then (TS*) has a solution by Theorems 9.4, 9.6, and 9.7. Conversely, if (TS*) has a solution $(w_1, w_2, \ldots, w_n, t)$, then it follows easily from the definition of ceilings and floors that $(w_1, w_2, \ldots, w_n, t)$ is a solution of (TS), and hence, f is a threshold function. $\square$

Example 9.12. *Consider again the function* $f = x_1 x_2 \vee x_1 x_3 x_4 \vee x_2 x_3 x_4$, *as in Example 9.11. The unique ceiling of* f *is the point* $(1,0,1,0)$, *and its floors are the* *points* $(1,1,0,0)$ *and* $(0,1,1,1)$. *Thus, the system* (TS*) *associated with* f *reads*

$$
\begin{array}{rcrcrcrcl}
w_1 & & & + & w_3 & & & \le & t \\
w_1 & + & w_2 & & & & & \ge & t+1 \\
& & w_2 & + & w_3 & + & w_4 & \ge & t+1 \\
w_1 & = & w_2 & \ge & w_3 & = & w_4 & \ge & 0.
\end{array}
$$

This system is equivalent to the system (TS*) *in Example 9.11.*   □

9.4.3 The general case

We have so far only handled the special case of the threshold recognition problem in which the input is the complete DNF of a positive function. Let us now drop this assumption, and let us assume that the input function is given by an arbitrary Boolean expression.

In this case, a generic approach for solving the threshold recognition problem can be sketched as follows [698]:

(a) Generate the prime implicants of f.
(b) Use Theorem 1.21 to check whether f is monotone. If not, then f is not a threshold function. Otherwise, convert f into a positive function by performing the change of variables: $y_i \leftarrow \overline{x}_i$ for all negative variables x_i.
(c) Use the procedure THRESHOLD to decide whether the resulting function is a threshold function.

Note that steps (b) and (c) of this procedure are easy in the sense that their complexity is polynomial in the number of prime implicants of f (that is, in the size of the output of step (a)). However, we know from Chapter 4 that step (a) is difficult even when f is in disjunctive normal form, and that its output may actually be exponentially large in the size of f. One may, therefore, wonder whether alternative, more efficient lines of attack could be devised.

An answer to this question was provided by Peled and Simeone [735], who established that the general version of the threshold recognition problem is likely to be significantly harder than the positive case.

Theorem 9.18. THRESHOLD RECOGNITION *is co-NP-complete even when its input is expressed as a DNF of degree 3.*

Proof. From the proof of Theorem 1.30 in Section 1.11, and from the observation that the quadratic DNF $\gamma = y_1 y_2 \vee y_3 y_4$ does not represent a threshold function, we can immediately conclude that THRESHOLD RECOGNITION is NP-hard when restricted to DNFs of degree 3. Thus, the only point requiring some attention is the claim that the associated decision problem is in co-NP.

To see that this is the case, let f be an arbitrary input function on $\mathcal{B}^n$, denote its false points by $\{X^1, X^2, \ldots, X^p\}$ and its true points by $\{Y^1, Y^2, \ldots, Y^m\}$. If f is not a threshold function, then (as in the proof of Theorem 9.14) the system (9.4)–(9.8) has a feasible solution in the variables u_i $(i = 1, 2, \ldots, p)$ and v_j $(j = 1, 2, \ldots, m)$.

Because this system has rational coefficients and involves $n+2$ equations, standard results about linear programming problems imply that (9.4)–(9.8) has a rational solution in which at most $n+2$ variables take a nonzero value, and whose size is polynomially bounded in n (see, e.g., [812]; in geometric terms, this is also a consequence of Caratheodory's theorem; see [199, 788]). Let $(U,V) \in \mathbb{R}^{p+m}$ be such a solution, with $I = \{i : u_i > 0, i = 1,2,\ldots,p\}$, $J = \{j : v_j > 0, j = 1,2,\ldots,m\}$, $|I| \leq (n+2)$, and $|J| \leq (n+2)$. Then, the points X^i $(i \in I)$ and Y^j $(j \in J)$, together with the coefficients u_i $(i \in I)$ and v_j $(j \in J)$, constitute a polynomial-size certificate of nonthresholdness for f. This implies that THRESHOLD RECOGNITION is in co-NP.   □

The bound on the degree of the input DNF is sharp in Theorem 9.18: Indeed, the prime implicants of a quadratic DNF can be generated in polynomial time (see Section 5.8), so that the generic recognition procedure sketched at the beginning of this subsection applies. In particular, the case of nonmonotone quadratic DNFs can be efficiently reduced to the positive case, which we discuss in greater detail in Section 9.7.

Application 9.7. (Integer programming.) *The aggregation problem for a system of linear inequalities in 0–1 variables can be stated as follows. Given a system of inequalities*

$$\sum_{j=1}^{n} a_{ij} x_j \leq b_i \quad (i = 1,2,\ldots,m), \tag{9.13}$$

is there a single inequality in $(x_1, x_2, \ldots, x_n)$, say,

$$\sum_{j=1}^{n} w_j x_j \leq t \tag{9.14}$$

such that (9.13) and (9.14) have the same set of solutions over $\mathcal{B}^n$? To establish the link between the aggregation problem and the threshold recognition problem, we rely on some of the concepts that have been introduced in Chapter 1, Section 1.13.6. Remember that the resolvent of the system (9.13) is the Boolean function $f(x_1, x_2, \ldots, x_n)$ whose false points are exactly the 0–1 solutions of (9.13) (see Section 1.13.6). Then, the aggregation problem is simply asking whether f is a threshold function, and Theorem 9.18 implies that the aggregation problem is NP-hard, even for systems of generalized covering inequalities (see Theorem 1.39). However, when (9.13) happens to be a system of set-covering inequalities, meaning that $a_{ij} \in \{-1,0\}$ and $b_i = -1$ $(i = 1,2,\ldots,m, \ j = 1,2,\ldots,n)$, then f is a negative function and its prime implicants are readily available. Hence, in this special case, the procedure THRESHOLD provides an efficient solution of the aggregation problem (Peled and Simeone [735]). See also Application 9.12 in Section 9.7 for related considerations.   □

9.5   Prime implicants of threshold functions

In the previous section, we tackled the problem of computing a structure of a threshold function when the function is expressed in DNF. We now turn to the opposite question; namely, given a separating structure, how can we generate the prime implicants of the corresponding threshold function? For the sake of simplicity, we consider only the case of positive functions; by virtue of Theorem 9.6, this assumption does not entail any loss of generality. Note also that, as an immediate consequence of Theorem 9.9, all results in this section carry over mutatis mutandis to the prime implicates of threshold functions.

We start with a simple characterization of the prime implicants of a threshold function in terms of a separating structure of the function.

Theorem 9.19. *Let $f(x_1, x_2, \ldots, x_n)$ be a positive threshold function with separating structure $(w_1, w_2, \ldots, w_n, t)$, where $w_j \geq 0$ for $j = 1, 2, \ldots, n$. The elementary conjunction $\bigwedge_{j \in P} x_j$ is a prime implicant of f if and only if $\sum_{j \in P} w_j > t$ and $\sum_{j \in P \setminus \{i\}} w_j \leq t$ for all $i \in P$.*

Proof. We know that $\bigwedge_{j \in P} x_j$ is a prime implicant of f if and only the point $X^P \in \mathcal{B}^n$, defined by $x_j^P = 1$ for $j \in P$ and $x_j^P = 0$ for $j \notin P$, is a minimal true point of f. This is trivially equivalent to the conditions given in the statement.   □

We now present an algorithm to generate all prime implicants (or, more precisely, all minimal true points) of a positive threshold function $f(x_1, x_2, \ldots, x_n)$ described by a structure $(w_1, w_2, \ldots, w_n, t)$. We assume that the variables of f have been permuted in such a way that $w_1 \geq w_2 \geq \cdots \geq w_n \geq 0$ and, to rule out the trivial cases where f is constant on $\mathcal{B}^n$, we also assume that $0 \leq t < \sum_{i=1}^{n} w_i$.

For $k = 1, 2, \ldots, n$, we denote by T_k the set of all points $(y_1^*, y_2^*, \ldots, y_k^*) \in \mathcal{B}^k$ such that f has a minimal true point of the form $(y_1^*, y_2^*, \ldots, y_n^*) \in \mathcal{B}^n$ for an appropriate choice of $(y_{k+1}^*, y_{k+2}^*, \ldots, y_n^*) \in \mathcal{B}^{n-k}$. Thus, T_n contains exactly the minimal true points of f. We also let $T_0 = \{()\}$, where $()$ is the "empty" vector. (Observe that this convention is coherent with the previous definition, since we have assumed that T_n is nonempty.)

Now, the prime implicant generation algorithm recursively generates $T_1, T_2, \ldots, T_n$: The next result explains how T_{k+1} can be efficiently generated when T_k is at hand.

Theorem 9.20. *Let $f(x_1, x_2, \ldots, x_n)$ be a positive threshold function with structure $(w_1, w_2, \ldots, w_n, t)$, where $w_1 \geq w_2 \geq \cdots \geq w_n \geq 0$; let $0 \leq k \leq n$; and let $(y_1^*, y_2^*, \ldots, y_k^*)$ be a point in T_k. Then,*

(1) $(y_1^*, y_2^*, \ldots, y_k^*, 1) \in T_{k+1}$ if and only if $\sum_{i=1}^{k} w_i y_i^* \leq t$;

(2) $(y_1^*, y_2^*, \ldots, y_k^*, 0) \in T_{k+1}$ if and only if $\sum_{i=1}^{k} w_i y_i^* + \sum_{i=k+2}^{n} w_i > t$.

Proof. Let $Y^* = (y_1^*, y_2^*, \ldots, y_k^*)$, and consider assertion (1). If $(Y^*, 1)$ is in T_{k+1}, then f has a minimal true point of the form $(Y^*, 1, Z^*) \in \mathcal{B}^n$, and hence, $(Y^*, 0, \ldots, 0) \in \mathcal{B}^n$ is a false point of f, meaning that $\sum_{i=1}^k w_i y_i^* \leq t$.

Conversely, assume now that $\sum_{i=1}^k w_i y_i^* \leq t$. Since $Y^* \in T_k$, the point $(Y^*, 1, \ldots, 1) \in \mathcal{B}^n$ is a true point of f, and hence, $\sum_{i=1}^k w_i y_i^* + \sum_{i=k+1}^n w_i > t$. Let $r \geq k+1$ be the smallest index such that $\sum_{i=1}^k w_i y_i^* + \sum_{i=k+1}^r w_i > t$. Define $y_{k+1}^* = \cdots = y_r^* = 1$, $y_{r+1}^* = \cdots = y_n^* = 0$, and $X^* = (Y^*, y_{k+1}^*, \ldots, y_n^*)$. Then, X^* is a minimal true point of f, and hence, $(Y^*, 1) \in T_{k+1}$ as required.

Consider now assertion (2), and assume that $(Y^*, 0)$ is in T_{k+1}. Then $(Y^*, 0, 1, \ldots, 1) \in \mathcal{B}^n$ is a true point of f; hence, $\sum_{i=1}^k w_i y_i^* + \sum_{i=k+2}^n w_i > t$.

Conversely, assume that $\sum_{i=1}^k w_i y_i^* + \sum_{i=k+2}^n w_i > t$. There are two cases:

- If $\sum_{i=1}^k w_i y_i^* \leq t$, let $r \geq k+2$ be the smallest index such that $\sum_{i=1}^k w_i y_i^* + \sum_{i=k+2}^r w_i > t$. Define $y_{k+1}^* = 0$, $y_{k+2}^* = \cdots = y_r^* = 1$, $y_{r+1}^* = \cdots = y_n^* = 0$, and let $X^* = (Y^*, y_{k+1}^*, \ldots, y_n^*)$. Then, X^* is a minimal true point of f, and hence, $(Y^*, 0) \in T_{k+1}$.
- If $\sum_{i=1}^k w_i y_i^* > t$, then $X^* = (Y^*, 0, \ldots, 0) \in \mathcal{B}^n$ is a true point of f. On the other hand, since $Y^* \in T_k$, there exists a minimal true point of f of the form $Z^* = (Y^*, y_{k+1}^*, \ldots, y_n^*)$. By minimality of Z^*, we conclude that $X^* = Z^*$, and hence, $(Y^*, 0) \in T_{k+1}$. □

Theorem 9.20 leads to the algorithm displayed in Figure 9.4. We illustrate this algorithm on a small example.

Example 9.13. *We apply procedure* MINTRUE *to the threshold function f represented by the separating structure* $(w_1, w_2, \ldots, w_5, t) = (5, 4, 3, 2, 1, 8)$. *Since* $\sum_{j=1}^5 w_j = 15 > 8$, *we start with* $T_0 = \{()\}$.

In order to generate T_1, *we use Theorem 9.20 with* $k = 0$ *and* $Y^* = ()$. *Since* $\sum_{i=1}^k w_i y_i^* = 0 \leq 8$ *and* $\sum_{i=1}^k w_i y_i^* + \sum_{i=k+2}^n w_i = 10 > 8$, *we obtain* $T_1 = \{(1), (0)\}$.

Procedure MINTRUE$(w_1, w_2, \ldots, w_n, t)$
Input: The separating structure $(w_1, w_2, \ldots, w_n, t) \in \mathbb{Q}^{n+1}$ of a threshold function f,
    with $w_1 \geq w_2 \geq \cdots \geq w_n \geq 0$.
Output: The set T of minimal true points of f.

begin
    if $\sum_{i=1}^n w_i \leq t$ **then return** $T := \emptyset$
    else if $t < 0$ **then return** $T := \{(0, \ldots, 0)\}$
    else begin
            $T_0 := \{()\}$;
            for $j := 1$ **to** n **do** use Theorem 9.20 to generate T_j;
            return $T := T_n$;
        end
end

Figure 9.4. Procedure MINTRUE.

Next, we let $k = 1$ in Theorem 9.20. When $Y^* = (1)$, we have $\sum_{i=1}^{k} w_i y_i^* = 5 \leq 8$ and $\sum_{i=1}^{k} w_i y_i^* + \sum_{i=k+2}^{n} w_i = 11 > 8$. Thus, the points $(1,0)$ and $(1,1)$ are in T_2. On the other hand, when $Y^* = (0)$, $\sum_{i=1}^{k} w_i y_i^* = 0 \leq 8$ and $\sum_{i=1}^{k} w_i y_i^* + \sum_{i=k+2}^{n} w_i = 6 \leq 8$. Hence, $(0,1)$ is in T_2, and we conclude that $T_2 = \{(1,0),(1,1),(0,1)\}$.

Continuing in this way, we successively produce

$$T_3 = \{(1,0,1),(1,1,0),(0,1,1)\}$$

$$T_4 = \{(1,0,1,1),(1,0,1,0),(1,1,0,0),(0,1,1,1)\}$$

$$T_5 = \{(1,0,1,1,0),(1,0,1,0,1),(1,1,0,0,0),(0,1,1,1,0)\}.$$

The set T_5 contains the complete list of minimal true points of f.   □

In the next statement, the term *(arithmetic) operations* denotes elementary operations, such as additions, subtractions, multiplications, comparisons, performed on numbers of size polynomially bounded in the size of the input.

Theorem 9.21. *Procedure* MINTRUE *is correct and can be implemented to perform $O(nm)$ arithmetic operations, where n is the number of variables, and m is the number of minimal true points of the input function.*

Proof. Theorem 9.20 implies that MINTRUE is correct.

To establish the complexity bound, it is useful to picture a binary tree $T(f)$ of height n, whose root is the "empty" point $()$, and whose vertices at height k are the elements of T_k ($k = 1, 2, \ldots, n$). The parent of vertex $(y_1^*, y_2^*, \ldots, y_{k+1}^*) \in T_{k+1}$ is vertex $(y_1^*, y_2^*, \ldots, y_k^*) \in T_k$. Note that, since the tree $T(f)$ has m leaves, it has $O(nm)$ vertices.

For an efficient implementation of MINTRUE, we do not explicitly record the components of vertex $(y_1^*, y_2^*, \ldots, y_k^*) \in T_k$, but only the quadruplet of labels $(k, y_k^*, \sum_{i=1}^{k} w_i y_i^*, \sum_{i=k+2}^{n} w_i)$. The root is labeled by the quadruplet $(0, *, 0, \sum_{i=2}^{n} w_i)$.

Now, the procedure MINTRUE builds $T(f)$ recursively, visiting every vertex of $T(f)$ exactly once in the process. Note that, for each element Y^* of T_k, testing the conditions in Theorem 9.20 and computing the labels associated with the children of Y^* requires a constant number of operations. Hence, the labels associated with T_{k+1} can be generated from those associated with T_k in time $O(|T_k|)$, and this implies that all minimal true points can be listed in total time $O(\sum_{k=1}^{n} |T_k|) = O(nm)$.   □

Procedure MINTRUE is a version of a procedure described by Hammer and Rudeanu [460]. For related work, see, for instance, Granot and Hammer [410]; Bradley, Hammer and Wolsey [148]; Lawler, Lenstra, and Rinnooy Kan [605], and so on.

For the sake of simplicity, we have described MINTRUE as a breadth-first search traversal of $T(f)$, but it should be obvious that it can also be implemented as a depth-first search procedure, possibly allowing reduction of storage requirements (details are left to the reader).

Procedure MINTRUE runs in polynomial total time, since its complexity is polynomially bounded in the size of its input and of its output (see Appendix B). In fact, the minimal true points can even be generated with polynomial delay if MINTRUE is implemented as a depth-first search procedure. In the worst case, however, the number of minimal true points to be generated could be exponentially large in the encoding size of the input structure.

Example 9.14. *For $n \geq 1$, consider the structure $(w_1, w_2, \ldots, w_n, t) = (1, 1, \ldots, 1, \lfloor \frac{n}{2} \rfloor)$ and the corresponding threshold function f_n. Then, the encoding size of the structure is $O(n)$. But f_n has $m(n) = \binom{n}{\lfloor \frac{n}{2} \rfloor}$ minimal true points, and $m(n)$ is not bounded by any polynomial in n.* $\square$

Application 9.8. (Integer programming.) *Consider the knapsack constraints (see Application 9.5)*

$$\sum_{i=1}^{n} w_i x_i \leq t \tag{9.15}$$

$$(x_1, x_2, \ldots, x_n) \in \mathcal{B}^n \tag{9.16}$$

and their continuous relaxation

$$\sum_{i=1}^{n} w_i x_i \leq t \tag{9.17}$$

$$0 \leq x_i \leq 1 \ (i = 1, 2, \ldots, n), \tag{9.18}$$

where we assume that $w_i \geq 0$ for $i = 1, 2, \ldots, n$.

The 0–1 solutions of (9.15)–(9.16) are the false points of a positive threshold function $f(x_1, x_2, \ldots, x_n)$, with structure $(w_1, w_2, \ldots, w_n, t)$. We call the convex hull of these 0–1 points a threshold (or knapsack) polyhedron. From Section 1.13.6, we know that, if $C_k = \bigwedge_{j \in P(k)} x_j$, $k = 1, 2, \ldots, m$, denote the prime implicants of f, then each of the inequalities

$$\sum_{i \in P(k)} x_i \leq |P(k)| - 1 \ (k = 1, 2, \ldots, m) \tag{9.19}$$

defines a valid inequality for the corresponding threshold polyhedron, meaning that every point in the threshold polyhedron satisfies the inequalities (9.19). Moreover, the solution set of the system (9.17)–(9.19) is, in general, strictly smaller than the solution set of (9.17)–(9.18). As a consequence, inequalities of the form (9.19) have been successfully used in cutting-plane algorithms for the solution of large-scale 0–1 linear programming problems. Each constraint of such a problem is then considered individually in order to generate the corresponding inequalities (9.19).

The investigation of the relationship between the facets of threshold polyhedra and the prime implicant inequalities (9.19) was initiated by Balas [42], Hammer,

Johnson, and Peled [444], and Wolsey [922], and further developed in numerous publications (see, e.g., Balas and Zemel [46]; Weismantel [904]; Zemel [935], etc.). Their practical use in 0–1 programming was first convincingly demonstrated by Crowder, Johnson, and Padberg [245]. We refer the reader to Nemhauser and Wolsey [707] or Wolsey [924] for more information on this topic.

We also note that, more recently, a number of researchers have examined efficient procedures to translate knapsack systems of the form (9.15)–(9.16) into equivalent Boolean DNF equations, possibly involving additional variables. This line of research, in the spirit of Chapter 2, Section 2.3, opens the possibility of relying on purely Boolean techniques (such as satisfiability solvers) to handle 0–1 linear optimization problems; see, for instance, Bailleux, Boufkhad, and Roussel [41]; Eén and Sörensson [290]; or Manquinho and Roussel [667].   □

Application 9.9. (Integer programming.) *In certain applications, it may be advantageous to substitute an initial separating structure by one with smaller weights and/or threshold value, but which defines the same threshold function. This type of transformation gives rise to a variety of coefficient reduction problems.*

To illustrate, consider for instance the following system of inequalities, defining the continuous relaxation of a particular knapsack problem:

$$10x_1 + 8x_2 + 7x_3 + 6x_4 \leq 22 \qquad (9.20)$$

$$0 \leq x_j \leq 1 \ (j = 1,2,3,4). \qquad (9.21)$$

It is easily seen that the inequality (9.20) has exactly the same 0–1 solutions as

$$2x_1 + x_2 + x_3 + x_4 \leq 3 \qquad (9.22)$$

(that is, both inequalities define the same threshold function $f(x_1,x_2,x_3,x_4) = x_1x_2x_3 \vee x_1x_2x_4 \vee x_1x_3x_4$), but some fractional solutions of (9.20)–(9.21) are cut off by the inequality (9.22): For instance, $(x_1^*,x_2^*,x_3^*,x_4^*) = (\frac{2}{3},\frac{2}{3},\frac{2}{3},\frac{2}{3})$ satisfies (9.20)–(9.21) but violates (9.22).

Even though it is not true that a reduction of the coefficient sizes always implies a strengthening of the inequality, this is, nevertheless, often the case. Coefficient reduction is therefore of interest in branch-and-bound and cutting-plane algorithms for 0–1 linear programming problems (see Bradley, Hammer, and Wolsey [148]; Nemhauser and Wolsey [707]; Williams [910], etc.). Similar issues also arise in electrical engineering; see, for instance, Muroga [698].

A possible approach to coefficient reduction problems goes as follows: Given the initial separating structure $(w_1,w_2,\ldots,w_n,t)$, generate the maximal false points $X^1,X^2,\ldots,X^p$ and the minimal true points $Y^1,Y^2,\ldots,Y^m$ of the corresponding threshold function f (as usual, we assume that f is positive). Then, in view of Theorem 9.13, a "reduced" structure of f can be found by solving the optimization *problem*

$$minimize\ g(w_1, w_2, \ldots, w_n, t) \qquad\qquad (9.23)$$

$$subject\ to\ \sum_{i=1}^{n} w_i\, x_i^j \leq t \quad (j = 1, 2, \ldots, p) \qquad\qquad (9.24)$$

$$\sum_{i=1}^{n} w_i\, y_i^j \geq t + 1 \ (j = 1, 2, \ldots, m) \qquad\qquad (9.25)$$

$$w_i \geq 0 \qquad (i = 1, 2, \ldots, n), \qquad\qquad (9.26)$$

where $g(w_1, w_2, \ldots, w_n, t)$ could be any of a variety of objective functions, such as: $\sum_{i=1}^{n} w_i$, *or* t, *or* $\sum_{i=1}^{n} w_i + t$, *or* $\max(w_1, w_2, \ldots, w_n)$, *and so on.*

Note, however, that the optimal solution of (9.23)–(9.26) depends on the choice of the objective function g. This is related to the fact that, in general, the solution set of (9.24)–(9.26) has no componentwise minimum element. □

Example 9.15. *The separating structures* $(W^*, t^*) = (13, 7, 6, 6, 4, 4, 4, 3, 2, 24)$ *and* $(V^*, t^*) = (13, 7, 6, 6, 4, 4, 4, 2, 3, 24)$ *define the same threshold function f. But no solution of the system (9.24)–(9.26) associated with f is componentwise smaller than both* (W^*, t^*) *and* (V^*, t^*) *(see Exercise 7 at the end of this chapter).* □

9.6 Chow parameters of threshold functions

We have had several opportunities to discuss the concept of Chow parameters (see, e.g., Sections 1.6, 1.13, and 8.2). Historically, the motivation to introduce this concept stemmed from the observation that the Chow parameters of threshold functions display numerous remarkable properties; Dertouzos [269], Dubey and Shapley [279], and Winder [920] present a wealth of information about this early stream of research.

Recall Definition 1.14 from Section 1.6.

Definition 9.5. *The* Chow parameters *of a Boolean function f on* $\mathcal{B}^n$ *are the* $n + 1$ *integers* $(\omega_1(f), \omega_2(f), \ldots, \omega_n(f), \omega(f))$, *where* $\omega(f)$ *is the number of true points of f and* $\omega_i(f)$ *is the number of true points* Y^* *of f such that* $y_i^* = 1$

$$\omega_i(f) = |\{Y^* \in \mathcal{B}^n \mid f(Y^*) = 1 \text{ and } y_i^* = 1\}|, \ i = 1, 2, \ldots, n.$$

When no confusion can arise, we sometimes drop the symbol f from the notation $\omega_i(f)$ or $\omega(f)$. Note that $(\omega_1, \omega_2, \ldots, \omega_n) = \sum_{j=1}^{\omega} Y^j$, where $Y^1, Y^2, \ldots, Y^\omega$ are the true points of f. We should also mention that many variants of Definition 9.5 have been used in the literature (see, e.g., [920] and Section 9.6.2). These variants give rise to different scalings of the Chow parameters, while preserving their main features.

9.6.1 Chow functions

Definition 9.6. *A Boolean function f is a* Chow function *if no other function has the same Chow parameters as f.*

Example 9.16. *The function $f = x_1 x_2 \vee \overline{x_1}\,\overline{x_2}$ is not a Chow function, since it has the same Chow parameters as $g = x_1\overline{x_2} \vee \overline{x_1}x_2$, namely, $(\omega_1, \omega_2, \omega) = (1,1,2)$.* $\square$

With Definition 9.6 at hand, we are now ready to state Chow's fundamental result (see Chow [194]; Muroga [698] also credits Tannenbaum [857] for this result).

Theorem 9.22. *Every threshold function is a Chow function.*

Proof. Consider a threshold function f on $\mathcal{B}^n$, and a function g on $\mathcal{B}^n$ having the same Chow parameters as f. We must show that $f = g$.

Let us denote by $Y^1, Y^2, \ldots, Y^\omega$ the true points of f, and by $X^1, X^2, \ldots, X^k$, $Y^{k+1}, Y^{k+2}, \ldots, Y^\omega$ the true points of g, where $\omega = \omega(f) = \omega(g)$, $0 \le k \le \omega$, and $X^1, X^2, \ldots, X^k$ are false points of f. Since f and g have the same Chow parameters,

$$\sum_{j=1}^{\omega} Y^j = \sum_{j=1}^{k} X^j + \sum_{j=k+1}^{\omega} Y^j,$$

or, equivalently,

$$\sum_{j=1}^{k} Y^j = \sum_{j=1}^{k} X^j. \tag{9.27}$$

Now, if $k \ge 1$, then (9.27) contradicts the fact that f is asummable. Hence, we conclude that $k = 0$, meaning that f and g have the same true points, and that $f = g$. $\square$

This result shows that every threshold function is uniquely identified by its Chow parameters. Chow parameters have therefore been used as convenient identifiers for cataloging threshold functions; see Muroga [698] for a table of threshold functions up to five variables; Muroga, Toda, and Kondo [701] or Winder [917] for functions of six variables; Winder [919] for functions of seven variables; and Muroga, Tsuboi, and Baugh [702] (cited in [698]) for functions of eight variables.

Observe that all points occurring in equation (9.27) are distinct. This motivates the introduction of yet another concept.

Definition 9.7. *A Boolean function is* weakly asummable *if, for all $k \ge 1$, there do not exist k distinct false points of f, say, $X^1, X^2, \ldots, X^k$, and k distinct true points of f, say, $Y^1, Y^2, \ldots, Y^k$, such that $\sum_{i=1}^{k} X^i = \sum_{i=1}^{k} Y^i$.*

Clearly, every asummable (that is, threshold) function is weakly asummable. Moreover, the proof of Theorem 9.22 actually establishes that every weakly asummable function is a Chow function. Yajima and Ibaraki [929] (see also Winder [920]) proved that the converse implication holds as well, namely:

$$(k > 2)$$
$$\text{threshold} \quad \Rightarrow \quad k\text{-asummable} \quad \Rightarrow \text{2-asummable}$$
$$\Updownarrow \qquad\qquad\qquad\qquad\qquad\qquad \Updownarrow \qquad \Rightarrow k\text{-monotone}$$
$$\text{asummable} \Rightarrow \text{weakly assumable} \Rightarrow \text{completely} \qquad (k \geq 1)$$
$$\Updownarrow \qquad\qquad\qquad \text{monotone}$$
$$\text{Chow}$$

Figure 9.5. A hierarchy of Boolean functions: Enlarged version.

Theorem 9.23. *A Boolean function is weakly asummable if and only if it is a Chow function.*

Proof. We only have to show that, if a Boolean function is not weakly asummable, then it is not a Chow function. Let $X^1, X^2, \ldots, X^q$ denote the false points of a function f, and let $Y^1, Y^2, \ldots, Y^p$ denote its true points. If f is not weakly asummable, then we can assume, without loss of generality, that $\sum_{i=1}^{k} X^i = \sum_{i=1}^{k} Y^i$ for some $k \geq 1$. Let g be the Boolean function whose true points are exactly $X^1, X^2, \ldots, X^k, Y^{k+1}, Y^{k+2}, \ldots, Y^p$. The functions f and g are distinct, but they have the same Chow parameters. Hence, f is not a Chow function. $\square$

There exist Chow functions that are not threshold, and the function constructed in the proof of Theorem 9.15 is completely monotone but not a Chow function. On the other hand, Yajima and Ibaraki [929] showed that Chow functions are completely monotone (we leave the proof of this assertion as an end-of-chapter exercise). Thus, we obtain the hierarchy displayed in Figure 9.5 (compare with Figure 9.2 in Section 9.3).

Chow's theorem has been more recently revisited by O'Donnell and Servedio [717], who established a "robust" generalization of it: Namely, they proved that if f is a threshold function, if g is an arbitrary function, and if $(\omega_1(f), \omega_2(f), \ldots, \omega_n(f), \omega(f))$ is "close" to $(\omega_1(g), \omega_2(g), \ldots, \omega_n(g), \omega(g))$ in some appropriate norm, then the functions f and g are also "close" in the norm $\sum_{X \in \mathcal{B}^n} |f(X) - g(X)|$ (if we replace "close" by "equal" in this statement, then we obtain exactly Theorem 9.22). Based on this result, O'Donnell and Servedio proposed a fast algorithmic version of Chow's theorem, which allows them to efficiently construct an approximate representation of a threshold function given its Chow parameters (an extension of this problem is mentioned in Application 9.10 hereunder). We refer to [717] for details and applications in learning theory; see also Matulef et al. [677] for additional far-reaching extensions of Chow's theorem.

9.6.2 Chow parameters and separating structures

It is natural to expect some sort of relationship between the Chow parameters of a threshold function and the separating structures defining the function, since both types of coefficients somehow provide a "measure" of the "influence" of each variable on the function (remember the discussion of power indices in Section 1.13.3). This relationship is most naturally expressed in terms of the so-called *modified Chow parameters* of the function, which were introduced in Section 1.13.3.

Definition 9.8. *The* modified Chow parameters *of a Boolean function* $f(x_1, x_2, \ldots, x_n)$ *are the* $(n + 1)$ *numbers* $(\pi_1, \pi_2, \ldots, \pi_n, \pi)$, *defined as* $\pi = \omega - 2^{n-1}$ *and* $\pi_k = 2\omega_k - \omega$ *for* $k = 1, 2, \ldots, n$, *where* $(\omega_1, \omega_2, \ldots, \omega_n, \omega)$ *are the Chow parameters of* f.

Since there is a bijective correspondence between Chow parameters and modified Chow parameters, Theorem 9.22 implies that every threshold function is uniquely determined by its modified Chow parameters, or by its Chow parameters, or by any of its separating structures.

The following statements display formal analogy with Theorems 9.6, 9.7, and 9.9, but they hold for *arbitrary* (not necessarily threshold) functions:

Theorem 9.24. *If* $f(x_1, x_2, \ldots, x_n)$ *is a Boolean function with modified Chow parameters* $(\pi_1, \pi_2, \ldots, \pi_n, \pi)$, *then, for all* $i, j \in \{1, 2, \ldots, n\}$,

(1) *if* f *is positive in* x_i *and* f *depends on* x_i, *then* $\pi_i > 0$;
(2) *if* f *is negative in* x_i *and* f *depends on* x_i, *then* $\pi_i < 0$;
(3) *if* f *does not depend on* x_i, *then* $\pi_i = 0$;
(4) *if* $x_i \succ_f x_j$, *then* $\pi_i > \pi_j$;
(5) *if* $x_i \approx_f x_j$, *then* $\pi_i = \pi_j$;
(6) *the modified Chow parameters of* f^d *are* $(\pi_1, \pi_2, \ldots, \pi_n, -\pi)$;
(7) *if* $f^d \le f$, *then* $\pi \ge 0$;
(8) *if* $f^d \ge f$, *then* $\pi \le 0$.

Proof. Let $(\omega_1, \omega_2, \ldots, \omega_n, \omega)$ denote the Chow parameters of f. Fix $i \in \{1, 2, \ldots, n\}$, and let $A = \{X \in \mathcal{B}^n : f(X) = 1, x_i = 1\}$, $B = \{X \in \mathcal{B}^n : f(X) = 1, x_i = 0\}$. So, $|A| = \omega_i$ and $|B| = \omega - \omega_i$. If f is positive in x_i, then the mapping $m(X) = X \vee e_i$ is one-to-one on B, and $m(B) \subseteq A$. Hence, $|B| \le |A|$ and $\pi_i \ge 0$. Moreover, if f depends on x_i, then $|B| < |A|$, and hence $\pi_i > 0$. This establishes assertion (1); assertions (2) and (3) are proved in a similar way.

Assertions (4) and (5) are a restatement of Theorem 8.4 in Section 8.2.

By definition of duality, $f^d(X) = 1$ if and only if $f(\overline{X}) = 0$. It follows directly that $\omega(f^d) = 2^n - \omega$, and hence, that $\pi(f^d) = \omega(f^d) - 2^{n-1} = 2^{n-1} - \omega = -\pi$. Similarly, for $i = 1, 2, \ldots, n$,

$$\omega_i(f^d) = |\{X : f^d(X) = 1, x_i = 1\}|$$
$$= |\{X : f(X) = 0, x_i = 0\}|$$
$$= |\{X : x_i = 0\}| - |\{X : f(X) = 1, x_i = 0\}|$$
$$= 2^{n-1} - (\omega - \omega_i),$$

and hence,

$$\pi_i(f^d) = 2\omega_i(f^d) - \omega(f^d)$$
$$= 2(2^{n-1} - \omega + \omega_i) - (2^n - \omega)$$
$$= 2\omega_i - \omega$$
$$= \pi_i.$$

This proves assertion (6). As for (7), observe that $f^d \leq f$ implies $\omega(f^d) \leq \omega$, and hence, $\pi \geq 0$. A similar reasoning yields (8). □

In the case of threshold functions, how much further does the analogy go between weights and modified Chow parameters? First, it can be informally stated that, for a threshold function with separating structure $(w_1, w_2, \ldots, w_n, t)$, the vectors $(w_1, w_2, \ldots, w_n)$ and $(\pi_1, \pi_2, \ldots, \pi_n)$ often turn out to be "roughly" proportional. This, in spite of the fact that, as the following example shows, proportionality can become quite rough when the separating structure is picked arbitrarily.

Example 9.17. *The threshold function with structure* $(w_1, w_2, w_3, t) = (1, 1, 1, 1)$ *has modified Chow parameters* $(\pi_1, \pi_2, \pi_3) = (2, 2, 2)$, *so that* (w_1, w_2, w_3) *is exactly proportional to* (π_1, π_2, π_3). *But* $(50, 50, 1, 50)$ *and* $(50, 33, 18, 50)$ *are two other structures of the same function, for which proportionality with* (π_1, π_2, π_3) *is much more approximative!* □

Based on the previous example, one may be tempted to go one step further and to conjecture that every threshold function admits a separating structure whose weights are proportional to the modified Chow parameters $\pi_1, \pi_2 \ldots \pi_n$ of the function. Or, in other words, that every such function has a structure of the form $(\pi_1, \pi_2, \ldots, \pi_n, t)$, for some suitable choice of t. This conjecture is easily disproved, however.

Example 9.18. *The function* $f(x_1, x_2, x_3, x_4, x_5) = x_1 x_2 \vee x_1 x_3 x_4 \vee x_1 x_3 x_5 \vee x_2 x_3 x_4 x_5$ *is a threshold function with separating structure* $(4, 3, 2, 1, 1, 6)$, *and its modified Chow parameters are* $(\pi_1, \pi_2, \pi_3, \pi_4, \pi) = (10, 6, 4, 2, 2, -4)$. *But this function has no structure of the form* $(10, 6, 4, 2, 2, t)$, *for any t (otherwise, $14 \leq t$, since $(1, 0, 1, 0, 0)$ is a false point of f, and $t < 14$, since $(0, 1, 1, 1, 1)$ is a true point of f). A similar reasoning also shows that f has no structure of the form* $(\omega_1, \omega_2, \omega_3, \omega_4, \omega_5, t) = (11, 9, 8, 7, 7, t)$. □

Notwithstanding this dispiriting news, Dubey and Shapley [279] observed that, in some sense, the vector of modified Chow parameters actually is proportional to the "average" of the vector of weights.

Theorem 9.25. *Let $w_1, w_2, \ldots, w_n$ be fixed nonnegative numbers and $W = \sum_{j=1}^n w_j$, let t be a random variable uniformly distributed on $[0, W]$, and let $f(x_1, x_2, \ldots, x_n)$ be the (random) threshold function with structure*

Table 9.1. Modified Chow parameters for Example 9.19

t	π_1	π_2	π_3	π_4	π_5
0, 14	1	1	1	1	1
1, 13	2	2	2	2	0
2, 12	3	3	3	1	1
3, 11	5	5	3	1	1
4, 10	7	5	3	3	1
5, 9	8	6	4	4	2
6, 8	9	7	5	3	1
7	10	6	6	2	2
Total	80	64	48	32	16

$(w_1, w_2, \ldots, w_n, t)$. Then, for $i = 1, 2, \ldots, n$, the expected value of the (random) modified Chow parameter π_i is equal to $2^{n-1} w_i / W$. □

Proof. Fix $i \in \{1, 2, \ldots, n\}$. By Theorem 1.37 in Section 1.13.3, the expected value of π_i is nothing but the expected number of swings of f for i. Now, if $X^* \in \mathcal{B}^n$ and $x_i^* = 0$, then,

$$\mathrm{Prob}(X^* \text{ is a swing of } f \text{ for } i) = \mathrm{Prob}\left(\sum_{j=1}^n w_j x_j^* \le t < \sum_{j=1}^n w_j x_j^* + w_i\right)$$

$$= \frac{w_i}{W} \text{ (since } t \text{ is uniformly distributed).}$$

Hence,

$$E[\pi_i] = \sum_{\{X \in \mathcal{B}^n : x_i = 0\}} \frac{w_i}{W} = 2^{n-1} \frac{w_i}{W}.$$

The same result holds, with the same proof, if the weights $w_1, w_2, \ldots, w_n$ are assumed to be nonnegative integers and if t is uniformly distributed on $\{0, 1, \ldots, W - 1\}$. Dubey and Shapley [279] illustrate this point with the following example.

Example 9.19. *Let* $(w_1, w_2, w_3, w_4, w_5) = (5, 4, 3, 2, 1)$ *and consider all threshold functions with separating structures* $(w_1, w_2, w_3, w_4, w_5, t)$, *where* t *can take any value in the set* $\{0, 1, \ldots, 14\}$. *The modified Chow parameters of these 15 functions are displayed in Table 9.1. The average value of* π_1 *for this set of functions is* $\frac{80}{15} = 2^{n-1} \frac{w_1}{W}$. *Note, however, that there is no single choice of the threshold* t *for which the vector of modified Chow parameters is exactly proportional to* $(w_1, w_2, \ldots, w_n)$. □

Additional theoretical results describing the relation between weights and Chow parameters of a threshold function, as well as algorithms allowing us to reconstruct a threshold function from the (approximate) knowledge of its Chow parameters, can be found in Alon and Edelman [17]; Aziz, Paterson, and Leech [39]; O'Donnell and Servedio [717], and so on.

We conclude this section with a discussion of interesting related issues arising in political science.

Application 9.10. (Political science, game theory.) *Picture a federation of states administered by a legislature in which each state, independently of its size, is represented by exactly one legislator. The legislature makes its decisions according to a weighted majority voting scheme, whereby every legislator carries a (possibly different) number of votes. In order to embody the "one man one vote" principle in the functioning of this legislature, the U.S. Supreme Court has ruled that "the voting power detained by each legislator ought to be proportional to the size of the constituency that he or she represents." The question is now: How is this principle to be put into practice?*

Apportionment problems of this nature, far from being only theoretical, actually arose in several U.S. elected bodies in the 1960s. They led John F. Banzhaf III [52] to propose the indices now bearing his name as adequate measures of voting power; his proposal was eventually adopted by several official bodies (see Dubey and Shapley [279], Felsenthal and Machover [329], or Lucas [631] for details on this story).

If we accept the principle that the Banzhaf index of a legislator can be equated with his share of voting power, then the Supreme Court decree can be mathematically reformulated as follows: Denote by $1, 2, \ldots, n$ the members of the legislature, and assume that member i represents a state of size s_i $(i = 1, 2, \ldots, n)$. Also, let w_i be the weight of legislator i in the voting system $(i = 1, 2, \ldots, n)$, and let $t + 1$ be the required number of votes for a resolution to pass in the legislature (we assume $w_1, w_2, \ldots, w_n$ and t to be integer). In other words, the weighted majority voting rule used by the legislature is described by the threshold function $f(x_1, x_2, \ldots, x_n)$ with structure $(w_1, w_2, \ldots, w_n, t)$.

Now, recall from Definition 1.34 in Section 1.13.3 that the ith (normalized) Banzhaf index of f is the quantity $\beta_i = \frac{\pi_i}{\sum_j \pi_j}$, where π_i is the ith modified Chow parameter of f $(i = 1, 2, \ldots, n)$.

Putting these facts together, we come to the conclusion that, according to the Supreme Court's interpretation of the "one man, one vote" principle, the weights $(w_1, w_2, \ldots, w_n)$ and the threshold t should be chosen in such a way that the vector $(\beta_1, \beta_2, \ldots, \beta_n)$ of Banzhaf indices be equal to the vector $\frac{1}{S}(s_1, s_2, \ldots, s_n)$ of relative population sizes, where $S = \sum_{i=1}^{n} s_i$ is the total population size.

As Example 9.17 shows, it is generally not sufficient to let $(w_1, w_2, \ldots, w_n) = (s_1, s_2, \ldots, s_n)$ to abide by the Supreme Court decree; see also Application 9.11 hereunder and the computations relative to the distribution of power in the European Union Council reported in Algaba, Bilbao, Fernández Garcia, and López [15]; Bilbao [79]; Bilbao, Fernández, Jiménez Losada, and López [80];

Bilbao, Fernández, Jiménez, and López [81]; Laruelle and Widgrén [600]; Leech [607], and so on.

Even more interestingly, the above mathematical model makes it very clear that the one man, one vote principle, as embodied in the Court decree and further interpreted in terms of Banzhaf indices, cannot always be implemented in real-world situations. Indeed, for fixed n, the number of possible realizations of the vector $\frac{1}{S}(s_1, s_2, \ldots, s_n)$ is infinite, whereas the number of Banzhaf vectors $(\beta_1, \beta_2, \ldots, \beta_n)$ is obviously finite (since the number of threshold functions of n variables is finite). So, for most distributions of population sizes, there exists no allocation of weights $(w_1, w_2, \ldots, w_n)$ that implies a distribution of power equal to $\frac{1}{S}(s_1, s_2, \ldots, s_n)$. In such cases, the need arises again to give an operational meaning to the one man, one vote principle. How can this be achieved?

One (rather intriguing) possibility raised by Papayanopoulos [729] would be to assign exactly s_i votes to legislator i, and to let the threshold t vary randomly between 0 and S (namely, the threshold would be drawn randomly in $[0, S]$ whenever the legislature is to vote). By virtue of Theorem 9.25, this would provide legislator i with an expected share of power proportional to the size of his or her constituency. Unfortunately, even though this solution may sound quite attractive to a mathematically inclined political scientist, it is doubtful that it will be adopted by any real-world legislature in the foreseeable future!

Another, more realistic way out of the dilemma has been actually implemented by some county supervisorial boards in the State of New York. In these bodies, the one man, one vote principle has been translated as follows: The voting weights $w_1, w_2, \ldots, w_n$ and the threshold t should be specified in such a way that the Banzhaf vector $(\beta_1, \beta_2, \ldots, \beta_n)$ be "as close as possible" to the population distribution $\frac{1}{S}(s_1, s_2, \ldots, s_n)$, or, in other words, so as to minimize the distance (in some appropriate norm) between $(\beta_1, \beta_2, \ldots, \beta_n)$ and $\frac{1}{S}(s_1, s_2, \ldots, s_n)$. This interpretation of the one man, one vote principle gives rise to an interesting, but hard, combinatorial optimization problem; see Alon and Edelman [17]; Aziz, Paterson, and Leech [39]; Lucas [632]; McLean [639]; O'Donnell and Servedio [717]; Papayanopoulos [727, 728, 729]; Laruelle and Widgrén [600]; or Leech [607, 608] for more information and related applications. □

9.6.3 Computing the Chow parameters

We now turn our attention to the algorithmic problem of computing the Chow parameters of a threshold function. As might be expected, the complexity of this problem depends very much on the format of its input. For instance, since threshold functions are 2-monotone (Theorem 9.7), the results in Chapter 7 and Chapter 8 (in particular, Application 8.5 in Section 8.2) imply that the Chow parameters of a threshold function can be computed in polynomial time when the list of prime implicants of the function is available. On the other hand, the problem becomes more difficult when the input function is described by a separating structure.

Indeed, the following result due to Garey and Johnson [371], in conjunction with Theorem 9.24, shows that it is already NP-complete to decide whether a modified Chow parameter vanishes or not (compare with Theorem 1.32).

Theorem 9.26. *Deciding whether a threshold function depends on its last variable is NP-complete when the function is described by a separating structure.*

Proof. The problem is obviously in NP. Now, recall that the following SUBSET SUM problem is NP-complete [371]: Given $n + 1$ positive integers $(w_1, w_2, \ldots, w_n, t)$, is there a point $X^* \in \mathcal{B}^n$ such that $\sum_{j=1}^{n} w_j x_j^* = t$?

With an arbitrary instance $(w_1, w_2, \ldots, w_n, t)$ of SUBSET SUM, we associate the threshold function $f(x_1, x_2, \ldots, x_{n+1})$ with structure $(w_1, w_2, \ldots, w_n, \frac{1}{2}, t)$. It is clear that f depends on its last variable x_{n+1} if and only if $(w_1, w_2, \ldots, w_n, t)$ is a Yes instance of SUBSET SUM.   □

Prasad and Kelly [754] actually proved that, for a threshold function given by a separating structure, computing Banzhaf indices – or, equivalently, Chow parameters – is #P-complete; compare with Theorem 1.38. (A similar observation was already formulated by Garey and Johnson [371] for Shapley-Shubik indices; see also Deng and Papadimitriou [268] and Matsui and Matsui [676].)

As a remarkable illustration of the occurrence of "dummy" variables in weighted majority systems, we mention a well-known story among political scientists (see, e.g., [150, 329]).

Application 9.11. (Political science, game theory.) *In 1958, the European Economic Community had six member-states, namely, Belgium, France, Germany, Italy, Luxembourg, and the Netherlands. Its Council of Ministers relied on a weighted majority decision rule with voting weight 4 for France, Germany and Italy, weight 2 for Belgium and the Netherlands, and weight 1 for Luxembourg. The threshold was set to $t = 11$. With these rules, it is readily seen that Luxembourg actually had no voting power at all, since the outcome of the vote was always determined regardless of the decision made by Luxembourg.*   □

The previous story, as well as the apportionment problem described in Application 9.10, build a strong case for "practically efficient" procedures for the computation of Chow parameters. Theorem 9.27 describes a simple *dynamic programming* (i.e., recursive) algorithm for the computation of ω, the number of true points of f.

Theorem 9.27. *If $f(x_1, x_2, \ldots, x_n)$ is a threshold function given by the integral structure $(w_1, w_2, \ldots, w_n, t)$, then the number of true points of f can be computed in $O(nt)$ arithmetic operations.*

Proof. We assume for the sake of simplicity that $w_1, w_2, \ldots, w_n$ and t are positive (only minor adaptations are required in the general case). For $j = 0, 1, \ldots, n$ and

$s = 0, 1, \ldots, t$, define $p(j,s)$ to be the number of points $X^* \in \mathcal{B}^n$ such that $x^*_{j+1} = \ldots = x^*_n = 0$ and $\sum_{j=1}^{n} w_j x^*_j = s$. In particular, $p(n,s)$ is the number of points such that $\sum_{j=1}^{n} w_j x^*_j = s$, and hence,

$$\omega = 2^n - \sum_{s=0}^{t} p(n,s). \tag{9.28}$$

The numbers $p(j,s)$ satisfy the recursions

$$p(j,s) = p(j-1,s) \text{ for } j = 1,2,\ldots,n; s = 1,2,\ldots,w_j - 1, \tag{9.29}$$

and

$$p(j,s) = p(j-1,s) + p(j-1,s-w_j) \text{ for } j = 1,2,\ldots,n; s = w_j, w_j+1,\ldots,t. \tag{9.30}$$

Indeed, when $s < w_j$, then $x^*_j = 0$ in all solutions X^* of $\sum_{j=1}^{n} w_j x^*_j = s$; Equation (9.29) follows from this observation. On the other hand, when $s \geq w_j$, then x^*_j can be either 0 or 1, and this gives rise to the two terms in Equation (9.30).

Note also that the initial conditions $p(j,0) = 1$ for $j = 0,1,\ldots,n$, and $p(0,s) = 0$ for $s = 1,2,\ldots,t$, must hold.

Equations (9.29) and (9.30), together with these initial conditions, can be used to fill in the $(n+1) \times (t+1)$ matrix with elements $p(j,s)$. This only requires $O(nt)$ arithmetic operations, and the theorem follows from (9.28).   □

Since the complexity of the algorithm in Theorem 9.27 increases polynomially with the value of the threshold t, we conclude that the number of true points of a threshold function can be computed in *pseudo-polynomial* time (which is the next best thing to a genuine polynomial algorithm).

Suppose next that we want to compute all $n + 1$ Chow parameters $(\omega_1, \omega_2, \ldots, \omega_n, \omega)$ of a threshold function f. Observe that $\omega_1 = \omega(f_{|x_1=1})$, where $f_{|x_1=1}$ is the threshold function with structure $(w_2, w_3, \ldots, w_n, t - w_1)$. It follows that in order to compute $(\omega_1, \omega_2, \ldots, \omega_n, \omega)$, we only need to apply the previous algorithm to $f_{|x_1=1}, f_{|x_2=1}, \ldots, f_{|x_n=1}$ and f. Hence, all Chow parameters can be computed in $O(n^2 t)$ operations.

Dynamic programming algorithms similar to the algorithm described in Theorem 9.27 are classical tools for the solution of knapsack problems (see [671]). Such algorithms have been proposed for the computation of power indices by Lucas [631]; see Matsui and Matsui [676]. Uno [878] showed that the Banzhaf indices (or Chow parameters) of all players can actually be computed in $O(nt)$ operations by eliminating redundant operations. Klinz and Woeginger [574] describe a dynamic programming algorithm with complexity $O(n^2 1.415^n)$. See also Pesant and Quimper [740] or Trick [869] for related work in the context of *constraint programming*.

Pseudo-polynomial algorithms based on the consideration of *generating functions* have been proposed for the computation of Shapley-Shubik indices by Mann

and Shapley [661] (who credit Cantor), and for Banzhaf indices by Brams and
Affuso [150]; see also Algaba et al. [15]; Bilbao [79]; Bilbao et al. [80, 81];
Fernández, Algaba, Bilbao, Jiménez, Jiménez and López [330]; Leech [607];
Papayanopoulos [729] for related work, extensions, and applications in various
political settings.

Finally, we refer to Crama and Leruth [236]; Crama, Leruth, Renneboog, and
Urbain [237]; Cubbin and Leech [246]; Gambarelli [368]; Leech [606, 608, 609]
for different approaches to the computation of Banzhaf indices in the framework
of corporate finance applications.

9.7 Threshold graphs

In this section, we specialize some of the above results to the case of graphic (that
is, purely quadratic and positive) functions. Recall that such a function can be identified with an undirected graph. More precisely, if $f(x_1, x_2, \ldots, x_n) = \bigvee_{(i,j) \in E} x_i x_j$,
we denote by G_f the graph (V, E), where $V = \{1, 2, \ldots, n\}$. Conversely, if
$G = (V, E)$ is an arbitrary graph, we define the Boolean function f_G by the
expression $\bigvee_{(i,j) \in E} x_i x_j$.

Definition 9.9. *A graph G is a* threshold *graph if the Boolean function f_G is
threshold. We say that $(w_1, w_2, \ldots, w_n, t)$ is a* separating structure *of G if it is a
separating structure of f_G.*

Example 9.20. *The function $f(x_1, x_2, x_3, x_4) = x_1 x_2 \vee x_1 x_3 \vee x_1 x_4 \vee x_2 x_3$ is
graphic, and the associated graph G_f is shown in Figure 9.6. Since f is a threshold function, G_f is a threshold graph. A separating structure for G_f is for instance
$(3, 2, 2, 1, 3)$.* □

Threshold graphs were introduced by Chvátal and Hammer [201] and, independently, by Henderson and Zalcstein [487]. Most of this section is based
on [201].

The central question we want to address is: Which graphic functions are threshold, or, equivalently, which graphs are threshold? In view of the correspondence

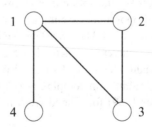

Figure 9.6. Graph G_f for Example 9.20.

between false points of f and stable sets of G_f (recall Application 1.13.5), we immediately obtain a first, trivial characterization.

Theorem 9.28. *A graph $G = (V, E)$ is a threshold graph if and only if there exists a structure $(w_1, w_2, \ldots, w_n, t)$ such that, for every subset S of vertices*

$$S \text{ is stable in } G \text{ if and only if } \sum_{i \in S} w_i \leq t.$$

Proof. This is a mere reformulation of Definition 9.9. $\qquad\square$

Recall that for a graph $G = (V, E)$ and a vertex $i \in V$, we denote by $N(i)$ the *neighborhhood* of i, that is, the set $N(i) = \{j \in V : (i, j) \in E\}$. We say that i is *isolated* if $N(i) = \emptyset$, and i is *dominating* if $N(i) = V \setminus \{i\}$. Note that isolated vertices of G correspond to inessential (dummy) variables of f_G.

Theorem 9.29. *For a graphic function $f(x_1, x_2, \ldots, x_n)$, the following statements are equivalent:*

(a) *f is a threshold function.*
(b) *f is a regular function.*
(c) *There is a permutation $(\sigma(1), \sigma(2) \ldots, \sigma(n))$ of $\{1, 2, \ldots, n\}$ such that, for every $i \in \{1, 2, \ldots, n\}$, $\sigma(i)$ is either isolated or dominating in the subgraph of G_f induced by $\{i, i+1, \ldots, n\}$.*

Proof. The implication (a) $\Longrightarrow$ (b) follows from Theorem 9.7.

We prove the implication (b) $\Longrightarrow$ (c) by induction on n. The implication certainly holds for $n = 1$. So, assume that $n > 1$, and let f be a regular graphic function with $x_1 \succeq_f x_2 \succeq_f \cdots \succeq_f x_n$. We first claim that either G_f has an isolated vertex, or vertex 1 is dominating in G_f.

Indeed, assume that G_f has no isolated vertex, and consider an arbitrary vertex j in $\{2, 3, \ldots, n\}$. Since j is not isolated, there exists a vertex $k \in N(j)$. If $k = 1$, then we are done. Otherwise, since $x_1 \succeq_f x_k$, Theorem 8.5 in Section 8.2 implies that $N(k) \setminus \{1\} \subseteq N(1) \setminus \{k\}$. Thus, $j \in N(1) \setminus \{k\}$, and the claim is proved.

Now, we can define $\sigma(1)$ as follows: If G_f has an isolated vertex, say i, then we let $\sigma(1) = i$; otherwise, we let $\sigma(1) = 1$.

Let H be the subgraph of G_f obtained by deleting $\sigma(1)$ from G_f. Thus, $H = G_g$, where g is the restriction of f obtained by fixing $x_{\sigma(1)}$ to 0. By Theorem 8.7, g is a regular graphic function, and the proof is complete by induction.

Finally, we use again induction on n to prove the implication (c) $\Longrightarrow$ (a). If $n = 1$, then the implication trivially holds. Otherwise, without loss of generality, assume that 1 is either an isolated or a dominating vertex of G_f. Let H be the induced subgraph of G_f obtained by deleting 1 from G_f. Then $H = G_g$, where g is the restriction of f to $x_1 = 0$.

Since condition (c) holds for H, the induction hypothesis implies that g is a threshold function. Let $(w_2, w_3, \ldots, w_n, t)$ be a separating structure for g. Using

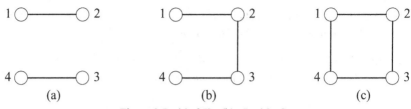

Figure 9.7. (a): $2K_2$, (b): P_4, (c): C_4.

Theorem 9.6 and the arguments in the proof of Theorem 9.4, we can assume that $w_i > 0$ for all $i = 2, 3, \ldots, n$, and that $t > 0$.

Now, it is easy to see that $(w_1, w_2, \ldots, w_n, t)$ is a structure for f, where $w_1 = 0$ if vertex 1 is isolated in G_f, and $w_1 = t$ if vertex 1 is dominating in G_f. Hence, f is a threshold function. $\qquad\square$

The graphs $2K_2$, P_4, and C_4 are represented in Figure 9.7.

Theorem 9.30. *A graph is a threshold graph if and only if it has no induced subgraph isomorphic to $2K_2$, P_4, or C_4.*

Proof. As a consequence of Theorem 9.3, every induced subgraph of a threshold graph is threshold. Moreover, it is easy to check that $2K_2$, P_4, and C_4 are not threshold. Therefore, a threshold graph cannot have any of these graphs as an induced subgraph.

To prove the converse statement, assume that G is not threshold. Hence, by Theorem 9.29, f_G is not regular. This means that there are two variables, say, x_i and x_j, that are not comparable in the strength preorder associated with f_G. From Theorem 8.5, it follows that $N(i) \setminus \{j\} \not\subseteq N(j) \setminus \{i\}$ and $N(j) \setminus \{i\} \not\subseteq N(i) \setminus \{j\}$. Let $k \in N(i) \setminus N(j)$, and let $\ell \in N(j) \setminus N(i)$, where i, j, k, ℓ are all distinct. Then, $\{i, j, k, \ell\}$ must induce $2K_2$ or P_4 or C_4. $\qquad\square$

Several characterizations of threshold graphs rely on the concept of degree sequence, which we define next.

Definition 9.10. *Let G be a graph on $\{1, 2, \ldots, n\}$, and let $d_i = deg(i)$ be the degree of vertex i, for $i = 1, 2, \ldots, n$. If $(\pi(1), \pi(2), \ldots, \pi(n))$ is a permutation of $\{1, 2, \ldots, n\}$ such that $d_{\pi(1)} \geq d_{\pi(2)} \geq \cdots \geq d_{\pi(n)}$, then we say that $d(G) = (d_{\pi(1)}, d_{\pi(2)}, \ldots, d_{\pi(n)})$ is the degree sequence of G. A degree sequence is called threshold if it is the degree sequence of at least one threshold graph.*

Theorem 9.31. *If G is a threshold graph and H is a graph such that $d(H) = d(G)$, then H is isomorphic to G.*

Proof. Consider a threshold graph G with degree sequence $d(G) = (d_1, d_2, \ldots, d_n)$, and let H be another graph with $d(H) = d(G)$. We assume, without loss of generality, that G and H have the same vertex–set, say, $\{1, 2, \ldots, n\}$, and that d_i is the degree of vertex i in both G and H, for $i = 1, 2, \ldots, n$. We now prove by induction on n that H is isomorphic to G.

If $n = 1$, this is trivial. If $n > 1$, then, by Theorem 9.29, there is a vertex $i \in \{1, 2, \ldots, n\}$ such that either $d_i = n - 1$ or $d_i = 0$. Recall that $G \setminus i$ is the subgraph obtained by deleting i from G, and let $d(G \setminus i) = \hat{d}$. Then, $\hat{d}$ is a threshold degree sequence and, by induction, $G \setminus i$ is (up to isomorphism) the unique graph with this degree sequence. In particular, since $d(H \setminus i) = \hat{d}$, we conclude that $G \setminus i$ is isomorphic to $H \setminus i$. From this, it easily follows that G is isomorphic to H. $\square$

Theorem 9.31 is closely related to Theorem 9.22: Indeed, the Chow parameters of a threshold graphic function can be explicitly expressed as a function of the corresponding degree sequence (see Exercise 19 at the end of the chapter).

An interesting corollary of this result is that all the information concerning the "thresholdness" of a graph is embodied in its degree sequence. In other words, it must be possible to decide whether a graph G is a threshold graph by simply examining its degree sequence. As a matter of fact, a careful reading of the proof of Theorem 9.31 indicates how to decide, in $O(n^2)$ operations, whether a sequence $(d_1, d_2, \ldots, d_n)$ of nonnegative integers is a threshold sequence. This result is not best possible: Threshold sequences and threshold graphs can be recognized in time $O(n)$; see Golumbic [398] or Mahadev and Peled [645] for details.

The foregoing observations can also be derived from an analytical characterization of threshold sequences due to Hammer, Ibaraki, and Simeone [442]. Before we state this result, we first recall a classical theorem of Erdős and Gallai [313] (see also [71]).

Theorem 9.32. *A sequence $(d_1, d_2, \ldots, d_n)$ with $d_1 \geq d_2 \geq \cdots \geq d_n \geq 0$ is a degree sequence if and only if $\sum_{i=1}^{n} d_i$ is even and, for $r = 1, 2, \ldots, n$,*

$$\sum_{i=1}^{r} d_i \leq r(r-1) + \sum_{i=r+1}^{n} \min\{r, d_i\}. \tag{9.31}$$

For $r = 1, 2, \ldots, n$, we call (9.31) the rth Erdős-Gallai inequality.

Example 9.21. *The sequence $d = (2, 2, 2, 2)$ is the degree sequence of the cycle C_4. For this sequence, (9.31) becomes*

$$2r \leq r(r-1) + r(4-r) \text{ when } r = 1, 2,$$

and

$$2r \leq r(r-1) + 2(4-r) \text{ when } r = 3, 4.$$

$\square$

Notice that, in the previous example, all of the Erdős-Gallai inequalities are satisfied as strict inequalities. This need not be the case in general.

Theorem 9.33. *The degree sequence* $d = (d_1, d_2, \ldots, d_n)$ *is a threshold sequence if and only if equality holds in the* r^{th} *Erdős-Gallai inequality associated with d, for all* $r \in \{1, 2, \ldots, n\}$ *such that* $r - 1 \leq d_r$.

We refer to the paper by Hammer, Ibaraki, and Simeone [442] for a proof of this result, and we simply illustrate it on a small example.

Example 9.22. *The degree sequence of the graph* G_f *described in Example 9.20 is* $(3, 2, 2, 1)$. *It is easy to check that* (9.31) *holds as equality for all* $r \in \{1, 2, 3, 4\}$ *such that* $r - 1 \leq d_r$, *that is, for* $r = 1, 2, 3$. $\qquad\qquad \square$

Many additional results on threshold graphs can be found in Golumbic [398] or Mahadev and Peled [645]. These books also describe applications of threshold graphs to integer programming, mathematical psychology, personnel scheduling and synchronization of parallel processes. We briefly discuss two of these applications.

Application 9.12. (Integer programming.) *A system of* set packing inequalities *is a system of the form*

$$\sum_{j=1}^{n} a_{kj} x_j \leq 1, \quad k = 1, 2, \ldots, m, \tag{9.32}$$

where $A = (a_{kj})$ *is an* $m \times n$ *matrix with 0–1 elements. As observed by Chvátal and Hammer [201], the aggregation problem (see Application 9.7) for set packing inequalities can be translated into the problem of recognizing threshold graphs.*
   Indeed, let us associate a graph $G(A) = (V, E)$ *with the system* (9.32), *where* $V = \{1, 2, \ldots, n\}$, *and* $E = \{ (i, j) : a_{ki} = a_{kj} = 1 \text{ for some } k \in \{1, 2, \ldots, m\} \}$. *It is easy to see that the system* (9.32) *has the same 0–1 solutions as the system*

$$x_i + x_j \leq 1, \quad (i, j) \in E. \tag{9.33}$$

   Now, the 0–1 solutions of (9.33) *are precisely the characteristic vectors of the stable sets of* $G(A)$. *Hence, by Theorem 9.28, there exists a single linear inequality having the same 0–1 solutions as* (9.32) *(or* (9.33)*) if and only if* $G(A)$ *is a threshold graph. In particular, this proves that the aggregation problem is polynomially solvable for set packing inequalities.*
   When (9.32) *is not equivalent to a single linear inequality, one can push the investigation a bit further and ask instead: What is the smallest integer* δ *for which there exists a system of* δ *linear inequalities*

$$\sum_{j=1}^{n} w_{kj} x_j \leq t_k, \quad k = 1, 2, \ldots, \delta, \tag{9.34}$$

such that (9.32) *and* (9.34) *have the same set of 0–1 solutions (Chvátal and Hammer [201], Neumaier [709])? We denote this value by* $\delta(A)$ *and call it the threshold dimension of A. Similarly, for a graph* $G = (V, E)$, *we can define the threshold dimension of G as* $\delta(G) := \delta(A(G))$, *where* $A(G)$ *is the coefficient matrix of the system (9.33) (notice that this definition is coherent in the sense that* $\delta(A) = \delta(G(A))$ *for every 0–1 matrix A).*

The threshold dimension has an interesting graph-theoretic interpretation. Of course, a graph $G = (V,E)$ is threshold if and only if $\delta(G) = 1$. But, more generally, it can also be shown that $\delta(G)$ is the smallest δ for which there exist δ threshold graphs $G_k = (V,E_k)$ $(k = 1,2,\ldots,\delta)$ satisfying $E_1 \cup E_2 \cup \cdots \cup E_\delta = E$ (this is left as an exercise to the reader).

Chvátal and Hammer [201] proved that computing $\delta(G)$ is NP-hard. Yannakakis [933] refined this result by proving that, for every fixed $k \geq 3$, it is NP-complete to decide whether $\delta(G) \leq k$. For many years, and in spite of a flurry of research on this topic, it remained unknown whether testing $\delta(G) \leq 2$ was NP-hard or not. Finally, the question was settled by Ma [641] who provided a polynomial-time algorithm for the recognition of graphs with threshold dimension 2. Other polynomial algorithms for this problem were later proposed by Raschle and Simon [779] and Sterbini and Raschle [847]. We refer the reader to the original papers or to Mahadev and Peled [645] for additional information.   □

Application 9.13. (Mathematical psychology, social choice.) *Let S denote a set of individuals, and let P denote a set of propositions, and assume that each individual declares either to "agree" or to "disagree" with each proposition. For instance, the individuals may be citizens and the propositions may be items in an opinion poll, or the individuals may be college students and the propositions may be math problems that the students can either solve or not, and so forth.*

We would like to map all individuals and all propositions to a common linear scale (e.g., from "left" to "right" or from "hard" to "easy") in such a way that an individual agrees with all propositions following it and disagrees with all propositions preceding him on the scale. Such a scale is called a Guttman *scale. More precisely, assume that the data of the problem are described by the bipartite graph* $H = (V,A)$, *where* $V = S \cup P$ *and*

$$A = \{(s,p) \in S \times P : s \text{ agrees with } p \}.$$

A Guttman scale for H is a mapping g from V to $\mathbb{R}$ *such that, for each* $s \in S$ *and* $p \in P$, $(s,p) \in A$ *if and only if* $g(s) < g(p)$.

Of course, not every bipartite graph admits a Guttman scale. To obtain a full characterization, consider the graph $G = (V,E)$, *where*

$$E = \{(s,t) \in S \times S : s \neq t\} \cup A.$$

The following result is due to Cozzens and Leibowitz [223].

Theorem 9.34. *The graph H has a Guttman scale if and only if G is a threshold graph.*

Proof. If G is a threshold graph, then it has a separating structure with threshold t, with weight $w(s)$ for vertex $s \in S$ and with weight $a(p)$ for vertex $p \in P$. We can now construct a Guttman scale g as follows. For $p \in P$, let $g(p) = a(p)$. For $s \in S$, consider the largest value $a(p^*)$ such that $w(s) + a(p^*) \leq t$, and define $g(s) = a(p^*)$. It is easy to check that g is a valid Guttman scale.

Conversely, if G is not a threshold graph, then by Theorem 9.30 it has four vertices, say, 1, 2, 3, 4, such that $(1,2) \in E$, $(3,4) \in E$, $(1,3) \notin E$ and $(2,4) \notin E$ (cf. Figure 9.7). We can assume, without loss of generality, that vertices 1 and 4 are in P, and vertices 2 and 3 are in S (indeed, 1 and 3 are not both in S, since they are not linked; hence, we can assume that $1 \in P$; then, $2 \in S$, etc.). Then, if g is a Guttman scale, there holds

$g(4) \leq g(2) < g(1)$, since 2 agrees with 1 but 2 does not agree with 4,

$g(1) \leq g(3) < g(4)$, since 3 agrees with 4 but 3 does not agree with 1,

and we reach a contradiction. □

We refer to the paper by Cozzens and Leibowitz [223] for additional information on the connections between Guttman scales and threshold graphs. □

Other connections between threshold functions and graph properties have been explored in several papers. For instance, Benzaken and Hammer [67] characterized *domishold* graphs: A graph $G = (V, E)$ is domishold if there exists a structure $(w_1, w_2, \ldots, w_n, t)$ such that, for every $S \subseteq V$,

$$S \text{ is dominating in } G \text{ if and only if } \sum_{i \in S} w_i \leq t.$$

Hammer, Maffray, and Queyranne [451] investigated *cut-threshold* graphs in which subsets of edges (or vertices) corresponding to cuts are characterized by a similar threshold-type property. We refer again to Mahadev and Peled [645] for more information on such graph classes.

9.8 Exercises

1. A Boolean function $f(x_1, x_2, \ldots, x_n)$ is a *ball* if there exist $(w_1, w_2, \ldots, w_n, t) \in \mathbb{R}^n$ such that, for all $(x_1, x_2, \ldots, x_n) \in \mathcal{B}^n$,

$$f(x_1, x_2, \ldots, x_n) = 0 \text{ if and only if } \sum_{i=1}^{n} (w_i - x_i)^2 \leq r^2.$$

   Prove that a function is a ball if and only if it is a threshold function (Hegedűs and Megiddo [481]).

2. (a) Show that, if a Boolean function can be represented by a DNF of degree k, then it is a polynomial threshold function of degree k in the sense of Section 9.1.

   (b) Show that the parity function $f(x_1, x_2, \ldots, x_n) = x_1 \oplus x_2 \oplus \ldots \oplus x_n$ is not a polynomial threshold function of degree k for any $k < n$ (Wang and Williams [897]).

3. Prove that every threshold function has an integral separating structure (Theorem 9.5).

4. Prove Theorem 9.6.

5. Derive Theorem 9.14 from the strong duality theorem of linear programming.

6. Consider the Boolean function f on $\mathcal{B}^{10}$ defined as follows: $f(x_1, x_2, \ldots, x_{10}) = 1$ if and only if

$$x_1 + x_2 + \ldots + x_{10} \geq 7$$

and

$$34x_1 + 29x_2 + 9x_3 + 7x_4 + 5x_5 + 5x_6 + 4x_7 + 3x_8 + 3x_9 + x_{10} \geq 50.$$

Prove that f is regular, but that f is not a threshold function.

7. Consider the separating structures $(W^*, t^*) = (13, 7, 6, 6, 4, 4, 4, 3, 2, 24)$ and $(V^*, t^*) = (13, 7, 6, 6, 4, 4, 4, 2, 3, 24)$, as in Example 9.15.

   (a) Show that (W^*, t^*) and (V^*, t^*) define the same self-dual threshold function f.

   (b) Observe that $Y^1 = (1,0,1,1,0,0,0,0,0,0)$, $Y^2 = (1,0,0,1,1,0,0,0,0,1)$, $Y^3 = (0,1,1,0,1,1,1,0,0)$, $Y^4 = (0,1,0,1,1,1,1,0,0)$, and $Y^5 = (0,0,1,1,0,1,1,1,1)$ are minimal true points of f, and that $X^1 = (1,1,0,0,0,1,0,0,0)$, $X^2 = (1,1,0,\ 0,0,0,1,0,0)$, $X^3 = (0,1,1,1,0,0,0,1,1)$, and $X^4 = (0,0,1,1,1,1,1,0,0)$ are maximal false points of f. Observe also that $14Y^1 + 3Y^2 + 15Y^3 + 12Y^4 + 11Y^5 - 8X^1 - 8X^2 - 10X^3 - 29X^4 = (1,1,1,1,1,1,1,1,4)$. Conclude that, in every solution of the system (9.24)–(9.26) such that $w_9 \leq 2$, there holds $\sum_{i=1}^{n} w_i \geq 49$, and hence, no solution of (9.24)–(9.26) is simultaneously smaller than both (W^*, t^*) and (V^*, t^*).

8. Prove that the following decision problem is co-NP-complete: Given a separating structure $(w_1, w_2, \ldots, w_n, t)$, decide whether the threshold function represented by this structure is self-dual. (Compare with Theorem 9.9.)

9. Use the lower bound (9.1) in Theorem 9.11 to prove the following: For every $k > 2$ and for n large enough, there exists a threshold function of n variables, so that any integral separating structure representing f involves a weight of magnitude at least $2^{n/k}$. (Compare with the – much stronger – lower bound in Theorem 9.12.)

10. Prove that any oracle algorithm for the threshold recognition problem must perform, in the worst case, an exponential number of queries on the oracle. (An *oracle algorithm* is an algorithm that can only gain information about the input function through queries of the form: "Is X^* a true point of the function?")

11. Let S_1 be the solution set of the system (9.17)–(9.18), and let S_2 be the solution set of the system (9.18)–(9.19). Show that no inclusion relation holds in general between S_1 and S_2.

12. In 1973, the voting weights of the nine members of the Council of Ministers of the European Economic Community were 10, 10, 10, 10, 5, 5, 3, 3, and 2, respectively. The threshold was 40 votes. Show that this voting procedure is equivalent to the procedure defined by the smaller weights 6, 6, 6, 6, 3, 3, 2, 2, 1 with threshold 24. Compute the Banzhaf indices of the nine states.

13. Prove that every Chow function is completely monotone.

14. Let f and g be two functions on $\mathcal{B}^n$. Prove that, if f is a positive threshold function and $\omega_i(f) = \omega_i(g)$ for $i = 1, 2, \ldots, n$, then either $f = g$ or $\omega(f) < \omega(g)$.

15. Prove that a graph $G = (V, E)$ is threshold if and only if there exist $(n+1)$ numbers $a_1, a_2, \ldots, a_n$ and q such that, for all i, j in V,

$$(i, j) \in E \text{ if and only if } a_i + a_j > q. \tag{9.35}$$

16. Show that, if G is a threshold graph and the numbers $a_1, a_2, \ldots, a_n, q$ satisfy (9.35), then $(a_1, a_2, \ldots, a_n, q)$ is not necessarily a separating structure of G.

17. A positive Boolean function $f(x_1, x_2, \ldots, x_n) = \bigvee_{P \in \mathcal{E}} \left(\bigwedge_{j \in P} x_j \right)$ is r-uniform if $|P| = r$ for all $P \in \mathcal{E}$. We say that a r-uniform function has property (T) if there exist $(n+1)$ numbers $a_1, a_2, \ldots, a_n$ and q such that, for all $P \subseteq \{1, 2, \ldots, n\}$ with $|P| = r$,

$$P \in \mathcal{E} \text{ if and only if } \sum_{i \in P} a_i > q.$$

Prove that

(a) if f is uniform and threshold, then f has property (T); if f is uniform and has property (T), then f is regular (Golumbic [398]); and

(b) the reverse of both implications in (a) may fail for 3-uniform functions (Reiterman, Rödl, Šiňajová, and Tůma [784]).

18. Let $G = (V, E)$ be a threshold graph on the vertex-set $V = \{1, 2, \ldots, n\}$, and let $(d_1, d_2, \ldots, d_n)$ be the degree sequence of G, where d_i is the degree of vertex i $(i = 1, 2, \ldots, n)$. Prove that

(a) $K = \{i \in V : i - 1 \leq d_i\}$ is a maximum clique of G;

(b) $V \setminus K$ is a stable set of G.

19. If G is a threshold graph, express the Chow parameters of f_G as a function of the degree sequence of $d(G)$. (B. Simeone, private communication.)

20. Show that the threshold dimension of a graph G is the smallest value of δ for which there exist δ threshold graphs $G_k = (V, E_k)$ $(k = 1, 2, \ldots, \delta)$ satisfying $E_1 \cup E_2 \cup \cdots \cup E_\delta = E$.

Question for thought

21. Let $k(n)$ be the smallest integer k such that every k-asummable function of n variables is a threshold function. It is known that $k(n) \geq \lfloor \sqrt{n} \rfloor$, and that $k(15) > \lfloor \sqrt{15} \rfloor$ (Muroga [698]). What else can be said about $k(n)$?

22. Is it possible to recognize threshold functions through an entirely combinatorial procedure, that is, without resorting to the solution of the system (TS) as in Theorem 9.16, or by developing a specialized combinatorial algorithm for its solution?

23. If $f(x_1, x_2, \ldots, x_n)$ is a positive Boolean function, denote by $\delta(f)$ (respectively, $\rho(f)$) the smallest number m such that f is the disjunction of m threshold (respectively, regular) functions.
    (a) Show that $\delta(f)$ and $\rho(f)$ can take any value between 1 and $\lfloor n/2 \rfloor$.
    (b) Is it true that, for every pair of integers (d, r) with $d \geq r$, there exists a positive function f with $\delta(f) = d$ and $\rho(f) = r$? (See Neumaier [709].)

Read-once functions

Martin C. Golumbic and Vladimir Gurvich

10.1 Introduction

In this chapter, we present the theory and applications of read-once Boolean functions, one of the most interesting special families of Boolean functions. A function f is called *read-once* if it can be represented by a Boolean expression using the operations of conjunction, disjunction, and negation in which every variable appears exactly once. We call such an expression a *read-once expression* for f. For example, the function

$$f_0(a,b,c,w,x,y,z) = ay \vee cxy \vee bw \vee bz$$

is a read-once function, since it can be factored into the expression

$$f_0 = y(a \vee cx) \vee b(w \vee z)$$

which is a read-once expression.

Observe, from the definition, that read-once functions must be monotone (or unate), since every variable appears either in its positive or negative form in the read-once expression (see Exercise 1 at the end of the chapter). However, we will make the stronger assumption that a read-once function is positive, simply by renaming any negative variable $\bar{x}_i$ as a new positive variable x'_i. Thus, every variable will be positive, and we may freely rely on the results presented earlier (in particular, in Chapters 1 and 4) on positive Boolean functions.

Let us look at two simple functions,

$$f_1 = ab \vee bc \vee cd$$

and

$$f_2 = ab \vee bc \vee ac.$$

Neither of these is a read-once function; indeed, it is impossible to express them so that each variable appears only once. (Try to do it.) The functions f_1 and f_2 illustrate the two types of forbidden functions that characterize read-once functions, as we

448

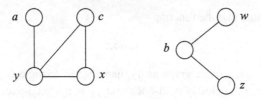

Figure 10.1. The co-occurrence graph of $f_0 = ay \vee cxy \vee bw \vee bz$.

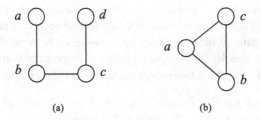

(a)   (b)

Figure 10.2. The co-occurrence graphs of (a): f_1, and (b): f_2, f_3.

will see. We begin by defining the co-occurrence graph of a positive Boolean function.

Let f be a positive Boolean function over the variable set $V = \{x_1, x_2, \ldots, x_n\}$. The *co-occurrence graph* of f, denoted $G(f) = (V, E)$, has vertex set V (the same as the set of variables), and there is an edge (x_i, x_j) in E if x_i and x_j occur together (at least once) in some prime implicant of f. In this chapter, we often regard a prime implicant as the set of its literals. Formally, let $\mathcal{P}$ denote the collection of prime implicants of f. Then,

$$(x_i, x_j) \in E \iff x_i, x_j \in P \text{ for some } P \in \mathcal{P}.$$

Figures 10.1 and 10.2 show the co-occurrence graphs of f_0, f_1, f_2.

We denote by P_4 the graph consisting of a chordless path on 4 vertices and 3 edges, which is the graph $G(f_1)$ in Figure 10.2 (see also Appendix A). A graph is called P_4-*free* if it contains no induced subgraph isomorphic to P_4. The P_4-free graphs are also known as *cographs* (for "complement reducible graphs"); we will have more to say about them in Section 10.4.

Since we have observed that f_1 is not read-once, and since its co-occurrence graph is P_4, it would be reasonable to conjecture that *the co-occurrence graph of a read-once function must be P_4-free*. In fact, we will prove this statement in Section 10.3. This is not enough, however. In order to characterize read-once functions in terms of graphs, we will need a second property called *normality*.[1]

---

[1] The property of normality is sometimes called *clique-maximality* in the literature. It also appears in the definition of *conformal hypergraphs* in Berge [71] and is used in the theory of acyclic hypergraphs.

To see this, note that the function

$$f_3 = abc$$

has the same co-occurrence graph as f_2, namely, the triangle $G(f_2) = G(f_3)$ in Figure 10.2, yet f_3 is clearly read-once and f_2 is not read-once. This example illustrates the motivation for the following definition.

A Boolean function f is called *normal* if every clique of its co-occurrence graph is contained in a prime implicant of f.

In our example, f_2 fails to be normal, since the triangle $\{a, b, c\}$ is not contained in any prime implicant of f_2. This leads to our second necessary property of read-once functions, namely, that *a read-once function must be normal*, which we will also prove in Section 10.3. Moreover, a classical theorem of Gurvich [422, 426] shows that combining these two properties characterizes read-once functions.

Theorem 10.1. *A positive Boolean function f is read-once if and only if its co-occurrence graph $G(f)$ is P_4-free and f is normal.*

A new proof of this theorem will be given in Section 10.3 as part of Theorem 10.6.

Read-once functions first appeared explicitly in the literature in the papers of Chein [190] and Hayes [479] that gave exponential time recognition algorithms for the family (see the historical notes at the end of this chapter). Gurvich [422, 425, 426] gave the first characterization theorems for read-once functions; they are presented in Section 10.3. Several authors have subsequently discovered and rediscovered these and a number of other characterizations. Theorem 10.1 also provides the justification for the polynomial time recognition algorithm of read-once functions by Golumbic, Mintz, and Rotics [401, 402], presented in Section 10.5. In particular, we will show how to factor read-once functions using the properties of P_4-free graphs.

Read-once functions have been studied in computational learning theory, where they have been shown to constitute a class that can be learned in polynomial time. Section 10.6 will survey some of these results. Additional applications of read-once functions are presented in Section 10.7.

Before turning our full attention to read-once functions, however, we review a few properties of the dual of a Boolean function and prove an important result on positive Boolean functions that will be useful in subsequent sections.

10.2 Dual implicants

In this section, we first recall some of the relationships between the prime implicants of a function f and the prime implicants of its dual function f^d in the case of positive Boolean functions. All of these properties were presented in Chapter 1 and Chapter 4. We then present a characterization of the subimplicants of the dual of a positive Boolean function, due to Boros, Gurvich, and Hammer [121]. This result will be used later in the proof of one of the characterizations of read-once functions.

10.2.1 Implicants and dual implicants

The dual of a Boolean function f is the function f^d defined by

$$f^d(X) = \overline{f(\overline{X})},$$

and an expression for f^d can be obtained from any expression for f by simply interchanging the operators $\wedge$ and $\vee$ as well as the constants 0 and 1. In particular, given a DNF expression for f, this exchange yields a CNF expression for f^d. This shows that the dual of a read-once function is also read-once.

The process of transforming a DNF expression of f into a DNF expression of f^d is called *DNF dualization*; its complexity for positive Boolean functions is still unknown, the current best algorithm being quasi-polynomial [347]; see Chapter 4.

Let $\mathcal{P}$ be the collection of prime implicants of a positive Boolean function f over the variables $x_1, x_2, \ldots, x_n$, and let $\mathcal{D}$ be the collection of prime implicants of the dual function f^d. We assume throughout that all of the variables for f (and hence for f^d) are essential. We use the term "dual (prime) implicant" of f to mean a (prime) implicant of f^d. For positive functions, the prime implicants of f correspond precisely to the set of minimal true points $minT(f)$, and the dual prime implicants of f correspond precisely to the set of maximal false points $maxF(f)$; see Sections 1.10.3 and 4.2.1.

Theorem 4.7 states that the implicants and dual implicants of a Boolean function f, viewed as sets of literals, have pairwise nonempty intersections. In particular, this holds for the prime implicants and the dual prime implicants. Moreover, the prime implicants and the dual prime implicants are minimal with this property, that is, for every proper subset S of a dual prime implicant of f, there is a prime implicant P such that $P \cap S = \emptyset$.

In terms of hypergraph theory, the prime implicants $\mathcal{P}$ form a clutter (namely, a collection of sets, or hyperedges, such that no set contains another set), as does the collection of dual prime implicants $\mathcal{D}$.

Finally, we recall the following properties of duality to be used in this chapter and which can be derived from Theorems 4.1 and 4.19.

Theorem 10.2. *Let f and g be positive Boolean functions over $\{x_1, x_2, \ldots, x_n\}$, and let $\mathcal{P}$ and $\mathcal{D}$ be the collections of prime implicants of f and g, respectively. Then the following statements are equivalent:*

(i) *$g = f^d$.*

(ii) *For every partition of $\{x_1, x_2, \ldots, x_n\}$ into sets A and $\overline{A}$, there is either a member of $\mathcal{P}$ contained in A or a member of $\mathcal{D}$ contained in $\overline{A}$, but not both.*

(iii) *$\mathcal{D}$ is exactly the family of minimal transversals of $\mathcal{P}$.*

(iv) $\mathcal{P}$ *is exactly the family of minimal transversals of* $\mathcal{D}$.

(v) (a) *For all* $P \in \mathcal{P}$ *and* $D \in \mathcal{D}$, *we have* $P \cap D \neq \emptyset$; *and*

(b) *For every subset* $B \subseteq \{x_1, x_2, \ldots, x_n\}$, *there exists* $D \in \mathcal{D}$ *such that* $D \subseteq B$ *if and only if* $P \cap B \neq \emptyset$ *for every* $P \in \mathcal{P}$.

We obtain from Theorem 10.2(v) the following characterization of dual implicants.

Theorem 10.3. *A set of variables B is a dual implicant of the function f if and only if* $P \cap B \neq \emptyset$ *for all prime implicants P of f.*

10.2.2 The dual subimplicant theorem

We are now ready to present a characterization of the subimplicants of the dual of a positive function, due to Boros, Gurvich, and Hammer [121]. This characterization is interesting on its own and also provides a useful tool for proving other results.

Let f be a positive Boolean function over the variables $V = \{x_1, x_2, \ldots, x_n\}$, and let f^d be its dual. As before, $\mathcal{P}$ and $\mathcal{D}$ denote the prime implicants of f and f^d, respectively. We assume throughout that all of the variables of f (and f^d) are essential.

A subset T of the variables is called a *dual subimplicant* of f if T is a subset of a dual prime implicant of f, that is, if there exists a prime implicant D of f^d such that $T \subseteq D$. A *proper dual subimplicant* is a nonempty proper subset of a dual prime implicant.

Example 10.1. *Let* $f = x_1 x_2 \vee x_2 x_3 x_4 \vee x_4 x_5$. *Its dual is* $f^d = x_1 x_3 x_5 \vee x_1 x_4 \vee x_2 x_4 \vee x_2 x_5$. *The proper dual subimplicants of* f *are the pairs* $\{x_1, x_3\}, \{x_3, x_5\}, \{x_1, x_5\}$ *and the five singletons* $\{x_i\}$, $i = 1, \ldots, 5$. $\qquad\square$

We will make use below of the following consequence of Theorem 10.3:

Remark 10.1. Let T be a subset of the variables $\{x_1, x_2, \ldots, x_n\}$. If T is a proper dual subimplicant of f, then there exists a prime implicant $P \in \mathcal{P}$ such that $P \cap T = \emptyset$. $\qquad\square$

Let T be a subset of the variables. Our goal will be to determine whether T is contained in some $D \in \mathcal{D}$, namely, whether T is a dual subimplicant. We define the following sets of prime implicants of f, with respect to the set T:

$$\mathcal{P}_0(T) = \{P \in \mathcal{P} \mid P \cap T = \emptyset\},$$

and, for all $x \in T$,

$$\mathcal{P}_x(T) = \{P \in \mathcal{P} \mid P \cap T = \{x\}\}.$$

Note that by Theorem 10.3, $\mathcal{P}_0(T)$ is empty if and only if T is a dual implicant, and by Remark 10.1, $\mathcal{P}_0(T)$ is nonempty when T is a proper dual subimplicant. The remaining prime implicants in $\mathcal{P}$, which contain two or more variables of T, will not be relevant for our analysis. (We may omit the parameter T from our notation when it is clear which subset is meant.)

A *selection* $\mathcal{S}(T)$, with respect to T, consists of one prime implicant $P_x \in \mathcal{P}_x(T)$ for every $x \in T$. A selection is called *covering* if there is a prime implicant $P_0 \in \mathcal{P}_0(T)$ such that $P_0 \subseteq \bigcup_{x \in T} P_x$. Otherwise, it is called *noncovering*. (See Example 10.2.)

We now present the characterization of the dual subimplicants of a positive Boolean function from [121].

Theorem 10.4. *Let f be a positive Boolean function over the variable set $\{x_1, x_2, \ldots, x_n\}$, and let T be a subset of the variables. Then T is a dual subimplicant of f if and only if there exists a noncovering selection with respect to T.*

Proof. Assume that T is a dual subimplicant of f, and let $D \in \mathcal{D}$ be a prime implicant of f^d for which $T \subseteq D$. For any variable $x \in T$, the subset $D \setminus \{x\}$ is a proper subset of D, and therefore, by Remark 10.1 (or trivially, if $D = \{x\}$), there exists a prime implicant $P_x \in \mathcal{P}$ such that $P_x \cap (D \setminus \{x\}) = \emptyset$. Since $P_x \cap D \neq \emptyset$ by Theorem 10.3, we have $\{x\} = P_x \cap D = P_x \cap T$, that is, $P_x \in \mathcal{P}_x(T)$.

If $\mathcal{S} = \{P_x | x \in T\}$ were a covering selection, then there would exist a prime implicant $P_0 \in \mathcal{P}_0(T)$ such that $P_0 \subseteq \bigcup_{x \in T} P_x$. But this would imply

$$P_0 \cap D \subseteq \left(\bigcup_{x \in T} P_x \right) \cap D = \bigcup_{x \in T} (P_x \cap D) = T,$$

which, together with $P_0 \cap T = \emptyset$, would give $P_0 \cap D = \emptyset$, contradicting Theorem 10.3. Thus, the selection $\mathcal{S}$ we have constructed is a noncovering selection with respect to T. (Note that in the special case when $T = D$, we would have $\mathcal{P}_0(T)$ empty, and any selection would be noncovering.)

Conversely, suppose there exists a noncovering selection $\mathcal{S} = \{P_x | x \in T\}$, where $P_x \in \mathcal{P}_x(T)$. Since $\mathcal{S}$ is noncovering, we have for all $P_0 \in \mathcal{P}_0(T)$ that

$$P_0 \not\subseteq \bigcup_{x \in T} P_x.$$

Let B be defined as the complementary set

$$B = \left(\{x_1, x_2, \ldots, x_n\} \setminus \bigcup_{x \in T} P_x \right) \cup T.$$

Clearly, for any prime implicant $P_0 \in \mathcal{P}_0(T)$, we have $P_0 \cap B \neq \emptyset$, since $\mathcal{S}$ is non-covering. Moreover, by definition, all other prime implicants $P \in \mathcal{P} \setminus \mathcal{P}_0(T)$ intersect T, and therefore, they intersect B, since $T \subseteq B$. Thus, we have shown that $P \cap B \neq \emptyset$ for all $P \in \mathcal{P}$, implying that B is a (not necessarily prime) dual implicant.

Let $D \in \mathcal{D}$ be a dual prime implicant such that $D \subseteq B$. From the definition of B, it follows that $P_x \cap B = \{x\}$ for all $x \in T$. But each P_x intersects D, since P_x is a prime implicant and D is a dual prime implicant, which, together with the fact that $D \subseteq B$, implies that $P_x \cap D = \{x\}$. Hence, $T \subseteq D$, proving that T is a dual subimplicant. $\qquad\square$

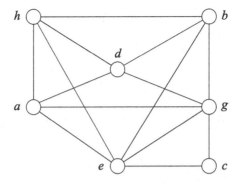

Figure 10.3. The co-occurrence graph for Example 10.2.

We will often apply Theorem 10.4 in its contrapositive form or in its dual form, as follows.

Remark 10.2. A subset T is not a dual subimplicant of f if and only if every selection with respect to T is a covering selection.   □

Remark 10.3. We may also apply Theorem 10.4 to subimplicants of f and dual selections, where the roles of $\mathcal{P}$ and $\mathcal{D}$ are reversed in the obvious manner.   □

Example 10.2. *Consider the positive Boolean function*

$$f = adg \vee adh \vee bdg \vee bdh \vee eag \vee ebg \vee ecg \vee eh,$$

whose co-occurrence graph is shown in Figure 10.3.

(i) *Let $T = \{b,c,h\}$. We have*

$$\mathcal{P}_0(T) = \{adg,eag\}, \quad \mathcal{P}_b(T) = \{bdg,ebg\}, \quad \mathcal{P}_c(T) = \{ecg\},$$
$$\mathcal{P}_h(T) = \{adh,eh\}.$$

The selection $S = \{bdg,ecg,eh\}$ is noncovering since $\{a,d,g\},\{a,e,g\} \nsubseteq \{b,c,d,e,g,h\}$; hence, by Theorem 10.4, T is a dual subimplicant.

(ii) *Now let $T' = \{a,b,g\}$. We have*

$$\mathcal{P}_0(T') = \{eh\}, \quad \mathcal{P}_a(T') = \{adh\}, \quad \mathcal{P}_b(T') = \{bdh\}, \quad \mathcal{P}_g(T') = \{ecg\}.$$

There is only one possible selection $S' = \{adh,bdh,ecg\}$ and S' is a covering selection since $\{e,h\} \subseteq \{a,b,c,d,e,g,h\}$. Hence, by Remark 10.2, T' is not a dual subimplicant.

It can be verified that T is contained in the dual prime implicant $abch$, and that, to extend T' to a dual implicant, it would be necessary to add either e or h; however, neither $abeg$ nor $abgh$ are prime (since $abe,bgh \in \mathcal{D}$), see Exercise 5 at the end of the chapter.   □

The problem of recognizing whether a given subset T is a dual subimplicant of a positive function f given by its complete DNF was shown to be NP-complete by Boros, Gurvich, and Hammer [121]. However, they point out that Theorem 10.4 can be applied in a straightforward manner to answer this recognition problem in $O(n|f|^{1+\min\{|T|,|\mathcal{P}_0(T)|\}})$ time, where $|f|$ denotes the number of literals in the complete DNF of f. This becomes feasible for very small and very large values of $|T|$, such as $2, 3, n-2, n-1$. Specifically, by applying this for every pair $T = \{x_i, x_j\}, 1 \le i < j \le n$, we obtain the following:

Theorem 10.5. *The co-occurrence graph $G(f^d)$ of the dual of a positive Boolean function f can be determined in polynomial time, when f is given by its complete DNF. The complexity of determining all the edges of $G(f^d)$ is at most $O(n^3|f|^3)$.*

Proof. Consider a given pair $T = \{x_i, x_j\}$. We observe the following:

(1) If either $\mathcal{P}_{x_i}$ or $\mathcal{P}_{x_j}$ is empty, then there is no possible selection (covering or noncovering). Hence, Theorem 10.4 implies that x_i and x_j are not contained together in a dual prime implicant and, therefore, are not adjacent in $G(f^d)$.
(2) If both $\mathcal{P}_{x_i}$ and $\mathcal{P}_{x_j}$ are nonempty, but $\mathcal{P}_0$ is empty, then there is a selection and every selection will be noncovering. Hence, Theorem 10.4 implies that $\{x_i, x_j\}$ is a dual subimplicant, and so x_i and x_j are adjacent in $G(f^d)$.
(3) If all three sets $\mathcal{P}_0, \mathcal{P}_{x_i}$ and $\mathcal{P}_{x_j}$ are nonempty, then we may have to check all possible $O(|f|^2)$ selections before knowing whether there is a noncovering selection.

We leave a detailed complexity analysis as an exercise for the reader. □

Example 10.3. *Let us calculate $G(f^d)$ for the function $f = abc \vee bde \vee ceg$, as illustrated in Figure 10.4.*
The pair (a,b) is not an edge: Indeed, we have in this case $\mathcal{P}_a = \emptyset$, so a and b are not adjacent in $G(f^d)$. Similarly, $(a,c),(b,d),(c,g),(d,e),(e,g)$ are also nonedges.
The pair (b,c) is an edge: In this case, both $\mathcal{P}_b$ and $\mathcal{P}_c$ are nonempty, but $\mathcal{P}_0$ is empty, so b and c are adjacent in $G(f^d)$. Similarly, $(b,e),(c,e)$ are also edges.

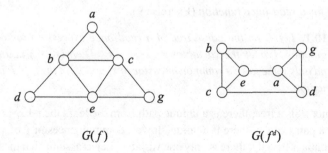

$G(f)$ $G(f^d)$

Figure 10.4. The co-occurrence graphs of f and f^d in Example 10.3.

The pair (a,e) is an edge: In this case, as in the previous one, both $\mathcal{P}_a \neq \emptyset$ and $\mathcal{P}_e \neq \emptyset$, but $\mathcal{P}_0 = \emptyset$, so a and e are adjacent in $G(f^d)$. Similarly, $(b,g),(c,d)$ are also edges.

The pair (a,d) is an edge: In this case, $\mathcal{P}_a = \{abc\}$, $\mathcal{P}_d = \{bde\}$, $\mathcal{P}_0 = \{ceg\}$. Since $\{c,e,g\} \not\subseteq \{a,b,c,d,e\}$, we conclude that a and d are adjacent in $G(f^d)$. Similarly, $(a,g),(d,g)$ are also edges.

Notice what happens if we add an additional prime implicant bce to the function f in this example. Consider the function $f' = abc \vee bde \vee ceg \vee bce$. Then ad is not a dual subimplicant of f' although it was of f. Indeed, there is still only one selection $\{abc,bde\}$, but now it is covering, since it contains bce. By symmetry, neither ag nor dg are dual subimplicants of f'. $\qquad\square$

10.3 Characterizing read-once functions

In this section, we present the mathematical theory underlying read-once functions due to Gurvich [422, 425, 426] and rediscovered by several other authors; see [293, 294, 548, 696]. The algorithmic aspects of recognizing and factoring read-once functions will be presented in Section 10.5.

Recall from Section 10.1 that a *read-once expression* is a Boolean expression in which every variable appears exactly once. A *read-once Boolean function* is a function that can be transformed (i.e., factored) into a read-once expression over the operations of conjunction and disjunction. We have also assumed read-once functions to be positive.

A positive Boolean expression, over the operations of conjunction and disjunction, may be represented as a (rooted) *parse tree* whose leaves are labeled by the variables $\{x_1,x_2,\ldots,x_n\}$, and whose internal nodes are labeled by the Boolean operations $\wedge$ and $\vee$. The parse tree represents the computation of the associated Boolean function according to the given expression, and each internal node is the root of a subtree corresponding to a part of the expression; see Figure 10.5. (A parse tree is a special type of combinational circuit, as introduced in Section 1.13.2.) If the expression is read-once, then each variable appears on exactly one leaf of the tree, and there is a unique path from the root to the variable.

We begin by presenting a very useful lemma relating a read-once expression to the co-occurrence graph of the function. It also shows that *the read-once expression is unique for a read-once function* (Exercise 9).

Lemma 10.1. *Let T be the parse tree of a read-once expression for a positive Boolean function f over the variables $x_1,x_2,\ldots,x_n$. Then (x_i,x_j) is an edge in $G(f)$ if and only if the lowest common ancestor of x_i and x_j in the tree T is labeled $\wedge$ (conjunction).*

Proof. Since T is a tree, there is a unique path from the leaf labeled x_i to the root. Thus, for a pair (x_i,x_j), there is a unique lowest common ancestor v of x_i and x_j.

The lemma is trivial if there is only one variable. Let us assume that the lemma is true for all functions with fewer than n variables, and prove the result by induction.

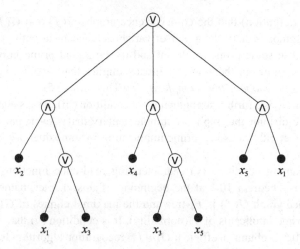

Figure 10.5. The parse tree of the expression $x_2(x_1 \vee x_3) \vee x_4(x_3 \vee x_5) \vee x_5 x_1$.

Let $u_1,\ldots,u_r$ be the children of the root of T, and for $k = 1,\ldots,r$, let T_k be the subexpression (subtree) rooted at u_k, denoting its corresponding function by f_k. Note that the variables at the leaves of T_k are disjoint from the leaves of T_l for $k \neq l$, since the expression is read-once.

If the root of T is labeled $\vee$, then $f = f_1 \vee \cdots \vee f_r$ and the graph $G(f)$ will be the disjoint union of the graphs $G(f_k)$ $(k = 1,\ldots,r)$, since multiplying out each of the expressions T_k will yield disjoint prime implicants of f. Thus, x_i and x_j are adjacent in $G(f)$ if and only if they are in the same T_k and adjacent in $G(f_k)$ and, by induction, if and only if their lowest common ancestor (in T_k and hence in T) is labeled $\wedge$ (conjunction).

If the root is labeled $\wedge$, then $f = f_1 \wedge \cdots \wedge f_r$ and the graph $G(f)$ will be the join of the graphs $G(f_k)$, $(k = 1,\ldots,r)$. That is, every vertex of the subgraph $G(f_k)$ is adjacent to every vertex of the subgraph $G(f_l)$ for $k \neq l$, since multiplying out each expression T_k and then expanding the entire expression T will put every pair of variables from different subtrees into some (perhaps many) prime implicants. Therefore, if x_i and x_j are on leaves of different subtrees, then they are connected in $G(f)$, and their lowest common ancestor is the root of T that is labeled $\wedge$. If x_i and x_j are on leaves of the same subtree, then again by induction, (x_i, x_j) is an edge in $G(f_k)$ if and only if the lowest common ancestor of x_i and x_j is labeled $\wedge$ (conjunction). $\square$

We are now ready to present and prove the main characterization theorem of read-once functions. We describe briefly what will be shown in our Theorem 10.6.

We saw in Theorem 10.2 that for any positive Boolean function f, every prime implicant P of f and every prime implicant D of its dual f^d must have at least one variable in common. This property is strengthened in the case of read-once functions, by condition (iv) in Theorem 10.6, which claims that f is read-once if and only if this common variable is unique. Moreover, this condition immediately implies (by definition) that the co-occurrence graphs $G(f)$ and $G(f^d)$ have no edges in common; otherwise, a pair of variables adjacent in both graphs would be contained in some prime implicant and in some dual prime implicant. This is condition (iii) of our theorem and already implies that *recognizing read-once functions has polynomial-time complexity* (by Theorem 10.5).

Condition (ii) is a further strengthening of condition (iii). It says that in addition to being edge-disjoint, the graphs are complementary, that is, every pair of variables appear together either in some prime implicant or in some dual prime implicant, but not both.

The remaining condition (v) characterizing read-once functions is the one mentioned as Theorem 10.1 at the beginning of this chapter, namely, that the co-occurrence graph $G(f)$ is P_4-free and the maximal cliques of $G(f)$ are pre-cisely the prime implicants of f (normality). It is condition (v) that will be used in Section 10.5 to obtain an efficient $O(n|f|)$ recognition algorithm for read-once functions.

Example 10.4. *The function*

$$f_4 = x_1x_2 \vee x_2x_3 \vee x_3x_4 \vee x_4x_5 \vee x_5x_1,$$

whose co-occurrence graph $G(f_4)$ is the chordless 5-cycle C_5, is normal but $G(f_4)$ is not P_4-free. Hence, f_4 is not a read-once function. Its dual

$$f_4^d = x_1x_2x_4 \vee x_2x_3x_5 \vee x_3x_4x_1 \vee x_4x_5x_2 \vee x_5x_1x_3,$$

whose co-occurrence graph $G(f_4^d)$ is the clique (complete graph) K_5, which is P_4-free, is not a normal function. $\square$

Theorem 10.6. *Let f be a positive Boolean function over the variable set $\{x_1, x_2, \ldots, x_n\}$. Then the following conditions are equivalent:*

(i) *f is a read-once function.*
(ii) *The co-occurrence graphs $G(f)$ and $G(f^d)$ are complementary, that is, $G(f^d) = \overline{G(f)}$.*
(iii) *The co-occurrence graphs $G(f)$ and $G(f^d)$ have no edges in common, that is, $E(G(f)) \cap E(G(f^d)) = \emptyset$.*
(iv) *For all $P \in \mathcal{P}$ and $D \in \mathcal{D}$, we have $|P \cap D| = 1$.*
(v) *The co-occurrence graph $G(f)$ is P_4-free and f is normal.*

Proof. (i) $\Longrightarrow$ (ii): Assume that f is a read-once function, and let T be the parse tree of a read-once expression for f. By interchanging the operations $\vee$ and $\wedge$, we obtain the parse tree T^d of a read-once expression for the dual f^d. By Lemma 10.1, (x_i, x_j) is an edge in $G(f)$ if and only if the lowest common ancestor of x_i and x_j in the tree T is labeled $\wedge$ (conjunction). Similarly, (x_i, x_j) is an edge in $G(f^d)$ if and only if the lowest common ancestor of x_i and x_j in the tree T^d is labeled $\wedge$ (conjunction). It follows from the foregoing construction that $G(f)$ and $G(f^d)$ are complementary.

(ii) $\Longrightarrow$ (iii): Trivial.

(iii) $\Longleftrightarrow$ (iv): As noted in the discussion above, by definition, the co-occurrence graphs $G(f)$ and $G(f^d)$ have no edges in common if and only if $|P \cap D| \leq 1$ for every prime implicant P of f and every prime implicant D of its dual f^d. However, for any positive Boolean function, we have $|P \cap D| \geq 1$ by Theorem 10.2(v), which proves the equivalence.

(iv) $\Longrightarrow$ (v): We first prove that the function f is normal (Claim 1), and then that the graph $G(f)$ is P_4-free (Claim 3). We may assume both conditions (iii) and (iv) since we have already shown that they are equivalent.

Claim 1. *The function f is normal, that is, every clique of $G(f)$ is contained in a prime implicant of f.*

The claim is certainly true for any clique of size one, since we assume that all variables are essential, and it is true for any clique of size two, by the definition of the co-occurrence graph $G(f)$. Let us consider the smallest value k ($k \geq 3$) for which the claim fails, that is, there exists a clique $K = \{x_1, \ldots, x_k\}$ of $G(f)$ that is not a subimplicant of f. We denote the subcliques of K of size $k-1$ by $K_i = K - \{x_i\}$, $i = 1, \ldots, k$.

By our assumption of k being smallest possible, each set K_i is a subimplicant of f, so each is contained, respectively, in a prime implicant $P_i \in \mathcal{P}$, which we can express in the form

$$P_i = K_i \cup A_i,$$

where $K \cap A_i = \emptyset$, since K is not a subimplicant.

In addition, each variable $x_i \in K$ is contained in a dual prime implicant $D_i \in \mathcal{D}$, which we can express in the form

$$D_i = \{x_i\} \cup B_i,$$

where $K \cap B_i = \emptyset$, by our assumption (iv). Applying (iv) further, we note that

$$|P_i \cap D_j| = |(K_i \cup A_i) \cap (\{x_j\} \cup B_j)| = 1$$

for all i, j. In the case of $i \neq j$, since $x_j \in K_i$, this implies

$$A_i \cap B_j = \emptyset \quad (\forall i \neq j). \tag{10.1}$$

In the case of $i = j$, we obtain

$$|A_i \cap B_i| = 1,$$

since the common variable cannot belong to K. This enables us to define

$$y_i = A_i \cap B_i \quad (i = 1, \ldots k). \tag{10.2}$$

Moreover, $y_i \neq y_j$ for $i \neq j$ by (10.1).

We now apply Theorem 10.4 (the dual subimplicant theorem). Consider a pair $T = \{x_i, x_j\}$ ($1 \leq i < j \leq k$). Since (x_i, x_j) is an edge of $G(f)$, by assumption (iii), it is not an edge of $G(f^d)$ and, hence, not a dual subimplicant. By Theorem 10.4, this implies that every selection $\mathcal{S}$ with respect to T must be a covering selection.

Now, $\mathcal{S} = \{P_i, P_j\}$ is a selection for $T = \{x_i, x_j\}$ since $P_i \cap \{x_i, x_j\} = \{x_j\}$ and $P_j \cap \{x_i, x_j\} = \{x_i\}$. Therefore, there exists a prime implicant P_0 such that $P_0 \cap \{x_i, x_j\} = \emptyset$ and $P_0 \subseteq P_i \cup P_j$. Thus, $P_0 \subseteq (K \setminus \{x_i, x_j\}) \cup A_i \cup A_j$.

Since, $1 = |P_0 \cap D_i| = |P_0 \cap B_i|$, it follows from (10.2) that $y_i \in P_0$. Similarly, $1 = |P_0 \cap D_j| = |P_0 \cap B_j|$, so $y_j \in P_0$. Thus, (y_i, y_j) is an edge in $G(f)$. In fact, since i and j were chosen arbitrarily, the set $Y = \{y_1, \ldots, y_k\}$ is a clique in $G(f)$.

Now, we apply Theorem 10.4 to the dual function f^d, as suggested in Remark 10.3. Since the clique K is not a subimplicant of f, every dual selection $\mathcal{S}'$ with respect to K must be a covering dual selection. In particular, $\mathcal{S}' = \{D_1, \ldots, D_k\}$ is such a selection since $D_i = \{x_i\} \cup B_i$ intersects K only in x_i. Therefore, there exists a dual prime implicant D_0 satisfying $D_0 \cap K = \emptyset$ and $D_0 \subseteq \bigcup_{x_i \in K}(\{x_i\} \cup B_i)$, or

$$D_0 \subseteq \bigcup_{x_i \in K} B_i. \tag{10.3}$$

For each i, we have $1 = |D_0 \cap P_i| = |D_0 \cap (K_i \cup A_i)|$. It therefore follows from (10.1), (10.2), and (10.3) that $D_0 \cap P_i = \{y_i\}$. Moreover, since i was chosen arbitrarily, $Y = \{y_1, \ldots, y_k\} \subseteq D_0$, implying that Y is a clique in $G(f^d)$. This is a contradition to (iii), since Y cannot be both a clique in $G(f)$ and a clique in $G(f^d)$. This proves Claim 1.

Claim 2. If $(x_1, x_2), (x_2, x_3) \in E(G(f))$ and $(x_1, x_3) \notin E(G(f))$, then $(x_1, x_3) \in E(G(f^d))$.

Suppose that (x_1, x_3) is not an edge of $G(f^d)$. Choose prime implicants

$$\{x_1, x_2\} \cup A_{12}, \{x_2, x_3\} \cup A_{23} \in \mathcal{P}$$

and dual prime implicants

$$\{x_1\} \cup B_1, \{x_3\} \cup B_3 \in \mathcal{D}.$$

By our assumptions,

$$\{x_1, x_2, x_3\} \cap (A_{12} \cup A_{23} \cup B_1 \cup B_3) = \emptyset.$$

By condition (iv), we have

$$|(\{x_1, x_2\} \cup A_{12}) \cap (\{x_1\} \cup B_1)| = 1 \Longrightarrow |A_{12} \cap B_1| = 0 \tag{10.4}$$

$$|(\{x_2, x_3\} \cup A_{23}) \cap (\{x_3\} \cup B_3)| = 1 \Longrightarrow |A_{23} \cap B_3| = 0 \tag{10.5}$$

and

$$|(\{x_1, x_2\} \cup A_{12}) \cap (\{x_3\} \cup B_3)| = 1 \Longrightarrow |A_{12} \cap B_3| = 1 \tag{10.6}$$

$$|(\{x_2, x_3\} \cup A_{23}) \cap (\{x_1\} \cup B_1)| = 1 \Longrightarrow |A_{23} \cap B_1| = 1. \tag{10.7}$$

From (10.6) and (10.7), we can define

$$y_1 = A_{12} \cap B_3$$

$$y_3 = A_{23} \cap B_1$$

and from (10.4) and (10.5),

$$y_1 \neq y_3.$$

On the one hand, because we have assumed that $\{x_1, x_3\}$ is not a subimplicant of the dual f^d, by Theorem 10.4, we claim that every selection with respect to $\{x_1, x_3\}$ is covering. Now,

$$S = \{\{x_1, x_2\} \cup A_{12}, \{x_2, x_3\} \cup A_{23}\}$$

is such a selection, so there exists a prime implicant

$$P_0 \subseteq \{x_2\} \cup A_{12} \cup A_{23}.$$

By condition (iv), (10.4), and (10.5), we have

$$|P_0 \cap (\{x_1\} \cup B_1)| = 1 \Longrightarrow P_0 \cap (\{x_1\} \cup B_1) = y_3$$

and

$$|P_0 \cap (\{x_3\} \cup B_3)| = 1 \Longrightarrow P_0 \cap (\{x_3\} \cup B_3) = y_1.$$

Hence, $\{y_1, y_3\} \subseteq P_0$ and (y_1, y_3) is an edge of $G(f)$, that is,

$$(y_1, y_3) \in E(G(f)). \tag{10.8}$$

On the other hand, since we have also assumed that $\{x_1, x_3\}$ is not a subimplicant of the original function f, we again apply Theorem 10.4, this time in its dual form, by claiming that every dual selection with respect to $\{x_1, x_3\}$ is covering. Now,

$$S' = \{\{x_1\} \cup B_1, \{x_3\} \cup B_3\}$$

is such a dual selection, so there exists a dual prime implicant

$$D_0 \subseteq B_1 \cup B_3.$$

By condition (iv), we have

$$|D_0 \cap (\{x_2, x_3\} \cup A_{23})| = 1 \Longrightarrow D_0 \cap (\{x_2, x_3\} \cup A_{23}) = y_3$$

and

$$|D_0 \cap (\{x_1, x_2\} \cup A_{12})| = 1 \Longrightarrow D_0 \cap (\{x_1, x_2\} \cup A_{12}) = y_1.$$

Hence, $\{y_1, y_3\} \subseteq D_0$ and (y_1, y_3) is an edge of $G(f^d)$, that is,

$$(y_1, y_3) \in E(G(f^d)). \tag{10.9}$$

Finally, combining the conclusions of (10.8) and (10.9), we have a contradiction, since $G(f)$ and $G(f^d)$ cannot share a common edge. *This proves Claim 2.*

Claim 3. *The graph $G(f)$ is P_4-free.*

Suppose $G(f)$ has a copy of P_4 with edges $(x_1, x_2), (x_2, x_3), (x_3, x_4)$ and nonedges $(x_2, x_4), (x_4, x_1), (x_1, x_3)$. By Claim 2, we have $(x_1, x_3), (x_2, x_4)$ are edges in $G(f^d)$. Choose prime implicants

$$\{x_1, x_2\} \cup A_{12}, \ \{x_3, x_4\} \cup A_{34} \in \mathcal{P}$$

and dual prime implicants

$$\{x_1, x_3\} \cup B_{13}, \{x_2, x_4\} \cup B_{24} \in \mathcal{D}.$$

By repeatedly using condition (iv), it is simple to verify that the sets

$$\{x_1, x_2, x_3, x_4\}, \quad A_{12} \cup A_{34}, \quad B_{13} \cup B_{24} \tag{10.10}$$

are pairwise disjoint.

Since $\{x_1, x_4\}$ is not a subimplicant of f, Theorem 10.4 implies that the dual selection

$$S' = \{\{x_1, x_3\} \cup B_{13}, \{x_2, x_4\} \cup B_{24}\}$$

with respect to $\{x_1, x_4\}$ must be covering. So there exists a dual prime implicant $D_0 \in \mathcal{D}$ satisfying $D_0 \subseteq S'$, where

$$S' = (\{x_1, x_3\} \cup B_{13}) \cup (\{x_2, x_4\} \cup B_{24})$$

and $x_1, x_4 \notin D_0$. By the pairwise disjointness of the sets in (10.10), we have

$$S' \cap (\{x_1, x_2\} \cup A_{12}) = \{x_1, x_2\},$$

so

$$D_0 \cap (\{x_1, x_2\} \cup A_{12}) = \{x_2\}.$$

Hence, $x_2 \in D_0$.

In a similar manner, we can show that

$$D_0 \cap (\{x_3, x_4\} \cup A_{34}) = \{x_3\}.$$

Hence, $x_3 \in D_0$.

Thus, we have shown $x_2, x_3 \in D_0$, implying that (x_2, x_3) is an edge of $G(f^d)$, a contradiction to condition (iii). *This proves Claim 3.*

(v) $\Longrightarrow$ (i): Let us assume that f is normal and that $G = G(f)$ is P_4-free. We will show how to construct a read-once formula for f recursively. In order to prove this implication, we will use the following property of P_4-free graphs (cographs) which we will prove in Section 10.4, Theorem 10.7.

Claim 4. *If a graph G is P_4-free, then its complement $\overline{G}$ is also P_4-free; moreover, if G has more than one vertex, precisely one of G and $\overline{G}$ is connected.*

The function is trivially read-once if there is only one variable. Assume that the implication (v) $\Rightarrow$ (i) is true for all functions with fewer than n variables.

By Claim 4, one of G or $\overline{G}$ is disconnected. Suppose G is disconnected, with connected components $G_1, \ldots, G_r$ partitioning the variables of f into r disjoint sets. Then the prime implicants of f are similarly partitioned into r collections $\mathcal{P}_i, (i = 1, \ldots, r)$, defining positive functions $f_1, \ldots, f_r$, respectively, where $G_i = G(f_i)$ and $f = f_1 \vee \cdots \vee f_r$. Clearly, $G(f_i)$ is P_4-free because it is an induced subgraph of $G(f)$, and each f_i is normal for the same reason. Therefore, by induction, there is a read-once expression F_i for each i, and combining these, we obtain a read-once expression for f given by $F = F_1 \vee \cdots \vee F_r$.

Now suppose that $\overline{G}$ is disconnected, and let $H_1,\ldots,H_r$ be the connected components of $\overline{G}$, again partitioning the variables into r disjoint sets. Define $G_i = \overline{H}_i$. We observe that every vertex x_i of G_i is adjacent to every vertex x_j of G_j for $i \neq j$, so each maximal clique of $G(f)$ consists of a union of maximal cliques of $G_1,\ldots,G_r$. Moreover, since f is normal, the maximal cliques are precisely the prime implicants. It now follows that by restricting f to the variables of G_i, we obtain a normal function f_i whose co-occurrence graph $G(f_i) = G_i$ is P_4-free, and $f = f_1 \wedge \cdots \wedge f_r$. Therefore, by induction, as before, there is a read-once expression F_i for each i, and combining these, we obtain a read-once expression for f given by $F = F_1 \wedge \cdots \wedge F_r$. $\square$

Example 10.5. *Let us again consider the function*

$$f_0 = ay \vee cxy \vee bw \vee bz,$$

whose co-occurrence graph $G(f_0)$ was shown in Figure 10.1. Clearly, f_0 is normal, and $G(f_0)$ is P_4-free and has two connected components $G_1 = G_{\{a,c,x,y\}}$ and $G_2 = G_{\{b,w,z\}}$. Using the arguments presented after Claim 4 above, we can handle these components separately, finding a read-once expression for each and taking their disjunction.

For G_1, we note that its complement $\overline{G_1}$ is disconnected with two components, namely, an isolated vertex $H_1 = \{y\}$ and $H_2 = \overline{G_1}_{\{a,c,x\}}$ having two edges; we can handle the components separately and take their conjunction. The complement $\overline{H_2}$ has an isolate $\{a\}$ and edge (c,x) which we combine with disjunction. Finally, complementing (c,x) gives two isolates which are combined with conjunction. Therefore, the read-once expression representing G_1 will be $y \wedge (a \vee [c \wedge x])$.

For G_2, we observe that its complement $\overline{G_2}$ has an isolate $\{b\}$ and edge (w,z), which we combine with conjunction, giving $b \wedge (w \vee z)$. So the read-once expression for f_0 is

$$f_0 = [y \wedge (a \vee [c \wedge x])] \vee [b \wedge (w \vee z)].$$

$\square$

10.4 The properties of P_4-free graphs and cographs

The recursive construction of a read-once expression, which we just saw illustrated at the end of the last section in Example 10.5, was based on the special properties of P_4-free graphs and, in particular, the use of Claim 4. We present these structural and algorithmic properties in this section.

The *complement reducible* graphs, or *cographs*, can be defined recursively as follows:

(1) A single vertex is a cograph.
(2) The union of disjoint cographs is a cograph.
(3) The join of disjoint cographs is a cograph,

where the *join* of disjoint graphs $G_1, \ldots, G_k$ is the graph G with $V(G) = V(G_1) \cup \cdots \cup V(G_k)$ and $E(G) = E(G_1) \cup \cdots \cup E(G_k) \cup \{(x, y) \mid x \in V(G_i), y \in V(G_j),$ for all $i \neq j\}$. An equivalent definition can be obtained by substituting for (3) the rule

(3′) the complement of a cograph is a cograph;

see Exercise 15 at the end of the chapter.

The building of a cograph G from these rules can be represented by a rooted tree T that records its construction, where

(a) the leaves of T are labeled by the vertices of G;
(b) if G is formed from the disjoint cographs $G_1, \ldots, G_k$ $(k > 1)$, then the root r of T has as its children the roots of the trees of $G_1, \ldots, G_k$; moreover,
(c) the root r is labeled 0 if G is formed by the *union* rule (2), and labeled 1 if G is formed by the *join* rule (3).

Among all such constructions, there is a canonical one whose tree T is called the *cotree* and satisfies the additional property that

(d) on every path, the labels of the internal nodes alternate between 0 and 1.

Thus, the root of the cotree is labeled 1 if G is connected and labeled 0 if G is disconnected; an internal node is labeled 0 if its parent is labeled 1, and vice versa. A subtree T_u rooted at an internal node u represents the subgraph of G induced by the labels of its leaves, and vertices x and y of G are adjacent in G if and only if their least common ancestor in the cotree is labeled 1.

Notice that the recursive application of rules (1)–(3) follows a *bottom-up* viewpoint of the construction of G. An alternate *top-down* viewpoint can also be taken, as a recursive decomposition of G, where we repeatedly partition the vertices according to either the connected components of G (union) or the connected components of its complement (join).

One can recognize whether a graph G is a cograph by repeatedly decomposing it this way, until the decomposition either fails on some component H (both H and $\overline{H}$ are connected) or succeeds, reaching all the vertices. The *cotree* is thus built top-down as the decomposition proceeds.[2]

The next theorem gives several characterizations of cographs.

Theorem 10.7. *The following are equivalent for an undirected graph G:*

(i) *G is a cograph.*
(ii) *G is P_4-free.*
(iii) *For every subset X of vertices $(|X| > 1)$, either the induced subgraph G_X is disconnected or its complement $\overline{G}_X$ is disconnected.*

---

[2] This latter viewpoint is a particular case of modular decomposition [358] that applies to arbitrary graphs, and any modular decomposition algorithm will produce a cotree when given a cograph, although such general algorithms [430, 636] are more involved than is necessary for cograph recognition.

In particular, any graph G for which both G and $\overline{G}$ are connected, must contain an induced P_4. This claim appears in Seinsche [817]; independently, it was one of the problems on the 1971 Russian Mathematics Olympiad, and seven students gave correct proofs; see [366]. The full version of the theorem was given independently by Gurvich [422, 423, 425] and by Corneil, Lerchs, and Burlingham [213], where further results on the theory of cographs were developed. Note that it is impossible for both a graph G and its complement $\overline{G}$ to be disconnected; see Exercise 7.

It is rather straightforward to recognize cographs and build their cotree in $O(n^3)$ time. The first linear $O(n+e)$ time algorithm for recognizing cographs appears in Corneil, Perl, and Stewart [214]. Subsequently, other linear time algorithms have appeared in [154, 155, 431]; a fully dynamic algorithm is given in [826], and a parallel algorithm was proposed in [251].

Proof of Theorem 10.7. (iii) $\implies$ (i): This implication follows immediately from the top-down construction of the cotree, as we just discussed.

(i) $\implies$ (ii): Let T be the cotree of G, and for vertex $x \in V(G)$, let p_x denote the path in T from the leaf labeled x to the root of the tree.

Suppose that G contains an induced P_4 with edges $(a,b),(b,c),(c,d)$. Since c and d are adjacent in G, their least common ancestor in T is an internal node u labeled 1. Consider the path p_a. Since both p_c and p_d must meet p_a in an internal node labeled by a 0, it follows that (i) they meet p_a in the same internal node, say v, and (ii) v is an ancestor of u.

Let us consider p_b. Now, p_b meets p_a in an internal node z labeled 1. If z is above v, then the least common ancestor of b and d will be z, which is labeled 1, contradicting the fact that b and d are nonadjacent in G. Furthermore, $z \neq v$, since they have opposite labels, which implies that z must lie below v on p_a. However, in this case, the least common ancestor of b and c will be v, which is labeled 0, contradicting the fact that b and c are adjacent in G. This proves the implication.

(ii) $\implies$ (iii): Assume that G is P_4-free, thus $\overline{G}$ is also P_4-free, since P_4 is self-complementary. Suppose that there is an induced subgraph H of G such that both H and its complement $\overline{H}$ are connected. Clearly, they are also P_4-free and can contain neither an isolated vertex nor a universal vertex (one that is adjacent to all other vertices).

We will construct an ordering $a_1, a_2, \ldots, a_n$ of $V(H)$ such that, for *odd*-indexed vertices a_{2j-1},

$$(a_i, a_{2j-1}) \in E(H), \text{for all } i < 2j-1,$$

and, for *even*-indexed vertices a_{2j},

$$(a_i, a_{2j}) \in E(\overline{H}), \text{for all } i < 2j.$$

In this case, a_n will either be an isolated vertex if n is even, or a universal vertex if n is odd, a contradiction.

Choose a_1 arbitrarily. Since a_1 cannot be universal in H, there is a vertex a_2 such that $(a_1, a_2) \in E(\overline{H})$. Since H is connected, there is a path in H from a_1

to a_2. Consider the shortest such path. It consists of exactly two edges of H, say, $(a_1,a_3),(a_2,a_3) \in E(H)$, since H is P_4-free.

By a complementary argument, since $\overline{H}$ is connected and P_4-free, there is a shortest path in $\overline{H}$ from a_2 to a_3 consisting of exactly two edges of $\overline{H}$, say, $(a_2,a_4),(a_3,a_4) \in E(\overline{H})$. Now we argue that $(a_1,a_4) \in E(\overline{H})$, since otherwise, H would have a P_4.

We continue constructing the ordering in the same manner. Assume we have $a_1,a_2,\ldots,a_{2j}$; we will find the next vertices in the ordering.

(**Find** a_{2j+1}). There is a shortest path in H from a_{2j-1} to a_{2j} consisting of exactly two edges of H, say, $(a_{2j-1},a_{2j+1}),(a_{2j},a_{2j+1}) \in E(H)$. Note that a_{2j+1} has not yet been seen in the ordering, since none of the a_i is adjacent to a_{2j}. We argue, for all $i < 2j - 1$, that $(a_i,a_{2j+1}) \in E(H)$, since otherwise, H would have a P_4 on the vertices $\{a_i,a_{2j-1},a_{2j+1},a_{2j}\}$. Thus, we have enlarged our ordering by one new vertex.

(**Find** a_{2j+2}). There is a shortest path in $\overline{H}$ from a_{2j} to a_{2j+1} consisting of exactly two edges of $\overline{H}$, say, $(a_{2j},a_{2j+2}),(a_{2j+1},a_{2j+2}) \in E(\overline{H})$. Now we argue, for all $i < 2j$, that $(a_i,a_{2j+2}) \in E(\overline{H})$, since otherwise, $\overline{H}$ would have a P_4 on the vertices $\{a_i,a_{2j},a_{2j+2},a_{2j+1}\}$. Thus, we have enlarged our ordering by another new vertex.

Eventually, this process orders all vertices, and the last one a_n will be either isolated or universal, giving the promised contradiction.   □

10.5 Recognizing read-once functions

Given a Boolean function f, can we efficiently determine whether f is a read-once function? This is known as the recognition problem for read-once functions, which we define as follows:

READ-ONCE RECOGNITION
Input: A representation of a positive Boolean function f by its list of prime implicants, namely, its complete DNF expression.
Output: A read-once expression for f, or "failure" if there is none.

Chein [190] and Hayes [479] first introduced read-once functions and provided an exponential-time recognition algorithm for the family. Peer and Pinter [734] also gave an exponential-time factoring algorithm for read-once functions, whose nonpolynomial complexity is due to the need for repeated calls to a routine that converts a DNF representation to a CNF representation, or vice-versa. We have already observed in Section 10.3 that combining Theorem 10.5 with condition (iii) of Theorem 10.6 implies that *recognizing read-once functions has polynomial-time complexity*, although without immediately providing the read-once expression.

In this section, we present the polynomial-time recognition algorithm due to Golumbic, Mintz, and Rotics [400, 401, 402] and analyze its computational

---

Procedure GMR READ-ONCE RECOGNITION(f)

Step 1: Build the co-occurrence graph $G(f)$.

Step 2: Test whether $G(f)$ is P_4-free. If so, construct the cotree T for $G(f)$. Otherwise, exit with "failure."

Step 3: Test whether f is a normal function, and if so, output T as the read-once expression. Otherwise, exit with "failure."

Figure 10.6.   Procedure GMR READ-ONCE RECOGNITION.

complexity. The algorithm is described in Figure 10.6. It is based on condition (v) of Theorem 10.6, that a function is read-once if and only if its co-occurrence graph is P_4-free (namely, is a cograph) and the function is normal. That is, we first test whether $G(f)$ is P_4-free and construct its cotree T, then we test whether f is normal. Passing both tests assures that f is read-once. Moreover, T will provide us with the read-once expression, see Remark 10.4.

Remark 10.4. The reader has no doubt noticed that the cotree of a P_4-free graph is very similar to the parse tree of a read-once expression. On the one hand, when a function is read-once, its parse tree is identical to the cotree of its co-occurrence graph: Just switch the labels $\{0,1\}$ to $\{\vee,\wedge\}$. On the other hand, a cotree always generates a read-once expression that represents "some" Boolean function g. Thus, the question to be asked is:

Given a function f, although $G(f)$ may be P_4-free and, thus, has a cotree T, will the read-once function g represented by T be equal to f or not? (In other words, $G(g) = G(f)$ and, by construction, the maximal cliques of $G(g)$ are precisely the prime implicants of g, so will these also be the prime implicants of f?)

The function $f = ab \vee bc \vee ac$ is a negative example; its graph is a triangle and $g = abc$.

The answer to our question lies in testing normality, that is, comparing the prime implicants of g with those of f, and doing it efficiently. $\square$

The main result of this section is the following:

Theorem 10.8. *[400, 401, 402] Given the complete DNF formula of a positive Boolean function f on n variables, the* GMR *procedure solves the* READ-ONCE RECOGNITION *problem in time* $O(n|f|)$, *where* $|f|$ *denotes the length of the DNF expression.*

Proof. **(Step 1.)** The first step of the GMR procedure is building the graph $G(f)$. If an arbitrary positive function f is given by its DNF expression, that is, as a list of its prime implicants $\mathcal{P} = \{P_1, \ldots, P_m\}$, then the edge set of $G(f)$ can be found in $O(\sum_{i=1}^{m} |P_i|^2)$ time. It is easy to see that this is at most $O(n|f|)$.

(Step 2.) As we saw in Section 10.4, the complexity of testing whether the graph $G(f)$ is P_4-free and providing a read-once expression (its cotree T) is $O(n+e)$, as first shown in [214]. This is at worst $O(n^2)$ and is bounded by $O(n|f|)$. (A straightforward application of Theorem 10.7 would yield complexity $O(n^3)$).

(Step 3.) Finally, we show that the function f can be tested for normality in $O(n|f|)$ time by a novel method, due to [400] and described more fully in [401, 402, 685][3]. As in Remark 10.4, we denote by g the function represented by the cotree T; we will verify that $g = f$.

Testing normality

We may assume that $G = G(f)$ has successfully been tested to be P_4-free, and that T is its cotree. We construct the set of maximal cliques of G recursively, by traversing the cotree T from bottom to top, according to Lemma 10.2 below. For a node x of T, we denote by T_x the subtree of T rooted at x, and we denote by g_x the function represented by T_x. We note that T_x is also the cotree representing the subgraph G_X of G induced by the set X of labels of the leaves of T_x.

First, we introduce some notation. Let $X_1, X_2, \ldots, X_r$ be disjoint sets, and let C_i be a set of subsets of X_i ($1 \le i \le r$). We define the *Cartesian sum* $C = C_1 \otimes \cdots \otimes C_r$, to be the set whose elements are unions of individual elements from the sets C_i (one element from each set). In other words,

$$C = C_1 \otimes \cdots \otimes C_r = \{C_1 \cup \cdots \cup C_r \mid C_i \in C_i, 1 \le i \le r\}.$$

For a cotree T, let $C(T)$ denote the set of all maximal cliques in the cograph corresponding to T. From the definitions of cotree and cograph, we obtain:

Lemma 10.2. *Let G be a P_4-free graph and let T be the cotree of G. Let h be an internal node of T and let $h_1, \ldots, h_r$ be the children of h in T.*

(1) *If h is labeled with 0, then $C(T_h) = C(T_{h_1}) \cup \cdots \cup C(T_{h_r})$.*
(2) *If h is labeled with 1, then $C(T_h) = C(T_{h_1}) \otimes \cdots \otimes C(T_{h_r})$.*

The following algorithm calculates, for each node x of the cotree, the set $C(T_x)$ of all the maximal cliques in the cograph defined by T_x. It proceeds bottom up, using Lemma 10.2, and also keeps at each node x:

$s(T_x)$: The number of cliques in $C(T_x)$. This number is equal to the number of prime implicants in g_x.
$L(T_x)$: The total length of the list of cliques at T_x, namely, $L(T_x) = \sum\{|C| : C \in C(T_x)\}$, which represents the total length of the list of prime implicants of g_x.

A global variable L maintains the overall size of the clique lists as they are being built. (In other words, L is the sum of all $L(T_x)$ taken over all x on the frontier as we proceed bottom up.)

---

[3] In [401] only a complexity bound of $O(n^2 k)$ was claimed, where k is the number of prime implicants; however, using an efficient data structure and careful analysis, it has been shown in [402], following [685], that the method can be implemented in $O(n|f|)$. For the general case of a positive Boolean function given in DNF form, it is possible to check normality in $O(n^3 k)$ time using the results of [538]; see Exercise 13 at the end of the chapter.

Procedure CHECKING NORMALITY(f)

Step 3a: Initialize k to be the number of terms (clauses) in the DNF representation of f. For every leaf a of T, set $\mathcal{C}(T_a) = \{a\}$ and set $s(T_a) = 1$, $L(T_a) = 1$, and $L = n$.

Step 3b: Scan T from bottom to top, at each internal node h reached, let $h_1, \ldots, h_r$ be the children of h and do:

(1) If h is labeled with 0:
- set $s(T_h) = s(T_{h_1}) + \cdots + s(T_{h_r})$
- if $s(T_h) > k$ stop, and claim that f is not normal; otherwise,
- set $L(T_h) = L(T_{h_1}) + \cdots + L(T_{h_r})$
- L remains unchanged
- set $\mathcal{C}(T_h) = \mathcal{C}(T_{h_1}) \cup \cdots \cup \mathcal{C}(T_{h_r})$

(2) If h is labeled with 1:
- set $s(T_h) = s(T_{h_1}) \times \cdots \times s(T_{h_r})$
- if $s(T_h) > k$ stop, and claim that f is not normal; otherwise,
- set $L(T_h) = \Sigma\{|C_1| + \cdots + |C_r| \mid (C_1, \ldots, C_r) \in \mathcal{C}(T_{h_1}) \times \cdots \times \mathcal{C}(T_{h_r})\}$
- set $L \leftarrow L + L(T_h) - [L(T_{h_1}) + \cdots + L(T_{h_r})]$
- if $L > |f|$ stop, and claim that f is not normal; otherwise,
- set $\mathcal{C}(T_h) = \mathcal{C}(T_{h_1}) \otimes \cdots \otimes \mathcal{C}(T_{h_r})$

Step 3c: Let y be the root of T, and let $\mathcal{C}(T_y)$ be the set of maximal cliques of the cograph, obtained by the preceding step.

- If $s(T_y) \neq k$ or if $|L| \neq |f|$ stop, and claim that f is not normal.
- Otherwise, compare the set $\mathcal{C}(T_y)$ with the set of prime implicants (from the DNF) of f, using radix sort as described in the proof. If the sets are equal, claim that f is normal. Otherwise, claim that f is not normal.

Figure 10.7. Procedure CHECKING NORMALITY.

The steps of the normality-checking procedure are given in Figure 10.7. This procedure correctly tests normality because it tests whether the maximal cliques of the cograph are precisely the prime implicants of f.

Complexity analysis

The purpose of comparing $s(T_h)$ with k at each step is simply a speedup mechanism to assure that the number of cliques never exceeds the number of prime implicants. Similarly, calculating $L(T_h)$, that is, $|g_h|$ and comparing L with $|f|$ at each step assures that the overall length of the list of cliques will never exceed the sum of the lengths of the prime implicants. (Note that we precompute L, and test against $|f|$ *before* we actually build a new set of cliques.)

For efficiency, we number the variables $\{x_1, x_2, \ldots, x_n\}$, and maintain both the prime implicants and the cliques as lists of their variables. Then, each collection of cliques $\mathcal{C}(T_x)$ is maintained as a list of such lists. In this way, constructing $\mathcal{C}(T_h)$ in Step 3b(1) can be done by concatenating the lists $\mathcal{C}(T_{h_1}), \ldots, \mathcal{C}(T_{h_r})$, and constructing $\mathcal{C}(T_h)$ in Step 3b(2) can be done by creating a new list of cliques by repeatedly taking r (sub)cliques, one from each set $\mathcal{C}(T_{h_1}), \ldots, \mathcal{C}(T_{h_r})$ and concatenating these r (disjoint) lists of variables.

Thus, the overall calculation of $C(T_h)$ takes at most $O(|f|)$ time. Since the number of internal nodes of the cotree is less than n, the complexity of Steps 3a and 3b is $O(n|f|)$.

It remains to compare the list of the prime implicants of f with the list of the maximal cliques $C(T_y)$, where y is the root of T. This can be accomplished using radix sort in $O(nk)$ time. Initialize two $k \times n$ bit matrices $\mathbf{P}$ and $\mathbf{C}$ filled with zeros. Each prime implicant P_i is traversed (it is a list of variables), and for every $x_j \in P_i$ we assign $\mathbf{P}_{i,j} \leftarrow 1$, thus, converting it into its characteristic vector, which will be in row i of $\mathbf{P}$. Similarly, we traverse each maximal clique C_i and convert it into its characteristic vector, which will be in row i of $\mathbf{C}$. It is now a straightforward procedure to lexicographically sort the rows of these two matrices and compare them in $O(nk)$ time.

This concludes the proof, since the complexity of each step is bounded by $O(n|f|)$. $\square$

Of course, the form in which a function f is given influences the computational complexity of recognizing whether it is read-once. For example, if f is initially represented by an arbitrary Boolean expression, we are required to pay a preprocessing expense to test that f is positive and to transform f into its DNF expression in order to apply the GMR procedure. The same would be true if f were to be given as a BDD. This preprocessing could be exponential in the size of the original input.

Actually, for a general (nonmonotone) DNF expression ψ, Theorem 1.30 (Section 1.11) implies that it is NP-hard to decide whether ψ represents a read-once function, and Aizenstein et al. [13] proved that this decision problem is in co-NP, but the question remains open for BDDs.

As for positive expressions (other than DNFs), the problem is co-NP-complete. More precisely, we are now going to show that it is co-NP-complete to decide whether a positive Boolean function given by an arbitrary positive Boolean expression is read-once.

In the remainder of this section, we let g_0 be the positive Boolean function defined by the quadratic DNF formula $\phi_0 = x_1 y_1 \vee \ldots \vee x_n y_n$.

Lemma 10.3. *(Gurvich and Khachiyan [429]) When h is a positive function defined by a CNF formula θ on the variables $x_1, y_1, \ldots, x_n, y_n$, it is co-NP-complete to verify the equality $g_0 \vee h = g_0$. Moreover, the problem remains co-NP-complete under the additional conditions that h has no linear implicants and no quadratic implicants.*

Proof. Let θ' be any CNF such that no variable appears more than three times in θ'. It is NP-complete to decide whether θ' is satisfiable; see [371, 932].

Now, replace $\bar{x}_i$ by y_i in θ' for all $i \in \{1, 2, \ldots, n\}$, denote by θ the resulting (positive) CNF, and denote by h the Boolean function represented by θ. It is easy to see that θ' is satisfiable if and only if $g_0 < h$ or, equivalently, if and only if $g_0 \vee h \neq g_0$. Moreover, if the number of clauses of θ' is large enough (say, at least 7), then θ has no implicants of degree smaller than three. $\square$

Remark 10.5. Recall that verifying the equality of two Boolean functions defined by positive DNF and CNF expressions, respectively, is exactly POSITIVE DNF DUALIZATION, which is not co-NP-complete unless every problem of co-NP can be solved in quasi-polynomial time; see Section 4.4.2. Yet, verifying the similar identity $g_0 \vee h = g_0$ appears harder. $\square$

Lemma 10.4. *Let h be a positive function without linear implicants. If the function $f = g_0 \vee h$ is read-once, then f is quadratic.*

Proof. Since h has no linear implicants, $x_i y_i$ is a prime implicant of $f = g_0 \vee h$ for each $i \in \{1, 2, \ldots, n\}$.

If f is read-once, let T be its associated parse tree. By definition, the leaves of T are labeled by the variables $x_1, y_1, \ldots, x_n, y_n$, each of which appears at most once, and in fact, exactly once, since $x_i y_i$ is a prime implicant of f for each $i \in \{1, 2, \ldots, n\}$. All other nodes of T are labeled by $\vee$ and $\wedge$.

For each $i \in \{1, 2, \ldots, n\}$, let us consider in T two paths p_i and r_i from the root v_0 to the leaves labeled by x_i and y_i, respectively, and denote by v_i the last common vertex of these two paths. Obviously, v_i is a $\wedge$-vertex, since $x_i y_i$ is a prime implicant of f. For the same reason, vertex v_i is of degree three in T: The corresponding three edges lead towards x_i, y_i, and v_0. Moreover, for the same reason, paths p_i and r_i have no other $\wedge$-vertices.

Since $i \in \{1, 2, \ldots, n\}$ was chosen arbitrarily, we conclude that every path in T from the root to a leaf contains exactly one $\wedge$-vertex, and that this vertex is of degree three. This easily implies that every prime implicant of f is quadratic. $\square$

The read-once functions constructed as in the previous lemma are more completely characterized in Exercise 21.

Remark 10.6. It is easy to demonstrate that the condition on h is essential in Lemma 10.4. Let us consider, for example, the function

$$h = (x_1 \vee x_2 \vee \ldots \vee x_n \vee y_1)(x_1 \vee x_2 \vee \ldots \vee x_n \vee y_2) \ldots (x_1 \vee x_2 \vee \ldots \vee x_n \vee y_n).$$

Obviously, the corresponding function $g_0 \vee h = x_1 \vee x_2 \vee \ldots \vee x_n \vee (y_1 y_2 \ldots y_n)$ is read-once, but it contains the prime implicant $(y_1 y_2 \ldots y_n)$, which is not quadratic when $n > 2$. Yet, in this case h has n linear prime implicants, namely, $x_1, x_2, \ldots, x_n$. $\square$

Lemma 10.5. *Let h be a positive function without linear or quadratic implicants. The function $f = g_0 \vee h$ is read-once if and only if $g_0 \vee h = g_0$.*

Proof. The "if" part is obvious, since function g_0 is read-once, while the "only if" part follows immediately from the previous lemma.   □

Remark 10.7. Again, it is easy to demonstrate that the assumptions on h are essential. For instance, when $n = 5$ and

$$h = (x_1 \vee y_2 \vee x_3 \vee x_4)(x_1 \vee y_2 \vee y_3 \vee y_4)(y_1 \vee x_2 \vee x_3 \vee x_4)(y_1 \vee x_2 \vee y_3 \vee y_4),$$

we find

$$g_0 \vee h = (x_1 \vee y_2)(x_2 \vee y_1) \vee (x_3 \vee x_4)(y_3 \vee y_4) \vee x_5 y_5,$$

so that $g_0 \vee h$ is read-once but distinct from g_0.   □

Now we are ready to prove the desired result.

Theorem 10.9. *For a Boolean function f given by a positive $\vee$-$\wedge$ expression, it is co-NP-complete to decide whether f is read-once.*

Proof. NP-hardness immediately follows from Lemmas 10.3 and 10.5. It remains to show that the decision problem is in co-NP. This will follow from Theorem 10.6: f is read once if and only if every prime implicant P of f and prime implicant D of f^d have exactly one variable in common. Hence, to disprove that f is read-once, it is sufficient (and necessary) to exhibit dual prime implicants P_0 and D_0 with at least two common variables. Furthermore, to verify that P_0 is a prime implicant of f, it is sufficient to check that

   (i) f is true if all variables of P_0 are true, while all others are false.
   (ii) f is false if all variables of P_0 but one are true, while all others are false.

This can be checked in polynomial time.

Similarly, we can check that D_0 is a prime implicant of f^d. To do so, it is enough to dualize the expression of f by swap of $\vee$ and $\wedge$ (see Theorem 1.3 in Section 1.3).   □

Remark 10.8. The recognition problem remains in co-NP when the function f is given by any polynomially computable representation (or polynomial oracle) and is guaranteed to be positive. Moreover, Aizenstein et al. [13] showed that the problem remains in co-NP even without assumption of the positivity of f.   □

Remark 10.9. Interestingly, the same arguments (three lemmas and theorem) prove that it is a co-NP-complete problem to recognize whether a positive Boolean formula, $\phi_0 \vee \theta$, defines a *quadratic* Boolean function. Indeed, the corresponding Boolean function $g_0 \vee h$ is quadratic if and only if $g_0 \vee h = g_0$, provided h has no implicants of degree less than three.   □

Exercise 27 at the end of the chapter raises some related open questions regarding the complexity of recognizing a read-once function depending on the representation of the function. For example, we may be fortunate to receive f as a very compact expression, yet not know how to take advantage of this. When might it be possible to efficiently construct the co-occurrence graph of a Boolean function and test normality for forms other than a positive DNF representation?

10.6 Learning read-once functions

I've got a secret. It's a Boolean function f. Can you guess what it is? You can ask me questions like: "What is the value of f at the point X?" Can you figure out my mystery function with just 20 questions?

The answer, of course, is *yes*, 20 questions are enough if the number of variables is at most 4. Otherwise, the answer is *no*. If there are n variables, then there will be 2^n independent points to be queried before you can "know" the function.

Suppose I give you a clue: The function f is a positive Boolean function. Now can you learn f with fewer queries?

Again the answer is *yes*. The extra information given by the clue allows you to ask fewer questions in order to learn the function. For example, in the case $n = 4$, first try $(1,1,0,0)$. If the answer is true, then you immediately know that $(1,1,1,0)$, $(1,1,0,1)$ and $(1,1,1,1)$ are all true. If the answer is false, then $(1,0,0,0)$, $(0,1,0,0)$ and $(0,0,0,0)$ are all false. Either way, you asked one question and got four answers. Not bad. Now if you query $(0,0,1,1)$, you will similarly get two or three more free answers. In the worst case, it could take 10 queries to learn the function (rather than 16 had you queried each point).

Learning a Boolean function in this manner is sometimes called EXACT LEARN-ING WITH QUERIES; see Angluin [21]. It receives as input an oracle for a Boolean function f, that is, a "black box" that can answer a query on the value of f at a given Boolean point in constant time. It then attempts to learn the value of f at all 2^n points and outputs a Boolean expression that is logically equivalent to f.

If we know something extra about the structure of the function f, then it may be possible to reduce the number of queries required to learn the function. We saw this earlier in our example with the clue (that the mystery function was positive). However, even for positive functions, the number of queries needed to learn the function remains exponential.

The situation is much better for read-once functions. In this case, the number of required queries can be reduced to a polynomial number, and the unique read-once formula can be produced, provided we "know" that the function is read-once. Thus, the read-once functions constitute a very natural class of functions that can be learned efficiently, and, for this reason, they have been extensively studied within the computational learning theory community.

For our purposes, we define the problem as follows:

> **Procedure** AHK READ-ONCE EXACT LEARNING(f)
>
> Step 0: Check whether f is a constant function, using the oracle: If $f(\mathbf{1}) = 0$ then f is constant 0; if $f(\mathbf{0}) = 1$ then f is constant 1.
> Step 1: Use the oracle to construct the co-occurrence graph $G(f)$.
> Step 2: Build a cotree T for $G(f)$ ("knowing" a priori that it must be P_4-free and thus will succeed).
> Step 3: Immediately output T as the read-once expression ("knowing" a priori that f is normal).

Figure 10.8. Procedure AHK READ-ONCE EXACT LEARNING.

READ-ONCE EXACT LEARNING

Input: A black-box oracle to evaluate f at any given point, where f is known a priori to be a positive read-once function.

Output: A read-once factorization for f.

Remark 10.10. There is a subtle but significant difference between the EXACT LEARNING problem and the RECOGNITION problem. With recognition, we have a DNF expression for f and must determine whether it represents a read-once function. With exact learning, we have an oracle for f whose correct usage relies upon the a priori assumption that the function to be learned is read-once. So the input assumptions are different, but the output goal in both cases is a correct read-once expression for f. Also, when measuring the complexity of recognition, we count the algorithmic operations; when measuring the complexity of exact learning, we must count both the operations implemented by the algorithm and the number of queries to the oracle.   □

As we saw in Section 10.5, the GMR recognition procedure: (1) uses the DNF expression to construct the co-occurrence graph $G(f)$, then (2) tests whether $G(f)$ is P_4-free and builds a cotree T for it, and (3) uses T and the original DNF formula to test whether f is normal; if so, T is the read-once expression.

In contrast to this, Angluin, Hellerstein, and Karpinski [22] give the exact learning algorithm in Figure 10.8.

The main difference between AHK exact learning and GMR recognition that concerns us will be Step 1, that is, *how to construct $G(f)$ using an oracle*. We outline the solution through a series of exercises at the end of the chapter.

(A) In a greedy manner, we can determine whether a subset $U \subseteq X$ of the variables contains a prime implicant, and find one when the answer is positive. Exercise 16 gives such a routine FIND-PI-IN(U), which has complexity $O(n)$ plus $|U|$ queries to the oracle. A similar greedy algorithm FIND-DUALPI-IN(U) will find a dual prime implicant contained in U.

(B) An algorithm FIND-ESSENTIAL-VARIABLES is developed in Exercises 17, 18, and 19 that not only finds the set Y of essential variables[4] but also, in the process, for each variable x_i in Y, generates a prime implicant $P[i]$ and a dual

---

[4] We have generally assumed throughout this chapter that all of the variables for a Boolean function f (and hence for f^d) are essential. However, in the exact learning problem, we may wish to drop this assumption and then need to find the set of essential variables.

prime implicant $D[i]$ containing x_i. This algorithm uses FIND-PI-IN and FIND-DUALPI-IN and can be implemented to run in $O(n^2)$ time using $O(n^2)$ queries to the oracle.

(C) Finally, we construct the co-occurrence graph $G(f)$ based on the following Lemma (whose proof is proposed as Exercise 14):

Lemma 10.6. *Let f be a nonconstant read-once function over the variables $N = \{x_1, x_2, \ldots, x_n\}$. Suppose that D_i is a dual prime implicant containing x_i but not x_j, and that D_j is a dual prime implicant containing x_j but not x_i. Let $R_{i,j} = (N \setminus (D_i \cup D_j)) \cup \{x_i, x_j\}$. Then (x_i, x_j) is an edge in the co-occurrence graph $G(f)$ if and only if $R_{i,j}$ contains a prime implicant.*

We obtain $G(f)$ using the oracle in the following way: For each pair of essential variables x_i and x_j,

C.1: if $x_i \in D[j]$ or $x_j \in D[i]$, then (x_i, x_j) is *not* an edge of $G(f)$;

C.2: otherwise, construct $R_{i,j}$ from $D[i]$ and $D[j]$ and test whether $R_{i,j}$ contains a prime implicant using just one query to the oracle, namely, is $f(X^{R_{i,j}}) = 1$? If so, then (x_i, x_j) is an edge in $G(f)$; otherwise, it is not an edge.

Complexity

The computational complexity of the procedure is determined as follows. Step 0 requires two queries to the oracle. Step 1 constructs the co-occurrence graph $G(f)$ by first calling the algorithm FIND-ESSENTIAL-VARIABLES (Part B) to generate $P[i]$ and $D[i]$ for each variable x_i in $O(n^2)$ time using $O(n^2)$ queries, then it applies Lemma 10.6 (Part C) to determine the edges of the graph. Step C.1 can be done in the same complexity as Step B; however, Step C.2 uses $O(n^3)$ time and $O(n^2)$ queries, since, for each pair i, j, we have $O(n)$ operations and 1 query. Step 2, building the cotree T for $G(f)$ takes $O(n^2)$ time using one of the fast cograph algorithms of [154, 214, 431], and Step 3 takes no time at all.

To summarize, the overall complexity using the method of Angluin, Hellerstein and Karpinski [22] will be $O(n^3)$ time and $O(n^2)$ queries. However, in an unpublished manuscript [250], Dahlhaus subsequently reported an alternative to Step C.2 using only $O(n^2)$ time. (Further generalizations by Raghavan and Schach [774] lead to the same time bound.)

The main result, therefore, is the following:

Theorem 10.10. *The READ-ONCE EXACT LEARNING problem can be solved with the AHK procedure in $O(n^2)$ time, using $O(n^2)$ queries to the oracle.*

Proof. The correctness of the AHK exact learning procedure follows from Lemma 10.6, Exercises 17–19, and Remark 10.4.   ☐

Remark 10.11. If a lying, deceitful, cunning adversary were to place a *non*-read-once function into our "black-box" query oracle, then the exact learning method described here would give an incorrect identification answer, since the "a priori read-once" assumption is vital for the construction of $G(f)$. (See the discussion in Exercise 28 concerning what might happen if such an oracle were to be applied to a non-read-once function.)   □

Further topics relating computational learning theory with read-once functions may be found in [13, 22, 162, 484, 396, 397, 482, 749, 774, 838, 884, etc.].

10.7 Related topics and applications of read-once functions

In this section, we briefly mention three topics related to read-once functions and application areas in which they play an interesting role.

10.7.1 The readability of a Boolean function

Suppose a given function f is not a read-once function. In this case, we may still want to obtain an expression that is logically equivalent to f and that has a small number of repetitions of the variables. The notion of the readability of a Boolean function is used to capture this notion.

We call a Boolean expression *read-m* if each variable appears at most m times in the expression. A Boolean function f is defined to be a *read-m function* if it has an equivalent read-m expression. Finally, the *readability* of f is the smallest number m such that f is a read-m function.

The definition of readability does not require the function to be positive. Thus, characterizing read-m Boolean functions and characterizing positive read-m Boolean functions appear to be separate questions.

As noted earlier in Section 10.5, recognizing whether a nonmonotone DNF represents a read-once function is NP-hard. The same result holds for recognizing whether a nonmonotone DNF represents a read-m function when $m > 1$. (This follows again from Theorem 1.30.)

To the best of our knowledge, the complexity of recognizing read-m functions given by an irredundant positive DNF is open for all fixed $m \geq 2$. Golumbic, Mintz, and Rotics therefore proposed in [401] to investigate restrictions of the general problem to special cases of positive Boolean functions f identified by the structure of the co-occurrence graph $G(f)$. As a first step in this direction, they showed the following result:

Theorem 10.11. *[401] Let f be a positive Boolean function. If f is a normal function and its co-occurrence graph $G(f)$ is a partial k-tree, then f is a read-2^k function and a read-2^k expression for f can be obtained in polynomial ($O(n^{k+1})$) time.*

Notice that if $G(f)$ is a tree, then f would immediately be normal. Therefore, in the case of $k = 1$, Theorem 10.11 reduces to the following:

Corollary 10.1. *Let f be a positive Boolean function. If $G(f)$ is a tree, then f is a read-twice function.*

10.7.2 Factoring general Boolean functions

Factoring is the process of deriving a parenthesized Boolean expression or *factored form* representing a given Boolean function. Since, in general, a function will have many factored forms, the problem of factoring Boolean functions into shorter, more compact, logically equivalent expressions is one of the basic operations in the early stages in designing logic circuits. Generating an optimum factored form (a shortest length expression) is an NP-hard problem. Thus, heuristic algorithms have been developed in order to obtain good factored forms.

An exception to this, as we have already seen, are the read-once functions. For a read-once function f, the read-once expression is unique, it can be determined very efficiently; moreover, it is the shortest possible expression for f. According to [734], read-once functions account for a significant percentage of functions that arise in real circuit applications. Some smaller or specifically designed circuits may indeed be read-once functions, but most often they will not even be positive functions. Nevertheless, we can use the optimality of factoring read-once functions as part of a heuristic method.

Such an approach for factoring general Boolean functions has been described in [399, 686], and is based on graph partitioning. Their heuristic algorithm is recursive and operates on the function and its dual to obtain the better factored expression. As a special class, which appears in the lower levels of the recursive factoring process, are the read-once functions.

The original function f is decomposed into smaller components, for example, $f = f_1 \vee f_2 \vee f_3$, and when a component is recognized to be read-once, a special purpose subroutine (namely, the GMR procedure of Section 10.5) is called to factor that read-once component efficiently and optimally. Their method has been implemented in the SIS logic synthesis environment, and an empirical evaluation indicates that the factored expressions obtained are usually significantly better than those from previous fast algebraic factoring algorithms and are quite competitive with previous Boolean factoring methods, but with lower computation costs (see [685, 686]).

10.7.3 Positional games

We introduce here the notions of normal, extensive, and positional game forms, and then show their relationship to read-once functions.

Definition 10.1. *Given three finite sets* $S^1 = \{s_1^1, s_2^1, ..., s_{m_1}^1\}$, $S^2 = \{s_1^2, s_2^2, ..., s_{m_2}^2\}$, *which are interpreted as the sets of strategies of the players* 1 *and* 2, *and* $X = \{x_1, x_2, ..., x_k\}$, *which is interpreted as the set of outcomes, a* game form *(of two players) is a mapping* $g : S^1 \times S^2 \to X$, *which assigns an outcome* $x(s^1, s^2) \in X$ *to every pair of strategies* $s^1 \in S^1$, $s^2 \in S^2$.

A convenient representation of a game form is a matrix $M = M(g)$ whose rows are labeled by S^1, whose columns are labeled by S^2, and whose elements are labeled by X. For example,

$$M_1 = \begin{bmatrix} x_1 & x_2 \\ x_2 & x_1 \end{bmatrix}.$$

Each outcome $x \in X$ may appear several times in $M(g)$, because g may not be injective. We can interpret $M(g)$ as "a game in normal form in which the payoff is not specified, yet."

Definition 10.2. *Two strategies s_1^i and s_2^i of player i, where $i = 1$ or 2, are called* equivalent *if for every strategy s^{3-i} of the opponent, we have $g(s_1^i, s^{3-i}) = g(s_2^i, s^{3-i})$; in other words, if in matrix $M(g)$, the rows ($i = 1$) or the columns ($i = 2$) corresponding to the strategies s_1^i and s_2^i, are equal.*

We will restrict ourselves by studying the game forms *without equivalent strategies*.

Definition 10.3. *Given a read-once function f, we can interpret its parse tree (or read-once formula) $T(f)$ as an extensive game form (or game tree) of two players. The leaves $X = \{x_1, x_2, ..., x_k\}$ of T are the final positions or outcomes. The internal vertices of T are the internal positions. The game starts at the root of T and ends in a final position $x \in X$. Each path from the root to a final position (leaf) is called a* play. *If an internal node v is labeled by $\vee$ (respectively, by $\wedge$), then it is the turn of player 1 (respectively, player 2) to move in v. This player can choose any vertex that is a child of v in T.*

A strategy *of a player is a mapping which assigns a move to every position in which this player has to move. In other words, a strategy is a plan of how to play in every possible situation.*

Any pair of strategies s^1 of player 1 and s^2 of player 2 define a play $p(s^1, s^2)$ and an outcome $x(s^1, s^2)$ that would appear if both players implement these strategies.

Two strategies s_1^i and s_2^i of player i, where $i = 1$ or 2, are called equivalent *if for every strategy s^{3-i} of the opponent the outcome is the same, that is, if $x(s_1^i, s^{3-i}) = x(s_2^i, s^{3-i})$. By suppressing all but one (arbitrary) strategy from every class of equivalent strategies, we obtain two reduced sets of strategies, denoted by $S^1 = \{s_1^1, s_2^1, ..., s_{m_1}^1\}$ and $S^2 = \{s_1^2, s_2^2, ..., s_{m_2}^2\}$.*

The mapping $g : S^1 \times S^2 \to X$, which assigns the outcome $x(s^1, s^2) \in X$ to every pair of strategies $s^1 \in S^1$, $s^2 \in S^2$, defines a game form, which we call the normal form *of the corresponding extensive game form.*

Note that such a mapping $g = g(T)$ may be not injective because different pairs of strategies may generate the same play.

We call a game form g positional *if it is the normal form of an extensive game form, that is, if $g = g(T(f))$ for a read-once function f.*

Example 10.6. *In the extensive game form defined by the read-once formula $((x_1 \vee x_2)x_3 \vee x_4)x_5$, each player has three strategies, and the corresponding normal game form is given by the following (3×3)-matrix:*

$$M_2 = \begin{bmatrix} x_1 & x_3 & x_5 \\ x_2 & x_3 & x_5 \\ x_4 & x_4 & x_5 \end{bmatrix}.$$

The game form given by the matrix

$$M_3 = \begin{bmatrix} x_1 & x_1 \\ x_2 & x_3 \end{bmatrix}$$

is also generated by a read-once formula, namely, by $x_1 \vee x_2 x_3$. $\qquad\square$

Our aim is to characterize the positional game forms.

Definition 10.4. *Let us consider a game form* g *and the corresponding matrix* $M = M(g)$. *We associate with* M *two DNFs, representing two Boolean functions* $f_1 = f_1(g) = f_1(M)$ *and* $f_2 = f_2(g) = f_2(M)$, *respectively, by first taking the conjunction of all the variables in each row (respectively, each column) of* M, *and then taking the disjunction of all these conjunctions for all rows (respectively, columns) of* M.

We call a game form g *(as well as its matrix* M*)* tight *if the functions* f_1 *and* f_2 *are mutually dual.*

Example 10.7. *Matrix* M_2 *of Example 10.6 generates the functions* $f_1(M_2) = x_1 x_3 x_5 \vee x_2 x_3 x_5 \vee x_4 x_5$ *and* $f_2(M_2) = x_1 x_2 x_4 \vee x_3 x_4 \vee x_5$. *These functions are mutually dual, thus the game form is tight. Matrix* M_3 *is also tight, because its functions* $f_1(M_3) = x_1 \vee x_2 x_3$ *and* $f_2(M_3) = x_1 x_2 \vee x_1 x_3$ *are mutually dual. However,* M_1 *is not tight, because its functions* $f_1(M_1) = f_2(M_1) = x_1 x_2$ *are not mutually dual.* $\qquad\square$

Remark 10.12. It is proven in [421] that a normal game form (of two players) is Nash-solvable (that is, for an arbitrary payoff the obtained game has at least one Nash equilibrium in pure strategies) if and only if this game form is tight. $\square$

Theorem 10.12. *Let* f *be a read-once function;* $T = T(f)$, *the parse tree of* f *interpreted as an extensive game form;* $g = g(T)$, *its normal form;* $M = M(g)$, *the corresponding matrix; and* $f_1 = f_1(M)$, $f_2 = f_2(M)$, *the functions generated by* M. *Then,* $f_1 = f$ *and* $f_2 = f^d$.

Proof. By induction. For a trivial function f the claim is obvious. If $f = f' \vee f''$, then $f_1 = f_1' \vee f_1''$ and $f_2 = f_2' \wedge f_2''$. If $f = f' \wedge f''$, then $f_1 = f_1' \wedge f_1''$ and $f_2 = f_2' \vee f_2''$. The theorem follows directly from the definition of strategies. $\square$

Definition 10.5. *We call a game form* $g : S^1 \times S^2 \to X$ *(as well as the corresponding matrix* M*)* rectangular *if every outcome* $x \in X$ *occupies a rectangular array in* M, *that is, if the following property holds:* $g(s_1^1, s_1^2) = g(s_2^1, s_2^2) = x$ *implies* $g(s_1^1, s_2^2) = g(s_2^1, s_1^2) = x$.

For example, matrices M_2 and M_3 above are rectangular, while M_1 is not.

Theorem 10.13. *A game form g and its corresponding matrix M are rectangular if and only if every prime implicant of $f_1(M)$ and every prime implicant of $f_2(M)$ have exactly one variable in common.*

Proof. Obviously, any two such prime implicants must have *at least* one common variable because every row and every column in M intersect, that is, row s^1 and column s^2 always have a common outcome $x = g(s^1, s^2)$. Let us suppose that they have another common outcome, namely, that there exist strategies s_i^1 and s_j^2 such that $g(s^1, s_j^2) = g(s_i^1, s^2) = x' \neq x$. Then, $g(s^1, s^2) = x$; thus, g is not rectangular.

Conversely, let us assume that g is not rectangular, that is, $g(s_1^1, s_1^2) = g(s_2^1, s_2^2) = x$, while $g(s_1^1, s_2^2) = x' \neq x$. Then row s_1^1 and column s_2^2 have at least two outcomes in common, namely, x and x'.   □

Theorem 10.14. *(Gurvich [423, 424]). A normal game form g is positional if and only if it is tight and rectangular.*

Proof. The normal form g corresponding to an extensive game form $T(f)$ is tight in view of Theorem 10.12, and g is rectangular in view of Theorem 10.13 and Theorem 10.6(iv).

Conversely, if g is tight and rectangular, then, by definition, $f_1(g)$ and $f_2(g)$ are dual. Further, according to Theorem 10.13, every prime implicant of $f_1(g)$ and every prime implicant of $f_2(g)$ have exactly one variable in common. Hence, by Theorem 10.6(iv), $f_1(g)$ and $f_2(g)$ are read-once; thus, g is positional.   □

Remark 10.13. In [423], this theorem is generalized for game forms of n players. The criterion is the same: A game form is positional if and only if it is tight and rectangular. The proof is based on the cotree decomposition of P_4-free graphs; see Sections 10.3, 10.5.   □

10.8 Historical notes

We conclude this chapter with a few brief remarks about the history of read-once functions. It is important to distinguish between

(A) the algorithms to verify read-onceness based on the parse tree decomposition, or, in other words, the $\vee$-$\wedge$ *disjoint* decomposition, and

(B) the criteria of read-onceness based on "rectangularity" of the pair f and f^d, or P_4-freeness and normality of f.

In fact, (A) is at least 20 years older than (B). The oldest reference we know is by Kuznetsov [592], in 1958. Kuznetsov claims that the parse tree decomposition is well defined (i.e., it is unique), and he also says a few words on how to get it; De Morgan's formulae are mentioned, too. This implies (A), though read-onceness is not mentioned explicitly in this paper.

In his 1978 doctoral thesis, Gurvich [423] remarked that the parse tree decomposition is a must for any minimum $\vee$-$\wedge$ formula for f, in both the monotone and general cases. However, a bit earlier, Michel Chein's short paper [190] based on his doctoral thesis of 1967 may be the earliest one mentioning "read-once" functions. J. Kuntzmann (Chein's thesis advisor) raised the question a few years earlier in the first edition (1965) of his book "Algèbre de Boole" [589], mentioning a problem called "dédoublement de variables," and in the second edition (1968) he cites Chein's work.

What Chein does (using our notation) is to look at the bipartite graph $B(f) = (\mathcal{P}, V, E)$, where $\mathcal{P}$ is the set of prime implicants, V is the set of variables, and edges represent containment, that is, for all $P \in \mathcal{P}, v \in V$,

$$(P, v) \in E \iff v \in P.$$

The reader can easily verify that $B(f)$ is connected if and only if the graph $G(f)$ is connected.

Chein's method is to check which of $B(f)$ or $B(f^d)$ is disconnected (failing if both are connected) and continuing recursively. An exponential price is paid for dualizing. Peer and Pinter [734] do something quite similar.

By contrast, as the reader also now knows, the polynomial-time algorithm of Golumbic, Mintz, Rotics similarly acts on $G(f)$ and $G(f^d)$, but $G(f^d)$ is gotten for free, without dualizing, thanks to the fact that $G(f^d)$ equals the graph complement of $G(f)$ (by Theorem 10.6), paying only an extra low price to check for normality.

Finally, to clarify complexities using our notation: Clearly, building $B(f^d)$ involves dualization of f; however, building $G(f^d)$ can be done in polynomial time for any positive Boolean function (i.e., without any dualization). The implication is that one can compute a unique read-once decomposition for any (positive) read-once Boolean function in polynomial time; see also Ramamurthy's book [777].

To summarize, testing read-onceness and obtaining a parse tree decomposition is just an extreme case of representing f by a minimum length $\vee$-$\wedge$ formula. The parse tree decomposition implies (A) and has been known since 1958 [592], whereas (B) has been known since 1977 [422, 423] and been rediscovered independently several times thereafter [293, 294, 548, 696]. Dominique de Werra has described it as "an additional interesting example of rediscovery by people from the same scientific community. It shows that the problem has kept its importance and [those involved] have good taste."

10.9 Exercises

1. Prove that a Boolean function f for which some variable appears in its positive form x in one prime implicant and in its negative form $\overline{x}$ in another prime implicant cannot be a read-once function.
2. Verify Remark 10.1; namely, if T is a proper dual subimplicant of f, then there exists a prime implicant of f, say, P, such that $P \cap T = \emptyset$.

3. Consider the positive Boolean function

$$f = x_1 x_2 \vee x_1 x_5 \vee x_2 x_3 \vee x_2 x_4 \vee x_3 x_4 \vee x_4 x_5.$$

   (a) Draw the co-occurrence graph $G(f)$. Prove that f is not a read-once function.
   (b) Let $T = \{x_1, x_4\}$. What are the sets $\mathcal{P}_0, \mathcal{P}_{x_1}, \mathcal{P}_{x_4}$? Prove that T is a dual subimplicant of f by finding a noncovering selection.
   (c) Let $T' = \{x_3, x_4, x_5\}$. What are the sets $\mathcal{P}'_0, \mathcal{P}'_{x_3}, \mathcal{P}'_{x_4}, \mathcal{P}'_{x_5}$? Prove that T' is not a dual subimplicant of f.

4. Consider the function $f = ab \vee bc \vee cd$. Verify that $\{a, d\}$ is not a dual subimplicant.

5. Verify that the function

$$f = adg \vee adh \vee bdg \vee bdh \vee eag \vee ebg \vee ecg \vee eh$$

   in Example 10.2 is not a normal function. Find the collection $\mathcal{D}$ of dual prime implicants of f. Is f^d normal?

6. Let f be a positive Boolean function over the variable set $\{x_1, x_2, ..., x_n\}$, and let T be a subset of the variables. Prove the following:
   (a) T is a dual prime implicant if and only if $\mathcal{P}_0 = \emptyset$ and there is a nonempty selection S for T (i.e., $\mathcal{P}_{x_i} \neq \emptyset$ for every $x_i \in T$).
   (b) T is a dual super implicant (i.e., $D \subset T$ for some dual prime implicant $D \in \mathcal{D}$) if and only if $\mathcal{P}_0 = \emptyset$ and $\mathcal{P}_{x_i} = \emptyset$ for some $x_i \in T$ (i.e., no selection S is possible).

7. Prove that for any graph G, $\overline{G}$ must be connected if G is disconnected.

8. Give a direct proof (using the dual subimplicant theorem) of the implication (iii) $\implies$ (ii) of Theorem 10.4; namely, if $G(f)$ and $G(f^d)$ do not share a common edge, then $G(f)$ and $G(f^d)$ are complementary graphs.

9. Using Lemma 10.1, prove that the read-once expression is unique for a read-once function (up to commutativity of the operations $\vee$ and $\wedge$).

10. Verify that the function $f = abc \vee bde \vee ceg$ from Example 10.3 is not normal, though its three prime implicants correspond to maximal cliques of the co-occurrence graph $G(f)$; see Figure 10.4. Verify that $G(f)$ contains an induced P_4. How many P_4's does it contain?

11. Consider two functions:

$$f_1 = x_1 x_3 x_5 \vee x_1 x_3 x_6 \vee x_1 x_4 x_5 \vee x_1 x_4 x_6 \vee x_2 x_3 x_5 \vee x_2 x_3 x_6 \vee x_2 x_4 x_5 \vee x_2 x_4 x_6$$

   and

$$f_2 = x_1 x_3 x_5 \vee x_1 x_3 x_6 \vee x_1 x_4 x_5 \vee x_1 x_4 x_6 \vee x_2 x_3 x_5 \vee x_2 x_3 x_6 \vee x_2 x_4 x_5.$$

   Verify that they generate the same co-occurrence graph G, which is P_4-free, and that all prime implicants of f_1 and f_2 correspond to maximal cliques of G; yet, f_1 is normal, while f_2 is not. Find the cotree for G and the read-once expression for f_1.

12. Give an example of a pair of functions g and f with same co-occurrence graph $G = G(g) = G(f)$, which is P_4-free, and where the number of prime implicants of g and f are equal; yet, g is normal and thus read-once, while f is not. (Hint: Combine nonnormal functions seen in this chapter whose graphs are P_4-free.)

13. Prove that for a positive Boolean function given by its complete DNF expression, it is possible to check normality in $O(n^3 k)$ time, where n is the number of essential variables, and k is the number of prime implicants of the function. (Hint: Use the results of [538].)

14. Prove Lemma 10.6: Let f be a nonconstant read-once function over the variables $N = \{x_1, x_2, \ldots, x_n\}$. Suppose that D_i is a dual prime implicant containing x_i but not x_j, and that D_j is a dual prime implicant containing x_j but not x_i. Let $R_{i,j} = (N \setminus (D_i \cup D_j)) \cup \{x_i, x_j\}$. Then, (x_i, x_j) is an edge in the co-occurrence graph $G(f)$ if and only if $R_{i,j}$ contains a prime implicant. (Hint: Use (iv) of Theorem 10.6, or see reference [22].)

15. Prove that the recursive definition of cographs based on rules $(1), (2), (3)$ in Section 10.4 is equivalent to the alternative definition using rules $(1), (2), (3')$.

16. Let f be a positive Boolean function over the variables $N = \{x_1, x_2, \ldots, x_n\}$, and let $U \subseteq N$.
    (a) Prove that the following greedy algorithm FIND-PI-IN(U) finds a prime implicant $P \subseteq U$ of f, if one exists, and can be implemented to run in $O(n)$ time using $|U|$ membership queries. (We denote by e_U the characteristic vector of U, where $(e_U)_i = 1$ for $x_i \in U$, and $(e_U)_i = 0$ otherwise.)
    Algorithm FIND-PI-IN(U)
    Step 1: Verify that $f(e_U) = 1$.
             Otherwise, exit with no solution, since U contains no prime implicant.
    Step 2: Set $S \leftarrow U$.
    Step 3: For all $x_i \in U$, **do**
             if $f(e_{S \setminus \{x_i\}}) = 1$ then $S \leftarrow S \setminus \{x_i\}$
             end-do
    Step 4: Set $P \leftarrow S$ and output P.
    (b) Write an analogous dual **Algorithm** FIND-DUALPI-IN(U) to find a dual prime implicant $D \subseteq U$ of f, if one exists.

17. The next three exercises are due to [22].
    Prove the following: Let f be a nonconstant read-once function, and let Y be a nonempty subset of its variables. Then Y is the set of essential variables of f if and only if for every variable $x_i \in Y$, x_i is contained in a prime implicant of f that is a subset of Y, and x_i is contained in a dual prime implicant of f that is a subset of Y.

18. Let f be a read-once function over the set of variables $N = \{x_1, x_2, \ldots, x_n\}$. Prove the following: If S is a prime implicant of f containing the variable

x_i, then $(N \setminus S) \cup \{x_i\}$ contains a dual prime implicant of f, and any such dual prime implicant contains x_i. Dually, if T is a dual prime implicant of f containing the variable x_i, then $(N \setminus T) \cup \{x_i\}$ contains a prime implicant of f, and any such prime implicant contains x_i.

19. Let f be a read-once function over the set of variables $N = \{x_1, x_2, \ldots, x_n\}$. Using Exercises 16, 17, and 18, prove that the following algorithm finds the set Y of essential variables and can be implemented to run in $O(n^2)$ time using $O(n^2)$ membership queries. In the process, for each variable x_i in Y, it generates a prime implicant $P[i]$ and a dual prime implicant $D[i]$ containing x_i.

    Algorithm FIND-ESSENTIAL-VARIABLES
    Step 1: Set $P[i] \leftarrow D[i] \leftarrow \emptyset$ for $i = 1, \ldots, n$.
    Step 2: Set $W \leftarrow P \leftarrow$ FIND-PI-IN(N), and
    for each $x_j \in P$, set $P[j] \leftarrow P$.
    Step 3: While there exists $x_i \in N$ such that exactly one of $P[i]$ and $D[i]$ is $\emptyset$, **do**
    (3a:) if $D[i] = \emptyset$, then set $D \leftarrow$ FIND-DUALPI-IN$((N \setminus P[i]) \cup \{x_i\})$, and for each $x_j \in D$, set $D[j] \leftarrow D$, and set $W \leftarrow W \cup D$.
    (3b:) if $P[i] = \emptyset$, then set $P \leftarrow$ FIND-PI-IN$((N \setminus D[i]) \cup \{x_i\})$, and for each $x_j \in P$, set $P[j] \leftarrow P$, and set $W \leftarrow W \cup P$.
    end-do
    Step 4: Set $Y \leftarrow W$ and output Y.

20. Give a counter example to show that the statement in Exercise 17 may fail when f is a positive Boolean function but is not read-once. Show that for an arbitrary positive Boolean function f, identifying the set of essential variables may require an exponential number of calls on a membership oracle.

21. Let f be a read-once positive function of $2n$ variables with n prime implicants $x_i y_i$ for $i \in N = \{1, \ldots, n\}$. Prove that there is a partition $N = I_1 \cup \ldots \cup I_k$ such that $f = \bigvee_{j=1}^{k} \mu_j \nu_j$, where, for $j \in \{1, \ldots, k\}$, μ_j and ν_j are elementary disjunctions, each containing exactly one of x_i, y_i for each $i \in I_j$, and no other variables. (See Lemma 10.4.)

22. (From Lisa Hellerstein.) Consider the function

$$f_1 = x_1 \vee x_2 \vee \ldots \vee x_n$$

and the class of functions $\mathcal{F} = \{f_A\}$, where A is an element in $\{0, 1\}^n$ having at least two 1's, and

$$f_A(X) = 1 \iff f_1(X) = 1 \text{ and } X \neq A.$$

(a) Prove that the functions f_A are not monotone.
(b) Prove that determining that a function is equal to f_1 and not some f_A requires querying all possible A's, and there are $\Theta(2^n)$ of them.

23. Prove directly that the normal form of any extensive game form is rectangular. In other words, if two pairs of strategies (s_1^1, s_1^2) and (s_2^1, s_2^2) result in the same play p, that is, $p(s_1^1, s_1^2) = p(s_2^1, s_2^2) = p$, then (s_1^1, s_2^2) and (s_2^1, s_1^2) also result in the same play, that is, $p(s_1^1, s_2^2) = p(s_2^1, s_1^2) = p$.

24. Verify that the following two game forms are tight:

$$M_4 = \begin{bmatrix} x_1 & x_2 & x_1 & x_2 \\ x_3 & x_4 & x_4 & x_3 \\ x_1 & x_4 & x_1 & x_5 \\ x_3 & x_2 & x_6 & x_2 \end{bmatrix},$$

$$M_5 = \begin{bmatrix} x_1 & x_1 & x_2 \\ x_1 & x_1 & x_3 \\ x_2 & x_4 & x_2 \end{bmatrix}.$$

Questions for thought

25. To what extent is Lemma 10.1 true for all expressions, that is, not just the read-once formula and the DNF formula of prime implicants?

26. The polynomial time complexity given in Theorem 10.5 can (almost certainly) be improved by a more careful choice of data structures. In this direction, what is the complexity of calculating $\mathcal{P}_0$ and P_{x_i} for all x_i? Consider using bit vectors to represent sets of variables.

27. What can be said about the complexity of recognizing read-once functions if the input formula is not a DNF, but some other type of representation, such as a BDD or an arbitrary Boolean expression? In such a case, we might have to pay a high price to convert the formula into a DNF or CNF and use the GMR method of Section 10.5. When is there an efficient alternative way to build the co-occurrence graph $G(f)$ directly from a representation of f that is different from the DNF or CNF expression? What assumptions must be made regarding f? When can normality also be tested?

    It is shown in [13] that if ψ is a nonmonotone DNF expression, the read-once recognition problem is co-NP-complete. Furthermore, as we saw in Theorem 10.9, the problem remains co-NP-complete even for arbitrary positive expressions. How does this impact the answer?

28. What would happen if we attempted to apply the read-once oracle learning method to a positive function f that was *not* read-once? In other words, in the building of the co-occurrence graph (Step 1), how did we rely upon the read-once assumption? Would the oracle fail, in which case we would know that f is not read-once, or would it produce some other graph? What graph would we get? When would it still yield the correct co-occurrence graph $G(f)$? If so, we can easily test whether it is a cograph, but how can we test whether the function is normal? For example, consider what would happen

for the functions f_1 and f_2 of Section 10.1. Could the oracle generate all prime implicants? What would be the complexity?

29. The two game forms M_4 and M_5 in Exercise 24 represent the normal form of some extensive games on graphs that have no terminal positions, and their cycles are the outcomes of the game. Find two graphs that generate M_4 and M_5.

11

Characterizations of special classes by functional equations

Lisa Hellerstein

The previous chapters covered a number of different classes of Boolean functions and provided a variety of characterizations of those classes. Some of those characterizations were in terms of functional equations or inequalities, such as the characterization of Horn functions by the inequality $f(XY) \le f(X) \vee f(Y)$ in Chapter 6. This chapter presents similar characterizations of other Boolean function classes.

This chapter also presents general results on characterizations of Boolean function classes by functional equations. Some important classes of Boolean functions can be characterized by a single simple functional equation. Other classes can be characterized by an infinite set of functional equations, but not by any finite set. Finally, some classes cannot be characterized even by an infinite set of functional equations.

Ekin, Foldes, Hammer, and Hellerstein [305] were the first to systematically study the characterization of Boolean functions by functional equations and similar logical expressions. Related results and characterizations, and extensions to non-Boolean classes of functions, appeared in a number of papers (cf. [748, 485, 334, 751, 340, 217, 218, 219, 220]); several of these papers point out the connections between equational characterizations of Boolean functions and Post's classical description of the classes of Boolean functions closed under compositions (see [753, 752]).

Except where otherwise noted, the results in this chapter are from Ekin et al. [305].

11.1 Characterizations of positive functions

To help motivate what follows, we begin with some simple characterizations.

Recall from Section 1.10 that for two points $X = (x_1, x_2, \ldots, x_n)$ and $Y = (y_1, y_2, \ldots, y_n)$ in $\mathcal{B}^n$, we write $X \le Y$ if $x_i \le y_i$ for all $i = 1, 2, \ldots, n$. Let $X \vee Y$ denote $(x_1 \vee y_1, \ldots, x_n \vee y_n)$, the bitwise disjunction of X and Y. Let $X \wedge Y$ (also written XY) and $\overline{X}$ similarly denote the bitwise conjunction of X and Y and the bitwise negation of X, respectively.

By Theorem 1.20, a Boolean function f on $\mathcal{B}^n$ is positive if and only if $f(X) \leq f(Y)$ for all $X, Y \in \mathcal{B}^n$ such that $X \leq Y$. The following theorem gives two other characterizations of positive functions:

Theorem 11.1. *A Boolean function f on $\mathcal{B}^n$ is positive if and only if the following inequality is satisfied for all $X, Y \in \mathcal{B}^n$:*

$$f(X) \leq f(X \vee Y) \tag{11.1}$$

or, equivalently, if and only if the following inequality is satisfied for all $X, Y \in \mathcal{B}^n$:

$$f(XY) \leq f(X). \tag{11.2}$$

Proof. We prove that the statement holds for the first inequality. The second is proved similarly.

Let f be a Boolean function defined on $\mathcal{B}^n$. Since for all $X, Y \in \mathcal{B}^n$, $X \leq X \vee Y$, if f is positive then f satisfies $f(X) \leq f(X \vee Y)$.

Conversely, suppose f satisfies $f(X) \leq f(X \vee Y)$. Consider $V, W \in \mathcal{B}^n$ such that $V \leq W$. Since $V \vee W = W$, f satisfies $f(V) \leq f(V \vee W) = f(W)$. Therefore, f is positive. $\qquad\square$

From the foregoing, it is easy to show that the class of negative Boolean functions is characterized by the inequalities

$$f(X \vee Y) \leq f(X)$$

and

$$f(X) \leq f(XY),$$

which are opposite to the inequalities given for positive functions.

We will show below that similar functional equations and inequalities characterize other interesting classes of Boolean functions.

11.2 Functional equations

In this section, we formally define what it means to characterize a class of Boolean functions using functional equations or inequalities.

We first give preliminary definitions and notation. Let $m, n > 0$. A Boolean expression ϕ on $\mathcal{B}^m$ can be interpreted as representing a function from $(\mathcal{B}^n)^m$ to $\mathcal{B}^n$, as follows.

Definition 11.1. *Let $\phi(x_1, \ldots, x_m)$ be a Boolean expression. Let $n \geq 1$, and let $Y_1 \ldots, Y_m$ be elements of $\mathcal{B}^n$. We define $\phi(Y_1, \ldots, Y_m)$ to be the vector obtained by applying ϕ componentwise to the entries of $Y_1, \ldots, Y_m$. More formally, letting $Y_i = (y_{i1}, \ldots, y_{in})$, for $1 \leq i \leq m$, we define $\phi(Y_1, \ldots, Y_m)$ to be equal to*

$$(\phi(y_{1,1}, \ldots, y_{m,1}), \phi(y_{1,2}, \ldots, y_{m,2}), \ldots, \phi(y_{1,n}, \ldots, y_{m,n})).$$

The expression $\phi(Y_1,\ldots,Y_m)$ *thus represents a function from* $(\mathcal{B}^n)^m$ *to* $\mathcal{B}^n$. *We call this function the interpretation of* ϕ *in* $\mathcal{B}^n$.

Example 11.1. *Let*

$$\phi_1(Y_1,Y_2,Y_3) = Y_1Y_2 \vee \overline{Y}_1Y_3,$$

$$\phi_2(Y_1) = 0.$$

Let $Y_1 = (1,0), Y_2 = (0,1)$ *and* $Y_3 = (1,1)$. *Then,*

$$\phi_1(Y_1,Y_2,Y_3) = \phi_1((1,0),(0,1),(1,1))$$
$$= (1,0)(0,1) \vee \overline{(1,0)}(1,1)$$
$$= (0,0) \vee (0,1)(1,1)$$
$$= (0,1)$$

and $\phi_2(1,0) = (0,0)$. □

As is standard with functions taking a single vector-valued input, we write, for example, $\phi_2(1,0)$ rather than $\phi_2((1,0))$.

Given a Boolean function g on $\mathcal{B}^n$ and a Boolean expression $\phi(Y_1,\ldots,Y_m)$, the expression $g(\phi(Y_1,\ldots,Y_m))$ denotes the composition of ϕ, interpreted in $\mathcal{B}^n$, and g. This composite function is a map from $(\mathcal{B}^n)^m$ to $\mathcal{B}$.

We now give a formal definition of a functional equation.

Definition 11.2. *A functional equation in the variables* $Y_1,\ldots,Y_m$ *and the function symbol* f *is an equation of the form*

$$h_1(f(\tau_1(Y_1,\ldots,Y_m)),\ldots,f(\tau_s(Y_1,\ldots,Y_m)))$$
$$= h_2(f(\tau'_1(Y_1,\ldots,Y_m)),\ldots,f(\tau'_t(Y_1,\ldots,Y_m))), \quad (11.3)$$

where $m,s,t \geq 1$, h_1 *is a Boolean expression on* $\mathcal{B}^s$, h_2 *is a Boolean expression on* $\mathcal{B}^t$, *and each* τ_i *and* τ'_i *is a Boolean expression on* $\mathcal{B}^m$.

We refer to the variables $Y_1,\ldots,Y_m$ as the *vector variables* of the equation. Functional inequalities are defined analogously to functional equations.

Example 11.2. *Consider the functional equation*

$$h_1(f(\tau_1(Y_1,Y_2)),f(\tau_2(Y_1,Y_2))) = h_2(f(\tau'_1(Y_1,Y_2))),$$

where $h_1(x_1,x_2) = x_1 \vee x_2$, $h_2(x_1) = x_1$, $\tau_1(x_1,x_2) = x_1$, $\tau_2(x_1,x_2) = x_1 \vee x_2$, *and* $\tau'_1(x_1,x_2) = x_1 \wedge x_2$.
We write this more succinctly as

$$f(Y_1) \vee f(Y_1 \vee Y_2) = f(Y_1Y_2).$$

□

Consider a functional equation $C = D$ in the variables $Y_1,\ldots,Y_m$ and the function symbol f, as in Equation 11.3. By replacing the function symbol f in C by a particular function g on $\mathcal{B}^n$ (for some $n \geq 0$), and interpreting the τ_i in C in $\mathcal{B}^n$, we obtain an expression representing a Boolean function on $(\mathcal{B}^n)^m$. We denote this function by $C_g(Y_1, \ldots, Y_m)$. The function D_g is defined analogously.

Example 11.3. *Let*

$$C = f(Y_1 \vee Y_3) \vee \overline{f(Y_2)}.$$

Let g be the function on $\mathcal{B}^2$ such that $g(x_1, x_2) = x_1 x_2$. Then,

$$C_g(Y_1, Y_2, Y_3) = g(Y_1 \vee Y_3) \vee \overline{g(Y_2)}.$$

The value of $C_g((0,1),(1,0),(0,0))$ can be computed as follows:

$$C_g((0,1),(1,0),(0,0)) = g((0,1) \vee (0,0)) \vee \overline{g(1,0)} = g(0,1) \vee \overline{g(1,0)} = 0 \vee \overline{0} = 1.$$

$\square$

We say that a particular Boolean function g on $\mathcal{B}^n$ *satisfies* a functional equation $C = D$ in the variables $Y_1, \ldots, Y_n$ and function symbol f, if for all $Y_1, \ldots, Y_m \in \mathcal{B}^n$,

$$C_g(Y_1, \ldots, Y_m) = D_g(Y_1, \ldots, Y_m).$$

Otherwise, we say that g *falsifies* the equation.

Example 11.4. *Consider the functional equation*

$$f(Y_1 \vee Y_3) \vee \overline{f(Y_2)} = f(Y_1 Y_2).$$

Let C denote the left-hand side of this equation and D the right-hand side. Note that C is the same as in the previous example.

Also as in the previous example, let g be the function on $\mathcal{B}^2$ such that $g(x_1, x_2) = x_1 x_2$, and let $Y_1 = (0,1), Y_2 = (1,0), Y_3 = (0,0)$. We showed that $C_g(Y_1, Y_2, Y_3) = 1$. For the same values of the Y_i's,

$$D_g(Y_1, Y_2, Y_3) = g((0,1)(1,0)) = g(0,0) = 0.$$

Thus g falsifies the above equation. $\square$

Definition 11.3. *A (possibly infinite) set I of functional equations characterizes a class $\mathcal{K}$ of Boolean functions if $\mathcal{K}$ consists precisely of the Boolean functions that satisfy all equations in I.*

Our primary focus is on characterization by functional equations of the form $C = D$. However, it is sometimes more convenient to consider characterizations by functional inequalities $C \leq D$.

Theorem 11.2. *The following two equations each characterize the same set of Boolean functions as the functional inequality $C \leq D$:*

- $C \vee D = D$.
- $CD = C$.

Proof. Follows directly from the fact that C and D are both Boolean-valued.   □

Example 11.5. *By Theorem 11.1, the inequality $f(X) \leq f(X \vee Y)$ characterizes the class of positive Boolean functions. Therefore, so do either of the following functional equations:*

$$f(X) \vee f(X \vee Y) = f(X \vee Y)$$

and

$$f(X)f(X \vee Y) = f(X).$$

□

An interesting alternative to using functional equations or inequalities is to instead use relations called *Boolean constraints*. These relations were introduced by Pippenger, who showed that a class of Boolean functions can be characterized by functional equations if and only if it can be characterized by a set of Boolean constraints [748] (cf. Exercise 4).

11.3 Characterizations of particular classes

In this section, we present and discuss functional equations and inequalities characterizing some important classes of Boolean functions.

11.3.1 Horn functions

In Chapter 6, Corollary 6.2, the following inequality was shown to characterize the class of Horn functions:

$$f(XY) \leq f(X) \vee f(Y). \tag{11.4}$$

This inequality for Horn functions is very similar to the inequality $f(XY) \leq f(X)f(Y)$. The latter inequality characterizes the positive functions. This can be shown by combining the inequality $f(XY) \leq f(X)$, previously shown to characterize positive functions (in Theorem 11.1), with the equivalent inequality $f(XY) \leq f(Y)$.

Recall that a Boolean function f is co-Horn if the function $g(X) = f(\overline{X})$ is Horn. From the inequality characterizing Horn functions, it is easy to show that the following inequality characterizes the co-Horn functions:

$$f(X \vee Y) \leq f(X) \vee f(Y). \tag{11.5}$$

11.3.2 Linear functions and related classes

In Chapter 1 (Definition 1.12), the degree of a DNF ϕ was defined to be the maximum degree (number of literals) in any term of ϕ. We now define the degree of a Boolean function.

Definition 11.4. *The* degree *of a Boolean function* f *is the degree of the complete DNF of* f. *Equivalently, it is the maximum degree of any prime implicant of* f. *A Boolean function is called* linear *if its degree is at most 1.*

If a Boolean function is representable by a DNF of degree 1, then all of its prime implicants have degree 1. Therefore, a Boolean function is linear if and only if it can be represented by a DNF of degree at most 1.

We discuss functions of degree $k \geq 2$ in the next section.

Polar functions were defined in Chapter 5, Section 5.3. A Boolean function is polar if it is representable by a DNF in which no term contains both a complemented and an uncomplemented variable. Equivalently, a Boolean function f is polar if $f = g \vee h$ for some positive function g and some negative function h.

Submodular functions were defined in Chapter 6, Section 6.9 to be the functions satisfying the inequality

$$f(X) \vee f(Y) \geq f(X \vee Y) \vee f(XY). \tag{11.6}$$

Supermodular functions are defined by reversing the inequality for submodular functions.

Definition 11.5. *A Boolean function is* supermodular *if it satisfies the inequality*

$$f(X) \vee f(Y) \leq f(X \vee Y) \vee f(XY). \tag{11.7}$$

In fact, the class of supermodular functions is identical to the class of polar functions.

Theorem 11.3. *A Boolean function* f *is polar if and only if it is supermodular.*

Proof. Suppose f is a polar function on $\mathcal{B}^n$. Let $f = g \vee h$, where g is positive and h is negative. Suppose $X, Y \in \mathcal{B}^n$ are such that $f(X) \vee f(Y) = 1$. Assume, without loss of generality, that $f(X) = 1$. Then, X satisfies either g or h, or both. If X satisfies g, then $X \vee Y$ must also satisfy g because g is positive, and hence, $f(X \vee Y) = 1$. If X satisfies h, then XY must satisfy h because h is negative, and hence, $f(XY) = 1$. Therefore, f is supermodular.

Conversely, suppose f is a supermodular function on $\mathcal{B}^n$. Define the following sets:

$$S = \{X \in \mathcal{B}^n \mid f(X) = 1 \text{ and for all } Y \in \mathcal{B}^n, X \leq Y \Rightarrow f(Y) = 1\},$$
$$T = \{X \in \mathcal{B}^n \mid f(X) = 1 \text{ and for all } Y \in \mathcal{B}^n, Y \leq X \Rightarrow f(Y) = 1\}.$$

Let g be the function on $\mathcal{B}^n$ such that $g(X) = 1$ if and only if $X \in S$, and let h be the function on $\mathcal{B}^n$ such that $h(X) = 1$ if and only if $X \in T$. Clearly g is positive, h is negative, and $g \vee h \leq f$. We will show that $f = g \vee h$. Suppose not. Then, there exist points $P, Q, R \in \mathcal{B}^n$ such that $f(Q) = 1$, $f(P) = f(R) = 0$, and $P \leq Q \leq R$. Define $Z = P \vee \overline{Q}R$. Since $P \leq Q \leq R$, $ZQ = P$ and $Z \vee Q = R$. But then $f(Z) \vee f(Q) = 1$ and $f(ZQ) \vee f(Z \vee Q) = 0$, contradicting that f is supermodular. Therefore $f = g \vee h$, and thus, f is polar. $\qquad\square$

The inequalities characterizing polar and submodular functions yield an equation characterizing linear functions.

Theorem 11.4. *A Boolean function is linear if and only if it satisfies the functional equation*

$$f(X) \vee f(Y) = f(X \vee Y) \vee f(XY).$$

Proof. A Boolean function is linear if and only if it is polar, Horn, and co-Horn. In Chapter 6, Section 6.9, it was shown that a Boolean function is submodular if and only if it is both Horn and co-Horn. Using these two facts, Theorem 11.4 follows immediately from Theorem 11.3 and the functional equation for submodular functions.   □

11.3.3   Quadratic and degree k functions

Quadratic functions were defined previously in Chapter 5 as the Boolean functions representable by DNFs of degree at most 2. By Theorem 5.1, if a function is quadratic, then all its prime implicants have degree at most 2. Hence the quadratic functions are precisely the functions of degree at most 2, in the sense of Definition 11.4.

In Chapter 1, Section 1.11, we defined $\mathcal{F}_k$ to be the class of Boolean functions representable by DNFs of degree at most k. For $k = 1$ and $k = 2$, $\mathcal{F}_k$ is also the class of functions of degree at most k. However, for $k \geq 3$, $\mathcal{F}_k$ is not the class of Boolean functions of degree at most k (in the sense of Definition 11.4). For example, the function $f(x_1, x_2, x_3, x_4, x_5) = x_1 x_2 \overline{x_3} \vee x_3 x_4 x_5$ is representable by the given DNF, which has degree 3, but it has a prime implicant of degree greater than 3, namely $x_1 x_2 x_4 x_5$.

As mentioned in Section 5.3.2 of Chapter 5, an early functional characterization of quadratic Boolean functions was given by Schaefer [807]. This characterization was rediscovered (in a slightly different form) by Ekin et al. [305], and we give their proof here.

Theorem 11.5. *Quadratic Boolean functions are characterized by the inequality*

$$f(XY \vee XZ \vee YZ) \leq f(X) \vee f(Y) \vee f(Z). \tag{11.8}$$

Proof. Suppose f is quadratic. Let Q, R, S be points such that

$$f(Q) \vee f(R) \vee f(S) = 0.$$

We will show that

$$f(QR \vee RS \vee QS) = 0. \tag{11.9}$$

Let P be a prime implicant of f. The prime implicant P contains at most two literals, and $Q, R,$ and S must each falsify at least one literal of P. Therefore, there exists a literal z of P that is falsified by at least two of $Q, R,$ and S. Without loss of generality, assume that Q and R both falsify z. Then, whether z is complemented or not, $QR \vee RS \vee QS$ also falsifies z, and hence, P as well. This implies (11.9) and completes the proof of inequality (11.8) for quadratic functions.

Conversely, suppose that f is not quadratic, that is, some prime implicant P of f has degree at least three. Then P can be written as

$$P_1 P_2 P_3,$$

where each factor P_i is an elementary conjunction with at least one variable, but no two of the three factors P_1, P_2, P_3 have a common variable. Define elementary conjunctions

$$R_1 = P_1 P_3, \quad R_2 = P_2 P_3, \quad R_3 = P_1 P_2.$$

Since P is a prime implicant, none of these R_i is an implicant of f, namely, there are points X, Y, Z such that

$$R_1(X) = R_2(Y) = R_3(Z) = 1$$

$$f(X) = f(Y) = f(Z) = 0.$$

These points violate (11.8).   □

Although linear and quadratic functions can be characterized by a functional equation, we will show in Section 11.4 that for $k > 2$, there is no set of functional equations that characterizes the functions of degree at most k. However, by generalizing the equation for quadratic functions, we obtain the following result for positive functions:

Theorem 11.6. *Let f be a positive Boolean function and let $k \geq 2$. Then f has degree at most k if and only if f satisfies the inequality*

$$f\left(\bigvee_{i=1}^{k+1} \bigwedge_{j \neq i} Y_j\right) \leq f(Y_1) \vee \ldots \vee f(Y_{k+1}). \tag{11.10}$$

Proof. Let f be defined on $\mathcal{B}^n$. First, we show that if f is of degree at most k, then (11.10) holds. Suppose

$$f(Y_1) = \ldots = f(Y_{k+1}) = 0$$

for some $Y_1, \ldots, Y_{k+1} \in \mathcal{B}^n$. Let P be a prime implicant of f. Then P contains at most k literals. Each Y_i must falsify at least one literal of P, and hence, there exists a literal z of P that is falsified by at least two of $Y_1, \ldots, Y_{k+1}$. Without loss of generality, assume Y_1 and Y_2 falsify z. Since P is positive, z is an uncomplemented literal. Thus, the variable z takes the value 0 in Y_1 and Y_2. Then, z also takes the value 0 in

$$\bigvee_{i=1}^{k+1} \bigwedge_{j \neq i} Y_j,$$

and hence, so does P. It follows that the left hand side of Equation (11.10) is 0.

Conversely, suppose that some prime implicant P of f has degree at least $k+1$. Then, P can be written as

$$P_1 \ldots P_{k+1},$$

where each factor P_i is an elementary conjunction with at least one variable, but no two factors have a common variable. For each $i = 1, \ldots, k+1$, let

$$R_i = \bigwedge_{j \neq i} P_i.$$

Since P is a prime implicant, there are points $Y_1, \ldots, Y_{k+1}$ such that

$$R_1(Y_1) = \cdots = R_{k+1}(Y_{k+1}) = 1$$

$$f(Y_1) = \cdots = f(Y_{k+1}) = 0.$$

These points violate (11.10).   □

For $k \geq 2$, the positive functions of degree at most k can be characterized by the inequality for positive functions together with the inequality in Theorem 11.10.

For arbitrary (i.e., not necessarily positive) Boolean functions, the inequality in Theorem 11.6 is a sufficient but not necessary condition for the function to have degree at most k.

11.4 Conditions for characterization

Having given explicit characterizations of a number of particular classes of Boolean functions, we now address the following general question: Which classes of Boolean functions can be characterized by a set of functional equations? Our answer to this question involves two operations on Boolean functions, *identification of variables* and *addition of inessential variables*.

Definition 11.6. *Let f be a Boolean function on B^n. Let $m \leq n$ and let r : $\{1, \ldots, n\} \rightarrow \{1, \ldots, m\}$ be a surjective function. We say that the Boolean function g on B^m defined by $g(x_1, \ldots, x_m) = f(x_{r(1)}, \ldots, x_{r(n)})$ is produced from f by* identification of variables. *We call r the* identification map *that produces g from f. If r is a bijection, we say that g is obtained from f by* permutation *of variables.*

Let $J = \{(x_{r(1)}, \ldots, x_{r(n)}) \mid (x_1, \ldots, x_m) \in B^m\}$. Let s be the bijection from J to B^m such that for all $(x_1, \ldots, x_m) \in B^m$, $s(x_{r(1)}, \ldots, x_{r(n)}) = (x_1, \ldots, x_m)$. We call s the vector map *associated with r. Clearly, for all $X \in J$, $f(X) = g(s(X))$.*

Definition 11.7. *Let f be a Boolean function on B^n. Let $k > 0$. Then, the function g on B^{n+k} defined by $g(x_1, \ldots, x_{n+k}) = f(x_1, \ldots, x_n)$ is said to be produced from g by* addition of inessential variables.

If f and g are such that g is produced from f by identification map r, and ϕ is a Boolean formula representing f, then one can produce a formula representing g by simply replacing each variable x_i in g by $x_{r(i)}$.

Example 11.6. *Let* $f(x_1,x_2,x_3) = x_1x_2 \vee x_1\overline{x}_3$. *Let* $r : \{1,2,3\} \to \{1,2\}$ *be such that* $r(1) = r(3) = 2$ *and* $r(2) = 1$. *Then* $g(x_1,x_2) = x_1x_2 \vee x_2\overline{x}_2 = x_1x_2$ *is produced from* f *by the identification map* r. *The function* $h(x_1,x_2,x_3,x_4) = x_1x_2$ *can be produced from* g *by addition of inessential variables. The function* $h'(x_1,x_2,x_3,x_4) = x_1x_4$ *can be produced from* h *by identification of variables (in fact, by permutation of variables).* ☐

The importance of the operations of identification of variables and addition of inessential variables can be seen in the following theorem:

Theorem 11.7. *If a class* $\mathcal{K}$ *of Boolean functions can be characterized by a set of functional equations, then* $\mathcal{K}$ *is closed under identification of variables and addition of inessential variables.*

Proof. Let $C = D$ be a functional equation. Let f be a Boolean function on $\mathcal{B}^n$ that satisfies $C = D$.

Consider a Boolean function f' that is produced from f by addition of inessential variables. Clearly, f' also satisfies $C = D$.

Now consider a Boolean function f' that is produced from f by identification of variables using an identification map r. Let $J = \{(x_{r(1)},\ldots,x_{r(n)}) \mid (x_1,\ldots,x_m) \in \mathcal{B}^m\}$. Let $s : J \to \mathcal{B}^m$ be the vector map associated with r. Consider any $X \in J$. Clearly, $f(X) = f'(s(X))$. Since $J \subseteq \mathcal{B}^n, C_f(X) = D_f(X)$, and hence, $C_{f'}(s(X)) = D_{f'}(s(X))$. Since s is surjective, it follows that f' satisfies $C = D$. ☐

We can use Theorem 11.7 to prove that certain classes of Boolean functions cannot be characterized by functional equations.

Theorem 11.8. *The following classes of functions do not have a characterization by a set of functional equations:*

  (a) *Monotone functions.*
  (b) *Functions of degree at most* k, *for all* $k \geq 3$.
  (c) *Shellable functions.*
  (d) *Regular functions.*
  (e) *Read-once functions.*

Proof. We show that each of these classes is not closed under identification of variables.

Monotone functions: Let $f(x_1,x_2,x_3,x_4) = x_1\overline{x}_2 \vee \overline{x}_3x_4$, and apply the identification map $r : \{1,2,3,4\} \to \{1,2\}$ such that $r(1) = 1, r(2) = 2, r(3) = 1$, and $r(4) = 2$ to yield $f'(x_1,x_2) = x_1\overline{x}_2 \vee \overline{x}_1x_2$. The function f' is neither positive nor negative in x_1 and x_2, and hence, it is not monotone (i.e., not unate).

Functions of degree at most k, *for all* $k \geq 3$: Let $f(x_1,\ldots,x_{2k}) = x_1x_2x_3\ldots x_k \vee \overline{x}_{k+1}\ldots\overline{x}_{2k}$, and apply the identification map $r : \{1,\ldots,2k\} \to \{1,\ldots,2k-1\}$ such that $r(i) = i$ for all $i < 2k$, and $r(2k) = 1$. The resulting function has

$x_2 x_3 \ldots x_k \overline{x}_{k+1} \ldots \overline{x}_{2k-1}$ as a prime implicant, and hence, it is not of degree at most k.

Shellable functions: (The following proof was provided by Yves Crama.) The function $f = x_1 x_2 \vee x_1 x_3 x_5 \vee x_2 x_3 x_5 \vee x_3 x_4 x_5$ is shellable, since it is represented by the orthogonal DNF $\phi = x_1 x_2 \vee x_1 \overline{x}_2 x_3 x_5 \vee \overline{x}_1 x_2 x_3 x_5 \vee \overline{x}_1 \overline{x}_2 x_3 x_4 x_5$.

Now, identify variables x_4 and x_5 in f. This yields the function $g = x_1 x_2 \vee x_1 x_3 x_4 \vee x_2 x_3 x_4 \vee x_3 x_4 x_4 = x_1 x_2 \vee x_3 x_4$, which is not shellable. Therefore, the class of shellable functions is not characterizable by functional equations.

Regular functions: The function $f(x_1, x_2, x_3, x_4, x_5, x_6) = x_4 x_6 \vee x_5 x_6 \vee x_2 x_4 x_5 \vee x_3 x_4 x_5 \vee x_1 x_2 x_3 x_5 \vee x_1 x_2 x_3 x_6$ is regular because (using the notation from Chapter 8) $x_1 \prec_f x_2 \approx_f x_3 \prec_f x_4 \prec_f x_5 \prec_f x_6$. Applying the identification map $r : \{1, \ldots, 6\} \to \{1, \ldots, 5\}$ such that $r(1) = 1, r(2) = 2$, and $r(i) = i - 1$ for all $i \geq 3$, yields the function

$$f'(x_1, x_2, x_3, x_4, x_5, x_6) = x_3 x_5 \vee x_4 x_5 \vee x_2 x_3 x_4 \vee x_1 x_2 x_4 \vee x_1 x_2 x_5.$$

The function f' is not regular because

$$f'(0,0,1,0,1) = 1$$
$$f'(0,1,0,0,1) = 0$$
$$f'(1,1,0,1,0) = 1$$
$$f'(1,0,1,1,0) = 0,$$

meaning that x_2 and x_3 are not comparable.

Read-once functions: Left as an end-of-chapter exercise for the reader (Exercise 2).   □

Surprisingly, closure under identification of variables and addition of inessential variables is not just a necessary condition for a class of functions to have a characterization by functional equations; it is also a sufficient condition.

Theorem 11.9. *Let $\mathcal{K}$ be a class of Boolean functions that is closed under identification of variables and addition of inessential variables. Then $\mathcal{K}$ can be characterized by a (possibly infinite) set of functional equations.*

Proof. This result was first shown by Ekin et al. [305]. The following version of the proof uses simplifications due to Pippenger [748].

Let $\mathcal{G}$ be the set of Boolean functions not in $\mathcal{K}$. For each $g \in \mathcal{G}$, we will construct a functional equation I_g such that I_g is falsified by g and satisfied by every function in $\mathcal{K}$. The set of equations $\{I_g \mid g \in \mathcal{G}\}$ clearly characterizes $\mathcal{K}$.

Let $g \in \mathcal{G}$ be defined on $\mathcal{B}^m$. The construction of I_g is as follows: Let $t = 2^m$. Let A be the $t \times m$ binary matrix whose rows are the t binary vectors of length m, listed in lexicographic order. Let $A_1, \ldots, A_t$ denote the rows of A. Let $col(A)$ denote the set of column vectors of A. All the columns are distinct.

For $i \in \{1,\dots,t\}$ let h_i be the Boolean function on $\mathcal{B}^t$ such that, for all $(x_1,\dots,x_t) \in \mathcal{B}^t$, $h_i(x_1,\dots,x_t) = x_i$ if the transpose of $(x_1,\dots,x_t)$ is in $col(A)$, and $h_i(x_1,\dots,x_t) = 0$ otherwise. Similarly, for $i \in \{1,\dots,t\}$ let h_{t+i} be the Boolean function on $\mathcal{B}^t$ such that for all $(x_1,\dots,x_t) \in \mathcal{B}^t$, $h_{t+i}(x_1,\dots,x_t) = x_i$ if the transpose of $(x_1,\dots,x_t)$ is in $col(A)$, and $h_{t+i}(x_1,\dots,x_t) = 1$ otherwise. For $i \in \{1,\dots,2t\}$, let $\phi_i(x_1,\dots,x_t)$ be a Boolean expression representing h_i.

For all $n \geq 0$, define the function $h_n^i : (\mathcal{B}^n)^t \to \mathcal{B}^n$ as follows: For all $X_1,\dots,X_t \in \mathcal{B}^n$, $h_n^i(X_1,\dots,X_t) = (y_1,\dots,y_n)$ such that, for all $j \in \{1,\dots,n\}$, $y_j = h_i(X_1[j],X_2[j],\dots,X_t[j])$. That is, h_n^i is the function obtained by applying h_i componentwise to $X_1,\dots,X_t$. Thus, h_n^i is the interpretation of $\phi_i(x_1,\dots,x_t)$ in $\mathcal{B}^n$. Because n may not be equal to 1, we will write $\phi_i(X_1,\dots,X_t)$ rather than $\phi_i(x_1,\dots,x_t)$, to emphasize that the variables of ϕ_i are vector variables.

Let $H = \{h_1,\dots,h_{2t}\}$. Define a partition of H into two sets, H_0 and H_1 as follows:

$$H_0 = \{h_{kt+i} : i \in \{1,\dots,t\}, k \in \{0,1\}, \text{ and } g(A_i) = 0\},$$

$$H_1 = \{h_{kt+i} : i \in \{1,\dots,t\}, k \in \{0,1\}, \text{ and } g(A_i) = 1\}.$$

The desired equation I_g is defined to be

$$\bigvee_{h_i \in H_0} (f(\phi_i(X_1,\dots,X_t))) \vee \left(\bigvee_{h_i \in H_1} \overline{f(\phi_i(X_1,\dots,X_t))} \right) = 1. \qquad (11.11)$$

We show that I_g is falsified by g but satisfied by all functions in $\mathcal{K}$.

Let $C(X_1,\dots,X_t)$ denote the functional expression on the left-hand side of I_g, so I_g is $C(X_1,\dots,X_t) = 1$. For all $k \in \{0,1\}, i \in \{1,\dots,t\}$, $\phi_{kt+i}(A_1,\dots,A_t) = h_{kt+i}^n(A_1,\dots,A_t) = A_i$. It follows from the definitions of H_0 and H_1 that $C_g(A_1,\dots,A_t) = 0$. Therefore, g falsifies I_g.

We now show that any Boolean function f falsifying I_g is not a member of $\mathcal{K}$. Suppose f is a Boolean function on $\mathcal{B}^n$ that falsifies I_g. Then, for some $W_1,\dots,W_t \in (\mathcal{B}^n)^t$, $C_f(W_1,\dots,W_t) = 0$. Let W be the $t \times n$ matrix whose rows are $W_1,\dots,W_t$. Since $C_f(W_1,\dots,W_t) = 0$, it follows that, for all $k \in \{0,1\}, i \in \{1,\dots,t\}$,

$$f(\phi_{kt+i}(W_1,\dots,W_t)) = g(A_i). \qquad (11.12)$$

The column vectors of W are not necessarily all distinct. Let $col(W)$ denote the set of column vectors of W. Let $q = |col(W) \cap col(A)|$.

We first consider the case $q > 0$. For each column vector in $col(W) \cap col(A)$, choose a column of W that is equal to that column vector. Let $k_1,\dots,k_q$ be the indices of the chosen columns. Let $j_1,\dots,j_q$ be the indices of the columns of A that are equal to columns $k_1,\dots,k_q$ of W respectively. Let $j_{q+1},\dots,j_m$ be the indices of the remaining columns of A. Let $r : \{1,\dots,n\} \to \{1,\dots,q\}$ be such that for $i \in \{1,\dots,n\}$, $r(i) = d$ if column i of W equals column k_d of W (and hence, column j_d of A), and $r(i) = 1$ if column i of W is not in $col(A)$.

Let f' be the function produced from f by the identification map r. Let f_0' be produced from f' by addition of $m - q$ inessential variables. Let $p : \{1,\dots,m\} \to$

$\{1,\ldots,m\}$ be the bijection such that, for all $u \in \{1,\ldots,m\}$, $p(u) = j_u$. Let f'' be the function produced from f_0' by the identification map p.

Let $i \in \{1,\ldots,t\}$. For index c, let W_{ic} and A_{ic} denote the cth components of W_i and A_i respectively. Let $\rho = W_{ik_1}$. Then,

$$h_{\rho t+i}^n(W_1,\ldots,W_t) = W_{ik_{r(1)}},\ldots,W_{ik_{r(n)}} \tag{11.13}$$

because if column c of W is equal to a column of A, then column c of W is equal to column $k_{r(c)}$ of W, and otherwise $r(c) = 1$.

We now have

$$
\begin{aligned}
g(A_i) &= f(h_{\rho t+i}^n(W_1,\ldots,W_t)) \quad \text{by Equation (11.12)} \\
&= f(W_{ik_{r(1)}},\ldots,W_{ik_{r(n)}}) \quad \text{by Equation (11.13)} \\
&= f'(W_{ik_1},\ldots,W_{ik_q}) \quad \text{because for all } (x_1,\ldots,x_q) \in \mathcal{B}^q, \\
&\qquad\qquad f'(x_1,\ldots,x_q) = f(x_{r(1)},\ldots,x_{r(n)}) \\
&= f_0'(W_{ik_1},\ldots,W_{ik_q},A_{ij_{q+1}},A_{ij_{q+2}},\ldots,A_{ij_m}) \\
&\qquad\qquad \text{by addition of inessential variables to } f' \\
&= f_0'(A_{ij_1},\ldots,A_{ij_q},A_{ij_{q+1}},A_{ij_{q+2}},\ldots,A_{ij_m}) \\
&\qquad\qquad \text{since } W_{ik_1},\ldots,W_{ik_q} \text{ equal } A_{ij_1},\ldots,A_{ij_q} \text{ respectively} \\
&= f''(A_{i1},\ldots,A_{im}) \quad \text{by definition of } f'' \\
&= f''(A_i).
\end{aligned}
$$

Thus $g(A_i) = f''(A_i)$ for all $i \in \{1,\ldots,t\}$. Since the rows of A are the t elements of the domain of g, $f'' = g$.

The class $\mathcal{K}$ is closed under identification of variables and addition of inessential variables. If f were in $\mathcal{K}$, then g would be also, since g can be obtained from f by these operations. Therefore, f is not in $\mathcal{K}$, which is what we wanted to show.

It remains to consider the case $q = 0$. Let $i \in \{1,\ldots,n\}$. By Equation (11.12), for $\rho \in \{0,1\}$, $g(h_{\rho t+i}^m(A_1,\ldots,A_t)) = f(h_{\rho t+i}^n(W_1,\ldots,W_t))$. By the definitions of h_i and h_{t+i}, it follows that $g(A_{i1},\ldots,A_{in}) = f(0,\ldots,0) = f(1,\ldots,1)$. Since this is true for all $i \in \{1,\ldots,t\}$, g is a constant function. The constant function g can be produced from f by first applying the identification map $r : \{1,\ldots,n\} \to \{1\}$ such that $r(u) = 1$ for all $u \in \{1,\ldots,n\}$, and then adding $m-1$ inessential variables. As in the case $q > 0$, it follows immediately that f is not in $\mathcal{K}$.   □

Theorem 11.9 can be used to show that particular classes of functions have a characterization by functional equations. For example, we can prove the following result for the class of threshold functions.

Theorem 11.10. *The class of threshold functions can be characterized by a set of functional equations.*

Proof. By Theorem 11.9, it suffices to show that the class of threshold functions is closed under identification of variables and addition of inessential variables. Closure under addition of inessential variables is obvious.

We show closure under identification of variables. Suppose $f(x_1,\ldots,x_n)$ is a threshold function. Then, for some $w_1,\ldots,w_n$ and t in $\mathbb{R}$, $f(x_1,\ldots,x_n) = 0$ if and only if $\sum_i w_i x_i \le t$. If $f'(x_1,\ldots,x_m)$ is obtained from f using an identification map r, then $f'(x_1,\ldots,x_m) = 0$ if and only if

$$\sum_{i=1}^{m} \left(\sum_{1 \le j \le n, r(j)=i} w_j \right) x_i \le t.$$

Therefore, f' is a threshold function. $\square$

Similarly, it is easy to show that the class $\mathcal{F}_k$ of functions representable by DNFs of degree at most k has a characterization by functional equations (see also Exercise 3). This is in contrast to the result (cf. Theorem 11.8) that, for $k \ge 3$, the class of functions of degree k has no such characterization.

Note that the set of equations constructed in the proof of Theorem 11.9 consists of one equation I_g for each function g not in the set $\mathcal{K}$ being characterized. Since there are an infinite number of Boolean functions that are not threshold functions, Theorem 11.9 implies that there is an infinite set of functional equations characterizing the class of threshold functions. (See Exercise 4 for another way to construct a characterization of Boolean threshold functions by an infinite set of functional equations.) In the next section, we address the question of whether the class of threshold functions can be characterized by a *finite* set of functional equations.

Combining Theorems 11.7 and 11.9 yields the following:

Theorem 11.11. *A class $\mathcal{K}$ of functions can be characterized by a set of functional equations if and only if $\mathcal{K}$ is closed under identification of variables and addition of inessential variables.*

11.5 Finite characterizations by functional equations

Theorem 11.12. *If a class $\mathcal{K}$ of Boolean functions can be characterized by a finite set of functional equations, then it can be characterized by a single functional equation.*

Proof. Let $\{C_1 = D_1,\ldots,C_m = D_m\}$ be a finite set of functional equations. Without loss of generality, assume that these equations are over disjoint sets of variables. A function g satisfies all the equations in the above set if and only if it satisfies the equation $\bigwedge_{i=1}^{m}(C_i D_i \vee \overline{C}_i \overline{D}_i) = 1$. $\square$

When can a class of Boolean functions be characterized by a finite set of functional equations (and hence by a single one)? We begin by describing a necessary condition.

Definition 11.8. *Let $\mathcal{K}$ be a class of Boolean functions. Let g be a Boolean function on $\mathcal{B}^n$. A certificate of nonmembership of g in $\mathcal{K}$ is a subset $Q \subseteq \mathcal{B}^n$ such that for all Boolean functions f on $\mathcal{B}^n$, if $f \in \mathcal{K}$, then there exists $X \in Q$ such that $f(X) \neq g(X)$. A class $\mathcal{K}$ of Boolean functions has* constant-size certificates of nonmembership *if there exists an integer $c \geq 0$ such that for every Boolean function $g \notin \mathcal{K}$, there is a certificate Q of nonmembership of g in $\mathcal{K}$ such that $|Q| \leq c$.*

Example 11.7. *Let $g(x_1, x_2) = x_1 \overline{x}_2 \vee \overline{x}_1 x_2$. By Theorem 11.1, the positive functions are characterized by the functional inequality $f(X) \leq f(X \vee Y)$. If $X = (0, 1)$ and $Y = (1, 1)$, then $g(X) > g(X \vee Y)$. Therefore, $\{(0, 1), (1, 1)\}$ is a certificate of nonmembership of g in the class of positive functions.*

Since every Boolean function g on $\mathcal{B}^n$ that is not a positive function must falsify $f(X) \leq f(X \vee Y)$, for each such g, there exists a set $\{X, Y\} \subseteq \mathcal{B}^n$ that is a certificate of nonmembership of g in the class of positive functions. Therefore, the class of positive functions has constant-size certificates of nonmembership. $\square$

By generalizing Example 11.7 we easily obtain the following result (Hellerstein [485]):

Theorem 11.13. *Let $\mathcal{K}$ be a class of functions that can be characterized by a finite set of functional equations. Then $\mathcal{K}$ has constant-size certificates of nonmembership.*

Proof. Let Z be a finite set of functional equations characterizing $\mathcal{K}$. Let c be the maximum number of vector variables in any equation in Z. Let g be a Boolean function on $\mathcal{B}^n$ that is not in $\mathcal{K}$. Then g falsifies some functional equation $C(X_1, \ldots, X_m) = D(X_1, \ldots, X_m)$ in Z, where $m \leq c$. The two sides of the equation are Boolean expressions over elements of the form $f(\tau(X_1, \ldots, X_m)))$, where τ is a Boolean expression on $\mathcal{B}^m$. For fixed $\{Y_1, \ldots, Y_m\} \in \mathcal{B}^n$, the value of $g(\tau(Y_1, \ldots, Y_m)))$, for all τ appearing in the equation, determines whether $C_g(Y_1, \ldots, Y_m) \neq D_g(Y_1, \ldots, Y_m)$. Since g falsifies the functional equation $C(X_1, \ldots, X_m) = D(X_1, \ldots, X_m)$, there exist $\{Y_1, \ldots, Y_m\} \in \mathcal{B}^m$ such that $C_g(Y_1, \ldots, Y_m) \neq D_g(Y_1, \ldots, Y_m)$; the set of vectors $\tau(Y_1, \ldots, Y_m) \in \mathcal{B}^n$, for all τ appearing in the functional equation, constitute a certificate that g is not in $\mathcal{K}$. Since each such τ expresses one of the 2^{2^m} functions on $\mathcal{B}^m$, it follows that this certificate has size at most $2^{2^m} \leq 2^{2^c}$. $\square$

By Theorem 11.10, threshold functions can be characterized by a set of functional equations. However, Hellerstein [485] showed that they cannot be characterized by a *finite* set of functional equations. This is proved using the following result:

Theorem 11.14. *Threshold functions do not have constant-size certificates of nonmembership.*

Proof. Suppose for contradiction that the set of threshold functions has certificates of nonmembership of size c.

Then for each Boolean function g that is not a threshold function, there exists a certificate Q_g of nonmembership of g in the set of threshold functions, such that Q_g has size at most c. Let $S_g = \{X \in Q_g | g(X) = 1\}$, and let $T_g = \{X \in Q_g | g(X) = 0\}$.

Consider an arbitrary Boolean function g on $\mathcal{B}^n$ that is not a threshold function. If the convex hull of S_g does not intersect the convex hull of T_g, then, by standard separation theorems, (see, e.g., [788]), there exists a hyperplane separating the points in S_g from the points in T_g. In this case, there exists a threshold function f' such that $f'(X) = g(X)$ for all $X \in Q_g$. This contradicts that Q_g is a certificate of nonmembership of g in the set of threshold functions. Hence, the convex hulls of S_g and T_g intersect.

Let $X_1, \ldots, X_t$ be the elements of Q_g. Let M_g be the $t \times n$ matrix whose rows are $X_1, \ldots, X_t$. Let M_g' be the matrix obtained from M_g by deleting all columns j from M_g such that for some $j' < j$, column j' and column j of M_g' are equal. Let m be the number of columns of M_g', and let $\hat{X}_1, \ldots, \hat{X}_t$ be the rows of M_g' corresponding to rows $X_1, \ldots, X_t$ of M_g.

Let $\hat{S}_g = \{\hat{X}_i | X_i \in S_g\}$, and $\hat{T}_g = \{\hat{X}_i | X_i \in T_g\}$. Since S_g and T_g are disjoint, so are $\hat{S}_g$ and $\hat{T}_g$. Also, since the convex hulls of S_g and T_g intersect, the convex hulls of $\hat{S}_g$ and $\hat{T}_g$ intersect.

Since the convex hulls of $\hat{S}_g$ and $\hat{T}_g$ intersect, it follows from the proof of Theorem 9.14 in Chapter 9 that, for some $z > 0$, there exist z points $\hat{X}_{i_1}, .., \hat{X}_{i_z}$ in $\hat{S}_g$ (not necessarily distinct), and z points $\hat{X}_{j_1}, \ldots, \hat{X}_{j_z}$ in $\hat{T}_g$ (not necessarily distinct) such that

$$\hat{X}_{i_1} + \cdots + \hat{X}_{i_z} = \hat{X}_{j_1} + \cdots + \hat{X}_{j_z}, \tag{11.14}$$

and hence,

$$X_{i_1} + \cdots + X_{i_z} = X_{j_1} + \cdots + X_{j_z}. \tag{11.15}$$

Let z_g be the smallest such z. Note that z_g is completely determined by $\hat{S}_g$ and $\hat{T}_g$.

The columns of M_g' are all distinct. Since there are only 2^t different binary vectors of length t, it follows that $m \leq 2^t$. Because Q_g has size at most c, $t \leq c$, and hence, $m \leq 2^c$.

Therefore, over all possible Boolean functions g that are not threshold functions, there are a finite number of possible values for $\hat{S}_g$ and $\hat{T}_g$ and, hence, a finite number of possible values for z_g.

Let α be the maximum value of z_g over all Boolean functions g that are not threshold functions.

As mentioned in Chapter 9, Section 9.3, Winder showed that for every k there is a function that is k-asummable but not a threshold function [917, 915, 860]. Consider a function g that is α-asummable but not a threshold function. Since g is not a threshold function, it follows that, for $z = z_g$, there exist z points $X_{i_1}, \ldots, X_{i_z}$

in S_g (not necessarily distinct), and z points $X_{j_1},...,X_{j_z}$ in T_g (not necessarily distinct), such that Equation (11.15) holds. Since $z_g \leq \alpha$, g is α-summable, a contradiction.   □

By Definition 9.2, any k-summable function has a certificate of size at most $2k$ that it is k-summable. Thus, Theorem 11.4 generalizes Winder's result that, for any fixed k, k-asummability is not a sufficient condition for thresholdness; see Section 9.3. Informally, it says that any condition depending on only a constant number of points of the function cannot be a sufficient condition for thresholdness.

Returning to the question of characterization by functional equations, we now have the following theorem:

Theorem 11.15. *Threshold functions cannot be characterized by a finite set of functional equations.*

Proof. Follows immediately from Theorems 11.13 and 11.14.   □

Although the existence of constant-size certificates of nonmembership is a necessary condition for characterization of a class by a finite set of functional equations, it is not a sufficient condition.

Example 11.8. *Let g be a Boolean function on $\mathcal{B}^n$ such that g is not a monotone function. Then, there exists $k \in \{1,...,n\}$ such that g is neither positive nor negative in the variable x_k. It follows that there exist $X = (x_1,...,x_n)$ and $Y = (y_1,...,y_n)$ in $\mathcal{B}^n$ such that*

$$g(x_1,...,x_{k-1},0,x_{k+1},...,x_n) = 0, \tag{11.16}$$

$$g(x_1,...,x_{k-1},1,x_{k+1},...,x_n) = 1, \tag{11.17}$$

$$g(y_1,...,y_{k-1},1,y_{k+1},...,y_n) = 0, \tag{11.18}$$

$$g(y_1,...,y_{k-1},0,y_{k+1},...,y_n) = 1. \tag{11.19}$$

The four vectors in the above equations constitute a certificate of nonmembership of g in the class of monotone functions. Since such a set of four vectors exists for each non-monotone g, monotone functions have constant-size certificates of nonmembership. However, by Theorem 11.8, monotone functions cannot be characterized by any set (finite or infinite) of functional equations.   □

Ekin et al. [305] showed that a condition that is both necessary and sufficient can be obtained by considering *identification minors*.

Definition 11.9. *Let f be a Boolean function, and let g be a function that is produced from f by identification of variables. The function g is called an* identification minor *of f. We use the notation $g \preceq f$ to denote that g is an identification minor of f.*

Identification minors are a restricted case of the *Boolean minors* introduced by Wang and Williams [897] and Wang [896]. They are called *minors* because of their similarity to graph minors, which have been extensively studied in graph theory.

Definition 11.10. *Let K be a class of Boolean functions. A Boolean function g is called a* forbidden identification minor *of K if g is not an identification minor of any function $f \in K$.*

Example 11.9. *The function $f(x_1,x_2) = x_1\bar{x}_2 \vee \bar{x}_1 x_2$ is a forbidden identification minor of the class of positive functions.*   □

Definition 11.11. *Let K be a class of Boolean functions, and let Z be a set of forbidden identification minors of K. The set Z* characterizes K *if every function not in K has an identification minor in Z.*

Theorem 11.16. *Let K be a class of Boolean functions. Then K can be characterized by a finite set of functional equations if and only if K is closed under addition of inessential variables and can be characterized by a finite set of forbidden identification minors.*

Proof. Suppose K can be characterized by a finite set of functional equations. By Theorem 11.7, K must be closed under addition of inessential variables. We show now that it can be characterized by a finite set of forbidden identification minors.

Since K can be characterized by a finite set of functional equations, by Theorem 11.12 it can be characterized by a single functional equation $E = F$. Let $X_1,\ldots,X_m$ be the vector variables appearing in $E = F$. Suppose f is a Boolean function on B^n such that $n > 2^m$ and f does not satisfy $E = F$. Then there exist $V_1,\ldots,V_m \in B^n$ such that $E_f(V_1,\ldots,V_m) \neq F_f(V_1,\ldots,V_m)$. Consider the $m \times n$ matrix W with rows $V_1,\ldots,V_m$, in that order. Let n' be the number of distinct columns of W. Clearly, $n' \leq 2^m$. Consider an identification map $r : \{1,\ldots,n\} \to \{1,\ldots,n'\}$ such that $r(i) = r(j)$ if and only if columns i and j of W are equal. This map produces an identification minor f' of f defined on $B^{n'}$. Let s be the vector map corresponding to r. For $i \in \{1,\ldots,n\}$, $f(V_i) = f'(s(V_i))$. Therefore, $E_{f'}(s(V_1),\ldots,s(V_m)) \neq F_{f'}(s(V_1),\ldots,s(V_m))$.

Thus, for every f defined on B^n, with $n > 2^m$ and $f \notin K$, there exists f' defined on $B^{n'}$ with $n' \leq 2^m$, such that $f' \preceq f$ and $f' \notin K$. The set of all such f' is finite, and forms a set of forbidden identification minors that characterizes K.

Conversely, suppose K is closed under addition of inessential variables and can be characterized by a finite set of forbidden identification minors. Clearly, K is closed under identification of variables. Let $Z = \{g_1,\ldots,g_n\}$ be a set of forbidden identification minors characterizing K.

Referring to the proof of Theorem 11.9, consider the equations $I_{g_1},\ldots,I_{g_n}$. By Theorem 11.7, if a function f satisfies these equations, then so do all identification minors of f. Because $g_1,\ldots,g_n$ do not satisfy all these equations, it follows that

$g_1, \ldots, g_n$ are not identification minors of f. Hence, $f \in \mathcal{K}$. Conversely, by the proof of Theorem 11.9, if f belongs to $\mathcal{K}$ then f satisfies every I_{g_i}. Therefore, the equations $\{I_{g_1}, \ldots, I_{g_n}\}$ characterize $\mathcal{K}$. $\square$

As we showed in Example 11.8, for arbitrary classes of Boolean functions, having constant-size certificates of nonmembership is a necessary, but not sufficient, condition for the class to have a characterization by a finite set of functional equations. However, Hellerstein [485] showed that for classes closed under identification of variables and addition of inessential variables, the condition is both necessary and sufficient.

Theorem 11.17. *Let $\mathcal{K}$ be a class of Boolean functions that is closed under identification of variables and addition of inessential variables. Then $\mathcal{K}$ can be characterized by a finite set of functional equations if and only if $\mathcal{K}$ has constant-size certificates of nonmembership.*

Proof. Necessity was shown in Theorem 11.13.

To show sufficiency, suppose every Boolean function not in $\mathcal{K}$ has a certificate of nonmembership of size at most c, for some constant c.

Let g be a Boolean function on $\mathcal{B}^n$ that is not in $\mathcal{K}$. Let $Q = \{Q_1, \ldots, Q_k\}$ be a certificate of nonmembership of g in $\mathcal{K}$ such that $k \leq c$.

Consider the matrix A whose rows are $Q_1, \ldots, Q_k$. Let n' be the number of distinct column vectors appearing as columns of A. Clearly $n' \leq 2^k$. Without loss of generality, assume that the first n' columns of A are distinct. Let $r : \{1, \ldots, n\} \to \{1, \ldots, n'\}$ be such that for all $j \in \{1, \ldots, n'\}, r(j) = i$, where $1 \leq i \leq n'$ and the ith and jth columns of A are equal. Let g' be the function produced from g using the identification map r.

Now consider the function g'' derived from g' by adding $n - n'$ inessential variables to g'. For each $Q_i \in Q$, $g''(Q_i) = g(Q_i)$. Since Q is a certificate of nonmembership of g in $\mathcal{K}$, it is also a certificate of nonmembership of g'' in $\mathcal{K}$. Thus $g'' \notin \mathcal{K}$.

Since g'' can be produced from g' by addition of inessential variables, and since $\mathcal{K}$ is closed under addition of inessential variables, $g' \notin \mathcal{K}$. Therefore, g has an identification minor g' that is not in $\mathcal{K}$ such that g' is defined on $\mathcal{B}^{n'}$ for some $n' \leq 2^c$. This holds for each g not in $\mathcal{K}$. Let Z be the set of all such g'. The set Z consists of forbidden identification minors of $\mathcal{K}$ and characterizes $\mathcal{K}$.

Because there are only a finite number of functions defined on $\mathcal{B}^{n'}$, for all $n' \leq 2^c$, Z is a finite set. By Theorem 11.16, $\mathcal{K}$ can be characterized by a finite set of functional equations. $\square$

Hellerstein and Raghavan [486] showed that, for any k, the class of functions representable by DNFs having at most k terms has constant-sized certificates of nonmembership, and hence, by the above theorem, it has a characterization by a finite set of functional equations.

We observed earlier that the class $\mathcal{F}_k$ can be characterized by a set of functional equations. In contrast to this result, $\mathcal{F}_k$ cannot be characterized by a finite set of functional equations (see Exercise 3).

11.6 Exercises

1. Give a functional equation characterizing the class of elementary conjunctions.

2. Prove that the class of read-once functions cannot be characterized by a set of functional equations.

3. This exercise is based on a result of Bernard Rosell (personal communication). Let $k > 2$. Recall that the class $\mathcal{F}_k$ consists of functions representable by DNFs of degree at most k. Let $f(x_1, \ldots, x_n)$ be the function whose output is 1 if and only if at least $\frac{k+1}{2}$ of its inputs are 1 and at least $\frac{k+1}{2}$ of its inputs are 0.

   (a) Show that the given function f is not in $\mathcal{F}_k$.

   (b) Prove a lower bound on the size of any certificate of nonmembership of f in $\mathcal{F}_k$. Use this lower bound to show that $\mathcal{F}_k$ cannot be characterized by a finite set of functional equations.

4. In [748], Pippenger defined a *Boolean constraint* to be a pair (R, S), where R and S are each a set of binary column vectors of length m, for some $m \geq 0$. If A is an $m \times n$ binary matrix, and $f(x_1, \ldots, x_n)$ is a Boolean function on n variables, then let $f(A)$ denote the column vector produced by applying f to each row of A; namely, $f(A)$ is the length m column vector whose ith entry is $f(A[i,1], A[i,2], \ldots, A[i,n])$, for all entries i. We write $A \prec R$ if each column of A is a member of R. Function $f(x_1, \ldots, x_n)$ *satisfies* constraint (R, S) if for all $m \times n$ binary matrices A, $A \prec R$ implies that $f(A) \in S$.

   A set I of Boolean constraints *characterizes* a class $\mathcal{K}$ of Boolean functions if $\mathcal{K}$ consists precisely of the Boolean functions that satisfy all constraints in I.

   (a) Show that the following constraint characterizes the class of positive Boolean functions: $\left(\left\{ \begin{bmatrix} 0 \\ 1 \end{bmatrix}, \begin{bmatrix} 1 \\ 1 \end{bmatrix}, \begin{bmatrix} 0 \\ 0 \end{bmatrix} \right\}, \left\{ \begin{bmatrix} 0 \\ 1 \end{bmatrix}, \begin{bmatrix} 1 \\ 1 \end{bmatrix}, \begin{bmatrix} 0 \\ 0 \end{bmatrix} \right\} \right)$.

   (b) Give a constraint that characterizes the class of Horn functions.

   (c) By Theorem 9.14, a Boolean function is a threshold function if and only if it is k-asummable for every $k \geq 2$.

   Describe a constraint that characterizes the set of functions that are k-asummable, for fixed $k \geq 2$. Then construct an infinite set of constraints that characterizes the class of threshold functions.

   (d) Show that a class of Boolean functions can be characterized by a set of functional equations if and only if it can be characterized by a set of Boolean constraints. (See [748].)

5. Let $\mathcal{K}$ be a class of Boolean functions. A Boolean function g defined on $\mathcal{B}^n$ is a *minimal* forbidden identification minor of $\mathcal{K}$ if it is a forbidden identification minor of $\mathcal{K}$, and, for every identification minor g' of g, if g' is defined on $\mathcal{B}^{n'}$ and $n' < n$, then $g' \in \mathcal{K}$.

   (a) Prove that, if a Boolean function f is defined on $\mathcal{B}^5$, then f is not a minimal forbidden identification minor of the class of linear functions.

   (b) Give an example of a function defined on $\mathcal{B}^3$ that is a minimal forbidden identification minor of the class of linear functions.

Part III

Generalizations

12

Partially defined Boolean functions

Toshihide Ibaraki

12.1 Introduction

Suppose that a set of data points is at hand for a certain phenomenon. A data point is called a *positive example* if it describes a case that triggers the phenomenon, and a *negative example* otherwise. We consider the situation in which all data points are binary and have a fixed dimension; namely, they belong to $\mathcal{B}^n$.

Given a set of positive examples $T \subseteq \mathcal{B}^n$ and a set of negative examples $F \subseteq \mathcal{B}^n$, we call the pair (T, F) a *partially defined Boolean function* (pdBf) on $\mathcal{B}^n$. For a pdBf (T, F) on $\mathcal{B}^n$, a Boolean function $f : \mathcal{B}^n \to \mathcal{B}$ satisfying

$$T(f) \supseteq T \text{ and } F(f) \supseteq F$$

is called an *extension* of (T, F), where

$$T(f) = \{A \in \mathcal{B}^n \mid f(A) = 1\}, \tag{12.1}$$

$$F(f) = \{B \in \mathcal{B}^n \mid f(B) = 0\}. \tag{12.2}$$

If we associate n Boolean variables $x_j, j = 1, 2, \ldots, n$, with the components of points in $\mathcal{B}^n$, then extensions are Boolean functions of the variables $x_1, x_2, \ldots, x_n$.

As an example of a pdBf (T, F), let us assume that each point $A = (a_1, a_2, \ldots, a_n) \in T \cup F$ indicates the result of physical tests applied to a patient, where T denotes the set of results for patients diagnosed as positive, and F denotes the set of negative results. Each component a_j of a point A gives the result of the j-th test; for example, $a_1 = 1$ may indicate that blood pressure is "high," while $a_1 = 0$ indicates, "low"; $a_2 = 1$ may say that body temperature is "high," while $a_2 = 0$ says "low," and so on. An extension f of this pdBf (T, F) then describes how the diagnosis of the disease could be formulated for all possible patients. In other words, this Boolean function f contains all the details of the diagnosis. As extensions of a given pdBf (T, F) are not unique, in general, it is interesting and important to investigate how to build meaningful extensions from given pdBfs.

This line of approach to data analysis recently received increasing attention in statistics and in artificial intelligence under various names such as data mining, knowledge discovery, and knowledge acquisition (Agrawal, Imielinski, and Swami [8]; Crama, Hammer, and Ibaraki [233]; Fayyad et al. [321]; Mangasarian [659]; Mangasarian, Setiono, and Wolberg [660]; Mannila, Toivonen, and Verkamo [665]; Quinlan [770, 771]), reflecting the current trend that large amount of data sets are available in many applications. In addition to the diagnosis of diseases, applications include the analysis of sales records at retail shops, economic indices of countries and enterprises, stock market records, DNA sequences, geological data, and many others. Extraction of meaningful information from such data sets is considered very important. It may be interesting to observe that closely related approaches have also been proposed in the social science literature, where they are specifically applied to the analysis of small sets of qualitative data which do not lend themselves to classical statistical approaches; see, for instance, Flament [333] and Ragin [775].

We term the approach in this chapter *logical analysis of data* (LAD) to emphasize its logical aspects in statistics and in artificial intelligence. The study of LAD was initiated by Crama, Hammer, and Ibaraki [233] and has been elaborated in subsequent papers, such as those by Boros et al. [122, 128, 130, 131, 139, 140] and Bonates and Hammer [102]. More references are found in other sections of this chapter.

A large body of studies on pdBfs can be found in switching theory (Curtis [248]; Hu [512]; Kuntzmann [589]; McCluskey [634]; Mendelson [680]; Muroga [699]; Prather [755]; Roth [793]; Urbano and Mueller [879]). In this area, pdBfs are often called "incompletely specified Boolean functions," as the value of Boolean functions is usually specified in most points, except for some binary points called "don't cares," which never arise as input vectors because of circuit specification constraints. The main issue here is to exploit don't cares to simplify the resulting circuits. Various minimization techniques used for Boolean functions (as discussed in Section 3.3) have been generalized to minimize functions with don't cares. Some discussion in this direction will be given in Section 12.6; see Villa, Brayton, and Sangiovanni-Vincentelli [891] for more information.

Extensions of pdBfs are also closely related to problems studied in computational learning theory (see, e.g., Aizenstein et al. [13]; Angluin [21]; Anthony [26]; Anthony and Biggs [29]; Bshouty [161]; Kearns, Li and Valiant [560]; Pitt and Valiant [749]; Sloan, Szörényi and Turán [838]; Valiant [884], etc.); in fact, some relevant results on pdBfs were first obtained in learning theory.

We also note that psychologists rely on Boolean functions to model human concept learning from examples, as explained, for instance, by Feldman [326, 327]; see also Ganter and Wille [369] for a general mathematical framework of concept formation.

Finally, discriminant functions studied in pattern recognition have obvious resemblance with extensions, although statistical models and methods are usually considered in pattern recognition (e.g., Gnanadesikan [388]; Hand [466]), in contrast with the purely logical and combinatorial methods to be covered in this chapter.

Table 12.1. An example of pdBf (T, F)

		x_1	x_2	x_3	x_4	x_5	x_6	x_7	x_8
	$A^{(1)} =$	0	1	0	1	0	1	1	0
T	$A^{(2)} =$	1	1	0	1	1	0	0	1
	$A^{(3)} =$	0	1	1	0	1	0	0	1
	$B^{(1)} =$	1	0	1	0	1	0	1	0
F	$B^{(2)} =$	0	0	0	1	1	1	0	0
	$B^{(3)} =$	1	1	0	1	0	1	0	1
	$B^{(4)} =$	0	0	1	0	1	0	1	0

Example 12.1. *Consider a pdBf (T, F) as shown in Table 12.1. Extensions of this pdBf can be expressed by the following DNFs:*

$$f_1 = \bar{x}_1 x_2 \vee x_2 x_5$$
$$f_2 = \bar{x}_1 \bar{x}_5 \vee x_3 \bar{x}_7 \vee x_1 x_5 \bar{x}_7$$
$$f_3 = x_5 x_8 \vee x_6 x_7.$$

It can be verified that all these functions are indeed extensions of (T, F), that is, $f_k(A^{(i)}) = 1$ holds for $i = 1, 2, 3$ and $f_k(B^{(i)}) = 0$ holds for $i = 1, 2, 3, 4$. As we shall see later, this pdBf has many other extensions. ☐

As extensions are Boolean functions, they can be represented by DNFs, CNFs, and other Boolean expressions. We shall also discuss decision trees as a means of representing extensions in Section 12.2.5.

When choosing among many extensions of a given pdBf, we need some criteria to guide the choice. We emphasize the following two points: First,

- the simplicity of extensions,

which may reflect our general belief that the truth is simple and beautiful, or as a translation of Occam's razor principle. Simplicity can be measured, for example, by the sizes of representations such as DNFs, CNFs, and decision trees. The size of a "support set," to be discussed in Section 12.2.2, is another measure of simplicity. Second,

- embodiment in the extensions of structural knowledge concerning the phenomenon to be modeled.

For example, if high blood pressure is known to favor the appearance of a disease, then we expect the extension f to depend positively on the variable x_j associated with blood pressure.

In more general mathematical terms, we require the obtained extension to belong to a specified class of functions $\mathcal{C}$. The selected class $\mathcal{C}$ may arise not only from prior structural information, but also from the application that we have in mind for the resulting extensions. For example, if an extension f is Horn, then f can be dealt with by Horn rules; as discussed in Chapter 6, this allows us to benefit from numerous convenient mathematical properties of Horn rules.

In this chapter, we consider the following classes of functions:

(1) The class of all Boolean functions, $\mathcal{F}_{\mathrm{ALL}}$.
(2) The class of positive functions, $\mathcal{F}_+$ (defined in Sections 1.10 and 1.11).
(3) The class of monotone, or unate functions, $\mathcal{F}_{\mathrm{UNATE}}$ (defined in Section 1.10).
(4) The class of functions representable by a DNF of degree at most k, $\mathcal{F}_k$ (defined in Sections 1.4 and 1.11).
(5) The class of Horn functions, $\mathcal{F}_{\mathrm{HORN}}$ (discussed in Chapter 6).
(6) The class of threshold functions, $\mathcal{F}_{\mathrm{Th}}$ (discussed in Chapter 9).
(7) The class of decomposable functions, $\mathcal{F}_{F_0(S_0, F_1(S_1))}$ (defined in Section 12.3.6).
(8) The class of k-convex functions, $\mathcal{F}_{k\text{-CONV}}$ (discussed in Section 12.3.7).

For other classes of functions studied in the literature on pdBfs, see [139].

In dealing with real-world data, we should also be aware that the data may contain errors as well as missing bits. A missing bit is denoted by $*$, meaning that it can be either 0 or 1. We shall discuss in Sections 12.4 and 12.5 how to deal with these situations, and we shall introduce various problems associated with the extensions in such cases.

12.2 Extensions of pdBfs and their representations

12.2.1 Definitions

Given a Boolean function of n variables $f : \mathcal{B}^n \to \mathcal{B}$, let $T(f)$ denote its set of *true points* and $F(f)$ its set of *false points*, as defined by (12.1)–(12.2). Obviously $T(f) \cap F(f) = \emptyset$ and $T(f) \cup F(f) = \mathcal{B}^n$ hold. For two Boolean functions f and g on the same set of n variables, recall that we write $f \le g$ if $f(A) \le g(A)$ holds for all $A \in \mathcal{B}^n$, where we consider $0 < 1$ for $\mathcal{B} = \{0, 1\}$. As already defined in Section 12.1, a Boolean function $f : \mathcal{B}^n \to \mathcal{B}$ is an *extension* of a pdBf (T, F), where $T \subseteq \mathcal{B}^n$ and $F \subseteq \mathcal{B}^n$, if $T(f) \supseteq T$ and $F(f) \supseteq F$ hold.

A fundamental question raised in Section 12.1 can be stated as follows, where $\mathcal{C}$ denotes an arbitrary class of Boolean functions:

PROBLEM EXTENSION($\mathcal{C}$)
Instance: A pdBf (T, F).
Question: Does (T, F) have an extension in $\mathcal{C}$?

When the answer to the question is "yes," it is frequently required to output an extension in $\mathcal{C}$.

In this section, we consider the class $\mathcal{C} = \mathcal{F}_{\mathrm{ALL}}$. Other classes will be discussed in subsequent sections. The following theorem is immediate from the above definitions.

Theorem 12.1. *A pdBf (T, F) has an extension in $\mathcal{F}_{\text{ALL}}$ if and only if $T \cap F = \emptyset$. Hence, problem* EXTENSION($\mathcal{F}_{\text{ALL}}$) *can be solved in polynomial time.* $\square$

If a pdBf (T, F) satisfies $T \cap F = \emptyset$, then it has $2^{2^n - |T| - |F|}$ extensions. Define two extensions $f_{\min}$ and $f_{\max}$ by

$$T(f_{\min}) = T, \quad F(f_{\min}) = \mathcal{B}^n \setminus T, \qquad (12.3)$$

$$T(f_{\max}) = \mathcal{B}^n \setminus F, \quad F(f_{\max}) = F. \qquad (12.4)$$

Then, any extension f of (T, F) satisfies

$$f_{\min} \le f \le f_{\max},$$

that is, $f_{\max}$ maximizes $T(f)$ and $f_{\min}$ minimizes $T(f)$ among all extensions f of (T, F). Furthermore, all extensions of (T, F) form a finite lattice with respect to the operations $\vee$ and $\wedge$ between functions. The largest element of this lattice is $f_{\max}$, and its smallest element is $f_{\min}$. A remaining question is: Which extensions in the lattice are appropriate for the purpose of logical analysis of data?

12.2.2 Support sets of variables

For a subset $U \subseteq \mathcal{B}^n$ and $S \subseteq \{1, 2, \ldots, n\}$, we denote by $U|_S$ the *projection* of U to S. In other words, $U|_S = \{A|_S \mid A \in U\}$, where $A|_S = (a_j \mid j \in S)$ is the point obtained from A by considering only those components a_j with $j \in S$. For example, for $U = \{(1, 0, 1, 1), (0, 1, 1, 0), (0, 0, 0, 1)\}$ and $S = \{2, 3\}$, we have $U|_S = \{(0, 1), (1, 1), (0, 0)\}$. Given a pdBf (T, F) with $T, F \subseteq \mathcal{B}^n$, and a class $\mathcal{C}$ of Boolean functions, a subset $S \subseteq \{1, 2, \ldots, n\}$ is called a *support set* for class $\mathcal{C}$ if $(T|_S, F|_S)$ has an extension in class $\mathcal{C}$. In a sense, given a support set S, all variables $x_j, j \in \{1, 2, \ldots, n\} \setminus S$, are redundant because there is an extension in $\mathcal{C}$ that does not depend on them.

From the viewpoint of pursuing simple extensions, therefore, it is meaningful to consider small support sets. We say that a support set S is *minimal* if there is no other support set properly contained in S, and *minimum* if it minimizes $|S|$.

PROBLEM MIN-SUPPORT($\mathcal{C}$)
Instance: A pdBf (T, F) (where we assume that (T, F) has an extension in $\mathcal{C}$).
Output: A minimum support set S of (T, F) for class $\mathcal{C}$.

We first show that this problem for class $\mathcal{F}_{\text{ALL}}$ can be formulated as a *set covering problem*. Recall that the set covering problem is the following NP-hard optimization problem [371]:

PROBLEM SET COVER
Instance: An $m \times n$ 0–1 matrix Q.
Output: An n-dimensional 0–1 vector $y = (y_1, y_2, \ldots, y_n)^t$ that satisfies $Qy \ge 1$

and that minimizes $\sum_{j=1}^{n} y_j$, where $\mathbf{1}$ is the n-dimensional column vector of 1's.

For any two points $A, B \in \mathcal{B}^n$, define

$$\Delta(A, B) = \{j \in \{1, 2, \ldots, n\} \mid a_j \neq b_j\}. \tag{12.5}$$

Let us introduce 0–1 variables $y_j, j = 1, 2, \ldots, n$, to denote whether $j \in S$ (i.e., $y_j = 1$) or $j \notin S$ (i.e., $y_j = 0$). It is easy to see that $A|_S \neq B|_S$ holds for $S = \{j \mid y_j = 1\}$ if

$$\sum_{j \in \Delta(A,B)} y_j \geq 1. \tag{12.6}$$

Therefore, as a result of Theorem 12.1, problem MIN-SUPPORT($\mathcal{F}_{\text{ALL}}$) can be formulated as follows:

$$\text{minimize} \sum_{j=1}^{n} y_j$$

$$\text{subject to} \sum_{j \in \Delta(A,B)} y_j \geq 1, \quad A \in T, B \in F \tag{12.7}$$

$$y_j \in \{0, 1\}, \quad j \in \{1, 2, \ldots, n\}.$$

The relation between support sets and the set covering problem has been observed in various early papers (e.g., Kambayashi [547]; Kuntzmann [589]; Necula [704]). The preceding description follows the presentation by Crama, Hammer, and Ibaraki [233].

Example 12.2. *The set covering problem (12.7) corresponding to the pdBf in Example 12.1 is given as follows.*

$$\text{minimize} \sum_{j=1}^{8} y_j$$

$$\text{subject to } y_1 + y_2 + y_3 + y_4 + y_5 + y_6 \geq 1$$

$$y_2 + y_5 + y_7 \geq 1$$

$$y_1 + y_7 + y_8 \geq 1$$

$$y_2 + y_3 + y_4 + y_5 + y_6 \geq 1$$

$$y_2 + y_3 + y_4 + y_7 + y_8 \geq 1$$

$$y_1 + y_2 + y_6 + y_8 \geq 1$$

$$y_5 + y_6 \geq 1$$

$$y_1 + y_2 + y_3 + y_4 + y_7 + y_8 \geq 1$$
$$y_1 + y_2 + y_7 + y_8 \geq 1$$
$$y_2 + y_3 + y_4 + y_6 + y_8 \geq 1$$
$$y_1 + y_3 + y_4 + y_5 + y_6 \geq 1$$
$$y_2 + y_7 + y_8 \geq 1$$
$$y_1, y_2, \ldots, y_8 \in \{0,1\}.$$

This set of inequalities contains many redundant inequalities, and can be greatly simplified. As already observed in Chapter 1, Section 1.13, the constraints of a set covering problem can be associated with a CNF such that a 0–1 assignment of values to y satisfies the set covering constraints if and only if it satisfies all clauses of the CNF. In the current example, we obtain the CNF

$$\psi = (y_1 \vee y_2 \vee y_3 \vee y_4 \vee y_5 \vee y_6)(y_2 \vee y_5 \vee y_7)(\cdots)(y_2 \vee y_7 \vee y_8).$$

It is not difficult to see that the prime implicants of the function represented by ψ correspond exactly to the minimal support sets of (T, F) (see Chapter 4, Section 4.2). Applying this procedure, we conclude that there are eight minimal support sets for our example, namely,

$$S_1 = \{5,8\}, \ S_2 = \{6,7\},$$
$$S_3 = \{1,2,5\}, \ S_4 = \{1,2,6\}, \ S_5 = \{2,5,7\},$$
$$S_6 = \{2,6,8\}, \ S_7 = \{1,3,5,7\}, \ S_8 = \{1,4,5,7\}.$$

The first two sets, S_1 and S_2, are the only minimum support sets, and the following DNFs provide two extensions associated with S_1 and S_2, respectively.

$$\varphi_1 = x_5 x_8 \vee \bar{x}_5 \bar{x}_8$$
$$\varphi_2 = x_6 x_7 \vee \bar{x}_6 \bar{x}_7.$$

$\square$

Theorem 12.2. *Problem* MIN-SUPPORT$(\mathcal{F}_{\text{ALL}})$ *is NP-hard.*

Proof. We provide a reduction from SET COVER. Given an instance Q of SET COVER, we consider the following instance of MIN-SUPPORT$(\mathcal{F}_{\text{ALL}})$:

$$T = \{Q_i \mid i = 1,2,\ldots,m\},$$
$$F = \{(0,0,\cdots,0)\},$$

where Q_i denotes the i-th row of the 0–1 matrix Q. It is easy to see that the formulation (12.7) for MIN-SUPPORT$(\mathcal{F}_{\text{ALL}})$ is exactly the same as the original instance of SET COVER. This shows that SET COVER is reducible to MIN-SUPPORT$(\mathcal{F}_{\text{ALL}})$, and proves the theorem. $\square$

Problem SET COVER has been intensively studied in operations research, as it has a wide variety of applications. Even though it is NP-hard, branch-and-bound algorithms can solve fairly large instances of SET COVER exactly (Nemhauser and Wolsey [707]), and there are various heuristic algorithms that can find very good feasible solutions of large instances (Caprara, Fischetti, and Toth [169]; Yagiura, Kishida, and Ibaraki [927]). A theoretical analysis of simple greedy heuristics can be found in papers by Chvátal [198] and Lovász [623]. These algorithms can be used to solve the formulation (12.7) of MIN-SUPPORT($\mathcal{F}_{ALL}$) exactly or approximately. Other types of heuristic algorithms to find support sets are described in Boros et al. [137].

12.2.3 Patterns and theories of pdBfs

In this section, we consider methods of obtaining extensions with rather simple DNFs. A DNF will be considered "simple" if all its terms are short and the number of its terms is small.

We say that a term t *covers* $A \in \mathcal{B}^n$ if $t(A) = 1$ holds, where t is regarded as a function. Let us define a term t as a *pattern* of a pdBf (T, F) if it covers some point $A \in T$, but does not cover any point $B \in F$; that is, if $T(t) \cap T \neq \emptyset$ and $T(t) \cap F = \emptyset$. For the pdBf of Example 12.1,

$$\bar{x}_1 x_2 \bar{x}_3 x_4 \bar{x}_5 x_6 x_7 \bar{x}_8, \quad \bar{x}_1 x_2 \bar{x}_3 x_4, \quad \bar{x}_1 x_2$$

are some of the patterns which cover $A^{(1)} \in T$.

Let t be a pattern of (T, F). We say that t is a *prime pattern* if no pattern of (T, F) can be obtained by deleting some literals from t, that is, if $T(t') \cap F \neq \emptyset$ holds for every $t' \neq t$ that absorbs t. Continuing the above example, we can see that

$$\bar{x}_1 x_2, \quad \bar{x}_1 \bar{x}_5, \quad x_2 x_7, \quad x_2 \bar{x}_8, \quad \bar{x}_3 x_7, \quad x_4 x_7, \quad \bar{x}_5 x_7, \quad \bar{x}_5 \bar{x}_8, \quad x_6 x_7$$

are all the prime patterns that cover $A^{(1)} \in T$.

Patterns and prime patterns of (T, F) are closely related to the function f_{max} defined by (12.4), as shown by the next lemma.

Lemma 12.1. *Let (T, F) be a pdBf. A term t is a pattern (respectively, a prime pattern) of (T, F) if and only if t is an implicant (respectively, a prime implicant) of $f_{\max}$ that covers some point in T.*

Proof. Let t be a term of (T, F). The condition $T(t) \cap F = \emptyset$ is equivalent to $T(t) \subseteq T(f_{\max})$, which means, in turn, that t is an implicant of $f_{\max}$. The characterization of patterns follows directly from this observation.

If t is a prime pattern of (T, F), then, every term t' obtained from t by deleting some literals satisfies $T(t') \cap F(f_{\max}) \neq \emptyset$, meaning that t' is not an implicant of $f_{\max}$; hence, t is a prime implicant of $f_{\max}$. The converse statement is proved similarly. $\square$

Generating all prime patterns is a very important problem in logical analysis of data, and in the design of logic circuits. Some methods for this purpose are discussed in Section 12.6.

Now consider a DNF φ consisting only of patterns of a given pdBf (T, F) such that every $A \in T$ is covered by some patterns in φ. Such a DNF φ represents an extension of (T, F) and is called a *theory* of (T, F). If the patterns in a theory φ are all prime, then φ is called a *prime theory*. Every prime theory is a theory, but the converse does not hold in general. If a theory φ has the additional property that none of its patterns can be removed without sacrificing the covering condition of T, it is called an *irredundant* theory of (T, F). In general, there exist many irredundant theories of (T, F), and every such theory is minimal (but may not be minimum) in the sense of the number of terms. In a similar manner, we can define a *prime irredundant* theory of (T, F). A prime irredundant theory is minimal with respect to the length of each term as well as the number of terms. In subsequent sections, when no confusion arises, we may sometimes call "theory" the extension represented by a theory.

Example 12.3. *Consider again the pdBf of Example 12.1. It is easy to see that*

$$\bar{x}_1 x_2 \text{ is a prime pattern that covers } A^{(1)},$$

$$x_2 x_5 \text{ is a prime pattern that covers } A^{(2)},$$

$$x_3 x_8 \text{ is a prime pattern that covers } A^{(3)},$$

and the following DNF gives a prime theory:

$$\varphi = \bar{x}_1 x_2 \vee x_2 x_5 \vee x_3 x_8.$$

However, this theory is not irredundant, because the DNF φ' obtained from φ by removing $x_3 x_8$ still covers all points in T:

$$\varphi' = \bar{x}_1 x_2 \vee x_2 x_5.$$

This prime theory φ' is irredundant, as none of the patterns in φ' can be removed any longer.   □

At this point, let us note that not all extensions of a given pdBf (T, F) are theories. For the pdBf of Example 12.1, for instance, the extension f with $T(f) = T \cup \{(1, 1, 1, 1, 1, 1, 1, 1)\}$ and $F(f) = \mathcal{B}^n \setminus T(f)$ is not a theory, since any term that covers a point in T and $(1, 1, 1, 1, 1, 1, 1, 1)$ must cover some other points not in T (hence in $F(f)$). In most cases, only a very small fraction of all extensions are theories (and an even smaller fraction are prime theories). In this sense, theories (in particular, prime irredundant theories) define extensions of a given pdBf (T, F) that can be considered as simple in their DNF expressions. Indeed, the following statement holds:

Theorem 12.3. *Let f be an extension of the pdBf (T, F) and let $\varphi = \bigvee_{i=1}^{m} C_i$ be an arbitrary DNF expression of f. If f is not a theory, then there exists a proper subset $S \subset \{1, 2, \ldots, m\}$ such that $\bigvee_{i \in S} C_i$ is a theory of (T, F).*

Table 12.2. All basic theories of the pdBf in Table 12.1

$\varphi_1 = x_5 x_8 \vee \bar{x}_5 \bar{x}_8$	$\varphi_7^1 = \bar{x}_1 \bar{x}_5 \vee x_3 \bar{x}_7 \vee x_1 x_5 \bar{x}_7$	$\varphi_8^1 = \bar{x}_1 \bar{x}_5 \vee \bar{x}_4 \bar{x}_7 \vee x_1 x_4 x_5$
$\varphi_2 = x_6 x_7 \vee \bar{x}_6 \bar{x}_7$	$\varphi_7^2 = \bar{x}_1 \bar{x}_5 \vee x_3 \bar{x}_7 \vee x_1 \bar{x}_3 x_5$	$\varphi_8^2 = \bar{x}_1 \bar{x}_5 \vee \bar{x}_4 \bar{x}_7 \vee x_1 x_5 \bar{x}_7$
$\varphi_3^1 = \bar{x}_1 x_2 \vee x_2 x_5$	$\varphi_7^3 = x_3 \bar{x}_7 \vee \bar{x}_3 x_7 \vee x_1 x_5 \bar{x}_7$	$\varphi_8^3 = x_4 x_7 \vee \bar{x}_4 \bar{x}_7 \vee x_1 x_4 x_5$
$\varphi_3^2 = \bar{x}_1 \bar{x}_5 \vee x_2 x_5$	$\varphi_7^4 = x_3 \bar{x}_7 \vee \bar{x}_3 x_7 \vee x_1 \bar{x}_3 x_5$	$\varphi_8^4 = x_4 x_7 \vee \bar{x}_4 \bar{x}_7 \vee x_1 x_5 \bar{x}_7$
$\varphi_4 = \bar{x}_1 x_2 \vee x_2 \bar{x}_6$	$\varphi_7^5 = x_3 \bar{x}_7 \vee \bar{x}_5 x_7 \vee x_1 x_5 \bar{x}_7$	$\varphi_8^5 = \bar{x}_4 \bar{x}_7 \vee \bar{x}_5 x_7 \vee x_1 x_4 x_5$
$\varphi_5^1 = x_2 x_5 \vee x_2 x_7$	$\varphi_7^6 = x_3 \bar{x}_7 \vee \bar{x}_5 x_7 \vee x_1 \bar{x}_3 x_5$	$\varphi_8^6 = \bar{x}_4 \bar{x}_7 \vee \bar{x}_5 x_7 \vee x_1 x_5 \bar{x}_7$
$\varphi_5^2 = x_2 x_5 \vee \bar{x}_5 x_7$		
$\varphi_6^1 = x_2 \bar{x}_6 \vee x_2 \bar{x}_8$		
$\varphi_6^2 = x_2 \bar{x}_8 \vee \bar{x}_6 x_8$		

Proof. For every $A \in T$, there is a term $C_{i(A)}$, with $i(A) \in \{1, 2, \ldots, m\}$, such that $C_{i(A)}(A) = 1$. Since $C_{i(A)}(B) = 0$ for all $B \in F$, we see that $C_{i(A)}$ is a pattern of (T, F). Now, let $S = \{i(A) \mid A \in T\}$. Then, $\bigvee_{i \in S} C_i$ is a theory of (T, F) and since f itself is not a theory, it must be the case that S is a proper subset of $\{1, 2, \ldots, m\}$. $\square$

Another measure of simplicity addressed in Section 12.2.2 is the minimality of a support set S. Any theory of $(T|_S, F|_S)$ over a support set S is a theory of the original pdBf (T, F). From the view point of simplicity, it is desirable that the support set S be minimal and the theory be prime. Furthermore, it is easy to show that any irredundant theory of $(T|_S, F|_S)$ for a support set S is an irredundant theory of (T, F). Combining these concepts together, we call *basic theory* any prime irredundant theory of a pdBf (T, F) defined over a minimal support set. A basic theory displays simplicity with respect to both the size of its DNF expression and the size of the support set.

Example 12.4. *As all minimal support sets were listed in Example 12.2 for the pdBf (T, F) of Example 12.1, we are now able to obtain all basic theories by enumerating all prime patterns for each support set S_k. Table 12.2 gives all basic theories thus obtained, where φ_k^i indicates the i-th basic theory generated from a minimal support set S_k. (The superscript i is not indicated if S_k has only one basic theory.)*
   Note that the pdBf (T, F) has $|\mathcal{B}^8 \setminus (T \cup F)| = 2^8 - 7 = 249$ unspecified points, implying that it has 2^{249} extensions in $\mathcal{F}_{\text{ALL}}$. Table 12.2 shows that only 21 of these extensions are basic theories. $\square$

The preceding discussion can be symmetrically applied to the set F of a pdBf (T, F) (in other words, when we consider the pdBf (F, T) instead of (T, F)). An implicant t of $\bar{f}_{min}$ that covers at least one point $B \in F$ is called a *copattern* of (T, F). A copattern is a *prime copattern* if it is a prime implicant of $\bar{f}_{min}$. Cotheories, prime cotheories, irredundant cotheories and basic cotheories can then be defined from copatterns and prime copatterns in the same manner. For the pdBf (T, F) of Example 12.1, $x_5 \bar{x}_8$ is a prime copattern that covers $B^{(1)}, B^{(2)}, B^{(4)}$,

Table 12.3. Flat data set corresponding to the
pdBf in Table 12.1

x_1	x_2	x_3	x_4	x_5	x_6	x_7	x_8	x_9
0	1	0	1	0	1	1	0	1
1	1	0	1	1	0	0	1	1
0	1	1	0	1	0	0	1	1
1	0	1	0	1	0	1	0	0
0	0	0	1	1	1	0	0	0
1	1	0	1	0	1	0	1	0
0	0	1	0	1	0	1	0	0

and $\bar{x}_5 x_8$ is a prime copattern that covers $B^{(3)}$. Therefore $\varphi = x_5 \bar{x}_8 \vee \bar{x}_5 x_8$ is a prime cotheory. It is easy to see that this extension φ is also a basic cotheory.

In concluding this subsection, we briefly comment upon the history of the fundamental concepts of patterns and theories. In the context of LAD, the definitions of patterns and theories were formulated in Crama, Hammer, and Ibaraki [233], and their properties have been studied in several subsequent papers; see, for example, Boros et al. [115]. However, as patterns and prime patterns for pdBfs are natural generalizations of implicants and prime implicants for Boolean functions, similar concepts can be found in early references such as Mendelson [680]; Prather [755]; and Roth [793, 879]. For example, in [755, 793], patterns are discussed under the name of "basic cells," and prime patterns under the name of "maximal basic cells." The concept of theories is also introduced in these references.

There also exists an interesting relation between patterns, as defined in this section, and *association rules*, which are a basic concept used in *data mining*. In data mining, a data set is usually given as a "flat" list of data points without "output bit," rather than as a pair of sets consisting of positive and negative examples. The data set of Example 12.1, for instance, would be given as Table 12.3, after adding the attribute x_9 that indicates the outcome of each data point.

A property that holds among such data points is called an association rule if it can be described as an implication of the form: "if $x_2 = 0$ and $x_4 = 1$, then $x_9 = 1$ holds." More formally, an association rule takes the form,

$$(x_{j_1} = a_{j_1}, x_{j_2} = a_{j_2}, \ldots, x_{j_k} = a_{j_k}) \Longrightarrow x_l = a_l,$$

where $a_{j_1}, a_{j_2}, \ldots, a_{j_k}$ and a_l are either 0 or 1, respectively. It is further required that at least one data point should satisfy the rule, that no data point should violate the rule, and that no shorter rule (namely, consisting of a subset of $\{x_{j_1} = a_{j_1}, x_{j_2} = a_{j_2}, \ldots, x_{j_k} = a_{j_k}\}$ in its left-hand side) should exist.

It is not difficult to see that, if we give a special role to the conclusion variable x_l, and if we define the set of data points satisfying $x_l = 1$ (respectively, $x_l = 0$) as T (respectively, F), then the above association rule actually asserts that

$\bigwedge_{i \in P} x_{j_i} \bigwedge_{i \in N} \bar{x}_{j_i}$ is either a prime pattern (in case $a_l = 1$) or a prime copattern (in case $a_l = 0$) of the pdBf (T, F), where $P = \{i \mid a_{j_i} = 1\}$ and $N = \{i \mid a_{j_i} = 0\}$.

In the discussion of association rules in data mining, data sets are usually supposed to contain errors and missing parts. To cope with such situations, the concepts of support and confidence are introduced as essential constituents of association rules. We do not go into details, but refer to Agrawal, Imielinski, and Swami [8]; Fayyad et al. [321]; and Mannila, Toivonen, and Verkamo [665] for further discussion. In this chapter, we shall deal with errors and missing bits of data in Sections 12.4 and 12.5, respectively, from a slightly different viewpoint.

Remark 12.1. There is some confusion in the use of the terms "theory" and "cotheory" in application areas. In learning theory and data-mining, "theory" is often used as a synonym of "extension." But, here we use it to mean a special extension with certain properties. A cotheory is sometimes referred to as a "negative theory" to emphasize its role with respect to the set of negative examples F (in this case, the theory itself is called "positive theory"). We do not follow these conventions, so as to avoid a potential confusion with the concepts of positive and negative functions.   □

12.2.4 Roles of theories and cotheories

In this subsection, we discuss some properties of theories and cotheories. Most results in this section are based on Boros et al. [115]. Given a pdBf (T, F), let us define the following theory $\alpha_{(T,F)}$ and cotheory $\beta_{(T,F)}$:

$$\alpha_{(T,F)} = \bigvee_{t \in P(T,F)} t \tag{12.8}$$

$$\beta_{(T,F)} = \bigvee_{t \in coP(T,F)} t, \tag{12.9}$$

where $P(T, F)$ (respectively, $coP(T, F)$) denotes the set of all patterns (respectively, copatterns) of (T, F). The suffix (T, F) of α and β may be omitted if no confusion arises. In words, α (respectively, β) is the largest theory (respectively, cotheory) of (T, F). We can also define α and β by taking the disjunction of all prime patterns and prime copatterns of (T, F), respectively. The resulting theory and cotheory are equivalent to those defined by (12.8)–(12.9), in the sense that they define the same functions on $\mathcal{B}^n$. Thus, we may also say that α (respectively, β) is the largest prime theory (respectively, prime cotheory) of (T, F).

As an important property of α and β, we can show that every point in $\mathcal{B}^n$ is a true point of either α or β. (But note that $T(\alpha) \cap T(\beta)$ is not empty in general.)

Theorem 12.4. *For every pdBf (T, F) on $\mathcal{B}^n$, $T(\alpha) \cup T(\beta) = \mathcal{B}^n$.*

Proof. Take an arbitrary point $X \in \mathcal{B}^n$, and let $A \in T \cup F$ be the closest point to X in the sense of the Hamming distance, which is defined by

$$d(V,W) = |\{j = 1,2,\ldots,n \mid v_j \neq w_j\}| = \sum_{j=1}^{n} |v_j - w_j| \text{ for all } V,W \in \mathcal{B}^n.$$

Assume $A \in T$ without loss of generality. For a point $Y \in \mathcal{B}^n$, we use the notation $L(Y)$ to denote the set of all literals in its minterm (e.g., if $Y = (1,0,1,1)$, we have $L(Y) = \{x_1, \bar{x}_2, x_3, x_4\}$). Then let t be the term consisting of all literals in $L(X) \cap L(A)$ (e.g., if $X = (0,0,1,1,1)$ and $A = (1,0,0,1,1)$, then $t = \bar{x}_2 x_4 x_5$). This term t satisfies $t(A) = 1$ by definition. Moreover, $t(B) = 0$ holds for all $B \in F$, since $d(X,A) \leq d(X,B)$ and $A \neq B$ imply that at least one literal in t does not coincide with $L(B)$. Thus, t is a pattern of (T,F), and hence, $\alpha(X) = 1$ holds, which establishes the theorem. $\square$

Example 12.5. *Consider the following pdBf* (T,F)*:*

$$T = \{(1,0,0),(1,1,1)\},$$
$$F = \{(0,0,0),(0,0,1),(0,1,1)\}.$$

This pdBf is illustrated in Figure 12.1. It has the following:

   Patterns: $x_1, x_1 x_2, x_1 \bar{x}_2, x_1 x_3, x_1 \bar{x}_3, x_1 x_2 x_3, x_1 \bar{x}_2 \bar{x}_3,$

   Copatterns: $\bar{x}_1, \bar{x}_1 x_2, \bar{x}_1 \bar{x}_2, \bar{x}_1 x_3, \bar{x}_1 \bar{x}_3, \bar{x}_1 x_2 x_3, \bar{x}_1 \bar{x}_2 x_3, \bar{x}_2 x_3, \bar{x}_1 \bar{x}_2 \bar{x}_3,$

   Prime patterns: $x_1,$

   Prime copatterns: $\bar{x}_1, \bar{x}_2 x_3.$

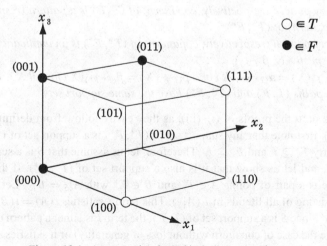

Figure 12.1. An example of pdBf in 3-dimensional space.

Therefore, the functions α and β, when represented by the disjunctions of all prime patterns and prime copatterns, can be written as:

$$\alpha = x_1, \quad \beta = \bar{x}_1 \vee \bar{x}_2 x_3.$$

This implies that

$$T(\alpha) = \{(1,0,0),(1,1,1),(1,1,0),(1,0,1)\},$$

$$T(\beta) = \{(0,0,0),(0,0,1),(0,1,1),(0,1,0),(1,0,1)\}.$$

Note that the point $(1,0,1)$ belongs to both $T(\alpha)$ and $T(\beta)$. □

For a pdBf (T,F), let us now define

$$T^* = F(\beta) = \{X \in \mathcal{B}^n \mid \beta_{(T,F)}(X) = 0\}, \tag{12.10}$$

$$F^* = F(\alpha) = \{X \in \mathcal{B}^n \mid \alpha_{(T,F)}(X) = 0\}. \tag{12.11}$$

Namely, T^* (respectively, F^*) is the set of points at which all cotheories of (T,F) evaluate to 0 (respectively, all theories evaluate to 0). Obviously, $T^* \supseteq T$ and $F^* \supseteq F$ hold.

Example 12.6. *For the pdBf (T,F) of Example 12.5, we obtain*

$$T^* = \{(1,0,0),(1,1,1),(1,1,0)\},$$

$$F^* = \{(0,0,0),(0,0,1),(0,1,1),(0,1,0)\}.$$ □

The two pdBfs (T,F) and (T^*,F^*) are mathematically very close, as evidenced by Lemma 12.2.

Lemma 12.2. *For a given pdBf (T,F), let (T^*,F^*) be the pdBf as defined above. Then,*

- (i) *every pattern (respectively, copattern) of (T,F) is a pattern (respectively, copattern) of (T^*,F^*);*
- (ii) *every pattern (respectively, copattern) of (T^*,F^*) is an implicant of $\alpha_{(T,F)}$ (respectively, $\beta_{(T,F)}$);*
- (iii) *$\alpha_{(T,F)}(X) = \alpha_{(T^*,F^*)}(X)$ and $\beta_{(T,F)}(X) = \beta_{(T^*,F^*)}(X)$ for all $X \in \mathcal{B}^n$;*
- (iv) *the pdBfs (T,F) and (T^*,F^*) have the same support sets.*

Proof. We omit the proofs of (i)–(iii), as they easily follow from definitions. To prove (iv), first note that any support set of (T^*,F^*) is a support set of (T,F) by the property $T^* \supseteq T$ and $F^* \supseteq F$. Therefore, let us assume that S is a support set of (T,F), and let us show that it is also a support set of (T^*,F^*). If this is not true, there is a pair of points $A \in T^*$ and $B \in F^*$ with $A|_S = B|_S$. Let t be the term consisting of all literals in $L(A|_S)$. This term t satisfies $t(A) = t(B) = 1$ by definition. Since S is a support set of (T,F), the term t is either a pattern of (T,F) (we ignore the case of copattern without loss of generality) or it satisfies $t(X) = 0$ for all points $X \in T \cup F$. If t is a pattern of (T,F), this implies that $t(B) = 0$

by the definition of F^*, leading to a contradiction. In the other case, take a point $C \in T \cup F$ for which the Hamming distance $d(A|_S, C|_S)$ is minimized among all points in $T \cup F$. We assume $C \in T$ without loss of generality. Then let t' be the term consisting of all literals in $L(A|_S) \cap L(C|_S)$. This t' satisfies $t'(A) = 1$ and $t'(Y) = 0$ for all $Y \in F$ by construction (use the argument based on the Hamming distance in the proof of Theorem 12.4); that is, t' is a pattern of (T, F). Then $t'(B) = 0$ follows from the definition of F^*, again contradicting the assumption $t(B) = 1$ since t' involves only a subset of the literals in t. □

It may be a lengthy procedure to generate all the elements in T^* and F^* from a given pdBf (T, F). We can, however, state the next theorem for membership testing.

Theorem 12.5. *For a given pdBf (T, F), the membership in T^* (or in F^*) can be tested in polynomial time.*

Proof. We consider only the membership in T^*, since the other case is similar. Let $X \in \mathcal{B}^n$ be a point not in $T \cup F$. Then $X \notin T^*$ if and only if there is a copattern t of (T, F) satisfying $t(X) = 1$. Let this t cover $B \in F$, and let $t_{(X,B)}$ be the term consisting of all the literals in $L(X) \cap L(B)$. By definition, $t_{(X,B)} \le t$ holds and $t_{(X,B)}$ is also a copattern of (T, F). This argument implies that the condition $X \notin T^*$ holds if and only if $t_{(X,B)}$ is a copattern for some $B \in F$. This test can be conducted in time polynomial in the input length $n(|T| + |F|)$. □

In view of the symmetric relation between theory and cotheory, it may be interesting to give special consideration to those theories whose complement is a cotheory: We say that a theory φ is a *bi-theory* of (T, F) if $\overline{\varphi}$ is a cotheory of (T, F) (more precisely, if $\overline{\varphi}$ is equivalent to some cotheory of (T, F)). For the pdBf of Example 12.5, we see that $\varphi = x_1$ is a bi-theory, since its complement $\overline{\varphi} = \bar{x}_1$ is a cotheory. There is another bi-theory $\varphi' = x_1 x_2 \vee x_1 \bar{x}_3$, and this exhausts all bi-theories for this example.

It is natural to ask whether every pdBf has a bi-theory, assuming of course that it has an extension. The next theorem answers this question.

Theorem 12.6. *If a pdBf has an extension, then it has at least one bi-theory.*

Proof. Consider a pdBf (T, F). For a point $X \in \mathcal{B}^n$ and a set $U \subseteq \mathcal{B}^n$, let $d(X, U) = \min_{Y \in U} d(X, Y)$ where d denotes the Hamming distance. To prove the lemma constructively, define the Boolean function f by

$$f(X) = \begin{cases} 1 & \text{if } d(X, T) \le d(X, F), \\ 0 & \text{otherwise.} \end{cases}$$

It is easy to see that if (T, F) has an extension, that is, if $T \cap F = \emptyset$, then f is an extension of (T, F).

For each $X \in T(f)$, let t_X be the term consisting of all the literals in $L(X) \cap L(A)$ for some $A \in T$ satisfying $d(X, A) = d(X, T)$. Then define

$$\varphi = \bigvee_{X \in T(f)} t_X.$$

Similarly, for each $Y \in F(f)$, let t_Y be the term consisting of all the literals in $L(Y) \cap L(B)$ for some $B \in F$ satisfying $d(Y, B) = d(Y, F)$, and define

$$\psi = \bigvee_{Y \in F(f)} t_Y.$$

It follows from these definitions that φ is a theory of (T, F) and ψ is a cotheory of (T, F). We are going to show that φ represents f and ψ represents $\bar{f}$ (and hence $\bar{\varphi}$), which together imply that φ is a bi-theory. For simplicity, we only prove the statement about φ, since the other statement is analogous.

From the definition of φ, it follows immediately that $\varphi(X) = 1$ holds for all $X \in T(f)$. To show that $\varphi(Y) = 0$ holds for all $Y \in F(f)$, choose $X \in T(f)$ and $Y \in F(f)$ arbitrarily. Let $d(X, A) = d(X, T)$ hold for $A \in T$ and $d(Y, B) = d(Y, F)$ hold for $B \in F$. Then we have

$$d(X, A) = d(X, T) \leq d(X, F) \leq d(X, B) \text{ and}$$

$$d(Y, B) = d(Y, F) < d(Y, T) \leq d(Y, A).$$

For these A and B, define $L_{AB} = L(A) \setminus L(B)$ and $L_{BA} = L(B) \setminus L(A)$ (namely, L_{AB} is the set of literals in $L(A)$ whose complements are in $L(B)$, and L_{BA} is defined similarly). Then $d(X, A) \leq d(X, B)$ implies $|t_X \cap L_{AB}| \geq |L_{AB}|/2$, where $t_X \cap L_{AB}$ denotes the set of literals in both t_X and L_{AB}. Similarly, $d(Y, B) < d(Y, A)$ implies $|t_Y \cap L_{BA}| > |L_{BA}|/2 = |L_{AB}|/2$. Therefore, there is at least one literal in t_X whose complement is in t_Y, and hence, $t_X(Y) = 0$ holds. As t_X was an arbitrary term in φ, this proves $\varphi(Y) = 0$. $\square$

With regard to bi-theories, the sets (T^*, F^*) defined by (12.10)–(12.11) can be characterized as follows (Boros et al. [115]): $X \in T^*$ if and only if $\varphi(X) = 1$ holds for all bi-theories φ, and $Y \in F^*$ if and only if $\varphi(Y) = 0$ holds for all bi-theories φ.

Bi-theories play an important role in logical analysis of data, as the classifications which they produce are justified both by examples from T and by examples from F. However, not much is known about the complexity of their recognition and generation. We refer the reader to Boros et al. [115] for additional details.

12.2.5 Decision trees

Decision trees were introduced in Chapter 1, Section 1.12.3, as a means of representing Boolean functions. Recall that a decision tree is a binary rooted tree, in which each intermediate node has exactly two children corresponding to the assignments $x_j = 0$ and $x_j = 1$ for a chosen variable x_j, and each leaf node carries

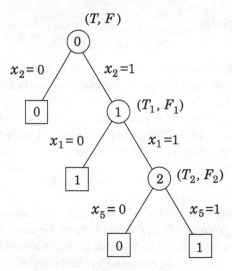

Figure 12.2. An example of a decision tree.

Procedure TREE-DNF

For each leaf node with value 1, construct the term corresponding to the path from the root to the leaf node, and take the disjunction of all such terms.

Figure 12.3. Procedure TREE-DNF.

the function value 0 or 1 for the assignment defined by the unique path from the root (the top node numbered 0) to the leaf node under consideration. Figure 12.2 shows an example of a decision tree, where intermediate nodes are drawn as circles and leaves are drawn as squares. For example, the rightmost bottom node (with assignment 1) indicates that the function value for the assignment $x_2 = 1$, $x_1 = 1$, and $x_5 = 1$ (along the rightmost path) is 1. To know the function value for a given data point $A = (0, 1, 0, 1, 0, 1, 1, 0)$, for example, we start from the root and follow the branch $x_2 = 1$ (since $a_2 = 1$) to the intermediate node 1. Then from node 1 we follow the branch $x_1 = 0$ (since $a_1 = 0$) to arrive at a leaf node with value 1. This tells us that $f(A) = 1$ for the function f represented by this decision tree.

Given a decision tree representing a Boolean function f, the above explanation entails that a DNF of f can be constructed by the procedure in Figure 12.3.

In this procedure, the term corresponding to a path is defined by including the literal x_j (respectively, $\bar{x}_j$) if the assignment $x_j = 1$ (respectively, $x_j = 0$) occurs along the path.

Example 12.7. *As the decision tree in Figure 12.2 has two leaf nodes with value 1, the following DNF is obtained by the procedure* TREE-DNF.

$$\varphi = \bar{x}_1 x_2 \vee x_1 x_2 x_5.$$

This can be simplified to

$$\varphi' = \bar{x}_1 x_2 \vee x_2 x_5,$$

which shows that the tree in Figure 12.2 represents the function f_1 of Example 12.1. The DNF φ' is actually one of the prime theories obtained in Example 12.3 for the pdBf of Example 12.1.

As observed in Section 1.12.3, a decision tree yields a DNF of $\bar{f}$ as well, by applying the procedure TREE-DNF *to all leaf nodes with value 0 (instead of those with value 1). For the above example, we obtain*

$$\bar{\varphi} = \bar{x}_2 \vee x_1 x_2 \bar{x}_5 = \bar{x}_2 \vee x_1 \bar{x}_5. \qquad \square$$

Let us now consider the problem of constructing a decision tree that represents an extension of a given pdBf (T, F). Similarly to the case of DNFs, there are many such decision trees, and it is desirable to obtain a "simple" one. The simplicity of decision trees may be measured by their number of nodes, or by their height. Exact minimization is, however, intractable; for example, it is known that finding a decision tree with the minimum number of nodes is NP-hard (Hyafil and Rivest [514]). Therefore, various heuristic algorithms have been proposed to obtain approximately minimum decision trees. When applied to a pdBf (T, F), most of these heuristics fit in the generic scheme described in Figure 12.4, where we assume that $T \cap F = \emptyset$.

The procedure PDBF DECISION TREE yields a decision tree $D = D(T, F)$ for every pdBf (T, F). The tree D can be viewed as representing a Boolean function

---

Procedure PDBF DECISION TREE

Start with the rooted tree consisting of a single node numbered 0, which is unprocessed, and which is associated with the original pdBf (T, F).

Repeat the following **Branching step** as long as there remains an unprocessed node:

(Branching step) Select an unprocessed node numbered k, associated with a pdBf (T_k, F_k), and process it according to the following rules.

1. If $T_k = \emptyset$, then node k becomes a leaf node with value 0.
2. If $F_k = \emptyset$, then node k becomes a leaf node with value 1.
3. If $T_k \neq \emptyset$ and $F_k \neq \emptyset$, then determine a branching variable x_{j_k} such that x_{j_k} does not take a constant value in $T_k \cup F_k$, and generate two children k_0 and k_1 corresponding to $x_{j_k} = 0$ and $x_{j_k} = 1$, respectively, for which the associated pdBfs (T_{k_0}, F_{k_0}) and (T_{k_1}, F_{k_1}) are defined as follows.

$$T_{k_0} = \{A \in T_k \mid a_{j_k} = 0\},$$
$$F_{k_0} = \{B \in F_k \mid b_{j_k} = 0\},$$
$$T_{k_1} = \{A \in T_k \mid a_{j_k} = 1\},$$
$$F_{k_1} = \{B \in F_k \mid b_{j_k} = 1\}.$$

Figure 12.4. Procedure PDBF DECISION TREE.

f_D, which is an extension of (T, F). It is interesting to observe that every extension produced in this way is a bi-theory, as introduced in Section 12.2.4:

Theorem 12.7. *For every pdBF* (T, F), *if* D *is a decision tree produced by the procedure* PDBF DECISION TREE, *and if* f_D *is the extension of* (T, F) *represented by* D, *then* f_D *is a bi-theory of* (T, F).

Proof. A DNF φ_D of the function f_D can be constructed by the procedure TREE-DNF. It is easy to see that each term of φ is a pattern of (T, F), and hence, the DNF φ_D is a theory of (T, F).

Similarly, the DNF ψ_D obtained by applying the procedure TREE-DNF to the leaf nodes of D with value 0 is a cotheory of (T, F). Since ψ_D represents $\overline{f_D}$, we conclude that f_D is a bi-theory. $\qquad\qquad\square$

Note that this result provides an alternative proof of Theorem 12.6. It is illustrated by Example 12.7. Additional connections between decision trees and bi-theories are established in Boros et al. [115].

When using the procedure PDBF DECISION TREE, the rule applied for choosing a branching variable at each intermediate node is crucial and determines the properties of the resulting decision tree, including its size. Different heuristic methods rely on different ways to select the branching variable. As a representative example, we describe now the "information theoretic" rule proposed by Quinlan [770] in a popular algorithm called ID3.

For a pdBf (T_k, F_k), let $p = |T_k|$ and $q = |F_k|$, and define the entropy of (T_k, F_k) by

$$I(p,q) = -\frac{p}{p+q} \log_2 \frac{p}{p+q} - \frac{q}{p+q} \log_2 \frac{q}{p+q}.$$

If a branching variable x_j yields the pdBfs (T_{k_0}, F_{k_0}) and (T_{k_1}, F_{k_1}), then their average entropy becomes

$$E(x_j) = \frac{p_0 + q_0}{p+q} I(p_0, q_0) + \frac{p_1 + q_1}{p+q} I(p_1, q_1).$$

where $p_0 = |T_{k_0}|, q_0 = |F_{k_0}|, p_1 = |T_{k_1}|$ and $q_1 = |F_{k_1}|$. This means that the amount of information gained by the decomposition based on the selction of x_j is

$$gain(x_j) = I(p,q) - E(x_j). \tag{12.12}$$

In ID3, the variable x_j that maximizes $gain(x_j)$ among all the remaining unfixed variables is selected as the branching variable at node k.

Example 12.8. *Let us apply the procedure ID3 to the pdBf of Example 12.1. We first apply rule 3 to the original pdBf* (T, F). *In determining the branching variable that maximizes* $gain(x_j)$, *we can choose the variable that minimizes* $E(x_j)$, *since* $I(p,q)$ *is constant for all* x_j *in* (12.12). *In order to illustrate the computation of*

$E(x_1)$, *observe that the following pdBfs* (T', F') *and* (T'', F'') *result when we fix* x_1 *to 0 and to 1, respectively,*

$$T' = \{A^{(1)}, A^{(3)}\}$$
$$F' = \{B^{(2)}, B^{(4)}\}$$
$$T'' = \{A^{(2)}\}$$
$$F'' = \{B^{(1)}, B^{(3)}\}.$$

Thus we have $p_0 = |T'| = 2, q_0 = |F'| = 2, p_1 = |T''| = 1, q_1 = |F''| = 2$, *and hence,*

$$E(x_1) = \frac{4}{7}(-\frac{2}{4}\log\frac{2}{4} - \frac{2}{4}\log\frac{2}{4}) + \frac{3}{7}(-\frac{1}{3}\log\frac{1}{3} - \frac{2}{3}\log\frac{2}{3}) = 0.77.$$

Similarly, we obtain

$$E(x_2) = 0.46,$$
$$E(x_3) = E(x_4) = E(x_6) = E(x_7) = E(x_8) = 0.77,$$
$$E(x_5) = 0.98.$$

Therefore, at the root, x_2 *minimizes* $E(x_j)$ *and is chosen as the branching variable.*

Now the two pdBfs (T_0, F_0) *and* (T_1, F_1) *that result by fixing* x_2 *to 0 and to 1, respectively, are given by*

$$T_0 = \emptyset,$$
$$F_0 = \{B^{(1)}, B^{(2)}, B^{(4)}\},$$
$$T_1 = \{A^{(1)}, A^{(2)}, A^{(3)}\},$$
$$F_1 = \{B^{(3)}\}.$$

As the pdBf (T_0, F_0) *corresponding to* $x_2 = 0$ *satisfies* $T_0 = \emptyset$, *we obtain a leaf node with value 0 by rule 1 of the branching step in procedure* PDBF DECISION TREE. *For the pdBf* (T_1, F_1) *corresponding to* $x_2 = 1$, *we again apply rule 3 to select a branching variable from among* $x_1, x_3, x_4, \ldots, x_8$; *in this case,* x_1 *is selected.*

Repeating this procedure, we eventually obtain the decision tree of Figure 12.2, in which the pdBf (T_2, F_2) *of node 2 does not depend on* x_1, x_2, *and is given by*

$$T_2 = \{A^{(2)}\}$$
$$F_2 = \{B^{(3)}\}. \qquad \square$$

Other types of selection rules for branching variables have also been proposed. The successful software C4.5 and its successor C5.0 by Quinlan [771, 772], for example, use a rule based on the *gain-ratio* in place of the above *gain* criterion. Another important addition included in these algorithms is the operation of "pruning," which is applied after a decision tree is constructed. This operation is performed on each intermediate node in order to test whether it is more beneficial to retain the node or to prune it into a leaf node, according to some statistical criterion. The resulting decision tree usually features a more robust behavior on new input samples.

Before closing this section, we briefly compare two representations of extensions of pdBfs, by DNFs and by decision trees, respectively. Generally speaking, if an extension f has a small decision tree, it tends to have a small DNF, and vice versa, since both representations are closely related as explained earlier in this section. A decision tree is visually appealing, while a DNF may be more convenient for the purpose of understanding the logical content of f. For certain function classes, such as $\mathcal{F}_+$, $\mathcal{F}_{\text{HORN}}$, and $\mathcal{F}_k$, it is easier to check whether a function belongs to the class when it is represented by a DNF.

The size of a support set is also positively correlated with that of a decision tree, but not always exactly. Recall that a support set is a set of variables which is required to represent an extension. On the other hand, heuristic minimization of a decision tree, such as performed by ID3, is based on choosing an appropriate branching variable at each node, independently of the choices at other nodes. As a result of this difference, minimization of a support set does not generally coincide with minimization of a decision tree.

12.3   Extensions within given function classes

In this section we consider the problem EXTENSION($\mathcal{C}$), defined in Section 12.2.1, for the function classes $\mathcal{F}_+$, $\mathcal{F}_{\text{UNATE}}$, $\mathcal{F}_{\text{HORN}}$, $\mathcal{F}_{\text{Th}}$, $\mathcal{F}_{F_0(S_0, F_1(S_1))}$, $\mathcal{F}_k$, and $\mathcal{F}_{k\text{-CONV}}$. We discuss necessary and sufficient conditions for the existence of such extensions, and the computational complexity of finding an extension in the class when there is one. The results are mainly borrowed from papers by Boros, Ibaraki, and Makino [139] and Crama, Hammer, and Ibaraki [233], which also consider other classes of functions.

12.3.1   Positive extensions

Let us first consider the class of positive functions $\mathcal{F}_+$, defined in Sections 1.10 and 1.11. For this class, we obtain (Zuev [939]):

Theorem 12.8. *A pdBf* (T, F) *has an extension* $f \in \mathcal{F}_+$ *if and only if there exists no pair* (A, B) *with* $A \in T$ *and* $B \in F$ *such that* $A \leq B$. *This condition can be checked in polynomial time.*

Proof. Necessity. Assume that (T, F) has a positive extension f, and let $A \in T$, $B \in F$. Then, $f(A) = 1$ and $f(B) = 0$, and the positivity of f rules out that $A \leq B$.

Sufficiency. Define a Boolean function $f_{\min}^+$ by

$$T(f_{\min}^+) = \{C \in \mathcal{B}^n \mid C \geq A \text{ holds for some } A \in T\},$$
$$F(f_{\min}^+) = \mathcal{B}^n \setminus T(f_{\min}^+).$$

It is clear that f_{min}^+ is a positive function. Furthermore, $T(f_{min}^+) \cap F = \emptyset$ holds by the assumption on T and F. Therefore, f_{min}^+ is a positive extension of (T, F).

Finally, the condition in the theorem statement can be checked by directly comparing all pairs (A, B) with $A \in T$ and $B \in F$. This can be done in $O(n|T||F|)$ time, which is polynomial in the input length $n(|T| + |F|)$.   □

The positive extension f_{min}^+ defined in the proof minimizes the set $T(f)$ among all positive extensions f of the pdBf (T, F). It is not difficult to show that f_{min}^+ is in fact the unique minimum positive extension of (T, F). We can also define f_{max}^+ dually:

$$F(f_{max}^+) = \{C \in \mathcal{B}^n \mid C \leq B \text{ holds for some } B \in F\},$$
$$T(f_{max}^+) = \mathcal{B}^n \setminus F(f_{max}^+).$$

This function f_{max}^+ is the unique extension that maximizes the set $T(f)$ among all positive extensions f. Any positive extension f of a pdBf (T, F) satisfies

$$f_{min}^+ \leq f \leq f_{max}^+,$$

and all positive extensions form a lattice under the operations $\vee$ and $\wedge$ between functions. This is a sublattice of the lattice of all extensions of a pdBf (T, F) introduced in Section 12.2.1.

Assume now that a pdBf (T, F) has positive extensions. We say that a set $S^+ \subseteq \{1, 2, \ldots, n\}$ is a *positive support set* for (T, F) if $(T|_{S^+}, F|_{S^+})$ has a positive extension, and we define

$$\Delta^+(A, B) = \{j \in \{1, 2, \ldots, n\} \mid a_j = 1, b_j = 0\}$$

(compare with $\Delta(A, B)$ of (12.5) in Section 12.2.2). Then the problem of finding a minimum positive support set can be formulated as the following set covering problem:

$$\text{minimize} \sum_{j=1}^{n} y_j$$

$$\text{subject to} \sum_{j \in \Delta^+(A,B)} y_j \geq 1, \quad A \in T, B \in F$$

$$y_j \in \{0, 1\}, \quad j \in \{1, 2, \ldots, n\}.$$

The next theorem can be proved similarly to Theorem 12.2 [233].

Theorem 12.9. *Problem* MIN-SUPPORT$(\mathcal{F}_+)$ *is NP-hard.*   □

12.3.2 Monotone (unate) extensions

As defined in Section 1.10, a Boolean function f is called *monotone* (or *unate*) if f is either positive or negative in each of its variables. In finding an extension

$f \in \mathcal{F}_{\text{UNATE}}$ of a pdBf (T, F), therefore, it is also required to know the polarity (either positive or negative) of each variable x_j. We first show that this problem can be formulated as a 0–1 integer programming problem, by adapting the argument used for support sets in Sections 12.2.2 and 12.3.1

Introduce two new 0–1 variables y_j and z_j for each $j \in \{1, 2, \ldots, n\}$, where $y_j = 1$ implies that variable x_j appears positively in a unate extension, while $z_j = 1$ implies that x_j appears negatively in this extension. The assignment $y_j = z_j = 1$ is prohibited, and $y_j = z_j = 0$ indicates that the extension does not depend on x_j. Define

$$\Delta^+(A, B) = \{j \in \{1, 2, \ldots, n\} \mid a_j = 1, b_j = 0\},$$
$$\Delta^-(A, B) = \{j \in \{1, 2, \ldots, n\} \mid a_j = 0, b_j = 1\}.$$

Then problem EXTENSION($\mathcal{F}_{\text{UNATE}}$) has a solution for a given pdBf (T, F) if and only if the following problem has a feasible solution:

$$\sum_{j \in \Delta^+(A, B)} y_j + \sum_{j \in \Delta^-(A, B)} z_j \geq 1, \quad A \in T, B \in F \tag{12.13}$$

$$y_j + z_j \leq 1, \quad j \in \{1, 2, \ldots, n\} \tag{12.14}$$

$$y_j \in \{0, 1\}, \ z_j \in \{0, 1\}, \quad j \in \{1, 2, \ldots, n\}. \tag{12.15}$$

Furthermore, since $y_j = 1$ or $z_j = 1$ implies that j is used in the resulting support set of $f \in \mathcal{F}_{\text{UNATE}}$, problem MIN-SUPPORT($\mathcal{F}_{\text{UNATE}}$) can be formulated as the 0–1 programming problem obtained by considering the objective function

$$\text{minimize} \sum_{j=1}^{n} y_j + \sum_{j=1}^{n} z_j$$

together with the constraint set (12.13)–(12.15).

For practical purposes, the above problems may be solved by existing integer programming algorithms, and heuristic algorithms may be developed to solve large problem instances. However, the constraint set (12.13)–(12.15) is more complicated than the set covering constraints used in Sections 12.2.2 and 12.3.1, due to the presence of the additional constraints (12.14). Therefore, the following theorem (due to [233]) should not come as a surprise.

Theorem 12.10. *Problem* EXTENSION($\mathcal{F}_{\text{UNATE}}$) *is NP-complete.*

Proof. The problem is obviously in the class NP, since it is straightforward to check whether any assignment of 0–1 values to the variables (y_j, z_j) satisfies the constraints (12.13)–(12.14).

To prove that EXTENSION($\mathcal{F}_{\text{UNATE}}$) is NP-complete, we provide a reduction from the following NP-complete problem:

DNF EQUATION
Instance: A DNF expression $\phi(X)$ on the variables $X = (x_1, x_2, \ldots, x_n)$.
Question: Is the equation $\phi(X) = 0$ consistent?

(see Chapter 2 and Appendix B). Given an instance $\phi(X)$ of DNF EQUATION, we construct a pdBf (T, F) such that the corresponding 0–1 problem (12.13)–(12.14) has a feasible solution if and only if the equation $\phi(X) = 0$ is consistent. For this purpose, let $t_1, t_2, \ldots, t_m$ denote the terms of the DNF ϕ, and let $C_i \subseteq \{x_1, \bar{x}_2, \ldots, x_n, \bar{x}_n\}$ be the set of literals that appear in t_i. We define

$$T = \{A^i \in \mathcal{B}^{n+m} \mid i = 1, 2, \ldots, m\},$$
$$F = \{B^i, D^i \in \mathcal{B}^{n+m} \mid i = 1, 2, \ldots, m\},$$

where, for $i, k = 1, 2, \ldots, m$ and $j = 1, 2, \ldots, n$,

$$a^i_j = 1 \text{ and } b^i_j = 0 \quad \text{if } x_j \in C_i,$$
$$a^i_j = 0 \text{ and } b^i_j = 1 \quad \text{if } \bar{x}_j \in C_i,$$
$$a^i_j = b^i_j = 0 \qquad \text{if } x_j, \bar{x}_j \notin C_i,$$
$$a^i_{n+i} = b^i_{n+i} = 1,$$
$$a^i_{n+k} = b^i_{n+k} = 0 \quad \text{if } k \neq i,$$
$$d^i_j = a^i_j,$$
$$d^i_{n+k} = 0.$$

For the pdBf (T, F), we obtain the following set of inequalities from (12.13)–(12.14):

(i) For $A^i \in T$ and $B^i \in F$ (for the same i),

$$\sum_{x_j \in C_i} y_j + \sum_{\bar{x}_j \in C_i} z_j \geq 1, \quad i \in \{1, 2, \ldots, m\}.$$

(ii) For $A^i \in T$ and $D^i \in F$ (for the same i),

$$y_{n+i} \geq 1, \quad i \in \{1, 2, \ldots, m\}.$$

(iii) $y_j + z_j \leq 1, \quad j \in \{1, 2, \ldots, n + m\}.$
(iv) Other inequalities.

From (ii) and (iii), we see that $y_{n+i} = 1$ and $z_{j+i} = 0$ must hold for all $i \in \{1, 2, \ldots, m\}$. This implies that the inequalities in (iv) are all redundant since any inequality in (iv) contains at least one variable y_{n+i} $(i = 1, 2, \ldots, m)$ in its left-hand side. Therefore, our problem EXTENSION($\mathcal{F}_{\text{UNATE}}$) becomes equivalent to deciding whether the constraints (i) and (iii) have a feasible 0–1 solution. It is now obvious that such a solution (Y, Z) exists if and only if the original Boolean equation has a solution X defined by $x_j = 1$ if $y_j = 1$, $x_j = 0$ if $z_j = 1$, and x_j arbitrary if $x_j = y_j = 0$. □

12.3.3 Degree-k extensions

A DNF φ is called a k-DNF if it has degree k, that is, if every term of φ contains at most k literals, where k is a given positive integer. We denote by $\mathcal{F}_k$ the class of Boolean functions which can be represented by a k-DNF. The following statement is an immediate corollary of Theorem 12.3 and of the definition of prime irredundant theories in Section 12.2.3:

Lemma 12.3. *If a pdBf (T, F) has an extension in $\mathcal{F}_k$, then it has a prime irredundant theory in $\mathcal{F}_k$.* $\square$

In view of this lemma, a pdBf (T, F) has an extension in $\mathcal{F}_k$ if and only if every point $A \in T$ is covered by a pattern of degree k or less. For a given $A \in T$, this property can be checked as follows. First, construct the minterm

$$t_A^* = \Big(\bigwedge_{j:a_j=1} x_j \Big)\Big(\bigwedge_{j:a_j=0} \bar{x}_j \Big),$$

and generate all terms t consisting of at most k literals chosen from the n literals in t_a^*. If at least one of these terms t satisfies $T(t) \cap F = \emptyset$, then t is the required pattern; otherwise, A is not covered by any pattern of degree k. A naive implementation of this procedure requires $O(n^k \times n|F|)$ time, which is polynomial when k is viewed as a constant.

Thus, we obtain:

Theorem 12.11. *The problem* EXTENSION($\mathcal{F}_k$) *can be solved in polynomial time when k is fixed. Similarly,* EXTENSION($\mathcal{F}_k^+$) *can be solved in polynomial time for every fixed k.*

Proof. The first part of the theorem follows from the above discussion. The statement about positive extensions can be shown similarly, by starting from $t_A^+ = \bigwedge_{j:a_j=1} x_j$ instead of t_A^*. $\square$

12.3.4 Horn extensions

Recall the characterization of a Horn function in Section 6.3: a Boolean function f is Horn if and only if $F(f) = F(f)^\wedge$ holds, where $U^\wedge$ denotes the *conjunction closure* of a set $U \subseteq \mathcal{B}^n$. This implies that any Horn extension f of a pdBf (T, F) satisfies $F(f) \supseteq F^\wedge$, and hence, $F^\wedge \cap T = \emptyset$ is a necessary condition for the existence of a Horn extension. The next theorem establishes that this condition is also sufficient:

Theorem 12.12. *A pdBf (T, F) has an extension $f \in \mathcal{F}_{\mathrm{HORN}}$ if and only if $F^\wedge \cap T = \emptyset$. This condition can be checked in polynomial time.*

Proof. The Boolean function f defined by $F(f) = F^\wedge$ is a Horn function, and it is an extension of (T, F) if and only if $F^\wedge \cap T = \emptyset$. This proves the first part of the theorem.

Table 12.4. A pdBf (T, F) with Horn extensions

		x_1	x_2	x_3	x_4	x_5	x_6	x_7	x_8	x_9
	$A^{(1)} =$	1	1	1	1	0	0	1	0	0
	$A^{(2)} =$	1	1	1	0	1	0	1	0	0
	$A^{(3)} =$	1	1	1	0	0	1	0	1	0
T	$A^{(4)} =$	0	0	1	0	0	0	1	0	0
	$A^{(5)} =$	1	0	0	0	0	0	1	0	0
	$A^{(6)} =$	0	1	1	0	0	0	0	0	1
	$A^{(7)} =$	1	1	0	0	0	0	0	0	1
	$A^{(8)} =$	1	1	1	1	1	1	0	0	0
	$B^{(1)} =$	1	1	1	1	0	0	1	1	0
F	$B^{(2)} =$	1	1	1	0	1	0	1	1	1
	$B^{(3)} =$	1	1	1	0	0	1	1	1	0
	$B^{(4)} =$	1	1	1	0	0	0	1	0	1

For the time complexity, note that condition $F^\wedge \cap T = \emptyset$ can be rewritten as

$$\bigwedge_{B \in F'} B \neq A, \quad \text{for all } F' \subseteq F \text{ and for all } A \in T.$$

For $A \in \mathcal{B}^n$, define

$$F_{\geq A} = \{B \in F \mid B \geq A\}.$$

For every $F' \subseteq F$, the condition $\bigwedge_{B \in F'} B = A$ implies that $B \geq A$ for all $B \in F'$ (i.e., $F' \subseteq F_{\geq A}$), and hence, that $\bigwedge_{B \in F_{\geq A}} B = A$ also holds. Therefore, the condition $F^\wedge \cap T = \emptyset$ is equivalent to

$$\bigwedge_{B \in F_{\geq A}} B \neq A, \quad \text{for all } A \in T, \tag{12.16}$$

which can be checked in $O(n|T||F|)$ time by scanning all $B \in F$ for each $A \in T$. $\square$

Example 12.9. *Consider the pdBf (T, F) defined in Table 12.4. It is easily checked that*

$$F_{\geq A^{(1)}} = \{B^{(1)}\},$$
$$F_{\geq A^{(2)}} = \{B^{(2)}\},$$
$$F_{\geq A^{(3)}} = \{B^{(3)}\},$$
$$F_{\geq A^{(4)}} = F_{\geq A^{(5)}} = \{B^{(1)}, B^{(2)}, B^{(3)}, B^{(4)}\},$$
$$F_{\geq A^{(6)}} = F_{\geq A^{(7)}} = \{B^{(2)}, B^{(4)}\},$$
$$F_{\geq A^{(8)}} = \emptyset,$$

and condition (12.16) *holds for all* $A^{(i)} \in T$. *Therefore, this pdBf has a Horn extension by Theorem 12.12.*   □

In general, a pdBf (T, F) may have many Horn extensions. Let $f_{\max}^{\text{HORN}}$ denote the Horn extension that maximizes $T(f)$ among all Horn extensions f. Then, it follows from the discussion before Theorem 12.12 that $f_{\max}^{\text{HORN}}$ is given by

$$F(f_{\max}^{\text{HORN}}) = F^{\wedge}$$
$$T(f_{\max}^{\text{HORN}}) = \mathcal{B}^n \setminus F^{\wedge},$$

and it is unique. On the other hand, there are generally many *minimal* Horn extensions, that is, Horn extensions f with minimal true set $T(f)$.

As observed in Chapter 6, DNFs of Horn functions have numerous special properties. Some of them can be generalized to Horn extensions of pdBfs. In particular, there are pdBfs (T, F) for which the number of prime implicants in $f_{\max}^{\text{HORN}}$ is exponential in the input length $n(|T| + |F|)$. There are algorithms for generating all prime implicants of $f_{\max}^{\text{Horn}}$, but none of them runs in polynomial time in its input and output length (Kautz, Kearns, and Selman [554]; Khardon [564]). It is known that this problem has a polynomial time algorithm if and only if there is a polynomial time algorithm (in its input and output length) to generate all prime implicants of the dual of a positive function (Kavvadias, Papadimitriou, and Sideri [558]). As discussed in Section 4.4, the complexity of the latter problem is still open. Observe that just finding *any* Horn DNF of $f_{\max}^{\text{HORN}}$ is not easier than finding all its prime implicants, since, from such a DNF, all prime implicants can be generated in polynomial total time (see Section 6.5). The complexity of this problem and other related problems, such as finding an irredundant DNF of $f_{\max}^{\text{HORN}}$ and finding a shortest DNF of $f_{\max}^{\text{HORN}}$, still remain to be studied.

On the other hand, DNFs of minimal (in the sense of $T(f)$) Horn extensions can be described in a canonical form, each of which is of polynomial length in the input length $n(|T| + |F|)$. To see this, let us introduce some notations. For a pdBf (T, F) with $T \cap F = \emptyset$, and for each $A \in T$,

$$I(A) = \{j \in \{1, 2, \ldots, n\} \mid a_j = 0 \text{ and } b_j = 1 \text{ for all } B \in F_{\geq A}\}$$

$$R(A) = \begin{cases} \{\bigwedge_{j=1}^{n} x_j\} & \text{if } A = (1, 1, \ldots, 1) \\ \{(\bigwedge_{j:a_j=1} x_j)\bar{x}_l \mid l \in I(A)\} & \text{if } A \neq (1, 1, \ldots, 1) \text{ and } I(A) \neq \emptyset \\ \emptyset & \text{if } A \neq (1, 1, \ldots, 1) \text{ and } I(A) = \emptyset. \end{cases}$$

Note that $R(A)$ is empty only if $A \neq (1, 1, \ldots, 1)$ and $I(A) = \emptyset$, in which case condition (12.16) implies that (T, F) has no Horn extension.

Now, when $R(A)$ is nonempty for all $A \in T$, we define a *canonical Horn* DNF for the pdBf (T, F) to be any DNF of the form

$$\varphi = \bigvee_{A \in T} t_A, \quad \text{where } t_A \in R(A).$$

In words, a canonical Horn DNF is obtained by choosing one term t_A from each set $R(A)$ and taking the disjunction of these terms over all $A \in T$. Note that each term $t_A \in R(A)$ satisfies $t_A(A) = 1$ and $t_A(B) = 0$ for all $B \in F$. Therefore, every canonical Horn DNF represents a Horn extension of (T, F), and its length is $O(n|T|)$.

Example 12.10. *Let us obtain $I(A)$ and $R(A)$ for all $A \in T$ of Example 12.9.*

$$
\begin{aligned}
I(A^{(1)}) &= \{8\}, & R(A^{(1)}) &= \{12347\bar{8}\} \\
I(A^{(2)}) &= \{8,9\}, & R(A^{(2)}) &= \{12357\bar{8}, 123579\} \\
I(A^{(3)}) &= \{7\}, & R(A^{(3)}) &= \{1236\bar{7}8\} \\
I(A^{(4)}) &= \{1,2\}, & R(A^{(4)}) &= \{\bar{1}37, \bar{2}37\} \\
I(A^{(5)}) &= \{2,3\}, & R(A^{(5)}) &= \{1\bar{2}7, 1\bar{3}7\} \\
I(A^{(6)}) &= \{1,7\}, & R(A^{(6)}) &= \{\bar{1}239, 23\bar{7}9\} \\
I(A^{(7)}) &= \{3,7\}, & R(A^{(7)}) &= \{12\bar{3}9, 12\bar{7}9\} \\
I(A^{(8)}) &= \{7,8,9\}, & R(A^{(8)}) &= \{123456\bar{7}, 123456\bar{8}, 123456\bar{9}\},
\end{aligned}
$$

where we employ a shorthand notation for terms; for example, $12347\bar{8}$ stands for $x_1x_2x_3x_4x_7\bar{x}_8$, and so on. Consequently, there are $1 \times 2 \times 1 \times 2 \times 2 \times 2 \times 2 \times 3 = 96$ canonical Horn DNFs, among which we find, for example,

$$
\varphi^{(1)} = 12347\bar{8} \vee 123579 \vee 1236\bar{7}8 \vee \bar{1}37 \vee 1\bar{3}7 \vee 23\bar{7}9 \vee 12\bar{7}9 \vee 123456\bar{7}
$$

$$
\varphi^{(2)} = 12347\bar{8} \vee 123579 \vee 1236\bar{7}8 \vee \bar{1}37 \vee 1\bar{3}7 \vee \bar{1}239 \vee 12\bar{3}9 \vee 123456\bar{9}.
$$
□

It can be proved that every minimal Horn extension has a canonical Horn DNF, but the converse is not always true; we refer the reader to Makino, Hatanaka, and Ibaraki [650] for details. In the above Example 12.10, $\varphi^{(2)}$ represents a minimal Horn extension, but $\varphi^{(1)}$ does not. It can be checked in polynomial time whether a canonical Horn DNF represents a minimal Horn extension or not [650]. Further properties of Horn extensions can be found in Ibaraki, Kogan, and Makino [518].

12.3.5 Threshold extensions

Because of their natural interpretation, threshold extensions of pdBf have been extensively studied in data mining and pattern recognition (see, e.g., Mangasarian [658]; Bradley, Fayyad, and Mangasarian [149]; Mangasarian, Setiono, and Wolberg [660]), in machine learning (see Matulef et al. [677]; O'Donnell and Servedio [717])), and in mathematical psychology (see Medina and Schwanen-flugel [679]; Smith, Murray, and Minda [843]; Wattenmaker et al. [901]).

The problem of deciding whether a pdBf admits a threshold extension is easily settled.

Theorem 12.13. *The pdBf (T,F) has a threshold extension if and only if the system of inequalities*

$$\sum_{j=1}^{n} w_j x_j \leq t \qquad \text{for all } X \in F, \tag{12.17}$$

$$\sum_{j=1}^{n} w_j x_j \geq t+1 \qquad \text{for all } X \in T, \tag{12.18}$$

has a solution $(w_1, w_2, \ldots, w_n, t)$. This condition can be checked in polynomial time.

Proof. The characterization follows immediately from the definition of threshold functions. The feasibility of the system of linear inequalities (12.17)–(12.18) can be checked in polynomial time (see, e.g., [76]).   □

Note however that, even when there exists a threshold extension, the solution of the system (12.17)–(12.18) does not immediately produce a DNF of the extension, but only a linear separating structure $(w_1, w_2, \ldots, w_n, t)$. Also, the existence of a threshold extension does not guarantee the existence of a threshold prime theory or of a threshold theory defined over a minimum cardinality support set.

Example 12.11. *Consider the pdBf given by*

$$T = \{(1,0,1,1), (1,1,0,0), (1,1,0,1), (1,1,1,0), (1,1,1,1)\},$$

$$F = \{(0,0,0,0), (0,0,0,1), (0,0,1,0), (0,1,0,0), (0,1,0,1),$$

$$(0,1,1,0), (1,0,0,0), (1,0,0,1), (1,0,1,0)\}.$$

This pdBf has four extensions, namely,

$$\psi_1 = x_1 x_2 \vee x_1 x_3 x_4,$$

$$\psi_2 = x_1 x_2 \vee x_1 x_3 x_4 \vee x_2 x_3 x_4,$$

$$\psi_3 = x_1 x_2 \vee x_1 x_3 x_4 \vee \bar{x}_2 x_3 x_4,$$

$$\psi_4 = x_1 x_2 \vee x_3 x_4.$$

Of these four extensions, only ψ_1 and ψ_2 are threshold. The unique prime theory of (T,F) is ψ_4, which is not threshold.

Similarly, the pdBf of Example 12.1 has several threshold extensions, but the extensions defined over the minimum cardinality support sets $S_1 = \{5,8\}$ and $S_2 = \{6,7\}$ (namely, φ_1 and φ_2 in Table 12.2) are not threshold.   □

12.3.6 Decomposable extensions

Consider a family of subsets $S_0, S_1, \ldots, S_k$, where $S_i \subseteq \{1, 2, \ldots, n\}$ for all i. In general, we allow S_i and S_j to intersect, that is, $S_i \cap S_j \neq \emptyset$, although the case of disjoint S_i's will be most interesting. We denote the *projection* of a vector of variables $X = (x_1, x_2, \ldots, x_n)$ to a set S as $X|_S = (x_j \mid j \in S)$.

Now, consider a Boolean function f on $\mathcal{B}^n$. We say that f is $F_0(S_0, F_1(S_1), \ldots, F_k(S_k))$-*decomposable* if there exist $(k+1)$ Boolean functions $g : \mathcal{B}^{|S_0|+k} \to \mathcal{B}$, and $h_i : \mathcal{B}^{|S_i|} \to \mathcal{B}$, for $i = 1, 2, \ldots, k$, such that f can be represented as the following composition of g and $h_i, i = 1, 2, \ldots, k$:

$$f(X) = g(X|_{S_0}, h_1(X|_{S_1}), \ldots, h_k(X|_{S_k})). \tag{12.19}$$

Here, $F_0(S_0, F_1(S_1), \ldots, F_k(S_k))$ is referred to as a *scheme* in which F_0 and F_i stand for some Boolean functions.

Decomposability of Boolean functions is an important topic in logic design [32, 248, 512], database theory [264], reliability and game theory [777], and other fields; we refer to Bioch [87] for a recent survey. In logic design, decompositions of partially defined Boolean functions received some attention in the foregoing early references, and enumerative type algorithms were proposed. Decomposability is also important from the viewpoint of logical analysis of data, since decompositions such as (12.19) reveal essential hierarchical logical structures in the underlying data sets. As the simplest decomposition scheme of this kind, we study in this section the decomposition scheme $F_0(S_0, F_1(S_1))$, where $\mathcal{F}_{F_0(S_0, F_1(S_1))}$ denotes the class of functions decomposable under this scheme. We also consider the class $\mathcal{F}^+_{F_0(S_0, F_1(S_1))}$ in which the functions g and h_1 are restricted to being positive in the decomposition $g(X|_{S_0}, h_1(X|_{S_1}]))$.

We first consider the problem EXTENSION($\mathcal{F}_{F_0(S_0, F_1(S_1))}$) for a given pair of sets S_0 and S_1. Let us define the structure graph $G_{(T,F)} = (V, E)$ by

$$V = V_0 \cup V_1,$$
$$E = E_F \cup E_T,$$
$$V_i = \{X|_{S_i} \mid X \in T \cup F\}, \quad i = 0, 1$$
$$E_T = \{(A|_{S_0}, A|_{S_1}) \mid A \in T\},$$
$$E_F = \{(B|_{S_0}, B|_{S_1}) \mid B \in F\}.$$

When displaying the graph $G_{(T,F)}$, we draw the edges in E_T as solid lines, and the edges in E_F as broken lines.

Example 12.12. *Consider the pdBf in Table 12.5 with* $S_0 = \{1, 2, 3\}$ *and* $S_1 = \{4, 5, 6\}$ *(ignore the column* h_1 *for the time being). The corresponding structure graph is shown in Figure 12.5.*   □

In view of Theorem 12.1, the pdBf (T, F) has an extension $f \in \mathcal{F}_{F_0(S_0, F_1(S_1))}$ if and only if there exists a function $h_1 : V_1 \to \mathcal{B}$ such that $T' \cap F' = \emptyset$, where

$$T' = \{(A|_{S_0}, h_1(A|_{S_1})) \mid A \in T\}$$
$$F' = \{(B|_{S_0}, h_1(B|_{S_1})) \mid B \in F\}.$$

Table 12.5. An example of pdBf having a
decomposition $g(X|_{S_0}, h_1(X|_{S_1}))$

	S_0	S_1	h_1
T	100	101	1
	011	110	0
F	011	010	1
	110	101	1
	100	110	0
	000	110	0
	000	010	1

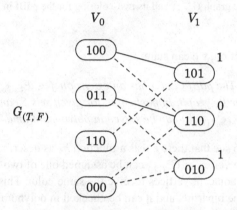

Figure 12.5. The structure graph $G_{(T,F)}$ of the pdBf in Example 12.12.

In terms of the graph $G_{(T,F)}$, this condition is described as follows:

Lemma 12.4. *The pdBf (T, F) has an extension $f \in \mathcal{F}_{F_0(S_0), F_1(S_1))}$ if and only if there exists a function $h_1 : V_1 \to \mathcal{B}$ such that, for every pair of edges $e = (X_0, X_1) \in E_T$ and $e' = (X_0', X_1') \in E_F$, either $X_0 \neq X_0'$ holds or $h_1(X_1) \neq h_1(X_1')$ holds.* $\square$

Example 12.13. *For the pdBf of Example 12.12, possible values of $h_1(X)$ ($X \in V_1$) are indicated in Table 12.5 and also beside the vertices in V_1, in Figure 12.5. It is easy to see that these values $h_1(X)$ satisfy the condition in Lemma 12.4, thus implying that the pdBf of Example 12.12 has an extension in $\mathcal{F}_{F_0(S_0), F_1(S_1))}$.* $\square$

In order to verify if there exists a function h_1 satisfying the condition of Lemma 12.4, let us construct the auxiliary graph $G_{(T,F)}^* = (V^*, E^*)$ as follows:

$$V^* = V_1$$

$$E^* = \{(X_1, X_1') \mid \text{there is a vertex } X_0 \in V_0 \text{ in } G_{(T,F)}$$

$$\text{such that } (X_0, X_1) \in E_T \text{ and } (X_0, X_1') \in E_F\}.$$

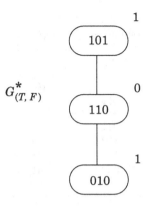

Figure 12.6. The graph $G^*_{(T,F)}$ and its two-coloring for the pdBf in Example 12.12.

With this construction, we can state:

Theorem 12.14. *The pdBf (T,F) has an extension $f \in \mathcal{F}_{F_0(S_0,F_1(S_1))}$ if and only if $G^*_{(T,F)}$ is a bipartite graph. In particular, for given sets S_0 and S_1, the problem* EXTENSION($\mathcal{F}_{F_0(S_0,F_1(S_1))}$) *can be solved in polynomial time.*

Proof. It is easy to see that there exists a function h_1 as described in Lemma 12.4 if and only if each vertex of $G^*_{(T,F)}$ can be assigned one of two colors, either 0 or 1, so that no two adjacent vertices receive the same color. This condition means that $G^*_{(T,F)}$ must be bipartite, and it can be checked in polynomial time. □

Example 12.14. *The auxiliary graph $G^*_{(T,F)}$ for the pdBf (T,F) of Example 12.12 is displayed in Figure 12.6. The colors satisfying the above condition are indicated beside the vertices. This construction illustrates how the h_1-values shown in Figure 12.5 were obtained.* □

We next turn to the class of positively decomposable functions $\mathcal{F}^+_{F_0(S_0,F_1(S_1))}$. In this case, we have to rely on Theorem 12.8 rather than on Theorem 12.1. Thus, let us define the positive structure graph $G^+_{(T,F)} = (V_0 \cup V_1, E_F \cup E_T \cup H_0 \cup H_1)$ for a given pdBf (T,F), by adding the following sets of *directed* arcs to the structure graph $G_{(T,F)} = (V_0 \cup V_1, E_F \cup E_T)$:

$$H_i = \{(X,X') \mid X,X' \in V_i \text{ and } X \leq X'\}, \quad i = 0,1.$$

The arcs (X,X') in $H_0 \cup H_1$ are drawn as solid arrows from X to X', respectively.

Example 12.15. *For the pdBf (T,F) of Table 12.6, assume that $S_0 = \{1,2\}$ and $S_1 = \{3,4,5\}$ are given (ignore the column h_1 temporarily). The positive structure graph $G^+_{(T,F)}$ is shown in Figure 12.7.* □

Table 12.6. An example of a pdBf
having a positive decomposition.

	S_0	S_1	h_1
	11	011	0
T	01	101	1
	01	110	1
	01	010	0
F	00	101	1
	10	110	1

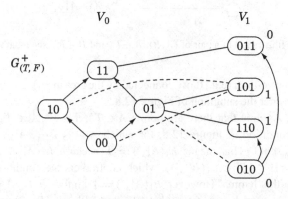

Figure 12.7. The positive structure graph $G_{(T,F)}^+$ of the pdBf in Example 12.15.

In view of Theorem 12.8, a positive decomposable extension of (T,F) exists if and only if there is a function $h_1 : V_1 \to \mathcal{B}$ such that

(i) for all $X_1, X_1' \in V_1$ with $X_1 \leq X_1'$, the inequality $h_1(X_1) \leq h_1(X_1')$ holds;

(ii) there is no pair of edges $e = (X_0, X_1) \in E_T$ and $e' = (X_0', X_1') \in E_F$ such that both inequalities $X_0 \leq X_0'$ and $h_1(X_1) \leq h_1(X_1')$ simultaneously hold.

The condition for the existence of such a function h_1 is expressed by the next lemma, where we let

$$T^* = \{A \in T \mid \text{there exists } B' \in F \text{ such that } B'|_{S_0} \geq A|_{S_0}\},$$

$$F^* = \{B \in F \mid \text{there exists } A' \in T \text{ such that } B|_{S_0} \geq A'|_{S_0}\}.$$

Lemma 12.5. *A pdBf (T,F) has an extension $f = g(X|_{S_0}, h_1(X|_{S_1})) \in \mathcal{F}_{F_0(S_0,F_1(S_1))}^+$ if and only if there is no pair of points $A \in T^*$ and $B \in F^*$ such that $A|_{S_1} \leq B|_{S_1}$.*

Proof. The general condition for the existence of a pair $A \in T^*$ and $B \in F^*$ such that $A|_{S_1} \leq B|_{S_1}$ is illustrated in Figure 12.8. The condition in the lemma asserts that the positive structure graph $G_{(T,F)}^+$ does not contain Figure 12.8 as a subgraph.

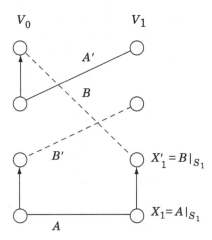

Figure 12.8. Illustration of a pair of (A, B), $A \in T^*$ and $B \in F^*$, such that $A|_{S_1} \leq B|_{S_1}$.

Note that some vertices (e.g., those connected by the arcs in H_i) may be contracted when we consider the subgraph of Figure 12.8.

Necessity. Assume that there are points $A \in T^*$, $B \in F^*$, $A' \in T$, $B' \in F$ satisfying the condition of Figure 12.8, for which $A|_{S_1} \leq B|_{S_1}$, $A|_{S_0} \leq B'|_{S_0}$ and $A'|_{S_0} \leq B|_{S_0}$ hold. This means $h_1(A|_{S_1}) = 1$ because $h_1(A|_{S_1}) = 0$ implies $(A|_{S_0}, h_1(A|_{S_1})) \leq (B'|_{S_0}, h_1(B'|_{S_1}))$, which contradicts the condition (ii) on h_1 stated before this lemma. However, $h_1(A|_{S_1}) = 1$ implies $h(B|_{S_1}) = 1$ by condition (i) on h_1, and hence, $(A'|_{S_0}, h_1(A'|_{S_1})) \leq (B|_{S_0}, h_1(B|_{S_1}))$, contradicting condition (ii) on h_1.

Sufficiency. If the subgraph of Figure 12.8 is not contained in $G^+_{(T,F)}$, then a positive function $h_1 : V_1 \to \{0, 1\}$ can be defined as follows:

$$h_1(X) = \begin{cases} 1 & \text{if some } A \in T^* \text{ satisfies } A|_{S_1} \leq X \\ 0 & \text{otherwise.} \end{cases} \tag{12.20}$$

It is straightforward to show that this function h_1 satisfies the above conditions (i) and (ii). □

Example 12.16. *It can be checked directly that the positive structure graph $G^+_{(T,F)}$ of Figure 12.7 for Example 12.15 does not contain the subgraph of Figure 12.8. The values of h_1, indicated in Figure 12.7 beside the vertices of V_1, are determined by (12.20). This assignment h_1 satisfies conditions (i) and (ii), as easily seen from Table 12.6, and we can conclude that the pdBf (T, F) has a positive extension $f = g(X|_{S_0}, h_1(X|_{S_1})) \in \mathcal{F}^+_{F_0(S_0, F_1(S_1))}.$* □

Since the condition of Lemma 12.5 can be checked in polynomial time by enumerating all possible subsets of eight vertices, we obtain the next theorem.

Table 12.7. Complexity results for decomposable extensions

	$\mathcal{F}_{\text{ALL}}$	$\mathcal{F}_+$
$F_0(S_0, F_1(S_1))$	P	P
$F_0(F_1(S_1), F_2(S_2))$	P	P
$F_0(S_0, F_1(S_1), F_2(S_2))$	NPC	P
$F_0(F_1(S_1), F_2(S_2), F_3(S_3))$	NPC	P
$F_0(S_0, F_1(S_1), \ldots, F_k(S_k)), k \geq 3$	NPC	NPC
$F_0(F_1(S_1), F_2(S_2), \ldots, F_k(S_k)), k \geq 4$	NPC	NPC

P: polynomial time, NPC: NP-complete

Theorem 12.15. *For given sets S_0 and S_1, problem* $\text{EXTENSION}(\mathcal{F}^+_{F_0(S_0, F_1(S_1))})$ *can be solved in polynomial time.* □

To conclude this section, we summarize in Table 12.7 the complexity status of problem $\text{EXTENSION}(\mathcal{C})$ for various decomposition schemes in $\mathcal{F}_{\text{ALL}}$ and $\mathcal{F}_+$, where $S_1, S_2, \ldots, S_k$ are given subsets. For $\mathcal{F}_+$, we require that the functions g and h_i in (12.19) should all be positive. In Table 12.7, a letter "P" indicates that the corresponding problem is solvable in polynomial time, and "NPC" means that it is NP-complete. Most of these results are due to Boros et al. [122], except the results for $\mathcal{C}^+_{F_0(S_0, F_1(S_1), \ldots, F_k(S_k))}$ with $k \geq 3$, and $\mathcal{C}^+_{F_0(F_1(S_1), F_2(S_2), \ldots, F_k(S_k))}$ with $k \geq 4$, which are proved by Makino, Yano, and Ibaraki [655]. The latter reference also considers cases in which some or all of the functions are restricted to be Horn. Further related results can be found in Ono, Makino, and Ibaraki [718].

12.3.7 k-convex extensions

The concept of k-convex function was introduced by Ekin, Hammer, and Kogan [306], and k-convex extensions were studied by the same authors in [307]. A Boolean function f is called k-*convex* for a given integer $k \geq 2$ if, for every pair of true points $A, C \in T(f)$ with Hamming distance $d(A, C) \leq k$, every point B located between A and C is also a true point of f. Here, we say that B is located between A and C if $d(A, B) + d(B, C) = d(A, C)$ holds. The class of k-convex functions is denoted $\mathcal{F}_{k\text{-CONV}}$.

The class $\mathcal{F}_{k\text{-CONV}}$ deserves attention in data analysis because k-convex functions can model situations in which the set of true points consists of a number of clusters that lie far apart (at distance larger than k) from each other.

Let us say that two terms s and t *conflict in h literals* if there are h variables, each of which appears in exactly one of the terms s and t as a positive literal, and in the other term as a negative literal. A k-convex function can be characterized as follows [306]:

Lemma 12.6. *For $k \geq 2$, a Boolean function f is k-convex if and only if every two prime implicants of f conflict in at least $k + 1$ literals.* $\square$

Example 12.17. *Consider a function f with two prime implicants,*

$$f = x_1 x_2 x_3 x_4 \vee \bar{x}_1 \bar{x}_2 \bar{x}_3.$$

Since the two prime implicants of f conflict in three literals, this function is 2-convex. In other words, $T(f)$ consists of two clusters represented by the two prime implicants, and any two points belonging to different clusters are at Hamming distance at least 3. $\square$

For a function f (which may not be k-convex), define the k-*convex envelope* of f to be the smallest k-convex majorant of f. The k-convex envelope of f is denoted by $[f]_k$. Thus, $[f]_k \in \mathcal{F}_{k\text{-CONV}}$, and $[f]_k \leq g$ for all $g \in \mathcal{F}_{k\text{-CONV}}$ such that $f \leq g$.

Ekin, Hammer and Kogan [306] introduced the k-convex envelope and proved that it always exists. In order to describe an algorithm to compute the k-convex envelope, we define as follows the *convex hull* of two terms s and t. Let y_j, $j = 1, 2, \ldots, n$, denote (positive or negative) literals, and assume that s and t are written as

$$s = \Big(\bigwedge_{j \in S_1} y_j\Big)\Big(\bigwedge_{j \in S_2} y_j\Big)\Big(\bigwedge_{j \in S_3} y_j\Big),$$

$$t = \Big(\bigwedge_{j \in S_1} \bar{y}_j\Big)\Big(\bigwedge_{j \in S_2} y_j\Big)\Big(\bigwedge_{j \in S_4} y_j\Big),$$

where S_1 denotes the set of indices of conflicting literals, S_2 the set of indices of common literals, and S_3 and S_4 (satisfying $S_3 \cap S_4 = \emptyset$) the sets of literals which appear only in s and only in t, respectively. The convex hull $[s, t]$ is defined as the conjunction of the common literals in s and t:

$$[s, t] = \bigwedge_{j \in S_2} y_j.$$

Given a DNF φ_0 of a function f, a DNF of its k-convex envelope $[f]_k$ is obtained by applying the following operation as long as possible.

If the current DNF φ contains two terms s and t conflicting in at most k literals, then remove s and t from φ, and add the new term $[s, t]$ to φ.

This algorithm terminates in polynomial time in the length of the DNF φ_0, since the number of terms decreases by one at each iteration.

Example 12.18. *Let us compute the 2-convex envelope of the following function:*

$$f = x_1 x_2 x_3 x_4 x_5 \vee x_1 x_2 x_3 x_4 x_6 \vee \bar{x}_1 \bar{x}_2 \bar{x}_3 x_5 x_6 \vee \bar{x}_1 \bar{x}_2 \bar{x}_3 x_4 \bar{x}_5 x_6 \vee \bar{x}_1 \bar{x}_2 \bar{x}_3 \bar{x}_4 \bar{x}_5 \bar{x}_6.$$

Taking the convex hull of the first two terms, which have no conflicting literal, we obtain

$$[x_1x_2x_3x_4x_5, x_1x_2x_3x_4x_6] = x_1x_2x_3x_4.$$

Similarly, from the third and fourth terms having one conflicting literal x_5, we obtain

$$[\bar{x}_1\bar{x}_2\bar{x}_3x_5x_6, \bar{x}_1\bar{x}_2\bar{x}_3x_4\bar{x}_5x_6] = \bar{x}_1\bar{x}_2\bar{x}_3x_6.$$

Finally, from this new term and the fifth term of f having one conflicting literal x_6, we obtain

$$[\bar{x}_1\bar{x}_2\bar{x}_3x_6, \bar{x}_1\bar{x}_2\bar{x}_3\bar{x}_4\bar{x}_5\bar{x}_6] = \bar{x}_1\bar{x}_2\bar{x}_3.$$

The resulting two terms conflict in three ($= k + 1$) literals, and thus we have obtained the 2-envelope of f:

$$[f]_2 = x_1x_2x_3x_4 \vee \bar{x}_1\bar{x}_2\bar{x}_3.$$

This is indeed a 2-convex function as already discussed in Example 12.17. $\square$

Now let (T, F) be a pdBf, and suppose we want to know whether (T, F) admits a k-convex extension. Let φ_T be the DNF consisting of all the minterms associated with the true points in T (φ_T is the minterm expression of $f_{\min}$; see (12.3)). Then from the preceding argument, the following theorem easily follows:

Theorem 12.16. *A pdBf (T, F) has an extension $f \in \mathcal{F}_{k\text{-CONV}}$ if and only if the k-envelope of φ_T satisfies $[\varphi_T]_k(B) = 0$ for all $B \in F$. This condition can be checked in polynomial time.*

Proof. Supppose that g is a k-convex extension of (T, F). Since $[\varphi_T] \leq g$, the definition of the k-convex envelope implies $[\varphi_T]_k \leq g$. Now, for all $B \in F$, $g(B) = 0$, and hence, $[\varphi_T]_k(B) = 0$.

The converse implication and the complexity statement are straightforward. $\square$

Remark 12.2. The problem EXTENSION($\mathcal{C}$) has also been extensively studied in computational learning theory, where it is usually called the *consistency problem*. This interest is motivated by the fact that a class $\mathcal{C}$ is not PAC learnable and not polynomially exact learnable with equivalence queries, if the consistency problem for $\mathcal{C}$ is NP-complete (provided, of course, $P \neq NP$); see, for example, Anthony [26] for details. For example, the consistency problem for the class of h-term DNF functions (namely, functions representable by a disjunction of at most h terms) was shown to be NP-complete by Pitt and Valiant [749]. For related topics, the reader is referred to Aizenstein et al. [13]; Angluin [21]; Bshouty [161]; Kearns, Li, and Valiant [560]; and Valiant [884], and so on. $\square$

12.4   Best-fit extensions of pdBfs containing errors

Real-world data sets represented as pdBfs (T, F) are prone to errors. Some points in $T \cup F$ may contain corrupted bits, some points may have been erroneously classified, and some attributes not included in the current data set may render it inconsistent. In this section, in order to cope with such situations, we allow an extension f "to make errors" in the sense that some points $A \in T$ may be classified in $F(f)$ $(f(A) = 0)$, and some points $B \in F$ may be classified in $T(f)$ $(f(B) = 1)$. However, we obviously want to minimize the magnitude of such errors. In order to state more precisely the resulting questions, let

$$w : T \cup F \to \mathcal{R}_+$$

be a *weighting function* that represents the importance of each data point in $T \cup F$. For a subset $U \subseteq T \cup F$, we let

$$w(U) = \sum_{A \in U} w(A).$$

Boros, Ibaraki, and Makino [139] introduced the following problem (see also Boros, Hammer, and Hooker [128]):

PROBLEM BEST-FIT($\mathcal{C}$)
Instance: A pdBf (T, F) and a weighting function w on $T \cup F$.
Output: A pdBf (T^*, F^*) (and an extension $f \in \mathcal{C}$ of (T^*, F^*)) with the following properties:

1. $T^* \cap F^* = \emptyset$ and $T^* \cup F^* = T \cup F$.
2. (T^*, F^*) has an extension in $\mathcal{C}$.
3. $w(T \cap F^*) + w(F \cap T^*)$ is minimized.

The conditions in this problem express that if we consider the points in $T \cap F^*$ and $F \cap T^*$ as erroneously classified, and if we change their classification accordingly, then the resulting pdBf (T^*, F^*) has an extension in the designated class $\mathcal{C}$. In case the weighting function w satisfies $w(A) = 1$ for all $A \in T \cup F$, the problem asks to minimize the number of erroneously classified points in $T \cup F$.

Clearly, problem BEST-FIT($\mathcal{C}$) contains problem EXTENSION($\mathcal{C}$) as a special case. Therefore, if EXTENSION($\mathcal{C}$) is NP-complete, then BEST-FIT($\mathcal{C}$) is NP-hard. Conversely, if BEST-FIT($\mathcal{C}$) is solvable in polynomial time, then so is EXTENSION($\mathcal{C}$). The next theorem indicates that BEST-FIT($\mathcal{C}$) is quite hard and is polynomially solvable only for very restrictive classes $\mathcal{C}$ (see [139] for additional results).

Theorem 12.17. *Problem* BEST-FIT($\mathcal{C}$) *can be solved in polynomial time for $\mathcal{C} = \mathcal{F}_{\text{ALL}}$ and $\mathcal{C} = \mathcal{F}_+$, but is NP-hard for $\mathcal{C} \in \{\mathcal{F}_{\text{UNATE}}, \mathcal{F}_{\text{Th}}, \mathcal{F}_{\text{HORN}}, \mathcal{F}_{F_0(S_0, F_1(S_1))}, \mathcal{F}_k\}$.*

Proof. We prove the polynomiality of BEST-FIT($\mathcal{C}$) for $\mathcal{F}_{\text{ALL}}$ and $\mathcal{F}_+$. Its NP-hardness for $\mathcal{F}_{\text{UNATE}}$ follows from Theorem 12.10. The results for other classes are omitted (see Boros, Ibaraki, and Makino [139]).

$\mathcal{C} = \mathcal{F}_{\text{ALL}}$: By Theorem 12.1, if (T, F) does not have an extension in $\mathcal{F}_{\text{ALL}}$, then $T \cap F \neq \emptyset$. The optimal pdBf (T^*, F^*) is obtained by reclassifying every point

$X \in T \cap F$ either into T^* or into F^*. Since both decisions carry the same weight $w(X)$, we can minimize $w(T^* \cap F) + w(F^* \cap T)$ by letting, for example,

$$T^* = T \setminus F, \quad F^* = F. \tag{12.21}$$

$\mathcal{C} = \mathcal{F}_+$: By Theorem 12.8, if the pdBf (T, F) does not have an extension in $\mathcal{F}_+$ then there are two points $A \in T$, $B \in F$ with $A \leq B$. Define a bipartite graph $H_{(T,F)} = (T \cup F, E)$ by

$$E = \{(A, B) \mid A \leq B, A \in T, B \in F\}.$$

This graph $H_{(T,F)}$ can be constructed from (T, F) in $O(n|T||F|)$ time. A *minimum vertex cover* of $H_{(T,F)}$ is a subset of vertices $U \subseteq T \cup F$ such that

(1) U is a vertex cover of $H_{(T,F)}$, that is, every edge $(A, B) \in E$ satisfies either $A \in U$ or $B \in U$, and
(2) $w(U)$ is minimum among all vertex covers.

Although the problem of finding a minimum vertex cover is NP-hard for general graphs, it is solvable in $O((|T| + |F|)^3)$ time for bipartite graphs (e.g., Ford and Fulkerson [341]; Kuhn [587]).

Let U be a minimum vertex cover of $H_{(T,F)}$. We can assume without loss of generality that U is a minimal cover, meaning that no proper subset of U is a vertex cover (this is certainly true if all weights w are strictly positive; otherwise, simply remove all redundant vertices from U).

Observe that for every positive Boolean function f, the set

$$W = (T \cap F(f)) \cup (F \cap T(f))$$

is a vertex cover of $H_{(T,F)}$. (Indeed, otherwise, there is an edge $(A, B) \in E$ such that $A \leq B, f(A) = 1$ and $f(B) = 0$, which contradicts the positivity of f.) This implies that

$$w(T \cap F(f)) + w(F \cap T(f)) \geq w(U) \tag{12.22}$$

for every positive function f.

Now define

$$T^* = (T \setminus U) \cup (F \cap U), \tag{12.23}$$
$$F^* = (T \cap U) \cup (F \setminus U). \tag{12.24}$$

We claim that the pdBf (T^*, F^*) has an extension in $\mathcal{F}_+$. Every such extension f satisfies

$$w(T \cap F(f)) + w(F \cap T(f)) = w(T \cap F^*) + w(F \cap T^*) = w(U),$$

and this, together with (12.22) implies that (T^*, F^*) provides an optimal solution of BEST-FIT($\mathcal{F}_+$). The total time required for the entire computation of (T^*, F^*) is $O(n|T||F| + (|T| + |F|)^3)$.

In order to prove the claim, assume that (T^*, F^*) does not have a positive extension. This means that there exist $A \in T^*$ and $B \in F^*$ such that $A \leq B$. We distinguish three cases, according to the definition (12.23)–(12.24) of (T^*, F^*):

(1) $A \in T \setminus U$ and $B \in F \setminus U$: Then, the edge (A, B) is in E, and this contradicts the assumption that U is a vertex cover of $H_{(T,F)}$.

(2) $A \in T \setminus U$ and $B \in T \cap U$: If there is an edge $(B, B') \in E$ with $B' \in F$ and $B \leq B'$, we have $A \leq B \leq B'$. Hence, (A, B') is an edge of $H_{(T,F)}$, and $B' \in U$, since U is a vertex cover and $A \notin U$. This shows that $U \setminus \{B\}$ is also a vertex cover contradicting the minimality of U.

(3) $A \in F \cap U$: If there is an edge $(A', A) \in E$ with $A' \in T$ and $A' \leq A$, we have $A' \leq A \leq B$. If $A' \in T \setminus U$, then, the same reasoning as in either (1) or (2) (with A' playing now the role of A) leads again to a contradiction. Thus, $A' \in U$, and we conclude that $U \setminus \{A\}$ is also a vertex cover of $H_{(T,F)}$, contradicting again the minimality of U.

This completes the proof of the claim and of the theorem.   □

Example 12.19. *Consider a pdBf* (T, F) *on* $\mathcal{B}^5$ *defined by*

$$T = \{(0,1,1,0,0), (0,1,0,1,0), (0,0,1,1,0), (0,0,1,0,1), (0,0,1,1,1)\},$$
$$F = \{(0,1,0,1,1), (1,1,0,1,0), (0,1,1,1,0), (0,0,1,1,1)\}.$$

The weighting function w *is given by* $w(A) = 1$ *for all* $A \in T \cup F$.
     Since $T \cap F = \{(0,0,1,1,1)\}$, *a best-fit extension in* $\mathcal{F}_{\mathrm{ALL}}$ *is obtained from the pdBf* (T^*, F^*) *defined by*

$$T^* = T \setminus \{(0,0,1,1,1)\},$$
$$F^* = F,$$

in view of (12.21).
     To solve BEST-FIT$(\mathcal{F}_+)$, *we then construct the bipartite graph* $H_{(T,F)}$ *of Figure 12.9. This graph has a minimum vertex cover*

$$U = \{(0,1,0,1,0), (0,1,1,1,0), (0,0,1,1,1)\},$$

as illustrated by the dark circles in the figure. Therefore, by (12.23)–(12.24) *in the above proof, we obtain*

$$T^* = (T \setminus U) \cup (F \cap U)$$
$$= \{(0,1,1,0,0), (0,0,1,1,0), (0,0,1,0,1), (0,0,1,1,1), (0,1,1,1,0)\},$$
$$F^* = (T \cap U) \cup (F \setminus U)$$
$$= \{(0,1,0,1,0), (0,1,0,1,1), (1,1,0,1,0)\}.$$

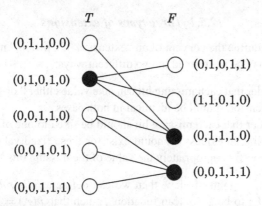

Figure 12.9. Bipartite graph $H_{(T,F)}$ for the pdBf in Example 12.19 (dark circles denote the vertices in a minimum vertex cover U).

It is easily checked that there is no pair (A, B) with $A \in T^*$ and $B \in F^*$ such that $A \leq B$, and hence, the pdBf (T^*, F^*) has an extension in $\mathcal{F}_+$ by Theorem 12.8.   □

The problem BEST-FIT was extensively studied by Boros, Ibaraki and Makino [139]. As the problem plays an important role in analyzing real-world data, efficient heuristic algorithms are necessary to deal with those classes $\mathcal{C}$ for which BEST-FIT$(\mathcal{C})$ is NP-hard. Some attempts in this direction have been made, for instance, in Boros et al. [131].

12.5 Extensions of pdBfs with missing bits

Not only do real-world data sets often contain erroneous data, but they may also turn out to be incomplete. By incomplete, we mean here that one or several bits may be missing from certain data points. Such bits simply may have been lost in the handling process, or may have been unavailable at the time of data collection, or intentionally omitted, for instance, because obtaining the bits is costly or dangerous. Let us denote each missing bit by $*$, and let

$$\mathcal{M} = \{0, 1, *\}.$$

We call *partially defined Boolean function with missing bits*, abbreviated as pBmb, any pair $(\tilde{T}, \tilde{F})$ consisting of a set of positive examples $\tilde{T} \subseteq \mathcal{M}^n$ and of a set of negative examples $\tilde{F} \subseteq \mathcal{M}^n$. Following the line of Boros, Ibaraki, and Makino [140], we introduce in the next subsection various types of extensions which are meaningful for pBmbs. Related complexity results are then discussed in Section 12.5.2.

12.5.1   Three types of extensions

When trying to define the concept of an "extension of a pBmb," missing bits "$*$" in data points may be interpreted in two different ways:

1. We consider that each missing bit can take value either 0 or 1, and the value of the extension should be identical in both cases.
2. We consider that each missing bit should be fixed to one of the two values 0 and 1, and an extension should exist for these fixed values. (Here, it is important to fix appropriately the value of the missing bits.)

If we take the first point of view, then we can define a *fully robust extension*[1] of a pBmb $(\tilde{T}, \tilde{F})$ to be a Boolean function f such that $f(A) = 1$ (respectively, $f(B) = 0$) for all $A \in \mathcal{B}^n$ (respectively, $B \in \mathcal{B}^n$) obtainable from a point $\tilde{A} \in \tilde{T}$ (respectively, $\tilde{B} \in \tilde{F}$) by fixing each missing bit to either 0 or 1. From the second point of view, we can define a *consistent extension* of $(\tilde{T}, \tilde{F})$ to be an extension of some pdBf (T', F') obtained from $(\tilde{T}, \tilde{F})$ by fixing all missing bits appropriately.

When a pBmb $(\tilde{T}, \tilde{F})$ admits a consistent extension, but no fully robust extension, then we may also take an intermediate view whereby we should fix a smallest possible number of missing bits so that the resulting pBmb has a fully robust extension. Such an extension is called a *most robust extension*.

To describe more precisely the above three problems, let us introduce some notations. For a set of points $\tilde{S} \subseteq \mathcal{M}^n$, let

$$AS(\tilde{S}) = \{(X, j) \mid X \in \tilde{S}, x_j = *\}.$$

For each subset $Q \subseteq AS(\tilde{S})$ and $\alpha \in \mathcal{B}^Q$, we interpret α as an assignment of values to the missing bits x_j, for all points X and all indices j such that $(X, j) \in Q$. The outcome of the assignment α to $\tilde{S}$ is denoted by $\tilde{S}^\alpha = \{X^\alpha \mid X \in \tilde{S}\}$, where

$$x_j^\alpha = \begin{cases} \alpha(X, j) & \text{if } (X, j) \in Q, \\ x_j & \text{otherwise.} \end{cases}$$

Example 12.20. *If*

$$\tilde{S} = \{X = (1, *, 0, 1), \ Y = (0, 1, *, *), \ Z = (1, 1, *, 0)\},$$

then

$$AS(\tilde{S}) = \{(X, 2), (Y, 3), (Y, 4), (Z, 3)\}.$$

If we consider $Q = \{(X, 2), (Y, 4)\}$ and the assignment $(\alpha(X, 2), \alpha(Y, 4)) = (1, 0) \in \mathcal{B}^Q$, then we obtain

$$\tilde{S}^\alpha = \{X^\alpha = (1, 1, 0, 1), \ Y^\alpha = (0, 1, *, 0), \ Z^\alpha = (1, 1, *, 0)\}. \qquad \square$$

---

[1] This extension is called a *robust extension* in [140].

We also use the following shorthand notations: For a given pBmb $(\tilde{T}, \tilde{F})$, we let

$$AS = AS(\tilde{T} \cup \tilde{F}),$$

and if $\tilde{S}$ is a singleton $\{X\}$, then we simply write $AS(X)$ for $AS(\tilde{S})$. Note that an assignment $\alpha \in \mathcal{B}^{AS}$ fixes all missing bits of the points in $\tilde{T} \cup \tilde{F}$.

Based on these definitions, we say that a Boolean function f is a *fully robust extension* of the pBmb $(\tilde{T}, \tilde{F})$ if the conditions

$$f(A^\alpha) = 1 \text{ for all } A \in \tilde{T}, \tag{12.25}$$

$$f(B^\alpha) = 0 \text{ for all } B \in \tilde{F}, \tag{12.26}$$

hold for all $\alpha \in \mathcal{B}^{AS}$. A Boolean function f is a *consistent extension* of $(\tilde{T}, \tilde{F})$ if there is an assignment $\alpha \in \mathcal{B}^{AS}$ for which (12.25)–(12.26) hold.

Various extension problems for partially defined Boolean functions with missing bits can now be defined as follows.

PROBLEM FRE($\mathcal{C}$) (fully robust extension)
Instance: A pBmb $(\tilde{T}, \tilde{F})$.
Question: Does $(\tilde{T}, \tilde{F})$ have a fully robust extension in $\mathcal{C}$? (When the answer is "yes," it is usually required to output one such extension.)

PROBLEM CE($\mathcal{C}$) (consistent extension)
Instance: A pBmb $(\tilde{T}, \tilde{F})$.
Question: Does $(\tilde{T}, \tilde{F})$ have a consistent extension in $\mathcal{C}$? (When the answer is "yes," it is usually required to output one such extension and the corresponding assignment $\alpha \in \mathcal{B}^{AS}$.)

PROBLEM MRE($\mathcal{C}$) (most robust extension)
Instance: A pBmb $(\tilde{T}, \tilde{F})$.
Question: Does $(\tilde{T}, \tilde{F})$ have a consistent extension in $\mathcal{C}$? If the answer is "yes," then output a subset $Q \subseteq AS$ and an assignment $\alpha \in \mathcal{B}^Q$ such that

(1) the pdBf $(\tilde{T}^\alpha, \tilde{F}^\alpha)$ has a fully robust extension in $\mathcal{C}$, and
(2) $|Q|$ is minimized among all (Q, α) satisfying condition (1).

As is obvious from these definitions, FRE($\mathcal{C}$) and CE($\mathcal{C}$) both contain EXTENSION($\mathcal{C}$) as a special case. MRE($\mathcal{C}$) is more general than FRE($\mathcal{C}$) and CE($\mathcal{C}$). Therefore, NP-hardness of EXTENSION($\mathcal{C}$) for a class $\mathcal{C}$ implies NP-hardness of FRE($\mathcal{C}$), CE($\mathcal{C}$), and MRE($\mathcal{C}$). Furthermore, if one of FRE($\mathcal{C}$) and CE($\mathcal{C}$) is NP-hard, then so is MRE($\mathcal{C}$). For polynomial solvability, these arguments can be reversed.

In the next section, we investigate the complexity of FRE($\mathcal{C}$), CE($\mathcal{C}$), and MRE($\mathcal{C}$) for some function classes $\mathcal{C}$ of interest.

Remark 12.3. There is yet another type of extension of a pBmb called *fully consistent extension*: A pBmb $(\tilde{T}, \tilde{F})$ is fully consistent in class $\mathcal{C}$ if for every assignment $\alpha \in \mathcal{B}^{AS}$ there is an extension $f \in \mathcal{C}$ of the pdBf $(\tilde{T}^{\alpha}, \tilde{F}^{\alpha})$. Note that the extensions may be different for different assignments $\alpha \in \mathcal{B}^{AS}$. Clearly, a pBmb $(\tilde{T}, \tilde{F})$ is fully consistent in class $\mathcal{C}$ if it has a fully robust extension in $\mathcal{C}$, but the converse may not be true. This type of extension was studied in Boros et al. [141, 142]. □

12.5.2 Complexity results

In this section, for two points $A, B \in \mathcal{M}^n$, we write $A \approx B$ if there exists an assignment $\alpha \in \mathcal{B}^{AS(\{A,B\})}$ such that $A^{\alpha} = B^{\alpha}$. We write $A \preceq B$ if $A^{\alpha} \leq B^{\alpha}$ holds for some assignment α. For example, $(0, *, 1, *) \approx (*, 1, 1, 0)$ and $(0, *, 0, *) \preceq (*, 1, 1, 0)$, but $(1, *, 1, *) \not\approx (0, 1, 1, 0)$ and $(0, *, 1, *) \not\preceq (*, 1, 0, *)$. For a point $A \in \mathcal{M}^n$, let A^1 denote the point in $\mathcal{B}^n$ obtained from A by fixing all missing bits $*$ to 1, and A^0 the point obtained from A by fixing all missing bits $*$ to 0.

Theorem 12.18. *A pBmb $(\tilde{T}, \tilde{F})$ has a fully robust extension if and only if there exists no pair (A, B) with $A \in \tilde{T}$ and $B \in \tilde{F}$ such that $A \approx B$. Hence,* $\mathrm{FRE}(\mathcal{F}_{\mathrm{ALL}})$ *can be solved in polynomial time.*

Proof. The necessary and sufficient condition is obvious from the definition of a fully robust extension. The condition $A \not\approx B$ is equivalent to the existence of an index j such that $a_j \neq b_j$, $a_j, b_j \in \{0, 1\}$. This can be checked in $O(n|\tilde{T}||\tilde{F}|)$ time by direct comparison of all points $A \in \tilde{T}$ and $B \in \tilde{F}$. □

The next lemma holds for the class $\mathcal{F}_+$ and for any subclass of $\mathcal{F}_+$.

Lemma 12.7. *A pBmb $(\tilde{T}, \tilde{F})$ has a fully robust extension in the class $\mathcal{C} \subseteq \mathcal{F}_+$ if and only if the pdBf (T^-, F^+) defined by*

$$T^- = \{A^0 \mid A \in \tilde{T}\}, \ F^+ = \{B^1 \mid B \in \tilde{F}\},$$

has an extension in $\mathcal{C}$.

Proof. If there is a fully robust extension $f \in \mathcal{C}$ of $(\tilde{T}, \tilde{F})$, then by definition, f is also an extension of the pdBf (T^-, F^+) since the latter is obtained from $(\tilde{T}, \tilde{F})$ by some assignment of values to missing bits.

To prove the converse, assume that (T^-, F^+) has an extension $g \in \mathcal{C}$. Then $A^0 \leq A^{\beta}$ holds for all $A \in \tilde{T}$ and all assignments $\beta \in \mathcal{B}^{AS(A)}$, and hence, $1 = g(A^0) \leq g(A^{\beta})$ implies $g(A^{\beta}) = 1$. Similarly, we obtain $g(B^{\beta}) = 0$ for all $B \in \tilde{F}$ and all $\beta \in \mathcal{B}^{AS(B)}$. This shows that g is a fully robust extension of $(\tilde{T}, \tilde{F})$. □

Example 12.21. *Consider the following pBmb* $(\tilde{T}, \tilde{F})$ *with* $n = 5$:

$$\tilde{T} = \{(0, 1, *, *, 0), (*, 1, 0, 1, 1)\},$$
$$\tilde{F} = \{(*, *, 1, 0, 1), (0, *, 1, *, 1)\}.$$

It is easily checked that $A \approx B$ *does not hold for any* $A \in \tilde{T}$, $B \in \tilde{F}$, *and hence, there is a fully robust extension of* $(\tilde{T}, \tilde{F})$ *in* $\mathcal{F}_{ALL}$. *Such a fully robust extension* f *is for example given by*

$$T(f) = \{(0, 1, 0, 0, 0), (0, 1, 0, 1, 0), (0, 1, 1, 0, 0), (0, 1, 1, 1, 0), (0, 1, 0, 1, 1),$$
$$(1, 1, 0, 1, 1)\}$$
$$F(f) = \mathcal{B}^5 \setminus T(f).$$

We next construct (T^-, F^+) *as in Lemma 12.7:*

$$T^- = \{(0, 1, 0, 0, 0), (0, 1, 0, 1, 1)\}$$
$$F^+ = \{(1, 1, 1, 0, 1), (0, 1, 1, 1, 1)\}.$$

Since $A \leq B$ *holds for* $A = (0, 1, 0, 0, 0) \in T^-$ *and* $B = (1, 1, 1, 0, 1) \in F^+$, (T^-, F^+) *does not have an extension in* $\mathcal{F}_+$ *by Theorem 12.8. Hence, by the previous lemma, the pBmb* $(\tilde{T}, \tilde{F})$ *does not have a fully robust extension in* $\mathcal{F}_+$.   □

A variant of Lemma 12.7 applies to consistent extensions:

Lemma 12.8. *A pBmb* $(\tilde{T}, \tilde{F})$ *has a consistent extension in the class* $\mathcal{C} \subseteq \mathcal{F}_+$ *if and only if the pdBf* (T^+, F^-) *defined by*

$$T^+ = \{A^1 \mid A \in \tilde{T}\}, \ F^- = \{B^0 \mid B \in \tilde{F}\}$$

has an extension in $\mathcal{C}$.

Proof. Assume first that there is a consistent extension $f \in \mathcal{C}$ of $(\tilde{T}, \tilde{F})$. That is, f is an extension of the pdBf $(\tilde{T}^\beta, \tilde{F}^\beta)$ for some assignment $\beta \in \mathcal{B}^{AS}$. Since f is positive and $A^\beta \leq A^1$, we see that $f(A^1) = 1$ holds for all $A \in \tilde{T}$. Similarly, $f(B^0) = 0$ for all $B \in \tilde{F}$. Therefore, f is an extension of (T^+, F^-).

The converse direction is obvious since (T^+, F^-) is obtained from $(\tilde{T}, \tilde{F})$ by an assignment.   □

We proved earlier that EXTENSION($\mathcal{C}$) is polynomially solvable for the classes $\mathcal{C} = \mathcal{F}_+, \mathcal{F}^+_{F_0(S_0, F_1(S_1))}, \mathcal{F}^+_k$, among others. The following theorem then immediately follows from Lemmas 12.7 and 12.8:

Theorem 12.19. *The problems* FRE($\mathcal{C}$) *and* CE($\mathcal{C}$) *are solvable in polynomial time for the classes* $\mathcal{C} = \mathcal{F}_+, \mathcal{F}^+_{F_0(S_0, F_1(S_1))},$ *and* $\mathcal{F}^+_k$.   □

Fully robust threshold extensions can also be identified in polynomial time.

Theorem 12.20. *The problem* FRE($\mathcal{F}_{\text{Th}}$) *can be solved in polynomial time.*

Proof. For a pBmb $(\tilde{T}, \tilde{F})$ on $\mathcal{M}^n$, consider the system of linear inequalities:

$$\sum_{(j:a_j=1)} w_j + \sum_{(j:a_j=*)} y_j \geq t+1 \text{ for all } A \in \tilde{T}, \tag{12.27}$$

$$\sum_{(j:b_j=1)} w_j + \sum_{(j:b_j=*)} z_j \leq t \quad \text{for all } B \in \tilde{F}, \tag{12.28}$$

$$y_j \leq w_j,\ y_j \leq 0 \qquad j=1,2,\ldots,n, \tag{12.29}$$

$$z_j \geq w_j,\ z_j \geq 0 \qquad j=1,2,\ldots,n. \tag{12.30}$$

We claim that this system has a feasible solution if and only if $(\tilde{T}, \tilde{F})$ has a fully robust threshold extension.

Let us assume first that (12.27)–(12.30) has a feasible solution (W, Y, Z, t). Then, for all $A \in \tilde{T}$ and for all $\alpha \in AS(A)$, we obtain from (12.27) and (12.29):

$$t+1 \leq \sum_{j:a_j=1} w_j + \sum_{j:a_j=*} y_j \leq \sum_{j:a_j^\alpha=1} w_j = \sum_{j=1}^n w_j a_j^\alpha. \tag{12.31}$$

Applying the same reasoning on $\tilde{F}$, we conclude that the structure (W, t) defines a fully robust threshold extension of $(\tilde{T}, \tilde{F})$.

Conversely, assume that $(\tilde{T}, \tilde{F})$ has a fully robust threshold extension f, and let (W, t) be a separating structure for f. Set $y_j = \min\{0, w_j\}$ and $z_j = \max\{0, w_j\}$ for $j = 1, 2, \ldots, n$. We claim that (W, Y, Z, t) is a feasible solution of the system (12.27)–(12.30). Indeed, for any $A \in \tilde{T}$, let us define α to be the assignment on $AS(A)$ which sets $a_j^\alpha = 1$ when $y_j = w_j$, and $a_j^\alpha = 0$ when $y_j = 0$. Since f is a fully robust extension, we have $f(A^\alpha) = 1$, and since f is threshold, we have $\sum_{j=1}^n w_j a_j^\alpha \geq t+1$. Thus,

$$\sum_{j=1}^n w_j a_j^\alpha = \sum_{j:a_j^\alpha=1} w_j = \sum_{j:a_j=1} w_j + \sum_{j:a_j=*} y_j, \tag{12.32}$$

and (12.27) is satisfied. The same reasoning holds for (12.28), and hence, (W, Y, Z, t) is a feasible solution of (12.27)–(12.30). $\square$

Finally, we establish another polynomially solvable case of the fully robust extension problem.

Theorem 12.21. *The problem* FRE($\mathcal{F}_{\text{HORN}}$) *is solvable in polynomial time.*

Proof. For a pBmb $(\tilde{T}, \tilde{F})$ and a point $A \in \tilde{T}$, define

$$\tilde{F}_{\geq A} = \{B \in \tilde{F} \mid B \geq A\}.$$

We claim that $(\tilde{T}, \tilde{F})$ has a fully robust extension in $\mathcal{F}_{\text{HORN}}$ if and only if, for every $A \in \tilde{T}$ such that $\tilde{F}_{\geq A} \neq \emptyset$,

there is an index j such that $a_j = 0$ and $b_j = 1$ hold for all $B \in \tilde{F}_{\geq A}$. (12.33)

This claim will prove the theorem, since condition (12.33) can be checked for all $A \in \tilde{T}$ in $O(n|\tilde{T}||\tilde{F}|)$ time.

To prove the claim, assume first that condition (12.33) holds. Consider a point $A \in \tilde{T}$ and the corresponding index j satisfying condition (12.33). Then, for all assignments $\alpha \in \mathcal{B}^{AS}$ and all $B \in \tilde{F}_{\geq a}$, we have $a_j^\alpha = 0$ and $b_j^\alpha = 1$. Therefore, the Horn term

$$t_A = (\bigwedge_{i:a_i=1} x_i) \, \bar{x}_j$$

satisfies $t_A(A^\alpha) = 1$ and $t_A(B^\alpha) = 0$ for all $\alpha \in \mathcal{B}^{AS}$ and all $B \in \tilde{F}_{\geq a}$. This term t_A also satisfies $t_A(B^\alpha) = 0$ for all $B \in \tilde{F} \setminus \tilde{F}_{\geq A}$ and all $\alpha \in \mathcal{B}^{AS}$; indeed, for all such B, there is some i such that $a_i = 1$ and $b_i = 0$ by the assumption that $B \not\geq A$. We conclude that the following Horn DNF represents a fully robust extension of $(\tilde{T}, \tilde{F})$:

$$\varphi = \bigvee_{A \in \tilde{T}} t_A. \tag{12.34}$$

Conversely, if condition (12.33) does not hold for some $A \in \tilde{T}$ with $\tilde{F}_{\geq A} \neq \emptyset$, then define an assignment $\alpha \in \mathcal{B}^{AS(\{A\} \cup \tilde{F}_{\geq A})}$ as follows: For all $(A,i) \in AS(A)$,

$$\alpha(A,i) = \begin{cases} \bigwedge_{B \in \tilde{F}_{\geq A}: b_i \neq *} b_i & \text{if there is a point } B \in \tilde{F}_{\geq A} \text{ such that } b_i \neq * \\ 1 & \text{otherwise,} \end{cases}$$

and for all for all $(B,i) \in AS(\tilde{F}_{\geq A})$,

$$\alpha(B,i) = \begin{cases} a_i^\alpha & \text{if } (A,i) \in AS(A), \\ a_i & \text{otherwise.} \end{cases}$$

Then, it can be checked that $(\tilde{F}_{\geq A})^\alpha = (\tilde{F}^\alpha)_{\geq A^\alpha} = \{B^\alpha \in \tilde{F}^\alpha \mid B^\alpha \geq A^\alpha\}$ satisfies

$$A^\alpha = \bigwedge_{B^\alpha \in (\tilde{F}^\alpha)_{\geq A^\alpha}} B^\alpha.$$

By condition (12.16) in the proof of Theorem 12.12, this implies that the pdBf $(\tilde{T}^\alpha, \tilde{F}^\alpha)$ does not have a Horn extension, and consequently, the pBmb $(\tilde{T}, \tilde{F})$ does not have a fully robust extension in $\mathcal{F}_{\text{HORN}}$. □

Example 12.22. *Let us consider the pBmb $(\tilde{T}, \tilde{F})$ of Example 12.21. For the two points in $\tilde{T}$, we obtain*

$$\tilde{F}_{\geq(0,1,*,*,0)} = \{(*,*,1,0,1), \, (0,*,1,*,1)\},$$

$$\tilde{F}_{\geq(*,1,0,1,1)} = \{(0,*,1,*,1)\}.$$

The condition (12.33) holds with $j = 5$ for $A = (0,1,*,*,0)$, and with $j = 3$ for $A = (*,1,0,1,1)$. Therefore, the DNF φ defined by (12.34), namely,

$$\varphi = x_2 \bar{x}_5 \vee x_2 x_4 x_5 \bar{x}_3,$$

is a Horn DNF representing a fully robust extension of $(\tilde{T}, \tilde{F})$.   □

In contrast with the previous positive results, Boros, Ibaraki, and Makino [140] also proved that, except for the special cases discussed in Theorems 12.18, 12.19, 12.20, and 12.21, all other variants of the problems FRE, CE, MRE are either NP-complete or NP-hard for the classes $\mathcal{F}_{ALL}, \mathcal{F}_{+}, \mathcal{F}_{UNATE}, \mathcal{F}_{Th}, \mathcal{F}_{HORN}, \mathcal{F}_{F_0(S_0, F_1(S_1))}$, and $\mathcal{F}_k$. We refer the reader to [140] for details and additional results.

12.6 Minimization with don't cares

In designing logic circuits of computers and other digital systems, Boolean theory has been extensively used to make the circuits efficient and economical. This has been discussed in several other chapters of this book, for instance, in Chapter 1, Section 1.13.2; in Chapter 2, Section 2.1; and, in particular, in Chapter 3, Section 3.3. In the process of logic design, complex logic functions are first decomposed into many small basic blocks, and each one is then realized as a logic circuit. This is illustrated in Figure 12.10 in which the central block realizes three Boolean functions f_1, f_2, f_3 of the variables x_1, x_2, x_3, x_4. Although each block has in general many outputs, for simplicity, we consider here the case of realizing a single function f.

In practical applications, there are usually many combinations of input values that can never be simultaneously observed and that are therefore called *don't care points*, or simply *don't cares*. For example, two physical lines associated with x_1 and x_2 may be used to represent the binary numbers "0" and "1" by a special coding scheme "0"= $(0, 1)$ and "1"= $(1, 0)$. Then, it is prohibited to use the combinations $(0, 0)$ or $(1, 1)$, meaning that we can ignore all inputs points X satisfying $(x_1, x_2) = (0, 0)$ or $(1, 1)$. In general, the values of input lines are mutually correlated, and these input values must satisfy many constraints. All input points X not satisfying such constraints are called don't cares, as we do not need to care about

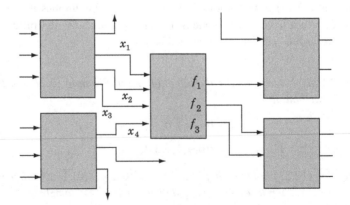

Figure 12.10. A basic block in logic circuits.

the output obtained for such input values when designing the logic circuit under consideration. This provides some freedom, which can be exploited in the design process. Since this aspect was not mentioned in Chapter 3, Section 3.3, we discuss it here very briefly (for more details, we refer the reader to the specialized literature; see, e.g., Umans, Villa and Sangiovanni-Vincentelli [877] or Villa, Brayton, and Sangiovanni-Vincentelli [891]).

In the terminology of this chapter, we can restate the logic synthesis problem as the problem of realizing some extension of a pdBf (T, F) (instead of a Boolean function) by a DNF ϕ, where all points in $\mathcal{B}^n \setminus (T \cup F)$ are interpreted as don't cares. Keeping in mind the difference between a Boolean function and a pdBf, we can accordingly adapt the discussion of logic minimization in Section 3.3. Namely, we want now to find an extension f of (T, F) having a shortest DNF ϕ, as measured either by $|\phi|$ (the number of literals) or by $||\phi||$ (the number of terms).

A main difference with the discussion in Chapter 3 is that here, we do not know the function f beforehand, but we have to select it among the extensions of (T, F). Note that, since our objective is to minimize the size of the DNF representation, Lemma 12.3 implies that there is no loss of generality in restricting our attention to prime irredundant theories (defined in Section 12.2.3). Therefore, we use prime patterns of (T, F) (instead of prime implicants of f in Chapter 3), and we aim to find a set of prime patterns that together cover T. The DNF ϕ defined as the disjunction of such prime patterns is a prime theory. If this prime theory minimizes $|\phi|$ (or $||\phi||$), then we deem it desirable from the point of view of circuit design.

In order to select an appropriate prime theory, the usual procedures require first to generate all prime patterns of the given pdBf (T, F). Since the prime patterns of (T, F) are among the prime implicants of the function $f_{\max}$ defined by (12.4) (see Lemma 12.1 in Section 12.2.3), we can proceed as described in Figure 12.11.

Note that $\psi_{f_{\max}}$ is explicitly available from F. The prime implicants of $f_{\max}$ can be generated, for instance, by (the dual version of) the procedure SD-DUALIZATION of Section 4.3.2.

Example 12.23. *Let us consider the pdBf in Table 12.1 of Section 12.1. First we construct $\psi_{f_{\max}}$ from $F = \{B^{(1)}, B^{(2)}, B^{(3)}, B^{(4)}\}$ as follows (we use the shorthand*

Procedure PRIME PATTERNS OF (T, F)

1. Construct the maxterm expression $\psi_{f_{\max}}$ (namely, the CNF expression of $f_{\max}$ introduced in Definition 1.11 of Section 1.4) by taking the conjunction of all maxterms associated with the points in $F = F(f_{\max})$.

2. Generate all prime implicants of $f_{\max}$ by one of the dualization methods discussed in Section 4.3 (see also Section 3.2.4).

3. Select the prime patterns of (T, F) among the prime implicants of $f_{\max}$.

Figure 12.11. Procedure PRIME PATTERNS OF (T, F).

notation $\bar{1}2$ for $\bar{x}_1 x_2$, etc.):

$$\psi_{f_{\max}} = (\bar{1} \vee 2 \vee \bar{3} \vee 4 \vee \bar{5} \vee 6 \vee \bar{7} \vee 8)(1 \vee 2 \vee 3 \vee \bar{4} \vee \bar{5} \vee \bar{6} \vee 7 \vee 8)$$
$$(\bar{1} \vee \bar{2} \vee 3 \vee \bar{4} \vee 5 \vee \bar{6} \vee 7 \vee \bar{8})(1 \vee 2 \vee \bar{3} \vee 4 \vee \bar{5} \vee 6 \vee \bar{7} \vee 8).$$

Expanding this CNF into a DNF, and then manipulating it as discussed in Section 4.3.2, we obtain the next DNF consisting of all prime implicants of $f_{\max}$:

$$\begin{aligned}
\varphi = {} & \bar{1}2^* \vee \bar{1}\bar{5}^* \vee \bar{1}8^* \vee 23^* \vee 2\bar{4}^* \vee 25^* \vee \bar{2}5 \vee 2\bar{6}^* \vee 27^* \vee 2\bar{8}^* \\
& \vee \bar{2}8 \vee 34 \vee 3\bar{4} \vee 35 \vee 36 \vee 3\bar{6}^* \vee \bar{3}7^* \vee 37^* \vee 38^* \vee \bar{4}5 \\
& \vee 4\bar{6}^* \vee \bar{4}6 \vee 47^* \vee 4\bar{7}^* \vee \bar{4}8^* \vee \bar{5}6 \vee \bar{5}7^* \vee 58^* \vee 5\bar{8}^* \vee 67^* \\
& \vee \bar{6}7^* \vee \bar{6}8^* \vee 78 \vee 1\bar{2}3 \vee 1\bar{2}4 \vee 1\bar{2}6 \vee 1\bar{2}7 \vee 13\bar{5}^* \vee 13\bar{8} \\
& \vee 145^* \vee 14\bar{8} \vee 156 \vee 15\bar{7}^* \vee 16\bar{8} \vee 1\bar{7}\bar{8} \vee 26\bar{8}^*.
\end{aligned}$$

*In this DNF, the terms marked with * are the prime patterns of (T,F), while the unmarked terms are not prime patterns. This example shows that the procedure may generate many prime implicants which are not prime patterns of (T,F).* $\quad\square$

Following the line of Section 3.3, the next step of logic minimization is to find a set of prime patterns that together cover the set T of true points. For the purpose of computing a theory ϕ which minimizes $|\phi|$ or $||\phi||$, the methods described in Section 3.3 (Quine-McCluskey method and its extensions) can be readily applied, if we simply replace the words "prime implicant" by "prime pattern." We illustrate this by continuing the foregoing example.

Example 12.24. *From the list of prime patterns marked in the DNF φ for Example 12.23, we choose a set of prime patterns which cover the set $T = \{A^{(1)}, A^{(2)}, A^{(3)}\}$ given in Table 12.1. It is easy to see that a single prime pattern cannot do this, and thus we must select at least two prime patterns. Even if we restrict ourselves to short prime implicants, there are many such sets, for example, $\{\bar{1}2, 25\}, \{\bar{1}2, 2\bar{6}\}, \ldots, \{67, \bar{6}\bar{7}\}$. The corresponding prime theories ϕ contain exactly two prime patterns of degree two and are minimum with respect to both norms $|\phi|$ and $||\phi||$. The extensions f_1 and f_3 given in Example 12.1 are two such minimum realizations of (T,F).* $\quad\square$

The preceding method based on generating all prime patterns seems reasonably efficient for those pdBfs (T,F) such that $T \cup F$ is not much smaller than $\mathcal{B}^n$ (which is often the case when don't cares are considered). But if the set $T \cup F$ is small, other methods that construct prime patterns directly from T may be more efficient. For example, Boros et al. [131] propose a naive method that first generates all terms of degree 1 one and picks up prime patterns from them, and then repeats the same for all terms of degree 2, and so on. This method can be used to obtain short prime patterns. Another approach is to apply, for each $A \in T$, a method to generate all prime patterns that cover A, by relying on the set covering characterization of

Table 12.8. Summary of complexity results obtained in this chapter

	EXT	MIN-SUPT	BEST-FIT	FRE	CE	MRE
$\mathcal{F}_{ALL}$	P	NPH	P	P	NPC	NPH
$\mathcal{F}_{+}$	P	NPH	P	P	P	NPH
$\mathcal{F}_{UNATE}$	NPC	NPH	NPH	NPC	NPC	NPH
$\mathcal{F}_{Th}$	P	NPH	NPH	P	NPC	NPH
$\mathcal{F}_{k}$	P	NPH	NPH	NPC	NPC	NPH
$\mathcal{F}_{HORN}$	P	NPH	NPH	P	NPC	NPH
$\mathcal{F}_{F_0(S_0,F_1(S_1))}$	P	NPH	NPH	NPC	NPC	NPH
$\mathcal{F}^{+}_{F_0(S_0,F_1(S_1))}$	P	NPH	NPH	P	P	NPH

P: polynomial time,   NPH: NP-hard,   NPC: NP-complete

patterns (see Exercise 2 of this chapter). This can be elaborated into an algorithm that runs with polynomial delay for the generation of all prime patterns, as discussed in Boros et al. [117].

12.7 Conclusion

In this chapter, we introduced partially defined Boolean functions (pdBfs) as fundamental models arising in various fields of applications, in particular, in logical analysis of data. We defined various problems and classified their computational complexity, with an emphasis on questions related to extensions of pdBfs. We summarize in Table 12.8 the main complexity results mentioned in this chapter. In this table, a letter P indicates that the corresponding problem is solvable in polynomial time, while NPH or NPC indicate that it is NP-hard or NP-complete, respectively. Also, EXT stands for EXTENSION and MIN-SUPT for MIN-SUPPORT.

As mentioned in the introduction of this chapter, acquiring or discovering meaningful information (or knowledge) from available data has recently received increased attention. The approach in this chapter may be regarded as a logical approach, since it is based solely on the consideration of pdBfs and of their extensions, viewed as Boolean functions having simple Boolean expressions. The performance of different approaches may be compared from several viewpoints, such as:

- accuracy of the performance of the obtained classification on new data points;
- ease of comprehension of the classification, and of the underlying knowledge unveiled by the approach;
- compactness of the representation of this knowledge, allowing its use for various purposes;
- efficiency of the computation of the classification.

It remains important to develop and to investigate better methods, possibly by combining existing approaches, so that they become more useful and more meaningful when applied to real-world situations.

12.8 Exercises

1. Prove Theorem 12.9, that is, prove that problem MIN-SUPPORT($\mathcal{F}_+$) is NP-hard.

2. Given a pdBf (T,F) on $\mathcal{B}^n$ and a point $A \in T$, let $t_A = z_1 z_2 \cdots z_n$ be the minterm of A (that is, $z_j = x_j$ if $a_j = 1$ and $z_j = \bar{x}_j$ otherwise). Define an $|F| \times n$ matrix Q by

$$Q_{ij} = \begin{cases} 1 & \text{if } b_j^{(i)} \neq a_j, \\ 0 & \text{otherwise,} \end{cases}$$

where $B^{(i)}$ is the i-th point in F. Then consider the following set covering constraints:

$$Qy \geq 1 \tag{12.35}$$

$$y \in \{0,1\}^n. \tag{12.36}$$

Show that y is a feasible solution of the system (12.35)–(12.36) if and only if the term $t = \bigwedge_{j:y_j=1} z_j$ is a pattern of (T,F) that covers A. Furthermore, y is a minimal solution of (12.35)–(12.36) (namely, no $y' \leq y$ with $y' \neq y$ is feasible) if and only if t is a prime pattern.

3. As a special case of $F_0(F_1(S_1), F_2(S_2))$-decomposability of a Boolean function f, where $S_1, S_2 \subseteq \{1, 2, \ldots, n\}$, let us consider a conjunctive decomposition of type

$$f(X) = h_1(X|_{S_1}) \wedge h_2(X|_{S_2}). \tag{12.37}$$

Define a bipartite graph $G_{(T,F)} = (V, E)$ by

$$V = V_1 \cup V_2,$$
$$E = E_F \cup E_T$$
$$V_i = \{X|_{S_i} \mid X \in T \cup F\}, \quad i = 1, 2$$
$$E_T = \{(X|_{S_1}, X|_{S_2}) \mid X \in T\},$$
$$E_F = \{(X|_{S_1}, X|_{S_2}) \mid X \in F\}.$$

Prove that (T,F) has an extension that is decomposable as in (12.37) if and only if $G_{(T,F)}$ has no four vertices A, B, C, D such that $A, C \in V_1$, $B, D \in V_2$, $(A, B) \in E_T$, $(C, D) \in E_T$, and $(C, B) \in E_F$.

4. Similarly to Exercise 3, consider now a disjunctive decomposition:

$$f(X) = h_1(X|_{S_1}) \vee h_2(X|_{S_2}),$$

and derive a necessary and sufficient condition for this type of decomposability.

5. Prove the second half of Theorem 12.11, that is, prove that problem EXTENSION($\mathcal{F}_k^+$) can be solved in polynomial time.

6. For a given graph $G = (V, E)$ with $V = \{1, 2, \ldots, n\}$, define the points $A^{(i,j)}$, $(i, j) \in E$, and $B^{(i)}$, $i \in V$, as follows:
   - $a_k^{(i,j)} = 1$ for $k \notin \{i, j\}$ and $a_k^{(i,j)} = 0$ for $k \in \{i, j\}$,
   - $b_k^{(i)} = 1$ for $k \neq i$ and $b_k^i = 0$ for $k = i$.

   Then define a pdBf (T, F) in $\mathcal{B}^n$ by

   $$T = \{A^{(i,j)} \mid (i, j) \in E\},$$
   $$F = \{B^{(i)} \mid i \in V\}.$$

   Show that

   $$\min(|T \cap F^*| + |F \cap T^*|) = \tau(G)$$

   holds, where the minimum is taken over all pdBfs (T^*, F^*) having an extension $f \in \mathcal{F}_{\text{HORN}}$, and $\tau(G)$ is the size of a minimum vertex cover in G. Knowing that the minimum vertex cover problem is NP-hard, prove that BEST-FIT($\mathcal{F}_{\text{HORN}}$) is NP-hard.

7. For each of the following conditions, construct a pBmb $(\tilde{T}, \tilde{F})$ satisfying it.
   a. $(\tilde{T}, \tilde{F})$ has a consistent extension in $\mathcal{F}_{\text{ALL}}$, but does not have a fully robust extension in $\mathcal{F}_{\text{ALL}}$.
   b. $(\tilde{T}, \tilde{F})$ has a fully robust extension in $\mathcal{F}_{\text{ALL}}$, but does not have a fully robust extension in $\mathcal{F}_+$.
   c. $(\tilde{T}, \tilde{F})$ has a consistent extension in $\mathcal{F}_+$, but does not have a fully robust extension in $\mathcal{F}_+$.

8. Consider the consistent extension problem CE($\mathcal{F}_{\text{ALL}}$) for a pBmb $(\tilde{T}, \tilde{F})$ such that each $A \in \tilde{T} \cup \tilde{F}$ has at most one missing bit. Recall that $(\tilde{T}, \tilde{F})$ has a consistent extension in $\mathcal{F}_{\text{ALL}}$ if and only if there is an assignment α such that $A^\alpha \neq B^\alpha$ holds for all pairs of $A \in \tilde{T}$ and $B \in \tilde{F}$. Show that the question of the existence of such an assignment can be formulated as a quadratic Boolean equation (or 2-SAT problem). Since quadratic equations are solvable in polynomial time, this proves that CE($\mathcal{F}_{\text{ALL}}$) is also solvable in polynomial time under the stated restriction.

13

Pseudo-Boolean functions

13.1 Definitions and examples

In Chapter 1, we defined a *pseudo-Boolean function* to be a mapping from $\mathcal{B}^n = \{0,1\}^n$ to $\mathbb{R}$. In other words, a pseudo-Boolean function is a real-valued function of a finite number of 0–1 variables. Identifying the Boolean symbols 0 and 1 (or T and F, *Yes* and *No*, etc.) with the corresponding integers, we see that pseudo-Boolean functions provide a proper generalization of Boolean functions. In fact, just as in the Boolean case, the deliberate ambiguity that results from this identification rarely causes any difficulties, but it is frequently the source of fruitful developments.

The systematic investigation of pseudo-Boolean functions, their theoretical properties, and their applications has been initiated by Hammer and Rudeanu in [460], building on previous ideas of Fortet [342, 343] and of Hammer, Rosenberg, and Rudeanu [458]. This field of research has given rise to countless subsequent publications over the last decades.

Since the element of $\{0,1\}^n$ are in one-to-one correspondence with the subsets of $N = \{1,2,\ldots,n\}$, every pseudo-Boolean function can also be viewed as a real-valued *set function* defined on $\mathcal{P}(N)$, the power set of $N = \{1,2,\ldots,n\}$. Set functions have been extensively studied because of their mathematical appeal and their presence in numerous fundamental models of mathematics and of applied sciences. By considering functions defined on $\{0,1\}^n$ rather than on $\mathcal{P}(N)$, however, the pseudo-Boolean approach provides an algebraic viewpoint, which sometimes carries clear advantages over the set-theoretic description. For instance, we mentioned in Chapter 1, Section 1.12.2, that every pseudo-Boolean function can be (uniquely) represented as a multilinear polynomial in its variables. This representation (and related ones) opens the door to algebraic and numerical manipulations of pseudo-Boolean functions that play a major role in many applications.

Another (voluminous) book would be required in order to discuss appropriately the enormous body of literature devoted to the investigation of pseudo-Boolean functions. Our intention in this chapter, therefore, is only to skim the surface of the topic and to briefly indicate some of the main research directions and techniques encountered in the field.

We now proceed with a description of a few representative problems arising in mathematics, computer science, and operations research, where pseudo-Boolean functions appear naturally and contribute to the analysis and the solution of area-specific problems.

Mathematics

Application 13.1. (Graph theory.) *As observed by Hammer and Rudeanu [460], many graph theoretic concepts can be easily formulated in the pseudo-Boolean language. We only give here a few examples.*

Let $N = \{1,2,\ldots,n\}$, and consider a graph $G = (N,E)$ with nonnegative weights $w : N \to \mathbb{R}^+$ on its vertices, and capacities $c : E \to \mathbb{R}^+$ on its (undirected) edges. For every $S \subseteq N$, the cut $(S, N \setminus S)$ is the set of edges having exactly one endpoint in S; the capacity of this cut is defined as $\sum_{(i,j) \in (S, N \setminus S)} c(i,j)$. The max-cut problem is to find a cut of maximum capacity in G. If $(x_1, x_2, \ldots, x_n)$ is interpreted as the characteristic vector of S, then the edge (i,j) has $i \in S$ and $j \notin S$ if and only if $x_i \overline{x}_j = 1$. Therefore, the max-cut problem is equivalent to the maximization of the quadratic pseudo-Boolean function

$$f(x_1, x_2, \ldots, x_n) = \sum_{(i,j) \in E} c(i,j)(x_i \overline{x}_j + \overline{x}_i x_j). \tag{13.1}$$

Recall that a stable set in G is a set $S \subseteq N$ such that no edge has both of its endpoints in S; the weight of S is $w(S) = \sum_{i \in S} w(i)$. The weighted stability problem is to find a stable set of maximum weight in G. If $(x_1, x_2, \ldots, x_n)$ denotes again the characteristic vector of S, then this is equivalent to maximizing the quadratic pseudo-Boolean function

$$f(x_1, x_2, \ldots, x_n) = \sum_{i=1}^{n} w(i) x_i - M \sum_{(i,j) \in E} x_i x_j \tag{13.2}$$

for a sufficiently large value of the penalty M (say, $M > \max_{1 \leq i \leq n} w(i)$).

Let us now assume that $w(i) = 1$ for $i = 1, 2, \ldots, n$. For every $A \subseteq N$, we denote by $\alpha_G(A)$ the stability number of the subgraph of G induced by A, that is, the size of a largest stable set of G contained in A. We can associate with G a pseudo-Boolean function f_{α_G} defined as follows: For each $X = (x_1, x_2, \ldots, x_n) \in \mathcal{B}^n$,

$$f_{\alpha_G}(X) = \alpha_G(supp(X)), \tag{13.3}$$

where $supp(X)$ denotes as usual the subset of N with characteristic vector X. Pseudo-Boolean functions defined in this way have been introduced in [66] in connection with the study of perfect graphs. To illustrate their interest, we mention for instance that G is the complement of a triangulated graph if and only if all the *coefficients in the multilinear polynomial representation of $f_{\alpha_G}(X)$ take values in $\{-1,0,1\}$ (see [66]).*   □

Application 13.2. (Linear algebra.) *Let V be a finite set of vectors over an arbitrary field and consider the set function $f : \mathcal{P}(V) \to \mathbb{R}$, where $f(T)$, $T \subseteq V$, is the rank of the matrix whose rows are the members of T. This rank function has two interesting properties that are further examined in Section 13.6. First, the function is monotone nondecreasing, that is,*

$$f(S) \leq f(T) \quad \text{whenever } S \subseteq T.$$

Second, the function is submodular, meaning that

$$f(S \cup T) + f(S \cap T) \leq f(S) + f(T) \quad \text{for all } S, T \subseteq V.$$

It is interesting to remark that both of these properties continue to hold for rank functions defined on subsets of elements of a matroid (see for instance Welsh [905]).   □

Computer science and engineering

Application 13.3. (Artificial intelligence, Maximum satisfiability.) *Expert systems are frequently described as systems of rules of the form*

$$(C_k(x_{i_1},\ldots,x_{i_{n_k}}) = 1) \Rightarrow (x_{j_k} = 1), \quad (k = 1,2,\ldots,m),$$

where the Boolean variables x_i, $i = 1,2,\ldots,n$, are associated with various control parameters, and where each C_k, $k = 1,2,\ldots,m$, is an elementary conjunction. Suppose that there is a real-valued penalty w_k for the violation of the k-th rule. Then, the total penalty incurred for the assignment of values $(x_1,x_2,\ldots,x_n)$ to the control variables is described by the pseudo-Boolean function

$$f = \sum_{k=1}^{m} w_k\, C_k(x_{i_1},\ldots,x_{i_{n_k}})\overline{x}_{j_k}.$$

This situation can be generalized as follows: Consider a CNF $\bigwedge_{k=1}^{m} C_k$, where each C_k is a Boolean clause (or elementary disjunction) of the form $C_k = (\bigvee_{i \in A_k} \overline{x}_i) \vee (\bigvee_{j \in B_k} x_j)$, and assume that a real weight w_k has been assigned to each clause C_k, for $k = 1,2,\ldots,m$. The weighted maximum satisfiability *(MAX SAT) problem is to find a point X^* in $\{0,1\}^n$ that maximizes the total weight of the satisfied clauses, that is, MAX SAT is the pseudo-Boolean optimization problem*

$$\text{maximize } \sum_{k=1}^{m} \{w_k \mid C_k(X) = 1\} \quad \text{subject to } X \in \mathcal{B}^n.$$

Clearly a clause C_k takes value 1 if and only if the term $(\prod_{i \in A_k} x_i)(\prod_{j \in B_k} \overline{x}_j)$ is equal to 0. Therefore, MAX SAT is equivalent to minimizing the pseudo-Boolean *function*

$$f = \sum_{k=1}^{m} w_k \left(\prod_{i \in A_k} x_i \right) \left(\prod_{j \in B_k} \overline{x}_j \right).$$

We refer the reader to Chapter 2, Section 2.11.4, for a more complete discussion of this well-known generalization of the Boolean satisfiability problem. □

Application 13.4. (Data mining, classification, learning theory.) *Consider a finite set* $\Omega^+ \subseteq \{0,1\}^n$ *of positive observations, and a finite set* $\Omega^- \subseteq \{0,1\}^n$ *of negative observations, such that* $\Omega^+ \cap \Omega^- = \emptyset$. *In order to distinguish the sets of positive and negative vectors, two families of elementary conjunctions* $C_1^+, \ldots, C_k^+$ *and* $C_1^-, \ldots, C_h^-$ *(called respectively positive and negative patterns) can be determined, such that for all* $X \in \Omega^+ \cup \Omega^-$,

$$C_i^+(X) = 1 \Rightarrow X \in \Omega^+ \quad (i = 1, \ldots k)$$
$$C_j^-(X) = 1 \Rightarrow X \in \Omega^- \quad (j = 1, \ldots h)$$

(see Chapter 12 for details). In Boros et al. [131], patterns have been used to define a family of discriminants, namely, pseudo-Boolean functions of the form

$$d(X) = \sum_{i=1}^{k} \alpha_i \, C_i^+(X) - \sum_{j=1}^{h} \beta_j \, C_j^-(X),$$

where the α_i*'s and the* β_j*'s are nonnegative reals, and* $\sum_{i=1}^{k} \alpha_i = \sum_{j=1}^{h} \beta_j = 1$. *An appropriate choice of the parameters* (α_i, β_j) *allows the construction of discriminants which take "high" values in positive observations, and "low" values in negative ones. We refer to [131] for details. See also Genkin, Kulikowski, and Muchnik [376] for other pseudo-Boolean models in data mining.* □

Application 13.5. (Computer vision.) *A fundamental problem in computer vision is to restore a "better" version of an initially blurred, or "noisy" image. Ideally, the restored image should be "similar" to the initial one but should display large "uniformly colored" regions with "crisp" transitions at boundaries between different colors.*

A basic formulation of the problem can be stated as follows: We are given a set $\mathcal{P} = \{1, 2, \ldots, n\}$ *of pixels, a set* $\mathcal{C} = \{1, 2, \ldots, C\}$ *of colors, an initial assignment* $c_0 : \mathcal{P} \to \mathcal{C}$ *of colors to pixels, and a so-called energy function* $E(c)$ *which measures the inadequacy of any new coloring* $c : \mathcal{P} \to \mathcal{C}$. *This energy function is to be minimized over all possible colorings* c.

Typically, the energy function takes the form

$$E(c) = \sum_{p \in \mathcal{P}} (c_0(p) - c(p))^2 + \sum_{(p,q) \in E} V(c(p), c(q)),$$

where E is a collection of "neighboring pixels." The first group of terms estimates the similarity between the initial coloring c_0 and the new coloring c, whereas the remaining terms penalize the assignment of distinct colors to neighboring pixels.

In the simplest (black-and-white) case, every pixel can take exactly one of two colors $(C = 2)$, so that each $c(p)$ can be viewed as a Boolean variable and $E(c)$ is a quadratic pseudo-Boolean function (by virtue of Theorem 13.1 hereunder, and because each term $V(c(p), c(q))$ depends on two Boolean variables only). In spite of its apparent simplicity, this binary model arises as a subproblem in the solution of more realistic formulations. We refer the reader to Boykov, Veksler, and Zabih [147] or to Kolmogorov and Rother [576] for more details on applications.   □

Operations research

Application 13.6. (0–1 linear programming.) *Consider the 0–1 linear programming problem*

$$\text{maximize} \quad z(x_1, x_2, \ldots, x_n) = \sum_{j=1}^{n} c_j x_j \tag{13.4}$$

$$\text{subject to} \quad \sum_{j=1}^{n} a_{ij} x_j = b_i, \quad i = 1, 2, \ldots, m \tag{13.5}$$

$$(x_1, x_2, \ldots, x_n) \in \{0, 1\}^n. \tag{13.6}$$

This fundamental problem of discrete optimization is equivalent to the unconstrained quadratic pseudo-Boolean optimization problem

$$\text{maximize} \quad f(x_1, x_2, \ldots, x_n) = \sum_{j=1}^{n} c_j x_j - M \sum_{i=1}^{m} (\sum_{j=1}^{n} a_{ij} x_j - b_i)^2$$

$$\text{subject to} \quad (x_1, x_2, \ldots, x_n) \in \{0, 1\}^n, \tag{13.7}$$

for a sufficiently large value of M.   □

Application 13.7. (Game theory.) *A game in characteristic form is a set function f defined on $\mathcal{P}(N)$, where $N = \{1, 2, \ldots, n\}$ is a finite set of players. The value of $f(S)$ is interpreted as the payoff that players in S can secure by acting together. It is usual to assume that $f(\emptyset) = 0$ and that f is monotone nondecreasing, that is, $f(S) \leq f(T)$ whenever $S \subseteq T$. Another frequent assumption is that f is superadditive, meaning that $f(S) + f(T) \leq f(S \cup T)$ whenever $S \cap T = \emptyset$. The economic interpretation of superadditivity is that players can achieve more value by cooperating than by acting separately.*

If f is viewed as a pseudo-Boolean function on $\mathcal{B}^n$, then its multilinear representation and its continuous extension f^c (see Sections 13.2 and 13.3) play an interesting role in this context. Indeed, several central concepts in game theory (such as imputations, core, Shapley value, Banzhaf index) have natural pseudo-Boolean interpretations, leading to interesting theoretical and algorithmic insights; see, for instance, [441, 456, 669, 719, 720] and Section 13.5.

Monotone nondecreasing functions such that $f(\emptyset) = 0$ and $f(N) = 1$ have also been examined in artificial intelligence under the name belief functions or Choquet capacities, where they are used to model uncertainty and subjective probabilities (see Chateauneuf and Jaffray [188]; Shafer [824, 825], etc.), and in multicriteria decision-making under the name fuzzy measures, as tools for the aggregation of interacting criteria (see Sugeno [851]; Grabisch [405]; Grabisch, Marichal, Mesiar, and Pap [406]; Marichal [668], etc.). They are discussed further in Section 13.6.2.

Application 13.8. (Production management and logistics.) *So-called fixed charge constraints of the form*

$$(y = 1) \quad \textit{if and only if} \quad (x_i = 1 \text{ for all } i \in A) \tag{13.8}$$

are encountered in many business decision problems, like capital budgeting, production planning, plant location, etc. Since constraint (13.8) simply expresses that $y = \prod_{i \in A} x_i$, pseudo-Boolean formulations of the associated problems often arise quite naturally by elimination of the y-variables. We briefly describe two models of this type.

A fundamental planning problem for flexible manufacturing systems (FMS) is the part selection problem. *A part-set containing n parts must be processed, one part at a time, on a single flexible machine. The machine can use different tools, numbered from 1 to m. Each part requires a specific subset of tools which have to be loaded in the tool magazine of the machine before the part can be processed: Say part i requires $T(i) \subseteq \{1, 2, \ldots, m\}$. The magazine features C tool slots. When loaded on the machine, tool j occupies s_j slots in the magazine ($j = 1, 2, \ldots, m$). The total number of tools required to process all parts can be much larger than C, so that it is sometimes necessary to change tools in order to process the complete part-set. Now, the part selection problem consists in determining the largest number of parts that can be produced without tool changes.*

This problem can be modeled as a pseudo-Boolean optimization problem in various ways. In the simplest model, a Boolean variable x_j indicates whether tool j is placed in the magazine or not ($j = 1, 2, \ldots, m$). Then, the part selection problem is

$$\textit{maximize} \quad f(x_1, x_2, \ldots, x_m) = \sum_{i=1}^{n} \left(\prod_{j \in T(i)} x_j \right) \tag{13.9}$$

$$\textit{subject to} \quad \sum_{j=1}^{n} s_j x_j \leq C, \tag{13.10}$$

$$(x_1, x_2, \ldots, x_m) \in \{0, 1\}^m, \tag{13.11}$$

where the product $\prod_{j \in T(i)} x_j$ takes value 1 only if part i can be processed by the selected tools. Different formulations and detailed discussions of this problem can be found, for instance, in [228, 238, 845].

For a second example, consider the classical simple facility location problem: Here, we must select an optimal subset of locations for some facilities (such as plants, warehouses, emergency facilities) in order to serve the needs of a set of users. Opening a facility in a given location i requires a fixed cost c_i, and delivering the service to user j from location i carries a cost d_{ji} ($j = 1, 2, \ldots, m$, $i = 1, 2, \ldots, n$).

Let us introduce a 0–1 variable x_i which indicates whether a facility is to be opened in location i ($i = 1, \ldots, n$). Two pseudo-Boolean functions can be defined: A function $c(X)$ to indicate the total fixed cost required to open a configuration $X = (x_1, x_2, \ldots, x_n)$, and a function $d(X)$ to indicate the optimal cost of serving the set of users from the corresponding locations. The optimal location problem (essentially) consists now in finding the minimum of the pseudo-Boolean function $c(X) + d(X)$. Detailed expressions of this function were first proposed by Hammer [434] and further examined in [70, 263, 394, 395, 900, etc.]. If we denote by $\pi(j) = (i_1(j), i_2(j), \ldots, i_n(j))$ a permutation of locations such that $d_{ji_1(j)} \le d_{ji_2(j)} \le \cdots \le d_{ji_n(j)}$, then the function to be minimized can be written as

$$f(X) = \sum_{i=1}^{n} c_i x_i + \sum_{j=1}^{m} \sum_{k=1}^{n} d_{ji_k(j)} x_{i_k(j)} \prod_{\ell < k} \overline{x}_{i_\ell(j)} + M \prod_{i=1}^{n} \overline{x}_i.$$

In this formulation, the last term involves a large penalty M; it is necessary to ensure that at least one facility is opened. □

13.2 Representations

Different application areas may rely on different descriptions of pseudo-Boolean functions. For instance, in game theory, the payoff of a coalition of players may be computed as the optimal value of an associated combinatorial optimization problem (see Bilbao [79]). In other models, the values assumed by a pseudo-Boolean function may be listed in a table, or computed by a black-box oracle.

One of the main impacts of the pseudo-Boolean viewpoint on the theory of set functions, however, is due to the existence of various *algebraic representations* of these functions. The properties of such algebraic representations are the main topic of this section.

13.2.1 Polynomial expressions, pseudo-Boolean normal forms and posiforms

The following representation theorem is stated in Hammer, Rosenberg, and Rudeanu [458] and in Hammer and Rudeanu [460] (where it is attributed to T. Gaspar).

Theorem 13.1. *For every pseudo-Boolean function f on $\mathcal{B}^n$, there exists a unique mapping $c \colon \mathcal{P}(N) \to \mathbb{R}$ such that*

$$f(x_1, x_2, \ldots, x_n) = \sum_{A \in \mathcal{P}(N)} c(A)\left(\prod_{i \in A} x_i\right). \tag{13.12}$$

Proof. For every point $X^* \in \mathcal{B}^n$, the expression

$$f(X^*)\left(\prod_{i \mid x_i^* = 1} x_i\right)\left(\prod_{j \mid x_j^* = 0} \overline{x}_j\right) \tag{13.13}$$

takes value $f(X^*)$ in the point X^*, and the value 0 in every other point of $\mathcal{B}^n$. Therefore,

$$f(x_1, x_2, \ldots, x_n) = \sum_{X^* \in \mathcal{B}^n} f(X^*)\left(\prod_{i \mid x_i^* = 1} x_i\right)\left(\prod_{j \mid x_j^* = 0} \overline{x}_j\right). \tag{13.14}$$

Replacing $\overline{x}_j$ by $(1 - x_j)$, expanding the products and using distributivity immediately yields a polynomial expression of the form (13.12).

Assume now that p_1 and p_2 are two different polynomial expressions of the form (13.12) with coefficients $c_1(A)$ and $c_2(A)$, $A \in \mathcal{P}(N)$, respectively. Let A^* be a subset of N such that $c_1(A^*) \neq c_2(A^*)$, and such that $c_1(A) = c_2(A)$ for all A with $|A| < |A^*|$. If X^* denotes the characteristic vector of A^*, then $p_1(X^*) - p_2(X^*) = c_1(A^*) - c_2(A^*) \neq 0$, so that p_1 and p_2 cannot both represent f. $\square$

Note that the polynomial (13.12) is linear in each of its variables: We say that it is *multilinear*.

Definition 13.1. *The expression in the right-hand side of* (13.12) *is the* (multilinear) *polynomial expression of f. The* degree *of f is the degree of this polynomial, namely, $\mathrm{degree}(f) = \max\{|A| : c(A) \neq 0\}$. We say that a pseudo-Boolean function is either* linear, *or* quadratic, *or* cubic *if its degree is at most 1, or 2, or 3, respectively.*

The set function $c \colon \mathcal{P}(N) \to \mathbb{R}$ is sometimes called the *Möbius transform* or the *mass function* associated with f (see for instance [407, 824]). In fact, it follows from the elementary theory of Möbius inversion for ordered sets that c can be computed as

$$c(A) = \sum_{S \subseteq A} (-1)^{|A| - |S|} f(e_S) \quad \text{for all } A \in \mathcal{P}(N),$$

where e_S denotes as usual the characteristic vector of S (see Aigner [12]). The bijective correspondence linking the functions f and c has been investigated in a broader context by various authors; see for instance Grabisch, Marichal, and Roubens [407].

The polynomial expression of a pseudo-Boolean function does not involve complemented variables. If we allow complementation, then we obtain a broader class of expressions.

Definition 13.2. *A* pseudo-Boolean normal form *(PBNF) is an expression ψ of the form*

$$\psi(x_1, x_2, \ldots, x_n) = b_0 + \sum_{k=1}^{m} b_k \left(\prod_{i \in A_k} x_i \right) \left(\prod_{j \in B_k} \bar{x}_j \right), \qquad (13.15)$$

where $b_0, b_1, \ldots, b_m$ are real coefficients, and $A_k \cap B_k = \emptyset$, $A_k \cup B_k \neq \emptyset$ for $k = 1, 2, \ldots, m$.

Every pseudo-Boolean function can be represented by (many) distinct PBNFs. For instance, the representation in equation (13.14) is a PBNF (called the *minterm PBNF*) of f that can be readily constructed from a table of values of f; see also Example 13.1 hereunder.

PBNFs with positive coefficients play a special role in many applications.

Definition 13.3. *The PBNF* (13.15) *is called a* posiform *if $b_k > 0$ for all $k = 1, \ldots, m$.*

Note that the sign of the free coefficient b_0 is unrestricted in a posiform. Hammer and Rosenberg [457] introduced posiforms and observed the following property:

Theorem 13.2. *Every pseudo-Boolean function can be represented by a* posiform.

Proof. Let us consider the polynomial representation (13.12) of a pseudo-Boolean function f. If $T = c\, x_{i_1}\, x_{i_2} \ldots x_{i_k}$ is a term of (13.12) with $c < 0$, then successive applications of the identity $x_{i_j} = 1 - \bar{x}_{i_j}$, for $j = k, k-1, \ldots$ down to 1, transform T into

$$T = c - c\,\bar{x}_{i_1} - c\,x_{i_1}\,\bar{x}_{i_2} - \cdots - c\,x_{i_1}\,x_{i_2} \cdots x_{i_{k-1}}\,\bar{x}_{i_k},$$

which is a posiform. Repeating this transformation for every negative term of (13.12) eventually produces a posiform of f. $\square$

Other posiforms representing the same pseudo-Boolean function would be obtained by applying in a different order the transformations described in the proof of Theorem 13.2.

Example 13.1. *The pseudo-Boolean function $f(x,y,z)$ defined by the table*

x y z	$f(x,y,z)$
0 0 0	3
0 0 1	1
0 1 0	0
0 1 1	−2
1 0 0	4
1 0 1	2
1 1 0	−5
1 1 1	6

admits the minterm PBNF

$$\mu = 3\,\overline{x}\,\overline{y}\,\overline{z} + \overline{x}\,\overline{y}z - 2\,\overline{x}yz + 4x\overline{y}\,\overline{z} + 2x\overline{y}z - 5xy\overline{z} + 6xyz.$$

Replacing each complemented variable $\overline{u}$ by $1-u$, we find the unique polynomial expression of f:

$$f = 3 + x - 3y - 2z - 6xy + 13xyz.$$

Replacing now the terms $-3y$ and $-2z$ by $-3 + 3\overline{y}$ and $-2 + 2\overline{z}$, respectively, and replacing the term $-6xy$ either by $-6 + 6\overline{x} + 6x\overline{y}$ or by $-6 + 6\overline{y} + 6\overline{x}y$, we obtain the posiform representations:

$$\psi_1 = -8 + x + 6\overline{x} + 3\overline{y} + 2\overline{z} + 6x\overline{y} + 13xyz$$

and

$$\psi_2 = -8 + x + 3\overline{y} + 6\overline{y} + 2\overline{z} + 6\overline{x}y + 13xyz,$$

which can be further simplified to

$$\psi_1' = -7 + 5\overline{x} + 3\overline{y} + 2\overline{z} + 6x\overline{y} + 13xyz$$

and

$$\psi_2' = -8 + x + 9\overline{y} + 2\overline{z} + 6\overline{x}y + 13xyz,$$

respectively. □

13.2.2 Piecewise linear representations

Hammer and Rosenberg [457] observed that every pseudo-Boolean function f can be expressed as the pointwise-minimum of a family of linear functions. To see this, consider an arbitrary posiform of f:

$$\psi(x_1, x_2, \ldots, x_n) = b_0 + \sum_{k=1}^{m} b_k \left(\prod_{i \in A_k} x_i \right) \left(\prod_{j \in B_k} \overline{x}_j \right) \tag{13.16}$$

where $A_k \cap B_k = \emptyset$, $A_k \cup B_k \neq \emptyset$ and $b_k > 0$ for $k = 1, 2, \ldots, m$. A *selector* for (13.16) is a vector $\sigma = (\sigma_1, \sigma_2, \ldots, \sigma_m)$ such that $\sigma_k \in A_k \cup B_k$ for $k = 1, 2, \ldots, m$. For every selector σ, the linear function

$$l^\sigma(x_1, x_2, \ldots, x_n) = b_0 + \sum_{\substack{k=1: \\ \sigma_k \in A_k}}^m b_k x_{\sigma_k} + \sum_{\substack{k=1: \\ \sigma_k \in B_k}}^m b_k (1 - x_{\sigma_k}) \qquad (13.17)$$

is a majorant of f, that is, $f(x_1, x_2, \ldots, x_n) \leq l^\sigma(x_1, x_2, \ldots, x_n)$ for all $(x_1, x_2, \ldots, x_n) \in \mathcal{B}^n$ (since the inequality holds termwise).

Theorem 13.3. *If $\mathcal{S}$ is the set of all selectors for (13.16), then*

$$f(x_1, x_2, \ldots, x_n) = \min_{\sigma \in \mathcal{S}} l^\sigma(x_1, x_2, \ldots, x_n) \quad \text{for all } (x_1, x_2, \ldots, x_n) \in \mathcal{B}^n. \quad (13.18)$$

Proof. The previous discussion implies that $f \leq \min_{\sigma \in \mathcal{S}} l^\sigma$ on $\mathcal{B}^n$.

To establish the reverse inequality, let X^* be a point in $\mathcal{B}^n$. We define a selector σ as follows. For $k = 1, 2, \ldots, m$, consider the value of $T_k^* = (\prod_{i \in A_k} x_i^*)(\prod_{j \in B_k} \overline{x_j^*})$. If $T_k^* = 1$, then σ_k can be an arbitrary index in $A_k \cup B_k$. If $T_k^* = 0$, then σ_k is either an index in A_k such that $x_{\sigma_k}^* = 0$ or an index in B_k such that $\overline{x_{\sigma_k}^*} = 0$. In all cases, it is easy to see that $f(X^*) = l^\sigma(X^*)$, and hence, equality holds in (13.18). $\qquad\square$

13.2.3 Disjunctive and conjunctive normal forms

An interesting representation of pseudo-Boolean functions is based on the use of elementary conjunctions and disjunctions, by analogy with classical representations of Boolean functions. Our discussion in this section is based on several papers by Foldes and Hammer [336, 337, 338], where additional information can be found. Closely related concepts are discussed by Cunninghame-Green [247]; Davio, Deschamps, and Thayse [259]; Grabisch et al. [406]; Marichal [668]; Störmer [849]; Sugeno [851], and so on.

Definition 13.4. *If f_1 and f_2 are two pseudo-Boolean functions on $\mathcal{B}^n$, their* disjunction *is the pseudo-Boolean function $f_1 \vee f_2$ defined as*

$$(f_1 \vee f_2)(X) = \max\{f_1(X), f_2(X)\} \text{ for all } X \in \mathcal{B}^n,$$

and their conjunction *is the function $f_1 \wedge f_2$ defined as*

$$(f_1 \wedge f_2)(X) = \min\{f_1(X), f_2(X)\} \text{ for all } X \in \mathcal{B}^n.$$

Clearly, if the functions f_1 and f_2 are Boolean, then disjunction and conjunction are simply the usual Boolean operators (and we sometimes omit to write the operator $\wedge$).

Definition 13.5. *A* (pseudo-Boolean) elementary conjunction *is an expression of the form*

$$p(X) = a + b\left(\bigwedge_{i \in A} x_i\right)\left(\bigwedge_{j \in B} \overline{x}_j\right), \tag{13.19}$$

where $a, b \in \mathbb{R}$, $b \geq 0$, *and* A, B *are subsets of indices with* $|A| + |B| \geq 1$ *and* $A \cap B = \emptyset$.

A (pseudo-Boolean) disjunctive normal form (DNF) *is a disjunction of elementary conjunctions that all have the same minimum, that is, an expression of the form*

$$f = \bigvee_{k=1}^{m}\left[a + b_k\left(\bigwedge_{i \in A_k} x_i\right)\left(\bigwedge_{j \in B_k} \overline{x}_j\right)\right], \tag{13.20}$$

where $b_1, b_2, \ldots, b_m \geq 0$, $|A_k| + |B_k| \geq 1$ *and* $A_k \cap B_k = \emptyset$ *for* $k = 1, 2, \ldots, m$. *We say that the right-hand side of* (13.20) *is a* DNF representation *or a* DNF expression *of the function* f.

Note that every constant function $p(X) = a$ is an elementary conjunction (with $b = 0$).

We know that Boolean functions can always be represented in disjunctive normal form. In the pseudo-Boolean case, we similarly obtain:

Theorem 13.4. *Every pseudo-Boolean function has infinitely many DNF representations.*

Proof. Let a be any constant such that $a \leq \min_{X \in \mathcal{B}^n} f(X)$. Then, f is represented by the DNF expression

$$\psi(x_1, \ldots, x_n) = \bigvee_{X^* \in \mathcal{B}^n}\left[a + (f(X^*) - a)\left(\bigwedge_{i \mid x_i^* = 1} x_i\right)\left(\bigwedge_{j \mid x_j^* = 0} \overline{x}_j\right)\right] \tag{13.21}$$

(compare with (13.14)). □

Example 13.2. *The pseudo-Boolean function*

$$f(x, y) = 6 + 3x - xy$$

attains its minimum value ($\min f(X) = 6$) *when* $(x, y) = (0, 0)$ *or when* $(x, y) = (0, 1)$. *Hence, using the construction* (13.21), *f can be expressed as*

$$f(x, y) = (6 + 2xy) \vee (6 + 3x\overline{y}),$$

or as

$$f(x, y) = (5 + 3xy) \vee (5 + \overline{x}y) \vee (5 + 4x\overline{y}) \vee (5 + \overline{x}\,\overline{y}),$$

or as

$$f(x,y) = (8x\,y) \vee (6\overline{x}\,y) \vee (9x\,\overline{y}) \vee (6\overline{x}\,\overline{y}),$$

and so on.   □

Pseudo-Boolean elementary disjunctions and conjunctive normal forms can be defined in a similar way. A *pseudo-Boolean elementary disjunction* is an expression of the form

$$a + b\left(\bigvee_{i \in A} x_i \vee \bigvee_{j \in B} \overline{x}_j \right), \tag{13.22}$$

where $a, b \in \mathbb{R}$, $b \geq 0$, and A, B are subsets of indices with $|A| + |B| \geq 1$, $A \cap B = \emptyset$. A *conjunctive normal form* (CNF) is a conjunction of elementary disjunctions that all have the same maximum M, namely, an expression of the form

$$f = \bigwedge_{k=1}^{m} \left[a_k + b_k \left(\bigvee_{i \in A_k} x_i \vee \bigvee_{j \in B_k} \overline{x}_j \right) \right], \tag{13.23}$$

where $b_k \geq 0$, $a_k + b_k = M$, $|A_k| + |B_k| \geq 1$ and $A_k \cap B_k = \emptyset$ for $k = 1, 2, \ldots, m$. The existence of CNF representations is shown similarly to that of DNFs.

Pseudo-Boolean implicants and implicates

An elementary conjunction (respectively, disjunction) p is an *implicant* (respectively, *implicate*) of a pseudo-Boolean function f if $p \leq f$ (respectively, $f \leq p$). An implicant p is called a *prime implicant* of f if f has no implicant p' such that $p \neq p'$ and $p \leq p'$. Prime implicates are similarly defined.

Let us establish some of the fundamental properties of pseudo-Boolean (prime) implicants and implicates.

Lemma 13.1. *Let f be a pseudo-Boolean function on $\mathcal{B}^n$, let*

$$p(X) = a + b\left(\bigwedge_{i \in A} x_i \right)\left(\bigwedge_{j \in B} \overline{x}_j \right) \tag{13.24}$$

be an elementary conjunction, let F_p denote the face

$$F_p = \{X \in \mathcal{B}^n \mid x_i = 1 \text{ for all } i \in A \text{ and } x_j = 0 \text{ for all } j \in B\}$$

if $b > 0$, and let $F_p = \mathcal{B}^n$ if $b = 0$. Let $f_{\min} = \min_{X \in \mathcal{B}^n} f(X)$ and $f_p = \min_{X \in F_p} f(X)$.

(i) *p is an implicant of f if and only if $a \leq f_{\min}$ and $a + b \leq f_p$.*
(ii) *If p is a prime implicant of f, then $a = f_{\min}$ and $a + b = f_p$.*

Proof. Note that the elementary conjunction $p(X)$ given by (13.24) takes value $a + b$ on F_p and value a elsewhere. Claim (i) follows immediately from these observations.

To prove Claim (ii), assume first that $b = 0$ and that $a < f_{min}$ (or equivalently in this case, $a + b < f_p$). Since $p'(X) = f_{min}$ is an implicant of f and since $p(X) = a < f_{min} = p'(X)$, we conclude that $p(X)$ is not prime.

So, let us assume from now on that $b > 0$. If $a < f_{min}$, let $0 < \epsilon \le \min(b, f_{min} - a)$, and define $p''(X) = (a+\epsilon) + (b-\epsilon)\left(\bigwedge_{i \in A} x_i\right)\left(\bigwedge_{j \in B} \overline{x}_j\right)$. There holds $p(X) \le p''(X) \le f(X)$ for all $X \in \mathcal{B}^n$, $p(X) \ne p''(X)$, and we conclude that $p(X)$ is not prime.

Finally, assume that $a + b < f_p$, let $0 < \eta \le f_p - (a + b)$, and define $p'''(X) = a + (b+\eta)\left(\bigwedge_{i \in A} x_i\right)\left(\bigwedge_{j \in B} \overline{x}_j\right)$. Here again, $p(X) \le p'''(X) \le f(X)$ for all $X \in \mathcal{B}^n$, and we conclude that $p(X)$ is not prime. $\square$

Theorem 13.5. *Every pseudo-Boolean function has an infinite number of implicants and implicates, and a finite number of prime implicants and prime implicates.*

Proof. We only discuss the case of implicants, as a similar reasoning applies for implicates. For every pseudo-Boolean function f, for every constant a strictly smaller than the minimum value of f, and for every sufficiently small constant b, the elementary conjunction (13.24) is an implicant of f. This shows that f has infinitely many implicants.

Let us say that an implicant of the form (13.24) is "tight" if $a = f_{min}$ and $a + b = f_p$, as in statement (ii) of Lemma 13.1. The lemma states that every prime implicant is tight. Moreover, the number of tight implicants is finite, since there is only a finite number of possible choices for the sets A and B in (13.24), and the value of a and b is fixed as soon as A and B are given. This proves the theorem. $\square$

Theorem 13.6. *For every implicant (respectively, implicate) p of a pseudo-Boolean function f, there is a prime implicant (respectively, implicate) p' of f such that $p \le p' \le f$ (respectively, $f \le p' \le p$).*

Proof. We concentrate on the claim concerning an implicant p. The proof of Lemma 13.1 actually implies that there is a tight implicant p' such that $p \le p' \le f$. Let us choose p' to be maximal with this property, that is, let us assume that there is no tight implicant q such that $p' \le q \le f$, $q \ne p'$ (this assumption is legitimate because the set of tight implicants is finite).

Now, if p' is not prime, then there exists another implicant p'' such that $p' \le p'' \le f$, $p' \ne p''$. But here again, the proof of Lemma 13.1 implies that p'' must be dominated by a tight implicant q such that $p'' \le q \le f$, contradicting the maximality of p'. $\square$

If ψ is a DNF expression of a pseudo-Boolean function f, then all elementary conjunctions appearing in ψ are implicants of f. Clearly, different DNFs may use very different sets of implicants. However, the prime implicants allow us to define a canonical DNF for each pseudo-Boolean function, thus extending the corresponding representation theory of Boolean functions. A similar situation arises for CNFs.

Theorem 13.7. *Every pseudo-Boolean function is the disjunction of its prime implicants and the conjunction of its prime implicates.*

Proof. This is an immediate consequence of Theorems 13.5 and 13.6. $\square$

Foldes and Hammer [336] propose an algorithm which produces all prime implicants of an arbitrary function expressed in DNF. Their algorithm is a generalization of the Boolean consensus method (see Chapter 3). It is also analogous to the consensus procedure for discrete functions described by Davio, Deschamps and Thayse [259].

13.3 Extensions of pseudo-Boolean functions

We denote by U^n the "solid" hypercube $U^n = [0, 1]^n$ spanned by $\mathcal{B}^n$.

Definition 13.6. *A (continuous) extension of the pseudo-Boolean function f : $\mathcal{B}^n \to \mathbb{R}$ is a function $g : U^n \to \mathbb{R}$ which coincides with f at the vertices of the hypercube, meaning that*

$$f(X) = g(X) \text{ for all } X \in \mathcal{B}^n.$$

Remark. The term "extension" was used with a different meaning in Chapter 12, where it applied to partially defined Boolean functions. On the other hand, the qualifier "continuous" is somewhat ambiguous in Definition 13.6, since this definition does not require that extensions be continuous in the standard sense for functions of real variables (namely, with respect to the Euclidean topology of $\mathbb{R}^n$); the word "continuous" only reminds us here that extensions are defined over a nondiscrete domain. Therefore, we generally use the short terminology "extension" in this chapter; this should hopefully cause no confusion. $\square$

Extensions of pseudo-Boolean functions find applications in optimization (see Section 13.4 hereunder), in reliability theory (see Section 1.13.4), in game theory (see, e.g., Alonso-Meijide et al. [18], Owen [719, 720]), or in multicriteria decision-making as illustrated by the next example.

Application 13.9. (Multicriteria decision making.) *Suppose that, in a particular decision problem, n relevant criteria $c_1, c_2, \ldots, c_n$ are defined and take value on a continuous $[0, 1]$ scale. Thus, $c_i(a)$ indicates the evaluation of a particular action a according to criterion c_i, and $(c_1(a), \ldots, c_n(a)) \in U^n$.*

A pseudo-Boolean function f on $\mathcal{B}^n$ can be used to model the importance of each subset of criteria: namely, for each $X \in \mathcal{B}^n$, the value $f(X)$ indicates the importance of the subset of criteria $\{c_i \mid x_i = 1\}$. Now, if g is an extension of

f on U^n, then $g(c_1(a),\ldots,c_n(a))$ can be interpreted as the global evaluation of action a. For instance, if $w_i \geq 0$ ($i = 1,2,\ldots,n$),

$$f(X) = \sum_{i=1}^{n} w_i\, x_i \quad \text{for all } X \in \mathcal{B}^n,$$

and

$$g(X) = \sum_{i=1}^{n} w_i\, x_i \quad \text{for all } X \in U^n,$$

then w_i can be viewed as the "weight" of criterion i in a simple additive weighing scheme.

Other classes of pseudo-Boolean functions and extensions can be used to model complex, nonlinear interactions among criteria (see for instance [405, 407, 668]). □

Of course, every pseudo-Boolean function f has infinitely many extensions. We now discuss some classes of extensions which have proved to be of special interest in various settings.

13.3.1 The polynomial extension

Definition 13.7. *When viewed as a mapping on* U^n, *the multilinear polynomial expression*

$$\sum_{A \in \mathcal{P}(N)} c(A) \left(\prod_{i \in A} x_i \right) \tag{13.25}$$

of a pseudo-Boolean function f *defines an extension of* f *that we call its polynomial extension and that we denote by* f^{pol}.

In game theory, f^{pol} is frequently called the *multilinear extension* of f; see Owen [719, 720].

More generally, if f is represented by the PBNF

$$\psi(x_1,x_2,\ldots,x_n) = b_0 + \sum_{k=1}^{m} b_k \left(\prod_{i \in A_k} x_i \right) \left(\prod_{j \in B_k} \bar{x}_j \right), \tag{13.26}$$

then the expression

$$\hat{\psi}(x_1,x_2,\ldots,x_n) = b_0 + \sum_{k=1}^{m} b_k \left(\prod_{i \in A_k} x_i \right) \left(\prod_{j \in B_k} (1 - x_j) \right) \tag{13.27}$$

provides an alternative representation of the polynomial extension f^{pol}. This easily follows from the observation that, if we expand all products in (13.27), then we obtain a polynomial, which, in view of Theorem 13.1, necessarily coincides with the multilinear polynomial expression of f.

Example 13.3. *Consider again the pseudo-Boolean function f introduced in Example 13.1, which can be represented by either of the expressions*

$$\phi = 3 + x - 3y - 2z - 6xy + 13xyz$$

or

$$\psi_1 = -8 + x + 6\overline{x} + 3\overline{y} + 2\overline{z} + 6x\overline{y} + 13xyz.$$

The expression

$$\hat{\psi}_1 = -8 + x + 6(1 - x) + 3(1 - y) + 2(1 - z) + 6x(1 - y) + 13xyz$$

represents the extension $f^{pol} = 3 + x - 3y - 2z - 6xy + 13xyz$ on U^3. □

The polynomial extension of f admits an interesting probabilistic interpretation.

Theorem 13.8. *Let f be a pseudo-Boolean function on $\mathcal{B}^n$. Assume that $x_1, x_2, \ldots, x_n$ are independent Bernoulli random variables, where x_i takes value 1 with probability p_i and value 0 with probability $1 - p_i$. Then, the expected value of f is equal to $f^{pol}(p_1, p_2, \ldots, p_n)$.*

Proof. Let f be given by (13.25) and denote by $E[u]$ the expectation of a random variable u. Then,

$$E[f(x_1, x_2, \ldots, x_n)] = \sum_{A \in \mathcal{P}(N)} c(A) E\left[\left(\prod_{i \in A} x_i\right)\right]$$

$$= \sum_{A \in \mathcal{P}(N)} c(A) \text{Prob}\left[\prod_{i \in A} x_i = 1\right]$$

$$= \sum_{A \in \mathcal{P}(N)} c(A) \left(\prod_{i \in A} p_i\right)$$

$$= f^{pol}(p_1, p_2, \ldots, p_n).$$

□

Example 13.4. *In Example 13.3, if each variable takes value 0 or 1 with probability $\frac{1}{2}$, then the expected value of f is $f^{pol}(\frac{1}{2}, \frac{1}{2}, \frac{1}{2}) = \frac{9}{8}$.* □

In the special case where f is a Boolean function, Proposition 13.8 has already been anticipated in our discussion of reliability theory, in Section 1.13.4 of Chapter 1. In this framework, the polynomial extension f^{pol} corresponds to the so-called *reliability polynomial*; see for instance Colbourn [205, 206], Ramamurthy [777].

13.3.2 Concave and convex extensions

Every pseudo-Boolean function f admits various concave and convex extensions that have been frequently examined in the optimization literature. A simple way to demonstrate the existence of such extensions is to observe that the piecewise linear representation (13.18) defines a concave real-valued function on $\mathbb{R}^n$, as the pointwise minimum of linear functions. Also, the function g defined by

$$g(x_1, x_2, \ldots, x_n) = f^{pol}(x_1, x_2, \ldots, x_n) + M \sum_{j=1}^{n} x_j (1 - x_j)$$

$$\text{for all } (x_1, x_2, \ldots, x_n) \in U^n$$

is an extension of f and is concave (respectively, convex) when M is a large enough positive (respectively, negative) number (for a quadratic function f, this was observed by Hammer and Rubin [459]; the general case was considered by Gianessi and Niccolucci [378] and by Kalantari and Rosen [544]).

The *concave envelope* of f, denoted f^{env}, is defined as the pointwise minimum of all concave extensions of f:

$$f^{env}(X) = \min\{g(X) \mid g \text{ is a concave extension of } f\} \quad \text{for all } X \in U^n.$$

Note that f^{env} is concave on U^n, as pointwise minimum of concave functions, and that it can be viewed as the smallest concave extension of f. The convex envelope of f would be similarly defined.

Another class of concave extensions has been introduced in Crama [227]. Suppose again that f is represented by the PBNF

$$\psi(x_1, x_2, \ldots, x_n) = b_0 + \sum_{k=1}^{m} b_k \left(\prod_{i \in A_k} x_i \right) \left(\prod_{j \in B_k} \overline{x}_j \right), \tag{13.28}$$

where $b_0, b_1, \ldots, b_m \in \mathbb{R}$, $A_k \cap B_k = \emptyset$, and $A_k \cup B_k \neq \emptyset$ for $k = 1, 2, \ldots, m$.

Then the function

$$\psi^{std}(x_1, x_2, \ldots, x_n) = b_0 + \sum_{k=1}^{m} b_k g_k(x_1, x_2, \ldots, x_n) \quad \text{for all } (x_1, x_2, \ldots, x_n) \in U^n$$

$$\tag{13.29}$$

where

$$g_k(x_1, x_2, \ldots, x_n) = \begin{cases} \min\left(\min(x_i \mid i \in A_k), \min(1 - x_j \mid j \in B_k) \right) & \text{if } b_k > 0 \\ \\ \max\left(0, 1 - |A_k| + \sum_{i \in A_k} x_i - \sum_{j \in B_k} x_j \right) & \text{if } b_k < 0 \end{cases}$$

is an extension of f and is concave: Indeed, $g_k(X) = \left(\prod_{i \in A_k} x_i \right) \left(\prod_{j \in B_k} \overline{x}_j \right)$ for all $X \in \mathcal{B}^n$, and each of the functions g_k is concave (respectively, convex) for

$b_k > 0$ (respectively, for $b_k < 0$). In [227], the function ψ^{std} is called the *standard extension* of f *associated with the* PBNF ψ.

The following facts will be useful:

Lemma 13.2. *Consider the PBNF* ψ *in* (13.28). *For* $k = 1, 2, \ldots, m$, *let* H_k *denote the polyhedron*

$$H_k = \{(X, y) \in U^{n+1} \mid y \le g_k(X)\} \; if \, b_k > 0$$

and

$$H_k = \{(X, y) \in U^{n+1} \mid y \ge g_k(X)\} \; if \, b_k < 0.$$

All vertices of H_k *are in* $\mathcal{B}^{n+1}$, *that is, they only have* 0–1 *components.*

Proof. The claim follows from the fact that, in both cases, the system of inequalities defining H_k is totally unimodular; this follows from Theorem 5.13 in Chapter 5 for the case where b_k is positive (see [47, 474, 786]); the other case is easily established by direct arguments. $\qquad\square$

The next lemma is found in Crama [227] (see also Hammer and Kalantari [445] and Hammer and Simeone [463]).

Lemma 13.3. *Consider the PBNF* ψ *in* (13.28). *If* ψ *consists of a single nonconstant term, that is, if* $\psi = b_1 \left(\prod_{i \in A_1} x_i \right) \left(\prod_{j \in B_1} \overline{x}_j \right)$, *then its standard extension* ψ^{std} *and its concave envelope* ψ^{env} *coincide on* U^n:

$$\psi^{env}(X) = \psi^{std}(X) = b_1 g_1(X) \quad for \, all \, X \in U^n.$$

Proof. Since ψ^{std} is concave, $\psi^{env} \le \psi^{std}$ on U^n. To establish the reverse inequality, let $X^* \in U^n$. Since the point $(X^*, g_1(X^*))$ is in H_1, it is a convex combination of vertices of H_1: That is, there exists a collection of 0–1 points $(X^r, y^r) \in H_1$ and of positive scalars λ_r ($r \in R$) such that $(X^*, g_1(X^*)) = \sum_{r \in R} \lambda_r (X^r, y^r)$ and $\sum_{r \in R} \lambda_r = 1$. Hence,

$$
\begin{aligned}
\psi^{std}(X^*) &= b_1 g_1(X^*) \\
&= \sum_{r \in R} \lambda_r b_1 y^r \\
&\le \sum_{r \in R} \lambda_r b_1 g_1(X^r) \quad \text{(since } (X^r, y^r) \in H_1) \\
&= \sum_{r \in R} \lambda_r \psi(X^r) \quad \text{(since } X^r \in \mathcal{B}^n \text{ by Lemma 13.2)} \\
&= \sum_{r \in R} \lambda_r \psi^{env}(X^r) \quad \text{(since } X^r \in \mathcal{B}^n) \\
&\le \psi^{env}\left(\sum_{r \in R} \lambda_r X^r\right) \quad \text{(by concavity of } \psi^{env}) \\
&= \psi^{env}(X^*).
\end{aligned}
$$

$\qquad\square$

Let us now introduce yet another class of concave extensions associated with the PBNF ψ. For $k = 1, 2, \ldots, m$, let p_k be any linear function such that $b_k \left(\prod_{i \in A_k} x_i \right) \left(\prod_{j \in B_k} \overline{x}_j \right) \le p_k(X)$ for all $X \in \mathcal{B}^n$. Then, the linear function

$$p(X) = b_0 + \sum_{k=1}^{m} p_k(X) \quad \text{for all } X \in U^n \tag{13.30}$$

is called a *paved upper-plane* of ψ (and of the function f represented by ψ). Clearly, a paved upper-plane is a linear majorant of f. Let now $\mathcal{P}$ denote the set of all paved upper-planes of f. The *paved upper-plane extension* of f associated with the PBNF ψ is the function ψ^{pup} defined by

$$\psi^{pup}(X) = \min_{p \in \mathcal{P}} p(X) \quad \text{for all } X \in U^n. \tag{13.31}$$

Our next result shows that, in spite of their very different definitions, ψ^{std} and ψ^{pup} turn out to be identical.

Theorem 13.9. *The standard extension ψ^{std} and the paved upper-plane extension ψ^{pup} associated with a same PBNF ψ coincide on U^n.*

Proof. Let $p(X)$ be a paved upper-plane of ψ given by (13.30). Since each term p_k $(k = 1, 2, \ldots, m)$ is a concave majorant of the corresponding term of ψ, it follows from Lemma 13.3 that $b_k g_k(X) \le p_k(X)$, and hence, $\psi^{std}(X) \le p(X)$ for all $X \in U^n$. So, $\psi^{std} \le \psi^{pup}$ on U^n.

To see that $\psi^{pup} \le \psi^{std}$ on U^n, fix $X^* \in U^n$ and consider the paved upper-plane $p(X)$ given by (13.30), where for each $k = 1, 2, \ldots, m$:

(a) $p_k = b_k x_i$ \qquad if $b_k > 0$ and $g_k(X^*) = x_i^*, i \in A_k$;
(b) $p_k = b_k(1 - x_j)$ \quad if $b_k > 0$ and $g_k(X^*) = 1 - x_j^*, j \in B_k$;
(c) $p_k = 0$ \qquad\qquad if $b_k < 0$ and $g_k(X^*) = 0$;
(d) $p_k = b_k(1 - |A_k| + \sum_{i \in A_k} x_i - \sum_{j \in B_k} x_j)$ if $b_k < 0$ and $g_k(X^*) > 0$.

(Apply an arbitrary tie-breaking rule to select the indices i and j if either (a) or (b) are ambiguous.) This construction is such that $p(X^*) = \psi^{std}(X^*)$, and hence, $\psi^{pup}(X^*) \le \psi^{std}(X^*)$. $\qquad\square$

Theorem 13.9 is due to Crama [227]. It generalizes a sequence of previous results by Hammer, Hansen, and Simeone [440]; Hansen, Lu, and Simeone [471]; Adams and Dearing [6], and so on, showing that the maximum of ψ^{std} and the maximum of ψ^{pup} coincide on U^n.

13.3.3 The Lovász extension

Consider again a pseudo-Boolean function f on B^n and its polynomial expression

$$f(X) = \sum_{A \in \mathcal{P}(N)} c(A) \left(\prod_{i \in A} x_i \right). \tag{13.32}$$

In this section, we assume for simplicity of notations that $f(0, 0, \ldots, 0) = 0$, that is, $c(\emptyset) = 0$.

Definition 13.8. *The Lovász extension of f is the extension f^L defined by*

$$f^L(X) = \sum_{A \in \mathcal{P}(N)} c(A) \min_{i \in A} x_i \quad \text{for all } (x_1, x_2, \ldots, x_n) \in U^n. \tag{13.33}$$

This extension was introduced by Lovász in [624]; see also [625]. Observe that, if $c(A) \geq 0$ for all $A \subseteq \{1, 2, \ldots, n\}$ such that $|A| \geq 2$, then f^L coincides with the standard extension associated with the polynomial representation of f, and it is concave. In general, however, f^L is neither concave nor convex on U^n, as illustrated by the next example.

Example 13.5. *The Lovász extension of* $f(x, y, z) = xy - xz$ *is the function* $f^L = \min(x, y) - \min(x, z)$, *which is neither concave nor convex on* U^3 *since*

$$\tfrac{1}{2} = \tfrac{1}{2} f^L(1, 1, 0) + \tfrac{1}{2} f^L(0, 1, 1) > f^L(\tfrac{1}{2}, 1, \tfrac{1}{2}) = 0,$$

and

$$-\tfrac{1}{2} = \tfrac{1}{2} f^L(1, 0, 1) + \tfrac{1}{2} f^L(0, 1, 1) < f^L(\tfrac{1}{2}, \tfrac{1}{2}, 1) = 0. \qquad \square$$

The following discussion provides a different perspective on the Lovász extension. For a set $A \subseteq \{1, 2, \ldots, n\}$, $A \neq \emptyset$, denote by $m(A)$ the smallest element in A: $m(A) = \min\{i \mid i \in A\}$. Let $S = \{X \in U^n \mid x_1 \leq x_2 \leq \ldots \leq x_n\}$ and observe that S is a *simplex*, that is, S is a full-dimensional convex bounded polyhedron with $n + 1$ vertices. Its vertices are exactly the points $(0, 0, \ldots, 0, 0, 0)$, $(0, 0, \ldots, 0, 0, 1)$, $(0, 0, \ldots, 0, 1, 1)$, ..., $(1, 1, \ldots, 1, 1, 1)$.

Consider now the restriction of f^L to the simplex S. This function, that we denote by f_S^L, is linear on S: Indeed, for all $X \in S$, Definition 13.8 yields

$$f^L(X) = f_S^L(X) = \sum_{A \in \mathcal{P}(N)} c(A) x_{m(A)}.$$

Even more, since f_S^L coincides with f at the $n + 1$ vertices of S, it follows that f_S^L actually is the *unique* linear extension of f on S.

This reasoning is easily generalized. For an arbitrary permutation π of $\{1, 2, \ldots, n\}$, let $S(\pi)$ be the simplex $S(\pi) = \{X \in U^n \mid x_{\pi(1)} \leq x_{\pi(2)} \leq \ldots \leq x_{\pi(n)}\}$ and let $f_{S(\pi)}^L$ be the restriction of f^L to $S(\pi)$. Then, $f_{S(\pi)}^L$ is the *unique* linear extension of f on $S(\pi)$. Moreover, since the cube U^n is covered by the family of simplices

$$\mathcal{S} = \{S(\pi) \mid \pi \text{ is a permutation of } \{1, 2, \ldots, n\}\},$$

it follows that f^L is the unique extension of f that is linear on every member of $\mathcal{S}$.

In order to obtain an analytical expression of the function $f_{S(\pi)}^L$, let us introduce the following notation: For $1 \leq k \leq n$, let

$$E^{\pi, k} = e_{\pi(k)} + e_{\pi(k+1)} + \ldots + e_{\pi(n)}.$$

We also let $E^{\pi, n+1} = (0, \ldots, 0)$, so that $E^{\pi, 1}, E^{\pi, 2}, \ldots, E^{\pi, n+1}$ are exactly the vertices of the simplex $S(\pi)$.

Theorem 13.10. *For every permutation* π *of* $\{1, 2, \ldots, n\}$ *and for every* $X \in S(\pi)$,

$$f_{S(\pi)}^L(X) = \sum_{k=1}^{n} (x_{\pi(k)} - x_{\pi(k-1)}) f(E^{\pi, k}), \qquad (13.34)$$

where $x_{\pi(0)} = 0$ *by convention.*

Proof. Since the right-hand side of (13.34) defines a linear function, it suffices to verify that this function coincides with f at every vertex of $S(\pi)$, which is true by construction. ☐

Equation (13.34) leads to the definition of f^L originally proposed by Lovász in [624, 625] (see also the end-of-chapter exercises). As observed by Singer [836], this approach to the construction of extensions can be further generalized by considering different coverings of U^n by collections of simplices.

13.4 Pseudo-Boolean optimization

We refer to the optimization of pseudo-Boolean functions over subsets of $\mathcal{B}^n = \{0,1\}^n$ as *pseudo-Boolean optimization* or *nonlinear* 0–1 *optimization*. This important field of research was popularized by Hammer and Rudanu [460], and is surveyed in [127, 469]. We mostly restrict ourselves here to a discussion of the *unconstrained maximization* problem

$$\text{maximize } f(X) \text{ subject to } X \in \mathcal{B}^n, \tag{13.35}$$

and we only mention a few fundamental results about it.

Remark. Some authors have recently started to use the term "pseudo-Boolean optimization problems" to designate 0–1 *linear* programming problems of the form (13.4)–(13.6), possibly subject to inequality constraints; see Eén and Sörensson [290]; Manquinho and Roussel [667], and so on. This usage is likely to create confusion with the classically accepted definition of pseudo-Boolean optimization problems, and we do not encourage it. ☐

Observe that the unconstrained problem (13.35) is NP-hard even when f is quadratic, since it subsumes several hard combinatorial problems, like max-cut, weighted stability, MAX 2-SAT, or 0–1 linear programming (see Section 13.1). We return to quadratic optimization in Section 13.6.1. On the other hand, problem (13.35) turns out to be easy when f is linear: Indeed, if

$$f(X) = \sum_{i=1}^{n} w_i x_i,$$

then the maximum of f is attained at any point $X^* \in \mathcal{B}^n$ such that

$$\begin{aligned} x_i^* &= 1 \quad \text{when } w_i > 0, \\ x_i^* &= 0 \quad \text{when } w_i < 0. \end{aligned} \tag{13.36}$$

13.4.1 Local optima

We start with a few definitions.

Definition 13.9. *Two points $X^*, Y^* \in \mathcal{B}^n$ are neighbors if they differ in exactly one component, that is, if they correspond to adjacent vertices of the unit hypercube.*

If f is a pseudo-Boolean function on $\mathcal{B}^n$, then $X^* \in \mathcal{B}^n$ is a local maximum of f if

$$f(X^*) \geq f(Y^*) \text{ for all neighbors } Y^* \text{ of } X^*.$$

Definition 13.10. *For $i = 1, 2, \ldots, n$, the i-th derivative of f is the pseudo-Boolean function*

$$\Delta_i f = f(x_1, \ldots, x_{i-1}, 1, x_{i+1}, \ldots, x_n) - f(x_1, \ldots, x_{i-1}, 0, x_{i+1}, \ldots, x_n). \quad (13.37)$$

Since $\Delta_i f$ does not depend on x_i, we may want to look at it as a function on $\mathcal{B}^n$ or on $\mathcal{B}^{n-1}$, as the context requires. It is easy to check that if the (unique) polynomial expression of f is written as

$$f(x_1, x_2, \ldots, x_n) = x_i \, g(x_1, \ldots, x_{i-1}, x_{i+1}, \ldots, x_n) + h(x_1, \ldots, x_{i-1}, x_{i+1}, \ldots, x_n), \quad (13.38)$$

where the polynomials g and h do not depend on x_i, then g is the (unique) polynomial expression of $\Delta_i f$. In other words, the polynomial expression of $\Delta_i f$ is obtained by writing the partial derivative $\frac{\partial f}{\partial x_i}$ of the polynomial expression of f with respect to x_i.

Fortet [343] and Hammer and Rudeanu [460] observed that the local maxima of a function are characterized by a system of implications involving its derivatives (compare with (13.36)).

Theorem 13.11. *If f is a pseudo-Boolean function on $\mathcal{B}^n$, then $X^* \in \mathcal{B}^n$ is a local maximum of f if and only if the following conditions hold for $i = 1, 2, \ldots, n$:*

$$\begin{aligned} x_i^* &= 1 \quad \text{when } \Delta_i f(X^*) > 0, \\ x_i^* &= 0 \quad \text{when } \Delta_i f(X^*) < 0. \end{aligned} \quad (13.39)$$

Proof. This is easily derived from (13.37) or from (13.38). □

Let now M_i be an arbitrary upper bound on $|\Delta_i f|$ (for instance, the sum of the absolute values of all coefficients in the polynomial representation of $\Delta_i f$). Then, it is easily seen that an equivalent characterization of the local maxima of f is given by the system of inequalities

$$M_i (x_i - 1) \leq \Delta_i f \leq M_i x_i, \quad \text{for } i = 1, 2, \ldots, n. \quad (13.40)$$

Thus, in principle, a local maximum of f could be obtained by finding a 0–1 solution of the system (13.40). This may be a difficult task in itself. It should be observed, however, that the system (13.40) is linear when f is quadratic and that it may lend itself to an easier treatment in this special case.

A local maximum of f can be found by any simple *local search procedure* starting from an arbitrary 0–1 point and moving from neighbor to neighbor as long as this improves the value of the function. Such algorithms tend to work very fast in practice; see, for instance, Boros, Hammer, and Tavares [136]; Boykov, Veksler, and Zabih [147]; Davoine, Hammer, and Vizvári [262]; Hansen and Jaumard [468]; Hvattum, Løkketangen, and Glover [513]; Lodi, Allemand, and Liebling [620], Merz and Freisleben [681], and so on.

From a theoretical perspective, however, things are not so nice. Indeed, it can be shown that in order to find a local maximum of a pseudo-Boolean function of n variables, such local search procedures may require a number of steps that grows exponentially with n (see Emamy-K. [311]; Hammer, Simeone, Liebling, and de Werra [464]; Hoke [496]; Tovey [866, 867, 868] for related investigations) or with the encoding size of the polynomial expression of f (see Schäffer and Yannakakis [806]). Moreover, Schäffer and Yannakakis [806] proved that computing a local maximum of a *quadratic* pseudo-Boolean function belongs to a class of hard (so-called PLS-complete), and likely intractable, local search problems (see also Pardalos and Jha [730]).

Finally, it should be observed that the value of f may be arbitrarily worse in a local maximum of f than in its global maximum (see Exercise 5 at the end of the chapter).

13.4.2 An elimination algorithm for global optimization

Hammer, Rosenberg and Rudeanu [458, 460] described a combinatorial variable elimination algorithm that finds a global maximum of a pseudo-Boolean function. The following streamlined version and an efficient implementation of this algorithm have been proposed by Crama, Hansen, and Jaumard [235].

Let $f_0(x_1, x_2, \ldots, x_n)$ be the function to be maximized. We can write

$$f_0(x_1, x_2, \ldots, x_n) = x_1 \Delta_1(x_2, x_3, \ldots, x_n) + h(x_2, x_3, \ldots, x_n),$$

where Δ_1 and h do not depend on x_1. As a slight extension of Theorem 13.11, it is easy to see that there exists a global maximum of f_0, say $(x_1^*, x_2^*, \ldots, x_n^*)$, with the property that

$$x_1^* = 1 \text{ if and only if } \Delta_1(x_2^*, x_3^*, \ldots, x_n^*) > 0. \tag{13.41}$$

This observation suggests a function $t_1(x_2, x_3, \ldots, x_n)$ defined as follows:

$$\begin{aligned} t_1(x_2, x_3, \ldots, x_n) &= \Delta_1(x_2, x_3, \ldots, x_n) \quad \text{if } \Delta_1(x_2, x_3, \ldots, x_n) > 0, \\ &= 0 \qquad\qquad\qquad\quad \text{otherwise.} \end{aligned} \tag{13.42}$$

Then, setting $f_1 = t_1 + h$, we have reduced the maximization of the original function f_0 in n variables to the maximization of f_1, which only depends on $n - 1$ variables: Indeed, if $(x_2^*, x_3^*, \ldots, x_n^*)$ is a maximum of f_1, then setting x_1^* to either 0 or 1 according to rules (13.41) yields a maximum of f_0.

Repeating n times this elimination process produces a sequence of pseudo-Boolean functions $f_0, f_1, \ldots, f_n$, where f_i depends on $n - i$ variables, and eventually allows us to determine a (global) maximum of f_0 by backtracking. (Note the analogy with the elimination techniques for the solution of Boolean equations presented in Chapter 2, Section 2.6, which originally inspired the development of this procedure.)

Assuming that f_0 is given in pseudo-Boolean normal form (13.15), the expensive step in the elimination process is to deduce a PBNF of f_{i+1} from a PBNF of f_i, for $i = 0,1,\ldots,n-1$. An efficient implementation of this step has been proposed in [235], where it is also proved that the elimination algorithm runs in polynomial time for a special class of pseudo-Boolean functions associated with graphs of *bounded tree-width*.

13.4.3 *Extensions and relaxations*

If g is an arbitrary extension of the pseudo-Boolean function f over the cube $U^n = [0,1]^n$, then $\max_{X \in U^n} g(X)$ is an upper bound for $\max_{X \in B^n} f(X)$. We now examine some properties of this bound for different families of extensions.

The polynomial extension

As observed by Rosenberg [789], the multilinear polynomial extension f^{pol} has the attractive feature that its maximum is attained at a vertex of the hypercube $[0,1]^n$ and hence, that this maximum coincides with the maximum of f.

Theorem 13.12. *For every pseudo-Boolean function f on B^n,*

$$\max_{X \in B^n} f(X) = \max_{X \in U^n} f^{pol}(X).$$

Proof. Let X^* denote a maximizer of f^{pol} on U^n and consider an arbitrary index $i \in \{1,2,\ldots,n\}$. Write f^{pol} as

$$f^{pol}(x_1,x_2,\ldots,x_n) = x_i \, g(x_1,\ldots,x_{i-1},x_{i+1},\ldots,x_n) + h(x_1,\ldots,x_{i-1},x_{i+1},\ldots,x_n),$$
$$(13.43)$$

where the polynomials g and h do not depend on x_i. The function

$$p(x_i) = x_i \, g(x_1^*,\ldots,x_{i-1}^*,x_{i+1}^*,\ldots,x_n^*) + h(x_1^*,\ldots,x_{i-1}^*,x_{i+1}^*,\ldots,x_n^*)$$

is linear in x_i, so that the maximum of $p(x_i)$ over $U = [0,1]$ is attained when $x_i = 0$ or when $x_i = 1$. Hence, if $0 < x_i^* < 1$, we can replace x_i^* by a 0–1 value without changing the value of f^{pol}. □

Note that Theorem 13.12 can alternatively be viewed as a corollary of Theorem 13.8: Indeed, for every point $(p_1, p_2, \ldots, p_n) \in U^n$, $f^{pol}(p_1, p_2, \ldots, p_n)$ is the expected value of f with respect to an appropriate probability distribution on B^n; hence, by well-known properties of the expectation, $\min_{X \in B^n} f(X) \le f^{pol}(p_1, p_2, \ldots, p_n) \le \max_{X \in B^n} f(X)$.

The proof of Theorem 13.12 actually implies that "rounding" a fractional point to a "better" 0–1 point can be performed efficiently. This result was already anticipated in earlier chapters of the book (see, e.g., Theorems 2.26, 2.27, and 2.28 in

Section 2.11.4), and was put to systematic use in Boros and Hammer [127], Boros and Prékopa [145], and so on.

Theorem 13.12 also suggests that continuous global optimization techniques can be applied to f^{pol} to compute the maximum of f. This approach has not proved computationally efficient in past experiments, but it remains conceptually valuable.

Linearization and concave extensions

A classical approach to pseudo-Boolean optimization consists in transforming the problem $\max\{f(X) : X \in \{0,1\}^n\}$ into an equivalent linear 0–1 programming problem by substituting a variable y_k for the kth monomial T_k of a PBNF representation, and by setting up a collection of linear constraints that enforce the equality $y_k = T_k$. More precisely, the following result can be traced to papers by Dantzig [256], Fortet [342, 343], and Glover and Woolsey [387]; see Hansen, Jaumard, Mathon [469] for additional references.

Theorem 13.13. *If the pseudo-Boolean function f is represented by the PBNF*

$$\psi(x_1, x_2, \ldots, x_n) = b_0 + \sum_{k=1}^{m} b_k \left(\prod_{i \in A_k} x_i \right) \left(\prod_{j \in B_k} \overline{x}_j \right), \qquad (13.44)$$

where $b_0, b_1, \ldots, b_m \in \mathbb{R}$, $A_k \cap B_k = \emptyset$, and $A_k \cup B_k \neq \emptyset$ for $k = 1, 2, \ldots, m$, then the maximum of f over $\mathcal{B}^n$ is equal to the optimal value of the 0–1 linear programming problem

$$\text{maximize} \quad b_0 + \sum_{k=1}^{m} b_k y_k \qquad (13.45)$$

$$\text{subject to} \quad y_k \leq x_i, \quad i \in A_k, k = 1, 2, \ldots, m, b_k > 0; \qquad (13.46)$$

$$y_k \leq 1 - x_j, \quad j \in B_k, k = 1, 2, \ldots, m, b_k > 0; \qquad (13.47)$$

$$1 - |A_k| + \sum_{i \in A_k} x_i - \sum_{j \in B_k} x_j \leq y_k, \quad k = 1, 2, \ldots, m, b_k < 0; \qquad (13.48)$$

$$x_i \in \{0, 1\}, \quad i = 1, 2, \ldots, n; \qquad (13.49)$$

$$y_k \in \{0, 1\}, \quad k = 1, 2, \ldots, m. \qquad (13.50)$$

Proof. In every optimal solution $(X^*, Y^*) \in \{0,1\}^{n+m}$ of (13.45)–(13.50), variable y_k^* takes value 1 if and only if $(\prod_{i \in A_k} x_i^*)(\prod_{j \in B_k} \overline{x}_j^*) = 1$. $\square$

This 0–1 linear model can be handled, in principle, by any algorithm for the solution of integer programming problems. The analysis of its facial structure has been been initiated by Balas and Mazzola [44, 45]. Its continuous relaxation, meaning the linear programming problem obtained after replacing the integrality requirements (13.49) and (13.50) by the weaker constraints $0 \leq x_i \leq 1$ $(i = 1, 2, \ldots, n)$ and $0 \leq y_k \leq 1$ $(k = 1, 2, \ldots, m)$, yields an easily computable upper bound W^{std}

on the maximum of f. It is easy to see that this upper bound is exactly the maximum over U^n of the concave standard extension ψ^{std} introduced in Section 13.3. Properties of the bound W^{std} have been investigated by Hammer, Hansen, and Simeone [440] and in a series of subsequent papers; see Crama [227] for a brief account and Section 13.6.1 for related considerations. Compare also with Theorem 2.26 and Theorem 2.28 in Section 2.11.4, where this relaxation was investigated in connection with the MAXIMUM SATISFIABILITY problem.

The Lovász extension

An analog of Rosenberg's Theorem 13.12 holds for the Lovász extension f^L.

Theorem 13.14. *For every pseudo-Boolean function f on B^n,*

$$\max_{X \in B^n} f(X) = \max_{X \in U^n} f^L(X).$$

Proof. This follows from Theorem 13.10, which shows that the Lovász extension is linear on every simplex $S(\pi)$: Hence, its maximum is necessarily attained at a vertex of B^n.   □

13.4.4 Posiform transformations and conflict graphs

In view of Theorem 13.2, every pseudo-Boolean optimization problem can be reduced to the optimization of a posiform

$$\psi = b_0 + \sum_{k=1}^{m} b_k T_k = b_0 + \sum_{k=1}^{m} b_k \left(\prod_{i \in A_k} x_i \right) \left(\prod_{j \in B_k} \overline{x}_j \right), \qquad (13.51)$$

where $A_k \cap B_k = \emptyset$, $A_k \cup B_k \neq \emptyset$, and $b_k > 0$ for $k = 1, 2, \ldots, m$. It turns out that both the minimization and the maximization of posiforms have natural connections with other fundamental combinatorial optimization problems.

First, Theorems 2.14 and 2.26 show that DNF equations and maximum satisfiability problems are easily expressed as posiform minimization problems. Conversely, a straightforward extension of Theorem 2.26 shows that *every* posiform minimization problem can be viewed as a maximum satisfiability problem: Indeed, minimizing a posiform ψ precisely consists in finding a point $X^* \in B^n$ that cancels (or "satisfies") as many terms as possible in ψ.

In this minimization setting, a useful remark is that, if (13.51) is an arbitrary posiform representation of a function f, then the free term b_0 is a *lower bound* on the global minimum of f (since the remaining terms are always nonnegative). In fact, for any function f, there always exists a posiform such that the free term b_0 is exactly equal to $\min_{X \in B^n} f(X)$ (we leave the proof of this claim as an exercise for the reader). Approaches to pseudo-Boolean minimization based on this observation have been developed for instance by Bourjolly, Hammer, Pulleyblank, and Simeone [146] and Hammer, Hansen, and Simeone [440]. The idea is here to "squeeze out" the highest possible constant b_0 by successive transformations of a posiform.

Let us now turn to the posiform maximization problem. As observed by Hammer [437, 465], this problem bears a fruitful relation to the maximum weighted stability problem described in Application 13.1. In order to discuss this relation, we first define the concept of *conflict graph* (conflict graphs were introduced in a slightly different framework in Chapter 5; see also [65, 69, 230, 461], etc.). Consider again the posiform (13.51), and assume for simplicity that $b_0 = 0$, as this assumption entails no loss of generality. We say that two terms T_k and T_ℓ *conflict* if $T_k T_\ell \equiv 0$ (that is, if a same variable appears both in T_k and T_ℓ, once complemented and once uncomplemented). Now, the conflict graph of ψ is the graph $G(\psi) = (V, E)$, where $V = \{1, 2, \ldots, m\}$, and where $(k, \ell) \in E$ if and only if T_k and T_ℓ conflict, for $k, \ell \in V$. We say that b_k is the *weight* of vertex k, for $k = 1, 2, \ldots, m$. Finally, we let $\alpha(\psi)$ denote the weight of a maximum weighted stable set in $G(\psi)$:

$$\alpha(\psi) = \max \left\{ \sum_{k \in S} b_k \mid S \text{ is a stable set of } G(\psi) \right\}.$$

Hammer [437, 465] proved:

Theorem 13.15. *For every posiform ψ on $\mathcal{B}^n$,*

$$\max_{X \in \mathcal{B}^n} \psi(X) = \alpha(\psi).$$

Proof. For any point $X^* \in \mathcal{B}^n$, let us observe first that the set

$$S(X^*) = \{ k \in \{1, 2, \ldots, m\} \mid T_k(X^*) = 1 \}$$

is a stable set of the graph $G(\psi)$. Indeed, no two terms in $S(X^*)$ can conflict, since otherwise, at least one of them would vanish at the point X^*. Hence,

$$\psi(X^*) = \sum_{k \in S(X^*)} b_k \leq \alpha(\psi) \quad \text{for all } X^* \in \mathcal{B}^n.$$

Conversely, if $S \subseteq V$ is a stable set of $G(\psi)$, then the terms associated with the vertices in S do not conflict, and thus all literals appearing in these terms can simultaneously be made equal to 1. In other words, for any stable set $S \subseteq V$, there exists a point $X^* \in \mathcal{B}^n$ such that $T_k(X^*) = 1$ for all $k \in S$. Applying this observation to a stable set S^* of maximum weight, we obtain

$$\alpha(\psi) = \sum_{k \in S^*} b_k = \sum_{k \in S^*} b_k T_k(X^*) \leq \psi(X^*) \leq \max_{X \in \mathcal{B}^n} \psi(X).$$

$\square$

So, every posiform maximization problem can be easily reduced to a graph stability problem. The converse statement is true as well, in view of the formulation (13.2) and of Theorem 13.2. In fact, another interesting transformation of the weighted stable set problem to posiform maximization can also be inferred from the following observations:

First, for a posiform ψ on $\mathcal{B}^n$ given by (13.51), consider an arbitrary variable x_i and define the sets

$$P_i = \{k \in \{1, 2, \ldots, m\} \mid i \in A_k\} \text{ and } N_i = \{k \in \{1, 2, \ldots, m\} \mid i \in B_k\}$$

(possibly $P_i = \emptyset$ or $N_i = \emptyset$). By definition, in the conflict graph $G(\psi)$, every vertex of P_i is linked to every vertex of N_i; in other words, the graph $H_i = (V_i, E_i)$ where $V_i = P_i \cup N_i$ and

$$E_i = \{(k, \ell) \in E \mid k \in P_i, \ell \in N_i\}$$

is a complete bipartite subgraph of $G(\psi)$. Moreover, $E = \bigcup_{i=1}^{n} E_i$, meaning that the edge-set of $G(\psi)$ is covered by the collection of complete bipartite graphs $H_1, H_2, \ldots, H_n$ associated with the variables of ψ.

Hammer [437] observed that this construction can be reversed and established the following result (recall that α_G denotes the weight of a maximum weighted stable set of G).

Theorem 13.16. *For every graph $G = (V, E)$ and vertex weights $w : V \to \mathbb{R}^+$, there exists a posiform ψ such that $G = G(\psi)$ and $\alpha_G = \max_{X \in \mathcal{B}^n} \psi(X)$.*

Proof. Consider any collection $H_1, H_2, \ldots, H_n$ of complete bipartite graphs covering the edges of G, and let $H_i = (P_i \cup N_i, E_i)$; thus, every edge of H_i has an endpoint in P_i and the other endpoint in N_i, and $E = \bigcup_{i=1}^{n} E_i$. If I is the set of isolated vertices of G, that is, $I = \{k \in V \mid \text{for all } e \in E, k \notin e\}$, and if I is nonempty, then assume, without loss of generality, that $P_n = I$ and $N_n = E_n = \emptyset$.

For $i = 1, 2, \ldots, n$, associate a variable x_i with the subgraph H_i, and for each $k \in V$ let

$$A_k = \{i \in \{1, 2, \ldots, n\} \mid k \in P_i\},$$
$$B_k = \{i \in \{1, 2, \ldots, n\} \mid k \in N_i\},$$
$$b_k = w(k).$$

With these definitions, if ψ is the posiform given by (13.51), then it is easy to check that $G = G(\psi)$. The equality $\alpha_G = \max_{X \in \mathcal{B}^n} \psi(X)$ follows from Theorem 13.15. $\square$

The relations between posiform maximization and weighted stability described in Theorems 13.15 and 13.16 have been exploited by several researchers. Ebenegger, Hammer, and de Werra [285], in particular, have proposed a specific posiform transformation technique leading to an algorithm called *struction* for the weigthed stability problem. Extensions and applications of struction to various classes of graphs have been investigated in [14, 453, 491], and so on. We refer the reader to these publications for more details.

13.5 Approximations

In this section, we briefly discuss the problem of approximating a pseudo-Boolean f on $\mathcal{B}^n$ by a "simpler" function. Hammer and Holzman [441] considered the specific version of this problem in which the objective is to find a function g of degree k, for a predetermined value of k, which minimizes the L_2-norm

$$\sum_{X \in \mathcal{B}^n} [f(X) - g(X)]^2. \tag{13.52}$$

When $k = 1$, g is the *best linear L_2-approximation* of f and we denote it by $L(f)$. Let us assume that f is represented by the polynomial expression (13.12). Then, in order to compute $L(f)$, it is sufficient to know how to compute the best linear approximation of a monomial. Indeed, $L(f)$ can be viewed as the projection of f on the subspace of linear functions, and hence, there holds

$$L(f) = \sum_{A \in \mathcal{P}(N)} c(A) L\left(\prod_{i \in A} x_i\right).$$

Hammer and Holzman [441] showed that

$$L\left(\prod_{i \in A} x_i\right) = \frac{1}{2^{|A|}} \left(1 - |A| + 2\sum_{i \in A} x_i\right) \quad \text{for all } A \subseteq N.$$

The best quadratic, cubic, and higher-order L_2-approximations can be derived by similar approaches; see also Ding, Lax, Chen, and Chen [273]; Ding, Lax, Chen, Chen, and Marx [274]; Grabisch, Marichal, and Roubens [407]; or Zhang and Rowe [936] for extensions of these results.

Important game-theoretical applications of best L_2-approximations consist in finding the Banzhaf indices of the players of a simple game, or the Shapley values of the players of an n-person characteristic function game. As shown in [441], these indices are simply the coefficients of best (weighted) linear L_2-approximations of the pseudo-Boolean functions describing the games.

Another application of these results allows the efficient determination of excellent heuristic solutions of unconstrained pseudo-Boolean optimization problems, as shown by Davoine, Hammer, and Vizvári [262]. Zhang and Rowe [936] discuss the relevance of pseudo-Boolean approximations for the development of evolutionary algorithms.

Finally, we note that (different types of) approximations of pseudo-Boolean functions are also of interest in the theory of probabilistic databases, where they can be used to track the most influential facts in the derivation of a conclusion; see Ré and Suciu [780].

13.6 Special classes of pseudo-Boolean functions

Many special classes of pseudo-Boolean functions can be defined by analogy with their Boolean counterparts: quadratic, monotone, supermodular, and so on.

13.6.1 Quadratic functions and quadratic 0-1 optimization

Quadratic pseudo-Boolean functions, or pseudo-Boolean functions of degree (at most) 2, have been the object of numerous investigations; surveys are provided by Boros and Hammer [127] and by Hammer and Simeone [463].

Quadratic 0–1 optimization, in particular, is an important special case of non-linear 0–1 optimization, both because numerous applications appear in this form (see Applications 13.1, 13.5, 13.6, etc.), and because the general case is easily reduced to it. This reduction can be performed in various ways. For instance, Theorems 13.2 and 13.15 suggest the following procedure: In order to maximize a pseudo-Boolean function f, produce a posiform of f, build the conflict graph G of this posiform, and formulate the weighted stability problem associated with G as a quadratic 0–1 maximization problem.

Another efficient transformation was proposed by Rosenberg [790]. It relies on the substitution of the product of any two variables by a new variable, and the addition of appropriate penalty terms which, at every optimal point, force the new variable to take the value of the product of the two substituted variables. More precisely:

Theorem 13.17. *Let f be a pseudo-Boolean function represented by the polynomial expression*

$$f(x_1, x_2, \ldots, x_n) = \sum_{k=1}^{m} c_k \left(\prod_{i \in A_k} x_i \right),$$

assume that $|A_1| \geq 2$, and select $j, \ell \in A_1$. Let y be a new 0-1 variable, different from $x_1, x_2, \ldots, x_n$, let M be a positive constant, and define

$$g(x_1, x_2, \ldots, x_n, y) = c_1 \left(\prod_{i \in A_1 \setminus \{j, \ell\}} x_i \right) y + \sum_{k=2}^{m} c_k \left(\prod_{i \in A_k} x_i \right)$$

$$- M(x_j x_\ell - 2x_j y - 2x_\ell y + 3y).$$

If M is large enough, then the maximum value of f over $\mathcal{B}^n$ is equal to the maximum value of g over $\mathcal{B}^{n+1}$.

Proof. Consider any point $(X^*, y^*) \in \mathcal{B}^{n+1}$. It is easy to check that the expression $x_j^* x_\ell^* - 2x_j^* y^* - 2x_\ell^* y^* + 3y^*$ is equal to 0 when $y^* = x_j^* x_\ell^*$, and is strictly positive otherwise.

Assume now that M is large (say, $M > |c_1|$). Then, $f(X^*) = g(X^*, y^*)$ for all $(X^*, y^*) \in \mathcal{B}^{n+1}$ such that $y^* = x_j^* x_\ell^*$, and $g(X^*, y^*) < f(X^*)$ for all other points in $\mathcal{B}^{n+1}$. The claim follows directly.   □

Note that, after applying the transformation described in Theorem 13.17, the degree of the first term of g is equal to $|A_1| - 1$. Thus, applying repeatedly this transformation eventually yields a function of degree 2 which has the same maximum value as f.

It is interesting to observe that this argument is analogous to the proof that every Boolean DNF equation is equivalent to a DNF equation of degree 3 (see Theorem 2.4 in Chapter 2). Actually, in many ways, it can be said that quadratic 0–1 optimization problems play the same fundamental role with respect to pseudo-Boolean optimization problems, as DNF equations of degree 3 (or 3-SAT problems) with respect to general DNF equations (or satisfiability problems).

Other transformations of pseudo-Boolean optimization problems to the quadratic case have been proposed and have been shown to be computationally effective by Buchheim and Rinaldi [164, 165].

Hammer, Hansen, and Simeone [440] showed that, for every quadratic pseudo-Boolean function f, one can efficiently construct a linear function

$$l(x_1, x_2, \ldots, x_n) = l_0 + \sum_{j=1}^{n} l_j x_j,$$

called the *roof dual* of f, that majorizes $f(x_1, x_2, \ldots, x_n)$ in every binary point and that has the following property of *strong persistency*: If l_j is strictly positive (respectively, negative), then x_j is equal to 1 (respectively, 0) in every maximizer of f. Thus, in some cases, strong persistency allows the determination of the optimal values of a subset of variables.

Note that the maximum of $l(X)$ over $\mathcal{B}^n$ is simply equal to $\rho(f) = l_0 + \sum_{j=1}^{n} \max(l_j, 0)$, and $\rho(f)$ provides an upper-bound on the maximum of f over $\mathcal{B}^n$. Hammer, Hansen, and Simeone [440] proved that $\rho(f)$ is exactly the optimal value W^{std} of the continuous relaxation of the 0–1 linear programming model (13.45)–(13.50) associated with the polynomial expression of f or with any posiform of f.

Moreover, the equality $\rho(f) = \max_{X \in \mathcal{B}^n} f(X)$ holds if and only if an associated quadratic Boolean function is consistent; therefore, the optimality of $\rho(f)$ can be tested in polynomial time (see Exercise 10 in Chapter 5).

The determination of the roof dual $l(X)$ was derived in [440] from the solution of the continuous relaxation of the model (13.45)–(13.50); Boros, Hammer, and Sun [125, 134] showed that the computation of the roof dual can be efficiently reduced to a maximum flow problem. We refer again to the survey by Boros and Hammer [127] for additional details, as well as to Boros, Crama, and Hammer [113, 114] or Boros, Lari, and Simeone [143] for extensions of roof duality theory.

The convex hull of the set of 0–1 solutions of (13.46)–(13.50) is called the *quadric polytope*, or *correlation polytope*. Its facial structure was investigated by Padberg [723] and by several other authors; see also Deza and Laurent [270] and Laurent and Rendl [601].

There is a huge number of papers discussing exact or heuristic optimization algorithms for quadratic pseudo-Boolean functions, and it is impossible to cite them all here. Among recent ones, let us only mention a variety of approaches by Billionnet and Elloumi [85]; Boros, Hammer, and Tavares [136]; Glover and Hao [386]; Gueye and Michelon [420]; Hansen and Meyer [472]; Lodi, Allemand, and

Liebling [620]; Merz and Freisleben [681]; Palubeckis [724], and so on, as well
as efficient implementations of the roof duality computations in the framework
of computer vision applications by Kolmogorov and Rother [576] and Rother,
Kolmogorov, Lempitsky, and Szummer [794].

13.6.2 Monotone functions

Definition 13.11. *A pseudo-Boolean function f on $\mathcal{B}^n$ is called* monotone
nondecreasing *if*

$$f(X) \leq f(Y) \text{ for all } X, Y \in \mathcal{B}^n \text{ such that } X \leq Y,$$

and it is called monotone nonincreasing *if*

$$f(X) \geq f(Y) \text{ for all } X, Y \in \mathcal{B}^n \text{ such that } X \leq Y.$$

As noted in Application 13.7, monotone nondecreasing functions such that
$f(0,\ldots,0) = 0$ and $f(1,\ldots,1) = 1$ have also been studied in the literature under
the names of Choquet capacities, belief functions, fuzzy measures, and so on.

Example 13.6. *The function $f_1(x,y) = 1 + 2x + 2y - xy$ is monotone nondecreasing, while $f_2(x,y) = 3 - y - xy$ is monotone nonincreasing.* $\qquad\square$

Just as in the case of functions of real variables, monotonicity properties can
be related to the signs of first-order derivatives.

Theorem 13.18. *The pseudo-Boolean function f is monotone nondecreasing if
and only if $\Delta_i f(X) \geq 0$ for all $X \in \mathcal{B}^n$ and for all $i = 1,2,\ldots,n$. It is monotone
nonincreasing if and only if $\Delta_i f(X) \leq 0$ for all $X \in \mathcal{B}^n$ and for all $i = 1,2,\ldots,n$.*

Proof. This is straightforward. $\qquad\square$

Extending Definition 13.11, we say that a function f is *monotone* if the sign of
$\Delta_i f$ is constant on $\mathcal{B}^n$ for each $i = 1,2,\ldots,n$ (Wilde and Sanchez-Anton [909]).
Maximizing a monotone function f on $\mathcal{B}^n$ is trivial if the sign of each first derivative is known: Indeed, a global maximum X^* is obtained by setting $x_i^* = 1$ if
$\Delta_i f(X) \geq 0$ on $\mathcal{B}^n$, and by setting $x_i^* = 0$ otherwise.

Note also that, for a function f given in polynomial form, the sign of each
derivative can be easily determined if we know beforehand that f is monotone
(Hammer [435]). However, *recognizing* whether a function is monotone is a hard
task in itself, as proved by Crama [226].

Theorem 13.19. *It is co-NP-complete to decide whether a pseudo-Boolean function expressed in polynomial form is monotone, even when the input is restricted
to cubic polynomials.*

Proof. The decision problem is in co-NP: Indeed, in order to establish that an instance f is not monotone, it suffices to exhibit two points $X^*, Y^* \in \mathcal{B}^n$ such that $\Delta_i f(X^*) > 0$ and $\Delta_i f(Y^*) < 0$.

To prove that the problem is co-NP-complete, we provide a transformation from the NP-complete SUBSET SUM problem, which can be stated as follows (see [371]): Given $n + 1$ positive integers $(w_1, w_2, \ldots, w_n, t)$, is there a point $X^* \in \mathcal{B}^n$ such that $\sum_{j=1}^n w_j x_j^* = t$?

With an arbitrary instance $(w_1, w_2, \ldots, w_n, t)$ of SUBSET SUM, we associate the linear function

$$r(x_1, x_2, \ldots, x_n) = \sum_{j=1}^n w_j x_j - t,$$

and the cubic function

$$f(x_1, x_2, \ldots, x_{n+1}) = (r^2(x_1, x_2, \ldots, x_n) - 1)x_{n+1} + 3C^2 \sum_{j=1}^n x_j,$$

where C is a large enough constant (say, $C = \sum_{j=1}^n w_j + t$). One easily verifies that $\Delta_i f \geq 0$ for $i = 1, 2, \ldots, n$, and that $\Delta_{n+1} f = r^2(x_1, x_2, \ldots, x_n) - 1$. Hence, f is not monotone if and only if there exists $X^* \in \mathcal{B}^n$ such that $r(X^*) = 0$, that is, if and only if the SUBSET SUM problem has a "Yes" answer. $\square$

The same argument shows that it is also co-NP-complete to decide whether a cubic function given in polynomial form is monotone nondecreasing or monotone nonincreasing. For quadratic polynomials, the problem is easy in view of Theorem 13.18.

Foldes and Hammer [337] have investigated monotone pseudo-Boolean functions expressed in disjunctive normal forms (see also Marichal [668], Sugeno [851]).

Example 13.7. *The functions in Example 13.6 can also be expressed as $f_1 = \bar{x} \vee 3\bar{x}y \vee 3x \vee 4xy$ and $f_2 = y \vee 3\bar{y} \vee 2\bar{x}y$.* $\square$

We have already seen that it is co-NP-complete to recognize whether a Boolean DNF is monotone (see Theorems 1.31 and 1.32 in Chapter 1). Since every Boolean DNF can be interpreted as a pseudo-Boolean DNF, it easily follows that recognizing monotone (nonincreasing or nondecreasing) pseudo-Boolean functions expressed in DNF is also co-NP-complete.

The following theorem generalizes another well-known result from the theory of Boolean functions (recall Theorem 1.21, and see Bioch [86] for an extension to the class of discrete functions).

Theorem 13.20. *For a pseudo-Boolean function f, the following conditions are equivalent:*

(i) f *is monotone nondecreasing.*
(ii) *Some DNF of f contains no complemented variables.*
(iii) *Some CNF of f contains no complemented variables.*

Proof. By monotonicity of the operators $\vee$ and $\wedge$, it is obvious that each of the properties (ii) and (iii) implies (i).

Assume now that f is monotone nondecreasing and let us show that this implies (ii) (the other case is similar). By Theorem 13.4, we know that f can be represented by a pseudo-Boolean DNF of the form $\psi = \bigvee_{k=1}^{m} p_k(X)$, where each term p_k has the form

$$p_k(X) = a + b_k \Big(\bigwedge_{i \in A_k} x_i \Big) \Big(\bigwedge_{j \in B_k} \overline{x}_j \Big),$$

where $b_k \geq 0$ for $k = 1, 2, \ldots, m$.

Suppose that some term of ψ contains at least one complemented variable, say, $B_1 \neq \emptyset$, and let

$$q_1(X) = a + b_1 \Big(\bigwedge_{i \in A_1} x_i \Big).$$

Since $p_1(X) \leq q_1(X)$, there holds

$$f(X) = \bigvee_{k=1}^{m} p_k(X) \leq q_1(X) \vee \bigvee_{k=2}^{m} p_k(X). \tag{13.53}$$

We claim that $q_1(X) \leq f(X)$ for all $X \in \mathcal{B}^n$. Indeed, assume that $q_1(X^*) > f(X^*)$ for some point $X^* \in \mathcal{B}^n$. Then, we define another point $Y^* \in \mathcal{B}^n$ as follows: $y_j^* = x_j^*$ for all $j \notin B_1$, and $y_j^* = 0$ for all $j \in B_1$. For this point Y^*,

$$p_1(Y^*) = q_1(X^*) > f(X^*) \geq f(Y^*)$$

(the last inequality holds because f is nondecreasing). But the conclusion $p_1(Y^*) > f(Y^*)$ is in contradiction with the definition of the DNF expression of f.

Thus, there holds $q_1(X) \leq f(X)$ for all X, and (13.53) leads to

$$f(X) = q_1(X) \vee \bigvee_{k=2}^{m} p_k(X).$$

Repeating this procedure for each term of the DNF ψ, we eventually conclude that the expression obtained by dropping all complemented literals from ψ is again a DNF of f (compare with Theorem 1.24 in Chapter 1). $\qquad\square$

Example 13.8. *Consider again the nondecreasing function f_1 already intro-duced in Example 13.6 and in Example 13.7. This function is represented by the DNF $\psi = 1 \vee 3y \vee 3x \vee 4xy$ and by the CNF $\phi = (3+x) \wedge (3+y) \wedge (1+3(x \vee y))$.* $\qquad\square$

13.6.3 Supermodular and submodular functions

Definition 13.12. *A pseudo-Boolean function* f *on* $\mathcal{B}^n$ *is supermodular if*

$$f(X) + f(Y) \leq f(X \vee Y) + f(X \wedge Y) \text{ for all } X, Y \in \mathcal{B}^n. \tag{13.54}$$

The function f *is submodular if* $(-f)$ *is supermodular, or equivalently if*

$$f(X) + f(Y) \geq f(X \vee Y) + f(X \wedge Y) \text{ for all } X, Y \in \mathcal{B}^n. \tag{13.55}$$

Supermodular and submodular functions arise in numerous contexts and have been thoroughly investigated in discrete mathematics, in combinatorial optimization, in algebra, in statistics, in game theory, in economics, in engineering, in artificial intelligence, and so on. We refer to Choquet [192], Edmonds [287] and Shapley [829] for early work, and to Fujishige [351], Iwata [523], Lovász [624, 625], McCormick [637], Narayanan [703], Nemhauser and Wolsey [707], Rosenmüller [791], Schrijver [814], and Topkis [865] for in-depth discussions and additional references.

Specific examples of supermodular functions were encountered earlier in this chapter. For instance, the objective function (13.9) in Application 13.8 is supermodular. (The reader can either try to check directly the conditions in Definition 13.12 or use Theorem 13.21 hereunder.) Note also that, if f is supermodular and $f(0, 0, \ldots, 0) = 0$, then f is superadditive in the sense of Application 13.7.

Examples of submodular functions have been provided in Application 13.1 (Equations (13.1) and (13.2)) and in Application 13.2. Submodular functions also arise in a variety of computer science models (data mining, see Application 13.4, Genkin, Kulikowski, and Muchnik [376]; computer vision, see Application 13.5, Boykov, Veksler, and Zabih [147], Kolmogorov and Zabih [577]; artificial intelligence, see Živný, Cohen, and Jeavons [938]).

As argued by Lovász in [624], supermodular functions share some of the characteristic features of concave *and* of convex functions on $\mathbb{R}^n$. In particular, similarly to convex functions, supermodular functions have nonnegative second derivatives (or, equivalently, in view of Theorem 13.18, nondecreasing first derivatives).

Theorem 13.21. *A pseudo-Boolean function* f *on* $\mathcal{B}^n$ *is supermodular if and only if*

$$\Delta_i \Delta_j f(X) \geq 0 \text{ for all } X \in \mathcal{B}^n \text{ and for all } i, j = 1, 2, \ldots, n. \tag{13.56}$$

The function f *is submodular if and only if*

$$\Delta_i \Delta_j f(X) \leq 0 \text{ for all } X \in \mathcal{B}^n \text{ and for all } i, j = 1, 2, \ldots, n. \tag{13.57}$$

Proof. We focus on the first statement, since the second one follows immediately by sign reversal.

Suppose first that f is supermodular and consider two indices $i < j$ (note that $\Delta_i \Delta_i f(X) \equiv 0$). In view of Definition 13.10,

$$
\begin{aligned}
\Delta_i \Delta_j f = {} & f(x_1, \ldots, x_{i-1}, 1, x_{i+1}, \ldots, x_{j-1}, 1, x_{j+1}, \ldots, x_n) \\
& - f(x_1, \ldots, x_{i-1}, 1, x_{i+1}, \ldots, x_{j-1}, 0, x_{j+1}, \ldots, x_n) \\
& - f(x_1, \ldots, x_{i-1}, 0, x_{i+1}, \ldots, x_{j-1}, 1, x_{j+1}, \ldots, x_n) \\
& + f(x_1, \ldots, x_{i-1}, 0, x_{i+1}, \ldots, x_{j-1}, 0, x_{j+1}, \ldots, x_n)
\end{aligned}
$$

for all $(x_1, \ldots, x_{i-1}, x_{i+1}, \ldots, x_{j-1}, x_{j+1}, \ldots, x_n) \in \mathcal{B}^{n-2}$. Letting

$$
X^* = (x_1, \ldots, x_{i-1}, 1, x_{i+1}, \ldots, x_{j-1}, 0, x_{j+1}, \ldots, x_n)
$$

and

$$
Y^* = (x_1, \ldots, x_{i-1}, 0, x_{i+1}, \ldots, x_{j-1}, 1, x_{j+1}, \ldots, x_n),
$$

we see that $\Delta_i \Delta_j f \geq 0$ holds as a consequence of (13.54).

Conversely, assume that (13.56) holds, and let $X^0, Y^0 \in \mathcal{B}^n$. We are going to establish that (13.54) holds for X^0, Y^0 by induction on the *Hamming distance* $d(X^0, Y^0)$ between X^0 and Y^0, where

$$
d(X, Y) = \sum_{i=1}^{n} |x_i - y_i|
$$

for all $X, Y \in \mathcal{B}^n$. When $d(X^0, Y^0) = 0$ or 1, the inequality (13.54) is trivially satisfied. For $d(X^0, Y^0) = 2$, it is a reformulation of (13.56), as follows from the first part of the proof. Assume now that $d(X^0, Y^0) \geq 3$, and assume without loss of generality that $X^0 = (0, 0, X^2)$ and $Y^0 = (1, 1, Y^2)$.

Introduce the point $U^0 = (0, 1, Y^2)$. There holds $d(X^0, U^0) = d(X^0, Y^0) - 1$, and hence, by induction,

$$
f(X^0) + f(U^0) \leq f(X^0 \vee U^0) + f(X^0 \wedge U^0). \tag{13.58}
$$

Moreover,

$$
d(X^0 \vee U^0, Y^0) = 1 + d(X^2 \vee Y^2, Y^2) \leq 1 + d(X^2, Y^2) = d(X^0, Y^0) - 1,
$$

hence, we obtain again by induction and after some easy computations:

$$
\begin{aligned}
f(X^0 \vee U^0) + f(Y^0) & \leq f(X^0 \vee U^0 \vee Y^0) + f\left((X^0 \vee U^0) \wedge Y^0\right) \\
& = f(X^0 \vee Y^0) + f(U^0). \tag{13.59}
\end{aligned}
$$

Adding (13.58) and (13.59) yields

$$
f(X^0) + f(Y^0) \leq f(X^0 \vee Y^0) + f(X^0 \wedge Y^0),
$$

and the proof is complete.   □

The sequence of Theorems 13.18 and 13.21 has been extended in Crama, Hammer, and Holzman [232] and Foldes and Hammer [339] to the characterization of functions with nonnegative derivatives of higher order (see also Choquet [192]).

Theorem 13.21 has several corollaries for a pseudo-Boolean function f given by its polynomial expression.

First, notice that f is linear if and only if all its second-order derivatives are identically zero. This implies that linear functions are exactly those pseudo-Boolean functions that are simultaneously supermodular and submodular; they are sometimes called "modular" in the literature.

Example 13.9. *A prime example of linear pseudo-Boolean function is provided by a probability measure on a finite set. Linearity is due to the defining identity*

$$\text{Prob}(A) = \sum_{j \in A} \text{Prob}(\{j\}) \ \text{ for all } A \subseteq \{1, 2, \ldots, n\},$$

whereas sub- and supermodularity appear clearly in the well-known inclusion-exclusion formula

$$\text{Prob}(A \cup B) = \text{Prob}(A) + \text{Prob}(B) - \text{Prob}(A \cap B). \qquad \square$$

Consider now the quadratic case. It follows from Theorem 13.21 that a quadratic function f is supermodular if and only all its quadratic terms have nonnegative coefficients (Nemhauser, Wolsey, and Fisher [708]). This property can easily be checked in polynomial time.

The second-order derivatives of cubic functions are linear functions. Hence, the minimum and maximum of these derivatives can be efficiently computed. This implies in turn that supermodular and submodular cubic functions can also be recognized in polynomial time. On the other hand, the following result was independently established by Crama [226] and by Gallo and Simeone [364].

Theorem 13.22. *It is co-NP-complete to decide whether a pseudo-Boolean function expressed in polynomial form is supermodular (or submodular), even when the input is restricted to polynomials of degree 4.*

Proof. The proof is similar to the proof of Theorem 13.19. We leave it as an end-of-chapter exercise to the reader. $\qquad \square$

An important connection between supermodularity and concavity was established by Lovász [624]. It relies on an elegant characterization of supermodular functions in terms of their Lovász extension (see Section 13.3.3, and remember that we have only defined the Lovász extension when $f(0, 0, \ldots, 0) = 0$).

Theorem 13.23. *A pseudo-Boolean function f such that $f(0, 0, \ldots, 0) = 0$ is supermodular if and only if its Lovász extension f^L is concave.*

Proof. We assume that f is defined on $\mathcal{B}^n$ and we use the same notations as in Section 13.3.3.

(If) Assume that f^L is concave and let $X, Y \in \mathcal{B}^n$. Observe that the points $X \wedge Y$ and $X \vee Y$ are in a same simplex $S(\pi) \in \mathcal{S}$ since $X \wedge Y \le X \vee Y$. Thus, we successively derive:

$$\tfrac{1}{2} f(X) + \tfrac{1}{2} f(Y) = \tfrac{1}{2} f^L(X) + \tfrac{1}{2} f^L(Y) \quad \text{(since } f^L \text{ is an extension of } f)$$
$$\le f^L(\tfrac{1}{2}(X+Y)) \quad \text{(by concavity of } f^L)$$
$$= f^L\big(\tfrac{1}{2}(X \vee Y) + \tfrac{1}{2}(X \wedge Y)\big)$$
$$= \tfrac{1}{2} f^L(X \vee Y) + \tfrac{1}{2} f^L(X \wedge Y) \quad \text{(by linearity of } f^L \text{ on } S(\pi))$$
$$= \tfrac{1}{2} f(X \vee Y) + \tfrac{1}{2} f(X \wedge Y) \quad \text{(since } f^L \text{ is an extension of } f).$$

This proves that f is supermodular.

(Only if) Assume that f is supermodular. Recall that, for an arbitrary permutation π of $\{1, 2, \dots, n\}$, $f_{S(\pi)}^L$ denotes the unique linear extension of f on $S(\pi)$ and that it can be expressed by Equation (13.34). By a slight abuse of notations, we look at $f_{S(\pi)}^L$ as being defined on $\mathbb{R}^n$, rather than on $S(\pi)$ only.

Consider now an arbitrary point $X \in U^n$, and assume that $x_{\pi^*(1)} \le x_{\pi^*(2)} \le \dots \le x_{\pi^*(n)}$, meaning that X is in the simplex $S(\pi^*)$ and $f^L(X) = f_{S(\pi^*)}^L(X)$. We are going to prove that, for every other permutation π,

$$f_{S(\pi^*)}^L(X) \le f_{S(\pi)}^L(X). \tag{13.60}$$

Observe that if (13.60) holds, then it follows that

$$f^L(X) = f_{S(\pi^*)}^L(X) = \min_{S(\pi) \in \mathcal{S}} f_{S(\pi)}^L(X), \tag{13.61}$$

and hence, f^L is concave because it is the pointwise minimum of (finitely many) linear functions.

In order to prove inequality (13.60), consider the smallest index j such that $x_{\pi(j)} > x_{\pi(j+1)}$. If j does not exist, then (13.60) holds as an equality since $X \in S(\pi) \cap S(\pi^*)$. Otherwise, define a permutation ρ by transposing j and $j+1$:

$$\rho(j) = \pi(j+1), \ \rho(j+1) = \pi(j), \text{ and } \rho(i) = \pi(i) \text{ for all } i \ne j, j+1.$$

Some computations show that

$$f_{S(\pi)}^L(X) - f_{S(\rho)}^L(X) = \sum_{k=1}^{n} \big[(x_{\pi(k)} - x_{\pi(k-1)}) f(E^{\pi,k}) - (x_{\rho(k)} - x_{\rho(k-1)}) f(E^{\rho,k}) \big]$$
$$= (x_{\pi(j)} - x_{\pi(j+1)})$$
$$\times \big[f(E^{\pi,j}) - f(E^{\pi,j+1}) - f(E^{\rho,j+1}) + f(E^{\pi,j+2}) \big].$$

Moreover, $E^{\pi,j} = E^{\pi,j+1} \vee E^{\rho,j+1}$ and $E^{\pi,j+2} = E^{\pi,j+1} \wedge E^{\rho,j+1}$. Therefore, supermodularity implies that

$$f_{S(\pi)}^L(X) - f_{S(\rho)}^L(X) \ge 0.$$

Repeating this argument at most n times eventually transforms π into a permutation ρ^* which sorts the components of X in nondecreasing order and such that

$$f_{S(\pi^*)}^L(X) = f_{S(\rho^*)}^L(X) \leq f_{S(\pi)}^L(X).$$

This establishes (13.60), and the proof is complete.   □

The proof of Theorem 13.23, in particular, Equation (13.61), shows that every supermodular function can be represented as the lower-envelope of linear (pseudo-Boolean) functions. Interestingly, supermodular functions can also be shown to be upper-envelopes of linear functions; this result is discussed in Rosenmüller [791], where it is used to characterize extreme rays of the cone of nonnegative supermodular functions.

Let us now turn to the problem of optimizing supermodular functions. Grötschel, Lovász, and Schrijver [414] were first to prove that supermodular functions can be maximized in polynomial time, even when the function can only be accessed via an oracle (that is, a black-box algorithm which returns the value $f(X)$ for every input $X \in \mathcal{B}^n$). Another proof of this result was provided by Lovász [624], as a direct consequence of Theorem 13.23, of the fact that concave functions can be maximized over convex sets in polynomial time, and of the observation that $\max_{X \in U^n} f^L(X) = \max_{X \in \mathcal{B}^n} f(X)$ (Theorem 13.14).

Strongly polynomial combinatorial algorithms for the maximization of supermodular functions were subsequently proposed by Iwata, Fleischer, and Fujishige [524] and Schrijver [813]; see also Fujishige [351] and Schrijver [814], as well as the surveys by Iwata [523] and McCormick [637].

When a supermodular function is given by its polynomial expression and is either quadratic or cubic, then its maximization can be reduced to a max-flow min-cut problem in an associated network (compare with Equation (13.1) in Section 13.1; see for instance Balinski [47], Billionnet and Minoux [84], Hansen and Simeone [474], Kolmogorov and Zabih [577], Picard and Ratliff [746], Rhys [786], Živný, Cohen, and Jeavons [938], and Section 13.6.4 hereunder for related considerations).

Finally, let us remark that even though the maximum of a supermodular (or the minimum of a submodular) function can be computed in polynomial time, the opposite optimization problems, namely, the maximization of a submodular (or the minimization of a supermodular) function is NP-hard; this follows easily, for instance, from the NP-hardness of the max-cut problem and of the weighted stability problem in graphs; see Application 13.1. However, a standard greedy procedure for the maximization of a submodular set function provides a $(1 - \frac{1}{e})$-approximation of the maximum; see Fisher, Nemhauser, and Wolsey [332, 708], Fujito [352], Nemhauser, and Wolsey [706], Wolsey [923], and so on. Goldengorin [393] reviews theoretical results about the structure of local and global maxima of submodular functions, and discusses specialized maximization algorithms.

13.6.4 Unimodular functions

Definition 13.13. *A pseudo-Boolean function is* almost-positive *if all its nonlinear terms (i.e., terms of degree at least 2) have nonnegative coefficients in its polynomial expression.*

Theorem 13.21 implies that almost-positive functions are supermodular, and that the converse relation holds for quadratic functions. It is well-known that the maximization of almost-positive functions can be performed efficiently, by reduction to the computation of a minimum cut in a network (Balinski [47]; Picard and Queyranne [745]; Picard and Ratliff [746]; Rhys [786]). This observation has prompted several researchers to investigate broader classes of functions for which the same property holds. In order to define these classes, we introduce the following *switching* operation. For a pseudo-Boolean function f on $\mathcal{B}^n$ and a subset S of $\{1, 2, \ldots, n\}$, we denote by f_S the function defined for all $(x_1, x_2, \ldots, x_n)$ in $\mathcal{B}^n$ by

$$f_S(x_1, x_2, \ldots, x_n) = f(y_1, y_2, \ldots, y_n), \text{ where } y_j = \overline{x}_j \text{ if } j \in S \text{ and } y_j = x_j \text{ if } j \notin S,$$

and we say that f_S is obtained from f by *switching* S. It is easy to see that the class of almost-positive functions is not closed under switching, and this motivates the next definition.

Definition 13.14. *A pseudo-Boolean function is* unate *if it can be obtained from an almost-positive function by switching a subset of its variables.*

Another extension of the class of almost-positive functions was introduced by Billionnet and Minoux [84].

Definition 13.15. *A posiform is* polar *if each of its terms involves either no complemented variables or no uncomplemented variables. A pseudo-Boolean function is* polar *if it has at least one polar posiform.*

Almost-positive functions are obviously polar. Moreover, Billionnet and Minoux [84] observed that the class of polar functions is properly included in the class of supermodular functions, and that both classes coincide when restricted to cubic functions. The maximization of polar functions is again reducible to a network min-cut problem; this follows, for instance, from the observation that the conflict graph of a polar posiform is bipartite, or from the special structure of the constraint matrix of the integer programming problem (13.45)–(13.50), which turns out to be totally unimodular for polar posiforms (as a consequence of Theorem 5.13). Since bipartiteness of the conflict graph and total unimodularity are preserved by switching operations, whereas polarity is not, we can define yet another class of functions.

Definition 13.16. *A pseudo-Boolean function is* unimodular *if it can be obtained from a polar function by switching a subset of its variables.*

Unimodular functions were introduced by Hansen and Simeone [474]. Their definition was directly stated in terms of total unimodularity, but the equivalence

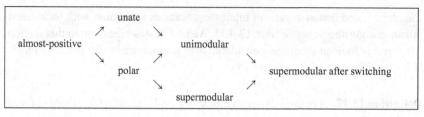

Figure 13.1. Classes related to unimodular and supermodular functions.

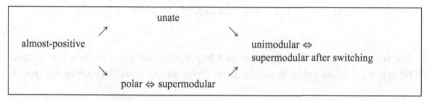

Figure 13.2. Classes related to unimodular and supermodular cubic functions.

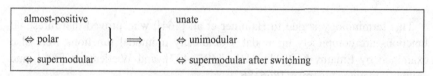

Figure 13.3. Classes related to unimodular and supermodular quadratic functions.

of both definitions was observed by Crama [226] and by Simeone, de Werra, and Cochand [835].

Clearly, almost-positive, unate, and polar functions are unimodular. Figure 13.1 summarizes the mutual relationships between several classes of functions. The simpler diagram obtained for cubic functions is displayed in Figure 13.2. For quadratic functions, the diagram shrinks even further, as shown in Figure 13.3.

Simeone, de Werra, and Cochand [835] proposed an efficient recognition algorithm for unate functions given in polynomial form. Crama [224, 226] described a polynomial-time algorithm that recognizes polar and unimodular functions; when a function f is unimodular, this algorithm produces a switching set S and a polar posiform of f_S.

13.6.5 Threshold and unimodal functions

Hammer, Simeone, Liebling, and de Werra [464] have introduced a hierarchy of pseudo-Boolean functions that generalize Boolean threshold functions (see

Chapter 9), and that also present interesting features in relation with local maximization algorithms (see Section 13.4.1). We briefly describe them in this section. All pseudo-Boolean functions considered here are assumed to be injective on B^n: If $X, Y \in B^n$ and $X \neq Y$, then $f(X) \neq f(Y)$.

Definition 13.17. *A pseudo-Boolean function f on B^n is called* threshold *if, for all $r \in \mathbb{R}^n$, there exist n weights $w_1(r), w_2(r), \ldots, w_n(r) \in \mathbb{R}$ and a threshold $t(r) \in \mathbb{R}$ such that, for all $(x_1, x_2, \ldots, x_n) \in B^n$,*

$$f(x_1, x_2, \ldots, x_n) \leq r \ \text{if and only if} \ \sum_{i=1}^{n} w_i(r) x_i \leq t(r).$$

So, for each value of r, there exists a hyperplane which separates the vertices of B^n where f takes value at most r from those where it takes value larger than r.

Definition 13.18. *A pseudo-Boolean function f on B^n is* unimax *if it has a unique local maximum in B^n. It is* completely unimodal *if, for each face F of B^n, the restriction of f to F is unimax.*

This terminology is due to Hammer et al. [464], who proved that threshold functions are completely unimodal. Completely unimodal functions were also examined by Emamy-K. [311], Hoke [496, 497], and Wiedemann [907], and unimax functions by Tovey [866, 867].

The main motivation for considering unimax functions is that local maximization algorithms could be expected to perform well for such functions. Indeed, if f is a unimax function, then the decision version of the maximization problem is in NP $\cap$ co-NP, since the global maximum of f is "well-characterized" [866]; based on this observation, it has been conjectured that unimax functions can be maximized in polynomial time. (Pardalos and Jha [730] proved that it is NP-hard to find the global maximum of a quadratic pseudo-Boolean function even when this global maximum is unique; however, this does not seem to have immediate consequences for unimax functions.)

When f is completely unimodal, Hammer et al. [464] proved that there always exists an increasing path of length at most n from any point $X \in B^n$ to the maximum of f. However, rather surprisingly, it has also been shown that simple local search procedures may perform an exponential number of steps before they reach a local (and global) maximum of a completely unimodal function; we refer to the above-mentioned references or to papers by Björklund, Sandberg, and Vorobyov [93, 94, 95] for related investigations and for applications in game theory and computer-aided verification.

Crama [226] proved that the recognition problem is NP-hard for threshold, completely unimodal, and unimax functions expressed in polynomial form. The question remains open, however, for quadratic unimax functions.

13.7 Exercises

1. Prove that the conjunction of two pseudo-Boolean elementary conjunctions is an elementary conjunction.

2. Show that condition (ii) in Lemma 13.1 does not completely characterize the prime implicants of a pseudo-Boolean function.

3. Consider the (simple) game associated with a Boolean function f, and let β_i denote the Banzhaf index of player i, as in Section 1.13.3. Show that $(\beta_1, \beta_2, \ldots, \beta_n)$ is proportional to the vector of first derivatives $(\Delta_1 f^{pol}(C), \Delta_2 f^{pol}(C), \ldots, \Delta_n f^{pol}(C))$ evaluated at the point $C = (\frac{1}{2}, \frac{1}{2}, \ldots, \frac{1}{2})$. (See Owen [720].)

4. (a) Show that every point $X \in U^n$ can be written in a unique way as a linear combination of the form

$$X = \sum_{k=1}^{K} \lambda_k X^k, \qquad (13.62)$$

where $\lambda_k > 0$ $(k = 1, 2, \ldots, K)$ and $X^1 \leq X^2 \leq \ldots \leq X^K$ are distinct points in $\mathcal{B}^n$.

(b) Show that the Lovász extension of a pseudo-Boolean function f can be expressed as

$$f^L(X) = \sum_{k=1}^{K} \lambda_k f(X^k) \quad \text{subject to (13.62)}.$$

5. Show that the value of a pseudo-Boolean function f may be arbitrarily worse in a local maximum than in the global maximum of f, even when f is assumed to be quadratic.

6. Show that the maximum of the max-cut function (13.1) is at least

$$\frac{1}{2} \sum_{1 \leq i < j \leq n} c(i, j).$$

Conclude that every graph contains a cut of capacity at least equal to half the sum of the edge capacities, and that such a cut can be found efficiently. (See Erdős [312] and Sahni and Gonzalez [798].)

7. Prove that every pseudo-Boolean function f on $\mathcal{B}^n$ has a posiform ψ of the form (13.51) such that $b_0 = \min_{X \in \mathcal{B}^n} f(X)$.

8. Prove that the optimal value of the linear relaxation of (13.45)–(13.50) is exactly the maximum of the concave standard extension ψ^{std} (13.29).

9. Show that the *hyperbolic* or *fractional programming* problem

$$\max_{X \in \mathcal{B}^n} f(X) = \frac{a_0 + \sum_{j=1}^{n} a_j x_j}{b_0 + \sum_{j=1}^{n} b_j x_j}$$

can be solved in polynomial time if

$$b_0 + \sum_{j=1}^{n} b_j x_j > 0 \quad \text{for all} \quad X \in \mathcal{B}^n,$$

but is NP-hard when this condition does not hold. (See Boros and Hammer [127]; Hammer and Rudeanu [460]; Hansen, Poggi de Aragão, and Ribeiro [473].)

10. Prove that it is co-NP-complete to decide whether a pseudo-Boolean function expressed in DNF is monotone.
11. Prove Theorem 13.22.
12. Prove that the concave envelope of a supermodular pseudo-Boolean function is its Lovász extension.
13. Show that the classes of almost-positive, supermodular, and polar functions are not closed under switching.
14. Establish all the implications displayed in Figures 13.1, 13.2 and 13.3, and show that they cannot be reversed.
15. Prove that threshold pseudo-Boolean functions are completely unimodal.
16. If f is a completely unimodal function on $\mathcal{B}^n$, prove that there always exists an increasing path of length at most n from any point $X \in \mathcal{B}^n$ to the maximum of f.
17. Prove that
    (a) it is NP-hard to decide whether a quadratic pseudo-Boolean function has a unique global maximum;
    (b) it is NP-hard to find the maximum of a quadratic pseudo-Boolean function even if we know that the global maximum is unique; (Pardalos and Jha [730]).

Question for thought

18. How difficult is it to recognize whether a quadratic pseudo-Boolean function is unimax?

Appendix A

Graphs and hypergraphs

This appendix proposes a short primer on graph and hypergraph theory. It sums up the basic concepts and terminology used in the remainder of the monograph. For (much) more information, we refer the reader to numerous excellent books dealing in-depth with this topic, such as Bang-Jensen and Gutin [51]; Berge [71, 72]; Brandstädt, Le, and Spinrad [152]; Golumbic [398]; Mahadev and Peled [645]; or Schrijver [814].

A.1 Undirected graphs

An *undirected graph*, or *graph* for short, is a pair of finite sets $G = (V, E)$ in which V is the set of *vertices* of the graph, and E is a set of unordered pairs of vertices called *edges* of the graph. Abiding by widespread conventions, we often use the notation (u, v), or even simply uv, for an edge $\{u, v\}$. Occasionally, we consider undirected graphs with *loops*, where a loop is an edge of the form (v, v) for $v \in V$ (we may view a loop as an edge of cardinality 1).

A graph can be represented as a diagram consisting of points (vertices) joined by lines (edges), as in Figure A.1.

When $e = (u, v)$ is an edge, we say that vertices u and v are *adjacent*, that u is a *neighbor* of v, that u and v are *incident* to e, that u and v are the *endpoints* of e, and so forth. The *neighborhood* of a vertex $u \in V$ is the set $N(u) = \{v \in V : (u, v) \in E\}$. The *degree* of u in G is the number of edges incident to u. We denote it by $deg_G(u)$ or simply $deg(u)$.

Two graphs (V, E) and (W, A) are *isomorphic* if there exists a bijection $\psi : V \to W$ such that, for all $u, v \in V$, $(u, v) \in E$ if and only if $(\psi(u), \psi(v)) \in A$. Intuitively, two graphs (V, E) and (W, A) are isomorphic if they can be represented by the same diagram.

The *complement* of the loopless $G = (V, E)$ is the graph $\overline{G} = (V, \overline{E})$ where $\overline{E} = \{(u, v) : u, v \in V, u \neq v, (u, v) \notin E\}$. So, the edges of $\overline{G}$ are exactly the nonedges of G.

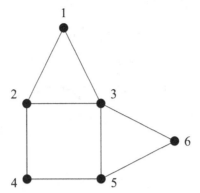

Figure A.1. Representation of a small graph.

A.1.1 Subgraphs

Let $G = (V, E)$ be a graph. A graph $H = (W, A)$ is a *subgraph* of G if $W \subseteq V$ and $A \subseteq E$. We say that H is the *subgraph* of G *induced by* W if A is exactly the set of edges of G that have both of their endpoints in W; namely, if $A = \{e \in E : e \subseteq W\}$. We sometimes denote by G_W the subgraph of G induced by W.

A subset of vertices $S \subseteq V$ is said to be a *stable set* (or an *independent set*) of G if S does not contain any edge of G. The subset S is a *clique* of G if every pair of vertices of S is an edge. It is a *transversal*, or a *vertex cover*, if every edge in E intersects S.

We denote by $\alpha(G)$ the maximum size of a stable set of G; by $\omega(G)$, the maximum size of a clique of G; and by $\tau(G)$, the minimum size of a vertex cover.

A subset of edges $M \subseteq E$ is called a *matching* of G if the edges in M are pairwise disjoint. A matching is *perfect* if it contains $\frac{1}{2}|V|$ edges, that is, if every vertex of G is incident to an edge of the matching.

A.1.2 Paths and connectivity

A graph can simply be viewed as a symmetric binary relation on its set of vertices. But the "pictorial" representation of a graph as a diagram of points (vertices) and lines (edges) naturally places the emphasis on topological notions like paths, cycles, or connectivity.

A *walk of length k* in a graph $G = (V, E)$ is a sequence

$$C = (v_1, e_1, v_2, e_2, v_3, \ldots, v_k, e_k, v_{k+1}) \tag{A.1}$$

in which $k \geq 0$, $v_1, v_2, \ldots, v_{k+1}$ are vertices, $e_1, e_2, \ldots, e_k$ are edges, and $e_i = (v_i, v_{i+1})$ for $i = 1, 2, \ldots, k$. It can also be denoted as $C = (v_1, v_2, v_3, \ldots, v_k, v_{k+1})$ or $C = (e_1, e_2, \ldots, e_k)$ when no confusion arises. The vertices v_1 and v_{k+1} are the *endpoints* of the walk C, and we say that they are *connected* by the walk. The walk is *closed* if $v_1 = v_{k+1}$.

The walk (A.1) is a *path* if all its vertices (and hence, all its edges) are distinct: $v_i \neq v_j$ for $1 \leq i < j \leq k+1$. The walk (A.1) is a *circuit* if is is closed ($v_1 = v_{k+1}$), if $v_1, v_2, \ldots, v_{k+1}$ are all distinct, and if $e_1, e_2, \ldots, e_k$ are all distinct.

A *connected component* of $G = (V, E)$ is a maximal subset $S \subseteq V$ such that, for all $u, v \in S$, u and v are the endpoints of a path in G. So, connected components are the equivalence classes of the equivalence relation "u and v are connected by a path." A graph is *connected* if it has a unique connected component.

A.1.3 Special classes of graphs

In this section, we introduce a few classes of graphs with special properties. More classes are defined in several chapters throughout the book.

First, we denote by P_n the path with vertex set $N = \{1, 2, \ldots, n\}$ and with edges $(i, i+1)$ for $i = 1, 2, \ldots, n-1$. Similarly, we denote by C_n the circuit with vertex set $N = \{1, 2, \ldots, n\}$ and with edges $(1, n)$ and $(i, i+1)$, $i = 1, 2, \ldots, n-1$.

The graph $G = (V, E)$ is *complete* if $E = \{(u, v) : u, v \in V\}$, that is, if V is a clique of G. We denote by K_n the complete graph on $N = \{1, 2, \ldots, n\}$.

The graphs P_4, C_4 and K_4 are represented in Figure A.2.

The graph $G = (V, E)$ is *bipartite* if there exists a partition of V into two subsets B, R (say, blue and red) such that every edge of G has one blue endpoint and one red endpoint, namely,

$$E \subseteq \{(u, v) : u \in B, v \in R\}. \tag{A.2}$$

The graph is called *complete bipartite* if E is exactly equal to the right-hand side of (A.2). For example, the graphs P_4 and C_4 are bipartite, and C_4 is complete bipartite (see Figure A.2). A *star* is a complete bipartite graph such that $|B| = 1$.

A graph is a *forest* if it contains no circuit. A *tree* is a connected forest. It is easy to see that, in a tree, there always exists a unique path between any pair of vertices. A *rooted tree* is a pair (T, r), where $T = (V, E)$ is a tree and $r \in V$ is a distinguished vertex called the *root* of T. A small rooted tree is shown in Figure A.3. Let

$$P = (v_1, e_1, v_2, e_2, v_3, \ldots, v_k, e_k, v_{k+1})$$

be a path in a rooted tree, with $v_1 = r$, and let v_j be one of the vertices in P, with $1 < j < k+1$. Then, we say that

- v_{j-1} is the (unique) *father* of v_j;

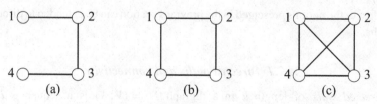

Figure A.2. (a): P_4, (b): C_4, (c): K_4.

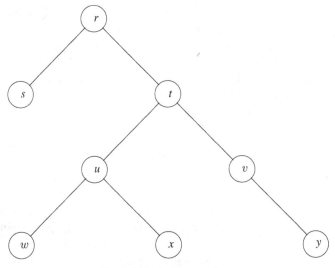

Figure A.3. A tree rooted at r.

- $v_1, v_2, \ldots, v_{j-1}$ are the *ancestors* of v_j;
- v_{j+1} is a *child* (not necessarily unique) of v_j;
- $v_{j+1}, v_{j+2}, \ldots, v_{k+1}$ are *successors* of v_j.

A.2 Directed graphs

A *directed graph*, or *digraph* for short, is a pair of finite sets $D = (V, A)$ where V is the set of *vertices* of the digraph and A is a collection of ordered pairs of vertices, called *arcs*. We think of every arc (u, v) as being directed from its *tail* u to its *head* v. A *loop* is an arc of the form (u, u).

The *outdegree* of vertex u is the number of arcs "leaving" u (that is, with tail u), and the *indegree* of vertex u is the number of arcs "entering" u (that is, with head u).

Every digraph can be obtained by *orienting* the edges of a graph $G = (V, E)$, that is, by replacing every edge $\{u, v\}$ of G by one (or both) of the arcs (u, v), (v, u). Conversely, by disregarding the orientation of its arcs, each digraph $D = (V, A)$ gives rise to the *underlying (undirected) graph* $G = (V, E)$, where $E = \{\{u, v\} \mid (u, v) \in A\}$.

Most of the notions presented in the previous section can be extended to directed graphs.

A.2.1 Directed paths and connectivity

A *directed walk* of length k in a digraph $D = (V, A)$ is a sequence $P = (v_1, a_1, v_2, a_2, v_3, \ldots, v_k, a_k, v_{k+1})$, where $k \geq 0$, $v_1, v_2, \ldots, v_{k+1}$ are vertices,

$a_1, a_2, \ldots, a_k$ are arcs, and $a_i = (v_i, v_{i+1})$ for $i = 1, 2, \ldots, k$. The directed walk P is *closed* if $v_1 = v_{k+1}$. It is a *directed path*, or *dipath*, if all its vertices (and hence, all its arcs) are distinct. It is a *cycle* if $k \geq 1$, $v_1, v_2, \ldots, v_k$ are all distinct, $v_1 = v_{k+1}$, and $a_1, a_2, \ldots, a_k$ are all distinct.

If there is a dipath from u to v in D, then we say that u is an *ancestor* of v, and that v is a *successor* of u.

A *strongly connected component*, or *strong component*, of $D = (V, A)$ is a maximal subset $S \subseteq V$ such that, for every pair u, v of distinct vertices in S, there is a directed path from u to v and a directed path from v to u in D. We say that D is *strongly connected* if V is its unique strong component. We simply say that D is *connected* if its underlying undirected graph is connected.

The *condensation* of digraph $D = (V, A)$ is the digraph $\hat{D} = (\hat{V}, \hat{A})$, where the elements of $\hat{V}$ are the strong components of D, and $(S_1, S_2) \in \hat{A}$ if there is at least one arc in D from some vertex of S_1 to some vertex of S_2. It is easy to see that $\hat{D}$ is an acyclic digraph.

A.2.2 Special classes of digraphs

An *arborescence rooted at r* is a pair (T, r), where $T = (V, A)$ is a digraph and $r \in V$ is a distinguished vertex of T such that, for every $v \in V$, there exists a unique directed path from r to v in D.

If (T, r) is a rooted (undirected) tree, where $T = (V, E)$, we obtain an arborescence $((V, A), r)$ as follows: For every edge $\{u, v\} \in E$, if u is the father of v, then we create the arc (u, v) in A (i.e., we orient every edge from father to son); this construction is illustrated in Figure A.4 for the tree of Figure A.3. Every arborescence arises in this way.

A digraph $D = (V, A)$ is *transitive* if the following implication holds for all $u, v, w \in V$:

$$(u, v) \in A \text{ and } (v, w) \in A \Rightarrow (u, w) \in A.$$

A *DAG* is a directed acyclic graph, that is, a directed graph without cycles. Every DAG D has at least one vertex with indegree 0, called a *source* of D, and at least one vertex with outdegree 0, called a *sink* or *leaf* of D. A *topological ordering* of a DAG $D = (V, A)$ is a bijection $\sigma : V \to \{1, 2, \ldots, n\}$ such that $\sigma(u) < \sigma(v)$ when $(u, v) \in A$. Every DAG has a topological ordering.

A.2.3 Transitive closure and transitive reduction

The *transitive closure* of a digraph $D = (V, A)$ is the smallest transitive digraph D^* that contains D as a subgraph; in other words, the transitive closure of D is the digraph $D^* = (V, A^*)$, where A^* contains all the arcs $(u, v) \in V \times V$ such that there is a directed path from u to v in D.

A *transitive reduction* of the digraph $D = (V, A)$ is any digraph $D' = (V, A')$ such that the transitive closure of D is equal to the transitive closure of D', and

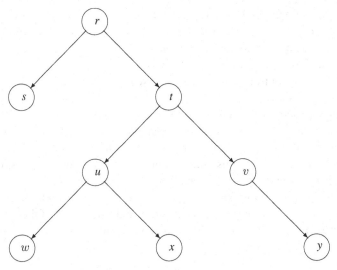

Figure A.4. An arborescence rooted at r.

such that the cardinality of A' is minimum with this property. If D is acyclic, then D has a unique transitive reduction.

A.3 Hypergraphs

A *hypergraph*, or *set system*, is a pair of sets $\mathcal{H} = (V, \mathcal{E})$, where V is the set of *vertices* of $\mathcal{H}$, and the elements of $\mathcal{E}$ are subsets of V called *edges* (or *hyperedges*) of the hypergraph.

Hypergraphs constitute a natural generalization of (undirected) graphs: Indeed, a graph is nothing but a hypergraph with edges of cardinality 2. As such, many of the concepts introduced for graphs can be extended (often in more than one way) to hypergraphs.

For instance, a subset of vertices is said to be *stable* in $\mathcal{H}$ if it does not contain any edge of $\mathcal{H}$, and it is a *transversal* of $\mathcal{H}$ if it intersects every edge of $\mathcal{H}$. A *matching* is a set of pairwise disjoint edges of $\mathcal{H}$.

A *clutter* (or *Sperner family*, or *simple hypergraph*) is a hypergraph $\mathcal{H} = (V, \mathcal{E})$ with the property that no edge is a subset of another edge: If $A \in \mathcal{E}, B \in \mathcal{E}$ and $A \neq B$, then $A \nsubseteq B$.

Appendix B

Algorithmic complexity

By and large, we assume that the readers of this book have at least some intuitive knowledge about algorithms and complexity. For the sake of completeness, however, we provide in this appendix an informal introduction to fundamental concepts of computational complexity: problems, algorithms, running time, easy and hard problems, etc. For a more thorough and rigorous introduction to this topic, we refer the reader to the classical monograph by Garey and Johnson [371], or to other specialized books like Aho, Hopcroft and Ullman [11], Papadimitriou [725], or Papadimitriou and Steiglitz [726]. Note that Cook et al. [211] and Schrijver [814] also provide gentle introductions to the topic, much in the spirit of this appendix.

In Section B.8, we propose a short primer on the complexity of list-generating algorithms; such algorithms are usually not discussed in basic textbooks on complexity theory, but they arise naturally in several chapters of our book.

B.1 Decision problems

Intuitively speaking, an *algorithmic* or *computational problem* is a generic question whose formulation contains a number of undetermined parameters. For instance, we can think of "addition" as the generic problem of adding two numbers (the numbers themselves must be specified before any specific computation can be performeed). Similarly, "solving quadratic equations" of the form $ax^2 + bx + c = 0$ is a problem which can be handled by an appropriate algorithm as soon as the numerical values of the parameters a, b, and c are known. In order to express these concepts more precisely, we need to explain how problems are stated and how their parameters are specified.

An *alphabet* is a finite set Σ, and a *word* on Σ is a finite (ordered) sequence of symbols from Σ. For instance, if Σ is the binary alphabet $B = \{0, 1\}$, then examples of words on Σ are: 000, 10100, 0011. If Σ consists of the Roman alphabet and of some usual typographical symbols, namely $\Sigma = \{a, b, c, \ldots, z, !, ;, -, \ldots\}$, then `do-not-disturb` or `dsfhuhf;;jseee` are examples of words on Σ. The *size* of a word W is the number of symbols in W; we denote it by $|W|$ and we denote by Σ^* the set of all words (of any size) on Σ.

When we fix an arbitrary alphabet Σ, the words in Σ^* can be used to encode many types of objects, such as a data set of numbers in binary format, or a text in natural language, or an algebraic equation, or a Boolean expression. Think of a word as the input string which is read by a computer program. Then, intuitively, a "problem" is a question that is asked about the input string: What is the largest number in the data set? Does the text contain the word `do-not-disturb`? Does the Boolean expression represent the constant **1**?

In particular, we say that a question about an input string is a *decision problem* if the answer to the question is either "Yes" or "No". More formally, a *decision problem* is simply defined as a subset Π of Σ^*. (Since decision problems are sets of words, they are also called *languages*.) In this context, an arbitrary word is called an *instance* (or *input*) of the problem. The word W is a "Yes-instance" for the decision problem Π if $W \in \Pi$, and it is a "No-instance" otherwise.

In our informal description of problems, we often use the following type of presentation:

PROBLEM Π
Instance: A word $W \in \Sigma^*$.
Question: Is W contained in Π?

To make things concrete, let us give some examples of problems:

QUADRATIC EQUATIONS
Instance: Three integers $a, b, c \in \mathbb{N}$.
Question: Does the equation $ax^2 + bx + c = 0$ have a solution in $\mathbb{R}$?

TREE
Instance: A graph $G = (V, E)$.
Question: Is G a tree?

HAMILTONIAN GRAPH
Instance: A graph $G = (V, E)$.
Question: Does G contain a Hamiltonian circuit, that is, does G contain a circuit that visits every vertex exactly once?

DNF EQUATION
Instance: A DNF expression $\phi(X)$.
Question: Is the equation $\phi(X) = 0$ consistent?

In the previous examples, we have (implicitly) assumed

$\Pi_{Quad} = \{W \in \Sigma^* : W$ represents a triplet $(a,b,c) \in \mathbb{N}^3$ such that

$$ax^2 + bx + c = 0 \text{ has a solution in } \mathbb{R}\},$$

$\Pi_{Tree} = \{W \in \Sigma^* : W$ represents a tree$\}$,

$\Pi_{Hamilton} = \{W \in \Sigma^* : W$ represents a Hamiltonian graph$\}$,

$\Pi_{DNF} = \{W \in \Sigma^* : W$ represents a DNF expression ϕ such that $\phi(X) \not\equiv 1\}$.

B.2 Algorithms

In order to solve a problem, we like to rely on an algorithm, that is, on a step-by-step procedure that describes how to compute a solution for each instance of the problem. Thus, for the problem QUADRATIC EQUATIONS, the algorithm may consist in computing the resolvent $\rho = b^2 - 4ac$, in testing whether $\rho \geq 0$, and in returning the answer either "Yes" or "No" depending on the outcome of the test.

More formally, algorithms (and computers) can be modelled in many different ways, such as Turing machines or random access machines (RAMs). A sketchy description of Turing machines [11, 371, 725] will suffice for our purpose. (In fact, we only need this description for the proof of Cook's theorem, in Section B.7 hereunder. So, the reader may choose to skip the following definitions in a first reading and to return to them later if necessary.)

A *one-tape Turing machine* $\mathcal{A}$ consists of

- a "processor," which is always in one of a finite number of "states"; the set of states, say Q, contains three distinguished states, namely, the "initial state" q_0, and the "final states" q_Y (for "Yes") and q_N (for "No");
- a single one-dimensional "tape," to be be viewed as memory space, which contains an infinite number of "cells" indexed by the integers in $\mathbb{Z}$; at any time, each cell of the tape can hold at most one symbol from the alphabet $\Sigma_0 = \Sigma \cup \{\Lambda\}$, where Λ is a special "blank" symbol;
- a "read-write (RW) head," which can move along the tape and scan any of its cells; when the RW head scans a cell, it can read the symbol marked in the cell and/or replace it by a new symbol;
- a "transition function" $T : Q \times \Sigma_0 \to Q \times \Sigma_0 \times \{-1, +1\}$, to be viewed as a primitive program.

A Turing machine operates on words of Σ^* according to the following recursive rules. Initially, the processor is in state q_0, the input word W is written in the adjacent cells $1, 2, \ldots, |W|$ of the tape (one symbol per cell), all the other cells contain the blank symbol Λ, and the RW head scans the leftmost symbol of W in cell 1. Suppose now that, at the start of iteration i, the processor is in state $q \in Q$, and the RW head scans cell $k \in \mathbb{Z}$, where k contains the symbol $\sigma \in \Sigma_0$. If $q = q_Y$ or $q = q_N$, then the machine ends its computation. Otherwise, let $T(q, \sigma) = (q', \sigma', m)$, where $m \in \{-1, +1\}$. Then, the processor changes its state from q to q', the RW head replaces the symbol σ by σ' in cell k, and the head moves to cell $k + m$ (that is, it moves either one step to the left or one step to the right). Iteration $i + 1$ can begin.

We say that the Turing machine $\mathcal{A}$ *accepts* the word $W \in \Sigma^*$ if it halts in state q_Y when applied to W. The set of words (that is, the language) accepted by $\mathcal{A}$ is, by definition, a decision problem $\Pi_{\mathcal{A}}$. Note that when $\mathcal{A}$ is applied to an input word that does not belong to $\Pi_{\mathcal{A}}$, then $\mathcal{A}$ may either halt in the state q_N, or it may go on computing forever. Since we are not fond of endless computations, we introduce one more concept: Namely, we say that the Turing machine $\mathcal{A}$ *solves* the decision problem Π if $\Pi = \Pi_{\mathcal{A}}$ and if $\mathcal{A}$ halts for all inputs $W \in \Sigma^*$. Thus, $\mathcal{A}$ returns the answer "Yes" when $W \in \Pi$, and it returns "No" otherwise.

We also note, for the record, that if a Turing machine $\mathcal{A}$ halts for all inputs $W \in \Sigma^*$, then it can be used to compute a function $f_{\mathcal{A}} : \Sigma^* \to \Sigma^*$, where $f_{\mathcal{A}}(W)$ is the word written on the tape when $\mathcal{A}$ halts, disregarding all blank symbols.

Despite its apparent simplicity, the Turing machine model is surprisingly powerful and can be used to simulate complex computations, such as those performed by real-world computers. Therefore, in the remainder of this appendix and throughout most of the book, we do not distinguish between "algorithms" and Turing machines unless the distinction is absolutely required. We refer again to the literature cited earlier for a discussion of the relation between Turing machines and other models of computation. Roughly speaking, however, the basic idea is that all these models are "essentially equivalent" from the point of view of their computational efficiency. Which brings us to our next topic...

B.3 Running time, polynomial-time algorithms, and the class P

The running time of a computer program on a given data set can be influenced by many factors, including the speed of the CPU, the skill of the programmer, the features of the programming language and of the compiler, and so on. But essentially, it is directly related to the number of elementary operations performed by the underlying algorithm and to the size of the data set. These observations motivate the following definitions.

Consider a problem Π and a Turing machine (or an algorithm) $\mathcal{A}$ that solves Π. For every $W \in \Sigma^*$, the *running time* of $\mathcal{A}$ on the input W is the number of iterations performed by $\mathcal{A}$ on W before it halts.

The *running time function* $R_{\mathcal{A}}(n)$ of $\mathcal{A}$ denotes the worst-case running time of $\mathcal{A}$ over all input words of size n, that is,

$$R_{\mathcal{A}}(n) = \max\{r : r \text{ is the running time of } \mathcal{A} \text{ on a word } W \in \Sigma^* \text{ such that } |W| = n\}.$$

The function $R_{\mathcal{A}}(n)$ is sometimes called the *time complexity function* of $\mathcal{A}$, or simply the *complexity* of $\mathcal{A}$.

Algorithm $\mathcal{A}$ *runs in polynomial time* if there exists a polynomial $p(n)$ such that $R_{\mathcal{A}}(n) \leq p(n)$ for all $n \in \mathbb{N}$. The complexity of a polynomial-time algorithm does not increase too fast with the size of the instances that it solves: We consider such an algorithm to be efficient.

The complexity class P contains the set of all problems that can be solved by a polynomial-time algorithm (or Turing machine):

P = {Π : Π is a decision problem and there is a polynomial-time algorithm
          that solves Π}.

The class P is of paramount importance in the theory of computation, so much so that, for combinatorial algorithmic problems, the qualifiers "solved in polynomial time," "well-solved," or "efficiently solved," have become quasi-synonymous. We refer again to [11, 371, 725, 726, 814] for a more thorough discussion.

By analogy with time complexity, one can also define the *space complexity* of a Turing machine $\mathcal{A}$ by reference to the number of cells scanned by the RW head until it halts. We do not make much use of this concept in the book.

B.4 The class NP

Another important complexity class is the class NP, where the initial "N" stands for "nondeterministic" and "P" stands for "polynomial." To understand its definition, consider again your favorite decision problem, say, DNF EQUATION as defined in Section B.1, and consider an instance ϕ of this problem, where

$$\phi = \overline{x}_1 x_2 \overline{x}_3 \vee x_1 \overline{x}_2 x_3 \vee \overline{x}_1 \overline{x}_2 x_4 \vee \overline{x}_1 x_3 \vee x_2 x_3 x_4 \vee x_4 x_5 x_6$$
$$\vee \overline{x}_4 \overline{x}_5 \overline{x}_6 \vee x_1 x_3 \overline{x}_4 \vee x_3 x_5 \overline{x}_6.$$

It may not be easy for you to decide whether the equation $\phi = 0$ is consistent or not. (Try!) But since we are nice people, we can provide some help: In fact, we can assure you of the existence of a solution, and we can even convince you easily that we are not lying. Indeed, $X^* = (1,0,0,1,0,0)$ is a solution.

Now, a crucial point in this example is that you do not need to know *how* we have found the solution in order to convince yourself of its correctness: We may have stumbled upon it by chance (nondeterministically), or guessed it otherwise. What matters is that, once you hold the candidate X^*, it is easy to check that the equation $\phi = 0$ is indeed consistent. (This situation is not as strange as it may initially appear; mathematicians, in particular, do not usually have to explain *how* they came up with the proof of a new theorem: Their professional community only requires that they be able to verify the validity of the alleged proof.)

Let us now generalize this idea. We say that a decision problem $\Pi \subseteq \Sigma^*$ is in the class NP if there exists a problem $\Pi' \in P$ and a polynomial $p'(n)$ such that, for every word $W \in \Sigma^*$, the following statements are equivalent:

(a) $W \in \Pi$, that is, W is a Yes-instance of Π.
(b) There exists a *certificate* $V \in \Sigma^*$ such that $|V| \leq p'(|W|)$ and such that $(V, W) \in \Pi'$.

To relate this formal definition to the previous discussion, note that for every Yes-instance $W \in \Pi$, there must exist a certificate V (in our previous example, a candidate solution X^*) that is reasonably short relative to W (this is ensured by the condition $|V| \leq p'(|W|)$), and such that checking the condition $W \in \Pi$ boils down to verifying that $(V, W) \in \Pi'$ (in our example, verifying that $\phi(X^*) = 0$). Moreover, the condition $(V, W) \in \Pi'$ must be testable in polynomial time; this is ensured by the assumption that $\Pi' \in P$.

It is easy to see that $P \subseteq NP$, meaning that every polynomially solvable problem is in NP. Indeed, if $\Pi \in P$, then it suffices to choose $\Pi' = \Pi$ and $p'(n) \equiv 0$ in the definition of NP (with V the empty string).

It is also quite obvious that the problem DNF EQUATION is in NP, just like HAMILTONIAN GRAPH and numerous other combinatorial problems (for example, any Hamiltonian circuit can be used to certify that a graph is Hamiltonian). To date, however, nobody has been able to devise a polynomial-time algorithm for DNF EQUATION or for HAMILTONIAN GRAPH; that is, nobody knows whether these problems are in P or in NP\P.

The vast majority of mathematicians and computer scientists actually believe that $P \neq NP$, but this famous conjecture has resisted all proof attempts (and there have been many) since the early 70s. To better appreciate this conjecture, it is useful to introduce the concepts of polynomial-time reductions and of NP-complete problems.

B.5 Polynomial-time reductions and NP-completeness

It is common practice in mathematics to establish that a problem Π can be viewed as a "special case" of another problem Π', and to solve Π by an algorithm originally designed for the more general problem Π'.

In our context, we say that a decision problem Π is (polynomially) *reducible* to a decision problem Π' if there is a polynomial-time algorithm $\mathcal{A}$ that computes, for any input word $W \in \Sigma^*$, another word $f_{\mathcal{A}}(W) = W' \in \Sigma^*$ such that

$$W \in \Pi \text{ if and only if } W' \in \Pi'.$$

The algorithm $\mathcal{A}$ that transforms any instance of Π into an instance of Π' is called a *polynomial-time reduction* of Π to Π'.

Note that, if Π is reducible to Π' and if Π' belongs to P, then Π also belongs to P. It is slightly less obvious, but equally true, that if Π is reducible to Π' and if Π' belongs to NP, then Π belongs to NP.

Now, a problem Π is called *NP-complete* if *every* problem in NP is reducible to Π. So, NP-complete problems can be viewed as the most general, or the hardest, problems in NP (they are "complete" in the sense that they "contain" every other problem of NP as a subproblem). It is not obvious, however, that NP-complete problems should actually exist. Cook's fundamental contribution was to demonstrate the existence of at least one natural NP-complete problem, namely, DNF

EQUATION [208] (see also Levin [610]). We sketch a proof of this seminal result later, in Section B.7.

Once we get hold of a first NP-complete problem Π, it becomes easier to establish that another problem Π' is also NP-complete: Indeed, to reach this conclusion, it suffices to prove that Π' is at least as hard as Π, or, more precisely, that Π is reducible to Π'. This type of reduction has been provided for thousands of decision problems, starting with the work of Cook [208] and Karp [550]; see also Ausiello et al. [36], Crescenzi and Kann [244], or Garey and Johnson [371]. Several examples of NP-completeness proofs are given in the book.

Note also that the existence of NP-complete problems has interesting consequences for the "P vs. NP" question stated above: Namely, to validate the conjecture that $P \neq NP$, it is sufficient to prove that at least one NP problem cannot be solved in polynomial time, and NP-complete problems are most natural candidates for this purpose. Moreover, the equality $P = NP$ holds if and only if at least one NP-complete problem happens to be polynomially solvable.

B.6 The class co-NP

The definition of the class NP in Section B.4 displays a striking asymmetry between "Yes-instances" and "No-instances" of decision problems. In fact, this apparent anomaly is well-grounded. Indeed, we have been able to argue that a problem like DNF EQUATION is in NP by observing that, when the DNF equation $\phi(X) = 0$ is consistent, any solution X^* provides a concise certificate of consistency (remember the small example in Section B.4). But when a DNF equation is *not* consistent, we may be hard put to provide a short proof of inconsistency.

As a consequence of this observation, we can introduce a new complexity class, to be called co-NP, by reversing the roles of "Yes-instances" and "No-instances" in the definition of the class NP. Equivalently, we say that a decision problem Π belongs to co-NP if and only if its complementary problem $(\Sigma^* \setminus \Pi)$ belongs to NP. Since $\Pi \in P$ trivially implies that $(\Sigma^* \setminus \Pi) \in P$, and since $P \subseteq NP$, we can also conclude that

$$P \subseteq NP \cap co\text{-}NP.$$

Problems in NP $\cap$ co-NP have short, polynomially verifiable certificates for both positive and negative instances. Therefore, these problems are sometimes called "well-characterized." Such problems are frequently known to belong to P as well, but the question of whether $P = NP \cap co\text{-}NP$ remains open.

Co-NP-complete problems can be defined by analogy with NP-complete problems, namely: Problem Π is co-NP-complete if and only if Π belongs to co-NP and every problem in co-NP is reducible to Π. Equivalently, Π is co-NP-complete exactly when its complementary problem $(\Sigma^* \setminus \Pi)$ is NP-complete.

Finally, we use the term *NP-hard* rather loosely to designate any problem Π (be it a decision problem or an optimization problem) that is at least as hard as every NP-complete problem in the sense that, if Π can be solved in polynomial time, then so can every NP-complete problem. In particular, NP-complete and co-NP-complete problems are NP-hard, as are certain problems that are not known to be either in NP or in co-NP.

B.7 Cook's theorem

In this section, we provide a proof of the following version of Cook's theorem [208]:

Theorem B.1. *The problem* DNF EQUATION *is NP-complete.*

Proof. We only sketch the main arguments of the proof, leaving aside some of the technical fine points, and we refer the reader to the specialized literature for details (see Cook's original paper or Garey and Johnson [371]).

The proof of the theorem heavily relies on the observation that the computations performed by a Turing machine can be "encoded" by the solution of a DNF equation, much in the same way that the output of a combinational circuit can be implicitly represented by the solution of a DNF equation (see Section 1.13.2). So, we start with a demonstration of this fact.

Consider an arbitrary decision problem Π, and suppose that Π is solved in polynomial time by a Turing machine $\mathcal{A}$. The complexity of $\mathcal{A}$ is bounded by a polynomial $p(n)$ for every instance of size $n \in \mathbb{N}$.

For simplicity, and without loss of generality, we assume that $\mathcal{A}$ works on the encoding alphabet $\mathcal{B} \cup \{\Lambda\}$, so that an input word of size n can be viewed as a point in $\mathcal{B}^n$.

We make the following claim:

Claim. For every $n \in \mathbb{N}$, there is an integer $m = O(p(n)^2)$ and a Boolean DNF $\phi(X, Y, z)$ (where $X \in \mathcal{B}^n$, $Y \in \mathcal{B}^m$, and $z \in \mathcal{B}$) with the property that, for every point $X^* \in \mathcal{B}^n$,

- (i) the DNF equation $\phi(X^*, Y, z) = 0$ has a unique solution $(X^*, Y^*, z^*) \in \mathcal{B}^{n+m+1}$, and
- (ii) when the Turing machine $\mathcal{A}$ operates on the input word X^*, the output of $\mathcal{A}$ is q_Y ("Yes") if $z^* = 1$, and the output of $\mathcal{A}$ is q_N ("No") if $z^* = 0$.

Moreover, the DNF ϕ can be constructed in time polynomial in n and $p(n)$.

Proof of the claim. For each fixed n, if the input point (or word) X is in $\mathcal{B}^n$, then the number of iterations performed by $\mathcal{A}$ is bounded by $p(n)$, so that the read-write head will be able to scan at most $p(n)$ cells of the tape until the Turing machine stops. More precisely, since the RW head initially scans cell 1, it can only scan the cells in $K = \{-p(n)+1, \ldots, p(n)\}$ until it stops.

Let us now introduce $p(n)(|Q|+8p(n))$ variables that completely describe the configuration of $\mathcal{A}$ in successive iterations: For $q \in Q$, $k \in K$, $t \in \{1, \ldots, p(n)\}$ and $\sigma \in \mathcal{B} \cup \{\Lambda\}$, we define

- variables $y_{q,t}^Q$: their intended meaning is that $y_{q,t}^Q = 1$ if $\mathcal{A}$ is in state q at iteration t;
- variables $y_{k,t}^H$, where $y_{k,t}^H = 1$ if the RW head scans cell k at iteration t;
- variables $y_{\sigma,k,t}^C$, where $y_{\sigma,k,t}^C = 1$ if σ is the symbol contained in cell k at iteration t.

For the variables $(y_{q,t}^Q, y_{k,t}^H, y_{\sigma,k,t}^C)$ to correctly describe the (uniquely defined) configuration of the Turing machine at every iteration t, there must hold

(a) $y_{q_0,1}^Q = 1$ (the machine is initially in state q_0) and, for all $q \in Q \setminus \{q_0\}$, $y_{q,1}^Q = 0$;

(b) $y_{1,1}^H = 1$ (the RW head initially scans cell 1) and, for all $k \neq 1$, $y_{k,1}^H = 0$;

(c) if $k \in \{1,2,\ldots,n\}$ and $\sigma = x_k$, then $y_{\sigma,k,1}^C = 1$; if $k \in K \setminus \{1,2,\ldots,n\}$ and $\sigma = \Lambda$, then $y_{\sigma,k,1}^C = 1$; for all other pairs (σ,k), $y_{\sigma,k,1}^C = 0$.

At every iteration, the variables describe a valid configuration resulting from a correct transition from the previous configuration, meaning that

(d) for all $t \in \{1,\ldots,p(n)\}$, for all $k \in K$, for all $q \notin \{q_Y,q_N\}$, for all σ, for $(q',\sigma',m) = T(q,\sigma)$, for all $q'' \neq q'$, for all $k' \neq k+m$, for all $k'' \neq k$, for all σ'',

if $y_{q,t}^Q = 1$ and $y_{k,t}^H = 1$ and $y_{\sigma,k,t}^C = 1$, then

$y_{q',t+1}^Q = 1, y_{q'',t+1}^Q = 0$ (the machine is in state q' at iteration $t+1$),

$y_{k+m,t+1}^H = 1, y_{k',t+1}^H = 0$ (the RW head scans cell $k+m$ at iteration $t+1$),

$y_{\sigma',k,t+1}^C = 1, y_{\sigma'',k,t+1}^C = 0$ (cell k contains the symbol σ' at iteration $t+1$),

$y_{\sigma'',k'',t+1}^C = y_{\sigma'',k'',t}^C$ (all cells other than cell k remain unchanged).

(These rules preserve the following property: At every iteration t, the machine is in a unique state, the RW head scans a unique cell, and each cell contains a unique symbol.)

(e) for all $t \in \{1,\ldots,p(n)-1\}$, for $q \in \{q_Y,q_N\}$ (if the machine has reached a halting state, then its configuration remains unchanged in subsequent iterations),

if $y_{q,t}^Q = 1$, then

$y_{q',t+1}^Q = y_{q',t}^Q$ for all $q' \in Q$,

$y_{k,t+1}^H = y_{k,t}^H$ for all $k \in K$,

$y_{\sigma,k,t+1}^C = y_{\sigma,k,t}^C$ for all $k \in K$ and for all σ.

(f) $z = y_{q_Y,p(n)}^Q$.

The conditions (a)–(f) are easily translated into a DNF equation $\phi(X,Y,z) = 0$. For instance, condition (c) can be written as

$$
\bigvee_{k=1}^{n}(y^C_{0,k,1}x_k \vee \overline{y}^C_{0,k,1}\overline{x}_k \vee y^C_{1,k,1}\overline{x}_k \vee \overline{y}^C_{1,k,1}x_k \vee y^C_{\Lambda,k,1}) \vee
$$

$$
\bigvee_{k \in K\setminus\{1,\dots,n\}}(y^C_{0,k,1} \vee y^C_{1,k,1} \vee \overline{y}^C_{\Lambda,k,1}) = 0.
$$

By construction, this equation has a unique solution $(X^*,Y^*,z^*) \in \mathcal{B}^{n+m+1}$ for every fixed $X^* \in \mathcal{B}^n$, and the values of Y^* and z^* in this solution describe the operations of the Turing machine on the input X^*. This establishes the claim.

We are now ready to conclude the proof of the theorem. Let Π be an arbitrary problem in NP. By definition, and with the same notations as in Section B.4, there is a problem $\Pi' \in P$ and a polynomial $p'(n)$ such that, for every instance $W \in \Sigma^*$, $W \in \Pi$ if and only if there exists a certificate $V \in \Sigma^*$ such that $|V| \le p'(|W|)$ and such that $(V,W) \in \Pi'$. Let $\mathcal{A}$ be a Turing machine that solves Π' in polynomial time.

For every fixed $n \in \mathbb{N}$, there is a Boolean DNF $\phi(X,Y,z)$ associated with $\mathcal{A}$ as in the proof of the claim. We can view every input word $X \in \mathcal{B}^n$ as consisting of two subwords V and W, with $V \in \mathcal{B}^r$ and $W \in \mathcal{B}^s$ for some fixed s and $r = p'(s)$. A word $W^* \in \mathcal{B}^s$ is a Yes-instance of Π if and only if there exists $V^* \in \mathcal{B}^r$ such that $X^* = (V^*,W^*) \in \Pi'$, or equivalently if and only if the equation

$$
\phi(V,W^*,Y,1) = 0
$$

is consistent. Observe that when the equation has a solution $(V^*,W^*,Y^*,1)$, the point V^* describes the certificate associated to W^*, and the point Y^* describes the steps of the verification of the certificate by $\mathcal{A}$. This completes the proof of Cook's theorem. $\square$

B.8 Complexity of list-generation and counting algorithms

In this book, we frequently investigate problems that are neither decision problems nor optimization problems, but problems of the following type: Given a binary relation $\Pi \subseteq \Sigma^* \times \Sigma^*$ and a word $W \in \Sigma^*$, we must generate all words $V \in \Sigma^*$ such that $(V,W) \in \Pi$. We say that this is the *list-generation problem associated with property* Π. Occasionally, we also consider the *counting problem associated with* Π, that is, the question of determining the number of words V such that $(V,W) \in \Pi$. To keep things reasonable, we further assume that the size of each "solution" V is polynomially-bounded in the size of the input W.

For example, if Π expresses the property "X is a solution of the DNF equation $\phi = 0$", and if the input string W encodes ϕ, then the associated counting problem asks for the number of solutions of the equation $\phi = 0$, and the list-generation problem consists in generating all solutions of the equation (these problems are considered in Sections 2.11.1 and 2.11.2, respectively). Similarly, if Π' expresses the property "C is a prime implicant of the function f", and if the input W encodes f (in some predetermined format), then the counting problem asks for the number of prime implicants of f, and the list-generation problem requires the production of all prime implicants of f (see Chapter 3).

We do not discuss the complexity of counting problems in detail here, as we encounter very few of them in this book. Let us simply say that we call #P-complete those counting problems that are "hardest" among a natural class of counting problems (essentially, among those counting problems such that the property $(V, W) \in \Pi$ can be verified in polynomial time). We refer to [371, 725, 883] for details.

By contrast, we find it necessary to discuss more formally the complexity of list-generation algorithms. The main difficulty here is that the *number* of solutions V satisfying the property $(V, W) \in \Pi$ may be *much* larger than the size $|W|$ of the input; to put it another way, the size of the output of a list-generation problem may be exponentially large in the size of its input, and hence, no polynomial-time algorithm can possibly exist for such a problem. Therefore, it makes sense to measure the complexity of list-generation algorithms as a function of their input size *and* of their output size. This notion has been formalized and used by many authors; early references include Read and Tarjan [781]; Valiant [883]; Lawler, Lenstra, and Rinnooy Kan [605]; and Johnson, Yannakakis, and Papadimitriou [538].

Consider a binary relation $\Pi \subseteq \Sigma^* \times \Sigma^*$ and the associated list-generation problem $\mathcal{L}_\Pi$. Let $\mathcal{A}$ be a list-generation algorithm for $\mathcal{L}_\Pi$, and suppose that, when running on the input W, $\mathcal{A}$ outputs the list $V_1, V_2, \ldots, V_m$, in that order. Note that the value of m depends on W but is independent of $\mathcal{A}$. We take it as a measure of the *output size* of $\mathcal{L}_\Pi$ for the instance $W \in \Sigma^*$ (remember that the size of each solution $V_1, V_2, \ldots, V_m$ has been assumed to be polynomially bounded in the size of W).

For $k = 1, \ldots, m$, we denote by $\tau(k)$ the running time required by $\mathcal{A}$ to output the first k elements of the list, that is, to generate $V_1, V_2, \ldots, V_k$. So, $\tau(m)$ is the total running time of $\mathcal{A}$ on W, and if we let $\tau(0) = 0$, then $\tau(k) - \tau(k - 1)$ is the time elapsed between the $(k - 1)$-st and the k-th outputs, for $k = 1, 2, \ldots, m$.

Following the terminology of Johnson, Yannakakis, and Papadimitriou [538], we say that

- $\mathcal{A}$ runs in *polynomial total time* if $\tau(m)$ is bounded by a polynomial in $|W|$ and m;
- $\mathcal{A}$ runs in *polynomial incremental time* if $\tau(k)$ is bounded by a polynomial in $|W|$ and k, for $k = 1, 2, \ldots, m$;
- $\mathcal{A}$ runs with *polynomial delay* if $\tau(k) - \tau(k - 1)$ is bounded by a polynomial in $|W|$, for $k = 1, 2, \ldots, m$.

Polynomial total time is, in a sense, the weakest notion of polynomiality that can be applied to $\mathcal{L}_\Pi$, since the running time of any algorithm for $\mathcal{L}_\Pi$ must grow at least linearly with m.

Polynomial incremental time captures the idea that the algorithm $\mathcal{A}$ outputs the solutions of $\mathcal{L}_\Pi$ sequentially and does not spend "too much time" between two successive outputs. Indeed, the definition implies that $\tau(k) - \tau(k-1)$ is polynomially bounded in $|W|$ and k, for all k. When generating the next element in the list, however, the algorithm may need to look at all previous outputs, and therefore, we allow $\tau(k)$ to depend on k as well as on the input size $|W|$.

Finally, an algorithm runs with polynomial delay when the time elapsed between two successive outputs is polynomial in the input size of the problem. This is a rather strong requirement, the strongest, in fact, among those discussed by Johnson, Yannakakis, and Papadimitriou [538].

In order to better understand the complexity of the list-generation problem $\mathcal{L}_\Pi$, it is also useful to grasp its relation with the following problem:

NEXT-GEN$_\Pi$

Instance: A word $W \in \Sigma^*$, and a set K of words such that $(V, W) \in \Pi$ for all $V \in K$.

Output: Either find a word $V \notin K$ such that $(V, W) \in \Pi$, or prove that no such word exists.

Clearly, if problem NEXT-GEN$_\Pi$ can be solved in polynomial time (meaning, in time polynomial in $|W|$ and $|K|$), then $\mathcal{L}_\Pi$ can be solved in polynomial incremental time: Indeed, starting from the empty list $K = \emptyset$, one can iteratively generate solutions of $\mathcal{L}_\Pi$ by solving a sequence of instances of NEXT-GEN$_\Pi$, until we can conclude that all solutions of $\mathcal{L}_\Pi$ have been generated.

Boros et al. [117] pointed out that, somewhat surprisingly, the converse relation also holds (see also Lawler et al. [605]). Namely, if algorithm $\mathcal{A}$ solves the list-generation problem $\mathcal{L}_\Pi$ in polynomial incremental time, then NEXT-GEN$_\Pi$ can be solved in polynomial time for every input (W, K) by a single run of $\mathcal{A}$ on the input W, which can be aborted after the generation of the first $|K| + 1$ solutions. Thus, investigating the complexity of NEXT-GEN$_\Pi$ provides valuable insights into the complexity of the list-generation problem $\mathcal{L}_\Pi$.

Appendix C

JBool: A software tool

Claude Benzaken and Nadia Brauner

C.1 Introduction

JBool is an application designed for teaching and illustrative purposes. It allows users to work with Boolean functions in disjunctive normal form (DNF) or in conjunctive normal form (CNF), and to easily manipulate the concepts described in this book or test conjectures on small-size examples. It is not an industrial software package, and it is not optimized to tackle large problems.

JBool can be downloaded freely from http://hdl.handle.net/2268/ 72714. The user interface is written in Java and the core engine for Boolean functions is written in ANSI C. The Java application requires a Java Runtime Environment (JRE) 1.3 or later, and binaries for the engine are available for the following platforms:

- Mac OS 10.3 and later
- Windows XP and later
- Linux x86

Source code is available, so the engine can be compiled for other platforms as well.

This appendix is organized as follows. First, the basic interface of the software is presented in Section C.2. The tools available to create, load, or save a function are described in Section C.3. The main functionalities of the software are then successively examined: Modify the elements of the edition (Section C.4), create several representations of the same function (Section C.5.1), apply various operators to the current function (Section C.5.2), perform operations on several functions and test properties of the current function (Section C.5). More details on all these functionalities can be found in the on-line help of the software.

C.2 Work interface

Figure C.1 displays the work interface of JBool. The main elements of this interface are described in the following sections.

C.2.1 Menu bar

When no Boolean function is selected, only [File], [Edit], and [Help] menus are visible in the menu bar. Other menus appear when a function is active.

- The [File] menu gives access to standard functionalities like New, Open, Save, and so on.
- The [Edit] menu contains classical commands like Cut, Copy, Paste, as well as some functionalities that change the function form.
- The [Presentation] menu contains items that produce an equivalent Boolean expression of the current function, like a dual form or an orthogonal form. In each case, a new name is created with structure `<Item name>(<function name>)`.

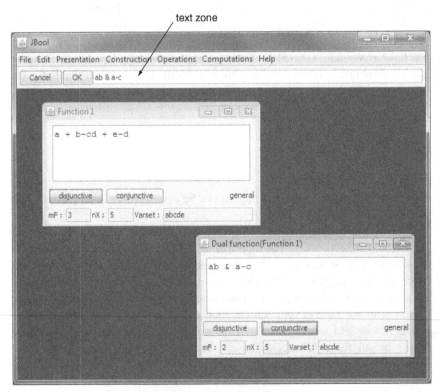

Figure C.1. JBool Interface.

- The [Construction] menu allows various constructions of new functions from the current one, like duplication of a function, restriction by assignment of values to literals, and so on. The same naming procedure is used for the new function as in the [Presentation] menu.
- The [Operations] menu allows the user to perform basic operations on pairs of functions, like disjunction or conjunction.
- The [Computation] menu allows the user to test properties (such as positivity, regularity, and so forth) of the current function.

C.2.2 Function windows

The Boolean functions are displayed in function windows. When a Boolean function is created, it appears in a new window with the default name *Function n*, where *n* is its sequence number. Some menu items open a new window associated with a new function, depending on the operation that has been performed. Then, several windows can be used simultaneously. The title bar of each window recalls the name of the operation used to create the function (for example, Dual function(Function 1)). When a window is selected, the corresponding function appears in the main text zone.

Each function window contains a general board that displays the following information: the number mF of terms or clauses, the number nX of variables, the variable set *Varset*, the normal form type (*disjunctive* or *conjunctive*), and a *general* or *positive* qualifier. Below the title bar, the text zone displays a normal form representation of the function, as explained in Section C.3.

C.2.3 Text zone

The *text zone* is located below the menu bar, as in Figure C.1. It is activated when a new function is created or loaded in the [File] menu. When a function window is selected, the corresponding Boolean function appears in the text zone. Each change in the text zone affects the corresponding function window when the [OK] button is selected.

C.3 Creating a Boolean function

C.3.1 Function syntax and presentation

The edition of a function is done only in the text zone. Each variable is represented by one character within the lowercase alphabet a to z or by an integer between 1 and 6. Thus one can use 32 variables in the set $\{a,\ldots,z\} \cup \{1,\ldots,6\}$. The software only allows the representation of functions as normal forms, either DNFs or CNFs. Terms or clauses are written as simple words, separated by "+" (representing the "or" operator) in a DNF, and separated by "&" (representing the "and" operator) in a CNF (when typing a function, one may input a space character instead of "+"

or "&"). Each word starts with the alphabetical list of positive literals, followed by the sign "-" and by the alphabetical list of negative (complemented) literals. Empty words are allowed. For instance, the DNF $(a \wedge \overline{b} \wedge \overline{f}) \vee (\overline{c}) \vee (\overline{d} \wedge e)$ is written as *a-bf + -c + e-d*. Similarly, the CNF $(a \vee \overline{b} \vee \overline{f}) \wedge (\overline{c}) \wedge (\overline{d} \vee e)$ is written as *a-bf & -c & e-d*.

An empty list ($mF = 0$) represents a a constant function (0 for a DNF, and 1 for a CNF) and is displayed as "F" (False) for a DNF and as "T" (True) for a CNF. (One may also simply type "T" or "F" in the text zone.)

All Boolean expressions are automatically simplified according to the absorption laws

$$x \wedge (x \vee y) = x, \quad x \vee (x \wedge y) = x.$$

For instance, the DNF expression $a \vee (a \wedge \overline{b} \wedge c)$ with the corresponding syntax *a + ac-b* is automatically simplified to the expression *a* (by absorption law).

C.3.2 Creation modes

There are four ways of creating a function: One can create a new empty function, generate a random function, calculate a threshold function from the definition of a separator, or load an existing function.

The [New] item in the [File] menu creates a new function whose default Boolean expression is F (False) in DNF, and T (True) in CNF. This function can be subsequently modified in the text zone.

The [Random...] item in the [File] menu opens a dialog box, as shown in Figure C.2, for generating a random expression. Six fields are displayed in the dialog box: the number of variables, an upper bound for the number of terms, the minimal degree (number of literals) of each term, a specification of uniform degree (all terms have equal degree), positivity of the function, and the conjunctive or disjunctive character of the normal form. Positivity here means that no negative literal appears in the expression. With the choices in Figure C.2, JBool might return the Boolean function $(d \vee \overline{e})(b \vee c \vee d \vee \overline{f})(\overline{a} \vee \overline{b} \vee \overline{e} \vee \overline{f})$.

The [Threshold...] item in the [File] menu opens a dialog box, as in Figure C.3, for generating a threshold function from the definition of a separator. First, the threshold value (which can be negative) and the number n of variables are required. Then a grid is opened with n boxes to be filled by integers (positive or negative) which are the weights of the n variables. The inequality corresponding to the example in Figure C.3 is $2a - b + 3c \geq 2$, and JBool returns the corresponding Boolean threshold function: $c \vee a\overline{b}$.

The [Open...] item in the [File] menu allows us to load a previously saved Boolean function. This item opens a dialog box for selecting a Boolean function file. Once opened, the Boolean function appears in the function window and in the text zone.

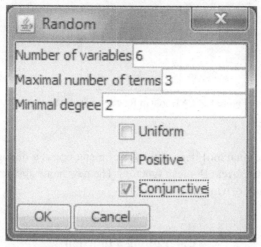

Figure C.2. Random function dialog.

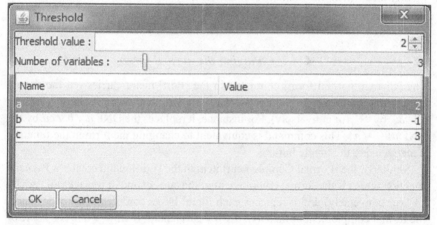

Figure C.3. Threshold function dialog.

C.3.3 Saving a function

The [Save As…] item in the [File] menu saves a Boolean function as a text file. This item opens a dialog box for entering the file name of the function. The [Save] item in the [File] menu saves an existing Boolean function. If the function does not exist yet, then this item opens a [Save As…] dialog.

Each Boolean function is saved in a text (*.txt*) file that only contains the function written in the JBool syntax. Figure C.4 shows an example of a Boolean function file.

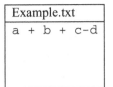

Figure C.4. A Boolean function file: Example.txt.

The [Rename function] item in the [File] menu opens a dialog for entering a new name for the current Boolean function. The new name appears in the title bar of the corresponding window.

C.4 Editing a function

The [Edit] menu contains tools for editing a function. Classical commands, like Cut, Copy, Paste are available. The menu also contains some functionalities that modify the form of the function: sort the terms by degree, change the normal form, modify the variable set, and so on. We next describe two of these operations.

C.4.1 Changing the normal form

The [Change normal form] command in the [Edit] menu carries out the formal transformation of the current normal form into the dual form: it simply replaces all "&" by "+" (or conversely). For instance, it replaces the DNF $a \vee \bar{b} \vee c\bar{d}$ by the CNF $a\bar{b}(c \vee \bar{d})$. This command is equivalent to changing the form in the function window using the toggle buttons.

Similarly, the [Formal Complement] item in the [Edit] menu creates a Boolean function which is the formal complement of the current one: it replaces all "&" by "+" (or conversely) and it replaces each literal by its complement. For instance, the formal complement of the DNF $a \vee \bar{b} \vee c\bar{d}$ is the CNF $\bar{a}b(\bar{c} \vee d)$. The name of the new function is Formal complement(< *function_name* >).

C.4.2 Modifying the variable set

The [Shift variables...] item in the [Edit] menu opens a dialog window for entering a number k. Each variable rank is then shifted by k. For instance, if $k = 2$, then the Boolean function $a \vee b\bar{c}$ becomes $c \vee d\bar{e}$. Notice that $32 - k$ has to be larger than the largest rank of the variables of the current function.

The [Add dummy] item in the [Edit] menu opens a dialog that asks for a subset A of variables. All variables in A are then added to the set of variables (Varset) of the current function. The [Delete dummy] item in the [Edit] menu deletes all variables on which the function does not effectively depend.

The [Compact...] item in the [Edit] menu deletes all dummy variables in the variable set (Varset) and replaces the rank of all the variables by the smallest possible rank. For instance, the Boolean function $e \vee j \vee hu$ becomes $a \vee c \vee bd$.

C.5 Operations on Boolean functions

This section briefly presents the main functionalities of the JBool software (we refer to the on-line help for details).

C.5.1 Equivalent presentations of a Boolean function

The [Presentation] menu contains commands that produce equivalent Boolean expressions of the current function, like a dual form, an orthogonal form, an irredundant form, the full list or prime implicants, or an irredundant list of prime implicants of the current function. For each item, a new function is created whose name is <Item Name>(<Function name>).

C.5.2 Constructions

The [Construction] menu allows various constructions of new functions from the current one, for instance, by duplication, dualization, or complementation of the current function; by assignment of values to subsets of literals; or by merging of variables. The new function can also be obtained by extracting terms of a given degree or by switching variables. In each case, a new function is created whose name is <Item Name>(<Function name>).

C.5.3 Operations on two Boolean functions

The possible operations are the disjunction and the conjunction of two Boolean function. The items in the [Operations] menu open a dialog with the list of all Boolean functions in use. One function must be selected in this dialog. Then, a new function is created by applying the chosen operation to the current function and to the function selected in the dialog.

C.5.4 Testing properties of a function

The [Computation] menu allows testing whether the current function is identically 1, monotone, 2-monotone, quadratic, pure Horn, disguised-Horn, or quasi-Horn-quadratic.

Bibliography

[1] P.A. Abdulla, P. Bjesse and N. Eén, Symbolic reachability analysis based on SAT-solvers, in: S. Graf and M. Schwartzbach, eds., *Tools and Algorithms for the Construction and Analysis of Systems*, Lecture Notes in Computer Science, Vol. 1785, Springer-Verlag, Berlin Heidelberg, 2000, pp. 411–425.

[2] J.A. Abraham, An improved algorithm for network reliability, *IEEE Transactions on Reliability* R-28 (1979) 58–61.

[3] D. Achlioptas and Y. Peres, The threshold for random k-SAT is $2^k \log 2 - O(k)$, *Journal of the American Mathematical Society* 17 (2004) 947–973.

[4] D. Achlioptas and G.B. Sorkin, Optimal myopic algorithms for random 3-SAT, *Proceedings of the 41st Annual IEEE Symposium on the Foundations of Computer Science*, IEEE, 2000, pp. 590–600.

[5] A. Adam, *Truth Functions and the Problem of Their Realization by Two-Terminal Graphs*, Akademiai Kiado, Budapest, 1968.

[6] W.P. Adams and P.M. Dearing, On the equivalence between roof duality and Lagrangian duality for unconstrained 0–1 quadratic programming problems, *Discrete Applied Mathematics* 48 (1994) 1–20.

[7] K.K. Aggarwal, K.B. Misra and J.S. Gupta, A fast algorithm for reliability evaluation, *IEEE Transactions on Reliability* R-24 (1975) 83–85.

[8] R. Agrawal, T. Imielinski and A. Swami, Mining association rules between sets of items in large databases, *International Conference on Management of Data* (SIGMOD 93), 1993, pp. 207–216.

[9] R. Agrawal, H. Mannila, R. Srikant, H. Toivonen and A.I. Verkamo, Fast discovery of association rules, in: U.M. Fayyad et al., eds., *Advances in Knowledge Discovery and Data Mining*, AAAI Press, Menlo Park, California, 1996, pp. 307–328.

[10] A.V. Aho, M.R. Garey and J.D. Ullman, The transitive reduction of a directed graph, *SIAM Journal on Computing* 1 (1972) 131–137.

[11] A.V. Aho, J.E. Hopcroft and J.D. Ullman, *The Design and Analysis of Computer Algorithms*, Addison-Wesley Publishing Company, Reading, MA, 1974.

[12] M. Aigner, *Combinatorial Theory*, Springer-Verlag, Berlin, Heidelberg, New York, 1979.

[13] H. Aizenstein, T. Hegedűs, L. Hellerstein and L. Pitt, Complexity theoretic hardness results for query learning, *Computational Complexity* 7 (1998) 19–53.

[14] G. Alexe, P.L. Hammer, V. Lozin and D. de Werra, Struction revisited, *Discrete Applied Mathematics* 132 (2003) 27–46.

[15] E. Algaba, J.M. Bilbao, J.R. Fernández Garcia and J.J. López, Computing power indices in weighted multiple majority games, *Mathematical Social Sciences* 46 (2003) 63–80.

[16] E. Allender, L. Hellerstein, P. McCabe, T. Pitassi and M.E. Saks, Minimizing disjunctive normal form formulas and AC^0 circuits given a truth table, *SIAM Journal on Computing* 38 (2008) 63–84.

[17] N. Alon and P.H. Edelman, The inverse Banzhaf problem, *Social Choice and Welfare* 34 (2010) 371–377.

[18] M. Alonso-Meijide, B. Casas-Méndez, M.J. Holler and S. Lorenzo-Freire, Computing power indices: Multilinear extensions and new characterizations, *European Journal of Operational Research* 188 (2008) 540–554.

[19] H. Andreka and I. Nemeti, The generalized completeness of Horn predicate-logic as a programming language, Research Report of the Department of Artificial Intelligence 21, University of Edinburgh, 1976.

[20] D. Angluin, Learning propositional Horn sentences with hints, Research Report of the Department of Computer Science 590, Yale University, 1987.

[21] D. Angluin, Queries and concept learning, *Machine Learning* 2 (1988) 319–342.

[22] D. Angluin, L. Hellerstein and M. Karpinski, Learning read-once formulas with queries, *Journal of the ACM* 40 (1993) 185–210.

[23] M.F. Anjos, An improved semidefinite programming relaxation for the satisfiability problem, *Mathematical Programming* 102 (2005) 589–608.

[24] M.F. Anjos, Semidefinite optimization approaches for satisfiability and maximum-satisfiability problem, *Journal on Satisfiability, Boolean Modeling and Computation* 1 (2005) 1–47.

[25] M. Anthony, *Discrete Mathematics of Neural Networks: Selected Topics*, SIAM Monographs on Discrete Mathematics and Applications, SIAM, Philadelphia, 2001.

[26] M. Anthony, Probabilistic learning and Boolean functions, in Y. Crama and P.L. Hammer, eds., *Boolean Models and Methods in Mathematics, Computer Science, and Engineering*, Cambridge University Press, Cambridge, 2010, pp. 197–220.

[27] M. Anthony, Neural networks and Boolean functions, in Y. Crama and P.L. Hammer, eds., *Boolean Models and Methods in Mathematics, Computer Science, and Engineering*, Cambridge University Press, Cambridge, 2010, pp. 554–576.

[28] M. Anthony, Decision lists and related classes of Boolean functions, in Y. Crama and P.L. Hammer, eds., *Boolean Models and Methods in Mathematics, Computer Science, and Engineering*, Cambridge University Press, Cambridge, 2010, pp. 577–595.

[29] M. Anthony and N. Biggs, *Computational Learning Theory*, Cambridge University Press, Cambridge, 1992.

[30] W.W. Armstrong, Dependency structures of database relationships, in: *IFIP-74*, North-Holland, Amsterdam, 1974, pp. 580–583.

[31] T. Asano and D.P. Williamson, Improved approximation algorithms for MAX SAT, Working paper, IBM Almaden Research Center, 2000. Preliminary version in the *Proceedings of the 11th ACM-SIAM Symposium on Discrete Algorithms*, 2000, pp. 96–105.

[32] R.L. Ashenhurst, The decomposition of switching functions, in: *Proceedings of the International Symposium on the Theory of Switching*, Part I, Harvard University Press, Cambridge, MA, 1959, pp. 75–116.

[33] B. Aspvall, Recognizing disguised NR(1) instances of the satisfiability problem, *Journal of Algorithms* 1 (1980) 97–103.

[34] B. Aspvall, M.F. Plass and R.E. Tarjan, A linear-time algorithm for testing the truth of certain quantified Boolean formulas, *Information Processing Letters* 8 (1979) 121–123.

[35] J. Astola and R.S. Stanković, *Fundamentals of Switching Theory and Logic Design: A Hands on Approach*, Springer, Dordrecht, The Netherlands, 2006.

[36] G. Ausiello, P. Crescenzi, G. Gambosi, V. Kann, A. Marchetti-Spaccamela and M. Protasi, *Complexity and Approximation*, Springer-Verlag, Berlin, 1999.

[37] G. Ausiello, A. D'Atri and D. Saccà, Minimal representation of directed hypergraphs, *SIAM Journal on Computing* 15 (1986) 418–431.

[38] A. Avidor, I. Berkovitch and U. Zwick, Improved approximation algorithms for MAX NAE-SAT and MAX SAT, in: T. Erlebach and G. Persiano, eds., *Approximation and*

Online Algorithms, Lecture Notes in Computer Science, Vol. 3879, Springer-Verlag, Berlin Heidelberg, 2006, pp. 27–40.

[39] H. Aziz, M. Paterson and D. Leech, Efficient algorithm for designing weighted voting games, in: *Proceedings of the 11th IEEE International Multitopic Conference*, IEEE Computer Society, 2007, pp. 1–6.

[40] J. Bailey, T. Manoukian and K. Ramamohanarao, A fast algorithm for computing hypergraph transversals and its application in mining emerging patterns, in: *Proceedings of the 3rd IEEE International Conference on Data Mining* Florida, USA, IEEE Computer Society, 2003, pp. 485–488.

[41] O. Bailleux, Y. Boufkhad and O. Roussel, A translation of pseudo-Boolean constraints to SAT, *Journal on Satisfiability, Boolean Modeling and Computation* 2 (2006) 191–200.

[42] E. Balas, Facets of the knapsack polytope, *Mathematical Programming* 8 (1975) 146–164.

[43] E. Balas and R. Jeroslow, Canonical cuts on the unit hypercube, *SIAM Journal on Applied Mathematics* 23(1972) 661–669.

[44] E. Balas and J.B. Mazzola, Nonlinear 0–1 programming: I. Linearization techniques, *Mathematical Programming* 30 (1984) 1–21.

[45] E. Balas and J.B. Mazzola, Nonlinear 0–1 programming: II. Dominance relations and algorithms, *Mathematical Programming* 30 (1984) 22–45.

[46] E. Balas and E. Zemel, Facets of the knapsack polytope from minimal covers, *SIAM Journal of Applied Mathematics* 34 (1978) 119–148.

[47] M.L. Balinski, On a selection problem, *Management Science* 17 (1970) 230–231.

[48] M.O. Ball and G.L. Nemhauser, Matroids and a reliability analysis problem, *Mathematics of Operations Research* 4 (1979) 132–143.

[49] M.O. Ball and J.S. Provan, Disjoint products and efficient computation of reliability, *Operations Research* 36 (1988) 703–715.

[50] H.-J. Bandelt and V. Chepoi, Metric graph theory and geometry: A survey, in: J.E. Goodman, J. Pach and R. Pollack, eds., *Surveys on Discrete and Computational Geometry: Twenty Years Later*, Contemporary Mathematics, Vol. 453, American Mathematical Society, Providence, RI, 2008, pp. 49–86.

[51] J. Bang-Jensen and G. Gutin, *Digraphs: Theory, Algorithms and Applications*, Springer-Verlag, London, 2000.

[52] J.F. Banzhaf, Weighted voting doesn't work: A mathematical analysis, *Rutgers Law Review* 19 (1965) 317–343.

[53] K. Barkaoui and M. Minoux, A polynomial-time algorithm to decide liveness of some basic classes of bounded Petri nets, in: *Application and Theory of Petri Nets 1992*, Lecture Notes in Computer Science, Vol. 616, Springer-Verlag, Berlin Heidelberg, 1992, pp. 62–75.

[54] R.E. Barlow and F. Proschan, *Statistical Theory of Reliability and Life Testing*, Holt, Rinehart and Winston, New York, 1975.

[55] R. Battiti and M. Protasi, Solving MAX-SAT with non-oblivious functions and history-based heuristics, in: D. Du, J. Gu and P.M. Pardalos, eds., *Satisfiability Problem: Theory and Applications*, DIMACS series in Discrete Mathematics and Theoretical Computer Science, Vol. 35, American Mathematical Society, 1997, pp. 649–667.

[56] R.J. Bayardo Jr. and J.D. Pehoushek, Counting models using connected components, in: *Proceedings of the 17th National Conference on Artificial Intelligence and 12th Conference on Innovative Applications of Artificial Intelligence*, Austin, TX, 2000, pp. 157–162.

[57] R.J. Bayardo Jr. and R.C. Schrag, Using CSP look-back techniques to solve exceptionally hard SAT instances, in: *Proceedings of the Second International Conference on Principles and Practice of Constraint Programming*, Lecture Notes in Computer Science, Vol. 1118, Springer, Berlin, 1996, pp. 46–60.

[58] R.J. Bayardo Jr. and R.C. Schrag, Using CSP look-back techniques to solve real-world SAT instances, in: *Proceedings of the Fourteenth National Conference on Artificial Intelligence*, Providence, RI, 1997, pp. 203–208.

[59] E. Benoist and J-J. Hebrard, Recognition of simple enlarged Horn formulas and simple extended Horn formulas, *Annals of Mathematics and Artificial Intelligence*, 37 (2003) 251–272.

[60] M. Ben-Or and N. Linial, Collective coin flipping, in: S. Micali, ed., *Randomness and Computation*, Academic Press, New York, 1990, pp. 91–115.

[61] C. Benzaken, Algorithmes de dualisation d'une fonction booléenne, *R.F.T.I.-Chiffres* 9 (1966) 119–128.

[62] C. Benzaken, Post's closed systems and the weak chromatic number of hypergraphs, *Discrete Mathematics* 23 (1978) 77–84.

[63] C. Benzaken, Critical hypergraphs for the weak chromatic number, *Journal of Combinatorial Theory B* 29 (1980) 328–338.

[64] C. Benzaken, From logical gates synthesis to chromatic bicritical clutters, *Discrete Applied Mathematics* 96–97 (1999) 259–305.

[65] C. Benzaken, S. Boyd, P.L. Hammer and B. Simeone, Adjoints of pure bidirected graphs, *Congressus Numerantium* 39 (1983) 123–144.

[66] C. Benzaken, Y. Crama, P. Duchet, P.L. Hammer and F. Maffray, More characterizations of triangulated graphs, *Journal of Graph Theory* 14 (1990) 413–422.

[67] C. Benzaken and P.L. Hammer, Linear separation of dominating sets in graphs, *Annals of Discrete Mathematics* 3 (1978) 1–10.

[68] C. Benzaken, P.L. Hammer and B. Simeone, Graphes de conflit des fonctions pseudo-booléennes quadratiques, in: P. Hansen and D. de Werra, eds., *Regards sur la Théorie des Graphes*, Presses Polytechniques Romandes, Lausanne, 1980, pp. 165–170.

[69] C. Benzaken, P.L. Hammer and B. Simeone, Some remarks on conflict graphs of quadratic pseudo-Boolean functions, *International Series of Numerical Mathematics* 55 (1980) 9–30.

[70] V.L. Beresnev, On a problem of mathematical standardization theory, *Upravliajemyje Sistemy* 11 (1973) 43–54 (in Russian).

[71] C. Berge, *Graphes et Hypergraphes,* Dunod, Paris, 1970. (*Graphs and Hypergraphs,* North-Holland, Amsterdam, 1973, revised translation.)

[72] C. Berge, *Hypergraphs,* North-Holland, Amsterdam, 1989.

[73] J. Berman and P. Köhler, Cardinalities of finite distributive lattices, *Mitteilungen aus dem Mathematischen Seminar Giessen* 121 (1976) 103–124.

[74] P. Bertolazzi and A. Sassano, An $O(mn)$ algorithm for regular set-covering problems, *Theoretical Computer Science* 54 (1987) 237–247.

[75] P. Bertolazzi and A. Sassano, A class of polynomially solvable set-covering problems, *SIAM Journal on Discrete Mathematics* 1 (1988) 306–316.

[76] D. Bertsimas and J. Tsitsiklis, *Introduction to Linear Optimization,* Athena Scientific, Paris, 1997.

[77] A. Bhattacharya, B. DasGupta, D. Mubayi and G. Turán, On approximate Horn minimization, manuscript, 2009.

[78] W. Bibel and E. Eder, Methods and calculi for deduction, in: D.M. Gabbay, C.J. Hogger and J.A. Robinson, eds., *Handbook of Logic in Artificial Intelligence and Logic Programming, Vol. 1, Logical Foundations*, Oxford Science Publications, Clarendon Press, Oxford, 1993, pp. 67–182.

[79] J.M. Bilbao, *Cooperative Games on Combinatorial Structures,* Kluwer Academic Publishers, Dordrecht, 2000.

[80] J.M. Bilbao, J.R. Fernández, A. Jiménez Losada and J.J. López, Generating functions for computing power indices efficiently, *Sociedad de Estadística e Investigación Operativa Top* 8 (2000) 191–213.

[81] J.M. Bilbao, J.R. Fernández, N. Jiménez and J.J. López, Voting power in the European Union enlargement, *European Journal of Operational Research* 143 (2002) 181–196.

[82] L.J. Billera, Clutter decomposition and monotonic Boolean functions, *Annals of the New York Academy of Sciences* 175 (1970) 41–48.

[83] L.J. Billera, On the composition and decomposition of clutters, *Journal of Combinatorial Theory* 11 (1971) 234–245.

[84] A. Billionnet and M. Minoux, Maximizing a supermodular pseudoboolean function: A polynomial algorithm for supermodular cubic functions, *Discrete Applied Mathematics* 12 (1985) 1–11.

[85] A. Billionnet and S. Elloumi, Using a mixed integer quadratic programming solver for the unconstrained quadratic 0–1 problem, *Mathematical Programming* 109 (2007) 55–68.

[86] J.C. Bioch, Dualization, decision lists and identification of monotone discrete functions, *Annals of Mathematics and Artificial Intelligence* 24 (1998) 69–91.

[87] J.C. Bioch, Decomposition of Boolean functions, in: Y. Crama and P.L. Hammer, eds., *Boolean Models and Methods in Mathematics, Computer Science, and Engineering*, Cambridge University Press, Cambridge, 2010, pp. 39–75.

[88] J.C. Bioch and T. Ibaraki, Generating and approximating non-dominated coteries, *IEEE Transactions on Parallel and Distributed Systems* 6 (1995) 905–914.

[89] J.C. Bioch and T. Ibaraki, Complexity of identification and dualization of positive Boolean functions, *Information and Computation* 123 (1995) 50–63.

[90] E. Birnbaum and E.L. Lozinskii, The good old Davis-Putnam procedure helps counting models, *Journal of Artificial Intelligence Research* 10 (1999) 455–477.

[91] Z.W. Birnbaum, On the importance of different components in a multicomponent system, in: P.R. Krishnaiah, ed., *Multivariate Analysis*-II, Academic Press, New York, 1969.

[92] Z.W. Birnbaum, J.D. Esary and S.C. Saunders, Multi-component systems and structures and their reliability, *Technometrics* 3 (1961) 55–77.

[93] H. Björklund, S. Sandberg and S. Vorobyov, Optimization on completely unimodal hyper-cubes, Technical Report TR-2002-018, Department of Information Technology, Uppsala University, Sweden May 2002.

[94] H. Björklund, S. Sandberg and S. Vorobyov, Complexity of model checking by iterative improvement: The pseudo-Boolean framework, in: M. Broy and A.V. Zamulin, eds., *Perspectives of System Informatics 2003*, Lecture Notes in Computer Science, Vol. 2890, Springer-Verlag, Berlin-Heidelberg, 2003, pp. 381–394.

[95] H. Björklund and S. Vorobyov, Combinatorial structure and randomized subexponential algorithms for infinite games, *Theoretical Computer Science* 349 (2005) 347–360.

[96] A. Björner, Homology and shellability of matroids and geometric lattices, in: N. White, ed., *Matroid Applications*, Cambridge University Press, Cambridge, 1992, pp. 226–283.

[97] A. Björner, Topological methods, in: R. Graham, M. Grötschel and L. Lovász, eds., *Handbook of Combinatorics*, Elsevier, Amsterdam, 1995, pp. 1819–1872.

[98] C.E. Blair, R.G. Jeroslow and J.K. Lowe, Some results and experiments in programming techniques for propositional logic, *Computers and Operations Research* 13 (1986) 633–645.

[99] A. Blake, *Canonical Expressions in Boolean Algebras*, Dissertation, Department of Mathematics, University of Chicago, 1937. Published by University of Chicago Libraries, 1938.

[100] B. Bollig, M. Sauerhoff, D. Sieling and I. Wegener, Binary decision diagrams, in: Y. Crama and P.L. Hammer, eds., *Boolean Models and Methods in Mathematics, Computer Science, and Engineering*, Cambridge University Press, Cambridge, 2010, pp. 473–505.

[101] B. Bollobás, C. Borgs, J. Chayes, J.H. Kim and D.B. Wilson, The scaling window of the 2-SAT transition, *Random Structures and Algorithms* 18 (2001) 201–256.

[102] T. Bonates and P.L. Hammer, Logical Analysis of Data: From combinatorial optimization to medical applications, *Annals of Operations Research* 148 (2006) 203–225.

[103] G. Boole, *An Investigation of the Laws of Thought*, Walton, London, 1854. (Reprinted by Dover Books, New York, 1954.)

[104] K.S. Booth, Boolean matrix multiplication using only $O(n^{\log_2 7} \log n)$ bit operations, *SIGACT News* 9 (Fall 1977) p. 23.

[105] E. Boros, Dualization of aligned Boolean functions, RUTCOR Research Report RRR 9-94, Rutgers University, Piscataway, NJ, 1994.

[106] E. Boros, Maximum renamable Horn sub-CNFs, *Discrete Applied Mathematics* 96–97
       (1999) 29–40.

[107] E. Boros and O. Čepek, Perfect $0, \pm 1$ matrices. *Discrete Mathematics* 165–166 (1997)
       81–100.

[108] E. Boros, O. Čepek and A. Kogan, Horn minimization by iterative decomposition, *Annals
       of Mathematics and Artificial Intelligence* 23 (1998) 321–343.

[109] E. Boros, O. Čepek, A. Kogan and P. Kučera, Exclusive and essential sets of implicates of
       Boolean functions, RUTCOR Research Report 10-2008, Rutgers University, Piscataway,
       NJ, 2008.

[110] E. Boros, O. Čepek, and P. Kučera, Complexity of minimizing the number of clauses and
       literals in a Horn CNF, manuscript, 2010.

[111] E. Boros, Y. Crama, O. Ekin, P.L. Hammer, T. Ibaraki and A. Kogan, Boolean normal
       forms, shellability and reliability computations, *SIAM Journal on Discrete Mathematics*
       13 (2000) 212–226.

[112] E. Boros, Y. Crama and P.L. Hammer, Polynomial-time inference of all valid implications
       for Horn and related formulae, *Annals of Mathematics and Artificial Intelligence* 1 (1990)
       21–32.

[113] E. Boros, Y. Crama and P.L. Hammer, Upper bounds for quadratic 01 maximization,
       Operations Research Letters 9 (1990) 7379.

[114] E. Boros, Y. Crama and P.L. Hammer, Chvátal cuts and odd cycle inequalities in quadratic
       0-1 optimization, *SIAM Journal on Discrete Mathematics* 5 (1992) 163–177.

[115] E. Boros, Y. Crama, P.L. Hammer, T. Ibaraki, A. Kogan and K. Makino, Logical Analysis
       of Data: Classification with justification, *Annals of Operations Research* (2011), to appear.

[116] E. Boros, Y. Crama, P.L. Hammer and M. Saks, A complexity index for satisfiability
       problems, *SIAM Journal on Computing* 23 (1994) 45–49.

[117] E. Boros, K.M. Elbassioni, V. Gurvich, L. Khachiyan and K. Makino, Dual-bounded
       generating problems: All minimal integer solutions for a monotone system of linear
       inequalities, *SIAM Journal on Computing* 31 (2002) 1624–1643.

[118] E. Boros, K.M. Elbassioni, V. Gurvich and K. Makino, Generating vertices of polyhe-
       dra and related monotone generation problems, in: D. Avis, D. Bremner and A. Deza,
       eds., *Polyhedral Computations*, CRM Proceedings and Lecture Notes, Vol. 48, Centre de
       Recherches Mathématiques and AMS (2009) pp. 15–44.

[119] E. Boros, K.M. Elbassioni and K. Makino, On Berge multiplication for monotone Boolean
       dualization, in: A. Luca et al., eds., *Proceedings of the 35th International Colloquium on
       Automata, Languages and Programming* (ICALP), Lecture Notes in Computer Science,
       Vol. 5125, Springer-Verlag, Berlin Heidelberg, 2008, pp. 48–59.

[120] E. Boros, S. Foldes, P.L. Hammer and B. Simeone, A restricted consensus algorithm for
       the transitive closure of a digraph, manuscript, in preparation, 2008.

[121] E. Boros, V. Gurvich and P.L. Hammer, Dual subimplicants of positive Boolean functions,
       Optimization Methods and Software 10 (1998) 147–156.

[122] E. Boros, V. Gurvich, P.L. Hammer, T. Ibaraki and A. Kogan, Decompositions of partially
       defined Boolean functions, *Discrete Applied Mathematics* 62 (1995) 51–75.

[123] E. Boros, V. Gurvich, L. Khachiyan and K. Makino, Dual-bounded generating problems:
       Partial and multiple transversals of a hypergraph, *SIAM Journal on Computing* 30 (2000)
       2036–2050.

[124] E. Boros, V. Gurvich, L. Khachiyan and K. Makino, Dual-bounded generating problems:
       Weighted transversals of a hypergraph, *Discrete Applied Mathematics* 142 (2004) 1–15.

[125] E. Boros and P.L. Hammer, A max-flow approach to improved roof duality in quadratic
       0–1 minimization, RUTCOR Research Report RRR 15-1989, Rutgers University, 1989.

[126] E. Boros and P.L. Hammer, A generalization of the pure literal rule for satisfiability
       problems, RUTCOR Research Report 20-92, Rutgers University, 1992.

[127] E. Boros and P.L. Hammer, Pseudo-Boolean optimization, *Discrete Applied Mathematics*
       123 (2002) 155–225.

[128] E. Boros, P.L. Hammer and J.N. Hooker, Predicting cause-effect relationships from incomplete discrete observations, *SIAM Journal on Discrete Mathematics* 7 (1994) 531–543.

[129] E. Boros, P.L. Hammer, T. Ibaraki and K. Kawakami, Polynomial time recognition of 2-monotonic positive Boolean functions given by an oracle, *SIAM Journal on Computing* 26 (1997) 93–109.

[130] E. Boros, P.L. Hammer, T. Ibaraki and A. Kogan, Logical analysis of numerical data, *Mathematical Programming* 79 (1997) 163–190.

[131] E. Boros, P.L. Hammer, T. Ibaraki, A. Kogan, E. Mayoraz and I. Muchnik, An implementation of logical analysis of data, *IEEE Transactions on Knowledge and Data Engineering* 12 (2000) 292–306.

[132] E. Boros, P.L. Hammer, M. Minoux and D.J. Rader Jr., Optimal cell flipping to minimize channel density in VLSI design and pseudo-Boolean optimization, *Discrete Applied Mathematics* 90 (1999) 69–88.

[133] E. Boros, P.L. Hammer and X. Sun, The DDT method for quadratic 0–1 minimization, RUTCOR Research Report 39-89, Rutgers University, 1989.

[134] E. Boros, P.L. Hammer and X. Sun, Network flows and minimization of quadratic pseudo-Boolean functions, RUTCOR Research Report 17-91, Rutgers University, 1991.

[135] E. Boros, P.L. Hammer and X. Sun, Recognition of q-Horn formulae in linear time, *Discrete Applied Mathematics* 55 (1994) 1–13.

[136] E. Boros, P.L. Hammer and G. Tavares, Local search heuristics for quadratic unconstrained binary optimization, *Journal of Heuristics* 13 (2007) 99–132.

[137] E. Boros, T. Horiyama, T. Ibaraki, K. Makino and M. Yagiura, Finding essential attributes from binary data, *Annals of Mathematics and Artificial Intelligence* 39 (2003) 223–257.

[138] E. Boros, T. Ibaraki and K. Makino, Boolean analysis of incomplete examples, in: R. Karlsson and A. Lingas, eds., *Algorithm Theory – SWAT'96*, Lecture Notes in Computer Science, Vol. 1097, Springer-Verlag, Berlin, 1996, pp. 440–451.

[139] E. Boros, T. Ibaraki and K. Makino, Error-free and best-fit extensions of partially defined Boolean functions, *Information and Computation* 140 (1998) 254–283.

[140] E. Boros, T. Ibaraki and K. Makino, Logical analysis of binary data with missing bits, *Artificial Intelligence* 107 (1999) 219–264.

[141] E. Boros, T. Ibaraki and K. Makino, Fully consistent extensions of partially defined Boolean functions, in: J. van Leeuwen, O. Watanabe, M. Hagiya, P.D. Mosses and T. Ito, eds., *Theoretical Computer Science - International Conference IFIP TCS 2000*, Lecture Notes in Computer Science, Vol. 1872, Springer, Berlin, 2000, pp. 257–272.

[142] E. Boros, T. Ibaraki and K. Makino, Variations on extending partially defined Boolean functions with missing bits, *Information and Computation* 180 (2003) 53–70.

[143] E. Boros, I. Lari and B. Simeone, Block linear majorants in quadratic 01 optimization, *Discrete Applied Mathematics* 145 (2004) 52–71.

[144] E. Boros and A. Prékopa, Closed form two-sided bounds for probabilities that at least r or exactly r out of n events occur, *Mathematics of Operations Research* 14 (1989) 317–342.

[145] E. Boros and A. Prékopa, Probabilistic bounds and algorithms for the maximum satisfiability problem, *Annals of Operations Research* 21 (1989) 109–126.

[146] J.-M. Bourjolly, P.L. Hammer, W.R. Pulleyblank and B. Simeone, Boolean-combinatorial bounding of maximum 2-satisfiability, in: O. Balci, R. Sharda, S. Zenios, eds., *Computer Science and Operations Research: New Developments in their Interfaces*, Pergamon Press, 1992, 23–42.

[147] Y. Boykov, O. Veksler and R. Zabih, Fast approximate energy minimization via graph cuts, *IEEE Transactions on Pattern Analysis and Machine Intelligence* 23 (2001) 1222–1239.

[148] G.H. Bradley, P.L. Hammer and L.A. Wolsey, Coefficient reduction for inequalities in 0–1 variables, *Mathematical Programming* 7 (1974) 263–282.

[149] P.S. Bradley, U.M. Fayyad and O.L. Mangasarian, Mathematical programming for data mining: Formulations and challenges, *INFORMS Journal on Computing* 11 (1999) 217–238.

[150] S.J. Brams and P.J. Affuso, Power and size: A new paradox, *Theory and Decision* 7 (1976) 29–56.

[151] A. Brandstädt, P.L. Hammer, V.B. Le and V.V. Lozin, Bisplit graphs, *Discrete Mathematics* 299 (2005) 11–32.

[152] A. Brandstädt, V.B. Le and J.P. Spinrad, *Graph Classes: A Survey*, SIAM Monographs on Discrete Mathematics and Applications, SIAM, Philadelphia, 1999.

[153] R.K. Brayton, G.D. Hachtel, C.T. McMullen, A.L. Sangiovanni-Vincentelli, *Logic Minimization Algorithms for VLSI Synthesis*, Kluwer Academic Publishers, Boston, 1984.

[154] A. Bretscher, D.G. Corneil, M. Habib and C. Paul, A simple linear time LexBFS cograph recognition algorithm (extended abstract), in: *Proceedings of the 29th International Workshop on Graph-Theoretic Concepts in Computer Science, WG2003*, Lecture Notes in Computer Science, Vol. 2880, Springer-Verlag, Berlin Heidelberg, 2003, pp. 119–130.

[155] A. Bretscher, D.G. Corneil, M. Habib and C. Paul, A simple linear time LexBFS cograph recognition algorithm, *SIAM Journal on Discrete Mathematics* 22 (2008) 1277–1296.

[156] F.M. Brown, *Boolean Reasoning: The Logic of Boolean Equations*, Kluwer Academic Publishers, Boston - Dordrecht - London, 1990.

[157] J. Bruck, Fourier transforms and threshold circuit complexity, in: Y. Crama and P.L. Hammer, eds., *Boolean Models and Methods in Mathematics, Computer Science, and Engineering*, Cambridge University Press, Cambridge, 2010, pp. 531–553.

[158] R. Bruni, On the orthogonalization of arbitrary Boolean formulae, *Journal of Applied Mathematics and Decision Sciences* 2 (2005) 61–74.

[159] R. Bruni and A. Sassano, A complete adaptive solver for propositional satisfiability, *Discrete Applied Mathematics* 127 (2003) 523–534.

[160] R.E. Bryant, Graph-based algorithms for Boolean function manipulation, *IEEE Transactions on Computers* 35 (1986) 677–691.

[161] N.H. Bshouty, Exact learning Boolean functions via the monotone theory, *Information and Computation* 123 (1995) 146–153.

[162] N. Bshouty, T.R. Hancock and L. Hellerstein, Learning boolean read-once formulas with arbitrary symmetric and constant fan-in gates, *Journal of Computer and System Sciences* 50 (1995) 521–542.

[163] N. Bshouty and C. Tamon, On the Fourier spectrum of monotone functions, *Journal of the Association for Computing Machinery* 43 (1996) 747–770.

[164] C. Buchheim and G. Rinaldi, Efficient reduction of polynomial zero-one optimization to the quadratic case, *SIAM Journal on Optimization* 18 (2007) 1398–1413.

[165] C. Buchheim and G. Rinaldi, Terse integer linear programs for Boolean optimization, *Journal on Satisfiability, Boolean Modeling and Computation* 6 (2009) 121–139.

[166] M. Buro and H. Kleine Büning, Report on a SAT competition, Report Nr. 110, Mathematik/Informatik, Universität Paderborn, 1992.

[167] W. Büttner and H. Simonis, Embedding Boolean expressions into logic programming, *Journal of Symbolic Computation* 4 (1987) 191–205.

[168] R. Cambini, G. Gallo and M.G. Scutellà, Flows on hypergraphs, *Mathematical Programming* 78 (1997) 195–217.

[169] A. Caprara, M. Fischetti and P. Toth, A heuristic method for the set covering problem, *Operations Research* 47 (1999) 730–743.

[170] C. Carlet, Boolean functions for cryptography and error-correcting codes, in: Y. Crama and P.L. Hammer, eds., *Boolean Models and Methods in Mathematics, Computer Science, and Engineering*, Cambridge University Press, Cambridge, 2010, pp. 257–397.

[171] C. Carlet, Vectorial Boolean functions for cryptography, in: Y. Crama and P.L. Hammer, eds., *Boolean Models and Methods in Mathematics, Computer Science, and Engineering*, Cambridge University Press, Cambridge, 2010, pp. 398–469.

[172] O. Čepek, Restricted consensus method and quadratic implicates of pure Horn functions, RUTCOR Research Report 31, Rutgers University, Piscataway, NJ September 1994.

[173] O. Čepek, *Structural properties and minimization of Horn Boolean functions*, Ph.D. thesis, RUTCOR, Rutgers University, Piscataway, NJ, October 1995.

[174] O. Čepek and P. Kučera, Known and new classes of generalized Horn formulae with polynomial recognition and SAT testing, *Discrete Applied Mathematics* 149 (2005) 14–52.

[175] O. Čepek and P. Kučera, On the complexity of minimizing the number of literals in Horn formulae, RUTCOR Research Report 11-2008, Rutgers University, Piscataway, NJ, 2008.

[176] S. Ceri, G. Gottlob and L. Tanca, *Logic Programming and Databases*, Springer-Verlag, Berlin Heidelberg, 1990.

[177] D. Chai and A. Kuehlmann, A fast pseudo-Boolean constraint solver, *IEEE Transactions on Computer-Aided Design of Integrated Circuits and Systems* 24 (2005) 305–317.

[178] S.T. Chakradhar, V.D. Agrawal and M.L. Bushnell, *Neural Models and Algorithms for Digital Testing*, Kluwer Academic Publishers, Boston - Dordrecht - London, 1991.

[179] A.K. Chandra, H.R. Lewis and J.A. Makowsky, Embedded implicational dependencies and their inference problem, in: *Proceedings of the 13th Annual ACM Symposium on the Theory of Computation*, ACM Press, New York, 1981, pp. 342–354.

[180] R. Chandrasekaran, Integer programming problems for which a simple rounding type algorithm works, in: W.R. Pulleyblank, ed., *Progress in Combinatorial Optimization*, Academic Press Canada, Toronto, 1984, pp. 101–106.

[181] V. Chandru, C.R. Coullard, P.L. Hammer, M. Montanez, and X. Sun. On renamable Horn and generalized Horn functions, *Annals of Mathematics and Artificial Intelligence* 1 (1990) 33–47.

[182] V. Chandru and J.N. Hooker, Extended Horn sets in propositional logic, *Journal of the ACM* 38 (1991) 205–221.

[183] V. Chandru and J.N. Hooker, Detecting embedded Horn structure in propositional logic, *Information Processing Letters* 42 (1992) 109–111.

[184] V. Chandru and J.N. Hooker, *Optimization Methods for Logical Inference*, John Wiley & Sons, New York etc., 1999.

[185] C.L. Chang, The unit proof and the input proof in theorem proving, *Journal of the ACM* 14 (1970) 698–707.

[186] C.-L. Chang and R.C. Lee, *Symbolic Logic and Mechanical Theorem Proving*, Academic Press, New York - San Francisco - London, 1973.

[187] M.T. Chao and J. Franco, Probabilistic analysis of a generalization of the unit-clause literal section heuristic for the k-satisfiability problem, *Information Science* 51 (1990) 289–314.

[188] A. Chateauneuf and J.Y. Jaffray, Some characterizations of lower probabilities and other monotone capacities through the use of Möbius inversion, *Mathematical Social Sciences* 17 (1989) 263–283.

[189] S.S. Chaudhry, I.D. Moon and S.T. McCormick, Conditional covering: Greedy heuristics and computational results, *Computers and Operations Research* 14 (1987) 11–18.

[190] M. Chein, Algorithmes d'écriture de fonctions Booléennes croissantes en sommes et produits, *Revue Française d'Informatique et de Recherche Opérationnelle* 1 (1967) 97–105.

[191] Y. Chen and D. Cooke, On the transitive closure representation and adjustable compression, in: *SAC06 – Proceedings of the 21st Annual ACM Symposium on Applied Computing*, Dijon, France, 2006, pp. 450–455.

[192] G. Choquet, Theory of capacities, *Annales de l'Institut Fourier* 5 (1954) 131–295.

[193] C.K. Chow, Boolean functions realizable with single threshold devices, in: *Proceedings of the IRE* 49 (1961) 370–371.

[194] C.K. Chow, On the characterization of threshold functions, in: *IEEE Symposium on Switching Circuit Theory and Logical Design*, 1961, pp. 34–48.

[195] F.R.K. Chung, R.L. Graham and M.E. Saks, A dynamic location problem for graphs, *Combinatorica* 9 (1989) 111–132.

[196] R. Church, Enumeration by rank of the elements of the free distributive lattice with 7 generators, *Notices of the American Mathematical Society* 12 (1965) 724.

[197] V. Chvátal, Edmonds polytopes and a hierarchy of combinatorial problems, *Discrete Mathematics* 4 (1973) 305–337.

[198] V. Chvátal, A greedy heuristic for the set-covering problem, *Mathematics of Operations Research*, 4 (1979) 233–235.

[199] V. Chvátal, *Linear Programming*, W.H. Freeman and Co., New York, 1983.

[200] V. Chvátal and C. Ebenegger, A note on line digraphs and the directed max-cut problem, *Discrete Applied Mathematics* 29 (1990) 165–170.

[201] V. Chvátal and P.L. Hammer, Aggregation of inequalities in integer programming, *Annals of Discrete Mathematics* 1 (1977) 145–162.

[202] V. Chvátal and B. Reed, Mick gets some (the odds are on his side), in: *Proceedings of the 33rd Annual IEEE Symposium on the Foundations of Computer Science*, IEEE, 1992, pp. 620–627.

[203] V. Chvátal and E. Szemerédi, Many hard examples for resolution, *Journal of the Association for Computer Machinery* 35 (1988) 759–788.

[204] E. Clarke, A. Biere, R. Raimi and Y. Zhu, Bounded model checking using satisfiability solving, *Formal Methods in System Design* 19 (2001) 7–34.

[205] C.J. Colbourn, *The Combinatorics of Network Reliability*, Oxford University Press, New York, 1987.

[206] C.J. Colbourn, Boolean aspects of network reliability, in: Y. Crama and P.L. Hammer, eds., *Boolean Models and Methods in Mathematics, Computer Science, and Engineering*, Cambridge University Press, Cambridge, 2010, pp. 723–759.

[207] M. Conforti, G. Cornuéjols and C. de Francesco, Perfect $0, \pm 1$ matrices, *Linear Algebra and its Applications* 43 (1997) 299–309.

[208] S.A. Cook, The complexity of theorem-proving procedures, in: *Proceedings of the Third ACM Symposium on the Theory of Computing,* 1971, pp. 151–158.

[209] S.A. Cook and D.G. Mitchell, Finding hard instances for the satisfiability problem: A survey, in: D. Du, J. Gu and P.M. Pardalos, eds., *Satisfiability Problem: Theory and Applications*, DIMACS series in Discrete Mathematics and Theoretical Computer Science, Vol. 35, American Mathematical Society, 1997, pp. 1–17.

[210] W.J. Cook, C.R. Coullard and Gy. Turán, On the complexity of cutting-plane proofs, *Discrete Applied Mathematics* 18 (1987) 25–38.

[211] W.J. Cook, W.H. Cunningham, W.R. Pulleyblank and A. Schrijver, *Combinatorial Optimization*, Wiley-Interscience, New York, 1998.

[212] D. Coppersmith and S. Winograd, On the asymptotic complexity of matrix multiplication, *SIAM Journal on Computing* 11 (1982) 472–492.

[213] D. Corneil, H. Lerchs and L. Burlingham, Complement reducible graphs, *Discrete Applied Mathematics* 3 (1981) 163–174.

[214] D. Corneil, Y. Perl and L. Stewart, A linear recognition algorithm for cographs, *SIAM Journal on Computing* 14 (1985) 926–934.

[215] G. Cornuéjols, *Combinatorial Optimization*, SIAM, Philadelphia, 2001.

[216] R.W. Cottle and A.F. Veinott, Polyhedral sets having a least element, *Mathematical Programming* 3 (1972) 238–249.

[217] M. Couceiro and S. Foldes, Definability of Boolean function classes by linear equations over GF(2), *Discrete Applied Mathematics* 142 (2004) 29–34.

[218] M. Couceiro and S. Foldes, On closed sets of relational constraints and classes of functions closed under variable substitutions, *Algebra Universalis* 54 (2005) 149–165.

[219] M. Couceiro and S. Foldes, Functional equations, constraints, definability of function classes, and functions of Boolean variables, *Acta Cybernetica* 18 (2007) 61–75.

[220] M. Couceiro and M. Pouzet, On a quasi-ordering on Boolean functions, *Theoretical Computer Science* 396 (2008) 71–87.

[221] O. Coudert, Two-level logic minimization: An overview, *Integration: The VLSI Journal* 17 (1994) 97–140.

[222] O. Coudert and T. Sasao, Two-level logic minimization, in: *Logic Synthesis and Verification*, S. Hassoun and T. Sasao, eds., Kluwer Academic Publishers, Norwell, MA, 2002, pp. 1–27.

[223] M.B. Cozzens and R. Leibowitz, Multidimensional scaling and threshold graphs, *Journal of Mathematical Psychology* 31 (1987) 179–191.

[224] Y. Crama, *Recognition and Solution of Structured Discrete Optimization Problems*, Ph.D. thesis, Rutgers University, Piscataway, NJ, 1987.

[225] Y. Crama, Dualization of regular Boolean functions, *Discrete Applied Mathematics* 16 (1987) 79–85.

[226] Y. Crama, Recognition problems for special classes of polynomials in 0–1 variables, *Mathematical Programming* 44 (1989) 139–155.

[227] Y. Crama, Concave extensions for nonlinear 0–1 maximization problems, *Mathematical Programming* 61 (1993) 53–60.

[228] Y. Crama, Combinatorial optimization models for production scheduling in automated manufacturing systems, *European Journal of Operational Research* 99 (1997) 136–153.

[229] Y. Crama, O. Ekin and P.L. Hammer, Variable and term removal from Boolean formulae, *Discrete Applied Mathematics* 75 (1997) 217–230.

[230] Y. Crama and P.L. Hammer, Recognition of quadratic graphs and adjoints of bidirected graphs, in: G.S. Bloom, R.L. Graham and J. Malkevitch, eds., *Combinatorial Mathematics: Proceedings of the Third International Conference*, Annals of the New York Academy of Sciences, Vol. 555, 1989, pp. 140–149.

[231] Y. Crama and P.L. Hammer, eds., *Boolean Models and Methods in Mathematics, Computer Science, and Engineering*, Cambridge University Press, Cambridge, 2010.

[232] Y. Crama, P.L. Hammer and R. Holzman, A characterization of a cone of pseudo-Boolean functions via supermodularity-type inequalities, in: P. Kall, J. Kohlas, W. Popp and C.A. Zehnder, eds., *Quantitative Methoden in den Wirtschaftswissenschaften*, Springer-Verlag, Berlin-Heidelberg, 1989, pp. 53–55.

[233] Y. Crama, P.L. Hammer and T. Ibaraki, Cause-effect relationships and partially defined Boolean functions, *Annals of Operations Research* 16 (1988) 299–326.

[234] Y. Crama, P.L. Hammer, B. Jaumard and B. Simeone, Product form parametric representation of the solutions to a quadratic Boolean equation, *RAIRO - Operations Research* 21 (1987) 287–306.

[235] Y. Crama, P. Hansen and B. Jaumard, The basic algorithm for pseudo-Boolean programming revisited, *Discrete Applied Mathematics* 29 (1990) 171–185.

[236] Y. Crama and L. Leruth, Control and voting power in corporate networks: Concepts and computational aspects, *European Journal of Operational Research* 178 (2007) 879–893.

[237] Y. Crama, L. Leruth, L. Renneboog and J.-P. Urbain, Corporate control concentration measurement and firm performance, in: J.A. Batten and T.A. Fetherston, eds., *Social Responsibility: Corporate Governance Issues*, Research in International Business and Finance (Volume 17), Elsevier, Amsterdam, 2003, pp. 123–149.

[238] Y. Crama and J.B. Mazzola, Valid inequalities and facets for a hypergraph model of the nonlinear knapsack and FMS part-selection problems, *Annals of Operations Research* 58 (1995) 99–128.

[239] J.M. Crawford and L.D. Auton, Experimental results on the crossover point in random 3-SAT, *Artificial Intelligence* 81 (1996) 31–57.

[240] N. Creignou, A dichotomy theorem for maximum generalized satisfiability problems, *Journal of Computer and System Sciences* 51 (1995) 511–522.

[241] N. Creignou and H. Daudé, Generalized satisfiability problems: Minimal elements and phase transitions, *Theoretical Computer Science* 302 (2003) 417–430.

[242] N. Creignou and H. Daudé, The SAT–UNSAT transition for random constraint satisfaction problems, *Discrete Mathematics* 309 (2009) 2085–2099.

[243] N. Creignou, S. Khanna and M. Sudan, *Complexity Classifications of Boolean Constraint Satisfaction Problems*, SIAM Monographs on Discrete Mathematics and Applications, SIAM, Philadelphia, 2001.

[244] P. Crescenzi and V. Kann, eds., *A compendium of NP optimization problems*, published electronically at http://www.nada.kth.se/~viggo/wwwcompendium/ (2005).

[245] H.P. Crowder, E.L. Johnson and M.W. Padberg, Solving large-scale zero–one linear programming problems, *Operations Research* 31 (1983) 803–834.

[246] J. Cubbin and D. Leech, The effect of shareholding dispersion on the degree of control in British companies: Theory and measurement, *The Economic Journal* 93 (1983) 351–369.

[247] R. Cunninghame-Green, *Minimax Algebra*, Lecture Notes in Economics and Mathematical Systems, Vol. 166, Springer, Berlin, 1979.

[248] H.A. Curtis, *A New Approach to the Design of Switching Circuits*, D. Van Nostrand, Princeton, NJ, 1962.

[249] S.L.A. Czort, *The Complexity of Minimizing Disjunctive Normal Form Formulas*, Master's thesis, University of Aarhus, 1999.

[250] E. Dahlhaus, Learning monotone read-once formulas in quadratic time, *Unpublished manuscript*, Department of Computer Science, University of Sydney, 1990.

[251] E. Dahlhaus, Efficient parallel recognition algorithms of cographs and distance hereditary graphs, *Discrete Applied Mathematics* 57 (1995) 29–44.

[252] V. Dahllöf, P. Jonsson and M. Wahlström, Counting models for 2SAT and 3SAT formulae, *Theoretical Computer Science* 332 (2005) 265–291.

[253] M. Dalal and D.W. Etherington, A hierarchy of tractable satisfiability problems, *Information Processing Letters* 44 (1992) 173–180.

[254] G. Danaraj and V. Klee, Which spheres are shellable? *Annals of Discrete Mathematics* 2 (1978) 33–52.

[255] E. Dantsin, A. Goerdt, E.A. Hirsch, R. Kannan, J. Kleinberg, Ch. Papadimitriou, P. Raghavan and U. Schöning, A deterministic $(2 - 2/(k + 1))^n$ algorithm for k-SAT based on local search, *Theoretical Computer Science* 289 (2002) 69–83.

[256] G.B. Dantzig, On the significance of solving linear programming problems with some integer variables, *Econometrica* 28 (1960) 30–44.

[257] A. Darwiche, New advances in compiling CNF to decomposable negation normal form, in: *Proceedings of the 16th European Conference on Artificial Intelligence*, Valencia, Spain, 2004, pp. 328–332.

[258] S.B. Davidson, H. Garcia-Molina and D. Skeen, Consistency in partitioned networks, *ACM Computing Surveys* 17 (1985) 341–370.

[259] M. Davio, J.-P. Deschamps and A. Thayse, *Discrete and Switching Functions*, McGraw-Hill, New York, 1978.

[260] M. Davis, G. Logemann and D. Loveland, A machine program for theorem-proving, *Communications of the ACM* 5 (1962) 394–397.

[261] M. Davis and H. Putnam, A computing procedure for quantification theory, *Journal of the Association for Computing Machinery* 7 (1960) 201–215.

[262] T. Davoine, P.L. Hammer and B. Vizvári, A heuristic for Boolean optimization problems, *Journal of Heuristics* 9 (2003) 229–247.

[263] P.M. Dearing, P.L. Hammer and B. Simeone, Boolean and graph theoretic formulations of the simple plant location problem, *Transportation Science* 26 (1992) 138–148.

[264] R. Dechter and J. Pearl, Structure identification in relational data, *Artificial Intelligence* 58 (1992) 237–270.

[265] E. de Klerk and J.P. Warners, Semidefinite programming relaxations for MAX 2-SAT and 3-SAT: Computational perspectives, in: P.M. Pardalos, A. Migdalas and R.E. Burkard, eds., *Combinatorial and Global Optimization*, Series on Applied Optimization, Volume 14, World Scientific Publishers, River Edge, NJ, 2002, pp. 161–176.

[266] E. de Klerk, J.P. Warners and H. van Maaren, Relaxations of the satisfiability problem using semidefinite programming, *Journal of Automated Reasoning* 24 (2000) 37–65.

[267] C. Delobel and R.G. Casey, Decomposition of a database and the theory of Boolean switching functions, *IBM Journal of Research and Development* 17 (1973) 374–386.

[268] X. Deng and C.H. Papadimitriou, On the complexity of cooperative solution concepts, *Mathematics of Operations Research* 19 (1994) 257–266.

[269] M.L. Dertouzos, *Threshold Logic: A Synthesis Approach*, M.I.T. Press, Cambridge, MA, 1965.

[270] M.M. Deza and M. Laurent, *Geometry of Cuts and Metrics*, Springer-Verlag, Berlin, 1997.

[271] I. Diakonikolas and R.A. Servedio, Improved approximation of linear threshold functions, in: *Proceedings of the 24th Annual IEEE Conference on Computational Complexity*, IEEE Computer Society, Los Alamitos, CA, 2009, pp. 161–172.

[272] G. Ding, Monotone clutters, *Discrete Mathematics* 119 (1993) 67–77.

[273] G. Ding, R.F. Lax, J. Chen and P.P. Chen, Formulas for approximating pseudo-Boolean random variables, *Discrete Applied Mathematics* 156 (2008) 1581–1597.

[274] G. Ding, R.F. Lax, J. Chen, P.P. Chen and B.D. Marx, Transforms of pseudo-Boolean random variables, *Discrete Applied Mathematics* 158 (2010) 13–24.

[275] C. Domingo, N. Mishra and L. Pitt, Efficient read-restricted monotone CNF/DNF dualization by learning with membership queries, *Machine Learning* 37 (1999) 89–110.

[276] G. Dong and J. Li, Mining border descriptions of emerging patterns from dataset pairs, *Knowledge Information Systems* 8 (2005) 178–202.

[277] W.F. Dowling and J.H. Gallier, Linear time algorithms for testing the satisfiability of propositional Horn formulae, *Journal of Logic Programming* 3 (1984) 267–284.

[278] D. Du, J. Gu and P.M. Pardalos, eds., *Satisfiability Problem: Theory and Applications*, DIMACS Series in Discrete Mathematics and Theoretical Computer Science, Vol. 35, American Mathematical Society, 1997.

[279] P. Dubey and L.S. Shapley, Mathematical properties of the Banzhaf power index, *Mathematics of Operations Research* 4 (1979) 99–131.

[280] O. Dubois, Counting the number of solutions for instances of satisfiability problems, *Theoretical Computer Science* 81 (1991) 49–64.

[281] O. Dubois, P. André, Y. Boufkhad and J. Carlier, SAT versus UNSAT, in: D.S. Johnson and M.A. Trick, eds., *Cliques, Coloring, and Satisfiability*, DIMACS Series in Discrete Mathematics and Theoretical Computer Science, Vol. 26, American Mathematical Society, 1996, pp. 415–436.

[282] O. Dubois, Y. Boufkhad and J. Mandler, Typical random 3-SAT formulae and the satisfiability threshold, in: *Proceedings of the Eleventh Annual ACM-SIAM Symposium on Discrete Algorithms*, 2000, pp. 126–127.

[283] O. Dubois and G. Dequen, A backbone-search heuristic for efficient solving of hard 3-SAT formulae, in: *Proceedings of the 17th International Joint Conference on Artificial Intelligence* (IJCAI'01), Seattle, Washington, 2001, pp. 248–253.

[284] P. Duchet, Classical perfect graphs, in: *Topics on Perfect Graphs*, North-Holland, Amsterdam, 1984, pp. 67–96.

[285] Ch. Ebenegger, P.L. Hammer and D. de Werra, Pseudo-Boolean functions and stability of graphs, *Annals of Discrete Mathematics* 19 (1984) 83–97.

[286] J. Ebert, A sensitive transitive closure algorithm, *Information Processing Letters* 12 (1981) 255–258.

[287] J. Edmonds, Submodular functions, matroids, and certain polyhedra, in: R. Guy, H. Hanani, N. Sauer and J. Schönheim, eds., *Combinatorial Structures and Their Applications*, Gordon and Breach, New York, 1970, pp. 69–87.

[288] J. Edmonds and D.R. Fulkerson, Bottleneck extrema, *Journal of Combinatorial Theory* 8 (1970) 299–306.

[289] N. Eén and N. Sörensson, An extensible SAT-solver, in: *Proceedings of the 6th International Conference on Theory and Applications of Satisfiability Testing*, 2003.

[290] N. Eén and N. Sörensson, Translating pseudo-Boolean constraints into SAT, *Journal on Satisfiability, Boolean Modeling and Computation* 2 (2006) 1–26.

[291] E. Einy, The desirability relation of simple games, *Mathematical Social Sciences* 10 (1985) 155–168.

[292] E. Einy and E. Lehrer, Regular simple games, *International Journal of Game Theory* 18 (1989) 195–207.

[293] T. Eiter, Exact transversal hypergraphs and application to Boolean μ-functions, *Journal of Symbolic Computation* 17 (1994) 215–225.

[294] T. Eiter, Generating Boolean μ-expressions, *Acta Informatica* 32 (1995) 171–187.

[295] T. Eiter and G. Gottlob, Identifying the minimal transversals of a hypergraph and related problems, *SIAM Journal on Computing* 24 (1995) 1278–1304.

[296] T. Eiter, T. Ibaraki and K. Makino, Double Horn functions, *Information and Computation* 144 (1998) 155–190.

[297] T. Eiter, T. Ibaraki and K. Makino, Computing intersections of Horn theories for reasoning with models, *Artificial Intelligence* 110 (1999) 57–101.

[298] T. Eiter, T. Ibaraki and K. Makino, Bidual Horn functions and extensions, *Discrete Applied Mathematics* 96 (1999) 55–88.

[299] T. Eiter, T. Ibaraki and K. Makino. On the difference of Horn theories, *Journal of Computer and System Sciences* 61 (2000) 478–507.

[300] T. Eiter, T. Ibaraki and K. Makino, Disjunction of Horn theories and their cores, *SIAM Journal on Computing* 31 (2001) 269–288.

[301] T. Eiter, P. Kilpelainen and H. Mannila, Recognizing renamable generalized propositional Horn formulas is NP-complete, *Discrete Applied Mathematics* 59 (1995) 23–31.

[302] T. Eiter, K. Makino and G. Gottlob, Computational aspects of monotone dualization: A brief survey, *Discrete Applied Mathematics* 156 (2008) 2035–2049.

[303] O. Ekin, *Special Classes of Boolean Functions*, Ph.D. Thesis, Rutgers University, Piscataway, NJ, 1997.

[304] O. Ekin Karaşan, Dualization of quadratic Boolean functions, *Annals of Operations Research* (2011), to appear.

[305] O. Ekin, S. Foldes, P.L. Hammer and L. Hellerstein, Equational characterizations of Boolean function classes, *Discrete Mathematics* 211 (2000) 27–51.

[306] O. Ekin, P.L. Hammer and A. Kogan, On connected Boolean functions, *Discrete Applied Mathematics* 96/97 (1999) 337–362.

[307] O. Ekin, P.L. Hammer and A. Kogan, Convexity and logical analysis of data, *Theoretical Computer Science* 244 (2000) 95–116.

[308] O. Ekin, P.L. Hammer and U.N. Peled, Horn functions and submodular Boolean functions, *Theoretical Computer Science* 175 (1997) 257–270.

[309] K.M. Elbassioni, On the complexity of monotone dualization and generating minimal hypergraph transversals, *Discrete Applied Mathematics* 156 (2008) 2109–2123.

[310] C.C. Elgot, Truth functions realizable by single threshold organs, in: *IEEE Symposium on Switching Circuit Theory and Logical Design*, 1961, pp. 225–245.

[311] M.R. Emamy-K., The worst case behavior of a greedy algorithm for a class of pseudo-Boolean functions, *Discrete Applied Mathematics* 23 (1989) 285–287.

[312] P. Erdős, On some extremal problems in graph theory, *Israel Journal of Mathematics* 3 (1965) 113–116.

[313] P. Erdős and T. Gallai, Graphen mit Punkten vorgeschriebenen Graden, *Mat. Lapok* 11 (1960) 264–274.

[314] P. Erdős and J. Spencer, *Probabilistic Methods in Combinatorics*, Akadémiai Kiadó, Budapest, 1974.

[315] B. Escoffier and V.Th. Paschos, Differential approximation of MIN SAT, MAX SAT and related problems, *European Journal of Operational Research* 181 (2007) 620–633.

[316] E. Eskin, E. Halperin and R.M. Karp, Efficient reconstruction of haplotype structure via perfect phylogeny, *Journal of Bioinformatics and Computational Biology* 1 (2003) 1–20.

[317] R. Euler, Regular (2,2)-systems, *Mathematical Programming* 24 (1982) 269–283.

[318] S. Even, A. Itai and A. Shamir, On the complexity of timetable and multicommodity flow problems, *SIAM Journal on Computing* 5 (1976) 691–703.

[319] R. Fagin, Functional dependencies in a relational database and propositional logic, *IBM Journal of Research and Development* 21 (1977) 534–544.

[320] R. Fagin, Horn clauses and database dependencies, *Journal of the ACM* 29 (1982) 952–985.

[321] U.M. Fayyad, G. Piatetsky-Shapiro, P. Smyth and R. Uthurusamy, *Advances in Knowledge Discovery and Data Mining*, The MIT Press, Cambridge, MA, 1996.

[322] T. Feder, Network flow and 2-satisfiability, *Algorithmica* 11 (1994) 291–319.

[323] T. Feder, *Stable Networks and Product Graphs*, Memoirs of the American Mathematical Society, Vol. 116, No. 555, Providence, RI, 1995.

[324] U. Feige, A threshold of $\ln n$ for approximating set cover, *Journal of the Association for Computing Machinery* 45 (1998) 634–652.

[325] U. Feige and M.X. Goemans, Approximating the value of two prover proof systems, with applications to MAX SAT and MAX DICUT, in: *Proceedings of the Third Israel Symposium on Theory of Computing and Systems*, Tel Aviv, Israel, 1995, pp. 182–189.

[326] J. Feldman, Minimization of Boolean complexity in human concept learning, *Nature* 407 (2000) 630–633.

[327] J. Feldman, An algebra of human concept learning, *Journal of Mathematical Psychology* 50 (2006) 339–368.

[328] V. Feldman, Hardness of approximate two-level logic minimization and PAC learning with membership queries, in: *Proceedings of the 38th ACM Symposium on Theory of Computing* (STOC) 2006, pp. 363–372.

[329] D.S. Felsenthal and M. Machover, *The Measurement of Voting Power: Theory and Practice, Problems and Paradoxes,* Edward Elgar, Cheltenham, UK, 1998.

[330] J.R. Fernández, E. Algaba, J.M. Bilbao, A. Jiménez, N. Jiménez and J.J. López, Generating functions for computing the Myerson value, *Annals of Operations Research* 109 (2002) 143–158

[331] M.J. Fischer and A.R. Meyer, Boolean matrix multiplication and transitive closure, in: *Proceedings of the 12th Annual IEEE Symposium on the Foundations of Computer Science*, IEEE, 1971, pp. 129–131.

[332] M.L. Fisher, G.L. Nemhauser and L.A. Wolsey, An analysis of approximations for maximizing submodular set functions - II, *Mathematical Programming Study* 8 (1978) 73–87.

[333] C. Flament, L'analyse booléenne de questionnaires, *Mathématiques et Sciences Humaines* 12 (1966) 3–10.

[334] S. Foldes, Equational classes of Boolean functions via the HSP Theorem, *Algebra Universalis* 44 (2000) 309–324.

[335] S. Foldes and P.L. Hammer, Split graphs, *Congressus Numerantium* 19 (1977) 311–315.

[336] S. Foldes and P.L. Hammer, Disjunctive and conjunctive normal forms of pseudo-Boolean functions, *Discrete Applied Mathematics* 107 (2000) 1–26.

[337] S. Foldes and P.L. Hammer, Monotone, Horn and quadratic pseudo-Boolean functions, *Journal of Universal Computer Science* 6 (2000) 97–104.

[338] S. Foldes and P.L. Hammer, Disjunctive analogues of submodular and supermodular pseudo-Boolean functions, *Discrete Applied Mathematics* 142 (2004) 53–65.

[339] S. Foldes and P.L. Hammer, Submodularity, supermodularity, and higher-order monotonicities of pseudo-Boolean functions, *Mathematics of Operations Research* 30 (2005) 453–461.

[340] S. Foldes and G.R. Pogosyan, Post classes characterized by functional terms, *Discrete Applied Mathematics* 142 (2004) 35–51.

[341] L.R. Ford and D.R. Fulkerson, *Flows in Networks*, Princeton University Press, Princeton, NJ 1962.

[342] R. Fortet, L'algèbre de Boole et ses applications en recherche opérationnelle, *Cahiers du Centre d'Etudes de Recherche Opérationnelle* 1 (1959) 5–36.

[343] R. Fortet, Applications de l'algèbre de Boole en recherche opérationnelle, *Revue Française de Recherche Opérationnelle* 4 (1960) 17–26.

[344] J. Franco, Probabilistic analysis of satisfiability algorithms, in: Y. Crama and P.L. Hammer, eds., *Boolean Models and Methods in Mathematics, Computer Science, and Engineering*, Cambridge University Press, Cambridge, 2010, pp. 99–159.

[345] J. Franco and M. Paull, Probabilistic analysis of the Davis-Putnam procedure for solving the satisfiability problem, *Discrete Applied Mathematics* 5 (1983) 77–87.

[346] L. Fratta and U.G. Montanari, A Boolean algebra method for computing the terminal reliability in a communication network, *IEEE Transactions on Circuit Theory* CT-20 (1973) 203–211.

[347] M. Fredman and L. Khachiyan, On the complexity of dualization of monotone disjunctive normal forms, *Journal of Algorithms* 21 (1996) 618–628.

[348] E. Friedgut, Sharp threshold of graph properties, and the k-SAT problem, *Journal of the American Mathematical Society* 12 (1999) 1017–1054 (with an appendix by J. Bourgain).

[349] A.M. Frieze and B. Reed, Probabilistic analysis of algorithms, in: M. Habib, C. McDiarmid, J. Ramirez-Alfonsin and B. Reed, eds., *Probabilistic Methods for Algorithmic Discrete Mathematics*, Springer, Berlin, 1998, pp. 36–92.

[350] A. Frieze and N.C. Wormald, Random k-SAT: A tight threshold for moderately growing k, *Combinatorica* 25 (2005) 297–305.

[351] S. Fujishige, *Submodular Functions and Optimization, Annals of Discrete Mathematics* Vol. 58, Elsevier, Amsterdam, 2005.

[352] T. Fujito, On approximation of the submodular set cover problem, *Operations Research Letters* 25 (1999) 169–174.

[353] D.R. Fulkerson, Networks, frames, blocking systems, in: G.B. Dantzig and A.F. Veinott Jr., eds., *Mathematics of the Decision Sciences - Part I*, American Mathematical Society, Providence, RI, 1968, pp. 303–334.

[354] M. Fürer and S.P. Kasiviswanathan, Algorithms for counting 2-SAT solutions and colorings with applications, *Algorithmic Aspects in Information and Management*, Lecture Notes in Computer Science, Vol. 4508, Springer-Verlag, Berlin, 2007, pp. 47–57.

[355] M.E. Furman, Application of a method of fast multiplication to the problem of finding the transitive closure of a graph, *Soviet Mathematics Doklady* 22 (1970) 1252.

[356] I.J. Gabelman, *The Functional Behavior of Majority (Threshold) Elements*, Ph.D. Dissertation, Department of Electrical Engineering, Syracuse University, NY, 1961.

[357] H.N. Gabow and R.E. Tarjan, A linear-time algorithm for a special case of disjoint set union, *Journal of Computer and System Sciences* 30 (1996) 209–221.

[358] T. Gallai, Transitiv orientierbare Graphen, *Acta Mathematica Academiae Scientiarum Hungaricae* 18 (1967) 25–66.

[359] H. Gallaire and J. Minker, eds., *Logic and Data Bases*, Plenum, New York, 1978.

[360] G. Gallo, C. Gentile, D. Pretolani and G. Rago, Max Horn sat and the minimum cut problem in directed hypergraphs, *Mathematical Programming* 80 (1998) 213–237.

[361] G. Gallo, G. Longo, S. Nguyen and S. Pallottino, Directed hypergraphs and applications, *Discrete Applied Mathematics* 42 (1993) 177–201.

[362] G. Gallo and M.G. Scutellà, Polynomially solvable satisfiability problems, *Information Processing Letters* 29 (1988) 221–227.

[363] G. Gallo and M.G. Scutellà, Directed hypergraphs as a modelling paradigm, *Rivista AMASES* 21 (1998) 97–123.

[364] G. Gallo and B. Simeone, On the supermodular knapsack problem, *Mathematical Programming Study* 45 (1989) 295–309.

[365] G. Gallo and G. Urbani, Algorithms for testing the satisfiability of propositional formulae, *Journal of Logic Programming* 7 (1989) 45–61.

[366] G. Galperin and A. Tolpygo, Moscow Mathematical Olympiads, in: A. Kolmogorov, ed., *Prosveschenie* (Education), Moscow, USSR, 1986, Problem 72 (in Russian).

[367] F. Galvin, Horn sentences, *Annals of Mathematical Logic* 1 (1970) 389–422.

[368] G. Gambarelli, Power indices for political and financial decision making, *Annals of Operations Research* 51 (1994) 165–173.

[369] B. Ganter and R. Wille, *Formal Concept Analysis - Mathematical Foundations*, Springer-Verlag, Berlin, 1999.

[370] H. Garcia-Molina and D. Barbara, How to assign votes in a distributed system, *Journal of the Association for Computer Machinery* 32 (1985) 841–860.

[371] M.R. Garey and D.S. Johnson, *Computers and Intractability: A Guide to the Theory of NP-Completeness*, W.H. Freeman, New York, 1979.

[372] M.R. Garey, D.S. Johnson and L. Stockmeyer, Some simplified NP-complete graph problems, *Theoretical Computer Science* 1 (1976) 237–267.

[373] M.A. Garrido, A. Márquez, A. Morgana and J.R. Portillo, Single bend wiring on surfaces, *Discrete Applied Mathematics* 117 (2002) 27–40.

[374] F. Gavril, Testing for equality between maximum matching and minimum node covering, *Information Processing Letters* 6 (1977) 199–202.

[375] F. Gavril, An efficiently solvable graph partition problem to which many problems are reducible, *Information Processing Letters* 45 (1993) 285–290.

[376] A. Genkin, C.A. Kulikowski and I.B. Muchnik, Set covering submodular maximization: An optimal algorithm for data mining in bioinformatics and medical informatics, *Journal of Intelligent and Fuzzy Systems* 12 (2002) 5–17.

[377] I. Gent, H. van Maaren and T. Walsh, eds., *SAT2000: Highlights of Satisfiability Research in the Year 2000*, IOS Press, Amsterdam, 2000.

[378] F. Giannessi and F. Niccolucci, Connections between nonlinear and integer programming problems, *Symposia Mathematica* XIX (1976) 161–176.

[379] R. Giles and R. Kannan, A characterization of threshold matroids, *Discrete Mathematics* 30 (1980) 181–184.

[380] P.C. Gilmore, A proof method for quantification theory: Its justification and realization, *IBM Journal of Research and Development* 4 (1960) 28–35.

[381] J.F. Gimpel, A method of producing a Boolean function having an arbitrarily prescribed prime implicant table, *IEEE Transactions on Electronic Computers* EC-14 (1965) 485–488.

[382] J.F. Gimpel, A reduction technique for prime implicant tables, *IEEE Transactions on Electronic Computers* EC-14 (1965) 535–541.

[383] A. Ginsberg, Knowledge-base reduction: A new approach to checking knowledge bases for inconsistency and redundancy, in: *Proceedings of the Seventh National Conference on Artificial Intelligence*, 1988, pp. 585–589.

[384] E. Giunchiglia, F. Giunchiglia and A. Tacchella, SAT-based decision procedures for classical modal logics, in: I. Gent, H. van Maaren and T. Walsh, eds., *SAT2000: Highlights of Satisfiability Research in the Year 2000*, IOS Press, Amsterdam, 2000, pp. 403–426.

[385] V.V. Glagolev, Some estimates of disjunctive normal forms of functions in the algebra of logic, in: *Problems of Cybernetics*, Vol. 19, Nauka, Moscow, 1967, pp. 75–94 (in Russian).

[386] F. Glover and J-K. Hao, Efficient evaluations for solving large 0-1 unconstrained quadratic optimisation problems, *International Journal of Metaheuristics* 1 (2010) 3–10.

[387] F. Glover and E. Woolsey, Converting the 0-1 polynomial programming problem to a 0-1 linear program, *Operations Research* 22 (1974) 180–182.

[388] R. Gnanadesikan, *Methods for Statistical Data Analysis of Multivariate Observations*, Wiley-Interscience, New York, 1977.

[389] M.X. Goemans and D.P. Williamson, New $\frac{3}{4}$-approximation algorithm for the maximum satisfiability problem, *SIAM Journal on Discrete Mathematics* 7 (1994) 656–666.

[390] A. Goerdt, A threshold for unsatisfiability, in: I.M. Havel and V. Koubek, eds., *Proceedings of the 17th International Symposium on Mathematical Foundations of Computer Science*, Lecture Notes in Computer Science, Vol. 629, Springer-Verlag, Berlin, 1992, pp. 264–274.

[391] G. Gogic, C. Papadimitriou and M. Sideri, Incremental recompilation of knowledge, *Journal of Artificial Intelligence Research* 8 (1998) 23–37.

[392] E. Goldberg and Y. Novikov, BerkMin: A fast and robust SAT solver, *Discrete Applied Mathematics* 155 (2007) 1549–1561.

[393] B. Goldengorin, Maximization of submodular functions: Theory and enumeration algorithms, *European Journal of Operational Research* 198 (2009) 102–112.

[394] B. Goldengorin, D. Ghosh and G. Sierksma, Equivalent instances of the simple plant location problem, SOM Research Report No. 00A54, University of Groningen, The Netherlands, 2000.

[395] B. Goldengorin, D. Ghosh and G. Sierksma, Branch and peg algorithms for the simple plant location problem, *Computers and Operations Research* 31 (2004) 241–255.

[396] S.A. Goldman, M.J. Kearns and R.E. Schapire, Exact identification of read-once formulas
       using fixed points of amplification functions, *SIAM Journal on Computing* 22 (1993)
       705–726.

[397] J. Goldsmith, R.H. Sloan, B. Szorenyi and G. Turán, Theory revision with queries: Horn,
       read-once, and parity formulas, *Artificial Intelligence* 156 (2004) 139–176.

[398] M.C. Golumbic, *Algorithmic Graph Theory and Perfect Graphs*, Academic Press,
       New York, 1980. Second edition: *Annals of Discrete Mathematics*, Vol. 57, Elsevier,
       Amsterdam, 2004.

[399] M.C. Golumbic and A. Mintz, Factoring logic functions using graph partitioning, in:
       Proceedings of the IEEE/ACM International Conference on Computer Aided Design,
       November 1999, pp. 195–198.

[400] M.C. Golumbic, A. Mintz and U. Rotics, Factoring and recognition of read-once functions
       using cographs and normality, in: *Proceedings of the 38th Design Automation Conference*,
       June 2001, pp. 109–114.

[401] M.C. Golumbic, A. Mintz and U. Rotics, Factoring and recognition of read-once functions
       using cographs and normality and the readability of functions associated with partial
       k-trees, *Discrete Applied Mathematics* 154 (2006) 1465–1477.

[402] M.C. Golumbic, A. Mintz and U. Rotics, An improvement on the complexity of factoring
       read-once Boolean functions, *Discrete Applied Mathematics* 156 (2008) 1633–1636.

[403] C.P. Gomes, B. Selman, N. Crato and H. Kautz, Heavy-tailed phenomena in satisfiability
       and constraint satisfaction problems, in: I. Gent, H. van Maaren and T. Walsh, eds.,
       SAT2000: Highlights of Satisfiability Research in the Year 2000, IOS Press, Amsterdam,
       2000, pp. 15–41.

[404] A. Goralcikova and V. Koubek, A reduct and closure algorithm for graphs, in: *Pro-
       ceedings of the 8th Symposium on Mathematical Foundations of Computer Science*
       (MFCS'79), Lecture Notes in Computer Science, Vol. 74, Springer-Verlag, Berlin, 1979,
       pp. 301–307.

[405] M. Grabisch, The application of fuzzy integrals in multicriteria decision making, *European
       Journal of Operational Research* 89 (1996) 445–456.

[406] M. Grabisch, J.-L. Marichal, R. Mesiar and E. Pap, *Aggregation Functions*, Cambridge
       University Press, Cambridge, 2009.

[407] M. Grabisch, J.-L. Marichal and M. Roubens, Equivalent representations of set functions,
       Mathematics of Operations Research 25 (2) (2000) 157–178.

[408] D. Granot and F. Granot, Generalized covering relaxations for 0–1 programs, *Operations
       Research* 28 (1980) 1442–1450.

[409] D. Granot, F. Granot and J. Kallberg, Covering relaxation for positive 0–1 polynomial
       programs, *Management Science* 25 (1979) 264–273.

[410] F. Granot and P.L. Hammer, On the use of Boolean functions in 0–1 programming,
       Methods of Operations Research 12 (1972) 154–184.

[411] F. Granot and P.L. Hammer, On the role of generalized covering problems, *Cahiers du
       Centre d'Etudes de Recherche Opérationnelle* 16 (1974) 277–289.

[412] J.F. Groote and J.P. Warners, The propositional formula checker HeerHugo, in: I. Gent,
       H. van Maaren and T. Walsh, eds., *SAT2000: Highlights of Satisfiability Research in the
       Year 2000*, IOS Press, Amsterdam, 2000, pp. 261–282.

[413] A. Grossi, Algorithme à séparation de variables pour la dualisation d'une fonction
       booléenne, *R.A.I.R.O.* 8 (B-1) (1974) 41–55.

[414] M. Grötschel, L. Lovász and A. Schrijver, The ellipsoid method and its consequences in
       combinatorial optimization, *Combinatorica* 1 (1981) 169–197.

[415] J. Gu, Efficient local search for very large-scale satisfiability problems, *SIGART Bulletin*
       3 (1992) 8–12.

[416] J. Gu, Local search for satisfiability (SAT) problems, *IEEE Transactions on Systems, Man
       and Cybernetics* 23 (1993) 1108–1129.

[417] J. Gu, Global optimization for satisfiability (SAT) problems, *IEEE Transactions on
       Knowledge and Data Engineering* 6 (1994) 361–381.

[418] J. Gu, P.W. Purdom, J. Franco and B.W. Wah, Algorithms for the satisfiability (SAT) problem: A survey, in: D. Du, J. Gu and P.M. Pardalos, eds., *Satisfiability Problem: Theory and Applications*, DIMACS series in Discrete Mathematics and Theoretical Computer Science, Vol. 35, American Mathematical Society, 1997. pp. 19–151.

[419] B. Guenin, Perfect and ideal 0, ±1 matrices, *Mathematics of Operations Research* 23 (1998) 322–338.

[420] S. Gueye and P. Michelon, A linearization framework for unconstrained quadratic (0–1) problems, *Discrete Applied Mathematics* 157 (2009) 1255–1266.

[421] V. Gurvich, Nash-solvability of positional games in pure strategies, *USSR Computer Mathematics and Mathematical Physics* 15(2) (1975) 74–87.

[422] V. Gurvich, On repetition-free Boolean functions, *Uspekhi Mat. Nauk.* 32 (1977) 183–184, (in Russian); translated as: On read-once Boolean functions, *Russian Mathematical Surveys* 32 (1977) 183–184.

[423] V. Gurvich, *Applications of Boolean Functions and Networks in Game Theory*, Ph.D. thesis, Moscow Institute of Physics and Technology, Moscow, USSR, 1978 (in Russian).

[424] V. Gurvich, On the normal form of positional games, *Soviet Mathematics Doklady* 25(3) (1982) 572–575.

[425] V. Gurvich, Some properties and applications of complete edge-chromatic graphs and hypergraphs, *Soviet Mathematics Doklady* 30(3) (1984) 803–807.

[426] V. Gurvich, Criteria for repetition-freeness of functions in the algebra of logic, *Soviet Mathematics Doklady* 43(3) (1991) 721–726.

[427] V. Gurvich, Positional game forms and edge-chromatic graphs, *Soviet Mathematics Doklady* 45(1) (1992) 168–172.

[428] V. Gurvich and L. Khachiyan On the frequency of the most frequently occurring variable in dual DNFs, *Discrete Mathematics* 169 (1997) 245–248.

[429] V. Gurvich and L. Khachiyan, On generating the irredundant conjunctive and disjunctive normal forms of monotone Boolean functions, *Discrete Applied Mathematics* 96 (1999) 363–373.

[430] M. Habib, F. de Montgolfier and C. Paul, A simple linear-time modular decomposition algorithm, in: *Proceedings of the 9th Scandinavian Workshop on Algorithm Theory - SWAT 2004*, Lecture Notes in Computer Science, Vol. 3111, Springer-Verlag, Berlin, 2004, pp. 187–198.

[431] M. Habib and C. Paul, A simple linear time algorithm for cograph recognition, *Discrete Applied Mathematics* 145 (2005) 183–197.

[432] M. Hagen, *Algorithmic and Computational Complexity Issues of MONET*, Ph.D. thesis, Friedrich-Schiller-Universität Jena, Germany, 2009.

[433] A. Haken, The intractability of resolution, *Theoretical Computer Science* 39 (1985) 297–308.

[434] P.L. Hammer, Plant location: A pseudo-Boolean approach, *Israel Journal of Technology* 6 (1968) 330–332.

[435] P.L. Hammer, A note on the monotonicity of pseudo-Boolean functions, *Zeitschrift für Operations Research* 18 (1974) 47–50.

[436] P.L. Hammer, Pseudo-Boolean remarks on balanced graphs, *International Series of Numerical Mathematics* 36 (1977) 69–78.

[437] P.L. Hammer, The conflict graph of a pseudo-Boolean function, Bell Laboratories, Technical Report, August 1978.

[438] P.L. Hammer, Boolean elements in combinatorial optimization, in: P.L. Hammer, E.L. Johnson and B. Korte, eds., *Discrete Optimization, Annals of Discrete Mathematics* Vol. 4, Elsevier, Amsterdam, 1979, pp. 51–71.

[439] P.L. Hammer and P. Hansen, Logical relations in quadratic 0–1 programming, *Revue Roumaine de Mathématiques Pures et Appliquées* 26 (1981) 421–429.

[440] P.L. Hammer, P. Hansen and B. Simeone, Roof duality, complementation and persistency in quadratic 0–1 optimization, *Mathematical Programming* 28 (1984) 121–155.

[441] P.L. Hammer and R. Holzman, Approximations of pseudo-Boolean functions: Applications to game theory, *ZOR - Methods and Models of Operations Research* 36 (1992) 3–21.

[442] P.L. Hammer, T. Ibaraki and B. Simeone, Threshold sequences, *SIAM Journal on Algebraic and Discrete Methods* 2 (1981) 39–49.

[443] P.L. Hammer, E.L. Johnson and U.N. Peled, Regular 0–1 programs, *Cahiers du Centre d'Etudes de Recherche Opérationnelle* 16 (1974) 267–276.

[444] P.L. Hammer, E.L. Johnson and U.N. Peled, Facets of regular 0–1 polytopes, *Mathematical Programming* 8 (1975) 179–206.

[445] P.L. Hammer and B. Kalantari, A bound on the roof duality gap, in: B. Simeone, ed., *Combinatorial Optimization*, Lecture Notes in Mathematics, Vol. 1403, Springer, Berlin, 1989, pp. 254–257.

[446] P.L. Hammer and A. Kogan, Horn functions and their DNFs, *Information Processing Letters* 44 (1992) 23–29.

[447] P.L. Hammer and A. Kogan, Optimal compression of propositional knowledge bases: complexity and approximation, *Artificial Intelligence* 64 (1993) 131–145.

[448] P.L. Hammer and A. Kogan, Graph based methods for Horn knowledge compression, in: *Proceedings of the 27th Hawaii International Conference on System Sciences*, IEEE Press, 1994, pp. 300–309.

[449] P.L. Hammer and A. Kogan, Quasi-acyclic propositional Horn knowledge bases: optimal compression, *IEEE Transaction on Knowledge and Data Engineering* 7(5) (1995) 751–762.

[450] P.L. Hammer and A. Kogan, Essential and redundant rules in Horn knowledge bases, *Decision Support Systems* 16 (1996) 119–130.

[451] P.L. Hammer, F. Maffray and M. Queyranne, Cut-threshold graphs, *Discrete Applied Mathematics* 30 (1991) 163–179.

[452] P.L. Hammer and N.V.R. Mahadev, Bithreshold graphs, *SIAM Journal on Applied Mathematics* 6 (1985) 497–506.

[453] P.L. Hammer, N.V.R. Mahadev and D. de Werra, The struction of a graph: Application to CN-free graphs, *Combinatorica* 5 (1985) 141–147.

[454] P.L. Hammer and S. Nguyen, APOSS – A partial order in the solution space of bivalent programs, in: N. Christofides, A. Mingozzi, C. Sandi, and P. Toth, eds., *Combinatorial Optimization*, John Wiley & Sons, Chichester, New York, 1979, pp. 93–106.

[455] P.L. Hammer, U.N. Peled and M.A. Pollatschek, An algorithm to dualize a regular switching function, *IEEE Transactions on Computers* C-28 (1979) 238–243.

[456] P.L. Hammer, U.N. Peled and S. Sorensen, Pseudo-Boolean functions and game theory I. Core elements and Shapley value, *Cahiers du Centre d'Etudes de Recherche Opérationnelle* 19, 1977, 159–176.

[457] P.L. Hammer and I.G. Rosenberg, Linear decomposition of a positive group-Boolean function, in: L. Collatz and W. Wetterling, eds., *Numerische Methoden bei Optimierung*, Vol. 2, Birkhauser, Basel, 1974, pp. 51–62.

[458] P.L. Hammer, I.G. Rosenberg and S. Rudeanu, On the determination of the minima of pseudo-Boolean functions (in Romanian), *Studii şi Cercetari Matematice* 14 (1963) 359–364.

[459] P.L. Hammer and A.A. Rubin, Some remarks on quadratic programming with 0–1 variables, *Revue Française d'Informatique et de Recherche Opérationnelle* 4 (1970) 67–79.

[460] P.L. Hammer and S. Rudeanu, *Boolean Methods in Operations Research and Related Areas,* Springer, Berlin, 1968.

[461] P.L. Hammer and B. Simeone, Quasimonotone Boolean functions and bistellar graphs, *Annals of Discrete Mathematics* 9 (1980) 107–119.

[462] P.L. Hammer and B. Simeone, Order relations of variables in 0 − 1 programming, in: C. Ribeiro, G. Laporte and S. Martello, eds., *Surveys in Combinatorial Optimization, Annals of Discrete Mathematics* Vol. 31, North-Holland, Amsterdam, 1987, pp. 83–111.

[463] P.L. Hammer and B. Simeone, Quadratic functions of binary variables, in: B. Simeone, ed., *Combinatorial Optimization*, Lecture Notes in Mathematics, Vol. 1403, Springer, Berlin, 1989, pp. 1–56.

[464] P.L. Hammer, B. Simeone, T. Liebling and D. de Werra, From linear separability to unimodality: A hierarchy of pseudo-Boolean functions, *SIAM Journal on Discrete Mathematics* 1 (1988) 174–184.

[465] A. Hamor (alias P.L. Hammer), Stories of the one-zero-zero-one nights: Abu Boul in Graphistan, in: P. Hansen and D. de Werra, eds., *Regards sur la Théorie des Graphes*, Presses Polytechniques Romandes, Lausanne, 1980.

[466] D.J. Hand, *Construction and Assessment of Classification Rules*, Wiley, Chichester, 1997.

[467] P. Hansen and B. Jaumard, Minimum sum of diameters clustering, *Journal of Classification* 4 (1987) 215–226.

[468] P. Hansen and B. Jaumard, Algorithms for the maximum satisfiability problem, *Computing* 44 (1990) 279–303.

[469] P. Hansen, B. Jaumard and V. Mathon, Constrained nonlinear 0–1 programming, *ORSA Journal on Computing* 5 (1993) 97–119.

[470] P. Hansen, B. Jaumard and M. Minoux, A linear expected-time algorithm for deriving all logical conclusions implied by a set of Boolean inequalities, *Mathematical Programming* 34 (1986) 223–231.

[471] P. Hansen, S.H. Lu and B. Simeone, On the equivalence of paved-duality and standard linearization in nonlinear 0–1 optimization, *Discrete Applied Mathematics* 29 (1990) 187–193.

[472] P. Hansen and C. Meyer, Improved compact linearizations for the unconstrained quadratic 0–1 minimization problem, *Discrete Applied Mathematics* 157 (2009) 1267–1290.

[473] P. Hansen, M.V. Poggi de Aragão and C.C. Ribeiro, Boolean query optimization and the 0–1 hyperbolic sum problem, *Annals of Mathematics and Artificial Intelligence* 1 (1990) 97–109.

[474] P. Hansen and B. Simeone, Unimodular functions, *Discrete Applied Mathematics* 14 (1986) 269–281.

[475] F. Harary, On the notion of balance of a signed graph, *Michigan Mathematics Journal* 2 (1954) 143–146.

[476] F. Harche, J.N. Hooker and G.L. Thompson, A computational study of satisfiability algorithms for propositional logic, *ORSA Journal on Computing* 6 (1994) 423–435.

[477] J. Håstad, On the size of weights for threshold gates, *SIAM Journal on Discrete Mathematics* 7 (1994) 484–492.

[478] J. Håstad, Some optimal inapproximability results, *Journal of the Association for Computing Machinery* 48 (2001) 798–859.

[479] J.P. Hayes, The fanout structure of switching functions, *Journal of the ACM* 22 (1975) 551–571.

[480] J.-J. Hebrard, Unique Horn renaming and unique 2-satisfiability, *Information Processing Letters* 54 (1995) 235–239.

[481] T. Hegedűs and N. Megiddo, On the geometric separability of Boolean functions, *Discrete Applied Mathematics* 66 (1996) 205–218.

[482] R. Heiman and A. Wigderson, Randomized vs. deterministic decision tree complexity for read-once Boolean functions, *Computational Complexity* 1 (1991) 311–329.

[483] I. Heller and C.B. Tompkins, An extension of a theorem of Dantzig, in: H.W. Kuhn and A.W. Tucker, eds., *Linear Inequalities and Related Systems*, Princeton University Press, Princeton, N.J., 1956, pp. 247–254.

[484] L. Hellerstein, Functions that are read-once on a subset of their variables, *Discrete Applied Mathematics* 46 (1993) 235–251.

[485] L. Hellerstein, On generalized constraints and certificates, *Discrete Mathematics* 226 (2001) 211–232.

[486] L. Hellerstein and V. Raghavan, Exact learning of DNF formulas using DNF hypothesis, *Journal of Computer and System Sciences* 70 (2005) 435–470.

[487] P.B. Henderson and Y. Zalcstein, A graph-theoretic characterization of the PV chunk class of synchronizing primitives, *SIAM Journal on Computing* 6 (1977) 88–108.

[488] L.J. Henschen, Semantic resolution for Horn sets, *IEEE Transactions on Computers* 25 (1976) 816–822.

[489] L.J. Henschen and L. Wos, Unit refutations and Horn sets, *Journal of the ACM* 21 (1974) 590–605.

[490] M. Herbstritt, *Satisfiability and Verification: From Core Algorithms to Novel Application Domains*, Suedwestdeutscher Verlag für Hochschulschriften, 2009.

[491] A. Hertz, On the use of Boolean methods for the computation of the stability number, *Discrete Applied Mathematics* 76 (1997) 183–203.

[492] E.A. Hirsch, New worst-case upper bounds for SAT, *Journal of Automated Reasoning* 24 (2000) 397–420.

[493] W. Hodges, Reducing first order logic to Horn logic, School of Mathematical Sciences, Queen Mary and Westfield College, London, 1985.

[494] W. Hodges, Logical features of Horn clauses, in: *Handbook of Logic in Artificial Intelligence and Logic Programming*, Vol. 1, Oxford University Press, 1993, pp. 449–503.

[495] A.J. Hoffman and J.B. Kruskal, Integral boundary points of convex polyhedra, in: H.W. Kuhn and A.W. Tucker, eds., *Linear Inequalities and Related Systems*, Princeton University Press, Princeton, N.J., 1956, 223–246.

[496] K. Williamson Hoke, Completely unimodal numberings of a simple polytope, *Discrete Applied Mathematics* 20 (1988) 69–81.

[497] K. Hoke, Extending shelling orders and a hierarchy of functions of unimodal simple polytopes, *Discrete Applied Mathematics* 60 (1995) 211–217.

[498] J.N. Hooker, A quantitative approach to logical inference, *Decision Support Systems* 4 (1988) 45–69.

[499] J.N. Hooker, Generalized resolution and cutting planes, *Annals of Operations Research* 12 (1988) 217–239.

[500] J.N. Hooker, Resolution vs. cutting plane solution of inference problems: Some computational experience, *Operations Research Letters* 7 (1988) 1–7.

[501] J.N. Hooker, Resolution and the integrality of satisfiability problems, *Mathematical Programming* 74 (1996) 1–10.

[502] J.N. Hooker, *Logic-Based Methods for Optimization: Combining Optimization and Constraint Satisfaction*, John Wiley & Sons, New York, 2000.

[503] J.N. Hooker, Optimization methods in logic, in: Y. Crama and P.L. Hammer, eds., *Boolean Models and Methods in Mathematics, Computer Science, and Engineering*, Cambridge University Press, Cambridge, 2010, pp. 160–194.

[504] J.N. Hooker and V. Vinay, Branching rules for satisfiability, *Journal of Automated Reasoning* 15 (1995) 359–383.

[505] H.H. Hoos and T. Stützle, Towards a characterisation of the behaviour of stochastic local search algorithms for SAT, *Artificial Intelligence* 112 (1999) 213–232.

[506] H.H. Hoos and T. Stützle, SATLIB: An online resource for research on SAT, in: I. Gent, H. van Maaren and T. Walsh, eds., *SAT2000: Highlights of Satisfiability Research in the Year 2000*, IOS Press, Amsterdam, 2000, pp. 283–292.

[507] H.H. Hoos and T. Stützle, Local search algorithms for SAT: An empirical evaluation, *Journal of Automated Reasoning* 24 (2000) 421–481.

[508] H.H. Hoos and T. Stützle, *Stochastic Local Search: Foundations and Applications*, Morgan Kaufmann Publishers, San Francisco, CA, 2005.

[509] A. Horn, On sentences which are true of direct unions of algebras, *Journal of Symbolic Logic* 16 (1951) 14–21.

[510] I. Horrocks and P.F. Patel-Schneider, Evaluating optimized decision procedures for propositional modal $K_{(m)}$ satisfiability, in: I. Gent, H. van Maaren and T. Walsh, eds., *SAT2000: Highlights of Satisfiability Research in the Year 2000*, IOS Press, Amsterdam, 2000, pp. 427–458.

[511] S.-T. Hu, *Threshold Logic*, University of California Press, Berkeley - Los Angeles, 1965.

[512] S.-T. Hu, *Mathematical Theory of Switching Circuits and Automata*, University of California Press, Berkeley - Los Angeles, 1968.

[513] L.M. Hvattum, A. Løkketangen and F. Glover, Adaptive memory search for Boolean optimization problems, *Discrete Applied Mathematics* 142 (2004) 99–109.

[514] L. Hyafil and R.L. Rivest, Constructing optimal binary decision trees is NP-complete, *Information Processing Letters* 5 (1976) 15–17.

[515] T. Ibaraki, T. Imamichi, Y. Koga, H. Nagamochi, K. Nonobe and M. Yagiura, Efficient branch-and-bound algorithms for weighted MAX-2-SAT, Technical Report 2007-011, Department of Applied Mathematics and Physics, Graduate School of Informatics, Kyoto University, May 2007.

[516] T. Ibaraki and T. Kameda, A theory of coteries: Mutual exclusion in distributed systems, *IEEE Transactions on Parallel and Distributed Systems* 4 (1993) 779–794.

[517] T. Ibaraki, A. Kogan and K. Makino, Functional dependencies in Horn theories, *Artificial Intelligence* 108 (1999) 1–30.

[518] T. Ibaraki, A. Kogan and K. Makino, Inferring minimal functional dependencies in Horn and q-Horn theories, *Annals of Mathematics and Artificial Intelligence*, 38 (2003) 233–255.

[519] J.P. Ignizio, *Introduction to Expert Systems: The Development and Implementation of Rule-Based Expert Systems*, McGraw-Hill, New York, 1991.

[520] J.R. Isbell, A class of simple games, *Duke Mathematical Journal* 25 (1958) 423–439.

[521] A. Itai and J.A. Makowsky, Unification as a complexity measure for logic programming, *Journal of Logic Programming* 4 (1987) 105–117.

[522] K. Iwama, CNF satisfiability test by counting and polynomial average time, *SIAM Journal on Computing* 18 (1989) 385–391.

[523] S. Iwata, Submodular function minimization, *Mathematical Programming* Ser. B 112 (2008) 45–64.

[524] S. Iwata, L. Fleischer and S. Fujishige, A combinatorial, strongly polynomial-time algorithm for minimizing submodular functions, in: *Proceedings of the 32nd ACM Symposium on Theory of Computing*, 2000, pp. 97–106.

[525] S. Janson, Y.C. Stamatiou and M. Vamvakari, Bounding the unsatisfiability threshold of random 3-SAT, *Random Structures and Algorithms* 17 (2000) 103–116.

[526] B. Jaumard, *Extraction et Utilisation de Relations Booléennes pour la Résolution des Programmes Linéaires en Variables 0-1*, Thèse de doctorat, Ecole Nationale Supérieure des Télécommunications, Paris, France, 1986.

[527] B. Jaumard, P. Marchioro, A. Morgana, R. Petreschi and B. Simeone, An $O(n^3)$ on-line algorithm for 2-satisfiability, *Atti Giornate di Lavoro AIRO*, Pisa, 1988, pp. 391–399.

[528] B. Jaumard, P. Marchioro, A. Morgana, R. Petreschi and B. Simeone, On-line 2-satisfiability, *Annals of Mathematics and Artificial Intelligence* 1 (1990) 155–165.

[529] B. Jaumard and M. Minoux, An efficient algorithm for the transitive closure and a linear worst-case complexity result for a class of sparse graphs, *Information Processing Letters* 22 (1986) 163–169.

[530] B. Jaumard and B. Simeone, On the complexity of the maximum satisfiability problem for Horn formulas, *Information Processing Letters* 26 (1987) 1–4.

[531] B. Jaumard, B. Simeone and P.S. Ow, A selected Artificial Intelligence bibliography for Operations Researchers, *Annals of Operations Research* 12 (1988) 1–50.

[532] B. Jaumard, M. Stan and J. Desrosiers, Tabu search and a quadratic relaxation for the satisfiability problem, in: D.S. Johnson and M.A. Trick, eds., *Cliques, Coloring, and Satisfiability,* DIMACS Series in Discrete Mathematics and Theoretical Computer Science, Vol. 26, American Mathematical Society, 1996, pp. 457–477.

[533] R.G. Jeroslow, *Logic-Based Decision Support - Mixed Integer Model Formulation*, North-Holland, Amsterdam, 1989.

[534] R.G. Jeroslow and J. Wang, Solving propositional satisfiability problems, *Annals of Mathematics and Artificial Intelligence* 1 (1990) 167–187.

[535] J.H.R. Jiang and T. Villa, Hardware equivalence checking, in: Y. Crama and P.L. Hammer, eds., *Boolean Models and Methods in Mathematics, Computer Science, and Engineering*, Cambridge University Press, Cambridge, 2010, pp. 599–674.

[536] D.S. Johnson, Approximation algorithms for combinatorial problems, *Journal of Computer and System Sciences* 9 (1974) 256–278.

[537] D.S. Johnson and M.A. Trick, eds., *Cliques, Coloring, and Satisfiability*, DIMACS Series in Discrete Mathematics and Theoretical Computer Science, Vol. 26, American Mathematical Society, 1996.

[538] D.S. Johnson, M. Yannakakis and C.H. Papadimitriou, On generating all maximal independent sets, *Information Processing Letters* 27 (1988) 119–123.

[539] N.D. Jones and W.T. Laaser, Complete problems for deterministic polynomial time, *Theoretical Computer Science* 3 (1976) 105–117.

[540] S. Joy, J. Mitchell and B. Borchers, A branch and cut algorithm for MAX-SAT and weigthed MAX-SAT, in: D. Du, J. Gu and P.M. Pardalos, eds., *Satisfiability Problem: Theory and Applications*, DIMACS series in Discrete Mathematics and Theoretical Computer Science, Vol. 35, American Mathematical Society, 1997, pp. 519–536.

[541] S. Jukna, A. Razborov, P. Savický and I. Wegener, On P versus NP ∩ co-NP for decision trees and read-once branching programs, in: I. Prívara and P. Ruzicka, eds., *Mathematical Foundations of Computer Science 1997*, Lecture Notes in Computer Science, Vol. 1295, Springer-Verlag, Berlin-New York, 1997, pp. 319–326.

[542] J. Kahn, Entropy, independent sets and antichains: A new approach to Dedekind's problem, *Proceedings of the American Mathematical Society* 130 (2002) 371–378.

[543] J. Kahn, G. Kalai and N. Linial, The influence of variables on Boolean functions, in: *Proceedings of the 29th Annual IEEE Symposium on the Foundations of Computer Science*, IEEE, White Plains, NY, 1988, pp. 68–80.

[544] B. Kalantari and J.B. Rosen, Penalty formulation for zero-one nonlinear programming, *Discrete Applied Mathematics* 16 (1987) 179–182.

[545] A.P. Kamath, N.K. Karmarkar, K.G. Ramakrishnan and M.G.C. Resende, Computational experience with an interior point algorithm on the satisfiability problem, *Annals of Operations Research* 25 (1990) 43–58.

[546] A.P. Kamath, N.K. Karmarkar, K.G. Ramakrishnan and M.G.C. Resende, A continuous approach to inductive inference, *Mathematical Programming* 57 (1992) 215–238.

[547] Y. Kambayashi, Logic design of programmable logic arrays, *IEEE Transactions on Computers* C-28 (1979) 609–617.

[548] M. Karchmer, N. Linial, I. Newman, M. Saks and A. Wigderson, Combinatorial characterization of read-once formulae, *Discrete Mathematics* 114 (1993) 275–282.

[549] H. Karloff and U. Zwick, A 7/8-approximation algorithm for MAX 3SAT?, in: *Proceedings of the 38th Annual IEEE Symposium on the Foundations of Computer Science*, IEEE, 1997, pp. 406–415.

[550] R.M. Karp, Reducibility among combinatorial problems, in: R.E. Miller and J.W. Thatcher, eds., *Complexity of Computer Computations*, Plenum Press, New York, 1972, pp. 85–103.

[551] R.M. Karp, M. Luby and N. Madras, Monte-Carlo approximation algorithms for enumeration problems, *Journal of Algorithms* 10 (1989) 429–448.

[552] M. Karpinski, H. Kleine Büning and P.H. Schmitt, On the computational complexity of quantified Horn clauses, in: E. Börger, H. Kleine Büning and M.M. Richter, eds., *CSL '87, First Workshop on Computer Science Logic*, Lecture Notes in Computer Science, Vol. 329, Springer-Verlag, Berlin, 1988, pp. 129–137.

[553] S.A. Kauffman, *The Origins of Order: Self-Organization and Selection in Evolution*, Oxford University Press, New York, 1993.

[554] H.A. Kautz, M.J. Kearns and B. Selman, Horn approximations of empirical data, *Artificial Intelligence* 74 (1995) 129–145.

[555] H. Kautz and B. Selman, Knowledge compilation and theory of approximation, *Journal of the ACM* 43 (1996) 193–224.

[556] H. Kautz and B. Selman, Pushing the envelope: Planning, propositional logic, and stochastic search, in: *Proceedings of the 13th National Conference on Artificial Intelligence*, Portland, OR, 1996, pp. 1188–1194.

[557] H. Kautz, B. Selman and Y. Jiang, A general stochastic approach to solving problems with hard and soft constraints, in: D. Du, J. Gu and P.M. Pardalos, eds., *Satisfiability Problem: Theory and Applications*, DIMACS series in Discrete Mathematics and Theoretical Computer Science, Vol. 35, American Mathematical Society, 1997, pp. 573–586.

[558] D.J. Kavvadias, C.H. Papadimitriou and M. Sideri, On Horn envelopes and hypergraph transversals, in: K.W. Ng et al., eds., *Algorithms and Computation – ISAAC'93*, Lecture Notes in Computer Science, Vol. 762, Springer-Verlag, Berlin, 1993, pp. 399–405.

[559] D.J. Kavvadias and E.C. Stavropoulos, An efficient algorithm for the transversal hypergraph generation, *Journal of Graph Algorithms and Applications* 9 (2005) 239–264.

[560] M. Kearns, M. Li and L. Valiant, Learning Boolean functions, *Journal of the Association for Computing Machinery* 41 (1994) 1298–1328.

[561] H. Kellerer, U. Pferschy and D. Pisinger, *Knapsack Problems*, Springer-Verlag, Berlin-Heidelberg-New York, 2004.

[562] L. Khachiyan, E. Boros, K. Elbassioni and V. Gurvich, Generating all minimal integral solutions to AND-OR systems of monotone inequalities: Conjunctions are simpler than disjunctions, *Discrete Applied Mathematics* 156 (2008) 2020–2034.

[563] S. Khanna, M. Sudan and D.P. Williamson, A complete classification of the approximability of maximization problems derived from Boolean constraint satisfaction, in: *Proceedings of the 29th Annual ACM Symposium on the Theory of Computing*, 1997, pp. 11–20.

[564] R. Khardon, Translating between Horn representations and their characteristic models, *Journal of Artificial Intelligence Research* 3 (1995) 349–372.

[565] R. Khardon, H. Mannila and D. Roth, Reasoning with examples: Propositional formulae and database dependencies, *Acta Informatica* 36 (1999) 267–286.

[566] R. Khardon and D. Roth, Reasoning with models, *Artificial Intelligence* 87 (1996) 187–213.

[567] S. Khot, G. Kindler, E. Mossel and R. O'Donnell, Optimal inapproximability results for MAX-CUT and other 2-variable CSPs?, *SIAM Journal on Computing* 37 (2007) 319–357.

[568] P. Kilby, J.K. Slaney, S. Thibaux and T. Walsh, Backbones and backdoors in satisfiability, *AAAI Proceedings*, 2005, pp. 1368–1373.

[569] V. Klee and P. Kleinschmidt, Convex polytopes and related complexes, in: R. Graham, M. Grötschel and L. Lovász, eds., *Handbook of Combinatorics*, Elsevier, Amsterdam, 1995, pp. 875–917.

[570] H. Kleine Büning, On generalized Horn formulas and k-resolution, *Theoretical Computer Science* 116 (1993) 405–413.

[571] H. Kleine Büning and T. Lettmann, *Propositional Logic: Deduction and Algorithms*, Cambridge University Press, Cambridge, 1999.

[572] D. Kleitman, On Dedekind's problem: The number of monotone Boolean functions, *Proceedings of the American Mathematical Society* 21 (1969) 677–682.

[573] D. Kleitman and G. Markowsky, On Dedekind's problem: The number of isotone Boolean functions. II, *Transactions of the American Mathematical Society* 213 (1975) 373–390.

[574] B. Klinz and G.J. Woeginger, Faster algorithms for computing power indices in weighted voting games, *Mathematical Social Sciences* 49 (2005) 111–116.

[575] D.E. Knuth, *The Art of Computer Programming*, Volume 4, Fascicle 0, *Introduction to Combinatorial Algorithms and Boolean Functions*, Stanford University, Stanford, CA, 2008. http://www-cs-faculty.stanford.edu/ knuth/taocp.html

[576] V. Kolmogorov and C. Rother, Minimizing nonsubmodular functions with graph cuts - A review, *IEEE Transactions on Pattern Analysis and Machine Intelligence* 29 (2007) 1274–1279.

[577] V. Kolmogorov and R. Zabih, What energy functions can be minimized via graph cuts?, *IEEE Transactions on Pattern Analysis and Machine Intelligence* 26 (2004) 147–159.

[578] A.D. Korshunov, The number of monotone Boolean functions, *Problemy Kibernetiki* 38 (1981) 5–108 (in Russian).

[579] A.D. Korshunov, Families of subsets of a finite set and closed classes of Boolean functions, in: P. Frankl et al., eds., *Extremal Problems for Finite Sets*, János Bolyai Mathematical Society, Budapest, Hungary, 1994, pp. 375–396.

[580] A.D. Korshunov, Monotone Boolean functions, *Russian Mathematical Surveys* 58 (2003) 929–1001.

[581] S. Kottler, M. Kaufmann and C. Sinz, Computation of renameable Horn backdoors, in: *Proceedings of the 11th International Conference on Theory and Applications of Satisfiability Testing* (SAT 2008), Lecture Notes in Computer Science, Vol. 4996, Springer-Verlag, Berlin, 2008, pp. 154–160.

[582] R. Kowalski, *Logic for Problem Solving*, North-Holland, Amsterdam-New York, 1979.

[583] M. Krause and I. Wegener, Circuit complexity, in: Y. Crama and P.L. Hammer, eds., *Boolean Models and Methods in Mathematics, Computer Science, and Engineering*, Cambridge University Press, Cambridge, 2010, pp. 506–530.

[584] L. Kroc, A. Sabharwal and B. Selman, Leveraging belief propagation, backtrack search, and statistics for model counting, in: L. Perron and M.A. Trick, eds., *Integration of AI and OR Techniques in Constraint Programming for Combinatorial Optimization Problems*, Lecture Notes in Computer Science Vol. 5015, Springer-Verlag, Berlin Heidelberg, 2008, pp. 127–141.

[585] P. Kučera, On the size of maximum renamable Horn sub-CNF, *Discrete Applied Mathematics* 149 (2005) 126–130.

[586] W. Küchlin and C. Sinz, Proving consistency assertions for automotive product data management, in: I. Gent, H. van Maaren and T. Walsh, eds., *SAT2000: Highlights of Satisfiability Research in the Year 2000*, IOS Press, Amsterdam, 2000, pp. 327–342.

[587] H.W. Kuhn, The Hungarian method for solving the assignment problem, *Naval Research Logistics Quarterly* 2 (1955) 83–97.

[588] O. Kullmann, New methods for 3-SAT decision and worst-case analysis, *Theoretical Computer Science* 223 (1999) 1–72.

[589] J. Kuntzmann, *Algèbre de Boole*, Dunod, Paris, 1965. English translation: *Fundamental Boolean Algebra*, Blackie and Son Limited, London and Glasgow, 1967.

[590] W. Kunz and D. Stoffel, *Reasoning in Boolean Networks*, Kluwer Academic Publishers, Boston - Dordrecht - London, 1997.

[591] Z.A. Kuzicheva, Mathematical logic, in: A.N. Kolmogorov and A.P. Yushkevich, eds., *Mathematics of the 19th Century*, Volume 1, 2nd revised edition, Birkhaüser Verlag, Basel, 2001, pp. 1–34.

[592] A.V. Kuznetsov, Non-repeating contact schemes and non-repeating superpositions of functions of algebra of logic, in: *Collection of Articles on Mathematical Logic and its Applications to Some Questions of Cybernetics*, Proceedings of the Steklov Institute of Mathematics, Vol. 51, Academy of Sciences of USSR, Moscow, 1958, pp. 862–25.

[593] L. Lamport, The implementation of reliable distributed multiprocess systems, *Computing Networks* 2 (1978) 95–114.

[594] M. Langlois, D. Mubayi, R.H. Sloan and G. Turán, Combinatorial problems for Horn clauses, manuscript, 2008.

[595] M. Langlois, R.H. Sloan, B. Szörényi and G. Turán, Horn complements: Towards Horn-to-Horn belief revision, in: D. Fox and C.P. Gomes, eds., *Proceedings of the Twenty-Third AAAI Conference on Artificial Intelligence*, AAAI 2008, Chicago, Illinois, USA, 2008, pp. 466–471.

[596] M. Langlois, R.H. Sloan and G. Turán, Horn upper bounds and renaming, in: J. Marques-Silva and K.A. Sakallah, eds., *Proceedings of the 10th International Conference on Theory and Applications of Satisfiability Testing* – SAT 2007, Lisbon, Portugal, 2007, pp. 80–93.

[597] E. Lapidot, Weighted majority games and symmetry groups of games, M.Sc. thesis (in Hebrew), Technion, Haifa, Israel, 1968.

[598] E. Lapidot, The counting vector of a simple game, *Proceedings of the American Mathematical Society* 31 (1972) 228–231.

[599] T. Larrabee, Test pattern generation using Boolean satisfiability, *IEEE Transactions on Computer-Aided Design* 11 (1992) 4–15.

[600] A. Laruelle and M. Widgrén, Is the allocation of voting power among EU states fair?, *Public Choice* 94 (1998) 317–339.

[601] M. Laurent and F. Rendl, Semidefinite programming and integer programming, in: K. Aardal, G. Nemhauser and R. Weismantel, eds., *Discrete Optimization*, Elsevier, Amsterdam, 2005, pp. 393–514.

[602] M. Laurent and A. Sassano, A characterization of knapsacks with the max-flow-min-cut property, *Operations Research Letters* 11 (1992) 105–110.

[603] E.L. Lawler, Covering problems: Duality relations and a new method of solution, *SIAM Journal on Applied Mathematics* 14 (1966) 1115–1132.

[604] E.L. Lawler, *Combinatorial Optimization: Networks and Matroids*, Holt, Rinehart and Winston, New York, 1976.

[605] E.L. Lawler, J.K. Lenstra and A.H.G. Rinnooy Kan, Generating all maximal independent sets: NP-hardness and polynomial-time algorithms, *SIAM Journal on Computing* 9 (1980) 558–565.

[606] D. Leech, The relationship between shareholding concentration and shareholder voting power in British companies: A study of the application of power indices for simple games, *Management Science* 34 (1988) 509–528.

[607] D. Leech, Designing the voting system for the Council of the European Union, *Public Choice* 113 (2002) 437–464.

[608] D. Leech, Voting power in the governance of the International Monetary Fund, *Annals of Operations Research* 109 (2002) 375–397.

[609] D. Leech, Computation of power indices, Warwick Economic Research Papers, Number 644, The University of Warwick, 2002.

[610] L.A. Levin, Universal'nye zadachi perebora, *Problemy Peredachi Informatsii* 9 (1973) 115–116 (in Russian); translated as: Universal sequential search problems, *Problems of Information Transmission* 9 (1974) 265–266.

[611] M. Lewin, D. Livnat and U. Zwick, Improved rounding techniques for the MAX 2-SAT and MAX DI-CUT problems, in: *Integer Programming and Combinatorial Optimization (IPCO)*, Lecture Notes in Computer Science, Vol. 2337, Springer-Verlag, Berlin Heidelberg New York, 2002, pp. 67–82.

[612] H.R. Lewis, Renaming a set of clauses as a Horn set, *Journal of the ACM* 25 (1978) 134–135.

[613] C.M. Li and Anbulagan, Heuristics based on unit propagation for satisfiability problems, *Proceedings of the Fifteenth International Joint Conference on Artificial Intelligence*, Morgan Kaufmann, 1997, pp. 366–371.

[614] N. Linial and N. Nisan, Approximate inclusion-exclusion, *Combinatorica* 10 (1990) 349–365.

[615] N. Linial and M. Tarsi, Deciding hypergraph 2-colourability by H-resolution, *Theoretical Computer Science* 38 (1985) 343–347.

[616] M.O. Locks, Inverting and minimalizing path sets and cut sets, *IEEE Transactions on Reliability* R-27 (1978) 107–109.

[617] M.O. Locks, Inverting and minimizing Boolean functions, minimal paths and minimal cuts: Noncoherent system analysis, *IEEE Transactions on Reliability* R-28 (1979) 373–375.

[618] M.O. Locks, Recursive disjoint products, inclusion-exclusion, and min-cut approxima-
       tions, *IEEE Transactions on Reliability* R-29 (1980) 368–371.

[619] M.O. Locks, Recursive disjoint products: A review of three algorithms, *IEEE Transactions
       on Reliability* R-31 (1982) 33–35.

[620] A. Lodi, K. Allemand and T.M. Liebling, An evolutionary heuristic for quadratic 0–1
       programming, *European Journal of Operational Research* 119 (1999) 662–670.

[621] D.E. Loeb and A.R. Conway, Voting fairly: Transitive maximal intersecting families of
       sets, *Journal of Combinatorial Theory A* 91 (2000) 386–410.

[622] L. Lovász, Normal hypergraphs and the perfect graph conjecture, *Discrete Mathematics*
       2 (1972) 253–267.

[623] L. Lovász, On the ratio of optimal and integral and fractional covers, *Discrete Mathematics*
       13 (1975) 383–390.

[624] L. Lovász, Submodular functions and convexity, in: A. Bachem, M. Grötschel and
       B. Korte, eds., *Mathematical Programming – The State of the Art*, Springer-Verlag, Berlin,
       1983, pp. 235–257.

[625] L. Lovász, *An Algorithmic Theory of Numbers, Graphs and Convexity*, Society for
       Industrial and Applied Mathematics, Philadelphia, 1986.

[626] L. Lovász, *Lecture Notes on Evasiveness of Graph Properties*, Notes by Neal Young,
       Computer Science Department, Princeton University, January 1994.

[627] D.W. Loveland, *Automated Theorem-Proving: A Logical Basis*, North-Holland, Amster-
       dam, 1978.

[628] L. Löwenheim, Über das Auflösungsproblem im logischen Klassenkalkul, *Sitzungs-
       berichte der Berliner Mathematischen Gesellschaft* 7 (1908) 89–94.

[629] L. Löwenheim, Über die Auflösung von Gleichungen im logischen Gebietkalkul,
       Mathematische Annalen 68 (1910) 169–207.

[630] E. Lozinskii, Counting propositional models, *Information Processing Letters* 41 (1992)
       327–332.

[631] W.F. Lucas, Measuring power in weighted voting systems, in: *Case Studies in Applied
       Mathematics*, Mathematical Association of America, 1976, pp. 42–106. Also Chapter 9
       in: S.J. Brams, W.F. Lucas and P.D. Straffin, Jr., eds., *Political and Related Models*,
       Springer-Verlag, Berlin Heidelberg New York, 1983.

[632] W.F. Lucas, The apportionment problem, Chapter 14 in: S.J. Brams, W.F. Lucas and P.D.
       Straffin, Jr., eds., *Political and Related Models*, Springer-Verlag, Berlin Heidelberg New
       York, 1983.

[633] E.J. McCluskey, Minimization of Boolean functions, *Bell Systems Technical Journal* 35
       (1956) 1417–1444.

[634] E.J. McCluskey, *Introduction to the Theory of Switching Circuits*, McGraw-Hill,
       New York, 1965.

[635] E.J. McCluskey, *Logic Design Principles*, Prentice-Hall, Englewood Cliffs, New Jersey,
       1986.

[636] R.M. McConnell and J.P. Spinrad, Modular decomposition and transitive orientation,
       Discrete Mathematics 201 (1999) 189–241.

[637] S.T. McCormick, Submodular function minimization, in: K. Aardal, G.L. Nemhauser,
       R. Weismantel, eds., *Discrete Optimization*, Handbooks in Operations Research and
       Management Science, Vol. 12, Elsevier, Amsterdam, 2005, pp. 321–391.

[638] J.C.C. McKinsey, The decision problem for some classes of sentences without quantifiers,
       Journal of Symbolic Logic 8 (1943) 61–76.

[639] I. McLean, Don't let the lawyers do the math: Some problems of legislative districting in
       the UK and the USA, *Mathematical and Computer Modelling* 48 (2008) 1446–1454.

[640] G.F. McNulty, Fragments of first order logic, I: Universal Horn logic, *Journal of Symbolic
       Logic* 42 (1977) 221–237.

[641] T.-H. Ma, On the threshold dimension 2 graphs, Technical report, Institute of Information
       Sciences, Academia Sinica, Taipei, Republic of China, 1993.

[642] F.J. MacWilliams and N.J.A. Sloane, *The Theory of Error-Correcting Codes,* North-Holland, Amsterdam, The Netherlands, 1977.

[643] K. Maghout, Sur la détermination des nombres de stabilité et du nombre chromatique d'un graphe, *Comptes Rendus de l'Académie des Sciences de Paris* 248 (1959) 3522–3523.

[644] K. Maghout, Applications de l'algèbre de Boole à la théorie des graphes et aux programmes linéaires et quadratiques, *Cahiers du Centre d'Etudes de Recherche Opérationnelle* 5 (1963) 21–99.

[645] N.V.R. Mahadev and U. Peled, *Threshold Graphs and Related Topics, Annals of Discrete Mathematics* Vol. 56, North-Holland, Amsterdam, The Netherlands, 1995.

[646] D. Maier, Minimal covers in the relational database model, *Journal of the ACM* 27 (1980) 664–674.

[647] D. Maier, *The Theory of Relational Databases*, Computer Science Press, Rockville, MD, 1983.

[648] D. Maier and D.S. Warren, *Computing with Logic: Logic Programming with PROLOG*, Benjamin/Cummings Publishing Co., Menlo Park, CA, 1988.

[649] K. Makino, A linear time algorithm for recognizing regular Boolean function, *Journal of Algorithms* 43 (2002) 155–176.

[650] K. Makino, K. Hatanaka and T. Ibaraki, Horn extensions of a partially defined Boolean function, *SIAM Journal on Computing* 28 (1999) 2168–2186.

[651] K. Makino and T. Ibaraki, Interior and exterior functions of Boolean functions, *Discrete Applied Mathematics* 69 (1996) 209–231.

[652] K. Makino and T. Ibaraki, The maximum latency and identification of positive Boolean functions, *SIAM Journal on Computing* 26 (1997) 1363–1383.

[653] K. Makino and T. Ibaraki, A fast and simple algorithm for identifying 2-monotonic positive Boolean functions, *Journal of Algorithms* 26 (1998) 291–305.

[654] K. Makino and T. Ibaraki, Inner-core and outer-core functions of partially defined Boolean functions, *Discrete Applied Mathematics* 96–97 (1999) 307–326.

[655] K. Makino, K. Yano and T. Ibaraki, Positive and Horn decomposability of partially defined Boolean functions, *Discrete Applied Mathematics* 74 (1997) 251–274.

[656] J.A. Makowsky, Why Horn formulas matter in computer science: Initial structures and generic examples, *Journal of Computer and System Sciences* 34 (1987) 266–292.

[657] A.I. Malcev, *The Metamathematics of Algebraic Systems, Collected Papers: 1936–1967*, North Holland, Amsterdam, 1971.

[658] O.L. Mangasarian, Linear and nonlinear separation of patterns by linear programming, *Operations Research* 13 (1965) 444–452.

[659] O.L. Mangasarian, Mathematical programming in neural networks, *ORSA Journal on Computing* 5 (1993) 349–360.

[660] O.L. Mangasarian, R. Setiono and W.H. Wolberg, Pattern recognition via linear programming: Theory and applications to medical diagnosis, in: T.F. Coleman and Y. Li, eds., *Large-Scale Numerical Optimization*, SIAM Publications, Philadelphia, 1990, pp. 22–30.

[661] I. Mann and L.S. Shapley, Values of large games VI: Evaluating the Electoral College exactly, RM-3158, The Rand Corporation, Santa Monica, CA, 1962.

[662] H. Mannila and K. Mehlhorn, A fast algorithm for renaming a set of clauses as a Horn set, *Information Processing Letters* 21 (1985) 261–272.

[663] H.K. Mannila and J. Räihä, *Design of Relational Databases*, Addison-Wesley, Wokingham, 1992.

[664] H.K. Mannila and J. Räihä, Algorithms for inferring functional dependencies, *Data and Knowledge Engineering* 12 (1994) 83–99.

[665] H.K. Mannila, H. Toivonen and A.I. Verkamo, in: U.M. Fayyad and R. Uthurusamy, eds., *Efficient Algorithms for Discovering Association Rules, AAAI Workshop on Knowledge Discovery in Databases*, 1994, pp. 181–192.

[666] V. Manquinho and J.P. Marques-Silva, On using cutting planes in pseudo-Boolean optimization, *Journal on Satisfiability, Boolean Modeling and Computation* 2 (2006) 209–219.

[667] V.M. Manquinho and O. Roussel, The first evaluation of pseudo-Boolean solvers (PB'05), *Journal on Satisfiability, Boolean Modeling and Computation* 2 (2006) 103–143.

[668] J.-L. Marichal, On Sugeno integral as an aggregation function, *Fuzzy Sets and Systems* 114 (2000) 347–365.

[669] J.-L. Marichal, The influence of variables on pseudo-Boolean functions with applications to game theory and multicriteria decision making, *Discrete Applied Mathematics* 107 (2000) 139–164.

[670] J.P. Marques-Silva and K.A. Sakallah, GRASP: A search algorithm for propositional satisfiability, *IEEE Transactions on Computers* C-48 (1999) 506–521.

[671] S. Martello and P. Toth, *Knapsack Problems: Algorithms and Computer Implementations*, John Wiley & Sons, Chichester, New York, 1990.

[672] U. Martin and T. Nipkow, Boolean unification: The story so far, *Journal of Symbolic Computation* 7 (1989) 275–293.

[673] M. Maschler and B. Peleg, A characterization, existence proof and dimension bounds for the kernel of a game, *Pacific Journal of Mathematics* 18 (1966) 289–328.

[674] W.J. Masek, Some NP-complete set covering problems, MIT, Cambridge, MA, unpublished manuscript, August 1979.

[675] F. Massacci and L. Marraro, Logical cryptanalysis as a SAT problem: Encoding and analysis of the U.S. data encryption standard, in: I. Gent, H. van Maaren and T. Walsh, eds., *SAT2000: Highlights of Satisfiability Research in the Year 2000*, IOS Press, Amsterdam, 2000, pp. 343–376.

[676] T. Matsui and Y. Matsui, A survey of algorithms for calculating power indices of weighted majority games, *Journal of the Operations Research Society of Japan* 43 (2000) 71–86.

[677] K. Matulef, R. O'Donnell, R. Rubinfeld and R. Servedio, Testing halfspaces, in: *ACM-SIAM Symposium on Discrete Algorithms* (SODA), 2009, pp. 256–264.

[678] C. Maxfield, *Bebop to the Boolean Boogie: An Unconventional Guide to Electronics Fundamentals, Components, and Processes*, LLH Technology Publications, Eagle Rock, VA, 1995.

[679] D.L. Medin and P.J. Schwanenflugel, Linear separability in classification learning, *Journal of Experimental Psychology: Human Learning and Memory* 7 (1981) 355–368.

[680] E. Mendelson, *Boolean Algebra and Switching Circuits*, Schaum's Outline Series, McGraw-Hill, New York, 1970.

[681] P. Merz and B. Freisleben, Greedy and local search heuristics for unconstrained binary quadratic programming, *Journal of Heuristics* 8 (2002) 197–213.

[682] M. Minoux, The unique-Horn satisfiability problem and quadratic Boolean equations, *Annals of Mathematics and Artificial Intelligence* 6 (1992) 253–266.

[683] M. Minoux and K. Barkaoui, Deadlocks and traps in Petri nets as Horn satisfiability solutions and some related polynomially solvable problems, *Discrete Applied Mathematics* 29 (1990) 195–210.

[684] M. Minsky and S. Papert, *Perceptrons*, MIT Press, Cambridge, MA, 1969.

[685] A. Mintz, *Multi-Level Synthesis: Factoring Logic Functions Using Graph Partitioning Algorithms*, Ph.D. Thesis, Bar-Ilan University, Ramat Gan, Israel, 2000.

[686] A. Mintz and M.C. Golumbic, Factoring Boolean functions using graph partitioning, *Discrete Applied Mathematics* 149 (2005) 131–153.

[687] D. Mitchell, B. Selman and H. Levesque, Hard and easy distributions of SAT problems, in: *AAAI'92, Proceedings of the Tenth National Conference on Artificial Intelligence*, San Jose, CA, 1992, pp. 459–465.

[688] R. Miyashiro and T. Matsui, A polynomial-time algorithm to find an equitable homeaway assignment, *Operations Research Letters* 33 (2005) 235–241.

[689] M. Molloy, The probabilistic method, in: M. Habib, C. McDiarmid, J. Ramirez-Alfonsin and B. Reed, eds., *Probabilistic Methods for Algorithmic Discrete Mathematics*, Springer, Berlin, 1998, pp. 1–35.

[690] B. Monien and E. Speckenmeyer, Solving satisfiability in less than 2^n steps, *Discrete Applied Mathematics* 10 (1985) 287–295.

[691] I.D. Moon and S.S. Chaudhry, An analysis of network location problems, *Management Science* 30 (1984) 290–307.

[692] E.F. Moore, Counterexample to a conjecture of McCluskey and Paull, unpublished memorandum, Bell Telephone Laboratories, 1957.

[693] M.W. Moskewicz, C.F. Madigan, Y. Zhao, L. Zhang and S. Malik, Chaff: Engineering an efficient SAT solver, in: *Proceedings of the 38th Design Automation Conference (DAC'01)*, 2001, pp. 530–535.

[694] H.M. Mulder, The structure of median graphs, *Discrete Mathematics* 24 (1978) 197–204.

[695] H.M. Mulder and A. Schrijver, Median graphs and Helly hypergraphs, *Discrete Mathematics* 25 (1979) 41–50.

[696] D. Mundici, Functions computed by monotone Boolean formulas with no repeated variables, *Theoretical Computer Science* 66 (1989) 113–114.

[697] I. Munro, Efficient determination of the transitive closure of a directed graph, *Information Processing Letters* 1 (1971) 56–58.

[698] S. Muroga, *Threshold Logic and Its Applications*, Wiley-Interscience, New York, 1971.

[699] S. Muroga, *Logic Design and Switching Theory*, Wiley-Interscience, New York, 1979.

[700] S. Muroga, S. Takasu and I. Toda, Theory of majority decision elements, *Journal of the Franklin Institute* 271 (1961) 376–418.

[701] S. Muroga, M. Kondo and I. Toda, Majority decision functions of up to six variables, *Mathematics of Computation* 16 (1962) 459–472.

[702] S. Muroga, T. Tsuboi and C.R. Baugh, Enumeration of threshold functions of eight variables, Department of Computer Science, University of Illinois, Report n° 245, 1967. Excerpts in *IEEE Transactions on Computers* C-19 (1970) 818–825.

[703] H. Narayanan, *Submodular Functions and Electrical Networks*, Annals of Discrete Mathematics Vol. 54, Elsevier, Amsterdam, 1997.

[704] N.N. Necula, O metodă pentru reducerea numărului de variabile ale functiilor Booleene foarte slab definite, *Studii şi Cercetari Matematice* 24 (1972) 561–566.

[705] R.J. Nelson, Simplest normal truth functions, *Journal of Symbolic Logic* 20 (2) (1955) 105–108.

[706] G.L. Nemhauser and L.A. Wolsey, Maximizing submodular set functions: Formulations and analysis of algorithms, *Annals of Discrete Mathematics* 11 (1981) 279–301.

[707] G.L. Nemhauser and L.A. Wolsey, *Integer and Combinatorial Optimization*, Wiley-Interscience Series in Discrete Mathematics and Optimization, John Wiley & Sons, New York, 1988.

[708] G.L. Nemhauser, L.A. Wolsey and M.L. Fisher, An analysis of approximations for maximizing submodular set functions - I, *Mathematical Programming* 14 (1978) 265–294.

[709] A. Neumaier, Inklusions- und Abstimmungssyteme, *Mathematische Zeitschrift* 141 (1975) 147–158.

[710] I. Newman, On read-once boolean functions, in: M.S. Paterson, ed., *Boolean Function Complexity: Selected Papers from LMS Symposium, Durham, July 1990*, Cambridge University Press, 1992, pp. 24–34.

[711] T.A. Nguyen, W.A. Perkins, T.J. Laffey and D. Pecora, Knowledge base verification, *AI Magazine* 8 (1987) 69–75.

[712] R.G. Nigmatullin, A variational principle in the algebra of logic, in: *Discrete Analysis*, Vol. 10, Novosibirsk, 1967, pp. 69–89 (in Russian).

[713] N.J. Nilsson, *Principles of Artificial Intelligence*, Morgan Kaufmann Publishers, San Francisco, CA, 1980.

[714] N. Nisan and M. Szegedy, On the degree of Boolean functions as real polynomials, *Computational Complexity* 4 (1994) 301–313.

[715] N. Nishimura, P. Ragde and S. Szeider, Detecting backdoor sets with respect to Horn and binary clauses, *Seventh International Conference on Theory and Applications of Satisfiability Testing* – SAT04, 2004, Vancouver, Canada.

[716] R. O'Donnell, Some topics in analysis of Boolean functions, in: *Proceedings of the 40th ACM Annual Symposium on Theory of Computing* (STOC), 2008, pp. 569–578.

[717] R. O'Donnell and R.A. Servedio, The Chow parameters problem, in: *Proceedings of the 40th ACM Annual Symposium on Theory of Computing* (STOC), 2008, pp. 517–526.

[718] H. Ono, K. Makino and T. Ibaraki, Logical analysis of data with decomposable structures, *Theoretical Computer Science* 289 (2002) 977–995.

[719] G. Owen, Multilinear extensions of games, *Management Science* 18 (1972) 64–79.

[720] G. Owen, *Game Theory*, Academic Press, San Diego, 1995.

[721] P. Padawitz, *Computing in Horn Clause Theories*, Springer-Verlag, Berlin, 1988.

[722] M.W. Padberg, Perfect zero-one matrices, *Mathematical Programming* 6 (1974) 180–196.

[723] M.W. Padberg, The Boolean quadric polytope: Some characteristics, facets and relatives, *Mathematical Programming* 45 (1989) 139–172.

[724] G. Palubeckis, Iterated tabu search for the unconstrained binary quadratic optimization problem, *Informatica* 17 (2006) 279–296.

[725] C.H. Papadimitriou, *Computational Complexity*, Addison Wesley Publishing Co., Reading, MA, 1994.

[726] C.H. Papadimitriou and K. Steiglitz, *Combinatorial Optimization: Algorithms and Complexity*, Prentice Hall, Englewood Cliffs, NJ, 1982.

[727] L. Papayanopoulos, Computerized weighted voting reapportionment, in: *AFIPS Proceedings*, Vol. 50, 1981, pp. 623–629.

[728] L. Papayanopoulos, On the partial construction of the semi-infinite Banzhaf polyhedron, in: A.V. Fiacco and K.O. Kortanek, eds., *Semi-Infinite Programming and Applications*, Lecture Notes in Economics and Mathematical Systems, Vol. 215, Springer-Verlag, Berlin-Heidelberg-New York, 1983, pp. 208–218.

[729] L. Papayanopoulos, DD analysis: Variational and computational properties of power indices, Research Report 83-18, Graduate School of Management, Rutgers University, NJ, 1983.

[730] P.M. Pardalos and S. Jha, Complexity of uniqueness and local search in quadratic 0–1 programming, *Operations Research Letters* 11 (1992) 119–123.

[731] R. Paturi, P. Pudlák, M.E. Saks and F. Zane, An improved exponential-time algorithm for k-SAT, in: *Proceedings of the 39th Annual IEEE Symposium on the Foundations of Computer Science*, IEEE, 1998, pp. 628–637.

[732] M.C. Paull and E.J. McCluskey, Jr., Boolean functions realizable with single threshold devices, *Proceedings of the IRE* 48 (1960) 1335–1337.

[733] M.C. Paull and S.H. Unger, Minimizing the number of states in incompletely specified sequential switching functions, *IRE Transactions on Electronic Computers* EC-8 (1959) 356–367.

[734] J. Peer and R. Pinter, Minimal decomposition of Boolean functions using non-repeating literal trees, in: *Proceedings of the International Workshop on Logic and Architecture Synthesis, IFIP TC10 WD10.5*, Grenoble, 1995, pp. 129–139.

[735] U.N. Peled and B. Simeone, Polynomial-time algorithms for regular set-covering and threshold synthesis, *Discrete Applied Mathematics* 12 (1985) 57–69.

[736] U.N. Peled and B. Simeone, An $O(nm)$-time algorithm for computing the dual of a regular Boolean function, *Discrete Applied Mathematics* 49 (1994) 309–323.

[737] B. Peleg, A theory of coalition formation in committees, *Journal of Mathematical Economics* 7 (1980) 115–134.

[738] B. Peleg, Coalition formation in simple games with dominant players, *International Journal of Game Theory* 10 (1981) 11–33.

[739] L.S. Penrose, The elementary statistics of majority voting, *Journal of the Royal Statistical Society* 109 (1946) 53–57.

[740] G. Pesant and C.-G. Quimper, Counting solutions of knapsack constraints, in: L. Perron and M.A. Trick, eds., *Integration of AI and OR Techniques in Constraint Programming for Combinatorial Optimization Problems*, Lecture Notes in Computer Science, Vol. 5015, Springer-Verlag, Berlin-Heidelberg, 2008, pp. 203–217.

[741] R. Petreschi and B. Simeone, A switching algorithm for the solution of quadratic Boolean equations, *Information Processing Letters* 11 (1980) 193–198.

[742] R. Petreschi and B. Simeone, Experimental comparison of 2-satisfiability algorithms, *RAIRO Recherche Opérationnelle* 25 (1991) 241–264.

[743] C.A. Petri, *Introduction to General Net Theory of Processes and Systems*, Springer-Verlag, Berlin, 1980.

[744] S.R. Petrick, A direct determination of the irredundant forms of a boolean function from the set of prime implicants, Technical Report AFCRC-TR-56-110, Air Force Cambridge Research Center, Cambridge, MA, April 1956.

[745] J.-C. Picard and M. Queyranne, A network flow solution to some nonlinear 0–1 programming programs, with applications to graph theory, *Networks* 12 (1982) 141–159.

[746] J.-C. Picard and H.D. Ratliff, Minimum cuts and related problems, *Networks* 5 (1975) 357–370.

[747] E. Pichat, The disengagement algorithm or a new generalization of the exclusion algorithm, *Discrete Mathematics* 17 (1977) 95–106.

[748] N. Pippenger, Galois theory for minors of finite functions, *Discrete Mathematics* 254 (2002) 405–419.

[749] L. Pitt and L.G. Valiant, Computational limitations on learning from examples, *Journal of the Association for Computing Machinery* 35 (1988) 965–984.

[750] D. Plaisted and S. Greenbaum, A structure-preserving clause form translation, *Journal of Symbolic Computation* 2 (1986) 293–304.

[751] G.R. Pogosyan, Classes of Boolean functions defined by functional terms, *Multiple Valued Logic* 7 (2002) 417–448.

[752] R. Pöschel and I. Rosenberg, Compositions and clones of Boolean functions, in: Y. Crama and P.L. Hammer, eds., *Boolean Models and Methods in Mathematics, Computer Science, and Engineering*, Cambridge University Press, Cambridge, 2010, pp. 3–38.

[753] E.L. Post, *The Two-Valued Iterative Systems of Mathematical Logic*, Annals of Mathematics Studies Vol. 5, Princeton University Press, Princeton, NJ, 1941.

[754] K. Prasad and J.S. Kelly, NP-completeness of some problems concerning voting games, *International Journal of Game Theory* 19 (1990) 1–9.

[755] R.E. Prather, *Introduction to Switching Theory: A Mathematical Approach*, Allyn and Bacon, Inc., Boston, MA, 1967.

[756] D. Pretolani, *Satisfiability and Hypergraphs*, Ph.D. thesis, University of Pisa, Pisa, Italy, 1992.

[757] D. Pretolani, A linear time algorithm for unique Horn satisfiability, *Information Processing Letters* 48 (1993) 61–66.

[758] D. Pretolani, Efficiency and stability of hypergraph SAT algorithms, in: D.S. Johnson and M.A. Trick, eds., *Cliques, Coloring, and Satisfiability*, DIMACS Series in Discrete Mathematics and Theoretical Computer Science, Vol. 26, American Mathematical Society, 1996, pp. 479–498.

[759] J.S. Provan, Boolean decomposition schemes and the complexity of reliability computations, *DIMACS Series in Discrete Mathematics* Vol. 5, American Mathematical Society, 1991, pp. 213–228.

[760] J.S. Provan and M.O. Ball, Efficient recognition of matroid and 2-monotonic systems, in: R.D. Ringeisen and F.S. Roberts, eds., *Applications of Discrete Mathematics*, SIAM, Philadelphia, 1988, pp. 122–134.

[761] P. Pudlák, Lower bounds for resolution and cutting planes proofs and monotone computations, *Journal of Symbolic Logic* 62 (1997) 981–998.

[762] P.W. Purdom, Solving satisfiability with less searching, *IEEE Transactions on Pattern Analysis and Machine Intelligence* 6(4) (1984) 510–513.

[763] P.W. Purdom, A survey of average time analyses of satisfiability algorithms, *Journal of Information Processing* 13 (1990) 449–455.

[764] I.B. Pyne and E.J. McCluskey, Jr., An essay on prime implicant tables, *Journal of the Society for Industrial and Applied Mathematics* 9 (1961) 604–631.

[765] I.B. Pyne and E.J. McCluskey, Jr., The reduction of redundancy in solving prime implicant tables, *IRE Transactions on Electronic Computers* EC-11 (1962) 473–482.

[766] W.V. Quine, The problem of simplifying truth functions, *American Mathematical Monthly* 59 (1952) 521–531.

[767] W.V. Quine, Two theorems about truth functions, *Boletin de la Sociedad Matemática Mexicana* 10 (1953) 64–70.

[768] W.V. Quine, A way to simplify truth functions, *American Mathematical Monthly* 62 (1955) 627–631.

[769] W.V. Quine, On cores and prime implicants of truth functions, *American Mathematical Monthly* (1959) 755–760.

[770] J.R. Quinlan, Induction of decision trees, *Machine Learning* 1 (1986) 81–106.

[771] J.R. Quinlan, *C4.5: Programs for Machine Learning*, Morgan Kaufmann Publishers, 1993.

[772] J.R. Quinlan, *Data mining tools See5 and C5.0*, published electronically at http://www.rulequest.com/see5-info.html/ (2000).

[773] R. Raghavan, J. Cohoon and S. Sahni, Single bend wiring, *Journal of Algorithms* 7 (1986) 232–257.

[774] V. Raghavan and S. Schach, Learning switch configurations, in: *Proceedings of the Third Annual Workshop on Computational Learning Theory*, Morgan Kaufmann Publishers, San Francisco, CA, 1990, pp. 38–51.

[775] C.C. Ragin, *The Comparative Method: Moving Beyond Qualitative and Quantitative Strategies*, University of California Press, Berkeley-Los Angeles-London, 1987.

[776] S. Rai, M. Veeraraghavan and K.S. Trivedi. A survey of efficient reliability computation using disjoint products approach, *Networks* 25 (1995) 147–163.

[777] K.G. Ramamurthy, *Coherent Structures and Simple Games*, Kluwer Academic Publishers, Dordrecht, 1990.

[778] B. Randerath, E. Speckenmeyer, E. Boros, O. Čepek, P.L. Hammer, A. Kogan, K. Makino and B. Simeone, Satisfiability formulation of problems on level graphs, in: H. Kautz and B. Selman, eds., *Proceedings of the LICS 2001 Workshop on Theory and Applications of Satisfiability Testing* (SAT 2001), Boston, MA, *Electronic Notes in Discrete Mathematics* 9 (2001) pp. 1–9.

[779] T. Raschle and K. Simon, Recognition of graphs with threshold dimension two, *Proceedings of the 27th Annual ACM Symposium on the Theory of Computing*, Las Vegas, NE, 1995, pp. 650–661.

[780] C. Ré and D. Suciu, Approximate lineage for probabilistic databases, *Proceedings of the Very Large Database Endowment* 1 (2008) 797–808.

[781] R.C. Read and R.E. Tarjan, Bounds on backtrack algorithms for listing cycles, paths, and spanning trees, *Networks* 5 (1975) 237–252.

[782] I.S. Reed, A class of multiple error-correcting codes and the decoding scheme, *IRE Transactions on Information Theory* IT-4 (1954) 38–49.

[783] R. Reiter, A theory of diagnosis from first principles, *Artificial Intelligence* 32 (1987) 57–95.

[784] J. Reiterman, V. Rödl, E. Šiňajová and M. Tůma, Threshold hypergraphs, *Discrete Mathematics* 54 (1985) 193–200.

[785] M.G. Resende, L.S. Pitsoulis and P.M. Pardalos, Approximate solution of weighted MAX-SAT problems using GRASP, in: D. Du, J. Gu and P.M. Pardalos, eds., *Satisfiability Problem: Theory and Applications*, DIMACS series in Discrete Mathematics

and Theoretical Computer Science, Vol. 35, American Mathematical Society, 1997, pp. 393–405.

[786] J.M.W. Rhys, A selection problem of shared fixed costs and network flows, *Management Science* 17 (1970) 200–207.

[787] J.A. Robinson, A machine oriented logic based on the resolution principle, *Journal of the Association for Computing Machinery* 12 (1965) 23–41.

[788] R.T. Rockafellar, *Convex Analysis*, Princeton University Press, Princeton, 1970.

[789] I.G. Rosenberg, 0–1 optimization and non-linear programming, *Revue Française d'Automatique, d'Informatique et de Recherche Opérationnelle (Série Bleue)* 2 (1972) 95–97.

[790] I.G. Rosenberg, Reduction of bivalent maximization to the quadratic case, *Cahiers du Centre d'Etudes de Recherche Opérationnelle* 17 (1975), 71–74.

[791] J. Rosenmüller, Nondegeneracy problems in cooperative game theory, in: A. Bachem, M. Grötschel and B. Korte, eds., *Mathematical Programming – The State of the Art*, Springer-Verlag, 1983, pp. 391–416.

[792] D. Roth, On the hardness of approximate reasoning, *Artificial Intelligence* 82 (1996) 273–302.

[793] J.P. Roth, Algebraic topological methods for the synthesis of switching systems, *Transactions of the American Mathematical Society* 88 (1958) 301–326.

[794] C. Rother, V. Kolmogorov, V. Lempitsky and M. Szummer, Optimizing binary MRFs via extended roof duality, in: *IEEE Conference on Computer Vision and Pattern Recognition*, June 2007.

[795] S. Rudeanu, *Boolean Functions and Equations*, North-Holland, Amsterdam, 1974.

[796] S. Rudeanu, *Lattice Functions and Equations*, Springer-Verlag, Heidelberg, 2001.

[797] Y. Sagiv, C. Delobel, D.S. Parker and R. Fagin, An equivalence between relational database dependencies and a fragment of propositional logic, *Journal of the ACM* 28 (1981) 435–453.

[798] S. Sahni and T. Gonzalez, P-complete approximation problems, *Journal of the ACM* 23 (1976) 555–565.

[799] M. Saks, Slicing the hypercube, in: K. Walker, ed., *Surveys in Combinatorics*, Cambridge University Press, Cambridge, 1993, pp. 211–255.

[800] M.T. Salvemini, B. Simeone and R. Succi, Analisi del possesso integrato nei gruppi di imprese mediante grafi, *L'Industria* XVI(4) (1995) 641–662.

[801] E.W. Samson and B.E. Mills, Circuit minimization: Algebra and algorithms for new Boolean canonical expressions, Air Force Cambridge Research Center, Technical Report TR 54-21, 1954.

[802] T. Sang, F. Bacchus, P. Beame, H.A. Kautz and T. Pitassi, Combining component caching and clause learning for effective model counting, in: *SAT 2004 - The Seventh International Conference on Theory and Applications of Satisfiability Testing*, Vancouver, Canada, 2004, pp. 20–28.

[803] A.A. Sapozhenko, On the complexity of disjunctive normal forms obtained by the use of the gradient algorithm, in: *Discrete Analysis*, Vol. 21, Novosibirsk, 1972, pp. 62–71 (in Russian).

[804] T. Sasao, *Switching Theory for Logic Synthesis*, Kluwer Academic Publishers, Norwell, Massachusetts, 1999.

[805] M. Sauerhoff, I. Wegener and R. Werchner, Optimal ordered binary decision diagrams for read-once formulas, *Discrete Applied Mathematics* 46 (1993) 235–251.

[806] A.A. Schäffer and M. Yannakakis, Simple local search problems that are hard to solve, *SIAM Journal on Computing* 20 (1991) 56–87.

[807] T.J. Schaefer, The complexity of satisfiability problems, in: *Proceedings of the 10th Annual ACM Symposium on the Theory of Computing*, San Diego, CA, 1978, pp. 216–226.

[808] I. Schiermeyer, Pure literal lookahead: an $O(1.497^n)$ 3-satisfiability algorithm, in: *Proceedings of the Workshop on Satisfiability*, Siena, Italy, 1996, pp. 63–72.

[809]   J.S. Schlipf, F.S. Annexstein, J.V. Franco and R.P. Swaminathan, On finding solutions for extended Horn formulas, *Information Processing Letters* 54 (1995) 133–137.

[810]   L. Schmitz, An improved transitive closure algorithm, *Computing* 30 (1983) 359–371.

[811]   W.G. Schneeweiss, *Boolean Functions with Engineering Applications and Computer Programs,* Springer-Verlag, Berlin, New York, 1989.

[812]   A. Schrijver, *Theory of Linear and Integer Programming*, Wiley-Interscience Series in Discrete Mathematics and Optimization, John Wiley & Sons, Chichester, 1986.

[813]   A. Schrijver, A combinatorial algorithm minimizing submodular functions in strongly polynomial time, *Journal of Combinatorial Theory B* 80 (2000) 346–355.

[814]   A. Schrijver, *Combinatorial Optimization: Polyhedra and Efficiency*, Springer, Berlin, 2003.

[815]   M.G. Scutellà, A note on Dowling and Gallier's top-down algorithm for propositional Horn satisfiability, *Journal of Logic Programming* 8 (1990) 265–273.

[816]   J. Sebelik and P. Stepanek, Horn clause programs for recursive functions, in: K.L. Clark and S.-A. Tarnlund, eds., *Logic Programming*, Academic Press, 1982, pp. 325–340.

[817]   D. Seinsche, On a property of the class of n-colorable graphs, *Journal of Combinatorial Theory B* 16 (1974) 191–193.

[818]   B. Selman, H. Kautz and B. Cohen, Noise strategies for improving local search, in: *Proceedings of the Twelfth National Conference on Artificial Intelligence*, Seattle, WA, 1994, pp. 337–343.

[819]   B. Selman, H. Kautz and B. Cohen, Local search strategies for satisfiability testing, in: D.S. Johnson and M.A. Trick, eds., *Cliques, Coloring, and Satisfiability,* DIMACS Series in Discrete Mathematics and Theoretical Computer Science, Vol. 26, American Mathematical Society, 1996, pp. 521–531.

[820]   B. Selman, H. Levesque and D. Mitchell, A new method for solving hard satisfiability problems, in: *AAAI'92, Proceedings of the Tenth National Conference on Artificial Intelligence*, San Jose, CA, 1992, pp. 440–446.

[821]   P.D. Seymour, The forbidden minors of binary matroids, *Journal of the London Mathematical Society* Ser. 2, 12 (1976) 356–360.

[822]   P.D. Seymour, The matroids with the max-flow min-cut property, *Journal of Combinatorial Theory B* 23 (1977) 189–222.

[823]   P.D. Seymour, Decomposition of regular matroids, *Journal of Combinatorial Theory B* 28 (1980) 305–359.

[824]   G. Shafer, *A Mathematical Theory of Evidence*, Princeton University Press, Princeton, 1976.

[825]   G. Shafer, Perspectives on the theory and practice of belief functions, *International Journal of Approximate Reasoning* 4 (1990) 323–362.

[826]   R. Shamir and R. Sharan, A fully dynamic algorithm for modular decomposition and recognition of cographs, *Discrete Applied Mathematics* 136 (2004) 329–340.

[827]   C.E. Shannon, The synthesis of two-terminal switching circuits, *Bell System Technical Journal* 28 (1949) 59–98.

[828]   L.S. Shapley, Simple games: An outline of the descriptive theory, *Behavioral Science* 7 (1962) 59–66.

[829]   L.S. Shapley, Cores of convex games, *International Journal of Game Theory* 1 (1971) 11–26.

[830]   D.R. Shier and D.E. Whited, Algorithms for generating minimal cutsets by inversion, *IEEE Transactions on Reliability* R-34 (1985) 314–318.

[831]   I. Shmulevich, E.R. Dougherty and W. Zhang, From Boolean to probabilistic Boolean networks as models of genetic regulatory networks, in: *Proceedings of the IEEE* 90 (2002) 1778–1792.

[832]   I. Shmulevich and W. Zhang, Binary analysis and optimization-based normalization of gene expression data, *Bioinformatics* 18 (2002) 555–565.

[833]   B. Simeone, *Quadratic 0–1 Programming, Boolean Functions and Graphs*, Ph.D. thesis, University of Waterloo, Ontario, Canada, 1979.

[834] B. Simeone, Consistency of quadratic Boolean equations and the König-Egerváry property for graphs, *Annals of Discrete Mathematics* 25 (1985) 281–290.

[835] B. Simeone, D. de Werra and M. Cochand, Recognition of a class of unimodular functions, *Discrete Applied Mathematics* 29 (1990) 243–250.

[836] I. Singer, Extensions of functions of 0–1 variables and applications to combinatorial optimization, *Numerical Functional Analysis and Optimization* 7 (1984-85) 23–62.

[837] P. Slavík, A tight analysis of the greedy algorithm for set cover, *Journal of Algorithms* 25 (1997) 237–254.

[838] R.H. Sloan, B. Szörényi and G. Turán, Learning Boolean functions with queries, in: Y. Crama and P.L. Hammer, eds., *Boolean Models and Methods in Mathematics, Computer Science, and Engineering*, Cambridge University Press, Cambridge, 2010, pp. 221–256.

[839] R.H. Sloan, K. Takata and G. Turán, On frequent sets of Boolean matrices, *Annals of Mathematics and Artificial Intelligence* 24 (1998) 1–4.

[840] N.J.A. Sloane, *The On-Line Encyclopedia of Integer Sequences*, published electronically at http://www.research.att.com/~njas/sequences/ (2006).

[841] J.-G. Smaus, On Boolean functions encodable as a single linear pseudo-Boolean constraint, in: P. Van Hentenryck and L.A. Wolsey, eds., *Proceedings of the 4th International Conference on Integration of AI and OR Techniques in Constraint Programming for Combinatorial Optimization Problems* (CPAIOR 2007), Lecture Notes in Computer Science, Vol. 4510, Springer-Verlag, Berlin-Heidelberg, 2007, pp. 288–302. Full version available as: Technical Report 230, Institut für Informatik, Universität Freiburg, Germany, 2007.

[842] D.R. Smith, Bounds on the number of threshold functions, *IEEE Transactions on Electronic Computers* EC-15 (1966) 368–369.

[843] J.D. Smith, M.J. Murray, Jr. and J.P. Minda, Straight talk about linear separability, *Journal of Experimental Psychology: Learning, Memory, and Cognition* 23 (1997) 659–680.

[844] Z. Stachniak, Going non-clausal, in: *Fifth International Symposium on the Theory and Applications of Satisfiability Testing*, SAT 2002, Cincinnati, Ohio, 2002, pp. 316–322.

[845] K.E. Stecke, Formulation and solution of nonlinear integer production planning problems for flexible manufacturing sytems, *Management Science* 29 (1983) 273–288.

[846] P.R. Stephan, R.K. Brayton and A.L. Sangiovanni-Vincentelli, Combinational test generation using satisfiability, *IEEE Transactions on Computer-Aided Design of Integrated Circuits and Systems* 15 (1996) 1167–1176.

[847] A. Sterbini and T. Raschle, An $O(n^3)$ time algorithm for recognizing threshold dimension 2 graphs, *Information Processing Letters* 67 (1998) 255–259.

[848] R.R. Stoll, *Set Theory and Logic*, Dover Publications, New York, 1979.

[849] H. Störmer, *Binary Functions and their Applications*, Lecture Notes in Economics and Mathematical Systems, Vol. 348, Springer, Berlin, 1990.

[850] P.D. Straffin, *Game Theory and Strategy*, The Mathematical Association of America, Washington, 1993.

[851] M. Sugeno, Fuzzy measures and fuzzy integrals: a survey, in: M.M. Gupta, G.N. Saridis and B.R. Gaines, eds., *Fuzzy Automata and Decision Processes*, North-Holland, Amsterdam, 1977, pp. 89–102.

[852] R. Swaminathan and D.K. Wagner, The arborescence realization problem, *Discrete Applied Mathematics* 59 (1995) 267–283.

[853] O. Sykora, An optimal algorithm for renaming a set of clauses into the Horn set, *Computers and Artificial Intelligence* 4 (1985) 37–43.

[854] S. Szeider, Backdoor sets for DLL subsolvers, *Journal of Automated Reasoning* 35 (2005) 73–88.

[855] W. Szwast, On Horn spectra, *Theoretical Computer Science* 82 (1991) 329–339.

[856] K. Takata, A worst-case analysis of the sequential method to list the minimal hitting sets of a hypergraph, *SIAM Journal on Discrete Mathematics* 21 (2007) 936–946.

[857] M. Tannenbaum, The establishment of a unique representation for a linearly separable function, Lockheed, Technical Note n° 20, 1961.

[858] R.E. Tarjan, Depth first search and linear graph algorithms, *SIAM Journal on Computing* 1 (1972) 146–160.

[859] R.E. Tarjan, Amortized computational complexity, *SIAM Journal on Algebraic and Discrete Methods* 6 (1985) 306–318.

[860] A.D. Taylor and W.S. Zwicker, Simple games and magic squares, *Journal of Combinatorial Theory A* 71 (1995) 67–88.

[861] A.D. Taylor and W.S. Zwicker, *Simple Games: Desirability Relations, Trading, Pseudoweightings*, Princeton University Press, Princeton, NJ, 1999.

[862] A. Thayse, *Boolean Calculus of Differences*, Lecture Notes in Computer Science, Vol. 101, Springer-Verlag, Berlin-Heidelberg-New York, 1981.

[863] A. Thayse, *From Standard Logic to Logic Programming*, John Wiley & Sons, Chichester etc., 1988.

[864] P. Tison, Generalization of consensus theory and application to the minimization of Boolean functions, *IEEE Transactions on Electronic Computers* EC-16, No. 4 (1967) 446–456.

[865] D.M. Topkis, *Supermodularity and Complementarity*, Princeton University Press, Princeton, NJ, 1998.

[866] C.A. Tovey, Hill climbing with multiple local optima, *SIAM Journal on Algebraic and Discrete Methods* 6 (1985) 384–393.

[867] C.A. Tovey, Low order polynomial bounds on the expected performance of local improvement algorithms, *Mathematical Programming* 35 (1986) 193–224.

[868] C.A. Tovey, Local improvement on discrete structures, in: E. Aarts and J.K. Lenstra, eds., *Local Search in Combinatorial Optimization*, John Wiley & Sons, Chichester, 1997, pp. 57–89.

[869] M.A. Trick, A dynamic programming approach for consistency and propagation for knapsack constraints, *Annals of Operations Research* 118 (2003) 73–84.

[870] K. Truemper, Monotone decomposition of matrices, Technical Report UTDCS-1-94, 1994.

[871] K. Truemper, *Effective Logic Computation*, Wiley-Interscience, New York, 1998.

[872] G.S. Tseitin, On the complexity of derivations in propositional calculus, in: A.O. Slisenko, ed., *Studies in Constructive Mathematics and Mathematical Logic*, Part II, Consultants Bureau, New York, 1970, pp. 115–125. (Translated from the Russian).

[873] S. Tsukiyama, M. Ide, H. Ariyoshi and I. Shirakawa, A new algorithm for generating all the maximal independent sets, *SIAM Journal on Computing* 6 (1977) 505–517.

[874] J.D. Ullman, *Principles of Database and Knowledge-Base Systems, Vol. I: Classical Database Systems*, Computer Science Press, New York, 1988.

[875] J.D. Ullman, *Principles of Database and Knowledge-Base Systems, Vol. II: The New Technologies*, Computer Science Press, New York, 1989.

[876] C. Umans, The minimum equivalent DNF problem and shortest implicants, *Journal of Computer and System Sciences* 63 (2001) 597–611.

[877] C. Umans, T. Villa and A.L. Sangiovanni-Vincentelli, Complexity of two-level logic minimization, *IEEE Transactions on Computer-Aided Design of Integrated Circuits and Systems* 25 (2006) 1230–1246.

[878] T. Uno, Efficient computation of power indices for weighted majority games, NII Technical Report NII-2003-006E, National Institute of Informatics, Japan, 2003.

[879] R.H. Urbano and R.K. Mueller, A topological method for the determination of the minimal forms of a Boolean function, *IRE Transactions on Electronic Computers* EC-5 (1956) 126–132.

[880] A. Urquhart, Hard examples for resolution, *Journal of the Association for Computing Machinery* 34 (1987) 209–219.

[881] A. Urquhart, The complexity of propositional proofs, *Bulletin of Symbolic Logic* 1 (1995) 425–467.

[882] A. Urquhart, Proof theory, in: Y. Crama and P.L. Hammer, eds., *Boolean Models and Methods in Mathematics, Computer Science, and Engineering,* Cambridge University Press, Cambridge, 2010, pp. 79–98.

[883] L.G. Valiant, The complexity of enumeration and reliability problems, *SIAM Journal on Computing* 8 (1979) 410–421.

[884] L.G. Valiant, A theory of the learnable, *Communications of the ACM* 27 (1984) 1134–1142.

[885] A. Van Gelder, A satisfiability tester for non-clausal propositional calculus, *Information and Computation* 79 (1988) 1–21.

[886] A. Van Gelder and Y.K. Tsuji, Satisfiability testing with more reasoning and less guessing, in: D.S. Johnson and M.A. Trick, eds., *Cliques, Coloring, and Satisfiability,* DIMACS Series in Discrete Mathematics and Theoretical Computer Science, Vol. 26, American Mathematical Society, 1996, pp. 559–586.

[887] J. van Leeuwen, Graph algorithms, in: J. van Leeuwen, ed., *Handbook of Theoretical Computer Science: Algorithms and Complexity,* Volume A, The MIT Press, Cambridge, MA, 1990, pp. 525–631.

[888] H. Vantilborgh and A. van Lamsweede, On an extension of Dijkstra's semaphore primitives, *Information Processing Letters* 1 (1972) 181–186.

[889] Yu.L. Vasiliev, On the comparison of the complexity of prime irredundant and minimal DNFs, in: *Problems of Cybernetics,* Vol. 10, PhysMatGIz, Moscow, 1963, pp. 5–61 (in Russian).

[890] Yu.L. Vasiliev, The difficulties of minimizing Boolean functions using universal approaches, *Doklady Akademii Nauk SSSR,* Vol. 171, No. 1, 1966, pp. 13–16 (in Russian).

[891] T. Villa, R.K. Brayton and A.L. Sangiovanni-Vincentelli, Synthesis of multi-level Boolean networks, in: Y. Crama and P.L. Hammer, eds., *Boolean Models and Methods in Mathematics, Computer Science, and Engineering,* Cambridge University Press, Cambridge, 2010, pp. 675–722.

[892] H. Vollmer, *Introduction to Circuit Complexity: A Uniform Approach,* Springer, Berlin - New York, 1999.

[893] J. von Neumann and O. Morgenstern, *Theory of Games and Economic Behavior,* Princeton University Press, Princeton, NJ, 1944.

[894] B.W. Wah and Y. Shang, A discrete Lagrangian-based global-search method for solving satisfiability problems, in: D. Du, J. Gu and P.M. Pardalos, eds., *Satisfiability Problem: Theory and Applications,* DIMACS series in Discrete Mathematics and Theoretical Computer Science, Vol. 35, American Mathematical Society, 1997, pp. 365–392.

[895] D. Waltz, Understanding line drawings of scenes with shadows, in: P.H. Winston, ed., *The Psychology of Computer Vision,* McGraw-Hill, New York, 1975.

[896] C. Wang, Boolean minors, *Discrete Mathematics* 141 (1995) 237–258.

[897] C. Wang and A.C. Williams, The threshold order of a Boolean function, *Discrete Applied Mathematics* 31 (1991) 51–69.

[898] H. Wang, H. Xie, Y.R. Yang, L.E. Li, Y. Liu and A. Silberschatz, Stable egress route selection for interdomain traffic engineering: Model and analysis, in: *Proceedings of Thirteenth IEEE Conference on Network Protocols* (ICNP '05), Boston, 2005, pp. 16–29.

[899] S. Warshall, A theorem on Boolean matrices, *Journal of the ACM* 9 (1962) 11–12.

[900] A. Warszawski, Pseudo-Boolean solutions to multidimensional location problems, *Operations Research* 22 (1974) 1081–1096.

[901] W.D. Wattenmaker, G.I. Dewey, T.D. Murphy and D.L. Medin, Linear separability and concept learning: Context, relational properties, and concept naturalness, *Cognitive Psychology* 18 (1986) 158–194.

[902] I. Wegener, *The Complexity of Boolean Functions,* Wiley-Teubner Series in Computer Science, John Wiley & Sons, Chichester etc., 1987.

[903] I. Wegener, *Branching Programs and Binary Decision Diagrams: Theory and Applications*, SIAM Monographs on Discrete Mathematics and Applications, SIAM, Philadelphia, PA, 2000.

[904] R. Weismantel, On the 0–1 knapsack polytope, *Mathematical Programming* 77 (1987) 49–68.

[905] D.J.A. Welsh, *Matroid Theory*, London Mathematical Society Monographs, Vol. 8, Academic Press, New York, 1976.

[906] D.J.A. Welsh, Matroids: Fundamental concepts, in: R. Graham, M. Grötschel and L. Lovász, eds., *Handbook of Combinatorics*, Elsevier, Amsterdam, 1995, pp. 481–526.

[907] D. Wiedemann, Unimodal set-functions, *Congressus Numerantium* 50 (1985) 165–169.

[908] D. Wiedemann, A computation of the eighth Dedekind number, *Order* 8 (1991) 5–6.

[909] D.J. Wilde and J.M. Sanchez-Anton, Discrete optimization on a multivariable Boolean lattice, *Mathematical Programming* 1 (1971) 301–306.

[910] H.P. Williams, Experiments in the formulation of integer programming problems, *Mathematical Programming Studies* 2 (1974) 180–197.

[911] H.P. Williams, Linear and integer programming applied to the propositional calculus, *Systems Research and Information Sciences* 2 (1987) 81–100.

[912] H.P. Williams, Logic applied to integer programming and integer programming applied to logic, *European Journal of Operational Research* 81 (1995) 605–616.

[913] R. Williams, C. Gomes and B. Selman, Backdoors to typical case complexity, in: *Proceedings of the International Joint Conference on Artificial Intelligence* (IJCAI) 2003, pp. 1173–1178.

[914] J.M. Wilson, Compact normal forms in propositional logic and integer programming formulations, *Computers and Operations Research* 90 (1990) 309–314.

[915] R.O. Winder, More about threshold logic, in: *IEEE Symposium on Switching Circuit Theory and Logical Design*, 1961, pp. 55–64.

[916] R.O. Winder, Single stage threshold logic, in: *IEEE Symposium on Switching Circuit Theory and Logical Design*, 1961, pp. 321–332.

[917] R.O. Winder, *Threshold Logic*, Ph.D. Dissertation, Department of Mathematics, Princeton University, Princeton, NJ, 1962.

[918] R.O. Winder, Properties of threshold functions, *IEEE Transactions on Electronic Computers* EC-14 (1965) 252–254.

[919] R.O. Winder, Enumeration of seven-arguments threshold functions, *IEEE Transactions on Electronic Computers* EC-14 (1965) 315–325.

[920] R.O. Winder, Chow parameters in threshold logic, *Journal of the Association for Computing Machinery* 18 (1971) 265–289.

[921] P.H. Winston, *Artificial Intelligence*, Addison-Wesley, Reading, MA, 1984.

[922] L.A. Wolsey, Faces for a linear inequality in 0–1 variables, *Mathematical Programming* 8 (1975) 165–178.

[923] L.A. Wolsey, An analysis of the greedy algorithm for the submodular set covering problem, *Combinatorica* 2 (1982) 385–393.

[924] L.A. Wolsey, *Integer Programming*, Wiley-Interscience Series in Discrete Mathematics and Optimization, John Wiley & Sons, New York, 1998.

[925] L. Wos, R. Overbeek, E. Lusk and J. Boyle, *Automated Reasoning: Introduction and Applications*, Prentice-Hall, Englewood Cliffs, NJ, 1984.

[926] Z. Xing and W. Zhang, MaxSolver: An efficient exact algorithm for (weighted) maximum satisfiability, *Artificial Intelligence* 164 (2005) 47–80.

[927] M. Yagiura, M. Kishida and T. Ibaraki, A 3-flip neighborhood local search for the set covering problem, *European Journal of Operational Research* 172 (2006) 472–499.

[928] S. Yajima and T. Ibaraki, A lower bound on the number of threshold functions, *IEEE Transactions on Electronic Computers* EC-14 (1965) 926–929.

[929] S. Yajima and T. Ibaraki, On relations between a logic function and its characteristic vector, *Journal of the Institute of Electronic and Communication Engineers of Japan* 50 (1967) 377–384 (in Japanese).

[930] M. Yamamoto, An improved $\tilde{O}(1.234^m)$-time deterministic algorithm for SAT, in: X. Deng and D. Du, eds., *Algorithms and Computation - ISAAC 2005*, Lecture Notes in Computer Science, Vol. 3827, Springer-Verlag, Berlin-Heidelberg, 2005, pp. 644–653.

[931] S. Yamasaki and S. Doshita, The satisfiability problem for a class consisting of Horn sentences and some non-Horn sentences in propositional logic, *Information and Control* 59 (1983) 1–12.

[932] M. Yannakakis, Node-and edge-deletion NP-complete problems, in: *Proceedings of the 10th Annual ACM Symposium on Theory of Computing* (STOC) 1978, ACM, NY, USA, pp. 253–264.

[933] M. Yannakakis, The complexity of the partial order dimension problem, *SIAM Journal on Algebraic and Discrete Methods* 3 (1982) 351–358.

[934] M. Yannakakis, On the approximation of maximum satisfiability, *Journal of Algorithms* 17 (1994) 475–502.

[935] E. Zemel, Easily computable facets of the knapsack polytope, *Mathematics of Operations Research* 14 (1989) 760–764.

[936] H. Zhang and J.E. Rowe, Best approximations of fitness functions of binary strings, *Natural Computing* 3 (2004) 113–124.

[937] Yu.I. Zhuravlev, Set-theoretical methods in Boolean algebra, *Problems of Cybernetics* 8 (1962) 5–44 (in Russian).

[938] S. Živný, D.A. Cohen and P.G. Jeavons, The expressive power of binary submodular functions, *Discrete Applied Mathematics* 157 (2009) 3347–3358.

[939] Yu.A. Zuev, Approximation of a partial Boolean function by a monotonic Boolean function, *U.S.S.R. Computational Mathematics and Mathematical Physics* 18 (1979) 212–218.

[940] Yu.A. Zuev, Asymptotics of the logarithm of the number of threshold functions of the algebra of logic, *Soviet Mathematics Doklady* 39 (1989) 512–513.

[941] U. Zwick, Approximation algorithms for constraint satisfaction problems involving at most three variables per constraint, in: *SODA '98: Proceedings of the 9th Annual ACM-SIAM Symposium on Discrete Algorithms*, 1998, SIAM, Philadelphia, PA, pp. 201–210.

Index

2-Sᴀᴛ problem, *see* quadratic equation
3-Sᴀᴛ problem, *see* DNF equation, degree 3

absorption, 9, 26
  closure, 131
affine Boolean function, 110
algebraic normal form, *see* representation over
    GF(2)
aligned function, 398–399
almost-positive pseudo-Boolean function, 604
apportionment problem, 434
approximation algorithm, 117
arborescence, 613
artificial intelligence, 50–52, 68, 124, 174,
    273–274, 279–280, 511, 566, 569, 599
association rule, 521–522
asummable function, 414–417
  k-asummable
    definition, 414
  2-asummable
    complete monotonicity, 396
    definition, 395
    vs. threshold function, 416
  threshold function, 414
  weakly asummable function, 429
    Chow function, 430

backdoor set, 83
Banzhaf index, 57–58
  and pseudo-Boolean approximations, 593
  and strength preorder, 360
  definition, 57
  in reliability, 59
  of threshold functions, 349, 434, 436, 437
  raw, 57

BDD, *see* binary decision diagram (BDD)
belief function, 569
belt function, 159
bidirected graph, 206, 210
binary decision diagram (BDD), 46–49
  and orthogonal DNF, 48
  ordered (OBDD), 47
bipartite graph, 542, 549, 611
  and conflict codes, 216, 592
  and Guttman scale, 443
  and posiforms, 604
  complete, 611
  recognition, 219
black box oracle, *see* oracle algorithm
blocker, 179
Boolean equation
  complexity, 72–74, 104–111
  consistent, 67
  definition, 67, 73
  DNF, *see* DNF equation
  generating all solutions, 112
  inconsistent, 67
  parametric solutions, 113–115
Boolean expression
  definition, 10
  dual, 14
  equivalent, 12
  length, size, 12
  of a function, 10–13
  read-once, 448
  satisfiable, 68
  tautology, 68
  valid, 68
Boolean function, 3
  expression, representation, 10–13
  normal form representations, 15–19

bottleneck optimization, 180
branching procedures for Boolean equations,
   80–87
   branching on terms, 83
   branching rules, 82
   complexity, 104, 105

Choquet capacity, 569
Chow function, 429
   weakly asummable function, 430
Chow parameters
   and Banzhaf indices, 57
   and degree sequences, 441
   and essential variables, 31
   and reliability, 59
   and strength preorder, 355
   complexity, 61, 436
   definition, 24
   modified, 56, 66, 431
   of threshold functions, 428–438
circuit: combinational, logic, switching, 5,
   52–55, 69–71, 142, 174, 405, 456, 558
clause, 15
clique, 610
clutter, 60, 177, 451, 614
CNF, see conjunctive normal form (CNF)
co-Horn DNF, 207
co-Horn function, 309
   and Schaefer's theorem, 110
co-occurrence graph, 449, 455
   of read-once function, 458
   oracle algorithm, 475
coalition
   blocking, 182
   maximal losing, 56
   minimal winning, 56
   winning, 56
cograph, see P_4-free graph
coloring
   of graphs, 121, 220
   of hypergraphs, 71–72, 178–180
complementation (Boolean), 8
complete DNF, 27
   recognition, 134
completely monotone function, 391–397
   2-asummability, 396
   dual-comparability, 394–395
   recognition, 396
   vs. threshold function, 392
completely unimodal pseudo-Boolean function,
   606
computer vision, 567
concave envelope of a pseudo-Boolean function,
   581

condensation of a digraph, 613
conflict code, 216
conflict graph, 216–218, 591
   stable set, 591
conflicting terms, 216, 545, 591
conjunction
   Boolean, 8
   pseudo-Boolean, 574
conjunctive normal form (CNF), 14–19
   clause, 15
   definition, 15
   of a function, 15–19
   pseudo-Boolean, 576
connected component, 611
consensus
   Chvátal cut, 99
   closure, 131, 293
   derivation, 93
   of two conjunctions, 92
   unit consensus, 95, 97, 102
consensus procedure for Boolean equations,
   92–95
   and cutting-plane proofs, 99
   complexity, 104, 108
   hard examples, 105
   input consensus, 122
   linear consensus, 102, 122
consensus procedure for prime implicants,
   130–138
   disengagement order, 256
   input consensus, 289
   input disengagement, 255
   input prime consensus, 291
   prime implicant depletion, 287
   term disengagement, 137
   variable depletion, 135
constraint satisfaction problem, 108–111
constraint set, 108
control set, 83, 86
convex Boolean function, 545
convex envelope
   of a Boolean function, 546
   of a pseudo-Boolean function, 581
convex hull of terms, 546
Cook's theorem, 41, 72–74, 230, 622–624
correlation polytope, 595
coterie, 182
crossover point, 108
cube, 15
cut-threshold graph, 444
cutset, 181
cutting-plane proof, 99
   Chvátal closure, 98
   Chvátal cut, 98

complexity, 104
consensus cut, 99
hard examples, 105

data mining, 511, 521, 522, 538, 567
databases, 274–275, 593
Davis-Putnam rules, *see* DNF equation,
    Davis-Putnam rules
De Morgan's laws, 9
decision tree, 47–49
  complexity, 175
  construction, 48
  depth, 49, 175
  of a pdBf, 526
decomposable function, 540
degree
  of a Boolean function, 492
  of a DNF, 18
  of a polynomial threshold function, 407
  of a pseudo-Boolean function, 571
  of an elementary conjunction, 18
  of prime implicants, 163, 193
degree-k DNF, 41, 197
  functions representable by, 41
    characterization by functional equations,
      500, 506
    recognition, 43
degree-k extension, 535
degree-k function, 491–495
  characterization by functional equations, 494,
    496
degree-k pseudo-Boolean approximation, 593
disjunction
  Boolean, 8
  pseudo-Boolean, 574
disjunctive normal form (DNF), 14–19
  complete, 27
    recognition, 134
  definition, 15
  degree
    definition, 18
    typical, 163
  extremal size, 161
  irredundant, 27
  linear, 18
  mixed, 207
  monotone, 35
  negative, 35
  of a function, 15–19
  orthogonal, *see* orthogonal DNF
  polar, 207
  positive, 35
  prime, 27
  pseudo-Boolean, 575

quadratic, 18
random, 106
redundant, 27
term, 15
transformation into, 19–22
distributed computing, 182, 407
DNF, *see* disjunctive normal form (DNF)
DNF equation
  branching procedures, 80–87
  consensus procedure, 92–95, 135
  counting solutions, 111
  Davis-Putnam rules, 84, 91
  definition, 67, 73
  degree 3 (3-SAT), 73, 74, 105, 121
    random, 107
    reduction to, 77
  heuristics, 86, 102
  Horn relaxation, 86
  integer programming approaches, 95–100
  nonlinear programming approaches, 100–102
  preprocessing, 84–87
  quadratic relaxation, 87
  random, 106
  relative strength of procedures, 104
  relaxation schemes, 86
  rewriting rules, 84
  satisfiability problem, 74
  variable elimination, 87–91
domishold graph, 444
don't care points, 512, 558
dual expression, 14
dual function, 13
  mutually dual functions, 168, 183, 190
dual implicant, 169, 177, 308, 450, 458
dual subimplicant, 452
  recognition, 455
  theorem, 453
dual-comparable function, 170–174, 178–179,
    394
dual-major function, 170, 179, 180, 182,
    394, 411
dual-minor function, 170, 178, 180, 182,
    394, 411
duality
  and bottleneck optimization, 181
  and game theory, 13, 56, 182, 477–480
  and hypergraphs, 61, 177, 179
  and integer programming, 180
  and reliability theory, 181
  principle, 169
dualization
  algorithms, 183–189, 192–196
  Berge multiplication, 187
  by sequential distributivity, 186–189

dualization (*cont.*)
  complexity, 183–186, 189–192, 481
  double dualization, 141, 188
  equivalent problems, 191, 196
  Fredman-Khachiyan algorithm, 192–196, 308
  of Horn functions, 306–309
  of quadratic functions, 263–266
  of regular functions, 369–377
  of shellable functions, 336–338
  recursive algorithm, 197
  vs. identification, 196

electrical engineering, 5, 52–55, 69–71, 174,
  224–226, 405
elementary conjunction
  Boolean, 15
  pseudo-Boolean, 575
elementary disjunction
  Boolean, 15
  pseudo-Boolean, 576
elementary operations, 8
  properties, 9
Espresso, 69, 82
essential variable, 30
  of a read-once function, 474, 484
  recognition, 43, 436, 484
exclusive-or, 44
expert system, 50, 68, 124, 205, 273, 512, 566
expression, *see* Boolean expression
extension of a pdBf, 514–558
  best fit, 548
  bi-theory, 525
  convex extension, 545–547
  decision tree, 529
  decomposable extension, 539–545
  definition, 511
  degree-k extension, 535
  existence, 514
  Horn extension, 535–538
  largest, 515
  monotone extension, 532–534
  positive extension, 531
  smallest, 515
  theory, 519
  threshold extension, 538–539
  with errors, 547–551
  with missing bits, 551–558
    consistent, 552
    fully robust, 552, 553
    most robust, 552
extension of a pseudo-Boolean function, 578
  concave, 581–583, 589–590
  concave envelope, 581
  convex, 581–583

convex envelope, 581
Lovász extension, 583, 590
paved upper-plane, 583
polynomial, 579, 588–589
standard, 582, 590, 595

facility location, 570
false points, 3
  maximal, 38
Fourier expansion, *see* representation over
  the reals
functional equations
  addition of inessential variables, 495
  certificate of non-membership, 501, 505
  characterizable classes, 490, 495–506
  definition, 489
  finitely characterizable classes, 500–506
  for co-Horn functions, 491
  for degree-k positive functions, 494
  for Horn functions, 208, 278, 491
  for linear functions, 493
  for positive functions, 488
  for quadratic functions, 208, 493
  for submodular functions, 309, 492
  for supermodular functions, 492
  forbidden identification minors, 504
  identification of variables, 495
  non-characterizable classes, 496
fuzzy measure, 569

game theory
  characteristic form, 568, 593
  positional games, 477–480
  simple games, 7, 55–58, 182, 348, 354, 360,
    391, 406, 434–436, 593
    constant-sum, 182
    decisive, 182
    proper, 182
    strong, 182
greedy heuristic
  for Boolean equations, 86, 135
  for logic minimization, 156–159
  for set covering, 156
  for submodular optimization, 603
Guttman scale, 443

Hamming distance, 523
heuristics
  for Boolean equations, 86, 102
Horn DNF
  definition, 51, 270
  dual recognition, 308
  extended, 319

generalized, 320
irredundant and prime, 283, 294–296, 314
literal minimization: approximability, 301
literal minimization: complexity, 301
minimization, 297
polynomial hierarchies, 320–321
pure, 270
renamable, 235, 314–316
source sides minimization: complexity, 302
term minimization: approximability, 300–301
term minimization: complexity, 298–300
Horn equation, 86, 281–286
  and forward chaining, 284–285
  and unit literal rule, 281–284
  unique solution, 286
Horn function
  acyclic, 313–314
  and Schaefer's theorem, 110
  bidual, 311–312
  characteristic models, 280, 322
  characterization by functional equations,
    278, 491
  definition, 271
  double Horn, 312–313
  dual, 306
  dual recognition, 307
  dualization, 306–309
  false points, 277–280
  generation of prime implicants, 140, 286–292
  maximal minorant, 279, 321–323
  minimal majorant, 321–323
  properties of prime implicants, 292–296
  pure, 271
  recognition, 272
  renamable, 316, 317
Horn term
  definition, 269
  head, 270
  positive, 270
  pure, 270
  subgoals, 270
hypergraph, 60–61, 298
  coloring, 71, 178–180
  definition, 614
  directed, 275
  hierarchy, 320
  regular, 360
  stable sets, 7, 34
  transversal, 177, 179

ideal function, 399–400
identification minor, 503
identification of variables, 495
implicant

complexity of recognition, 138
definition, 26
dual, 169, 450
geometric interpretation, 31
pseudo-Boolean, 576
vs. true point, 39
implicate
  definition, 28
  dual, 169
  pseudo-Boolean, 576
  vs. false point, 39
implication graph, 212–216, 247
  and generation of prime implicants, 250–254
  and irredundant quadratic DNFs, 261–263
  Mirror property, 212
  solving quadratic equations, 239–240
incompletely specified function, see partially
    defined Boolean function
inessential variable, 30
  addition, 495
input consensus, 122, 255, 289, 291
integer programming
  MAX SAT, 117
  aggregation problem, 422, 442
  Boolean equations, 95–100
  Boolean formulations, 62–65, 180–181
  Boolean preprocessing, 72
  extended Horn equations, 319
  Horn equations, 276–277
  knapsack problem, 407, 426, 427
  logic minimization, 143, 156
  monotone pdBf extension, 533
  nonlinear, see pseudo-Boolean optimization
  pdBf support set, 532
  pseudo-Boolean formulations, 568–570
  regular set covering, 377–380, 388–390
  set packing, 442
  strength preorder, 359
  total unimodularity, 221

k-monotone function, 391–397
  dual, 394
  related properties, 430
Karnaugh map, 32
knapsack problem, 64, 407, 426, 427
Kőnig-Egerváry property, 72, 211, 223–224, 231

LE property, see lexico-exchange property
leader, 339, 359
left-shift, 362–365
  ceiling, 363, 420
  definition, 362, 420
  floor, 363, 420

level graph, 227
lexico-exchange property, 338–346
    dualization, 345–346
    orthogonal form, 345
    quadratic functions, 346–347
    recognition, 340–345
    regular functions, 359
    shellability, 339
lexicographic order
    of points in $\mathbb{R}^n$, 366, 374
    of sets, 336, 338, 362
linear equations over GF(2), 110
linear function
    characterization by functional equations, 493
linearly separable function, *see* threshold
        function
list-generating algorithms
    complexity, 625–626
    for prime implicants, 128–141
    for solutions of Boolean equations, 112
    for solutions of quadratic equations, 244
local maximum of pseudo-Boolean function, 606
    complexity, 587
    definition, 586
logic minimization, 141–159
    approximability, 156–159
    complexity, 150–156
    extremal number of terms, 161
    literal minimization, 142
    set covering formulation, 143–150
    term minimization, 142
    typical number of terms, 164
    with don't cares, 558–561
logical analysis of data, 512
Lovász extension, 583, 590
    and supermodular functions, 601

Möbius transform, 571
machine learning, 47, 52, 196, 473, 512, 522,
        538, 547, 567
majorant
    definition, 24
    Horn, 321–323
    regular, 380–390
    smallest positive, 66
    smallest regular, 380, 386
matched graph, *see* quadratic equation
matching, 610, 614
matroid, 197, 348, 566
MAX SAT problem, *see* maximum satisfiability
max-cut problem, 565, 603
max-quadratic pseudo-Boolean function, 226
maximum satisfiability, 115–121
    MAX 2-SAT, 249

approximability, 118–121
    complexity, 116
    integer programming formulation, 117
    pseudo-Boolean formulation,
        566, 590
maxterm, 17
    (Boolean) expression, 17
median graph, 243
membership problems
    complexity, 42
    definition, 40
    DNF, 40
    functional, 40
        certificate of non-membership, 501
min-cut problem, 603, 604
minorant
    definition, 24
    largest positive, 66, 166
    largest regular, 380, 382, 389
    maximal Horn, 279, 321–323
    regular, 380–390
minterm, 17
    (Boolean) expression, 17
    (pseudo-Boolean) PBNF, 572
models (of a Boolean equation), 51, 68, 273,
        279, 280
monomial, 15
monotone Boolean function, 34
    k-monotone, 392
    characterization by functional equations,
        496
    completely monotone, 392
monotone literal rules, 85
monotone pseudo-Boolean function, 566, 568,
        569, 596–598
    derivatives, 596
    DNF and CNF, 597
    recognition, 596
multicriteria decision-making, 569, 578
multilinear expression, extension, *see*
        polynomial expression, extension
mutually dual functions, 168, 183, 190

negation (Boolean), 8
negative function, 34
Negative Unit Literal Rule (NULR), 281
neural network, 406, 408
nonlinear programming
    for Boolean equations, 100–102
normal form of a pseudo-Boolean function
        (PBNF), 572
normal function, 450
    recognition, 468, 483

ODNF, *see* orthogonal DNF (ODNF)
oracle algorithm, 5
  for essential variables, 484
  for identification, 196
  for read-once learning, 473–476, 485
  for read-once recognition, 472
  for regular recognition, 369
  for supermodular optimization, 603
  for threshold recognition, 417, 445
orthogonal DNF (ODNF), 22, 48, 326
  algorithms, 326–330
  of a regular function, 361
  of a shellable DNF, 331–335
  reliability, 58
  vs. dual DNF, 329

P_4-free graph, 449
  complement, 462
  cotree, 464
  recognition, 465
  vs. cograph, 463–466
parametric solution of equations, 113–115
  quadratic equations, 246–249
  reproductive, 114
parity function, 44, 66, 162, 444
parse tree
  as game tree, 478
  of a positive expression, 456
  of a read-once expression, 456
partially defined Boolean function, 511
  basic theory, 520
  bi-theory, 525–526, 529
  co-pattern, 520
  co-theory, 520
  decision tree, 526–531
  extension, 511, 514–558
  irredundant theory, 519
  logic minimization, 558–561
  missing bits, 551
  pattern, 518, 567
  prime pattern, 518
  prime theory, 519
  support set, 515
  theory, 518–526
    definition, 519
pathset, 58, 181, 348
pattern (of a pdBf), *see* partially defined
    Boolean function, pattern
paved upper-plane, 583
  extension, 583
PBNF, *see* normal form of a pseudo-Boolean
    function
pdBF Decision Tree, 528
pdBf, *see* partially defined Boolean function

perceptron, 406
perfect matrix, 277
Petrick function, 144
phase transition for DNF equations, 108
piecewise linear representation of a
    pseudo-Boolean function, 574
pigeonhole formula, 105
polar function
  Boolean, 208, 492
  pseudo-Boolean, 604
polynomial delay: definition, 625
polynomial expression
  of a Boolean function, 45
  of a pseudo-Boolean function, 571
polynomial extension of a pseudo-Boolean
    function, 579
  optimization, 588–589
polynomial incremental time: definition, 625
polynomial total time: definition, 625
posiform of a pseudo-Boolean function, 572
  optimization, 590–592
positive function, 34
  recognition, 43
prime implicant
  definition, 26
  dual, 169, 177, 183
  essential, 146, 311, 314
  linear, 125
  number of, 129, 159–161
  of Horn functions, 292–296
  of positive functions, 35–38
  pseudo-Boolean, 576
  quadratic, 126
  redundant, 146, 314
  small degree, 125, 193
  vs. minimal pathset, 58
  vs. minimal true point, 39
  vs. minimal winning coalition, 56
prime implicant generation, 128–141
  by consensus, 130–138
  by double dualization, 141, 188
  by term disengagement, 137
  by variable depletion, 135
  complexity, 139, 141
  for Horn functions, 140, 286–292
  for quadratic functions, 140, 250–254
  for threshold functions, 423–428
  from transitive closure, 251–254, 310
  from true points, 128
  tractable classes, 140
prime implicate
  definition, 28
  dual, 169
  pseudo-Boolean, 576

prime implicate (*cont.*)
  vs. maximal false point, 39
  vs. maximal losing coalition, 56
problem definition
  $(T, F) \parallel \phi \parallel$-MINIMIZATION, 143
  $T \parallel \phi \parallel$-MINIMIZATION, 143
  $minT \parallel \phi \parallel$-MINIMIZATION, 143
  $\parallel \phi \parallel$-MINIMIZATION, 143
  ADD-j, 197
  BEST-FIT($\mathcal{C}$), 548
  BOOLEAN EQUATION, 73
  CE($\mathcal{C}$) (consistent extension), 553
  DNF DUALIZATION , 183
  DNF EQUATION, 73
  DNF MEMBERSHIP IN $\mathcal{C}$, 40
  DUAL RECOGNITION , 183
  DUALIZATION , 183
  EXTENSION($\mathcal{C}$), 514
  FRE($\mathcal{C}$) (fully robust extension), 553
  FORBIDDEN-COLOR GRAPH BIPARTITION, 221
  FUNCTIONAL MEMBERSHIP IN $\mathcal{C}$, 40
  HORN μ-MINIMIZATION, 298
  HORN DUAL RECOGNITION, 308
  IDENTIFICATION, 196
  IMPLICANT RECOGNITION, 138
  ISOTONY, 227
  MRE($\mathcal{C}$) (most robust extension), 553
  MIN-SUPPORT($\mathcal{C}$), 515
  POSITIVE DNF DUALIZATION, 190
  POSITIVE DUAL RECOGNITION , 190
  PRIME IMPLICANTS, 128
  QUADRATIC DNF DUALIZATION, 263
  READ-ONCE EXACT LEARNING, 474
  READ-ONCE RECOGNITION, 466
  REGULAR DUALIZATION, 369
  REGULARITY RECOGNITION, 365
  SATISFIABILITY, 74
  SET COVER, 515
  SHELLABILITY, 349
  THRESHOLD RECOGNITION, 417
procedure
  AHK READ-ONCE EXACT LEARNING, 474
  BRANCH (solving equations), 81
  CHECKING NORMALITY, 469
  CONSENSUS* (prime implicants), 131
  CONSENSUS (solving equations), 94
  DECISION TREE, 48, 528
  DISENGAGEMENT CONSENSUS, 138
  DUALREGO (regular dualization), 372
  DUALREG (regular dualization), 376
  ELIMINATE (solving equations), 89
  EXPAND* (Tseitin's procedure), 77
  EXPAND (Tseitin's procedure), 22

FK-DUALIZATION (Fredman and Khachiyan), 196
FORWARD CHAINING, 285
GMR READ-ONCE RECOGNITION, 467
IMPLICANT, 342
INPUT CONSENSUS, 289
INPUT DISENGAGEMENT CONSENSUS, 257
INPUT PRIME CONSENSUS, 291
LE-PROPERTY (lexico-exchange), 344
MINTRUE, 424
NULR (Negative Unit Literal Rule), 282
ORTHOGONALIZE., 328
PRIME IMPLICANT DEPLETION, 287
PRIME PATTERNS, 559
QUADRATIC PRIME IMPLICANTS, 254
RECOGNIZE DUAL, 194
REGCOVERO (regular set covering), 377
REGCOVER (regular set covering), 379
REGMINORO (largest regular minorant), 383
REGMINOR (largest regular minorant), 386
REGULAR (recognition), 368
SD-DUALIZATION (sequential distributive), 188
THRESHOLD (recognition), 418
TREE-DNF, 527
production planning, 569
projection of a Boolean function, 28
propositional logic, 5, 50–52, 68, 124, 273–274
pseudo-Boolean function
  almost-positive, 604
  approximation, 593
  completely unimodal, 606
  concave extension, 581–583
  continuous extensions, 578–585
  convex extension, 581–583
  definition, 564
  degree, 571
  derivative, 586, 596, 599
  DNF and CNF, 574–578
  linear, 601
  Lovász extension, 583–585
  modular, 601
  monotone, 596–598
  normal form (PBNF), 571–572
  paved upper-plane extension, 583
  piecewise linear representation, 573–574
  polynomial expression, 570–571
  polynomial extension, 579–580
  posiform, 572–573
  quadratic, 594–596
  representation, 570–578
  standard extension, 582
  submodular, 599–603
  supermodular, 599–605

threshold, 606
unate, 604
unimax, 606
unimodal, 605–606
unimodular, 604–605
pseudo-Boolean optimization
    concave extension, 589–590
    conflict graph, 590–592
    continuous relaxations, 588–590
    linearization, 589
    local optima, 585–587, 606
    Lovász extension, 590
    methods, 585–592
    models, 565–570
    polynomial extension, 588–589
    quadratic, 594–596
    rounding (Rosenberg's theorem), 101, 193,
        588, 590
    standard extension, 589–590
    standard extension bound, 590, 595
    supermodular function, 603–605
    transformation to the quadratic case, 594
    variable elimination, 587–588
psychology, 443, 512, 538

Q-Horn
    DNF, 317
    equation, 318
    function, 317
quadratic equation (2-SAT), 230–243
    Alternative Labeling algorithm, 234–235
    and Schaefer's theorem, 109
    definition, 204
    forced literal, 214
    generating all solutions, 244
    implication graph, 212–216, 239–240, 247
    Labeling algorithm, 231–234
    matched graph, 210–212, 236–239
    number of solutions, 61, 244
    on-line, 249
    parametric solution, 246–249
    random, 107, 241–243, 267–268
    reducibility to, 218–230
    set of solutions, 243
    Strong Components algorithm, 239–240
    Switching algorithm, 235–239
    twin literals, 215
    variable elimination, 230
    vs. 2-SATISFIABILITY, 205
quadratic function
    and Schaefer's theorem, 109
    and transitive closure, 210, 251–254
    characterization by functional equations, 208,
        493

definition, 203
    generation of prime implicants, 140, 250–254
    purely quadratic, 204
    shellability, 346–347
quadratic graph, 216–218
quadratic irredundant DNF
    and transitive reduction, 261–263
quadratic pseudo-Boolean function
    optimization, 594–596
    super- or submodular, 601
quadric polytope, 595

random Boolean function, 163
random DNF equation, 106
    crossover point, 108
    phase transition, 108
    quadratic, 241–243, 267–268
    threshold conjecture, 107
random DNF expression, 106
rank function (vector space, matroid), 566
read-m function, 476–477
read-once expression
    definition, 448
    parse tree, 456
    unicity, 456
read-once function
    P_4-freeness, 461
    characterization, 450, 458
    characterization by functional equations, 496
    co-occurrence graph, 456
    definition, 448
    learning with oracle, 473–476
    normality, 459
    positional game, 477–480
    recognition, 466–473
        arbitrary representation, 470–473
        complete DNF expression, 466–470
Reed-Muller expansion, see representation over
    GF(2)
regular function
    characterization by functional equations, 496
    definition, 354
    dualization, 369–377
    largest regular minorant, 380–390
    left-shifts, 362–365
    lexico-exchange property, 359
    maximal false points, 371, 375
        number of, 376
    prime implicant recognition, 367–368
    recognition, 365–369
    set covering, 377–380
    smallest regular majorant, 380–390
    threshold graph, 439
    vs. aligned, 400

regular function (*cont.*)
  vs. ideal, 400
  vs. weakly regular, 400
  Winder matrix, 367
regular hypergraph, 360
reliability polynomial, 59, 361, 580
reliability theory, 7, 34, 58–59, 181, 348,
      361–362, 406
renaming variables, *see* switching variables
representation by Boolean expressions, 10–13
  normal forms, 15–19
representation over GF(2), 44
  linear equations, 110
representation over the reals, 45
resolution principle, 92
resolvent of Boolean constraints, 62, 407, 422
restriction of a Boolean function, 28
roof dual, 595
  strong persistency, 595

satisfiability problem, *see* DNF equation
satisfiable expression, 68
Schaefer's theorem, 108–111
self-dual extension, 172, 402, 412
self-dual function, 13, 170, 179–180, 182–183,
      196, 199, 402, 412, 445
set covering problem, 63, 66, 143, 147, 150,
      156, 180, 299, 422, 515, 532, 562
  generalized, 62, 276, 422
  regular, 377–380, 388–390
set function, 564
shadow, 332, 337, 362
Shannon expansion, 29–30, 44, 88
shellable DNF, 330
  dualization, 336–338, 349
  orthogonal form, 331–335
  recognition, 335, 349
shellable function, 335
  characterization by functional equations, 496
shelling, 330
  recognition, 335
signed graph, 220
Sperner family, Sperner hypergraph, 60, 177,
      614
split graph, 220
spread of a function, 162
stability function, 7, 34, 60, 209, 402
stable set, 7, 34, 60, 71, 177, 197, 263, 439, 610,
      614
  and pseudo-Boolean optimization, 565,
      590–592, 594
standard extension of a pseudo-Boolean
      function, 582
  bound, 590, 595

strength relation, 351–362, 391–397
  k-monotonicity, 392
  and Chow parameters, 355
  and regular functions, 354
  and reliability theory, 361
  and simple games, 355, 360, 391
  complete monotonicity, 392
  leader, 359
  of dual function, 358
  of restrictions, 357, 393
  on subsets, definition, 391
  on variables, definition, 352
  recognition, 356, 366–369, 401
  Winder matrix, 366
strong component, 613
strongly connected digraph, 613
struction, 592
stuck-at fault, 69–71, 87
subcube, 31
submodular Boolean function, 208, 309–311
  characterization by functional equations, 309,
      492
submodular pseudo-Boolean function, 566,
      599–603
  derivatives, 599
  quadratic, 601
  recognition, 601
sum of disjoint products, *see* orthogonal DNF
superadditive pseudo-Boolean function, 568
supermodular Boolean function, 208
  characterization by functional equations, 492
supermodular pseudo-Boolean function,
      599–605, 608
  derivatives, 599
  Lovász extension, 601
  optimization, 603–605
  quadratic, 601
  recognition, 601
support set of a pdBf, 515–518
  minimum, 515
    complexity, 517
  positive, 532
swing, 57, 433
switching variables, 235–239, 314–316, 321,
      604–605
symmetric
  function, 65, 159, 361
  variables, 352, 356
system of equations (Boolean), 74–76

tautology, 68
term, 15
theory (of a pdBf), *see* partially defined Boolean
      function, theory

threshold conjecture, 107
threshold dimension, 443, 446
threshold function
  asummability, 414, 502
  certificate of non-membership, 501
  characterization by functional equations, 499,
    501–503
  Chow function, 429
  Chow parameters, 428–438
    computation, 435
    vs. separating structure, 430–435
  complete monotonicity, 392, 411
  definition, 404
  dual, 411
  generation of prime implicants, 423–428
  linear programming characterization, 413
  number of threshold functions, 412
  polynomial threshold function, 407
  prime implicant characterization, 423
  pseudo-Boolean, 606
  recognition, 417–422
    complexity, 418, 421
    oracle algorithm, 417, 445
  regularity, 410
  related graph classes, 438–444
  related properties, 430
  restrictions, 408
  separating structure, 404, 408–410
    cone, 409
    integral, 409
    random, 432, 435
    size of weights, 413
  shellability, 348
  vs. 2-asummability, 416
threshold graph, 438–444
  aggregation problem, 442
  degree sequence, 440–442, 446
  forbidden subgraphs, 440
  Guttman scale, 443
  threshold dimension, 443
threshold synthesis, 417
topological ordering of a digraph, 613
totally unimodular matrix

and Horn functions, 277
  pseudo-Boolean optimization, 604
  recognition, 221
transitive closure, 613
  and quadratic functions, 210
  by consensus, 255–261
transitive reduction, 613
  and irredundant quadratic DNFs, 261–263
transversal, 64, 177–181, 610, 614
tree, 611
  rooted, 611
triangulated graph, 346, 565
true points, 3
  minimal, 38
  number of, 23, 24, 61, 111, 163, 327, 436
truth table, 4, 143, 156, 158, 417
Tseitin's procedure, 19–22, 76

unate function
  Boolean, 34
  pseudo-Boolean, 604
unimax pseudo-Boolean function, 606
unimodal pseudo-Boolean function, 605–606
unimodular pseudo-Boolean function, 604–605
unique games conjecture, 121
unit literal rules, 85
  and extended Horn equations, 319
  and Horn equations, 281–284
unit resolution, 85

variable elimination
  complexity, 104, 106
  for Boolean equations, 87–91
  for pseudo-Boolean optimization, 587–588

weakly regular function, 397–398
weighted majority game, 56, 182, 348, 406, 407,
  434, 436

Zhegalkin polynomial, see representation over
  GF(2)